FLORA ZAMBESIACA

Flora terrarum Zambesii aquis conjunctarum

T0132985

VOLUME SEVEN: PART ONE

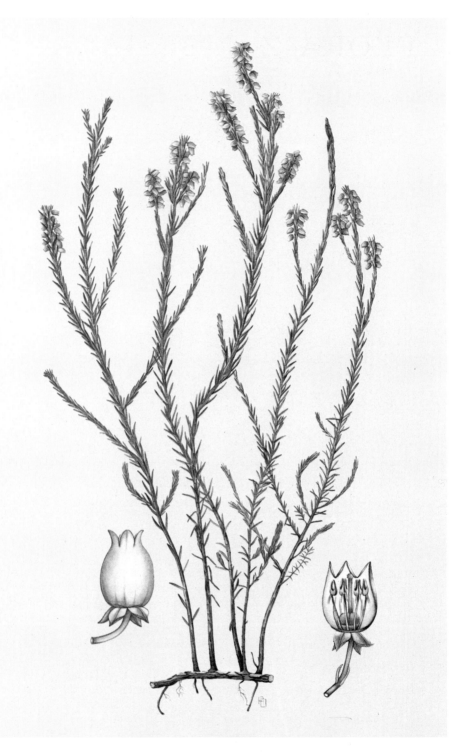

ERICA WHYTEANA

FLORA ZAMBESIACA

MOZAMBIQUE
MALAWI, ZAMBIA, ZIMBABWE
BOTSWANA

VOLUME SEVEN: PART ONE

Edited by
E. LAUNERT

on behalf of the Editorial Board:

J. P. M. BRENAN
Royal Botanic Gardens, Kew

E. LAUNERT
British Museum (Natural History)

E. J. MENDES
*Centro de Botânica, Instituto de Investigação
Científica Tropical, Lisboa*

H. WILD
University of Zimbabwe

Published by the Managing Committee on behalf of
the contributors to Flora Zambesiaca
1983

© *Flora Zambesiaca Managing Committee, 1983*

Printed in Great Britain by
Clark Constable (1982) Ltd., Edinburgh

ISBN 0 95076 82 00

CONTENTS

EDITOR'S PREFACE

The first part of volume VII follows, out of sequence, the publication of volume IV. The intervening volumes (volume V, comprising the *Compositae* and volume VI, comprising the *Rubiaceae*) will be published in due course. In order to make volume V more compact by devoting it entirely to the *Compositae* two related families, namely the *Valerianaceae* (95) and the *Dipsacaceae* (96), are included in volume VII. Moreover, for reasons given in the preface to volume IV the *Crassulaceae* are contained in this volume. The *Escalloniaceae* which were inadvertently omitted from volume IV are also included here. Political events in southern tropical Africa have led to numerous changes in place names in recent years. We have endeavoured to take account of these changes in this volume, but regretfully there was not sufficient time to implement the most recent changes in Zimbabwe. The changes in the geographical divisions of Mozambique are as follows: for SS (Sul do Save) read GI (Gaza and Inhambane); for LM (Lorenço Marques) read M (Maputo). Dr F. K. Kupicha, who contributed families 104–106 and 108 and critically read the manuscripts of other families, did this work in her capacity as Krukoff Curator of African Botany.

E. L.

LIST OF NEW NAMES AND TAXA
PUBLISHED IN THIS WORK

LIST OF FAMILIES INCLUDED IN
VOLUME VII, PART 1

64a. ESCALLONIACEAE

By B. Verdcourt

Trees or shrubs. Leaves simple, alternate, rarely subopposite or subverticillate, usually glandular-serrate; stipules usually absent or minute. Flowers hermaphrodite or less often dioecious or polygamous, mostly in terminal or axillary racemes, panicles, or cymes; in one genus epiphyllous. Sepals 4–5, mostly united at the base or rarely free, imbricate or valvate, often persistent. Petals 4–5, free or rarely connate into a short tube, imbricate or valvate. Disk annular or with lobes alternating with the stamens. Stamens (4)5(6), sometimes alternating with staminodes, perigynous, free; anthers 2-celled, opening by longitudinal slits. Ovary superior or inferior, syncarpous or apocarpous, 1–6-locular; ovules with axile or parietal placentation; ovules numerous; styles 1–6, free or ± joined. Fruit a capsule or berry. Seeds few to many, with small or large embryo and copious endosperm.

A rather small somewhat poorly defined family formerly included in Saxifragaceae with about 23 genera when defined in a broad sense but no more than 7 when dismembered into smaller families. For the purposes of the Flora Brexiaceae (*Brexia*) and Montiniaceae (*Grevea* & *Montinia*) have been considered worthy of separate family status; the remaining genus, *Choristylis*, is retained in Escalloniaceae although it is sometimes separated, together with *Itea*, as a small family Iteaceae* mainly characterized by extraordinary 2-porate subisopolar pollen grains.

Various species and hybrids of *Escallonia* are grown quite frequently in gardens in Salisbury and probably elsewhere e.g. Salisbury, Mt. Pleasant, fl. 11.i.1974, *Biegel* 4483 (SRGH) is a form of *E. rubra* (Ruiz & Pav.) Pers., (*E. punctata* DC.) which has numerous variants and enters into many hybrids.

CHORISTYLIS Harv.

Choristylis Harv. in Hook., Lond. Journ. Bot. 1: 19 (1842).

Shrubs. Leaves alternate, closely glandular-serrate, penninerved; stipules minute, linear. Flowers hermaphrodite or polygamous, small, in much-branched axillary panicles shorter than the leaves. Calyx-tube obconic, adnate to the ovary; lobes 5, subulate, distant, triangular at the base, puberulous, persistent. Petals 5, ovate-deltoid to narrowly triangular, perigynous, broad at the base and confluent with the epigynous disk, valvate, puberulous, persistent. Stamens 5, alternating with the petals, inserted at the margin of the disk, connivent; filaments subulate; anthers dorsifixed, hairy, small, ovoid, the cells separated by a thick connective. Ovary partly inferior, 2-locular; ovules numerous on axile placentas; styles 2, subulate, adherent or diverging; stigmas capitate. Capsule half-superior, 2-locular, dehiscing septicidally between the styles, many-seeded. Seeds irregularly obovoid or oblong-obovoid, slightly curved with a thick tesselated testa.

A monospecific genus restricted to E., Central and S. Africa and which presumably migrated northwards along the mountains, but seems to be absent from the more recent volcanic mountains in E. Africa.

Choristylis rhamnoides Harv. in Hook., Lond. Journ. Bot. **1**: 19 (1842); Thes. Cap. **2**: 15, t. 123 (1859); in F.C. **2**: 308 (1862). — Brenan in Mem. N.Y. Bot. Gard. **8**: 433 (1954). — Chapman, Veg. Mlanje Mts. Nyasaland: 62 (1962). — Boutique in Bull. Jard. Bot. Brux. **34**: 504 (1964). — Liben in F.C.B., Saxifragaceae: 2, t. 1 (1969). — Palmer & Pitman, Trees of S. Africa **1**: 658, figs. (1972). — Ross, Fl. Natal: 181 (1973). — Verdc. in F.T.E.A., Escalloniaceae: 3, t. 1 (1973). — Compton, Fl. Swaziland: 225 (1976). — Palmer, Field Guide Trees Southern Africa: 42 (1977). — Palgrave, Trees of Southern Africa: 202, map & fig. (1977). TAB. **1**. Type from S. Africa.

C. *shirensis* Bak.f.† in Trans. Linn. Soc., Lond. Ser. 2, Bot. **4**: 13, t. 3, figs. 1–6 (1894).

* E.g. in Willis "A Dictionary of Flowering Plants and Ferns", ed. 7: 584 (1966), where a very narrow view of the family Escalloniaceae is taken.

† A C. *virescens* Bak.f., referred to in the Index Kewensis as published on the same page seems to be fictitious.

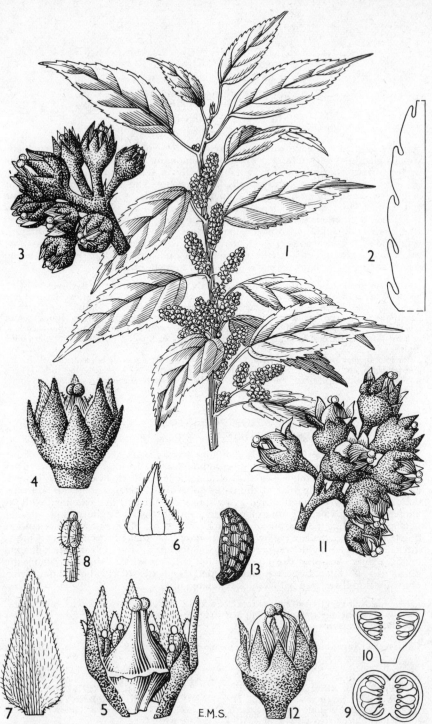

Tab. 1. CHORISTYLIS RHAMNOIDES. 1, habit (×⅔); 2, outline of leaf-margin (×6); 3, inflorescence (×6); 4, flower (×12); 5, same, opened out to show ovary (×12); 6, sepal (×24); 7, petal (×24); 8, stamen (×24); 9, diagrammatic transverse section of ovary; 10 diagrammatic longitudinal section of ovary; 11, infructescence (×4); 12, fruit (×6); 13, seed (×24). 1–10 from *Clements* 79; 11–13 from *Buchanan* 1432. From F.T.E.A.

— Brenan, T.T.C.L.: 196 (1949). Types: Malawi, Mt. Mulanje, *Whyte* 53 (BM, syntype; K) & Shire Highlands, *Buchanan* 158 (BM, syntype; K) & *Buchanan* 1468 (BM, syntype).

More or less erect or usually climbing shrub or small tree 1·8–10 m. tall or long; stems angular, often long and thin, sometimes trailing or hanging, diverging from the base, the young shoots often purplish, puberulous. Leaves petiolate; blades ovate, elliptic or oblong to ovate-lanceolate, 1·5–10·3(–12) cm. long, 0·9–5·5 cm. wide, acute to acuminate or rarely rounded at the apex, cuneate to unequally rounded at the base, glabrous or with sparse hairs on the main nerves beneath and often with domatia, somewhat glossy above, margins sharply and closely serrate; costa and main nerves impressed above, prominent beneath, the tertiary venation closely reticulate; petioles often purple, 0·5–2 cm long. Inflorescences 1–4·5 cm. long, 1–2·5 cm. wide, many-flowered, finely pubescent-tomentose, the flowers creamy-white or greenish-yellow, sometimes tinged red, sweetly scented; peduncles 0–1·2 cm. long; pedicels 1–2(–5) mm. long; bracts and bracteoles narrowly triangular, 0·5–1·5 mm. long, 0·2–0·5 mm. wide. Calyx-tube 1–1·5 mm. long, 1–1·5 mm. wide; lobes 0·5–2 mm. long, 0·3–0·5(–1) mm. wide at the base. Petals (1·6–) 2–3·3 mm. long, 1–2 mm. wide at the base, 3-nerved (eventually distinctly spreading in cultivated material recently seen). Styles yellowish, turning brown, 0·5–1 mm. long; stigmas 0·3–0·5 mm. in diameter, free or sometimes cohering for a time. Capsule campanulate to turbinate below, conical above, 3–5(–6) mm. long, 3 mm. wide, finely pubescent, slightly ribbed. Seeds brownish, ± 1 mm. long, 0·5 mm. wide.

Zimbabwe. C: Wedza Mt., fl. 18.ix.1964, *West* 4983 (K; LISC; SRGH). E: Umtali, Tsetsera Mts., Ranti North, fr. 3.x.1962, *Chase* 7819 (K; LISC; SRGH). **Malawi**. N: Nyika Plateau, fl. 2.ix.1964, *Robinson* 6244 (K; SRGH). S: Mt. Mulanje, Chambe basin, fr. 19.vi.1967, *Hilliard & Burtt* 4501 (K). **Mozambique.** Z: Gúrùe, fl. 20.ix.1944, *Mendonça* 2164 (LISC). MS: Manica, Serra Zuira, planalto Tsetsera, fr. 6.xi.1965, *Torre & Pereira* 12730 (LISC).

Also in Uganda, Tanzania, Burundi, Zaire, South Africa (Transvaal, Natal, Cape Province), Lesotho and Swaziland. Evergreen forest margins, riverine and valley forests, open hillsides (derived), sometimes in rock crevices and often by streams, 900–2300 m.

67. CRASSULACEAE

By R. Fernandes

Annual, biennial or perennial herbs, undershrubs or shrubs, usually ± succulent. Leaves opposite, verticillate or alternate, usually simple, sometimes compound, exstipulate, often ± thick and fleshy-succulent. Flowers (3)4–5(6–32)-merous, actinomorphic, usually bisexual, in axillary or terminal cymes often grouped in corymb- or panicle-like inflorescences, less often in racemes or spikes or solitary in the leaf-axil. Sepals free or ± united at the base, sometimes in a ± long tube. Petals as many as the sepals, free or ± connate, hypogynous. Stamens as many or twice as many as the petals; filaments free or more or less connate with the corolla-tube; anthers 2-celled, introrse, basifixed, dehiscing longitudinally. Ovary superior; carpels equal in number to the petals, free or united (up to the middle); ovules many or few to one in each carpel, anatropous, inserted on the adaxial suture; styles short, sometimes nearly absent or ± elongate; stigmas capitate. Fruit follicular, membranous or leathery, opening along the free part of the adaxial suture. Seeds mostly minute with smooth, rugose or tuberculate tegument and usually fleshy endosperm; embryo straight. Hypogynous scales (scale-like nectaries) usually present, small, applied to the base of the outer face of the carpels.

A family of about 33 genera occurring in tropical and temperate regions.

1. Stamens equal in number to the petals; flowers small to medium-sized, usually white; petals free or slightly connate at the base; leaves opposite - - - - - **1. Crassula**
 - Stamens twice as many as the petals; flowers fairly large, usually showy; petals united beyond the middle into a tube; leaves opposite or verticillate (rarely alternate) - - 2
2. Flowers 5-merous; shrubs or undershrubs - - - - - - **4. Cotyledon**
 - Flowers 4-merous - - - - - - - - - - - 3

3. Filaments connate with the corolla tube at least up to the middle of the latter (appearing to be inserted on the corolla tube at or above the middle); styles usually shorter than the follicles, often included; sepals usually not connate beyond the middle of calyx, the tube not or slightly swollen (appressed to the corolla tube); flowers not pendulous   -   **2. Kalanchoe**
   - Filaments free or apparently inserted below the middle of the corolla tube; styles longer than the follicles; sepals often connate beyond the middle into a swollen tube (not appressed to the corolla); flowers pendulous -   -   -   -   -   -   -   **3. Bryophyllum**

1. CRASSULA L.

Crassula L., Sp. Pl. 1: 282 (1753); Gen. Pl., ed. 5: 136 (1754).

Annual or perennial succulent herbs, sometimes with a tuber-like root, or undershrubs or shrubs with ± woody root-stock and succulent usually ± fleshy leaves. Leaves opposite, usually decussate, the lowermost frequently rosulate, free or ± connate in a sheath, usually simple, undivided and entire, thin to ± thick, flat, terete, semiterete, ovoid, etc. Flowers (3–4)5(6–9)-merous, isomerous, usually small and not showy, in cymes arranged either in dense subsessile or pedunculate axillary clusters sometimes forming thyrsoid inflorescences, or in corymb-like axillary or terminal loose or ± dense inflorescences, sometimes 1(2) flowers in the leaf-axils. Calyx usually shorter than the corolla, with the sepals free or slightly connate at the base, appressed to the corolla, ± succulent. Corolla usually white or whitish turning orange or brownish-red when dry, sometimes red or carmine, rarely bright yellow, persistent; petals erect or stellate, connate at the base into a ± short tube. Stamens free or with the lower part of the filaments connate with the corolla-tube, alternipetalous; anthers ovate or oblong, sometimes nearly circular. Carpels free or connate at the base, oblong or obovoid, attenuate towards or contracted into a ± short usually terminal style, sometimes the styles nearly absent and stigmas subdorsal, completely glabrous or papillose along the suture; ovules numerous or sometimes 1–4. Scales shorter than the carpels, hyaline or reddish-brown or rose, thin, narrowly to broadly spathulate or obovate or cuneate.

More than 300 species, the majority in South Africa, others in tropical Africa, Madagascar, Arabia, India, some of them almost worldwide.

1. Flowers 4-merous, axillary, usually solitary; pedicels 2–7 mm. long in flower, up to 12 mm. in fruit; leaves 5·5–20 × 1–5 mm., linear to oblong-spathulate; annual terrestrial plant of wet places or aquatic plant with rooting stem   -   -   -   -   -   **1. granvikii**
   - Flowers 5-merous; plants without the assemblage of characters above   -   -   -   2
2. Flowers in ± dense axillary clusters along the stem   -   -   -   -   -   3
   - Flowers axillary and solitary and/or in cymes not arranged in axillary clusters -   7
3. Carpels 2-seeded; flowers very small (usually not more than 2 mm. long); petals not mucronate; leaves sessile, small (not more than 12·5(15) × 2(2·5) mm.; plants moss- or lycopodium-like   -   -   -   -   -   -   -   -   -   -   4
   - Carpels 3-many-seeded; petals mucronate below the apex; leaves larger than above; plants neither moss- nor lycopodium-like   -   -   -   -   -   -   -   -   6
4. Anthers 0·25–0·3 mm. long; calyx usually less than 2 mm. long; nectary-scales spathulate to nearly flabellate; usually perennial herbs   -   -   -   -   -   **3. schimperi**
   - Anthers less than 0·25 mm.; calyx c. 2 mm. long, distinctly longer than the corolla; scales narrowly spathulate to linear-spathulate   -   -   -   -   -   -   5
5. Anthers minute, usually not more than 0·15 mm.; pedicels 1–3·5 mm. long, very unequal in each fascicle; cauline leaves linear or linear-lanceolate, obtuse or truncate and papillose at the apex; erect annual herb with narrowly winged stem, never rooting   -   **2. rhodesica**
   - Anthers a little longer than above; pedicels shorter; cauline leaves narrowly ovate to lanceolate, acute, not papillose but usually setose at the apex; usually perennial herb with a 4-angled stem, sometimes rooting   -   -   -   -   -   -   -   **3. schimperi**
6. Corolla not more than 2·0 mm. long; petals not inflexed at apex; follicles slightly attenuate, with apical stigmas   -   -   -   -   -   -   -   **13. cooperi** var. **subnodulosa**
   - Corolla 2·25–4 mm. long; petals with slightly inflexed apex; follicles obliquely ovoid with subsessile, finally sublateral stigmas   -   -   -   -   -   -   **17. nodulosa**
7. Leaves not more than 2(3) mm. broad, usually linear, either flat or semiterete or subtrigonous; corolla not more than 3·75(4) mm. long   -   -   -   -   -   8
   - At least some leaves more than 3 mm. broad   -   -   -   -   -   -   10
8. Perennial or annual, glabrous herb; stem leafy throughout; flowers axillary and solitary and in terminal cymes or sometimes in axillary cymes; pedicels up to 20 mm. long, elongating to 40(55) mm. in fruit; capillary; petals not mucronate   -   -   -   -   **4. expansa**
   - Shrubs or undershrubs, hairy or papillose at least on the branchlets; old branches leafless; flowers always in terminal inflorescences; pedicels not longer than 6 mm.; petals mucronate or submucronate   -   -   -   -   -   -   -   -   -   -   -   9

9. Erect shrub or undershrub 30–90 cm. high; leaves (8) 10–15·5(17) × 1–1·5(2) mm., acute, glabrous; pedicels 3–6 mm. long - - - - - - 11. *sarcocaulis* subsp. *rupicola*
 – Undershrub with prostrate stem and erect branches up to 11 cm. long, mat-forming; leaves 13–45 mm. long, obtuse, densely hispidulous; pedicels up to 1·5 mm. long   23. *zombensis*
10. Shrubs or sarmentose perennials; leaves not ciliate at the margin - - - - 11
 – Annual or perennial ± succulent herbs or undershrubs - - - - - 14
11. Leaves crenate-serrate, 2–4·5(6) × 1·2–2·5(3·5) cm., ovate or elliptic, the lower ones petiolate; plant sarmentose with a long stem (1 m. or more long), loosely branched, loosely leafy - - - - - - - - - - - - - 9. *sarmentosa*
 – Leaves entire, sessile; plants not sarmentose; stem ± branched, at least the young branches densely leafy - - - - - - - - - - - - - - 12
12. Leaves subtrigonous or very convex beneath, rather thick, 5–10·5(17·5) × 1–1·5(2·5) cm., densely and minutely papillose (grey-green, whitish or glaucous in colour); corolla (4·5)5–9(10) mm. long - - - - - - - - - 24. *heterotricha*
 – Leaves not as above, not more than 5 cm. long; corolla not more than 6·5(8) mm. long   13
13. Leaves obovate to obovate-spathulate, up to 5 × 3 cm., shortly acuminate or rounded at the apex, shining when alive, micaceous on drying; calyx up to 2·5 mm. long with broadly triangular sepals; shrub up to 3 m. or more tall - - - - - - 10. *ovata*
 – Leaves elliptic-linear to narrowly ovate, up to 3·5 × 0·1–0·8 cm., acute, neither shining nor micaceous; calyx (1)1·5–3·5(4) mm. long with oblong-lanceolate or lanceolate sepals; shrub up to 1·20 m. tall - - - - - - - 11. *sarcocaulis* subsp. *sarcocaulis*
14. Leaves ± scattered throughout the stem and branches, neither very thick nor very fleshy, not or indistinctly ciliate at the margin; stem usually slender and herbaceous - - 15
 – Leaves ± crowded below, often rosulate at the base of the floriferous axis, usually rather fleshy and ± thick, usually ciliate at the margin; undershrubs or succulent perennials   19
15. Petals (3)4–6 mm. long, lanceolate, acute; leaves ± petiolate, with glabrous, lineolate lamina - - - - - - - - - - - - - - 16
 – Petals usually shorter than above, ovate, broadly elliptic or oblong to obovate-oblong, obtuse or rounded at the apex; leaves sessile or shortly petiolate, not lineolate - - 17
16. Flowers in terminal corymb-like inflorescences; sepals 2–3(3·5) mm. long, usually less than half as long as corolla, oblong or linear, not hyaline at the margin - - 7. *alticola*
 – Flowers axillary and solitary towards the apices of stem and branches (sometimes few-flowered terminal cymes also present); sepals (2·5)3·5–5 mm. long, slightly shorter than the corolla, lanceolate to oblong-lanceolate, white-hyaline at the margin - 8. *alsinoides*
17. Flowers axillary, solitary and usually from the base to the apices of the stem and branches; corolla 3·5–4 mm. long; petals 1·5–2 mm. broad, suberect; plant completely glabrous
                                                    5. *maputensis*
 – Flowers only in terminal cymes or some flowers also axillary and solitary; corolla not longer than 3·5 mm.; petals narrower than above; plants hispid or puberulous - - 18
18. Pedicels 14–30(35) mm. long; flowers in terminal, lax cymes and also axillary and solitary; petals not mucronate; leaves 0·3–3·5(4·5) × 0·25–1·3 cm., usually ± petiolate, green
                                                    6. *fragilis*
 – Pedicels up to 0·5 mm. long; flowers in dense cymes either axillary or grouped in terminal corymbose inflorescences; petals mucronate; leaves 1·2–2·2 × 0·25–0·9 cm., sessile, whitish or greyish - - - - - - - - - - - - - 22. *leachii*
19. Floriferous axis leafless, only with 1–3(4) pairs of small bracts - - - - 20
 – Floriferous axis leafy from base to apex; cauline leaves often smaller than the basal ones, the upper bract-like; leaves ciliate at the margin - - - - - - - 24
20. Leaves usually not less than 5 times longer than broad, subtrigonous or very convex beneath, linear or oblong, whitish or greyish, uniformly papillose or hairy, not or indistinctly ciliate at the margin - - - - - - - - - 21
 – Leaves usually less than 5 times longer than broad, flattish, ± ciliate at the margin -   22
21. Stem prostrate and rooting with erect branches up to 11 cm. high; leaves 1·3–4·5 × 0·2–1 cm., very slightly connate; corolla 3·25–4·5 mm. long; undershrub forming dense mats
                                                    23. *zombensis*
 – Stem not rooting, up to 50 cm. (or more) long; leaves 5–10·5(17·5) × 1–1·5(2·5) cm., rather connate; corolla (4·5) 5–9(10) mm. long; an undershrub, but not as above   24. *heterotricha*
22. Floriferous axis and inflorescence — branches usually densely hispid; leaves usually hairy on both sides, rarely glabrous; inflorescence usually more than 5 cm. in diam.; sepals subobtuse, ciliate at margin and hispid outside - - - - - 21. *swaziensis*
 – Floriferous axis and inflorescence-branches ± appressed hairy (hairs retrorse) or glabrous; leaves usually glabrous on both sides; inflorescence not more than 5 cm. in diam.; sepals acute, ciliate at the margin and along the median line of back - - - - 23
23. Floriferous axis hairy from base to apex; leaves usually not more than twice as long as broad and obtuse or rounded at apex (if acute, then marginal cilia rather long); sepals as long as or longer than the half of corolla - - - - - - - 18. *globularioides*
 – Floriferous axis glabrous or sparsely hairy below the inflorescence; leaves usually narrower, acute, with very short marginal cilia; sepals shorter than half of corolla
                                                    19. *morrumbalensis*
24. Much-branched undershrub; old branches covered with the dried, persistent leaves; young

branches up to 5 cm. long, ending in dense leafy rosettes; leaves up to 1·2 × 1 cm., rounded
at the top, subglabrous   -   -   -   -   -   -   -   -   -   20. *nyikensis*
- Succulent perennial without the above assemblage of characters   -   -   -   25
25. Root not tuberous; stems slender, usually not more than 22 cm. long; leaves not more than
3(3·5) × 1 cm., oblong-lanceolate to suborbicular, usually not attenuate and often obtuse or
rounded at the apex; sheaths of leaves up to 2 mm. long (usually shorter)   12. *setulosa*
- Root tuberous; stems usually ± robust, up to 90 cm. tall; leaves oblong-lanceolate to linear,
relatively longer and narrower than above, 2·5–26·5 × (0·1–0·2)0·4–2·7(3·3) cm., ±
attenuate, acute; sheaths of lower leaves not less than (2)2·5 mm. long   -   -   26
26. Sheaths of the lower leaves usually not less than 1 cm. long (frequently longer, up to 2–3
cm. in strong plants), those of median and upper ones distinctly developed; sepals with a
long terminal obtuse papilla (obtuse marginal papillae sometimes also present); corolla
usually white or whitish   -   -   -   -   -   -   -   -   -   14. *vaginata*
- Sheaths of the lower leaves usually not longer than 1 cm., those of the median and upper
leaves very short to absent; sepals without a terminal obtuse papilla, serrulate to ciliate-
pectinate (denticules or cilia acute) at the margin; corolla usually red or carmine,
sometimes greenish-white   -   -   -   -   -   -   -   -   -   27
27. Corolla 2·5–3·5(4) mm. long; calyx 1–2·5(3) mm. long; sepals triangular- or ovate-
lanceolate, sometimes lanceolate, usually less than half as long as corolla; anthers ± 0·5
mm. long; sheaths of the lower leaves not longer than 0·5 cm.   -   -   16. *similis*
- Corolla (3)4–7 mm. long; calyx (2·5)3–5·5(6·5) mm. long; sepals lanceolate to linear-
lanceolate, usually more than half as long as corolla; anthers 0·5–0·75(1) mm. long; sheaths
of the lower leaves up to 0·8(1·2) cm. long   -   -   -   -   -   -   15. *alba*

1. **Crassula granvikii** Mildbr. in Notizbl. Bot. Gart. Berl. **8**: 227 (1922). — Hedberg in
Symb. Bot. Upsal. **15**, 1: 100 et 278–281, fig. 23 et 24 (1957). — Merxm. & al. in Ann.
Naturh. Mus. Wien **75**: 112 (1971). — Tölken in Contr. Bolus Herb. **8**: 88 (1977) in adnot.
— G. E. Wickens & Bywater in Kew Bull. **34**, 4: 631, t. 17 fig. A-B (1980). Type from
Kenya.
   Crassula vaillantii sensu Engl., Hochgebirgsfl. Trop. Afr.: 231 (1892). — Schönl. in
Trans. Roy. Soc. S. Afr. **17**, 3: 182 (1929) pro parte. — Toussaint in F.C.B. **2**: 572 (1951)
non (Willd.) Roth. (1827).
   Crassula vaillantii var. *kilimandscharica* Engl. in Bot. Jahrb. **39**: 468 (1907). — Berger
in Engl. & Prantl., Nat. Pflanzenfam. ed. 2, **18a**: 388 (1930). Type from Tanzania.
   Tillaea rivularis A. Peter in Abh. Ges. Wiss. Göttingen, N. Ser. **13**, 2: 80, fig. 14 (1928).
Syntypes from Tanzania.
   Tillaea repens A. Peter, tom. cit.: 80, fig. 15 (1928). Syntypes from Tanzania.
   Tillaea vaillantii var. *kilimandscharica* (Engl.) A. Peter, tom. cit.: 80 (1928). Type as for
Crassula vaillantii var. *kilimandscharica*.
   Tillaea aquatica sensu Hancock & Soundy in Journ. E. Afr. Ug. Nat. Hist. Soc. **36**: 181
(1931) non L. (1753)
   Crassula wrightiana Bullock in Kew Bull. **1932**: 487 (1932); op. cit. **1933**: 62 (1933); in
Hook., Ic. Pl., Ser. 5, 3: t. 3218 (1933). Type from Kenya.
   Crassula erubescens Bullock, tom. cit.: 488 (1932); tom. cit.: 63 (1933). — Bally in Journ.
E. Afr. Ug. Nat. Hist. Soc. **15**: 12 (1940). Type from Kenya.
   Crassula aquatica sensu Staner in Rev. Zool. Bot. Afr. **23**: 216 (1932). — Bullock, tom.
cit.: 62 (1933) non (L.) Schönl.
   Crassula rivularis (A. Peter) Hutch. & Bruce in Kew Bull. **1941**: 88 (1941). — Robyns,
Fl. Parc Nat. Alb. **1**: 229 (1948). — Toussaint, op. cit.: 571 (1951). — Cufod. in Bull. Jard.
Bot. Brux. **24**, Suppl.: 171 (1954). Type as for *Tillaea rivularis*.

   Aquatic or terrestrial completely glabrous herb, varying in habit according to
habitat. Stem 6–38 cm. long, either erect or ascending and simple, rooting at lower
nodes or at most nodes except for the apical ones, or stem widely trailing and rooting
at nodes and emitting short, erect branches throughout, these simple or with 1–2
branchlets towards the apex, fleshy, somewhat thick but flaccid, green, mauve or
purplish at the nodes; internodes up to 3·8 cm. long, the basal and median ones
longer than the leaves, the uppermost equalling them or shorter. Leaves
5·5–20 × 1–5 mm., linear or oblong to oblong spathulate, acute to obtuse, sessile and
connate at the base in a hyaline or purplish, 1–2 mm. long sheath, erect or erect-
spreading but the undermost patent or slightly reflexed, fleshy, somewhat thick and
rigid or nearly membranous on drying, light green, neither lineolate nor punctate
near the margin. Flowers 4-merous, axillary, solitary or sometimes 2–3 together;
pedicels 2–7 mm. long in flower, up to 12 mm. long in fruit, slender. Calyx 0·75–1·25
mm. long, green or yellow-green; sepals oblong, subacute to rounded at the apex,
connate at the base. Petals 1–1·75 mm. long, oblong or oblong-obovate, rounded at
the apex, white, sometimes with carmine-red midrib or completely pinkish-white to
carmine. Filaments 0·5–0·75 mm. long, carmine-red; anthers 0·15–0·25 mm. — long,

transversely ovate or subcircular, purplish. Follicles 1–1·25 mm., obovoid, rounded at the top, contracted into the style, carmine; styles c. 0·5 mm. long, reflexed. Seeds 1–2(4) per follicle, c. 0·5 to 1 mm. long, narrowly oblong, obtuse at the base and apex, minutely and densely tuberculate, the tubercules in longitudinal rows. Scales 0·5–0·75 mm. long, spathulate, truncate at the apex, purplish-hyaline.

Malawi. N: Rumphi District, Lake Kaulime, Nyika Plateau, 2250 m., 4.i.1959, *Richards* 10467 (K).
Also in Ethiopia, Uganda, Kenya, Tanzania, Rwanda, Burundi and Zaire. In damp places along permanent streams or in water, at high levels up to 3200 m. alt.

C. *granvikii* is described with 4-seeded carpels, whereas the type of C. *wrightiana* has 1-seeded carpels. However, we found that Tanzanian specimens referred to the latter had 1–2-seeded carpels. The species C. *erubescens* has been distinguished from C. *granvikii* by its larger leaves. We think that these three taxa should be sunk into a single species which has leaves of variable size according to the humidity of the habitat. The ovaries are initially 4-ovulate, but only 1 or 2 mature seeds are usually formed. C. *granvikii* and C. *erubescens* correspond to the land form, whereas C. *wrightiana* is the form characteristic of more moist or aquatic habitats.

2. **Crassula rhodesica** (Merxm.) Wickens & Bywater in Kew Bull. **34**, 4: 632, t. 18 fig. A–B (1980). — R. Fernandes in Bol. Soc. Brot., Sér. 2, **55**: 110 (1982). Type: Zimbabwe, Hartley Poole, *Hornby* 2911 (K; M, holotype; SRGH).
    Tillaea pentandra sensu Britten in F.T.A. **2**: 386 (1871) pro parte; sensu Hiern, Cat. Afr. Pl. Welw. **1**: 324–325 (1896) non Royle ex Edgew. (1846).
    Crassula campestris sensu Schönl. in S. Afr. Journ. Sci. **17**, 2: 188 (1921) non (Eckl. & Zeyh.) Endl. ex Walp. (1843).
    Crassula pharnaceoides subsp. *rhodesica* Merxm. in Mitt. Bot. Staatss. München, **1**, 3: 82 (1951). — Cufod. in Bull. Jard. Bot. Brux. **24**. Suppl.: 171 (1954). Type as above.
    Crassula pharnaceoides sensu Friedr. in Prodr. Fl. SW. Afr. **52**: 12 et 33 (1968) non Fisch. & Mey. (1841).
    Crassula campestris subsp. *pharnaceoides* (Fisch. & Mey.) Tölken in Contr. Bolus Herb. **8**, 1: 130–131 (1977) quoad syn. pro parte et specim. Namibiae.
    Crassula campestris subsp. *rhodesica* (Merxm.) R. Fernandes in Bol. Soc. Brot., Sér. 2, **52**: 168 (1978).

Annual, everywhere glabrous herb with slender root. Stems up to 22·5 cm. long, ± 1 mm. in diam., soft, distinctly 4-winged (at least when dried), whitish to pale green, the wings hyaline, simple or branched sometimes from the base (the branches ascending or erect, shorter than the main stem), leafy nearly from the base to the apex or leafless below; internodes usually longer than the leaves, the lower and median ones up to 14 mm. long. Leaves 5–8·5 mm. long, linear or linear-lanceolate, usually with nearly the same breadth from base to apex, the cauline and rameal ones blunt or truncate at the apex which is provided with ± numerous short papillae, entire, sessile and connate at the base into a sheath up to 0·5 mm. long, not spurred below their insertion with the stem; axillary-fascicles distinctly shorter than the subtending leaf. Flowers (4)5-merous in 2–7-flowered axillary fascicles nearly along all the stem and branches or sometimes flowers solitary; pedicels up to 3·5 mm. long, very unequal in length in each fascicle, filamentose. Calyx 1–2 mm. long, ± longer than the corolla, pale green or whitish; sepals oblong-lanceolate, attenuate-subulate to the apex which is acute or nearly so, very thin, stellate. Corolla 0·6–1 mm. long, white or pale pink, thinly membranous, hyaline to brownish on drying; petals ovate-lanceolate or ovate, folded above the middle and thus attenuate and cuspidate, nearly free, stellate. Filaments ± attaining the middle of the corolla; anthers c. 0·15 mm., subcircular, yellow. Follicles ± 0·75 mm., obliquely obovoid, contracted into the very short styles, pale green, nearly translucent, 2-seeded. Seeds c. 0·5 × 0·22 mm., oblong, obtuse at base and apex, shallowly longitudinally ridged, the ridges straight, not shining. Scales minute, narrowly spathulate.

Zambia. S: 8 km. S. of Kalomo, iii.1969, *Williamson* 1534 (SRGH). **Zimbabwe.** N: Gokwe, Sengwa Research Station, along the Sengwa road, 5.iv.1969, *N. Jacobsen* 541 (K; SRGH). W: Farm Besna Kobila, c. 1536 m., *Miller* 4193 (K; LISC; SRGH). C: Marandellas, Cave, 6.iv.1950, *Wild* 3317 (SRGH). S: Zimbabwe, 17.iv.1946,*Greatrex* (K; SRGH).
Also in Kenya, Tanzania, Angola and Namibia. On wet soil or rocks in Mopane woodland in shade.

3. **Crassula schimperi** Fisch. & Mey., Ind. Sem. Hort. Petropol. **8**: 56 (1841). — Walp., Repert. **2**: 254 (1843). — Cufod. in Bull. Jard. Bot. Brux. 24, Suppl: 171 (1954). Wickens

in Kew Bull. **36**, 4: 666 (1982). TAB. **2**. Type a cultivated plant at Hort. Bot. Petropol., from seeds collected in Ethiopia (K, isotype).
    Combesia schimperi (Fisch. & Mey.) Schweinf., Beitr. Fl. Aethiop.: 80 (1867).

Moss- or lycopodium-like perennial herbs or sometimes dwarf undershrubs, totally glabrous or sometimes papillose on leaves, calyces and extremities of stem and branches. Stems ± woody towards the base or herbaceous throughout, 4-gonous above, either slender or somewhat thick throughout or only below, leafy from base to apex or leafless on the older parts. Leaves not to ± succulent, soft to ± rigid, ovate to linear, obtuse or acute, sometimes setose at the apex, sessile, connate at the base into a short sheath, slightly thickened and inflexed at the margin, ± canaliculate above towards the apex, the apical ones ± dense. Flowers 5-merous, small, shortly pedicellate, in ± dense axillary fascicles shorter than to equalling or a little exceeding the subtending leaf, sometimes flowers solitary. Calyx slightly shorter or longer than thè corolla or subequalling it; sepals lanceolate, acute to rather attenuate and nearly subulate, thin to somewhat thick, nearly hyaline to whitish or pale green, shortly connate at the base. Corolla whitish, yellowish or rose, turning reddish-brown on drying; petals oblong or ovate-oblong, connate at base up to 0·5 mm., folded upwards and appearing cuspidate there. Anthers ovate, yellow. Follicles 0·75–1 mm. long, contracted or shortly attenuate into a very short style. Seeds 2 in each follicle. Scales flabellate to spathulate.

Sheath of the leaves usually 0·5 mm. or more long (up to 1·5–2 mm.); leaves up to 8 mm. long, somewhat spurred at the base; calyx (1·25)1·5–2 mm. long, usually distinctly longer than the corolla, with rather attenuate-subulate sepals; scales ± narrowly spathulate
                                                           subsp. *schimperi*
Sheath of the leaves usually less than 0·5 mm. long (often ± 0·25 mm.); leaves not or very slightly spurred at the base; calyx usually slightly longer to slightly shorter than the corolla, rarely distinctly longer than it; scales subflabellate to ± broadly spathulate
                                                           subsp. *transvaalensis*

Subsp. **schimperi**
    Tillaea pentandra Royle & Edgew. in Trans. Linn. Soc. **20**: 50 (1846). — Hook. & Thomson in Journ. Linn. Soc., Bot. **2**: 90 (1858). — Hook. f. in Journ. Linn. Soc., Bot., **7**: 192 (1864). — Britten in F.T.A. **2**: 386 (1871) pro parte. — C. B. Clarke in Hook., Fl. Br. India, **2**: 412 (1878). — Sacleux in Bull. Mus. Hist. Nat., Paris, **14**: 244 (1908). Type from India.
    Disporocarpa pentandra (Royle ex Edgew.). Aschers. in Schweinf., Beitr. Fl. Aethiop., Aufzähl.: 271 (1867) *comb. inval.*
    Crassula pentandra (Royle ex Edgew.) Schönl. in Engl. & Prantl, Nat. Pflanzenfam. **3**, 2a: 37 (1890). — Engl., Hochgebirgsfl. Trop. Afr.: 230 (1892) pro parte; in Ann. R. Ist. Bot. Rome, **9**: 251 (1902). — Penzig in Estr. Atti. Congr. Intern. Bot. Genova: 32 (1892). — Hutch. & Dalz., F.W.T.A. **1**, 1: 103 (1927). — Bally in Journ. E. Afr. Uganda Nat. Hist. Soc. **15**: 12 (1940). — Toussaint in F.C. **2**: 571 (1951). — Suesseng. & Merxm. in Mitt. Bot. Staatss. München, **1**, 3: 82 (1951). — Keay in F.W.T.A. ed. 2, **1**: 116 (1954). Type as for *Tillaea pentandra*.
    Crassula schimperi var. *schimperi* — Tölken in Journ. S. Afr. Bot. **41**, 2: 117 (1975); in Contr. Bolus Herb. **8**, 1: 133 (1977) quoad basion. et syn. pro parte.

Differs from subsp. *transvaalensis* by the longer leaf-sheaths, usually not less than 0·5 mm. but up to 1·5(2) mm. in some Indian plants; by the spurred underside of the leaf-base which is distinctly produced below the insertion on drying; by the calyx distinctly longer than the corolla (sometimes nearly twice as long); by the smaller anthers; and by the narrower nectary-scales which are usually narrowly spathulate.

Zambia. N: Mbala Distr., Itembwe Gap and Gorge, 1500 m., 7.i.1968, *Richards* 22862 (K).
    Also in Asia, Ethiopia, Kenya, Tanzania and perhaps elsewhere. In rock crevices and on ledges.

Subsp. **transvaalensis** (Kuntze) R. Fernandes in Bol. Soc. Brot., Sér. 2, **52**: 172 (1978); op. cit. **55**: 111 (1982). Type from S. Africa (Transvaal).
    Sedum transvaalense Kuntze, Rev. Gen. Pl. **3**, 2: 85 (1898). Type as above.
    Crassula transvaalensis (Kuntze) K. Schum. in Just's Jahresb. **26**: 347 (1900). — Schönl. in Ann. Bolus Herb. **2**: 66, t. 5 fig. 19 (1917); in S. Afr. Journ. Sci. **17**, 2: 188 (1921); in Trans. Roy. Soc. S. Afr. **17**, 3: 188 (1929). — Burtt Davy, F.P.F.T. **1**: 141 (1926). — Berger in Engl. & Prantl, Nat. Pflanzenfam. ed. 2, **18a**: 390 (1930). — Wilman, Check List Fl. Pl. Ferns Griqualand W.: 78 (1946). — Hutch., Botanist in S. Afr.: 286, 400 (1946). — Suesseng & Merxm. in Mitt Bot Staatss. München, **1**, 3: 83 (1951). — Friedr. in Prodr. Fl. SW. Afr. **52**: 36 (1968). — Jacobs., Das Sukk. Lexikon: 149 (1970).

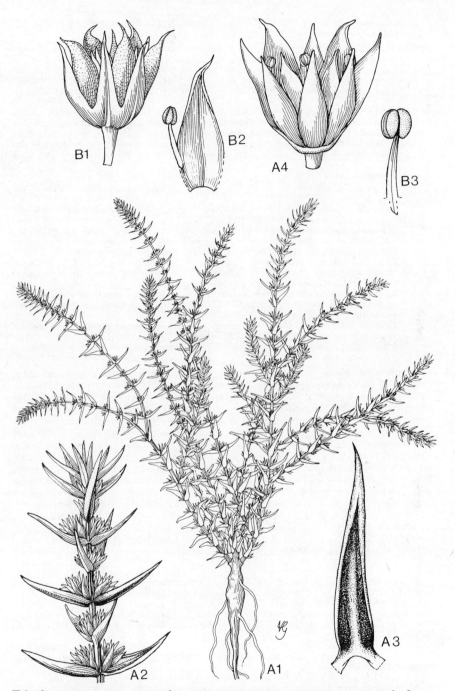

Tab. 2. CRASSULA SCHIMPERI subsp. TRANSVAALENSIS var. TRANSVAALENSIS. A, forma
TRANSVAALENSIS: A1, habit (× 1); A2, upper part of a flowering branch (× 4); A3, leaf
(× 12); A4, flower without frontal sepals and petals to show the follicles (× 18), from *Lemos
& Balsinhas* 250. B, forma ABBREVIATA. B1, flower (× 18); B2, petal with stamen (× 24);
B2, stamen (× 36), from *Müller* 2250.

— Guillarmod, Fl. Lesotho: 183 (1971). — Merxm. & al. in Ann. Naturh. Mus. Wien, **75**: 112 (1971). — J. H. Ross, Fl. Natal: 180 (1972). Type as above.

1. Stems usually erect or, if prostrate, then ± woody (thick and succulent when fresh), usually
   not rooting; leaves ± rigid, ± attenuate, often acute and ± setose at the apex   -   2
 - Stems usually ± prostrate and rooting, not very thick or very slender; leaves (not more than
   5(6) mm. long) soft or not very rigid, slightly attenuate, usually obtuse and not setose at
   the apex, often papillose; calyx 1–1·5 mm. long, subequalling to longer than the corolla
                                                                         var. *illecebroides*
2. Calyx 0·75–1·5 mm. long, shorter than to ± subequalling the corolla; stems erect, usually
   several often from a carrot-like, tuberous root, simple or branched at the base; internodes
   ± visible at least along the floriferous parts   -   -   -   var. *transvaalensis*
 - Calyx up to 1·8 mm. long, longer than the corolla; stems erect or sometimes prostrate, with a
   woody rootstock, rather branched; internodes usually not visible -   - var. *denticulata*

Var. **transvaalensis**. TAB. 2, fig. A.

    Thysantha subulata Hook., Ic. Pl. **6**: t. 590 (1843). Type from S. Africa (Orange Free State).

    Crassula subulata (Hook.) Harv. in Harv. & Sond., F.C. **2**: 352 (1862). — Engl., Hochgebirgsfl. Trop. Afr.: 230 (1892) non L. (1771). Type as above.

    Tillaea subulata (Hook.) Britten in F.T.A. **2**: 387 (1871). — Hiern, Cat. Afr. Pl. Welw. **1**: 325 (1896). — M. Wood, Handb. Fl. Natal: 46 (1907). Type as above.

    Crassula selago Dinter in Fedde, Repert. **16**: 243 (1919). Type from Namibia.

    Crassula schimperi var. *schimperi*. — Tölken in Journ. S. Afr. Bot. **41**, 2: 117 (1975) pro parte quoad syn.; in Contr. Bolus Herb. **8**, 1: 132–136 (1977) quoad syn. pro parte et specim. Afr. Austr., non Fisch. & Mey (1841).

    Crassula schimperi subsp. *transvaalensis* var. *transvaalensis* forma *transvaalensis* — R. Fernandes in Bol. Soc. Brot., Sér. 2, **52**: 173 (1978). Type as for *Crassula schimperi* subsp. *transvaalensis*.

    Crassula schimperi subsp. *transvaalensis* var. *transvaalensis* forma *abbreviata* R. Fernandes, tom. cit.: 174. Type from Zimbabwe.

    Most of the citations given above under *C. transvaalensis* refer to var. *transvaalensis*.

A perennial several- to many-stemmed herb frequently forming ± dense tufts, from a ± long descending tap-root. Stems 2–14(22) cm. long, slightly woody below, herbaceous or somewhat succulent upwards, slender, erect or diverging, usually straight, simple or branched usually near the base (the branches also erect or suberect, ± long), 4-angled, usually smooth or rarely shortly papillose, ± leafy all along, not rooting at lower nodes; internodes shorter than or sometimes the median ones subequalling the leaves and visible. Leaves decussate, (3)6–12·5(15) × (1)1·25–2(2·5) mm., linear-lanceolate or the lowermost ovate-lanceolate, ± attenuate and ± setaceous at the apex, usually not spurred at the base, pale green, whitish-green or glaucescent, slightly fleshy when fresh, ± rigid when dried, straight, erect and imbricate in the apical region and along the non-floriferous zones (densely so on sterile shoots), usually spreading or sometimes reflexed along the floriferous ones; sheath usually ± 0·25 mm. high. Pedicels 0·5–0·75 mm. long. Calyx 0·75–1·5 mm. long, usually slightly shorter than the corolla; sepals ± rigid, acute, pale green or whitish. Corolla 1·25–1·75 mm. long. Anthers c. 0·3 mm., oblong-ovate. Scales 0·3–0·4 × 0·25 mm., subflabellate-cuneate.

Zimbabwe. W: Victoria, Kyle National Park Game Reserve, 23.v.1971, *Grosvenor* 636 (SRGH). E: Inyanga, 21.ii.1946, *Wild* 825 (SRGH). S: N of Gutu on the Devule R., c. 1440 m., v.1966, *Goldsmith* 13/66 (SRGH). **Mozambique**. GI: Gaza, Limpopo, at 3 km. from Combomune to Mapuipanse, 28.viii.1969, *Correia & Marques* 1304 (LMU). M: Bela Vista, on road from Régulo Santaca to Catuane, 7.xii.1961, *Lemos & Balsinhas* 250 (BM; COI; K; LISC; LMA; PRE; SRGH).

Also in Angola, Namibia, Swaziland, Lesotho and S. Africa (Transvaal, Natal, Orange Free State and Cape Province). In crevices of rocks, on shallow soil on rocks or in sandy soils in grassland or savannas.

Many plants from Zimbabwe (Marandellas, Inyanga, Umtali, Bulawayo, Matopos) have shorter stems (2–5·5(10) cm. high), shorter leaves (up to 4(6) mm. long) and slightly smaller flowers than typical specimens of *C. transvaalensis* where the stems are usually 6–23 cm. high and the leaves 3–12·5(15) mm. long. Besides this, in some Zimbabwean plants the leaves as well as the sepals are papillose-scabrous, less acute and more shortly setose at the apex than usual. We had considered them as a form (forma *abbreviata* R. Fernandes, loc. cit.) but perhaps the above characters are conditioned by poor, drier soils and more sunny situations.

Var. **denticulata** (Brenan) R. Fernandes in Bol. Soc. Brot., Sér. 2, **52**: 175 (1978). Type: Malawi, Mulanje Mt., Litchenia Plateau, *Brass* 16784 (K, holotype; BR).
  Crassula pentandra var. *denticulata* Brenan in Mem. N.Y. Bot. Gard. **8**, 5: 434 (1954), excl. specim. Tibet. — Binns, H.C.L.M.: 41 (1968). Type as above.

A dwarf undershrub sometimes like a miniature juniper. Stems up to 20 cm. long, erect or prostrate, not rooting, rather thick and woody towards the base, usually very much branched. Leaves up to 9 mm. long, similar to those of var. *transvaalensis* but denser and all erect or those of floriferous regions not so much spreading, reddish-brown on drying, papillose-denticulate or smooth at the margins as well as the sepals; internodes short, concealed by the leaves; axillary fascicles rather densely leafy, usually equalling or a little exceeding the subtending leaf. Calyx 1·5–1·8 mm. long, slightly to distinctly longer than the corolla, with rather attenuate sepals which are completely reddish-brown or with whitish margins and reddish centre (on drying).

Malawi. N: Nyika Plateau, 2400 m., 14.iii.1961, *Robinson* 4521 (K). S: Mulanje Mt., base of W. Peak, 2010 m., 11.vi.1962, *Richards* 16633 (K).
Known only from Malawi: on shallow soil on rocks.

Var. **illecebroides** (Welw. ex Hiern) Rowley in Cactus et Succ. Journ. Gt. Brit. **40**, 2: 53 (1978). — R. Fernandes in Bol. Soc. Brot., Sér. 2, **55**: 111 (1982). Type from Angola.
  Tetraphyle muscosa (L.) Eckl. & Zeyh., Enum. Pl. Afr. Austr. **3**: 294 (1837) pro parte, non *Crassula muscosa* L. (1760).
  Crassula muscosa sensu Harv. in Harv. & Sond., F.C. **2**: 351 (1862) non L. (1760).
  Tillaea subulata var. *illecebroides* Welw. ex Hiern, Cat. Afr. Pl. Welw. **1**: 325 (1896). Type as for *Crassula schimperi* subsp. *transvaalensis* var. *illecebroides*.
  Sedum muscosum (L.) Kuntze, Rev. Gen. Pl. **3**, 2: 85 (1898) pro parte.
  Crassula filamentosa Schönl. in Ann. Bolus Herb. **2**: 63, fig. text. 13 et t. 5 fig. 16 (1917); in Trans. Roy. Soc. S. Afr. **17**, 3: 188 (1929). — Burtt Davy, F.P.F.T. **1**: 141 (1926). — Berger in Engl. & Prantl, Nat. Pflanzenfam. ed. 2, **18a**: 389 (1930). — ? Friedr. in Prodr. Fl. SW. Afr. **52**: 28 (1968). — Guillarmod, Fl. Lesotho: 181 (1971). — Schijff & Schoonr. in Bothalia **10**, 3: 489 (1971). — J. H. Ross, Fl. Natal: 179 (1972). Type from S. Afr. (Cape Prov.).
  Crassula campestris forma *laxa* Schönl., tom. cit.: 65 (1917) quoad *Schlechter* 6911.
  Crassula parvula sensu Schönl., tom. cit.: 66 pro parte et t. 5 fig. 18 (1917). — sensu Adamson in Adamson & Salter, Fl. Cape Penins.: 432 (1950). — sensu Guillarmod, Fl. Lesotho: 182 (1971) non (Eckl. & Zeyh.) Endl. ex Walp. (1843).
  Crassula schimperi sensu Suesseng. & Merxm. in Mitt. Bot. Staatss. München, **1**, 3: 82 (1951) non Fisch. & Mey (1841).
  Crassula pentandra sensu Brenan in Mem. N.Y. Bot. Gard. **8**, 5: 433 (1954); — Binns H.C.L.M.: 41 (1968) non. (Royle ex Edgew.) Schönl. (1890).
  Crassula pharnaceoides sensu Binns, loc. cit. (1968) non Fisch. & Mey (1841).
  Crassula schimperi var. *lanceolata* (Eckl. & Zeyh.) Tölken in Journ. S. Afr. Bot. **41**: 117 (1975); in Contr. Bolus Herb. **8**, 1: 136–139 (1977) pro parte, *comb. illegit.*
  Crassula schimperi subsp. *transvaalensis* var. *illecebroides* (Welw. ex Hiern) R. Fernandes in Bol. Soc. Brot., Sér. 2, **52**: 175 (1978). Type as for *Crassula schimperi* subsp. *transvaalensis* var. *illecebroides*.
  Crassula schimperi subsp. *transvaalensis* var. *illecebroides* forma *illecebroides*. — R. Fernandes, loc. cit. Type as above.
  Crassula schimperi subsp. *transvaalensis* var. *illecebroides* forma *filamentosa* (Schönl.) R. Fernandes, tom cit.: 177 (1978). Type as for *Crassula filamentosa*.

A perennial (or also annual?) herb with prostrate or ascending or sometimes suberect stems, normally rooting at the nodes sometimes throughout a great extent of the main stem and branches. Stems slender, often very thin, flexuous, sometimes somewhat woody below. Leaves 2–5(6) mm. long, ovate to lanceolate, arched at the base and ascending or spreading, slightly connate at the base (sheath 0·25 mm. or less high), usually obtuse and not setose at the apex, often shortly papillose, soft or slightly rigid on drying. Flowers usually 1–1·5 mm. long with the calyx slightly longer than the corolla or subequalling it; axillary fascicles half as long as to subequalling the subtending leaf. Scales broad, rounded at the apex, shortly stipitate.

Zimbabwe. E: Mutare, Engwa, 1830 m., 3.ii.1955, *E. M. & W.* 156 (BM; LISC; SRGH). S: Zimbabwe, viii.1927, *King in Eyles* 5037 (K; SRGH). **Malawi.** C: Dedza Mt., 2140–2230 m., 20.iii.1955, *E. M. & W.* 1090 (BM; LISC; SRGH). S: Zomba Distr., Zomba Plateau, 1430 m., 30.v.1946, *Brass* 16092 (K; SRGH). **Mozambique.** T: Macanga, Furancungo Mt., 1380–1420 m., 15.iii.1966. *Pereira, Sarmento & Marques* 1743 (LMU). GI: Gaza, Chibuto pr. Chaimite, 22.v.1948, *Torre* 7907 (LISC; LMU).

Also in Angola, Namibia and S. Africa. On rock crevices at high situations (up to 2130 m.), sometimes in savannas.

Most specimens of *C. schimperi* subsp. *transvaalensis* var. *illecebroides* from the F.Z. area belong to forma *illecebroides*. This has relatively thick stems and long leaves (5(6) mm.) compared with forma *filamentosa*. The latter, common in S. Africa, has very slender, sometimes almost filiform stems, and leaves only up to 3(4) mm. long. Plants with short stems resemble *C. campestris* subsp. *campestris* in habit. This, however, is an annual with neither prostrate nor rooting stems; it has longer pedicels, a relatively longer calyx with more attenuate sepals, smaller anthers and narrower nectary scales. Robust specimens of forma *illecebroides*, e.g. *Nobbs in Eyles* 1318 from Zimbabwe, Makoni (BR; K), are similar to weaker, prostrate specimens of var. *denticulata* (Malawi).

The name *Crassula lanceolata* (Eckl. & Zeyh.) Endl. ex Walp. (based on *Tetraphyle lanceolata* Eckl. & Zeyh., 1837) was placed in synonymy under *C. campestris* by Harvey in F.C. **2**: 351, 1862. Later Schönland (in Ann. Bolus Herb. **2**: 65, 1917) made *C. lanceolata* a form, forma *laxa*, of *C. campestris*. Recently, however, Tölken (loc. cit., 1975 & 1977) has considered it to be a variety of *C. schimperi*, identical with *C. filamentosa* Schönl. This treatment is nomenclaturally incorrect: at species level, the name *Tetraphyle lanceolata* has priority over *C. schimperi*; and even were this not so, at varietal rank var. *illecebroides* (Welw. ex Hiern) must be employed in preference to var. *lanceolata*.

The holotype of *C. lanceolata* is *Ecklon & Zeyher* 1874 (S). This plant differs from *C. filamentosa* in many respects; the distinction may be expressed by the following key:

Stems relatively short (up to 8 cm.); internodes shortening from stem base to apex (from 11 mm. to 1·25–1–0·25 mm. in one branch); leaves shortening from 7 to 3 mm. on the same branch; leaf apex terminating in a long hyaline seta; pedicels relatively long (up to 3·75 mm.) and slender; flowers stellate after anthesis, with spreading calyx and corolla; calyx 1·5–1·75 mm. long, exceeding corolla (c. 1 mm. long); sepals very attenuate, hyaline-setose; petals relatively attenuate and acute; anthers smaller than below; seeds elliptic, c. 0·4 × 0·25 mm.; nectary scales relatively long and narrow   -   -   -   -   -   *C. lanceolata*
Stems longer than above; internodes of a branch ± equal in length; leaves of a branch ± equal in length; leaf apex usually blunt, often papillose; pedicels shorter and thicker; flowers not stellate at anthesis with erect or suberect calyx and corolla; calyx equalling or only a little longer than corolla; sepals acute to slightly attenuate; petals less attenuate and acute than above; seeds oblong, relatively longer and narrower than above   -   -   *C. filamentosa*

Although the basal part of the type specimen of *C. lanceolata* cannot be studied in detail without causing damage some nodes at least have adventitious roots. However, the stem does not seem to be prostrate and the plant is apparently annual.

In its almost erect habit, leaf shape, inflorescence type, length and slenderness of pedicel, and size and shape of anthers, follicles, seeds and nectary scales, the type of *C. lanceolate* is identical with *C. campestris* subsp. *campestris*, differing vegetatively only in its few rooting nodes, longer internodes and longer more attenuate leaves. On the other hand, it presents one character state never observed in either *C. campestris* or *C. filamentosa*: the follicles are covered with minute, distinct tubercles on the part covering the lower seed. Similar tubercles were also found in *C. thunbergiana* (where plants with smooth follicles also exist), a S. African glabrous species with 2-seeded follicles in the same section (Sect. *Glomeratae* Harv.) as *C. campestris*. *C. thunbergiana* is an annual with erect stems like *C. campestris*, but with the stems sometimes rooting at the nodes. The leaves and flowers of *C. thunbergiana* are very different from those of *C. lanceolata*.

Three alternative hypotheses may be considered to interpret *C. lanceolata* as follows:

a) *C. lanceolata* is a form of *C. campestris*, its rooting nodes and its unusually long, soft stems and leaves having been determined by a shaded, moist environment — "*in nemoribus*" fide Ecklon & Zeyher.

b) *C. lanceolata* is a variety or a new subspecies of *C. campestris*, distinguished by its stem, leaf and carpel characters.

c) *C. lanceolata* is a hybrid between *C. campestria* and another parent, possibly *C. thunbergiana*. In the latter case nearly all the characters of *C. campestris* would be dominant.

In conclusion, therefore, we have no hesitation in considering that *C. lanceolata* and *C. schimperi* do not belong to the same species but more material of *C. lanceolata* from the locus classicus (K'aka-kamma, Uitenhage) is necessary to decide its exact relationships.

4. **Crassula expansa** Dryand. in Aiton, Hort. Kew. **1**: 390 (1789); op. cit., ed. 2, **2**: 196 (1811). — Willd. in L., Sp. Pl. ed. 4, **1**, 2: 1561 (1798). — Pers., Synops. **2**: 340 (1805). — Haw., Synops. Pl. Succ.: 59 (1812). — DC., Prodr. **3**: 389 (1828). — Harv. in Harv. & Sond., F.C. **2**: 354 (1862). — Britten & Bak. f. in Journ. Bot., Lond. **35**: 483 (1897); in Ann. S. Afr. Mus. **9**: 52 (1912). — M. Wood, Handb. Fl. Natal: 46 (1907). — Schönl. in Ann. Bolus Herb. **2**: 55, fig. text. 6, t. 3 fig. 8 (1917); in Trans. Roy. Soc. S. Afr. **17**, 3: 184 (1929). — Burtt Davy, F.P.F.T. **1**: 142 (1926). — Berger in Engl. & Prantl, Nat. Pflanzenfam. ed. 2, **18a**: 388 (1930). — Hutch., Botanist in S. Afr.: 312 (1946). — J. H. Ross, Fl. Natal: 179 (1972). — R. Fernandes in Bol. Soc. Brot., Sér. 2, **55**: 105 (1982). Type from S. Africa (Cape Prov.).

Sedum expansum (Dryand.) Kuntze, Rev. Gen. Pl. **3**, 2: 84 (1898). Type as above.
Crassula expansa subsp. *expansa* — Tölken in Journ. S. Afr. Bot. **41**, 2: 104 (1975); in
Contr. Bolus Herb. **8**, 1: 163 (1977). Type as above.

Perennial or (sometimes) annual, somewhat branched, completely glabrous herb,
sometimes rooting below. Stems slender, usually herbaceous or somewhat woody
towards the base. Leaves usually not more than 3 mm. broad, linear in outline,
semiterete (somewhat thick and rigid in the dry state) or linear-elliptic, elliptic or
narrowly spathulate and flat (submembranous on drying), attenuate towards the
acute apex, not or somewhat narrowing towards the base and there slightly connate
(sheath up to 0·5 mm. high), succulent, dark green to purplish. Flowers 5-merous,
solitary in the axils along the upper half of the stem and also in corymbose cymes at
the end of the branches and stem; pedicels capillary, at first subequalling the flower,
then elongating up to 40(55) mm. in the fruit. Calyx 1·25–2·5 mm. long, longer than
half to subequalling the corolla; sepals 0·5–0·8 mm. broad, oblong or oblong-
lanceolate, obtuse or sub-obtuse, slightly connate at the base, fleshy, green. Corolla
(1·75)2–3(4) mm. long, white; petals 1–1·25 mm. broad, ovate-oblong, slightly
contracted towards the obtuse apex, erect. Filaments 0·75–1·5 mm. long; anthers
0·25–0·4 mm., subcircular. Follicles 1·5–2·5 mm. (with the styles). Seeds 0·3–0·4
mm. long.
This description applies only to *C. expansa* sensu str.

Leaves up to 13 mm. long; pedicels up to 40(55) mm. long in fruit; plant usually diffuse,
    sometimes rooting at nodes -   -   -   -   -   -   -   - var. *expansa*
Leaves up to 34 mm. long; pedicels up to 20 mm. long in fruit; plant erect, not rooting
                                                                var. *longifolia*

Var. **expansa**
Crassula prostrata Thunb., Prodr. Pl. Cap.: 54 (1794); Fl. Cap., ed. Schultes: 282
(1823). Syntypes from S. Afr. (Cape Prov.).
Crassula parviflora E. Mey. ex Drège, Zwei Pfl. Docum.: 57, 66 & 134 (1843) *nom. nud.*
Crassula albicaulis Harv., op. cit.: 353 (1862). — Schönl., tom. cit.: 61 (1917); tom. cit.:
187 (1929). Type from S. Africa (Cape Prov.).

Zimbabwe: S. Bank of Zambesi, c. 960 m., xii.1903, *Rogers* 13021 (K).
Also in Natal, Transvaal and the Cape Prov. In wet situations amongst short grass, from
coastal regions nearly at sea-level inland up to at least c. 1440 m. alt.

The above specimen is a very poor one and its determination is somewhat doubtful.

Var. **longifolia** R. Fernandes in Bol. Soc. Brot., Sér. 2, **52**: 178, t. 1: (1978). Type: Mozambique, between Mangorro and Panda near the bridge on the Inhassume River, *Barbosa &
Lemos* 8520 (COI, holotype; K; LISC; LMA; PRE; SRGH).

Erect, up to 30 cm. high, sometimes branched from the base, the branches also
erect, seeming annual but somewhat woody near the base, not rooting. Leaves up to
34 × 2–3 mm., more than twice as long as in the type variety, linear, semiterete, rigid
on drying; internodes up to 35 mm. long. Flowers usually in terminal and axillary
somewhat dense cymes; fructiferous pedicels up to 20 mm. long (not so elongated in
fruit as in the type var.). Calyx 1·25–1·75 mm. long. Corolla 2–2·25 mm. long.
Follicles (with styles) 1·5–2 mm. long.

Mozambique. GI: Massinga, 11.ii.1898, *Schlechter* 12131 (BM; BR; K; P).
Known only from Mozambique. In the understorey of *Brachystegia* woodland.
Should perhaps be considered as a subspecies.

5. **Crassula maputensis** R. Fernandes in Bol. Soc. Brot., Sér. 2, **52**: 178, t. 2 (1978). Type:
Mozambique, Maputo, on the road Salamanga-Ponta do Ouro, *A. & R. Fernandes & A.
Pereira* 38 (COI, holotype; LMU).
Crassula expansa sensu Schönl. in Ann. Bolus Herb. **2**: 57 (1917) pro parte quoad
specim. *Schlechter* 11534 non Dryand. (1789).
Crassula sp. *sensu* Mogg in Macnae & Kalk, Nat. Hist. Inhaca Isl.: 145 (1958).

An annual (or also perennial), diffuse, somewhat fleshy, completely glabrous herb.
Stems up to 17(28) cm. long, procumbent, subdichotomously branched, rigid and
sometimes subligneous towards the base, rooting at the lower nodes; branches up to
18 cm. long, prostrate, ascending or erect, slender, herbaceous and rather leafy
towards the extremity; lower internodes up to 3·5 cm. long, the others successively

shorter. Leaves 0·6–2·5 × 0·15–0·7 cm., spathulate or oblong-spathulate to elliptic or oblong-elliptic, obtuse or nearly so at the apex, the narrowest ones nearly acute, entire, narrowing towards a subpetiolar base or rarely contracted below, slightly connate, flat, neither lineolate nor pellucid-punctate, dark green and membranous or nearly translucent on drying. Flowers 5-merous, solitary and axillary usually all along the stem and branches; pedicels capillary, 7–40 mm. long. Calyx 1·25–2 mm. long, up to half as long as corolla; sepals oblong-linear, obtuse, slightly connate at the base. Corolla 3·5–4 mm. long, white; petals 1·5–2 mm. broad, oblong-elliptic or obovate, concave, suberect. Stamens more than half as long as corolla; filaments 2–2·25 mm. long; anthers c. 0·25 mm. long, yellow. Follicles 2–2·5 mm. long, tapering into the c. 1 mm. long styles. Seeds 0·3–0·4 mm., longitudinally lineolate-ribbed.

Mozambique. M: between Boane and Porto Henrique, near Boane, 28.vi.1961, *Balsinhas* 487 (BM; K; LISC; LMA; LMU; PRE); Inhaca Island, Delagoa Bay, 31.viii.1959, *Watmough* 345 (BR; K; LISC; SRGH).

Known only from Mozambique, but may also occur in Natal. In shaded grassland, open forests and thickets, on wet sandy or clayey-sandy soils.

C. maputensis differs from its neighbour *C. expansa* by the broader, flat, obtuse leaves, narrowing distinctly towards the base; by the flowers disposed from below to above along the stem and branches, while in *C. expansa* they are placed only along the upper half; by the thinner pedicels; by the relatively shorter sepals, usually only ½ as long as corolla whereas in *C. expansa* they are equal ½–⅔ as long; by the broader petals which are less connate at the base and, on account of this, somewhat spreading (they are nearly erect in *C. expansa*); and by the follicles being longer and more tapering into the styles.

6. **Crassula fragilis** Bak. in Journ. Linn. Soc., Bot. **22**: 469 (1887). — Jacobs., Das Sukk. Lexikon: 142 (1970); R. Fernandes in Bol. Soc. Brot., Sér. 2, **55**: 106 (1982) non Schönl. (1929). Type from Madagascar.
    Crassula expansa subsp. *fragilis* (Bak.) Tölken in Journ. S. Afr. Bot. **41**, 2: 105 (1975); in Contr. Bolus Herb. **8**, 1: 169 (1977). — G. E. Wickens & Bywater in Kew Bull. **34**, 4: 635, t. 19 fig. A–B (1980). Type as above.

A perennial (or also annual ?), procumbent, delicate, succulent herb. Stems several from a short woody base, up to 25 cm. long, subdichotomously branched, sometimes from below, 4-gonous, slender, somewhat rigid and brownish below, herbaceous and green above, usually puberulous or hispid everywhere (hairs acute, hyaline, spreading, denser and longer above), or sometimes glabrous towards the base, or rarely papillose-puberulous; internodes variable in length, often longer than the leaves. Leaves 0·3–3·5(4·5) cm. long (incl. subpetiolar base) and 0·25–1·3 cm. broad, elliptic, spathulate or obovate to subcircular, acute or nearly so to obtuse or rounded at the apex, entire, cuneate and narrowing into a subpetiolar base, or subsessile, rarely distinctly petiolate, flat, subfleshy when fresh, membranaceous on drying, dark green, sometimes with a purple pattern, usually puberulous or hispidulous on both faces, rarely glabrous, slightly connate at the base. Flowers 5-merous, axillary and solitary, only one at each node, and also in terminal, lax cymes, sometimes in fascicles-like cymes at the end of ± short axillary branchlets; pedicels filiform, hispidulous, elongating in fruit up to 30(35) mm. Calyx 1·5–2·5(2·75) mm. long, green, hispidulous; sepals linear or linear-lanceolate, obtuse, connate at the base for 0·3–0·6 mm. Corolla 2·25–3·5 mm. long, white or pale pink, petals ovate or oblong-obovate, somewhat contracted towards the obtuse apex, smooth, with the median nerve hispid outside. Filaments 1·25–2·25 mm. long. Follicles (with styles) 2·25–3 mm. long; styles 0·3–1 mm. long. Seeds muriculate.

Leaves 0·8–3·5(4·5) × 0·25–1·3 cm., elliptic or spathulate to narrowly obovate, acute to obtuse,
    indistinctly petiolate or sessile; filaments 1·25–1·5 mm. long; follicles (with the styles)
    2·25–2·5 mm. long; styles 0·3–0·5 mm. long - - - - - - var. *fragilis*
Leaves-lamina up to 0·9 × 0·7 cm., broadly obovate to subcircular, rounded at the apex; petiole
    distinct, 1–5 mm. long; filaments ± 2·25 mm.; follicles c. 3 mm. long; styles c. 1 mm. long
                                                                    var. *suborbicularis*

Var. **fragilis.**
    Crassula furcata Schönl. in S. Afr. Journ. Sci. **17**, 2: 188 (1921) *nom. nud.*, non (Eckl. & Zeyh.) Endl. ex Walp. (1843).
    Crassula browniana Burtt Davy, F.P.F.T. **1**: 38, 141 (1926). — Schönl. in Trans. Roy. Soc. S. Afr. **17**, 3: 185 (1929). — Berger in Engl. & Prantl, Nat. Pflanzenfam. ed 2, **18a**: 388 (1930). — Suesseng. & Merxm. in Trans. Rhod. Sci. Assoc. **43**: 15 (1951). — Wild in

Rhod. Agr. Journ. **49**, 5: 286 (1952); Fl. Vict. Falls: 143 (1953). — Jacobs., Das Sukk. Lexikon: 140 (1970). — Schijff & Schoonr. in Bothalia **10**, 3: 489 (1971). — J. H. Ross, Fl. Natal: 179 (1972). Type from Transvaal.

Zambia. S: Victoria Falls, 10.iv.1955, *E. M.* & *W.* 1453 (BM; LISC; SRGH). **Zimbabwe.** N: Lomagundi, Farm Whindale below "The Cliffs", c. 1230 m., 18.ii.1968, *Jacobsen* 3386 (PRE; SRGH). W: Bulalima-Mangwe, about 10 km. N. of Marula on road to Mananda Dam, 20.iv.1972, *Grosvenor* 748 (COI; SRGH). C: Salisbury, Prince Edward Dam, c. 1600 m., 8.i.1952, *Wild* 3741 (K; LISC; SRGH). E: Umtali, cliffs of W. slope of Murakwa's Hill, Commonage, c. 1300 m., *Chase* 8156 (K; LISC; SRGH). S: Chibi, Nyoni Hills, 3.iii.1970, *Mavi* 1061 (K; PRE; SRGH). **Mozambique.** MS: ridge ascending Garuso, iv.1935, *Gilliland* 1958 (BM; K). M: 9 miles N. of Moamba, c. 96 m., 29.ix.1963, *Leach & Bayliss* 11746 (COI; K; LISC; SRGH).

Also in Angola, Swaziland, S. Africa (Natal, Transvaal and the Cape Prov.) and Madagascar. In humus-pockets in rock crevices, on termite mounds, in wet, shady situations in bush, tree savanna, riverine forests, etc.

Similar to *C. expansa* subsp. *expansa* but differing in the following characters: habit usually procumbent, with thicker, longer stems and branches; leaves thinner, always flat, larger and relatively broader (2·5–13 mm. broad as against up to 3 mm. in *C. expansa*); hydathodes scattered over the upper leaf surface (compared with the situation in *C. expansa* where they are in 1–2 marginal rows); pedicels shorter in fruit; calyx, filaments, anthers and follicles usually a little longer; indumentum present on stem, leaves, pedicels and calyx.

Specimens of *C. fragilis* from the African continent have larger anthers and longer styles than the holotype, *Baron* 3348 (K) which is from Madagascar. Thus, the mainland plants, corresponding with *C. browniana* Burtt Davy, have anthers and styles c. 0·5 mm. long, whereas in *Baron* 3348 they are c. 0·25 mm. long. We have seen only this one specimen from Madagascar. Should the differences just noted be constant, the continental plant would perhaps be considered as a distinct variety or form.

C. thorncroftii Burtt Davy and *C. woodii* Schönl. are placed as synonyms of *C. fragilis* by Tölken (loc. cit., 1977). However, from their descriptions, both are completely glabrous (even on the calyx) and have shorter pedicels than the latter. *C. woodii*, in addition, has smaller sepals and petals. Both taxa are placed by Schönland (tom. cit.: 1929) in a different section from *C. fragilis* (as *C. browniana*).

C. zimmermannii Engl. from Tanzania (Usambara) seems from description to be close to *C. fragilis*, but the type is apparently lost.

The specimen *Leach & Bayliss* 11746 has a stronger, not hispidulous but only shortly papillose-puberlous stem, relatively broader leaves and slightly larger flowers than most of the specimens.

Var. **suborbicularis** R. Fernandes in Bol. Soc. Brot., Sér. 2, **52**: 181 (1978). Type: Zimbabwe, Chipinge, Farfell Farm, N. of Gunguinyana Forest Reserve, alt. c. 1120 m., *Goldsmith* 167/62 (K; LISC; PRE, holotype; SRGH).

Differs from var. *fragilis* in having longer, denser indumentum, more rigid stems, smaller and relatively broader leaves more rounded at apex and thicker when dry, with a distinct petiole; and in its less numerous flowers by each plant on shorter pedicels, with longer corolla, filaments, follicles and styles.

Zimbabwe. E: Chipinge, Farfell Farm, N. of Gunguinyana Forest Reserve, c. 1120 m., viii.1962, *Goldsmith* 167/62 (K; LISC; PRE; SRGH).

Known only from Zimbabwe, in massive shale outcrop, in rock cracks on very thin soil, at shade. Only the type specimen seen.

7. **Crassula alticola** R. Fernandes in Bol. Soc. Brot., Sér. 2, **52**: 182 (1978). TAB. **3**. Type: Zimbabwe, Banti Forest, alt. c. 1980 m., *E. M. & W.* 217 (BM; LISC, holotype; SRGH).

A perennial herb. Stem up to 35 cm. long, prostrate or ascending, rooting at the lower nodes, simple or loosely branched above, sometimes with very short axillary branchlets, 4-gonous, glabrous towards the base, papillose-hairy in two opposite rows upwards; nodes usually distant. Leaves (1)1·5–3·7(5·0) cm. long, (with the petioles sometimes missing) and (0·7)1·2–1·8(2·0) cm. broad, elliptic to broadly elliptic or rhombic, shortly tapering below the middle into a very short petiole or petiolar flat base, slightly connate, glabrous except for a tuft of hyaline papillose hairs inserted on the middle of the margin of the sheath, dark green and brown-lineolate on drying, flat, spreading or reflexed. Flowers 5-merous in dense to loose corymb-like cymes at the end of the stem and branches, with the main axis and secondary inflorescence-branches papillose-pilose along two opposite sides, the other branches papillose all around; pedicels filiform, papillose (papillae short, obtuse), those of dichotomies up to 15 mm. long, the others shorter, not or slightly dilatate upwards.

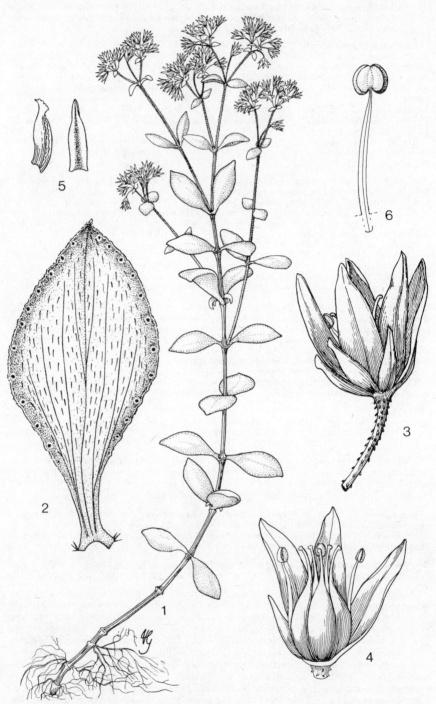

Tab. 3. CRASSULA ALTICOLA. 1, habit (×⅔); 2, leaf (×2); 3, flower (×8); 4, flower without frontal sepals and petals to show the follicles (×8); 5, sepals in ± dorsal and ventral view (×8); 6, stamen (×8), 1, 3–6 from *Phipps* 676, 2 from *Wild* 849.

Sepals 2–3(3·5) × 0·5–0·75 mm., usually less than half as long as corolla or sometimes up to 2/3 as long, oblong or linear, nearly acute or obtuse and slightly thickened-subtrigonous at the apex, slightly connate at the base, glabrous, not or narrowly hyaline-margined. Corolla 4–5·5 mm. long, white; petals 1–1·25 mm. broad, oblong-lanceolate, acute, nearly free. Filaments 3–3·25 mm. long; anthers c. 0·6 mm. long, purple. Follicles (with styles) 3–4 mm. long, usually contracted into the 1·25–1·5 mm. long styles. Seeds c. 0·5 mm. long, oblong-ellipsoid, minutely spinulose-tuberculate.

Zimbabwe. E: Melsetter, Chimanimani Mt., c. 2400 m., fl. 16.iii.1957, *Phipps* 676 (K; PRE; SRGH). **Mozambique.** MS: Manica, Serra Zuira, Tsetsera, on road to Vila Pery, alt. c. 2100 m., fl. 2.iv.1966, *Torre & Correia* 15562 (LISC; LMU; MO).
In damp places, crevices of rocks, stream banks in high mountains.

This species, which is rather common in the mountains of E. Zimbabwe between Inyanga and Melsetter, has been identified in herbaria either as *C. lineolata* Dryand. (most specimens), or as *C. alsinoides* (Hook. f.) Engl. or as *C. brachypetala* Drège ex Harv. Like these three taxa, *C. alticola* belongs to the *C. pellucida* L. complex (sect. *Anacampseroideae* Haw. subsect. *Fasciculares* Tölken); it differs from each of them by an assemblage of characters.
C. lineolata was recently sunk into *C. marginalis* Dryand. by Tölken (in Contr. Bolus Herb. **8**, 1: 187 (1977), a point of view that in our opinion needs further clarification. Furthermore, Tölken envisages the above taxa as having subspecific rank within *L. pellucida* (in Journ. S. Afr. Bot. **41**: 114, 1975). However, we consider that *C. pellucida, C. marginalis, C. brachypetala* and *C. alsinoides* differ from one another by several well-defined characters apart from those differential characters on which Tölken based his treatment, and that these fully justify their traditional recognition as separate species.
C. alticola resembles *C. brachypetala*, its nearest ally in the complex. The two can be separated as follows:

Indumentum composed of relatively short, broad and obtuse papillose hairs; leaves elliptic or rhombic, broadest in the middle, attenuate towards the base; calyx up to half as long as corolla; sepals linear or oblong, scarcely attenuate and ± obtuse at apex, not or indistinctly hyaline-margined; petals usually relatively narrow - - - - - *C. alticola*
Indumentum composed of relatively longer, narrower and acute papillose hairs; leaves cordate-ovate or cordate-lanceolate, broadest near the base, contracted into the petiole; calyx half as long as corolla or more; sepals (this includes var. *parvisepala* Schönl.) lanceolate, very attenuate and acute, sometimes subsetaceous at apex and very distinctly white-hyaline-margined - - - - - - - - - - - *C. brachypetala*

8. **Crassula alsinoides** (Hook. f.) Engl., Hochgebirgsfl. Trop. Afr.: 231 (1892); Pflanzenw. Ost-Afr. **C**: 189 (1895). — R. E. Fr., Wiss. Ergebn. Schwed. Rhod.-Kongo-Exped. **1**: 58 (1914). — Mildbr., Wiss. Ergebn. Zweit. Deutsche Z.-Afr.-Exped. **2**: 183 (1922). — De Wild., Pl. Bequart. **2**: 45 (1923). — Burtt Davy, F.P.F.T. **1**: 141 (1926). — Hutch. & Dalz., F.W.T.A. **1**, 1: 103 (1927). — Schönl. in Trans. Roy. Soc. S. Afr. **27**, 3: 197 (1929). — Gilliland in Journ. S. Afr. Bot. **4**: 94 (1938). — Bally in Journ. E. Afr. Uganda Nat. Hist. Soc. **15**: 12 (1940). — Hutch. & Bruce in Kew Bull. **1941**: 88 (1941). — Robyns, Fl. Parc Nat. Alb. **1**: 230 (1948). — J. Blake, Gard. E. Afr.: 179 (1950). — F. W. Andr., Fl. Pl. Anglo-Egypt. Sudan **1**: 76 (1950). — Toussaint, F.C.B. **2**: 570 (1951). — Cufod. in Bull. Jard. Bot. Brux. **24**, Suppl.: 170 (1954). — Keay in F.W.T.A., ed. 2, **1**, 1: 116 (1954). — Hedberg in Symb. Bot. Ups. **15**, 1: 99 et 278 (1957). — Binns, H.C.L.M.: 41 (1968). — Merxm. & al. in Ann. Naturh. Mus. Wien **75**: 113 (1971). — Greenway & Fitzgerald in Journ. E. Afr. Hist. Nat. Mus. **28**, 130: 5 (1972). — G. E. Wickens & Bywater in Kew Bull. **34**, 4: 636, t. 20 fig. D–E (1980). — R. Fernandes in Bol. Soc. Brot., Sér. 2, **55**: 101 (1982). Type from Fernando-Po.
Tillaea alsinoides Hook. f. in Journ. Linn. Soc., Bot. **7**: 192 (1864). — Schweinf. & Aschers. in Schweinf., Beitr. Fl. Aethiop., Aufzähl.: 271 (1867). — Britten in F.T.A. **2**: 387 (1871). Type as above.
Crassula nummulariifolia Bak. in Journ. Linn. Soc., Bot. **20**: 38 (1883). Type from Madagascar.
Crassula lineolata sensu J. H. Ross, Fl. Natal: 180 (1972), pro parte quoad specim. *Wood* 4566, non Dryand. (1789).
Crassula pellucida subsp. *alsinoides* (Hook. f.) Tölken in Journ. S. Afr. Bot. **41**, 2: 114 (1975); in Contr. Bolus Herb. **8**, 1: 193 (1977). Type as above.

A perennial, creeping herb, forming dense to lax mats. Stems up to 35(46) cm. long, prostrate or ascending, rooting at the lower nodes (roots very slender), simple or often the ± long prostrate part emitting short to ± long erect or ascending branches, these also rooting, mostly herbaceous-succulent (nearly translucent on drying), 4-gonous, glabrous except for two longitudinal lines of papillose hairs along two opposite sides, pale green; lower and median internodes up to 7·5 cm. long, rather longer than the leaves, the others successively shortening upwards (the uppermost shorter than the leaves). Leaves 0·5–2·8 cm. long (with the petiole) and 0·4–1·5 cm. broad, ovate-cuneate, triangular-ovate to elliptic or rhombic, rarely suborbicular, shortly acuminate, acute and sometimes apiculate at the apex, rarely (the suborbicular ones) obtuse, entire, sometimes minutely papillose-serrulate at the margin, cuneate or ± long attenuate into the petiole, glabrous, pale green (paler below), flat, succulent, usually thinly membranous on drying, sometimes longitudinally parallel-lineolate (the lines short, slender, brownish) and with a submarginal row of pellucid small dots; petiole ± 0·4 to ± 1·3 cm. long (delimitation from the lamina usually difficult), glabrous, dilatate below and connate with the opposite one in a sheath, provided with axillary fascicles of minute leaves or with short branchlets; sheath 0·5–1·5 mm. high, papillose-hairy at the margin. Flowers 5(6)-merous, usually solitary in the axils towards the extremities of the stem and branches, one for each node and alternating at the contiguous nodes, sometimes also in contracted terminal few-flowered inflorescences ± concealed by the upper leaves; pedicels 3–21 mm. long (the lowest ones the longest), filiform, ampliate-obconical below the calyx, usually glabrous. Calyx (2·75)4·5–5·5 mm. long, glabrous, green, usually more than half as long as but not exceeding the corolla; sepals (2·5)3·5–5 × 0·75–1(1·5) mm., lanceolate or oblong-lanceolate, attenuate towards the rather acute apex, connate at base for c. 0·5 mm., flat, with whitish-hyaline margins, sometimes brownish-lineolate along the median part. Petals (3)5–6 × 1·25–1·75 mm., oblong-lanceolate, acute, not mucronate, smooth on both faces, nearly free, usually white, rarely tinged with pink towards the apex outside. Filaments ± 3 mm. long; anthers c. 4 mm. long. Follicles (3)4–5·5 mm. long (with the styles), oblong, contracted into the styles, ± equalling the corolla-length, minutely pitted; styles ± 0·5 mm. to ± 0·75 mm. long. Seeds c. 0·5 mm. long, oblong, brownish, minutely tuberculate.

Zimbabwe. W: Coronation Vlei, xii.1940, *Farrar* 3655 (SRGH). E: Chipinge, Gunguinyana Forest Reserve, c. 1056 m., xi.1962, *Goldsmith* 224/62 (EA; K; LISC; PRE; SRGH). **Malawi.** N: Mugesse, Misuku, c. 1600 m., 12.xi.1952, *Williamson* 69 (BM). S: Zomba Plateau, edge of Mlunguzi Dam, 15.v.1967, *Salubeni* 808 (K; LISC; PRE; SRGH). **Mozambique.** MS: Gorongosa Mt. near Morombosi R. falls, 23.x.1945, *Pedro* 436 (K; LMA; PRE).

Also in Fernando Po, Cameroon, Sudan, Ethiopia, Somalia, Uganda, Kenya, Tanzania, Zaire, Swaziland, S. Africa (Natal, Transvaal and the Cape). In wet or moist forests of mountains up to 3000 m. In the F.Z. area *C. alsinoides* has glabrous pedicels, in contrast with the type (from Fernando Po) in which they are papillose. However, glabrous pedicels have also been found in specimens from Cameroon.

9. **Crassula sarmentosa** Harv. in Harv. & Sond., F.C. **2**: 348 (1862). — M. Wood, Handb. Fl. Natal: 46 (1907). — Schönl. in Engl. & Prantl, Nat. Pflanzenfam. **3**, 2 a: 36 (1890); in Trans. Roy. Soc. S. Afr. **17**, 3: 199 (1929). — Berger in Engl. & Prantl, op. cit. ed. 2, **18a**: 390 (1930). — Uhl in Amer. Journ. Bot. **35**: 698 (1948). — Suesseng. & Merxm. in Trans. Rhodes. Sci. Assoc. **43**: 15 (1951). — W. Wright, Wild Fl. S. Afr. (Natal): 55, fig. pag. 56 (1963). — Jacobsen, Das Sukk. Lexikon: 148, t. 48 fig. 1 (1970). — Merxm. & al. in Ann. Naturh. Mus. Wien **75**: 113 (1971). — J. H. Ross, Fl. Natal: 180 (1972). — Tölken in Contr. Bolus Herb. **8**, 1: 198 (1977). — R. Fernandes in Bol. Soc. Brot., Sér. 2, **55**: 111 (1982). Type from Natal.

A sarmentose everywhere glabrous perennial. Stem up to 1 m. or more long, trailing or climbing, simple or loosely branched, terete, reddish or with elongate whitish spots, leafless towards the base, sparsely leafy upwards; internodes up to 10 cm. long, longer to subequalling the leaves. Leaves 2–4·5(6) × 1·2–2·5(3·5) cm., ovate to elliptic, acute to acuminate at the apex, serrate at the margin with the teeth narrowly white-cartilaginous, the lower-ones petiolate, the uppermost subsessile and with axillary fascicles of minute leaves, flat, slightly fleshy, shining, green, sometimes tinged with red mainly the teeth, spreading; petiole usually not longer than 4 mm. somewhat thick, slightly connate at the base with the opposite one. Flowers 5-merous in terminal lax to ± dense paniculate inflorescences; peduncle terete, red; bracts lanceolate-linear, acute; pedicels 4–6 mm. long. Floral buds fusiform. Calyx 1·75–2·5 mm. high, equalling ⅓–¼ the length of the corolla; sepals c. 0·5 mm. wide at the base, lanceolate-linear, acute or obtusiuscule, entire, narrowly hyaline-margined, convex on the back, red or green. Corolla (4) 6–8 mm. long white or ± tinged with red; petals c. 1·5 mm. wide, lanceolate, ± attenuate towards the ±

acute apex, smooth, not dorsally mucronate below the apex, spreading, connate for c. 0·5 mm. at the base. Stamens with filaments 1·25–4·5 mm. long and suborbicular or lanceolate anthers 0·5–0·75 mm. long, at level of ⅓–½ of the corolla. Follicles c. 3 mm. long, very attenuate into the styles; styles 1·5–2·5 mm. long; stigmas at level of ¾ of corolla length.

Zimbabwe. C: Enterprise, Christian's rockery, vi.1932, *Eyles* 7011 (K; SRGH). E: Chipinga, Chirinda Forest, 31.iii.1950, *Hack* 146/50 (SRGH).
Also in S. Africa (Natal and the Cape). Possibly not native in the F.Z. area.

The two rather poor Zimbabwean specimens mentioned above differ from Natal plants by the slightly smaller corolla with more attenuate petals and the shorter filaments (1·25–2 mm. as against 3–4·5 mm.). In the Zimbabwean plants the anthers are lanceolate and c. 0·75 mm. long, in all flowers of the same inflorescence. Within gatherings from Natal we have found three different situations: a) plants in which the anthers of all flowers are identical to the Zimbabwean ones; b) plants in which all the anthers are smaller (c. 0·5 mm.) and subcircular, reaching a higher position in the flower; and c) plants where, in the same inflorescence, there is a mixture of long- and short-anthered flowers.

It is possible that some external cause (e.g. parasitism by insects, fungus, etc.) may have modified the size and shape of anthers and influenced filament elongation, but this cannot be ascertained from the available material. More observations are needed to understand the factors affecting stamen development. The few plants studied are also insufficient to allow any judgement about the value of differences in size and shape of petals.

For the present, therefore, we place the Zimbabwean plants with typical *C. sarmentosa.* var. *integrifolia* Tölken, with somewhat longer petioles and entire leaf margin which is known only from Natal.

10. **Crassula ovata** (Mill.) Druce in Rep. Bot. Exch. Cl. Brit. Isles **1916**: 617 (1917). — Tölken in Contr. Bolus Herb. **8**, 1: 211 (1977). — R. Fernandes in Bol. Soc. Brot., Sér. 2, **55**, 109 (1982). Described from a cultivated plant in Chelsea Physick Garden.
    Cotyledon ovata Mill., Gard. Dict. ed. **8**: no. 8 (1768), non Haw. (1812). Type as above. Neotype: *Tölken* 1772 (BOL), from S. Africa (Cape Prov.).
    Crassula argentea Thunb. in Nova Acta Acad. Caes. Leop.-Carol. **6**: 329 et 337 (1778); Prodr. Pl. Cap.: 56 (1794); Fl. Cap., ed. Schultes: 289 (1823). — L. f., Suppl. Sp. Pl.: 188 (1781). — Murray, Syst. Veg. ed. 14: 306 (1784). — Willd. in L., Sp. Pl. ed. 4, **1**, 2: 1549 (1798). — DC., Prodr. **3**: 383 (1828). — Schönl. in Ark. för Bot. **21A**, 16: 4 (1928); in Trans. Roy. Soc. S. Afr. **17**, 3: 201 (1929). — Berger in Engl. & Prantl, Nat. Pflanzenfam. ed. 2, **18a**: 391 (1930). — K. Gram in Kongel. Veter. -og Landb. **1941**: 40, fig. 1b, 2b, 3b, 4b (1941). — Uhl in Amer. Journ. Bot. **35**: 698 (1948). — Higgins in Journ. S. Afr. Bot. **24**: 110 (1958). — Merxm. & al. in Ann. Naturh. Mus. Wien **75**: 113 (1971). Type from S. Africa (Cape Prov.).
    Crassula portulacea Lam., Encycl. Méth. Bot. **2**: 172 (1786). — Pers., Synops. **2**: 338 (1805) "portulacacea". — DC., Hist. Pl. Grasses: t. 79 (1801); loc. cit. (1828). — Eckl. & Zeyh., Enum. Pl. Afr. Austr.: 295 (1837). — Harv. in Harv. & Sond., F.C. **2**: 337 (1862). — Schönl. in Engl. & Prantl, Nat. Pflanzenfam. **3**, 2a: 36 (1890). — K. Schumann & Rümpler, Die Sukk.: 91 (1892). — M. Wood, Handb. Fl. Natal: 46 (1907). — Nash in Addisonia, **3**: t. 109 (1918). — Pole Evans in Fl. Pl. S. Afr. **4**: t. 156 (1924). — Higgins, tom. cit.: 111 (1958). — Gledhill, East. Cape Veld Fl.: 125, t. 28 fig. 5 (1969). — Jacobsen, Das Sukk. Lexikon: 146, t. 46 fig. 2 (1970). — J. H. Ross, Fl. Natal: 180 (1972). Type a cultivated plant in the "Jardin du Roi" (Paris).
    Crassula obliqua Soland. in Ait., Hort. Kew. **1**: 393 (1789). — Willd., tom. cit.: 1553 (1798). — Pers., Synops. **2**: 338 (1805). — Haw., Synops. Pl. Succ.: 52 (1812). — Higgins loc. cit. (1958). — Jacobsen, Das Sukk. Lexikon: 145, t. 45 fig. 5 (1970) *nom. illegit.*
    Crassula articulata Zuccagni in Roem., Collect.: 136 (1806). Type unknown.
    Crassula nitida Schönl. in Rec. Albany Mus. **1**: 54 (1903). Type unknown.

A large, much branched, completely glabrous, very floriferous shrub, 0·6–3 m. or more high. Stem up to 20 cm. in diam. at the base, subterete, succulent, greyish; branches denuded, ± marked with the scars of the fallen leaves. Leaves 2–5 × 1–3 cm., obovate or broadly elliptical to obovate-spathulate, shortly acuminate and acute or subrounded at the top, entire and reddish at the sharply edged margin, attenuate towards a subpetiolar base, subconnate when young, free with age, fleshy, flattened, slightly concave above and slightly convex beneath, with nectary dots in one row on both sides near the margin (but dots also scattered on both surfaces), green and shining when fresh, dark brown and ± covered with a micaceous caducous peeling layer when dry, ± condensed at the extremities of the branchlets, longer than the internodes. Flowers 5-merous, arranged in terminal, ± dense to loose corymbose, pedunculate inflorescences; peduncles 1–4 cm. long; pedicels 4·5–8 mm. long. Calyx

$\pm$ 5 mm. in diam., campanulate to nearly saucer-shaped; sepals 1–1·5 mm. high and $\pm$ 1·5 mm. broad at the base, broadly triangular, acute, connate for $\pm$ 1 mm., separated by a rounded sinus, fleshy, their external surface similar to that of the leaves. Corolla white or faintly rose; petals (5)6·25–7·5(10) mm. long, oblong or lanceolate, keeled, acute, mucronate at the apex, slightly connate at the base, spreading. Stamen-filaments 4–5·5 mm. long; anthers fertile c. 0·5 mm. broad, transverse-oblong, or sterile c. 0·75 mm. long and oblong or lanceolate. Follicles 3–3·25 mm. long, obliquely oblong-ovoid; styles 2–3 mm. long, filiform. Scales broader than long.

Mozambique. M: 14 km. from Manhoca to Maputo R., *Correia & Marques* 2167 (LMU). Also in S. Africa (Natal and Cape Prov.). In open forests, on sandy-clayey grey soil.

The flowers of the specimen cited above have petals relatively broader than in S. African plants (6·25–7·5 × 2·5–3·5 mm. as against (5)7–10 × up to 2 mm.). The bracts are also broader, and the anthers sterile in all flowers examined. However, in the other Mozambican collection which we have seen (Maputoland Expedition no. 132) the anthers are fertile and the petals only 2(2·5) mm. broad. This suggests that increase in petal width may be correlated with anther-sterility.

C. ovata cannot be confused with *C. arborescens* (Mill.) Willd. which has broader petals than the former, because their flowering times are different (June–July for *C. ovata* in Mozambique, compared with November–December for *C. arborescens*), the leaf surface, when dry, has a different appearance (micaceous versus white-farinaceous) and *C. ovata* has relatively longer styles.

11. **Crassula sarcocaulis** Eckl. & Zeyh., Enum. Pl. Afr. Austr.: 295 (1837). Type from S. Africa (Cape Prov.).

Shrubs erect and $\pm$ branched. Old branches woody, leafless, glabrous, those of the current year succulent, shortly papillose or papillose hairy. Leaves sessile, acute at the apex, entire, with a row of marginal dots, slightly connate at the base, usually glabrous, sometimes papillose at the margin, erect or ascending or the lowermost spreading, $\pm$ caducous. Flowers 5-merous in corymb-like dense cymes at the end of the leafy branchlets; peduncles and cyme-branches glabrous or papillose. Calyx glabrous, pale green; sepals oblong or oblong-lanceolate or lanceolate, subacute, subobtuse or obtuse, glabrous. Corolla with ovate-oblong or oblong-ovate, obtuse or acute petals. Stamens slightly shorter than the corolla. Seeds minutely and densely spinulose-tuberculate.

Leaves (4)6–35 × (1)2–5·5(8) mm., narrowly ovate or elliptic to narrowly linear-elliptic, at least
  the upper ones flattish; corolla (4)4·5–6·5(8) mm. long, white or pinkish; anthers
  (0·5)0·75–1 mm. long; follicles 5–7·25 mm. long -   -   -   -   subsp. *sarcocaulis*
Leaves 10–15·5(17) × 1–1·5(2) mm., linear, terete or trigonous at least towards the apex; corolla
  3·25–4 mm. long, white; anthers 0·25–0·4 mm. long; follicles 2·5(3) mm. long
                                                      subsp. *rupicola* var. *mlanjiana*

Subsp. **sarcocaulis** — Tölken in Contr. Bolus Herb. **8**, 1: 252 (1977). — R. Fernandes in Bol. Soc. Brot., Sér. 2, **52**: t. 4 fig. a-d, t. 5 (1978); op. cit. **55**: 110 (1982).
    Crassula sarcocaulis Eckl. & Zeyh., loc. cit., sensu str. — Walp., Repert. **2**: 254 (1843). — Harv. in Harv. & Sond., F.C. **2**: 341 (1862) pro parte. — Schönl. in Trans. Roy. Soc. S. Afr. **17**, 3: 214 (1929) pro parte.
    Crassula parvisepala Schönl. in Journ. Linn. Soc., Bot. **31**: 549 (1897); in Rec. Albany Mus. **1**, 1: 61 (1903); in Trans. Roy. Soc. S. Afr. **17**, 3: 215 (1929) — Burtt Davy, F.P.F.T. **1**: 140 (1926). — Schijff & Schoonr. in Bothalia, **10**, 3: 489 (1971). — Jacobs., Das Sukk. Lexikon: 146 (1970). — Merxm. & al. in Ann. Naturh. Mus. Wien, **75**: 113 (1971). — J. H. Ross, Fl. Natal: 180 (1972). — Type from S. Africa (Transvaal).
    Crassula lignosa Burtt Davy, op. cit.: 38 & 140. Type from S. Africa (Transvaal).

Shrub (15)30–120 cm. tall. Old branches irregularly rugose, dark grey to nearly black, those of the current year dark brown or reddish-brown, 4-angular or narrowly 4-winged on drying, at first $\pm$ densely foliate, at fruiting time nearly leafless; internodes 1·7–3 mm. long, shortening successively towards the apex of branchlets. Leaves (4)6–35 × (1)2–5·5(8) mm., narrowly ovate or elliptic to narrowly linear-elliptic, rarely ovate, slightly asymmetrical, $\pm$ attenuate towards the usually very acute apex, entire or sometimes slightly undulate at margin on drying, with the marginal dots rarely conspicuous, fleshy, flat (at least the upper ones), green or sometimes dark brownish-green to blackish and often rigid rarely membranous in dry state, very caducous; sheath up to 1 mm. high. Inflorescences 2–5·5 cm. in diam.; pedicels 1·75–4 mm. long, glabrous, brown-reddish. Calyx (1)1·5–3·5(4) mm. long, usually shorter than the half of the corolla; sepals oblong-lanceolate or lanceolate, slightly carenate, submucronate near the apex, connate for up to 1 mm. at the base. Corolla (4–4·5)5–6·5(8) mm. long; petals ovate-oblong or -elliptic, somewhat carenate, subacute and submucronate below the apex, slightly reflexed, connate for ± 0·5–0·75 mm. at the base. Stamens slightly shorter than the corolla; filaments 3·75–4·5(6) mm. long; anthers (0·5)0·75–1 mm. long, dark purple. Follicles 5–7·25 mm. long, attenuate into the straight styles. Seeds c. 0·5 mm.

Zimbabwe. W: Matopos, Farm Besna Kobila, c. 1504 m., ii.1954, *Miller* 2205 (K; LISC; SRGH). E.: Inyanga, Ingangwe Fort, on granite ruins, c. 1920 m., 3.ix.1952, *Wild* 3854 (K; LISC; PRE; SRGH). S: Bikita, lower E. slope of granite Whaleback on W. bank of Turgwe R. at confluence with Dafana R., 110 m., 5.v.1969, *Biegel* 3017 (LISC; SRGH). **Mozambique.** MS: Manica, Serra Zuira, Tsetsera, on road to Vila Pery, c. 1800 m., 3.iv.1966, *Torre &* *Correia* 15650 (C; LISC; LMU).
Also in S. Afr. (Transvaal, Natal, Cape Prov.) and Swaziland. On rock clump-vegetation of high mountains up to 2400 m.
C. lignosa Burtt Davy, known only from the type which is a very poor specimen, is not distinct from *C. sarcocaulis* subsp. *sarcocaulis*, corresponding to its broad-leaved forms; the irregular, marginal crenations of the leaves referred to by Burtt Davy may be the result of an inapt desiccation. Affinities of *C. lignosa* with *C. lactea* Soland. as indicated by Burtt Davy or with *C. rubicunda* E. Mey. as suggested by Schönland seem to us inadmissible.

Subsp. **rupicola** Tölken in Journ. S. Afr. Bot. **41**, 2: 116 (1975); in Contr. Bolus Herb. **8**, 1: 254 (1977). — R. Fernandes in Bol. Soc. Brot., Sér. 2, **52**: t. 4 fig. e-i (1978); op. cit. **55**: 110 (1982). Type from Natal.
Crassula sarcocaulis sensu Schönl. in Trans. Roy. Soc. S. Afr. **17**, 3: 214 (1929). — Guillarmod, Fl. Lesotho: 183 (1961). — J. H. Ross, Fl. Natal: 180 (1972) non Eckl. & Zeyh. (1837).

Var. **mlanjiana** R. Fernandes in Bol. Soc. Brot., Sér. 2, **52**: 186, t. 4 fig. k-p (1978). Type: Malawi, Mulanje Mt., path from Tuchila Hut to head of Ruo Basin, *Brummitt* 9658 (K).
Crassula sarcocaulis sensu Brenan in Mem. N.Y. Bot. Gard. **8**, 5: 434 (1954). — Chapman, Veg. Mlanje Mt. Nyasal.: 66 (1962). — Binns, H.C.L.M.: 41 (1968) non Eckl. & Zeyh. (1837).

A very much branched shrub, 30–90 cm. high. Branches fastigiate, the oldest ones very woody, thick, naked, glabrous, with smooth or irregularly sulcate, blackish or grey bark, those of the current year very short, 4-gonous, slender, densely foliate, brownish, those of intermediate age also leafless but with raised nodes and ± marked scars of fallen leaves; internodes rather shorter than the leaves, the apical ones with the nodes nearly contiguous. Leaves (8)10–15·5(17) × 1–1·5(2) mm., linear, sub-terete on account of the reflexed margins and the convex upper face, trigonous towards the acute apex, entire, provided with a marginal row of minute, scattered, brown, circular glandular dots, flattish towards the base, canaliculate beneath, succulent, green, sometimes tipped with red, rigid on drying, straight or curved, very dense; sheaths very short, often ± overlapping. Cymes repeatedly dichotom-ous, surrounded at the base by the uppermost leaves; pedicels 3–6 mm. long, scabrous. Calyx 1·5–2 mm. long; sepals c. 0·75 mm. long. Corolla 3·25–4 mm. long; petals c. 2 mm. wide, oblong-ovate, obtuse, slightly thickened-mucronulate below the apex outside, white, spreading in the fully open flower. Filaments 2·5–3 mm. long; anthers 0·25–0·4 mm. long, purple. Follicles (with the styles) ± 2·5–(3) mm. long, equalling c. ⅔ of corolla, not or slightly attenuate into the indistinct styles, stellate-recurved above. Scales c. 0·6 mm. long, cuneate, truncate or subrounded at the apex.

Malawi. S: Mulanje Mt., Sambani Plateau, c. 2075 m., 12.v.1963, *Wild* 6217 (BR; K; LISC; SRGH).
Known only from Malawi. On rocky slopes of high mountains.
Typical subsp. *rupicola* is found in Lesotho and S. Africa (Transvaal, Orange Free State, Natal and the Cape Prov.)
Subsp. *rupicola* might perhaps be considered an independent species, since it differs from subsp. *sarcocaulis* in a variety of ways besides those mentioned in the key. The habit is different, being shorter and more richly branched with shorter fastigiate branchlets. The leaves form dense tufts at branchlet ends and are linear (up to 10 times longer than broad) rather than ovate or elliptic; they are subterete and trigonous towards the apex, not completely flat; and they dry a paler green than those of typical *C. sarcocaulis*, with more conspicuous marginal dots. The flowers of subsp. *rupicola* are usually smaller and have a proportionally longer calyx than subsp. *sarcocaulis*. The biology also seems to be different: subsp. *sarcocaulis* is reported to have an unpleasant odour, unlike subsp. *rupicola*; their flowering times are separate, at least in S. Africa

(Tölken, tom. cit.: 253 & 255, 1977). In the F.Z. area the flowering periods of two subspecies are much longer and overlap.

Var. *mlanjiana* differs from var. *rupicola* in having longer pedicels (up to 6 mm. compared with only up to 3 mm.) and therefore looser cymes; flowers more open, with almost spreading petals, rather than flowers with erect petals; shorter petals (up to 4 mm. as against up to 5 mm.); anthers less than, rather than exceeding, 0·5 mm. long; and shorter follicles (2·5(3) mm. long, rather than ± 5 mm., including styles) with styles less distinctly delimited and more stellate-spreading. Var. *mlanjiana* is thus more distinct from subsp. *sarcocaulis* in its characters of flower and fruit, than var. *rupicola*.

12. **Crassula setulosa** Harv. in Harv. & Sond., F.C. **2**: 347 (1862). — Schönl. in Trans. Roy. Soc. S. Afr. **17**, 3: 237 (1929). — Berger in Engl. & Prantl, Nat. Pflanzenfam. ed. 2, **18a**: 395 (1930). — Jacobs., Das Sukk. Lexikon: 148 (1970). — Tölken in Contr. Bolus Herb. **8**, 2: 340 (1977). Syntypes from S. Africa.

A perennial herb with the stem simple or branched from the base (thus seeming many-stemmed and caespitose), sometimes shortly stoloniferous, forming small dense cushions. Stems up to 22 cm. long (usually shorter, sometimes dwarfed), herbaceous or in stronger plants woody towards the base, erect or ascending or sometimes prostrate and rooting, usually leafy from base to apex or rarely with only a few pairs of leaves, hispid or retrorsely appressed papillose-hairy (hairs usually shorter and thinner than the marginal cilia of the leaves, or papillae short, obtuse). Basal leaves congested, subrosulate, up to 3(3·5) × 1 cm., obovate to oblong; cauline leaves up to 1·5 × 0·8 cm., rather smaller than the basal ones, oblong-lanceolate to ovate or sometimes broadly ovate to suborbicular, ± scattered to subimbricate, spreading to erect or reflexed, gradually modified into the bracts of the inflorescence, sometimes all the cauline leaves bract-like, rounded or obtuse to acute at the apex, entire, with a row of contiguous to ± scattered, rigid cilia at the margin (the cilia somewhat thick and ± long or short, white, acute or obtuse, straight or curved and retrorse); all leaves sessile and contracted or not into a broad base, connate for ± 0·5 to ± 2 mm., flattish, green and succulent when fresh, membranous and ± dark coloured on drying, all glabrous on both faces or at least the basal ones hispid or appressed hairy on both sides or only above, the hairs similar to those of stem; internodes longer to subequalling or shorter than the leaves. Flowers 5-merous in dense cymes forming terminal small subcapitate inflorescences or in corymbs up to 7 cm. in diam., sometimes with small cymes or 1–2 flowers in the axils of the uppermost leaves; inflorescence-branches hispid; pedicels up to 4 mm. long, somewhat thick. Calyx (1·5)2–2·75(3) mm. long, ½–⅔ as long as corolla, rarely subequalling it; sepals lanceolate or linear-lanceolate, acute to very attenuate, either glabrous except for a terminal papillose hair or with some cilia towards the apex (marginal and dorsal) or with a marginal row of long to short cilia and sometimes also ciliate all along the median line. Corolla (2)2·5–4(4·5) mm., white, rarely red; petals 0·75–1·5(2·25) mm. broad, oblong, oblanceolate to obovate, acute to obtuse or rounded at the apex, erect, smooth, without or with a mucro below the apex. Anthers 0·3–0·5(0·7) mm., subcircular to subrectangular. Follicles (1·5)2–3(3·5) mm. long (with the styles). Seeds c. 0·5 mm. long, subcostate-tuberculate.

A very polymorphic species in which several varieties, some of them certainly mere variations without taxonomical value, have been distinguished.

Most plants which we have seen from the Transvaal, as well as one from Zimbabwe (*Goldsmith* 11/72) have longer flowers, with relatively longer, narrower sepals and longer anthers, than the syntypes (*Zeyher* 650 and *Burke* 401). In addition, they are less rigidly hispid and have sepals glabrous or with a few papillose hairs or cilia at the apex (the type specimens have sepals long ciliate all round the margin). The rigid indumentum and small flowers of the syntypes of *C. setulosa* are possibly related to a drier habitat.

1. Plant not stoloniferous; stem up to 22 cm. long, hispid, the hairs ± dense, thin, acute; basal leaves up to 3(3·5) × 1 cm. - - - - - - - - - - - - - 2
 – Plant stoloniferous; stem shorter, with appressed, retrorse, papillose hairs; leaves smaller var. *rubra*
2. Corolla 2·5–4 mm. long; petals up to 1·5 mm. broad, acute to rounded at apex, with or without a mucro - - - - - - - - var. *setulosa* forma *setulosa*
 – Corolla 4–4·5 mm. long; petals c. 2·25 mm. broad, rounded at the apex, without a mucro var. *setulosa* forma *latipetala*

Var. **setulosa**. — Tölken in Contr. Bolus Herb. **8**, 2: 342 (1977). — R. Fernandes in Bol. Soc. Brot., Sér. 2, **55**: 112 (1982).
Forma **setulosa**.

Crassula setulosa Harv. sens. str. — Burtt Davy, F.P.F.T. **1**: 141 (1926). — Letty, Wild Fl. Transv.: 149, t. 75 fig. 1 (1962). — Guillarmod, Fl. Lesotho: 183 (1971). — Schijff & Schoonr. in Bothalia **10**, 3: 489 (1971).
?*Crassula scheppiggiana* Diels in Engl., Bot. Jahrb. **39**: 465 (1907), ab descr. — Burtt Davy, F.P.F.T. **1**: 141 (1926). Syntypes from Transvaal (B, †).
Crassula setulosa var. *lanceolata* Schönl. in Trans. Roy. Soc. S. Afr. **17**, 3: 237 (1929). Syntypes from Transvaal and O.F.S.
Crassula setulosa var. *ovata* Schönl., op. cit.: 238 (1929) pro max. parte (excl. *Bolus* 10533). Syntypes from Natal and Transvaal.
Crassula setulosa var. *ramosa* Schönl. in Rec. Albany Mus. **2**: 144 (1907); loc. cit. (1929). Type from Transvaal.
Crassula setulosa var. *robusta* Schönl., loc. cit. (1929). Syntypes from Transvaal.
Crassula setulosa var. *basutica* Schönl., tom. cit.: 239 (1929). — Guillarmod, loc. cit. (1971). Syntypes from Lesotho.
Crassula cooperi sensu Guillarmod, Fl. Lesotho: 181(1971) pro parte non Regel (1874).

Zimbabwe. C: Salisbury, 26.v.1943, *Craster* 24 (SRGH). E: Umtali, Vumba Mts., Chikwera Peak, c. 1650 m., 13.v.1956, *Chase* 6134 (BM; K; LISC; SRGH). **Malawi.** C: Dedza Mt., rocky outcrop, 1980–2050 m., 24.iv.1970, *Brummitt* 10117 (K). S: Shire Highlands, s.d., *Buchanan* 35 (K). **Mozambique.** MS: Tsetsera, 1950 m., 3.iii.1954, *Wild* 4491 (K; LISC; PRE; SRGH).
Also in Swaziland, Lesotho and S. Africa (Transvaal, Natal, Orange Free State and Cape Prov.). On rocks in woodland and dense mist forest, sometimes pendulous.

The specimens from Zimbabwe and Mozambique which we have included in forma *setulosa* have prostrate, rooting, sometimes pendulous stems, woody towards the base; leaves glabrous, except for the ciliate margin (the cilia being shorter and less acute than in plants of Transvaal); and petals obovate, up to 1·5 mm. broad. The sepal's, follicle's and anther's size is as in the type. These specimens from the F.Z. area may perhaps represent a distinct variety, but more material is needed before a decision can be made.

Forma **latipetala** R. Fernandes in Bol. Soc. Brot., Sér. 2, **52**: 188, t. 6 et 10 fig. a (1978). Type: Zimbabwe, Melsetter, Peni Mt., *Goldsmith* 11/72 (SRGH, holotype).

Zimbabwe. E: Melsetter, eastern slopes of Peni Mt., 1700 m., vi.1972, *Goldsmith* 11/72 (SRGH).

This form differs from the type mainly by its larger flowers with petals usually broader (c. 2·25 mm.). In many aspects (glabricity and shape of leaves, indumentum of stem and relatively ample inflorescences) it is similar to most plants from Zimbabwe and Mozambique, differing from them by the larger flowers with broader petals, and the larger anthers and follicles. In the length of sepals, petals, anthers and follicles it approaches most of the Transvaal plants seen.
Known only from the type. Herb forming dense clusters at the top of rocky places.

Var. **rubra** (N.E. Br.) Rowley in Cactus & Succ. Journ. Gt. Brit. **40**, 2: 53 (1978). — R. Fernandes in Bol. Soc. Brot., Sér. 2, **55**: 113 (1982). Type from S. Africa (Natal).
Crassula curta N.E. Br. in Kew Bull. **1895**: 144 (1895). M. Wood, Handb. Fl Natal: 46 (1907). Type from S. Africa (Natal).
Crassula curta var. *rubra* N.E. Br., tom. cit.: 145 (1895). Type as above.
Crassula setulosa var. *curta* (N.E. Br.) Schönl. in Trans. Roy. Soc. S. Afr. **17**, 3: 238 (1929). — Hutch., Botanist in S. Afr.: 286 (1946). — J. H. Ross, Fl. Natal: 180 (1972). — Tölken in Contr. Bolus Herb. **8**, 2: 348 (1977). Type as above.
Crassula cooperi sensu Guillarmod, Fl. Lesotho: 181 (1971) pro parte, non Regel (1874).

Mozambique. N: Ribáuè Mts., Mepáluè, c. 1600 m., 28.i.1964, *Torre & Paiva* 10316 (LISC).
Known also from S. Africa (Natal, Orange Free State and Cape Prov.) and Lesotho. On rocks.

We refer the specimen cited above to var. *rubra* only with some misgivings. It may be merely a very small form of *C. setulosa* var. *setulosa* the habit of which (stems very slender, up to 1·7 cm. long; inflorescences few-flowered) was determined by the dry environment ("*saxideserta*"). More material is needed to elucidate the situation.

13. **Crassula cooperi** Regel in Gartenfl. **23**: 36, t. 786 (1874). — K. Schum. & Rümpl., Die Sukk.: 94, fig. 53 (1892). — Burtt Davy, F.P.F.T. **1**: 141 (1926) pro parte — Schönl. in Trans. Roy. Soc. S. Afr. **17**, 3: 236 (1929) excl. *Dinter* 27. — Berger in Engl. & Prantl, Nat. Pflanzenfam. ed. 2, **18a**: 395 (1930). — Jacobsen, Das Sukk. Lexikon: 141 (1970). Type from S. Africa.
Crassula bolusii Hook. f. in Curtis, Bot. Mag., Ser. 3, **31**: t. 6194 (1875). Type from S. Africa (Cape Prov.).

Sedum regelii Kuntze, Rev. Gen. Pl. **3**, 2: 85 (1898) *nom. illegit.* Type as for *C. cooperi.*
Crassula exilis subsp. *cooperi* (Regel) Tölken in Journ. S. Afr. Bot. **41**, 2: 104 (1975); in
Contr. Bolus Herb. **8**, 2: 338 (1977). Type as above.

A perennial succulent herb with numerous stems forming tufts, from a vertical
woody rootstock. Stems up to 15 cm. long, ascending, very slender, herbaceous,
hispid with whitish, spreading, obtuse hairs up to 1 mm. long. Leaves crowded
towards the base of the stems forming cushions, scattered upwards, spreading, the
lower up to 2·6 × 0·25 cm., spathulate-oblanceolate, the median and upper ones
0·8–1·3 cm. long, elliptic-lanceolate, ± fleshy, pale green with dark spots on the
lamina, ciliate at the margin. Flowers shortly pedicellate in few-flowered cymes.
Calyx half as long as corolla, or more; sepals acute, ciliate at the margin. Petals ovate-
lanceolate, acute, with a short, acute, smooth mucro below the apex, stellately
spreading, smooth. Anthers 0·25–0·4 mm. long, suborbicular to suboblong. Follicles
± 1·5 mm. long, smooth, finally contracted into the terminal short styles. Scales
emarginate.

Var. **subnodulosa** R. Fernandes in Bol. Soc. Brot., Sér. 2, **52**: 188, t. 7 et t. 10 fig. b (1978).
    Type: Zimbabwe, Umtali, Maranka Reserve, *Chase* 2169 (BM, holotype; K; SRGH).

Herb 6–12 cm. high. Sterile rosettes of leaves present in the lowermost nodes with
the leaves obovate to suborbicular; lower cauline leaves up to 0·5 cm. broad, ovate to
oblong-spathulate, obtuse, the succeeding ones up to 2 × 0·32 cm., oblong-
spathulate to oblong, obtuse to acute, attenuate to a broad subpetiolar base, connate
for 0·5–2 mm., with the marginal cilia ± 0·5 to 0·7 mm. long, very thin, sometimes
nearly translucent and without dark spots on drying, glabrous on the lamina or with
some hairs towards the apex or hairy throughout (hairs longer on the upper face).
Cymes (sometimes reduced to one flower) shortly pedunculate or subsessile, usually
in verticillasters in the axils of the upper 1–6 pairs of leaves, the lower ones
sometimes shorter than the subtending leaves, the rest or all, longer than them;
pedicels 1–2 mm. long, glabrous or sparsely hispid. Calyx 1–1·5 mm. long. Corolla
1·75–2 mm. long.

Zimbabwe. N: Lomagundi, Whindale Farm, Mangula, alt. c. 1280 m., 18.i.1970, *Jacobsen*
4141 (PRE).
Not known from elsewhere. In humus pockets among dolomite, granite or limestone
boulders, on rocky outcrops in sheltered gullies and woodland.

Var. *subnodulosa* differs from var. *cooperi*, which is found in the Cape Prov., as follows:

Cymes terminal and axillary in the axils of the uppermost 1–6 leaf-pairs; calyx 1–1·5 mm. long;
    corolla 1·75–2 mm. long, white -   -   -   -   -   -   -   - var. *subnodulosa*
Cymes terminal only; calyx c. 2·25 mm. long; corolla 2·75–3(4) mm. long, rose-pink
                                                                    var. *cooperi*

The Zimbabwean variety may merit recognition as a distinct species, but more material is
needed to decide this question. *C. cooperi* is apparently its nearest ally but it also shows a strong
resemblance to *C. schmidtii* Regel (a species of uncertain origin) and to some forms of *C.
setulosa*, particularly var. *rubra*. It differs from the latter mainly in its non-stoloniferous stems,
narrower leaves, different inflorescences and acute, mucronate petals (var. *rubra* has obtuse
petals lacking a mucro). *C. cooperi* var. *subnodulosa* is also somewhat similar, in general aspect,
to *C. tabularis* Dinter, but the latter is a stronger plant with broader (up to 1·5 cm. broad) and
relatively shorter leaves which are thicker and have longer marginal cilia. *C. tabularis* also has
more numerous verticillasters, shorter pedicels, longer and more attenuate sepals, and larger
(2·5 mm. long) petals which are not so spreading and have a longer mucro than in var.
subnodulosa.

14. **Crassula vaginata** Eckl. & Zeyh., Enum. Pl. Afr. Austr. **3**: 298 (1837). — Walp., Repert.
    2: 254 (1843). — Harv. in Harv. & Sond., F.C. **2**: 341 (1862). — M. Wood, Handb. Fl.
    Natal: 46 (1907). — Schönl. in Engl., Bot. Jahrb. **43**: 359 (1909); in Trans. Roy. Soc. S.
    Afr. **17**, 3: 226 (1929). — Burtt Davy, F.P.F.T. **1**: 140 (1926). — Berger in Engl. & Prantl,
    Nat. Pflanzenfam. ed. 2, **18a**: 394 (1930). — Hutch., Botanist in S. Afr.: 404 (1946). —
    Letty, Wild Fl. Transv.: 149 (1962). — Riley, Fam. Fl. Pl. S. Afr.: 157 (1963). — J. H.
    Ross, Fl. Natal: 180 (1972). — Tölken in Contr. Bol. Herb. **8**, 2: 356 (1977). — R.
    Fernandes in Bol. Soc. Brot., Sér. 2, **55**: 115 (1928). — G. E. Wickens in Kew Bull. **36**, 4;
    667 (1982). TAB. **4** fig. B. Type from S. Africa (Cape Prov.).
        Crassula mannii Hook. f. in Journ. Linn. Soc., Bot. **7**: 193 (1864). — Hutch. & Dalz.,
    F.W.T.A. **1**, 1: 104 (1927). — Merxm. & al. in Ann. Naturh. Mus. Wien **75**: 113 (1971).
    Type from Cameroon.

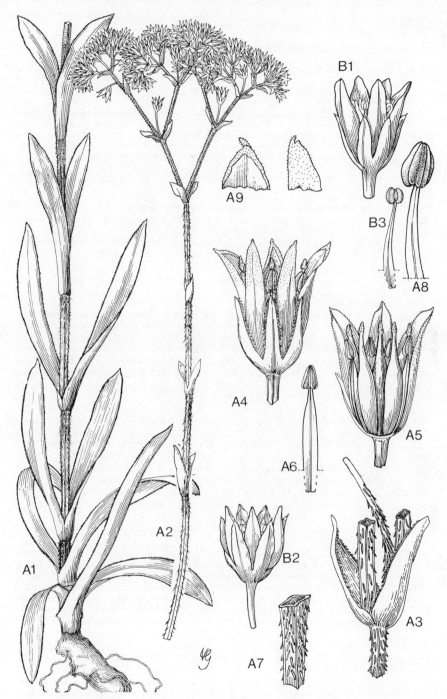

Tab. 4. A. — CRASSULA ALBA. A1, lower and median part of flowering stem (× ⅔); A2, upper part of flowering stem with inflorescence (× ⅔); A3, part of an inflorescence, showing two bracts, the lower part of two branches and pedicel of dichotomy (× 4); A4, flower (× 6); A5, flower without frontal sepals and petals to show the follicles (× 6); A6, stamen after dehiscence (× 8); A7, portion of stem (× 4); A8, stamen (× 10); A9, upper part of a petal in ventral and lateral view, from *Torre* 6693. B. — C. VAGINATA. B1, B2, flowers (× 6); B3, stamen (× 10), from *Torre* 10792.

Crassula abyssinica sensu Britten, F.T.A. **2**: 388 (1871) pro parte. — Engl., Hochgebirgsfl. Trop. Afr.: 231 (1892) pro parte. — Burtt Davy, loc. cit. — Gilliland in Journ. S. Afr. Bot. **4**: 91 (1938). — Toussaint in F.C.B. **2**: 569 (1951). Non A. Rich. (1847).

Crassula abyssinica var. *mannii* (Hook. f.) Engl., loc. cit. (1892). — Schönl., tom. cit.: 360 (1909). Type as for *Crassula mannii*.

Crassula abyssinica var. *vaginata* (Eckl. & Zeyh.) Engl., Pflanzenw. Ost. Afr. **C**: 189 (1895). — Hiern, Cat. Afr. Pl. Welw. **1**: 326 (1896). — De Wild. in Ann. Mus. Congo, Bot., Sér. 2, **2**: 20 (1900). Type as for *Crassula vaginata*.

Sedum vaginatum (Eckl. & Zeyh.) Kuntze, Rev. Gen. Pl. **3**, 2: 86 (1898). Type as above.

Crassula schweinfurthii De Wild., Ic. Sel. Hort. Then. **1**: 93, t. 22 (1900). Type a cultivated plant of Ethiopian origin.

Crassula abyssinica var. *angolensis* Schönl., loc. cit. (1909); tom. cit.: 227 (1929). Holotype from Angola.

Crassula abyssinica var. *nyikensis* Schönl., loc. cit. (1909). Type: Malawi, Nyika Plateau, alt. 1920–2240 m., *Whyte* 165 (B, holotype).

Crassula abyssinica var. *robusta* Schönl., loc. cit. (1909). Syntypes from Angola, Malawi and Kenya (lectotype: Malawi, *Buchanan* 340 (K)).

Crassula abyssinica var. *ovata* Schönl., loc. cit. (1909). Syntypes from Tanzania and Cameroon lectotype: Kilimandjaro, *Volkens* 1905 (K)).

Crassula abyssinica var. *transvaalensis* Schönl., tom. cit.: 226 (1929). Type from S. Africa (Transvaal).

Crassula alba sensu.: Berger in Engl. & Prantl, Nat. Pflanzenfam. ed. 2, **18a**: 394 (1930) pro parte. — Brenan in Mem, N.Y. Bot. Gard. **8**, 5: 434 (1954). — Cufod. in Bull. Jard. Bot. Brux. **24**, Suppl.: 169 (1954) pro parte. — Keay in F.W.T.A. ed. 2, **1**, 1: 116 (1954) pro parte; — Hedberg in Symb. Bot. Upsal. **15**, 1: 101 (1957) pro parte; — Chapman, Veg. Mlanje Mt. Nyasal.: 52 et 56 (1962); — Binns, H.C.L.M.: 41 (1968) non Forsskal. (1775).

Crassula retrorsa Hutch. in Kew Bull. **1933**: 420 (1933). — Jacobsen, Das Sukk. Lexicon: 147 (1970). Type from S. Africa (Transvaal).

A succulent, 8–90 cm. high perennial, forming clumps up to about 90 cm. in diameter. Stem arising from a fleshy to woody obconical or swollen tuber, erect, straight, usually simple, ± stout (up to 10 mm. in diameter at the base) or sometimes ± slender, terete, pale- or yellow-green below, bright red to reddish-brown above, usually crimson on drying, glabrous towards the base, papillose-hairy or shortly papillose upwards below the nodes along two ± short stripes alternating with the lamina of the leaves (the papillae usually very short, patent, obtuse or, if long, then hair-like), or completely glabrous, rarely papillose throughout. Leaves 2·5–26·5 × (0·1–0·2) 0·4–2·7(3·3) cm., sometimes up to 37 times longer than wide, the basal ones the longest and rosulate, oblong- to linear-lanceolate or linear, ± attenuate, the lower cauline leaves similar to the basal ones but shorter, the succeeding ones lanceolate or linear to oblong- or ovate-lanceolate, the uppermost sometimes bract-like, all acute or subacute at the apex, entire and with a row of short, contiguous, obtuse, white papillose cilia at the margin, sessile, ± long-connate at the base, clasping the stem below, then subappressed to it, finally erect or somewhat arched, flat, fleshy (but not very thick), pale to bright green when fresh, greenish-brown or dark brown when dry, usually glabrous on both faces or very rarely papillose-hairy beneath or on both surfaces; sheath (0·5)1–3 cm. long in the lower cauline leaves, shortening in the following; all internodes shorter than the leaves (sometimes the stem completely concealed by the sheaths and the clasping part of the laminas) or the upper ones subequalling or longer than them. Flowers 5-merous in a repeatedly dichotomous, usually very dense, subhemispherical, corymbose inflorescence, 1·5–22(30) cm. in diam.; branches ± papillose to nearly glabrous; pedicels 2–6 mm. long, glabrous or sparsely papillose. Calyx (1)1·5–3(4) mm. long, ⅓–½ (⅔) as long as corolla, rarely subequalling it, pale green; sepals ovate-lanceolate or ovate-elliptic to linear-lanceolate, not or ± attenuate, acute, with one terminal ± long obtuse papilla at the apex, usually entire and without marginal papillae or sometimes obtuse papillae present over a short to ± long extent of the margin as well as on the median nerve above. Corolla (2·25)2·5–3·5(4·5) mm. long, white, pinkish-white or cream-white, sometimes red-tipped, or yellow, rarely red; petals 0·75–1·5(2) mm. wide, elliptic-oblong to ovate-oblong or ovate, obtuse or nearly so, slightly thickened at the top, not or indistinctly mucronate below the apex, smooth. Filaments 1·5–2 mm. long; anthers (0·25)0·3–0·4(0·5) mm., subcircular. Follicles (incl. styles) 2–2·5(2·75) mm. long, contracted into the very short styles. Seeds c. 0·5 mm. long, oblong.

Zambia. N: Abercorn distr., Sunzu Hill, c. 2176 m., 28.iv.1936, *B. D. Burtt* 6145 (BM; K).
W: Mwinilunga, Lisombo R., 11.vi.1963, *Loveridge* 919 (K; SRGH). C: Serenje, 18.ii.1955,
Fanshawe 2086 (BR; K). E: Lundazi distr., Nyika Plateau, c. 2100 m., 3.i.1959, *Richards* 10428
(K; SRGH). **Zimbabwe.** E: Melsetter distr., Chimanimani Mts., behind the mountain hut, c.
1760 m., 10.iv.1967, *Grosvenor* 370 (K; LISC; PRE; SRGH). **Malawi.** N: Mzimba Distr.,
Viphya, 28.iii.1954, *G. Jackson* 1289 (BM; K). C: Dedza Distr., Chongoni, on slopes of Ciwau
near the Meteorological Station, 8.iii.1961, *Chapman* 1163 (K; SRGH). S: Zomba Mt., Forest
Nursery, 1830 m., *E. M. & W.* 756 (BM; LISC; SRGH). **Mozambique.** N: Massangulo Mt.,
41 mil. N. of Mandimba, c. 1440 m., 26.v.1961, *Leach & Rutherford-Smith* 11039 (K; LISC;
SRGH). T: Angónia, Monte Dómuè, c. 1700 m., 9.iii.1964, *Torre & Paiva* 11117 (LISC). MS:
Monte Chimanimane, 6.vi.1949, *Munch* 166 (SRGH).

Also in Ethiopia, Sudan, Kenya, Uganda, Tanzania, Swaziland, S. Afr. (Natal, Transvaal,
Orange Free State and the Cape) and from the Cameroons to Angola. In open grassland or
among rocks on mountain slopes up to 2680 m.

This species is similar to *C. alba* Forssk. (= *C. abyssinica* A. Rich.) The two can be
distinguished as follows (some characters not included in the key are mentioned here):

Indumentum relatively sparse, composed of relatively short papillae; sheaths of lower leaves 1
(rarely less)–3 cm. long; sepals usually up to half as long as corolla, not denticulate, either
entire or with obtuse marginal papillae; petals smooth and indistinctly mucronate; anthers
suborbicular, 0·3–0·5 mm. long - - - - - - - - - - *C. vaginata*
Indumentum relatively dense, composed of longer papillae; sheaths of lower leaves usually less
than 1 cm. long; sepals longer than ½–up to as long as corolla, the margins acutely
denticulate; petals sometimes subtuberculate-scabrous on the upper part of the outer
surface, distinctly mucronate; anthers oblong, 0·5–1 mm. long - - - *C. alba*

Many specimens of *C. vaginata* from S. Africa have broader and denser leaves than is usual in
the F.Z. area, as well as more frequently glabrous or subglabrous stems and sepals lacking
marginal papillae. However, similar leaves have been found in other regions, as in the
Cameroons (*C. mannii*), Sudan, Uganda, Tanzania (*C. abyssinica* var. *ovata*), Zaire and Angola;
and they occur here and there in the F.Z. area. Sepals without marginal papillae are also known
in both F.Z. specimens and ones from other tropical countries.

As can be seen from the synonymy many varieties have been distinguished. However, it seems
quite impossible to separate infraspecific taxa within *C. vaginata*, based on such characters as
stem stoutness, leaf size and shape, presence, absence or quantity of indumentum or occurrence
of marginal papillae on sepals, because they vary independently.

C. vaginata was described with "*flores flavi*". Unless it was noted by the collector, it is often
impossible to decide on the original petal colour; nearly all specimens have reddish- to blackish-
brown corollas, although they were ± white at anthesis. However, yellow corollas have never
been mentioned by collectors in the F.Z. area. *Torre & Correia* 15377 (LISC) from
Mozambique has red corollas.

The many specimens seen from Mulanje Mt. (Malawi) have a characteristic habit, being only
(2·5)4–10(12) cm. high, with the rosulate leaves narrowly linear and more persistent than usual,
and the cauline leaves similar. They seem to represent a dwarf altitudinal form of exposed
places and thin soil over rock. As intermediates occur at lower altitudes between them and more
robust plants of the normal form, it is advisable not to recognise them formally.

Several specimens from the Chimanimani Mts. region of Zimbabwe (E) and Mozambique
(MS), for example *Goodier* 177 (SRGH), *Phipps* 647 (K; SRGH), *Hall* 429 (SRGH) and
Pereira, Sarmento & Marques 1315 (LMU) are distinctly low-growing, with small, relatively
broad leaves connate for not more than 5 mm. at the base, and with a long, dense, stem
indumentum. They may be hybrids between *C. vaginata* and *C. setulosa*, of which the latter also
grows in this area.

15. **Crassula alba** Forssk., Fl. Aegypt.–Arab.: 60 (1775). — DC., Prodr. 3: 390 (1828). —
Schweinf. in Bull. Herb. Boiss. 4, App. 2: 197 (1896). — Berger in Engl. & Prantl, Nat.
Pflanzenfam, ed. 2, **18a**: 394 (1930). — Cufod. in Bull. Jard. Bot. Brux. 24, Suppl.: 169
(1954) pro parte; in Senckenb. Biol. **39**: 123 (1958). — Hedberg in Symb. Bot. Upsal. **15**,
1: 101 (1957), pro parte. — Keay in F.W.T.A., ed. 2, **2**: 116 (1954) quoad syn. pro parte.
— Tölken in Contr. Bolus Herb. **8**, 2: 362 (1977). — R. Fernandes in Bol. Soc. Brot.,
Sér. 2, **55**: 100 (1982). — G. E. Wickens in Kew Bull. **36**, 4: 669 (1982). TAB. **4** fig. A.
Type from Arabia.
  Crassula puberula R. Br. in Salt, Voy. Abyssinia, App. **4**: 64 (1814) *nom. nud.*
  Crassula rubicunda Drège, Zwei Pfl. Doc.: 155 (1843) *num. nud.*
  Globulea stricta Drège, Zwei Pfl. Doc.: 159 (1843) *nom. nud.*
  Crassula abyssinica A. Rich., Tent. Fl. Abyss. **1**: 309 (1848). — Schweinf. & Aschers. in
Schweinf., Beitr. Fl. Aeth., Aufzähl. 271 (1867). — Britten in F.T.A. **2**: 388 (1871) pro
parte. — Engl., Hochgebirgsfl. Trop. Afr.: 231 (1892) pro parte; in Ann. R. Ist. Bot.
Roma, **9**: 252 (1902). — Sacleux in Bull. Hist. Nat. Mus. Paris, **14**: 243 (1908). — Hutch.
& Bruce in Kew Bull. **1941**: 88 (1941). — Type from Ethiopia.
  Crassula rubicunda Drège ex Harv. in Harv. & Sond., F.C. **2**: 341 (1862). — Bak. in
Saunders Ref. Bot. **5**: t. 339 (1863). — M. Wood, Handb. Fl. Natal: 46 (1907). — Schönl.

in Rec. Albany Mus. **3**, 1: 57 (1914); in Trans. Roy. Soc. S. Afr. **17**, 3: 229 (1929). — Burtt
Davy, F.P.F.T. **1**: 140 (1926). — Dyer in Fl. Pl. S. Afr. **38**: t. 1520 (1967). — Jacobs., Das
Sukk. Lexikon: 147 (1970). — Guillarmod, Fl. Lesotho: 183 (1971). — Venter in Journ.
S. Afr. Bot. **37**, 2: 106 (1971). — J. H. Ross, Fl. Natal: 180 (1972). — Type from Natal.
    Crassula recurva N.E. Br. in Gard. Chron., Ser. 3, **8**: 684 (1890). — M. Wood, Natal.
Pl. **6**: t. 576 (1912). — Schönl. in Trans. R. Soc. S. Afr. **27**, 3: 231 (1929). — Jacobsen, Das
Sukk. Lexikon: 147 (1970). — J. H. Ross, loc. cit. (1972). Type a cultivated plant in Kew
Gardens of Zululandian origin.
    Crassula abyssinica var. *typica* Schönl. in Engl., Bot. Jahrb. **43**: 359 (1909).
    ?*Crassula ellenbeckiana* Schönl., op. cit. **39**: 361 (1909) — G. E. Wickens, tom. cit.: 666
(1982). Type from Ethiopia.
    Crassula abyssinica var.-Schönl. in S. Afr. Journ. Sci. **17**, 2: 188 (1921).
    Crassula milleriana Burtt Davy, op. cit.: 38 et 140 (1926). Type from Swaziland.
    Crassula stewartiae Burtt Davy, loc. cit. — Type from Swaziland.
    Crassula rubincunda var. *typica* Schönl., loc. cit.: 229 (1929). Type as for *C. rubicunda.*
    Crassula rubicunda var. *milleriana* (Burtt Davy) Schönl., tom. cit.: 230 (1929). Type as
for *C. milleriana.*
    Crassula rubicunda var. *hispida* Schönl., loc. cit. (1929). — Guillarmod, loc. cit.
Lectotype from Lesotho.
    Crassula rubicunda var. *subglabra* Schönl., tom. cit.: 231 (1929). Type from Natal.
    Crassula rubicunda var. *flexuosa* Schönl., loc. cit. (1929). Type from Transvaal.

A perennial or biennial succulent herb up to 90 cm. high with a tuberous root
producing one to many flowering stems, surrounded at base by dense rosulate leaves
which usually disappear at anthesis. Stem ± stout, erect, terete, simple, fleshy, from
base to apex ± covered all around with retrose, appressed, whitish, ± long hair-like
papillae (or papillae short, obtuse or acute, sometimes nearly bulliform), rarely
glabrous, fleshy, green or reddish-purple. Rosette-leaves up to 16(20) × 2·5 cm.,
oblong-lanceolate or lanceolate, attenuate towards the acute apex, entire and ciliate
or pectinate-ciliate at the margin (the cilia usually nearly contiguous, acute), sessile
and connate at the base, flat, fleshy but not very thick, usually glabrous, sometimes
papillose-pilose, green, sometimes with purple marks or ± purple beneath,
deflexed; cauline leaves similar to the basal ones but usually smaller, the lower longer
than the internodes, the upper successively shorter, becoming erect and ± clasping
the stem, the upper cauline leaves sometimes ovate-lanceolate or elliptic-oblong,
sometimes nearly bract-like; sheaths of lower cauline leaves 0·2–0·8(1·2) cm. long,
those of the upper ones very short. Flowers 5-merous, arranged in a terminal, ±
dense, up to 17(30) cm. in diam., corymbose, flat-topped inflorescence, with
papillose or hairy branches; pedicels up to 15 mm. long, glabrous or nearly so. Calyx
(2·5)3–5·5(6·5) mm. long, usually longer than ½ corolla to equalling it; sepals 1–1·25
mm. broad, lanceolate to linear-lanceolate, very attenuate, ± acute, serrulate or
denticulate to ciliate-pectinate at the margins (cilia acute, ± rigid), or entire, green
or reddish-purple. Corolla (4·5)5–7 mm. long; petals 1·25–2 mm. broad, oblong,
obtuse or subacute, erect but with ± reflexed or spreading tips, mucronate below the
slightly inflexed apex, sometimes minutely scabrid outside towards the apex, white,
greenish-white flushed with pink or bright to deep red. Filaments (2)2·5–3(3·5) mm.
long; anthers (0·5)–0·75(1) mm. long, rectangular. Follicles (2·75)3·5–5·5 mm. long
including the styles. Scales ± 0·5 mm. long, rectangular or subquadrate, truncate.

    Zimbabwe. W: Bulawayo, corolla whitish, v.1915 *Rogers* 13711 (K; Z). C. Selukwe Peak, c.
1615 m., corolla whitish 19.iii.1964, *Wild* 6433 (K; LISC; PRE; SRGH). S: Victoria, Kyle
National Park Game Reserve, near the base of Mtunumashawa Hill, corolla red, 21.v.1971,
Grosvenor 514 (SRGH). **Mozambique.** M: Bela Vista, Zitundo, Ponta do Ouro, corolla red,
3.x.1968, *Balsinhas* 1343 (LISC; LMA; PRE).
    Also in Arabia, Ethiopia, Uganda, Kenya, Tanzania, Swaziland, Lesotho and S. Africa
(Transvaal, Natal, Orange Free State and the Cape). In woodland and savannas.

    The type of *C. alba* is an Arabian specimen with white flowers; both flowers and fruits are
slightly smaller than those of most specimens from the F.Z. area. Thus in typical *C. alba* the
calyx is 2·5–3 mm. long, the corolla is 4–4·5 mm., the filaments are ± 2 mm., the anthers 0·5
mm. and the follicles (including styles) 3–3·5 mm. long, whereas in plants from the F.Z. area
and from S. Africa the calyx is 3·5–5·5(6) mm., the corolla is (4·5)5–7 mm., the filaments are
2·75–3·5 mm., the anthers are (0·5)0·75–1 mm. and follicles 3·5–5·5 mm. long. Flowers and
fruits as large as these, in which the corollas may be either white or red, are also found in Eritrea,
Ethiopia and elsewhere (some of these have been determined as *C. ellenbeckiana* at K). Apart
from their large flowers, plants from both the F.Z. area and S. Africa have sepals almost free at
the base and with longer marginal cilia than in specimens outside this region where the sepals

are often entire or have only very short cilia which are sometimes replaced by denticula. One could only decide whether the southern plants should be regarded as a distinct infraspecific taxon after detailed statistical studies on more material.

C. abyssinica (whose type has a red corolla but is identical in other important characters with the type of *C. alba*) has frequently been confused with *C. vaginata*. For a discussion on the differences between *C. alba* and *C. vaginata*, see under the latter species (no. 14).

16. **Crassula similis** Bak. f. in Bull. Herb. Boiss., Sér. 2, **3**: 814 (1903). — Burtt Davy, F.P.F.T. **1**: 140 (1926). — R. Fernandes in Bol. Soc. Brot., Sér. 2, **55**: 113 (1982). Type from S. Africa (Transvaal).

    Crassula wilmsii Diels in Engl., Bot. Jahrb. **39**: 464 (1907). — Burtt Davy, loc. cit. (1926). — Merxm. & al. in Ann. Naturh. Mus. Wien **75**: 113 (1971). Type from S. Africa (Transvaal).

    Crassula atrosanguinea Beauverd in Bull. Herb. Boiss., Sér. 2, **7**: 1013, fig. pag. 1014 (1907). — Barbey in Kew Bull. **1908**, App.: 84 (1908). Type a cultivated plant at "La Pierrière", (Chambéry, near Genève) originally from S. Africa (Transvaal).

    Crassula rubicunda var. *similis* (Bak. f.) Schönl. in Trans. Roy. Soc. S. Afr. **17**, 3: 230 (1929). Type as for *Crassula similis*.

    Crassula rubicunda var. *parvisepala* Schönl., tom. cit.: 231 (1929). Type as for *Crassula wilmisii*.

    Crassula rubicunda var. *lydenburgensis* Schönl., loc. cit. (1929). Type from S. Africa (Transvaal, Lydenburg).

    Crassula alba var. *parvisepala* (Schönl.) Tölken in Journ. S. Afr. Bot. **41**, 2: 93 (1975); in Contr. Bolus Herb. **8**, 2: 368 (1977). Type as for *Crassula wilmsii*.

Very similar to *C. alba* (No. 15) but differing vegetatively in the more numerous, denser and more persistent basal leaves which become coriaceous with age, the oldest ones drying and disintegrating to leave a proximal stub; in the stem indumentum which usually consists of very short, bulliform papillae; in the shorter (up to 0·5 cm. long) cauline leaf-sheaths; and in the usually shorter cauline leaves of which the uppermost rarely exceed the internodes. Distinguishing features of the reproductive plant include denser, hemispherical to subspherical, not flat-topped, inflorescences, and smaller flowers and fruits. Thus, calyx (1)1·25–2·5(3) mm. long, usually less than half as long as corolla, with triangular or ovate-lanceolate sepals less acute than in *C. alba* and marginal cilia absent or present and then small and few; corolla 2·5–3·5(4) mm. long, carmine or purple; stamens with filaments 1·75–2·25 mm. and anthers 0·5 mm. long. According to Tölken (tom. cit.: 365 et 368, 1977) the ovary in *C. similis* contains 8–10 ovules compared with 14–30 in *C. alba*.

Mozambique. M: Libombos Mts. near Namaacha, Ponduine Mt., 800 m., 22.ii.1955, *E. M. & W.* 515 (LISC; SRGH).

Also in S. Africa (Transvaal) and Swaziland. In the rupideserta.

17. **Crassula nodulosa** Schönl. in Rec. Albany Mus. **1**, 1: 56 (1903); tom. cit.: 64; in Trans. Roy. Soc. S. Afr. **17**, 3: 247 (1929). — Burtt Davy, F.P.F.T. **1**: 141 (1926). — Hutch., Botanist in S. Afr.: 400 et 673 (1946). — Wilman, Check List Fl. Pl. Ferns Griqual. — W.: 77 (1946). — Suesseng. & Merxm. in Trans. Rhodes. Sci. Assoc. **43**: 15 (1951). — Friedr. in Mitt. Bot. Staatss. München, **3**: 594 (1960); in Prodr. Fl. SW. Afr. **52**: 33 (1968). — Jacobs., Das Sukk. Lexikon: 145 (1970). — Guillarmod, Fl. Lesotho: 182 (1971). — R. Fernandes in Bol. Soc. Brot., Sér. 2, **55**: 108 (1982). TAB. 5. Type from S. Africa (Cape Prov.).

    Crassula enanthiophylla Bak. f. in Bull. Herb. Boiss., Sér. 2, **3**: 816 (1903). Type from S. Africa (Transvaal).

    Crassula elata N.E. Br. in Kew Bull. **1909**: 110 (1909). — Dinter in Fedde, Repert. **16**: 243 (1919). Type: Botswana, Palapye, *Lugard* 247 (K, holotype.).

    Crassula pectinata Conrath in Kew Bull. **1914**: 246 (1914). Type from S. Africa (Transvaal).

    Crassula avasimontana Dinter in Fedde, Repert. **19**: 148 (1923). Syntypes from Namibia.

    Crassula guchabensis Merxm. in Mitt. Bot. Staatss. München, **1**: 81 (1951). Type from Namibia.

    Crassula capitella Thunb. subsp. *nodulosa* (Schönl.) Tölken in Journ. S. Afr. Bot. **41**, 2: 100 (1975); in Contr. Bolus Herb. **8**, 2: 390 (1977). Type as for *Crassula nodulosa*.

    Crassula capitella subsp. *enanthiophylla* (Bak. f.) Tölken, loc. cit. (1975); tom. cit.: 392 (1977). Type as for *Crassula enanthiophylla*.

A biennial or perennial succulent herb with a tuberous root and a basal rosette of leaves up to 10 cm. in diameter, sometimes very dense and ± persisting at flowering time. Stem up to 60(80) cm. high, leafy below, floriferous above, usually solitary,

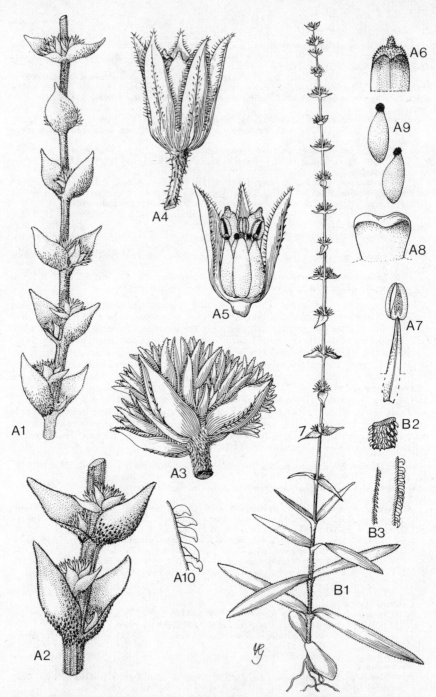

Tab. 5. CRASSULA NODULOSA. A. — var. LONGISEPALA. A1, part of a flowering stem (× ⅔); A2, the two lower nodes of A1 (× 1); A3, verticillaster (× 4); A4, flower (× 8); A5, flower without frontal sepals and petals to show the pistils (× 8); A6, upper part of a petal in ventral view (× 12); A7, stamen (× 12); A8, nectariferous scale (× 50); A9, follicles in dorsal and lateral view (× 12); A10, marginal cilia of a bract (× 25), all from *Simon* 765. B. — var. NODULOSA forma RHODESICA. B1, habit (× ⅔); B2, part of a stem (× 12); B3, marginal cilia of leaves (× 12), from *Müller* 610.

rarely 2–3 per root, simple, rarely with some slender short to ± long branches at the median cauline nodes or at the lower floral ones, terete, firm, slender to ± thick (up to 8(11) mm. in diameter at the base), whitish, greyish or fulvous-papillose-pilose (the hairs up to 0·75 mm. long, patent or subretrorse, ± dense, obtuse or acute, sometimes very short and bubble-like). Rosulate leaves ± dense, all spreading or the uppermost erect, up to 5(7) × 3·5 cm., obovate-spathulate to broadly obovate, ovate or suborbicular or oblong to oblanceolate or lanceolate, rounded, obtuse or acute at the apex, entire and pectinate-ciliate at the margin (the cilia up to 1 mm. long, retrorse, acute, rigid, whitish), sessile and somewhat connate at the base, flat, fleshy and green when fresh, firm and ± dark on drying, glabrous or ± densely to sparsely papillose-hairy on both faces or only on the lower one (sometimes glabrous leaves and hairy ones occur in different plants of the same gathering); lower cauline leaves similar to the uppermost basal ones but smaller and less obtuse or more acute, ± approximate to ± distant, subpatent to suberect, the median ones passing gradually into the leafy bracts; lower cauline internodes up to 6·5 cm. long, the following shortening successively into the floral ones. Flowers 5-merous in numerous (up to 34) verticillasters in the axils of opposite leafy bracts, forming a ± long, usually narrow, sometimes up to 4·3 cm. in diam., spike-like inflorescence, interrupted below, nearly continuous towards the apex; lower floral internodes up to 4 cm. long; individual cymes usually very condensed and subcapitate, up to 2·5 cm. in diameter, shortly pedunculate (apparently sessile); bracts similar to the upper cauline leaves, but shorter and relatively broader, the lower ones shorter or longer than the cymes, the others shorter, ± densely to sparsely papillose at least towards the base on the underside or glabrous everywhere; pedicels up to 3·5(4) mm. long, papillose-hairy or glabrous. Calyx (1·5)2–4(4·5) mm. long; sepals ovate-lanceolate to oblong-lanceolate, not to ± attenuate, acute, erect or slightly spreading, ciliate at the margin, connate for ± 0·75 mm. at base, usually hispidulous all over the dorsal surface or only along the median line, rarely completely glabrous, purplish. Corolla 2·25–4 mm. long, white or reddish; petals erect and subconnivent, ovate-oblong to oblong-elliptic or oblong, obtuse or sometimes subacute, with slightly inflexed apex, dorsally mucronate at the inflexion (the mucro ± up to 0·5 mm. long, conical, solid, acute or obtuse), smooth or ciliate at the margins, connate at the base, minutely tuberculate-scabrid (including the mucro) on the outer surface towards the apex. Filaments 0·75–2 mm. long; anthers ± 0·5 mm. to 0·75 mm. long, oblong-ovate. Follicles 1–1·5(2) mm. long, obliquely ovate-oblong, contracted into the very short or almost absent styles; stigmas thick, subsessile, lateral (at least at maturity). Scales up to 0·5 mm. long, as long as to shorter than broad, cuneate-spathulate to subrectangular, truncate or slightly emarginate, narrowly thickened at the upper margin. Seeds minutely tuberculate.

A very polymorphic species. In F.Z. area two main, rather constant infraspecific taxa can be distinguished.

1. Calyx (2·5)3–4(4·5) mm. long, distinctly longer than the corolla; sepals usually oblong-lanceolate, ± attenuate, very acute - - - - - - - var. *longisepala*
 - Calyx (1·5)2–3·5(3·75) mm. long, half as long as to subequalling the corolla; sepals ovate-lanceolate to lanceolate, not or scarcely attenuate, acute - - - - - - 2
2. Median cauline leaves usually not more than 3·5 times as long as broad, obovate or ovate to oblong, not or a little attenuate, usually obtuse or rounded at the apex
 var. *nodulosa* forma *nodulosa*
 - Median cauline leaves 2–6 × 0·3–0·7(1) cm., 3·5 or more times longer than broad, lanceolate to elliptic, ± attenuate, ± acute - - - - - var. *nodulosa* forma *rhodesica*

Var. **nodulosa**.
Forma **nodulosa**.

Botswana N: Kwebe Hills, *Lugard* 224 (K). **Zimbabwe.** C: Salisbury, Umvukwe Mt., 1680 m., 5.iii.1961, *Richards* 14576 (K). E: Inyangani Mt., 2000–2200 m., 21–22.iv.1934, *Humbert* 15788 (P). **Mozambique**. MS: Manica e Sofala, c. 1440 m., 8.x.1969, *Leach & Cannell* 14397 (SRGH).
Also in Angola, Namibia and S. Africa (Transvaal, Orange Free State and the Cape), and perhaps also in Uganda, Kenya and Tanzania. In grassland, bushveld in rocky situations, in mountains up to 2200 m.

Most of the synonyms given above under *C. nodulosa* apply to this taxon. The type of *C. elata*, however, consists only of the upper part of the stem and thus it is impossible to tell the shape of the median leaves.

C. nodulosa was treated by Tölken (loc. cit., 1975 & 1977) as a subspecies of *C. capitella*. We interpret it as an independent species differing from *C. capitella* as follows:

Perennial rarely biennial, with a tuberous base and usually ± swollen root; stems hairy or papillose; mucro of petals solid, conical, usually acute, tuberculate-scabrid, inserted near the apex of petals, the latter inflexed and obtuse; young follicles at first shortly attenuate with stigmas on very short styles, the stigmas becoming lateral at maturity; nectary-scales truncate or slightly emarginate, with thickened upper margin; seeds minutely tuberculate
C. nodulosa

Biennial rarely perennial without tuberous base and with several main roots; stems glabrous; appendage of petals hollow, broader than above, subcylindrical, obtuse, smooth, inserted further from the apex of the petals, the latter more inflexed than above and acute; young follicles having sessile lateral stigmas; nectary scales distinctly and more deeply emarginate than above, not thickened; seeds smooth   -   -   -   -   -   -   *C. capitella*

Tölken placed plants with a tuberculate-scabrid petal-appendage in *C. capitella* subsp. *enantiophylla*; this includes all specimens from the F.Z. area. According to his keys, subsp. *nodulosa* has a smooth petal-mucro. However, not only the type of *C. nodulosa* (*Adams* 28), but also many plants from the Transvaal and the Cape Prov. determined by Tölken as subsp. *nodulosa* have distinctly tuberculate-scabrid mucros. He refers to the appendage as "terminal", but this is inexact as the position is actually subterminal and only appears terminal by the inflexion of the petal apex.

The type of *C. enanthiophylla* is a plant with laxer cymes than typical *C. nodulosa*, a condition which we think was probably related to its shady habitat.

Many plants from the F.Z. area of var. *nodulosa* forma *nodulosa* have the calyx somewhat longer than is usual in material from other regions (Transvaal, Angola, etc.) thus being intermediate in this respect between var. *nodulosa* and var. *longisepala*.

Forma **rhodesica** R. Fernandes in Bol. Soc. Brot., Sér. 2, **52**: 189, t. 8, t.10 fig. et t. 12 fig. i–o (1978). TAB. **5** fig. B. Type: Zimbabwe, between Lundi and Tokwe, *Pole Evans* 19 (PRE). Also cultivated in Gardens of the Division of Pl. Industry, Union Dep. Agr., Pretoria, Nat. Herb. 14707 (K; PRE; also photos).

 Crassula rhodesica Mogg in sched., *nom. nud.*
 ?*Crassula albanensis* Schönl. in Rec. Albany Mus. **1**: 55 (1903); loc. cit. (1929). Type from S. Africa (Cape Prov.).

Zimbabwe. C: Macheke, c. 1600 m., xii.1919, *Eyles* 2019 (K; SRGH). E: Odzi, c. 1120 m., bank of dry stream in shade, 29.v.1936, *Eyles* 8562 (K). S: Chibi, Nyoni Mts., S. slopes, ± 1000 m., 20.iv.1967, *T. Müller* 610 (K; SRGH).

Known so far only from Zimbabwe, but possibly occurring elsewhere. Habitat as for forma *nodulosa*.

Apart from its lanceolate, attenuate and acute leaves, this form often has more slender, shorter stems than forma *nodulosa*, and smaller flowers (calyx usually not more than 2·5 mm. and anthers slightly less than 0·5 mm. long). The verticillasters are also less dense and smaller in diameter than in the type form and most specimens of var. *longisepala*. As far as can be ascertained by herbarium specimens this form is the only one of *C. nodulosa* present in certain regions (Ruwa, Rusape, Odzi, Victoria, Chibi). But in Marandellas, var. *longisepala* is also found.

The type of *C. albanensis* is rather similar to forma *rhodesica*. It differs mainly in the indumentum which consists of very short bulliform papillae. The same kind of indumentum is found sometimes in typical *C. nodulosa* (e.g. in the type of *C. guchabensis*) and in var. *longisepala*.

Var. **longisepala** R. Fernandes in Bol. Soc. Brot., Sér. 2, **52**: 190, t. 9, t. 10 fig. d, t. 11 et t. 12 fig. p–v (1978) TAB. **5** fig. A. Type: Zimbabwe, Inyanga, Rhodes Inyanga Hotel, Viaduct of Maroni R., c. 2080 m., *Simon* 756 (K, holotype; PRE; SRGH).

 ?*Crassula mariae* R.-Hamet in Bull. Herb. Boiss., Sér. 2, **8**: 717 (1908). Type: Mozambique, *Junod* s.n. (G?; Herb. R.-Hamet, ubi?).
 Crassula nodulosa sensu Gilliland in Journ. S. Afr. Bot. **4**: 91 (1938) non Schönl. (1903).

Zimbabwe. E: Umtali, Nuza Plateau, iii.1935, *Gilliland* 1799 (BM; K; PRE). S: Wedza, on slopes of Wedza Mt., c. 1824 m., xii.1961, *R. M. Davies* 2950 (SRGH). **Mozambique.** MS: Manica, Serra Zuira, Tsetsera, on road to Vila Pery, c. 1800 m., 3.iv.1966, *Torre & Correia* 15633 (LISC).

Known only from Zimbabwe and Mozambique. Habitat as for forma *nodulosa*.

This variety has the largest flowers of the species, with very attenuate, acute sepals exceeding the corolla, and the most robust plants with tall thick stems, large, broad leaves and broad, very condensed verticillasters. It is very frequent in the eastern province of Zimbabwe, mainly in Inyanga District and Umtali (Mt. Nuza). No plants of forma *rhodesica* have been collected in gatherings mixed with var. *longisepala*. However, some specimens (e.g. *Vereker* 15553 from

Enterprise and *Müller* 791 from Belingwe) have shorter calices and are transitional with typical *C. nodulosa*.

From its description, *C. mariae* agrees with var. *longisepala*. The "types" in the Barbey-Boissier Herbarium (G) do not include Junod's original plant, collected in (fide R.-Hamet, loc. cit.) in Mozambique, but two specimens cultivated in the gardens of "La Pierrière" at Chambéry and belonging to two different species. One of them seems to be near to *C. setulosa*, the other to *C. nodulosa*; but the poor axillary cymes, possibly modified by culture, with flowers in a very young state, do not permit sure identification.

18. **Crassula globularioides** Britten in F.T.A. **2**: 389 (1871). — Bak. f. in Trans. Linn. Soc., Ser. 2, Bot. **4**: 13 (1894). — Brenan in Mem. N.Y. Bot. Gard. **8**, 5: 434 (1954). — Chapman, Veg. Mlanje Mt. Naysal.: 22, 56 & 65 (1962). — Binns, H.C.L.M.: 41 (1968). — R. Fernandes in Bol. Soc. Brot., Sér. 2, **52**: t. 13 fig. a–e (1978). Type: Malawi, Mt. Chiradzura, *Meller* s.n. (K, holotype).
   Crassula sp. — Chapman, op. cit.: 44 (1962).
   Crassula globularioides subsp. *globularioides*. — Tölken in Contr. Bolus Herb. **8**, 2: 459 (1977).

A dwarf undershrub up to 13(20) cm. high, forming small compact mats, with a much branched woody rootstock. Branches short, up to 5 mm. in diam., leafless, blackish and with lower parts covered by the persistent, dry sheaths of the old leaves, reddish and leafy along the terminal 1·5–6·5(15) cm. Leaves up to 2·2 × 1·2 cm., oblong to oblong- or subcircular-obovate, usually obtuse at the apex, entire and ciliate at the margin (cilia rigid, white), slightly attenuate or contracted at the broad base, flat, ± succulent and green tinged red when fresh, turning somewhat rigid, thin and reddish-brown to nearly black on drying, usually glabrous on both sides, sessile, usually 4-ranked and ± condensed (rosettes up to 4·5 cm. in diameter, sometimes nearly spherical) or somewhat scattered and decussate along the stem but always longer than the internodes, erect or ± spreading. Flowers 5-merous in terminal, dichotomous, corymbose or sometimes dense and globular inflorescences, 1–5 cm. in diameter; peduncle 1–8 cm. long, erect, slender, retrorsely appressed papillose-hairy, with two pairs of opposite, connate at base, oblong, obtuse or subacute ciliate bracts, one pair between the base and the middle of the peduncle, the other below the inflorescence; pedicels up to 2·5 mm. long. Calyx (2)2·5–3 mm. long, half as long as corolla or more; sepals (1)1·5–1·75 mm. broad, triangular or lanceolate, attenuate towards the acute apex, keeled, ciliate at the margin and along the keel (cilia ± rigid, acute, spreading, white, ± scattered). Corolla 4–5 mm. long, white; petals oblong-obovate to broadly ovate-elliptical, obtuse to nearly acute, shortly mucronate below the apex, fleshy, smooth, erect or slightly spreading or reflexed at the top. Filaments 1·5–2·25 mm. long; anthers 0·5–0·75 mm. long. Follicles (with styles) 2·5–3 mm. long. Scales 0·4–0·6 × 0·3–0·4 mm., subrectangular, truncate or emarginate at the top, purple.

1. Leaves glabrous on both faces - - - - - - - - - 2
 – Leaves papillose-hairy beneath - - - - - - - forma *pilosa*
2. Leaves rather thin on drying, frequently acute; marginal cilia up to 1·75(2) mm. long, not very close, ± at right angles to the margin; sepals not or sparsely ciliate *longeciliata*
 – Leaves thicker, obtuse or rounded at the apex; marginal cilia shorter, very close, retrorse; sepals ciliate - - - - - - - - - forma *globularioides*

Forma **globularioides**.

Malawi. S: Mulanje Mt., c. 2080 m., 14.vii.1958, *Chapman* H/715 (K; SRGH).

Known only from Malawi (also present in Kenya and Tanzania?). On bare or lichen-covered rocks or among moss in sunny situations of high mountains at 1280–3040 m.

The specimen *Chapman* 719 (SRGH) is an abnormal plant with an elongated stem and scattered leaves; its habit was undoubtedly conditioned by the habitat, described as "on a big rock on a steep face along with moss, small ferns and *Streptocarpus*, in river bed, growing in an almost permanently cool, damp climate". This is the *Crassula* sp. referred to by Chapman, loc. cit. Despite its unusual appearance, the shape, size and glabricity of the leaves, the peduncle indumentum and both the shape and characteristic marginal cilia of the sepals, are typical of *C. globularioides* forma *globularioides*.

Forma **pilosa** R. Fernandes in Bol. Soc. Brot., Sér. 2, **52**: 191 (1978). Type: Malawi, Mulanje Mt., Ruo Gorge, *Newman & Whitmore* 631 (BM, holotype; SRGH).

Malawi. S: Mulanje Mt., west face of Gt. Ruo Gorge, c. 1280 m., 29.vii.1956, *Newman & Whitmore* 631 (BM; SRGH).

Known only from Malawi. On rocks.

Like forma *globularioides* but leaves ± sparsely papillose-hairy beneath, the hairs white, subappressed, retrorse.

Forma **longiciliata** R. Fernandes in Bol. Soc. Brot., Sér. 2, **52**: 192 (1978). Type: Malawi, Dedza, Giwaa, *Chapman* 1433 (LISC; SRGH, holotype).

Malawi. C: Dedza Distr., Chiwau Hill, Chongoni Forest Reserve, 4.x.1968, *Salubeni* 1140 (SRGH).
Known only from Malawi (Dedza). On rocks.

If the characters ascribed to this form prove to be constant and not merely caused by a more moist environment, it could perhaps be considered at subspecific level.

C. globularioides is very close to *C. swaziensis* Schönl., from which it differs in having smaller and usually more condensed rosettes of leaves; leaves not or only slightly contracted at the base and usually glabrous on both sides (in *C. swaziensis* they are usually hairy on both faces), thinner and less rigid on drying and with the cilia longer, stronger, contiguous, more recurved and rigid than in *C. swaziensis*; inflorescences smaller and denser, on shorter, more slender, appressed retrorse-hairy peduncles (these are hispid in *C. swaziensis*); sepals often more than half as long as corolla, broader, narrowing gradually from base to apex, acute, usually ciliate only at the margin and along the keel (with stronger, more scattered, not so longer cilia) while in *C. swaziensis* they are often less than half as long as corolla, elliptic or oblong, subobtuse, and dorsally hispid; petals not contracted at the middle, etc.

19. **Crassula morrumbalensis** R. Fernandes in Bol. Soc. Brot., Sér. 2, **55**: 97 (1982). Type: Mozambique, Mt. Morrumbala, *Torre* 4551 (LISC, holotype).

A perennial succulent herb. Stem procumbent up to 4 mm. thick, glabrous, rooting at some nodes, branched, leafless at the prostrate part but with leaves at the ascending extremity, emitting short erect branchlets lengthwise and sometimes also subsessile rosette of leaves. Branchlets up to 5 cm. long, sharply 4-angular, concave at the sides, thickened at nodes, covered with a papery ± detaching bark (or epidermis?) which is brown-blackish towards the base of branches and brown-reddish towards the apex, foliate; middle-internodes up to 8 mm. long, successively decreasing towards apex and base, those of rosettes very short. Leaves either oblong-elliptical, up to 24 × 5 mm., attenuate towards the base (± 2·5 mm. broad just above the sheath) and acute, or elliptical, up to 13 × 7 mm. and nearly acute at the apex, all entire, with acute, ciliate margins, (the cilia very short, obtuse, white, rather close, sometimes near touching, retrorse, those near the leaf-base somewhat longer (up to 0·3 mm. long), acute and ± scattered), glabrous on both surfaces, flat, decussate, erect and disposed in rosettes at the extremities of stem and branches, the other erect-spreading or spreading, neither very thick nor very rigid and brown-reddish (the older ones blackish) on drying; sheath up to 1·5 mm. long. Peduncles solitary at the end of the erect branchlets, surrounded at the base by the leaf-rosettes, slender, glabrous or very slightly pilose at the apex, (the hairs very short, retrorse, appressed, whitish), with two pairs of small, ciliate bracts, one pair between the base and the middle of peduncle, the other between the middle of the peduncle and the base of the inflorescence, sometimes only a pair present. Flowers 5-merous, disposed in small cymes forming dense corymbs up to 2 cm. in diam. at the end of the peduncles; branches of inflorescence somewhat thick, glabrous or sparsely-pilose, the hairs appressed, very short; pedicels up to 1·5 mm. long; bracts and bracteoles very shortly ciliate at the margin. Calyx 2–2·25 mm. long, obconical below the sepals; sepals 1·25–1·75 × 0·5–0·75 (at the base) mm., shorter than the half of corolla, oblong-lanceolate, subacute, apparently not or very slightly keeled, ciliate at the margin and along the median line of the back, the cilia very short, sparse. Corolla ± 4 mm. long, white; petals oblong-obovate, 1·25–1·5 mm. broad, obtuse, minutely scabrous-papillose at the upper ⅔ on the outside and with a very small mucro just below the apex, connate at the base for ca. 0·6 mm. Stamens shorter than the corolla; filaments 2–2·25 mm. long; anthers ± 0·6 mm. long, ovate before dehiscence, oblong after dehiscence. Follicles (with styles) 3·5–4 mm., long, tapering in the upper part towards the styles, smooth; styles ca. 0·5 mm. long, smooth; stigmas minutely capitate. Seeds ca. 0·5 mm. long, rounded at the top, slightly ribbed. Scales ca. 0·4 mm. long, subquadrate, very slightly emarginate.

Mozambique. Z: Massingire, Morrumbala Mt., 6.viii.1942, *Torre* 4551 (C; COI; K; LISC, holotype).
Known only from Mozambique. In savannas of mountain slopes.

20. **Crassula nyikensis** Bak. f. in Kew Bull. **1897**: 265 (1897). — Berger in Engl. & Prantl, Nat. Pflanzenfam. ed. 2, **18a**: 394, fig. 191 A–C (1930). — Binns, H.C.L.M.: 41 (1968). — Jacobs., Das Sukk. Lexikon: 145 (1970). Type: Malawi, Nyika Mt., *Whyte* 491 (K, holotype).

   Crassula whyteana Schönl. in Engl., Bot. Jahrb. **43**: 360 (1909). Type: Malawi, Nyika Plateau, Whyte s.n. (B? holotype; K, isotype).

A low, much branched undershrub. Old branches woody, covered by the persistent, dense, dried leaves of previous years; young branches up to 5 cm. long, ending in leaf-rosettes up to 2 cm. in diam., either barren or surrounding a flowering axis. Leaves of the rosettes up to 1·2 × 1 cm., densely imbricate, broadly obovate to subcircular, rounded at the apex, entire and ciliate at the margin (cilia contiguous, white), sessile and rather connate at the base, flattish, concave, fleshy and green when fresh, slightly rigid to submembranous and brownish, dark reddish-brown or nearly black on drying; older leaves spreading to finally subdeflexed, nearly black, the young ones erect. Flowering stems up to 10 cm. long, erect, shortly and subpatently papillose-hairy, leafy from the base to the inflorescence, the leaves ± dense, decussate, the undermost similar to those of rosettes and longer than the internodes, the others successively smaller and relatively narrower, longer than or subequalling the internodes, appressed or subreflexed, papillose on the outside towards the base. Flowers 5-merous, in dense, terminal corymbs, 1·5–3 cm. in diameter; inflorescence-branches papillose-pilose; pedicels up to 4 mm. long, sparsely papillose to glabrous. Calyx 2·25–2·6 mm.; sepals 1·5–2 mm. long, triangular-lanceolate, attenuate towards the acute apex, carinate, equalling or longer than the half of the corolla, glabrous, shortly and irregularly ciliate at the margin and sometimes along the median line, rarely without cilia. Corolla c. 3·5 mm. long, white; petals 1·5–1·75 mm. broad, obovate-elliptic, obtuse, not or indistinctly mucronate, smooth. Filaments c. 1·5 mm.; anthers c. 0·4 mm., subcircular. Follicles 2·25–2·5 mm. long (with the styles). Scales c. 0·5 × 0·5 mm., subrectangular-cuneate, rounded-truncate at the apex.

Malawi. N: on highest ridges of Nyika Mountains, 2400 m., vii.1896, *Whyte* 491 (K). Known only from Malawi. On bare gneiss and granite rocks.

Similar to *C. globularioides* Britten from which it differs by the smaller, thinner and more densely arranged leaves, the older ones not being caducous above the base but completely persistent on the old branches; by the smaller corolla, anthers and follicles; and, above all, by the flowering stem which is leafy from base to apex in contrast to *C. globularioides* where it is leafless and provided with only 1–2 pairs of small bracts.

As Tölken suggests (in Contr. Bolus Herb. **8**, 2: 460 (1977)), it is perhaps only an ecological variant of *C. globularioides*.

21. **Crassula swaziensis** Schönl. in Journ. Linn. Soc., Bot. **31**: 548 (1897); in Rec. Albany Mus. **1**, 1: 62 (1903). — Burtt Davy, F.P.F.T.: 140 (1926). Type from Swaziland.

An undershrub or perennial herb up to 20 cm. high (incl. the inflorescence) with a woody or succulent rootstock, emitting a short, little branched stem; lower part of branches leafless, covered with the persistent, dry, dark bases of fallen leaves. Leaves sessile, erect or erect-spreading, usually ± condensed at the end of branches, often in ± dense rosettes, or sometimes the pairs ± scattered over 4–7·5 cm., fleshy and often reddish when fresh, rather thick and rigid on drying, dark when glabrous or subglabrous, whitish to grey or yellowish-grey when hairy (hairs retrorse, appressed or subspreading, short, whitish, acute, thin), ciliate or indistinctly so at the margin. Flowers 5-merous, sessile to shortly pedicellate, in ± condensed cymes forming corymb-like (up to 11 cm. in diameter, usually rather smaller) or sometimes panicle-like (up to 15 × 14 cm.) inflorescences; peduncles 4·5–16 cm. long, erect, straight, terete, usually slender, with 1–3(4) pairs of small bracts, subappressed-retrorsely (towards the base) to subpatently (towards the apex) hairy, the hairs as in the leaves but longer and thinner; bracts 5–7(9) mm. long, oblong, slightly connate; pedicels 0–1·5 mm. long, hispid. Calyx 1·5–2·5(3) mm. long, up to ½ as long as corolla, ± densely whitish hispid everywhere outside; sepals 0·75–1 mm. broad, ovate-oblong or subelliptical, rarely triangular-oblong, subobtuse, usually not carinate, ciliate at the margin. Corolla (3)3·25–4·5(5) mm. long, campanulate, somewhat contracted just above the top of sepals, white or cream, flushed with pink when fresh, rusty-brown or orange when dry, then forming a distinct contrast in colour with the ± whitish calyx; petals oblong, oblong-spathulate, broadly elliptic to subovate, usually obtuse or sometimes roundish or subacute at the top, with a subterminal ± distinct mucro, often scabrid-papillose on both surfaces or only externally towards the upper half or smooth everywhere, nearly hyaline or scarious and concave near the base and there connate for 0·5–0·7 mm., spreading or reflexed above. Filaments 1–2·5 mm. long; anthers 0·5–0·75 mm. long. Follicles (1)1·5–3 mm. long (with the styles), attenuate or somewhat contracted into the distinct or obsolete styles. Scales 0·4–0·75 × 0·3–0·6 mm., obovate-oblong, cuneate or rectangular, obtuse, truncate or emarginate at the apex, pale to purple. Seeds ± 0·5 mm. long, oblong, papillose.

C. swaziensis is a very polymorphic species. We have distinguished the following infraspecific taxa in the area of Flora Zambesiaca:

1. Follicles (1)1·5–1·75 mm. long; styles obsolete with sessile or subsessile, sublateral stigma; corolla (2·75)3–3·75(4–4·5) mm. long; scales usually less than 0·4 mm. wide, often as long as or longer than twice the breadth - - - - - - - - subsp. *brachycarpa*
– Follicles 2–3 mm. long; styles usually distinct with terminal stigma; corolla 3–4·5(5) mm. long; scales 0·4–0·6 mm. wide, less than twice as long as broad - - - - - - 2
2. Petals up to 1·5 mm. broad, oblong, obtuse or sometimes acute, usually scabrid-papillose upwards - - - - - - - - - - - - - - - - - - 3
– Petals 2–2·5 mm. broad, broadly elliptic, with the apex rounded, smooth on both surfaces 4
3. Leaves glabrous or very sparsely pilose on both surfaces
                                 subsp. *swaziensis* var. *swaziensis* f. *swaziensis*
– Leaves ± pilose on both surfaces - - subsp. *swaziensis* var. *swaziensis* f. *argyrophylla*
4. Leaves glabrous on both surfaces - - - subsp. *swaziensis* var. *guruensis* f. *guruensis*
– Leaves pilose on both surfaces - - - subsp. *swaziensis* var. *guruensis* f. *brevipilosa*

Subsp. **swaziensis**. R. Fernandes in Bol. Soc. Brot., Sér. 2. **52**: 192 (1978); op. cit. **55**: 114 (1982).

Var. **swaziensis**.

A perennial herb, growing singly, in pairs or gregariously; old bark falling off in irregular flakes. Leaves 1·2–5(6) × 0·7–2·5(4·3) cm., obovate, obovate-spathulate or -cuneate to suborbicular, somewhat asymmetrical, usually rounded or obtuse or shortly acuminate or sometimes nearly acute at the top, entire and undulate or flat at the often ciliate margin, contracted below or attenuate towards the base and here broadly widened, flattish, concave above, caducous just above the connate base; internodes usually many times shorter than the leaves, sometimes elongate up to 2(4) cm. Calyx whitish hairy usually everywhere; sepals ovate-oblong or subelliptical. Corolla (3)3·5–4·5(5) mm. long; petals usually papillose-scabrid like the follicles.

Forma **swaziensis**. — R. Fernandes in Bol. Soc. Brot., Sér. 2, **52**: 193, t. 13 fig. f-i (1978). Type as for *C. swaziensis*.
    Crassula argyrophylla var. *swaziensis* (Schönl.) Schönl. in Trans. Roy. Soc. S. Afr. **17**, 3: 258 (1929). — Jacobs., Das Sukk. Lexikon: 140 (1970) *comb. illegit*.

In this form the leaves are either completely glabrous on both faces or only above or beneath; if pubescent then sparsely so.

Zimbabwe. E: Umtali, Tshakwe Mts., Burma Valley, Burma Farm, c. 1470 m., 21.viii.1955, *Chase* 5811 (BM; LISC; SRGH).
Also in Swaziland and S. Africa (Natal and Transvaal). In shallow soil on rocks.

Forma **argyrophylla** (Diels ex Schönl. & Bak. f.) R. Fernandes in Bol. Soc. Brot., Sér. 2, **52**: 193, t. 13 fig. j-r (1958). Syntypes from S. Africa (Transvaal).
    Crassula argyrophylla Diels ex Schönl. & Bak. f. in Journ. of Bot. **40**: 290 (1902). — Schönl., in Rec. Albany Mus. **1**, 1: 63 (1903), in S. Afr. Journ. Sci. **17**, 2: 188 (1921); pro parte quoad *Eyles* 811; loc. cit. (1929). — Engl. in Sitz.-Ber. Königl. Preuss. Akad. Wiss. **52**: 868 (1906). — Burtt Davy, F.P.F.T. **1**: 141 (1926). — Berger in Engl. & Prantl, Nat. Pflanzenfam. ed. 2, **18a**: 397 (1930). — Pole Evans in Fl. Pl. S. Afr. **19**: t. 754 (1939). — Hutch., Botanist in S. Afr.: 673 (1946). — Brenan in Mem. N.Y. Bot. Gard. **8**, 5: 435 (1954). — Letty, Wild Fl. Transv.: 149 (1962). — Binns, H.C.L.M.: 41 (1968). — Schields in Bull. A.S.P.S. **3**, 3: 88 (1968). — Jacobs., Das Sukk. Lexikon: 140 (1970). — Merxm. & al. in Ann. Naturh. Mus. Wien, **75**: 144 (1971). — J. H. Ross, Fl. Natal: 179 (1972). Syntypes as above.
    Crassula argyrophylla Diels in Engl., Bot. Jahrb. **39**: 465 (1907). Holotype from S. Africa (Transvaal), (*Wilms* 527).
    Crassula argyrophylla var. *ramosa* Schönl., loc. cit. (1929). — Jacobs., loc. cit. Type from S. Africa (Transvaal).
    Crassula globularioides subsp. *argyrophylla* (Diels ex Schönl. & Bak. f.) Tölken in Journ. S. Afr. Bot. **41**, 2: 106 (1975); in Contr. Bolus Herb. **8**, 2: 460 (1977). Syntypes as for *Crassula swaziensis* forma *argyrophylla*.

In this form the leaves, especially when young, are ± densely covered by an indumentum of short, retrorse, subappressed or subspreading white hairs. Old leaves or sometimes all leaves have a less dense indumentum, the colour of these being then ± dark, sometimes nearly black on drying. At the margin the leaves are ciliate, with a row of cilia usually not very distinct from the other hairs, spreading or retrorse, ± arched, not contiguous but often somewhat scattered between them. The sepals are also whitish hairy everywhere on the outside and ciliate at the margin.

Zimbabwe. C: Rusape, iv.1948, *Munch* 54 (K; SRGH). E: Inyanga, Nyangwe Fort, Inyanga Nat. Park, c. 2050 m., 2.vi.1968, *Plowes* 2897 (K; LISC; SRGH). **Malawi.** N: Nkhata Bay, Kamunga Rock, c. 1800 m., 16.iv.1967, *Pawek* 976 (SRGH). C: Chenga Hill, 9.ix.1846, *Brass* 17611 (K; SRGH)). **Mozambique.** T: cult. in a garden at Sintra (Portugal, 1966) from a shoot gathered at Angónia, Dómué Mt., 1964, by *Correia* (LISC). MS: Gorongosa, at the top of Nhandore Mt., c. 1840 m., 19.x.1965, *Torre & Pereira* 12426 C: (LISC). M: Libombos Mts., M' Ponduine Mt., Namaacha, 25.iv.1947, *Pedro & Pedrógão* 725 (LMA; LMU).

Also in Swaziland, Transvaal and Natal. In the rupideserta, *Brachystegia*-savanna, in leaf-mould, on granite- or quartzite-rocks, on quartzite-sand, etc. Frequent.

This is the most common form. Mozambican plants of Tete and Libombos Mts. and the specimen *Brass* 17611 of Malawi have smooth petals, whereas in most other specimens the petals are papillose-scabrid, as are the follicles. On the other hand, specimens *Correia* s.n. and *Brass* 17611 have narrower, subacute petals with longer mucro than the type of forma *argyrophylla*. The same plants, by their shorter follicles, approach subsp. *brachycarpa* R. Fernandes.

Var. **guruensis** R. Fernandes in Bol. Soc. Brot., Sér. 2, **52**: 194, t. 13 fig. s–v (1978). Type: Mozambique, Sururua (Gúruè) Namúli Mt., *Andrada* 1868 (COI, holotype; LISC).

It differs from var. *swaziensis* by the larger corymbs: by the calyx not so densely hispid, sometimes only with the sepals ciliate at the margin and along the median line; by the broader petals (2·0–2·5 mm. broad, whereas in var. *swaziensis* they are not broader than 1·5 mm.), elliptical and rounded at the apex and smooth on both faces.

Forma **guruensis**. — R. Fernandes in Bol. Soc. Brot., Sér. 2, **52**: 195 (1978). Type as above.

Mozambique. Z: Gúruè Mt., near the waterfalls of the river, 1600 m., 20.x.1944, *Mendonça* 2173 (COI; LISC; MO).
Known only from Mozambique. In Rupideserta, on rocks.
This form, which has leaves glabrous on both faces, is the most frequent in var. *guruensis*. In some aspects (glabrous leaves, sepals usually less hispid, smooth petals and follicles with very distinct styles) it approaches *C. globularioides*. Nevertheless, it differs from the latter by the leaves usually larger and thicker; the rather wider inflorescences; the linear, obtuse sepals shorter relative to the corolla than in *C. globularioides*; by the thinner and broader petals, etc. On account of both its characters and geographical situation, var. *guruensis* could perhaps be considered at subspecific rank.

Forma **brevipilosa** R. Fernandes in Bol. Soc. Brot., Sér. 2, **52**: 195 (1978). Type: Mozambique, Sururua (Gúruè), Namúli Mt., *Andrada* 1867 (COI, holotype; LISC).

This form is distinguished from the typical form by the leaves which are shortly appressed retrorse-pilose on both faces.

Mozambique. Z: Sururua (Gúruè), Namúli Mt., *Andrada* 1867 (COI; LISC).

Subsp. **brachycarpa** R. Fernandes in Bol. Soc. Brot., Sér. 2, **52**: 197, t. 13 fig. 1–9 (1978). Type: Malawi, Zomba Distr., Ntonya Mt., *Newman & Whitmore* (BM, holotype; BR; K; SRGH).
    Crassula argyrophylla sensu Schönl. in S. Afr. Journ. Sci. **17**, 2: 188 (1921) pro parte quoad *Teague* 255 ("a new species?"). — sensu Chapman, Veg. Mlanje Mt. Nyasal.: 32 (1962).

This taxon is separate from the typical subspecies by the usually smaller flowers with corolla (2·75)3–3·5(4–4·5) mm. long and filaments only 1–1·5 mm. long (not 2–2·5 mm.); by the shorter follicles (1)1·5–1·75 mm. long (not 2–3 mm.), neither attenuate nor subcontracted into the styles but with the styles absent or nearly so, the stigmas being sessile or subsessile and sublateral; and by the relatively longer and narrower scales. Plants with smooth petals and others with scabrid-papillose ones, as in subsp. *swaziensis*, are also found.

Zimbabwe. N: Mazoe, Iron Mask Hill, c. 1600 m., vi.1915, *Eyles* 617 (BM; K; SRGH). E: Umtali, Dora R., c. 960 m., 5.vi.1948, *Fisher* 1567 (K; PRE; SRGH). S: Belingwe E. Reserve, 7.vii.1953, *Wild* 4130 (K; LISC; PRE; SRGH). **Malawi.** S: Zomba Plateau, 13.viii.1960, *Leach* 10435 (K; SRGH).

Known only in Zimbabwe and Malawi. Among rocks or on dry open rock faces.

We think that this taxon is not merely a form whose smaller flowers and fruits were conditioned by the environment, because, not only are the scales usually longer than in the type, but also the leaves and inflorescences are not reduced, being sometimes of a large size. By the ensemble of its characters, it might perhaps be considered as an independent species.

Tölken (in Contr. Bolus Herb. **8**, 2: 460, 1977) reports for Malawi the subsp. *illichiana* (Engl.) Tölken of *C. globularioides*, no specimen being cited. *C. illichiana* Engl., of which we have seen one of the syntypes (*Buchwald* 174, K, from Tanzania (Usambara)), is in general aspect. rather distinct from *C. globularioides*, seeming to us more like *C. swaziensis* subsp. *swaziensis* var. *swaziensis*. However, on account of the slightly shorter calyx and longer carpels, perhaps it deserves to be considered as another variety of *C. swaziensis* of Tanzania and Uganda.

22. **Crassula leachii** R. Fernandes in Bol. Soc. Brot., Sér. 2, **55**: 95, t. 1 (1982). Type: Mozambique, Chimoio (Vila Pery), *Leach* 8135 (M; PRE, holotype; SRGH).

A perennial herb (or undershrub?). Main branches up to 23 cm. long and $\pm$ 2·5 mm. in diam., flexuose, herbaceous or slightly woody towards the base, the others shorter, more slender, ascending, all fleshy, terete, nearly of an equal diameter from base to top, foliate lengthwise, dark reddish, densely hispid by spreading or slightly retrorse whitish short hairs, ending in a peduncle; nodes not dilatate; internodes 0·4–2 cm. long, the uppermost the shortest. Leaves 1·2–2·2 × 0·25–0·9 cm., oblong to elliptic, obtuse at apex, entire and not ciliate at margin, not contracted, sessile and slightly connate at the base, spreading or erect-spreading, seeming to have been flat in living state, neither very thin (but not membranous) nor rigid on drying, with an indumentum on both faces as that of the branches and on account of this greyish or whitish. Flowers 5-merous, sessile or very shortly pedicellate, in condensate small cymes, forming 3-furcate, up to 2·8 cm. in diam. corymbs at the end of the main branches, or only a cyme at the end of the branchlets; peduncles 4·5–8 cm. long, slender, densely hispid, with 2–3 pairs of small (up to 0·5 cm. long), oblong, hispid bracts; pedicels absent or up to 0·5 mm. long, thick. Calyx 1·6–2 mm. long, equalling to slightly longer than the half of the corolla; sepals ovate to triangular, obtuse or nearly so, connate at base for $\pm$ 0·75 mm., hispid outside, ciliate at the margin. Corolla 3–3·25 mm. long, campanulate, white; petals oblong or obovate-oblong, obtuse or rounded at the apex, erect below, reflexed or spreading upwards, minutely scabrid-papillose on the outside at the upper ⅔ and with a small mucro just below the apex. Filaments 1–1·25 mm. long; anthers 0·4–0·6 mm., suborbicular, purple. Follicles c. 1·5 mm., not or shortly attenuate; stigmas subterminal, nearly sessile. Scales c. 0·5 mm. long, cuneate, truncate or slightly emarginate at the apex.

Mozambique. MS: c. 10 miles S. of Vila Pery, 12.vi.1961, *Leach*, 11107 (PRE; SRGH). Known only from Mozambique. On granite rocks.

This taxon was considered (in shedis) as *C. argyrophylla* (*C. swaziensis* var. *ramosa* Schönl.). We think that it is neither this entity (to which we do not attribute any taxonomical value) nor any infraspecific taxon of *C. swaziensis*, notwithstanding by its small flowers, fruits and indumentum it approaches subsp. *brachycarpa* of that species. If the rather longer and not woody branches and the scattered leaves of *C. leachii* were conditioned by a less dry and more shady habitat than the usual, then the same habitat conditions would determine also a looser inflorescence, larger flowers on longer pedicels and a laxer indumentum, but this is not the case. On the other hand, the shape of the leaves is different from those found in all specimens of *C. swaziensis* seen by us.

The specimen *Brummitt* 15147 from Malawi (Distr. Mangochi, Mangochi Mt., 20.xi.1977), of which we have seen only a fragment (in spirit) of the plant cultivated at Kew Gardens, is very similar to *C. leachii*, seeming to differ by the longer styles and more attenuate follicles. More material with a complete stem from wild plants is, however, necessary to elucidate the problem.

23. **Crassula zombensis** Bak. f. in Kew Bull. **1897**: 266 (1897). — Binns, H.C.L.M.: 41 (1968). Type: Malawi, Zomba Mt., *Whyte* s.n. (K).
    Crassula swaziensis var. *zombensis* (Bak. f.) R. Fernandes in Bol. Soc. Brot., Sér. 2, **52**: 196, t. 14 et t. 15 (1978).

A mat forming perennial herb or suffrutex with prostrate, woody, leafless, glabrous, branched main stem which emits erect or ascending branches. Branches inserted 0·7–3 cm. apart, up to 11 cm. long and 0·5 cm. in diam., subquadrangular or cylindric, often 2-forked or sometimes simple, with brown-scarious ± fissured bark, leafless or, when simple, leafy upwards, glabrescent, with the scars of the fallen leaves ring-shaped and pale, rather visible but not prominent; branchlets of the year usually the only leafy, 1·5–6 cm. long, herbaceous, grey or green-yellowish, hispidulous, the hairs short, whitish; internodes 0·2–1·5 cm., all ± shorter than the leaves, the apical of the leafy branchlets the shortest ones. Leaves 1·3–4·5 cm. long and 2–10 mm. broad, linear, 3-gonous or semiterete (convex beneath and canaliculate above in the living state), obtuse at the apex, entire, slightly attenuate towards the base, sessile and very shortly connate, not ciliate at the obtuse margins, ± covered with whitish, short, appressed or subspreading, subretrorse hairs, fleshy ± in living state, rigid on drying, erect or erect-spreading, somewhat arched; leaves of the old branches falling off together with the connate base. Flowers 5-merous, disposed in dense cymes, forming 4–7 branched corymb-like, terminal inflorescences, 2–3 cm. long and 3·5–5 cm. in diam.; peduncule 7·5–9 cm. long, slender, hispidulous, with 2–3 pairs of opposite bracts; pedicels up to 1·5 mm. long. Calyx 2·5 mm. long, densely hispidulous by whitish hairs; sepals c. 2 × 1 mm., oblong-triangular, convex, fleshy. Corolla 3·25–4·5 mm. long, stellate, white in living state, dark-brown on drying and forming a strong contrast in colour with the whitish calyx; petals 1·5–2·25 mm. broad, obovate, rounded at the top and with a subapical small mucro on the outside, smooth or nearly so, reflexed. Filaments ± 2·75 mm. long. Follicles 2·5–3·5 mm. long, rather attenuate, with distinct, divergent styles and apical stigmas. Scales 0·4 mm., subrectangular.

Malawi: Zomba Distr., Zomba Plateau, below road to summit opposite Malosa saddle, rocky slopes, 1900 m., 2.viii.1970, *Brummitt & Banda* 12377 (K). **Mozambique.** Z: Namúli Peaks, W. face, c. 1440 m., 26.vii.1962, *Leach & Schelpe* 11483 (K; LISC; SRGH). Also cultivated in a Garden at Sintra (Portugal) from *Torre & Correia* 14734 and 16030 (Gúruè, Mozambique).

Known only from Malawi and Mozambique. On rocks in dry situations.

24. **Crassula heterotricha** Schinz in Bull. Herb. Boiss. **2**: 203 (1894). — M. Wood, Handb. Fl. Natal: 46 (1907). — Jacobs., Das. Sukk. Lexikon: 143 (1970). — Venter in Journ. S. Afr. Bot. **37**, 2: 106 (1971). — J. H. Ross, Fl. Natal: 179 (1972). — R. Fernandes in Bol. Soc. Brot., Sér. 2, **55**: 106 (1982). Type from S. Africa (Natal).

   Crassula pallida sensu M. Wood, Natal Pl. **4**: t. 323 (1903) non Bak. (1874).
   Crassula perfoliata sensu Burtt Davy, F.P.F.T. **1**: 140 (1926). — Schönl. in Trans. Roy. Soc. S. Africa. **17**, 3: 224 (1929) pro parte, non L. (1753).
   Crassula perfoliata var. *heterotricha* (Schinz) Tölken in Journ. S. Afr. Bot. **41**, 2: 115 (1975); in Contr. Bolus Herb. **8**, 2: 503 (1977). Type as for *Crassula heterotricha*.

A succulent undershrub or shrub, 20–50 cm. (or more?) high. Stem sprawling with apical portion erect, ± branched, 6–25 mm. in diameter, woody, blackish, leafless and marked by the scars of the fallen leaves or with dried older leaves towards the base, leafy like the branches for the uppermost 3–22 cm. Leaves (3)5–10·5(17·5) × 1–1·5(2·5) cm., the median ones the longest, oblong or triangular in outline, usually nearly parallel-sided, obtuse, rounded or rarely acute at the apex, entire at the margin, sessile and connate-amplexicaul at the base, very fleshy, flat above, convex beneath or sometimes subtrigonous, grey-green or glaucous to whitish, closely covered everywhere (except towards the base on the upper face) by an indumentum of contiguous, very short, blunt, whitish papillae, decussate or in two opposite rows, usually very dense, spreading to erect-spreading, rarely arched at the base and then upright, straight (not falcate); internodes usually many times shorter than the leaves, the upper ones completely concealed by the sheaths of the leaves. Flowers 5-merous, usually in ± dense, subspheric, corymbose in-florescences, 3·5–7 cm. in diameter; peduncle 7·5–14 cm. long and up to 3·5 mm. in diameter below, erect, papillose as the leaves towards the base and patently or retrorsely hairy upwards (the hairs dense, thin, acute, fulvous or whitish), with 3–4 pairs of bracts, 3-forked at the apex; pedicels up to 3 mm. long, hispidulous. Calyx 3–4·5 mm. long, greyish, hairy; sepals c. 1 mm. broad at the base, triangular to linear-lanceolate, acute to subobtuse, connate for c. 1 mm. Corolla 5–9(10) mm. long, white; petals 1–1·5 mm. broad, oblong or linear-oblong, subobtuse to acute and mucronulate just below the apex, smooth or minutely papillose-scabrous on both faces, connate at base for c. 1·75 mm., reflexed at the top. Filaments 4–6 mm. long; anthers 1–1·25(1·5) mm. long when unopened, shorter after dehiscence. Follicles

5·5–8·5 mm. long (with the styles), attenuate above into the styles, minutely tuberculate everywhere or only along the suture. Scales $\pm$ 0·5 × 0·75 mm.

Zimbabwe. E: Umtali, Mandambisi Mt., 32 km. SW. of Umtali, 26.v.1949, *Chase* 1705 (BM; K; SRGH). S: Belingwe, Emberengwa (?), $\pm$ 1500 m., 7.vi.1965, *Leach & Bullock* 12886 (SRGH). **Mozambique.** M: Namaacha Falls, 3.vi.1960, *Leach* 9985 (K; SRGH).

Also in Swaziland and S. Africa (Natal, Transvaal and Cape Prov.). In rocky places.

C. heterotricha is very close to *C. perfoliata* L. and *C. falcata* Wendl. It differs from *C. perfoliata* in its habit (stem relatively shorter, usually prostrate and more branched, with shorter internodes than in *C. perfoliata* in which the stem is erect and few-branched); in its leaves, which are not so broad at the base and not constricted into the upper half, but either narrow gradually to the apex or have $\pm$ parallel sides, and have a flat, not canaliculate, upper surface; in the flowers and fruits, which are at least twice as long; and in petal shape (*C. heterotricha* has petals linear-oblong or oblong, subobtuse to acute and mucronate; *C. perfoliata* has petals obovate, obtuse or rounded and not so distinctly mucronate). *C. heterotricha* differs from *C. falcata* in having a prostrate stem; in leaf shape (not more than 2 cm. broad at base, and straight in *C. heterotricha*; 4 cm. or more broad, and falcate in *C. falcata*); in calyx shape and size (3–4·5 mm. long, half as long as corolla or more, sepals relatively longer and more acute in *C. heterotricha*; 2–3 mm. long, less than half as long as corolla in *C. falcata*), and in petal-shape and -colour (often acute, always mucronate, relatively narrow, white in *C. heterotricha*; subobtuse, not mucronate, red in *C. falcata*). Apart from these morphological differences, the three species seems to be separable on biological characters: *C. heterotricha* flowers from May to August, *C. perfoliata* from October to January and *C. falcata* from November to February (Tölken, tom. cit.: 501–503, 1977).

Specimens of *C. heterotricha* from the F.Z. area have smooth petals, 8–9(10) mm. long, and follicles tuberculate only along the suture, whereas those we have seen from Natal usually have smaller (5–7 mm. long), papillose-scabrous petals and follicles tuberculate throughout. Unless these differences are due to the wetter habitat in the F.Z. area, perhaps specimens from this region should be placed in a separate variety.

SPECIES DOUBTFULLY RECORDED

Crassula cf. **lasiantha** E. Mey.

The specimen *Junod* 65 from Delagoa Bay, referred to by Schinz (in Mém. Herb. Boiss. no. 37, 1900) as *C.* cf. *lasiantha* was not seen by us. *C. lasiantha* Drège ex. Harv. (not E. Mey.) is a species known only from the Cape (cf. Tölken in Contr. Bolus Herb. **8**, 1: 293, 1977).

Crassula orbicularis L.

Crassula rosularis Haw., Rev. Pl. Succ.: 13 (1821). — Brenan in Mem. N.Y. Bot. Gard. **8**, 5: 435 (1954).

At Kew there is a specimen "cultivated in New York Botanical Garden, said to have been collected by *Brass* in Vernay Nyasaland Expedition, 1946" which belongs to *C. orbicularis* L. However, we have not found any herbarium material of the corresponding wild collection, or any other specimens of this species from the F.Z. area. According to Tölken (tom. cit.: 325–326, 1977), *C. orbicularis* occurs only in Natal and the Cape Prov.

Crassula tenuicaulis Schönl.

Tölken (op. cit.: 172 (1977)) mentions this species for "northeaster Rhodesia" without citing any evidence. There are, however, in K and LD specimens, collected by *Nordlindh & Weimarck* (4905) in Zimbabwe (Makoni, ad villam Wick, 9.ii.1931) named by Tölken *C. tenuicaulis*. Examination of those specimens has revealed that they do not belong to *Crassula* at all but may represent a species of *Portulaca*. Whereas the flowers have 2 sepals, 4 petals and 4 stamens the fruit is a pyxidium. We have not seen any material from the F.Z. area which could be attributed to *C. tenuicaulis* which, so far as is known, is restricted in its distribution to S. Africa (Natal) and Lesotho.

SPECIES INSUFFICIENTLY KNOWN

Crassula sp. 1.

The specimen *Andrada* 1865 (COI) from Mozambique Rué-Rué (Gúruè), "a meia encosta do Námuli, solo esquelético da própria penedia do Nãmuli, vegetação rupícola por vezes com abundancia de *Aloe* sp., planta gorda, atingindo 1 m., flores brancas de jaspe, 12.viii.1949", comprises leafless fragments of branches and a "capsule" with detached, mostly broken leaves and inflorescences. The inflorescences suggest an affinity with sect. *Anacampseroideae* Haw., subsect. *Latifoliae* (DC.) Tölken. On the other hand, the leaves are very different from those of either *C. ovata* (Mill.) Druce or *C. arborescens* (Mill.) Willd., the two members of this subsection. As the leaves, flowers and branches of the specimen are all detached, we cannot be certain that they belong to one species. In these circumstances the specimen remains insufficient for determination.

2. KALANCHOE Adans.

Kalanchoe Adans., Fam. Pl. **2**: 248 (1763).

Biennial or perennial or sometimes annual succulent herbs often with subrosulate leaves and scape-like inflorescence-axes, sometimes undershrubs or shrubs. Leaves usually opposite and decussate, connate at the base, the lower more or less approximate, the uppermost bract-like, sessile or petiolate; lamina undivided or rarely pinnatifid, entire, crenate or serrate, usually flat, sometimes semiterete, fleshy-succulent, sometimes thin. Inflorescences terminal, panicle- or corymb-like or thyrsoid, composed of cymes, usually many-flowered. Flowers 4-merous, erect, $\pm$ pedicellate, medium-sized or $\pm$ large. Calyx shorter than or sometimes equalling the corolla-tube, completely green or lineolate with purple or red; sepals nearly free or more or less connate, rarely up to or beyond the middle. Corolla gamopetalous along at least the lower $\frac{2}{3}$; tube $\pm$ distinctly 4-angled, rounded and swollen near the base, usually constricted upwards; lobes 4, spreading or reflexed or erect and sometimes connivent, $\pm$ succulent, sometimes minutely papillose above, usually apiculate. Stamens 8, usually included; filaments $\pm$ connate with the corolla-tube, free above the middle of the tube; anthers ovate or oblong. Carpels 4, slightly connate at the base; styles usually shorter than the ovaries. Seeds numerous, oblong, with longitudinally rugose tegument. Scales semi-orbicular to linear, entire, crenulate or $\pm$ emarginate at the top.

A genus widely distributed in tropical Africa, Namibia, South Africa and Madagascar, with some species in Arabia, India, Ceylon, China, Indochina, Malaysia, Java and one species in tropical America.

1. Plants with an indumentum of simple hairs - - - - - - - - 2
 – Plants completely glabrous - - - - - - - - - 13
2. Hairs not capitate-glandular, short, dense, spreading - - - - - 3
 – Hairs capitate-glandular at least on the inflorescence - - - - - 4
3. Leaves undivided - - - - - - 9. *velutina* subsp. *chimanimanensis*
 – Leaves trisect or pinnatifid - - - . - - - - - 6. *laciniata*
4. Tube of corolla 30–34·5 mm. long; styles 17·5–20 mm. long, longer than the follicles, c. 3·5 mm. exserted - - - - - - - - - 10. *latisepala*
 – Tube of corolla not more than 14 mm. long; styles not more than 3·5 mm., shorter than the follicles, included - - - - - - - - - - 5
5. Leaves trisect or pinnatifid - - - - - - - 6. *laciniata*
 – Leaves undivided, with the margin entire, crenate or dentate, rarely lobed - - 6
6. Calyx-tube 5·5–8 mm. long; sepals thick, with inflexed margins, finally cylindric-canaliculate and with subulate apex at fruiting; leaves not broader than 2·7 cm. (usually less), linear to elliptic, canaliculate and subcylindric towards the apex - 8. *hametorum*
 – Calyx-tube shorter; sepals membranous or submembranous on drying, flattish (neither cylindric-canaliculate nor subulate); leaves relatively broader, flattish, not cylindric at the apex - - - - - - - - -. - - - - 7
7. Hairs of stem rather long, up to 2·5(3) mm.; leaves sessile, usually entire, linear-lanceolate to obovate, longer than 2·5 times the breadth; annual or biennial with 4-gonous stem (sometimes terete and with 4 raised longitudinal lines) - - - - 1. *lanceolata*
 – Hairs of stem 0·5–1(1·5) mm. long; leaves often petiolate (at least the lower ones) and without the assemblage of the other characters above; perennials with terete, smooth stem - - - - - - - - - - - - 8
8. At least the median leaves deeply and irregularly lobed to 3-lobed - - - 7. *lobata*
 – All leaves unlobed, usually with crenate or dentate margin - - - - - 9
9. Leaves usually densely glandular hairy on both faces; lamina, at least in median leaves, with length equalling or shorter than twice the breadth - - - - - - 10
 – At least the lower and median leaves glabrous, as long as or longer than twice the breadth; hairs of stem usually not more than 0·5 mm. long, hyaline - - - - 12
10. Hairs of indumentum rusty to tawny; leaves crenate; pedicels not distinctly thickened below the calyx; calyx (4)5–10·5 mm. long; corolla-lobes spreading to reflexed - 11
 – Hairs of indumentum hyaline to slightly tawny; leaves irregularly dentate to repand-dentate; pedicels rather thickened below the calyx; calyx not more than 5·5 mm. long; corolla lobes erect to connivent - - - - - - - - 2. *hirta*
11. Lower leaves condensate, distinct from the upper ones either in size or shape; upper internodes up to 29 cm. long, rather longer than the lower ones; calyx-tube 0·5–1(2) mm. long; sepals not or slightly acuminate - - - - - - 4. *lateritia*
 – Lower leaves not condensate, not or scarcely distinct from the upper ones; upper internodes not longer than 7·5 cm., not much longer than the lower ones; calyx-tube 1·4–2 mm. long; sepals rather acuminate - - - - - - - 5. *fernandesii*

12. Leaves rounded at apex; sepals up to 10 mm. long, usually very attenuate and acute; corolla up to 19(22) mm. long with lobes 4·5–7·5 × (2·5)3·5–5 mm. (if calyx only up to 5 mm. and corolla up to 16 mm. long, then anthers of the upper stamens exserted   -   - 3. *crenata*
  - Leaves acute; sepals usually not more than 5·5 mm., acute but not attenuate; corolla not more than 14(16) mm., long; all anthers included -   -   -   -   - 6. *laciniata*
13. Corolla not less than 30 mm. long; styles 15–25(30) mm. long, subequalling to longer than the follicles   -   -   -   -   -   -   -   -   -   -   - 14
  - Corolla shorter; styles not longer than 5·5 mm., shorter than the follicles   -   - 15
14. Corolla-tube 43–50 mm. long, straight; corolla-lobes 18–25 × 9·5–15 mm., spreading, white; styles 22–24 mm. long, included   -   -   -   -   -   - 11. *dyeri*
  - Corolla-tube 21·5–37·5(45) mm. long, curved; corolla-lobes 10–16 × 3·5–6(7·25) mm., yellow-orange to light red; styles 15–25(30) mm. long, very exserted   -   - 12. *elizae*
15. Nectariferous scales broader than long or not more than twice longer than broad; plants covered everywhere with a chalk-white powdery ± caducous clothing; cymes axillary forming terminal spike-like or thyrsoid inflorescences   -   -   -   -   - 16
  - Nectariferous scales several times longer than broad, linear, very narrow; plants without a chalk-white powdery clothing; cymes forming ± ample panicles or corymbs   -   - 18
16. Filaments 4 or 5 mm. long, inserted ± at the ¾ of corolla-tube; anthers all exserted; corolla-tube very contracted just below the limb   -   -   -   -   -   - 14. *luciae*
  - Filaments not more than 2 mm. long, inserted a little below the throat of corolla-tube; only the upper anthers completely exserted; corolla-tube not or slightly contracted at corolla-mouth   -   -   -   -   -   -   -   -   -   -   -   - 17
17. Corolla-tube 7–11 mm. long; follicles ± 7·5 mm. long; styles c. 0·5 mm. long or almost absent; scales broader than longer, 1·5 × 2·5–2·75 mm., entire   -   -   - 15. *wildii*
  - Corolla-tube 11·5–15 mm. long; follicles (12·5)13·5–15 mm. long; styles (1·5)2·5–3 mm. long; scales rectangular (1·75)2–3 × (1)1·25–2 mm., emarginate   -   - 16. *thyrsiflora*
18. Corolla 5–7(8) mm. long with the lobes veined dark purple or completely purplish; stem up to 11 cm. long (up to 20 cm. in cult. pl.); leaves condensed at base of the stem - 13. *humilis*
  - Corolla usually longer than 8 mm.; lobes neither veined with nor completely dark-purplish; stem usually taller   -   -   -   -   -   -   -   -   -   -   - 19
19. Calyx membranous to almost translucent on drying, usually lineolate with purple; sepals (3·25)3·5–10 mm. long, linear-lanceolate to lanceolate, very attenuate and acute
    3. *crenata*
  - Calyx not membranous, but somewhat thick and rigid on drying, not lineolate; sepals 1·5–3·5(5) mm. long, ovate, ovate-lanceolate or triangular, not or slightly attenuate -   20
20. Styles 2–4·5 mm. long; corolla-tube whitish-hyaline or pale salmon ± contrasting with the cinnamon-coloured or yellow-salmon corolla-lobes on drying; branches of the inflorescence arched below and spreading; leaves chartaceous, usually crenate and ± distinctly petiolate   -   -   -   -   -   -   -   -   - 21. *sexangularis*
  - Styles not more than 1·5 mm. long; tube and corolla-lobes ± of same colour on drying; branches of inflorescence erect or erect-spreading; leaves petiolate or sessile   -   - 21
21. At least the basal and median leaves distinctly petiolate, entire or rarely 3-lobed, thin on drying; corolla pink to brick, with the tube very twisted above the follicles after anthesis
    17. *rotundifolia*
  - Leaves sessile or the lower attenuate or contracted into a broad subpetiolar base, chartaceous or ± rigid on drying; corolla yellow, with the tube not or slightly twisted below the lobes   -   -   -   -   -   -   -   -   -   -   -   -   - 22
22. Corolla 13·5–16(17) mm. long; tube subconical after anthesis with faintly marked angles; lobes subconnivent, sometimes twisted; leaves up to 26 × 7 cm., usually dentate, crenate or lobed   -   -   -   -   -   -   -   -   -   -   -   - 19. *brachyloba*
  - Corolla usually smaller (not longer than 14 mm.); tube strongly 4-angled, contracted above the follicles; corolla-lobes erect but not connivent, or ± spreading   -   -   - 23
23. Leaves large, broadly ovate to suborbicular, entire, the lower up to 24·5 × 16·5 cm.; corolla 11·5–14 mm. long   -   -   -   -   -   -   -   -   -   - 18. *paniculata*
  - Leaves smaller and relatively narrower, usually oblong to oblong-spathulate, crenate to irregularly serrate-lobed, up to 10 × 2(2·8) cm.; corolla 8–11·5 mm. long   20. *leblancae*

1. **Kalanchoe lanceolata** (Forssk.) Pers., Synops. Sp. Pl. **1**: 446 (1805), "*Calanchoe*". — DC., Prodr. **3**: 395 (1828). — Haw. in Phil. Mag. **6**: 304 (1829). — Schweinf. in Bull. Herb. Boiss. **4**, App. 2: 202 (1896), "*Calanchoe*". — Sacleux in Bull. Mus. Nat. Hist. Nat., Paris **14**: 244 (1908). — R.-Hamet in Bull. Herb. Boiss., Sér. 2, **8**: 32 (1908). — Engl., Pflanzenw. Afr. **1**: 121, 122 et 136 (1910); op. cit. **3**: 285 (1915). — Fiori in N. Giorn. Bot. Ital. **19**: 446 (1912). — Schönl. in S. Afr. Journ. Sci. **2**: 187 (1921). — Burtt Davy, F.P.F.T. **1**: 144 (1926). — Hutch. & Dalz., F.W.T.A. **1**: 105 (1927). — Exell in Journ. Bot., Lond., **67**, Suppl. Polyp.: 161 (1928) pro parte. — Berger in Engl. & Prantl, Nat. Pflanzenfam. ed, 2, **18a**: 406 (1930). — Cufod. in Miss. Biol. Borana, **4**: 54 (1939). — É. François in Rev. Hort., N. Sér., **26**: 68 (1938). — Schwartz, Fl. Trop. Arab.: 79 (1939). — Hutch. & Bruce in Kew Bull. **1941**: 88 (1941). — Hutch., Botanist in S. Afr.: 459, 465 et 484 (1946). — R.-Hamet in Bull. Jard. Bot. Bruxelles, **19**: 437 (1949). — F.W. Andr., Fl. Pl. Anglo-Egypt. Sudan: 79 (1950). — J. Blake, Gard. E. Afr.: 180 (1950). — Toussaint in

F.C.B. **2**: 564 (1951). — Suesseng. & Merxm. in Trans. Rhodes. Sci. Assoc. **43**: 15 (1951). — Wild, Guide Fl. Vict. Falls: 143 (1953). — Brenan in Mem. N.Y. Bot. Gard. **8**: 435 (1954). — Cufod. in Bull. Jard. Bot. Brux. **24**, Suppl.: 168 (1954). — Keay, F.W.T.A. ed. 2, **1**: 118 (1954). — Chapman, Veg. Mlanje Mt. Nyasal.: 32 (1962). — Morton in Compt. Rend. 4. ème Réun. AETFAT: 292, map 2 (1962). — R.-Hamet & Marnier-Lapostolle in Arch. Mus. Nation. Hist. Nat. Paris., Sér. 7, **8**: 77, t. 26 fig. 86 et t. 27 fig. 87–88 (1964). — Cufod. in Webbia, **19**, 2: 728 (1965). — Binns, H.C.L.M.: 41 (1968). — Friedr. in Prodr. Fl. SW. Afr. **52**: 38 (1968). — Jacobs., Das Sukk. Lexikon: 253 (1970). — Greenway & Fitzgerald in E. Afr. Nat. Hist. Mus. **28**, no. 130: 5 (1972). — Raadts in Willdenowia, **8**: 139 (1977). — R. Fernandes in Bol. Soc. Brot., Sér. 2, **53**: 381 (1980). TAB. **6**. Type from Arabia (Yemen).

Cotyledon lanceolata Forssk., Fl. Aegypt.-Arab.: CXI et 89 (1775). — Vahl, Symb. Bot. **2**: 51 (1791). — Willd. in L., Sp. Pl. ed. 4, **2**, 1: 758 (1799). Type as above.

Kalanchoe pubescens R. Br. in Salt, Voy. Abyss., App. 4: LXIV (1814) *nom. nud.*, non Bak. (1887).

Verea lanceolata (Forsk.) Spreng., Syst. Veg. ed. 16, **2**: 260 (1825). Type as for *Kalanchoe lanceolata*.

Kalanchoe glandulosa Hochst. ex A. Rich., Tent. Fl. Abyss. **1**: 312 (1848). — Schweinf., Beitr. Fl. Aeth.: 81 (1867). — Aschers. & Schweinf. in Schweinf., op. cit., Aufzählung: 271 (1867). — Britten in F.T.A. **2**: 396 (1871). — C. B. Clarke in Hook., Fl. Br. India **2**: 414 (1878). — Engl., Hochgebirgsfl. Trop. Afr.: 233 (1892); Pflanzenw. Ost-Afr. **C**: 189 (1895); in Ann. R. Ist. Bot. Roma, **9**: 252 (1902). — Penzig in Estr. Atti Congr. Intern. Bot. Genova: 34 (1892). — Schweinf., Abyss. Pflanzen-Nam.: 57 (1893). — Cooke, Bombay Fl. **1**: 466 (1903). — Pax in Engl., Bot. Jahrb. **39**: 621 (1907). — Th. & H. Dur., Syll. Fl. Cong.: 193 (1909). — Rendle in Journ. of Bot. **70**: 90 (1932). Type from Ethiopia.

Kalanchoe ritchieana Dalziell in Hook. Journ. of Bot. **4**: 346 (1852). — Dalziell & Gibson, Bombay Fl.: 105 (1861). — Drury, Handb. Ind. Fl. **1**: 105 (1864). Type from India.

Meristostylus macrocalyx Klotzsch in Peters, Reise Mossamb., Bot. **1**: 269 (1861). — Harms in Engl. & Prantl, Nat. Pflanzenfam. ed. 2, **18a**: 404 (1930) in adnot. Type: Mozambique, Boror, *Peters* s.n. (B).

Kalanchoe modesta Kotschy & Peyr., Pl. Tinn.: 18 (1867). Type ?

Kalanchoe brachycalyx sensu Britten in F.T.A. **2**: 396 (1871) pro parte. — Engl., Hochgebirgsfl. Trop. Afr.: 233 (1892) pro parte. — Pichi-Sermolli in Miss. Lago Tana 7, 1: 47 (1951). — Cufod. in Bull. Jard. Bot. Brux., Suppl.: 166 (1954) non A. Rich. (1848).

Kalanchoe platysepala Welw. ex Britten in F.T.A. **2**: 393 (1871). — Engl., Pflanzenw. Ost-Afr. **C**: 189 (1895). — Hiern, Cat. Afr. Pl. Welw. **1**: 327 (1896). — R.-Hamet in Bull. Herb. Boiss., Sér. 2, **8**: 31 (1908). — Berger in Engl. & Prantl, Nat. Pflanzenfam. ed. 2, **18a**: 406 (1930). — Binns, H.C.L.M.: 41 (1968). Type from Angola.

Kalanchoe glandulosa var. *benguellensis* Engl., Hochgebirgsfl. Trop. Afr.: 233 (1892). — Hiern, Cat. Afr. Pl. Welw. **1**: 328 (err. 823) (1896). — Schinz in Bull. Herb. Boiss. **5**, App. 3: 99 (1897). — De Wild., Ann. Mus. Congo, Bot., Sér. 4, **1**: 179 (1903); Contr. Fl. Katanga: 65 (1921). — Engl. & Gilg in Warb., Kunene-Samb.-Exped. Baum: 242 (1903). — Th. & H. Dur., Syll. Fl. Cong.: 193 (1909). Type from Angola.

Kalanchoe pilosa Bak. in Kew Bull. **1895**: 289 (1895). Type: Zambia, Mwero Plateau, *Carson* 3 (K, holotype).

Kalanchoe crenata var. *collina* Engl., Pflanzenw. Ost-Afr. **C**: 189 (1895) pro parte.

Kalanchoe pentheri Schlecht. in Journ. of Bot. **35**: 341 (1897). — R.-Hamet in Bull. Herb. Boiss., Sér. 2, **8**: 32 (1908); in Bull. Soc. Bot. Fr. **57**: 22 (1910). — Schönl. in S. Afr. Journ. Sci. **17**, 2: 188 (1921). — Burtt Davy, F.P.F.T. **1**: 144 (1926). Type from S. Africa (Transvaal).

Kalanchoe glandulosa var. *rhodesica* Bak. f. in Journ. of Bot. **37**: 434 (1899). — Engl. in Sitz. Königl. Preuss. Akad. Wiss. **52**: 892 et 896 (1906). — Rendle in Journ. of Bot. **70**: 91 (1932). Type: Zimbabwe, Salisbury, *Rand* 465 (BM, holotype).

Kalanchoe glandulosa var. *tomentosa* Keissler in Ann. Naturh. Mus. Wien, **15**: 36 (1900).

Kalanchoe goetzei Engl. in Bot. Jahrb. **30**: 312 (1901). Type from Tanzania.

Kalanchoe laciniata sensu R.-Hamet in Bull. Herb. Boiss., Sér. 2, **8**: 18 (1908) quoad *Courbon* 217 (P) non L. (1753).

Kalanchoe hirta sensu Dinter, Deutsch-Südw.-Afr. Flora, Forst-und-landwirtsch. Fragm.: 70 (1909) non Harv. (1862).

Kalanchoe ellacombei N.E. Br. in Kew Bull. **1912**: 329 (1912). — R.E. Fr. in Wiss. Ergebn. Schwed. Rhod.-Kongo-Exped. **1**: 58 (1916). — Wild, Guide Fl. Vict. Falls: 143 (1953). Type: Zambia, Livingstone, on the N. bank of River Zambezi, *Ellacombe* (K, holotype).

Kalanchoe homblei De Wild. in Fedde, Repert. **12**: 298 (1913); in Ann. Soc. Sci. Brux. **38**: 12 (1914); op. cit. **40**: 88 (1921); Contr. Fl. Katanga: 65 (1921); op. cit., Suppl. **1**: 16 (1927). — R.-Hamet in Bull. Jard. Bot. Brux. **19**: 437 (1949). Type from Zaire (Katanga).

Kalanchoe homblei f. *reducta* De Wild. in Ann. Soc. Sci. Brux., loc. cit. (1914) et (1921). Type from Zaire.

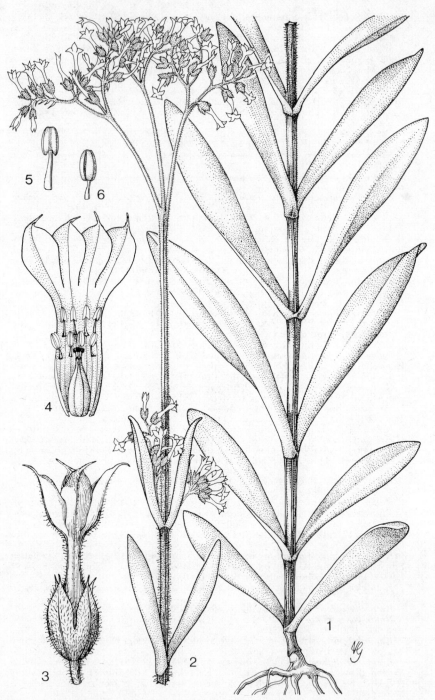

Tab. 6. KALANCHOE LANCEOLATA. 1, lower part of flowering stem (× ⅔), from *Teixeira* 982; 2, upper part of flowering stem (× ⅔); 3, flower (× 4); 4, flower opened to show stamens and pistils (× 4); 5, stamen of the upper verticil (× 8); 6, stamen of the lower verticil (× 8), 2–6 from *Pollhill & Paulo* 2103.

?Kalanchoe laciniata var. *brachycalyx* Chiov., Risult. Sci. Miss. Stef.-Paoli: 75 (1916), "*Calanchoe*".
Kalanchoe gregaria Dinter in Fedde, Repert. **18**: 433 (1922). Type from Namibia.
Kalanchoe diversa sensu Hutch. & Gillett in Kew Bull. **1941**: 88 (1941) non N.E. Br. (1902).
Kalanchoe lanceolata var. *lanceolata*. — Cufod. in Webbia **19**, 2: 729 (1965).
Kalanchoe lanceolata var. *glandulosa* (Hochst. ex A. Rich.) Cufod., tom. cit.: 730 (1965). — Jacobs., Das Sukk. Lexikon: 253 (1970). Type as for *Kalanchoe glandulosa*.

Annual or biennial succulent viscid herb, 11·5–150 cm. high (incl. the inflorescence). Stem erect, usually simple, rarely furcate or branched above the base, 3–15 mm. thick below, 4-gonous to narrowly winged or subterete and with 4 longitudinal raised lines, green or whitish on drying, glabrous or sparsely hairy below, more densely and with longer hairs above, the hairs slender, soft, spreading, up to 3 mm. long, minutely capitate-glandular; internodes 1–8·5 cm. long. Leaves 3–23(30) × 0·7–9 cm., obovate, obovate-oblong to narrowly oblong- or linear-lanceolate, obtuse to acute, entire or sinuate to crenate or sometimes irregularly shallow-lobed, sessile, connate (sheath up to 0·5 cm.) or the uppermost free, decurrent, glabrous to shortly glandular-hairy, succulent but not thickly fleshy, flat, papery to membranous and yellowish-green on drying. Flowers erect in ± long cymes grouped in panicle- or thyrse-like inflorescences up to 50 cm. long or sometimes reduced to the terminal group and then corymbose; internodes of inflorescences ± long, the lower ones up to 12·5 cm. long; branches of inflorescences up to 20 cm. long, erect, in the axils of leaf-like bracts, glandular-hairy; pedicels 1·5–3 mm. long, glandular-hairy. Calyx (4)5–7(9) mm. long, rounded at the base, green, glandular-hairy; tube less than ⅓ to slightly more than ½ of the total calyx length; sepals up to 3·5 mm. broad at base, ovate, ovate-lanceolate or elliptic-oblong, attenuate-cuspidate, pale green, glandular-hairy on both sides. Corolla-tube (8)9–14 mm. long, 4-gonous in the part included into the calyx, contracted above the ovary in a slender subcylindrical tubular portion, then suddenly dilatate below the limb, membranaceous, pale yellow or pale pink, pilose and matt outside, somewhat shining and glabrous inside, somewhat inflate and slightly rigid in fruit, splitting ready above the follicles and falling off; corolla-lobes 3–6 × 1·7–3·5(4) mm., obovate or broadly ovate, abruptly apiculate (apicule ± 0·75 mm.), rarely attenuate, pale yellow or greenish yellow to deep yellow, salmon, scarlet or deep orange. Stamens included, inserted above the middle of corolla-tube; filaments of lower stamens 0·5 mm., those of the upper ones 1·5 mm. long; anthers 0·5–0·75 mm. long, ovate, pale yellow, the upper ones c. 1·75–2 mm. below the mouth of corolla-tube. Follicles 5–8 mm. long, fusiform, attenuate above, pale; styles 0·5–0·75 mm. long. Scales 2·5–4·5 mm. long, linear. Seeds 0·5–0·75 mm. long, clavate, with longitudinal raised lines.

Caprivi Strip: all over the Caprivi, 15.iv.1946, *Krüger* s.no. (PRE). **Botswana**. N: Okavango swamps on Gwetshaa island, 3.v.1973, *Smith* 565 (SRGH). SW: 15 miles W. Ghanzi, 25.iv.1963, *Ballance* 633 (SRGH). SE: 75 miles W. NW. of Francistown on Maun road, near a tributary of R. Masupe, 960 m., 2.v.1957, *Drummond* 5292 (SRGH). **Zambia**. B: Gonya Falls, among rocks on river bank, c. 1088 m., 18.vii.1952, *Codd* 7119 (BM; K; PRE). N: Abercorn, lake Tanganyika, Mpulungu, rocks and very dry cliffs of Crocodile Island, 780 m., 12.iv.1957, *Richards* 11204 (BR; K; LISC). W: Kitwe, 2.vii.1955, *Fanshawe* 2359 (BR; EA; K; SRGH). C: Walamba, 23.v.1954, *Fanshawe* 1247 (K; SRGH). E: Lukusuzi Game Reserve, c. 1088 m., 6.v.1970, *Sayer* 167? (SRGH). S: Livingstone, Katambora, c. 960 m., 7.vii.1956, *Gilges* 622 (K; SRGH). **Zimbabwe**. N: Darwin, Mzarabani Tribal Trust Land, on peak of escarpment, 2.v.1972, *Mavi* 1386 (COI; K; SRGH). W: Matopos, near Maleme Dam, 16.v.1967, *Plowes* 2859 (K; LISC; SRGH). C: Gwelo, 13 miles E. of Gwelo, 3.vii.1930, *Hutchinson & Gillett* 3398 (BM; COI; K; LISC; SRGH). E: Umtali, 3 miles S. of Odzi, 12.vi.1968, *Plowes* 2898 (K; LISC; SRGH). S: Lundi River, 30.vi.1930, *Hutchinson & Gillett* 3251 (BM; K). **Malawi**. C: Ntchisi Distr., Ntchisi Mt., 1400 m., 5.viii.1946, *Brass* 17134 (K). S: Chikwawa Distr., Chikwawa, 2000 m., 2.x.1946, *Brass* 17887 (BR; K; PRE; SRGH). **Mozambique**. N: Niassa, Malema, 26.v.1937, *Torre* 1520 (COI; LISC). Z: Quelimane, Lugela-Mocuba, Namagoa Estate, viii.1946, *Faulkner* 75 (BR; COI; EA; K; P; SRGH). T: Tete, Songo, at the lower side of the Zanco, 17.iv.1972, *Macedo* 5204 (COI; LISC; LMA). MS: Beira, Chemba, 29 km. from Tambara to the plateau of "Serra Lupata" (Nhamalongo Mts.), c. 430 m., 14.v.1971, *Torre & Correia* 18405 (LISC; LMA; LMU). GI: Gaza, Caniçado, 17 km. from Massingir to the Singuédzi River, near the village of chief Bengo, 19.vii.1969, *Correia & Marques* 977 (LMU). M: Umbelúzi, 8.ix.1968, *Ferreira Marques* 56 (COI).

A very frequent species with a large distribution in tropical Africa (from Ethiopia and Somalia to Transvaal and Madagascar in the East; from Mali and Ghana to Namibia in the west; and widespread also in the central region), found also in Asia (Yemen and India). In various habitats: under trees and rocks in woodland, in savannas, on rock crevices of dry cliffs, on damp termite mounds, on river banks, etc.

Very variable as is reflected by the vast synonymy. Two varieties, based on degree of connation of calyx, were recently distinguished: var. *lanceolata* with the sepals somewhat more than half the total calyx-length, and var. *glandulosa* with the sepals equalling to slightly shorter than the connate portion. Considering that there is a nearly continuous variation in the extension of the connate part from less than ⅓ to more than ½ of total calyx-length, we prefer (at least at present) to ignore varieties based on this character. In the F.Z. area, however, plants with a short calyx-tube are the dominant ones.

The specimen *Andrada* 677 (LISU) from Mozambique, in its somewhat tawny indumentum, distinctly petiolate and shorter leaves, narrower sepals and shorter connate portion of calyx, is perhaps a hybrid between *K. lanceolata* and *K. lateritia* Engl.

'*K. pilosa* Bak. is excluded by Cufodontis (tom. cit.: 740, 741, 1965) from the synonymy of *K. lanceolata* on account of the hairs being not glandular. However, we have seen glands at the apex of the shortest hairs. The type is a delicate specimen with the calyx-tube subequalling the calyx-lobes (cf. R. Fernandes in Bol. Soc. Brot., Sér. 2, **53**: 391, 1980).

Kalanchoe diversa N.E. Br., described from Somalia, was considered by Cufodontis (tom. cit.: 729, 1965) as identical with *K. lanceolata* but as an independent species (Bull. Jard. Bot. Brux. **24**, Suppl.: 167 (1954). Its stem is not 4-angled and it has shorter hairs than in *K. lanceolata* (and only towards the extremity), the leaves are petiolate whereas in *K. lanceolata* they are sessile, and the sepals are free nearly to the base of calyx. It is a doubtful plant, which must not be regarded as a synonym of *K. lanceolata* (cf. R. Fernandes, tom. cit.: 392).

Kalanchoe brachycalyx A. Rich. is another taxon of doubtful position and must also be removed from *K. lanceolata's* synonymy (cf. R. Fernandes, tom. cit.: 384, 1980).

Kalanchoe floribunda Wight & Arn, (type from Ceylon does not seem to us identical with *K. lanceolata* as Cufodontis (tom. cit.: 728, 1965) asserts, but possibly to *K. crenata* (cf. R. Fernandes, tom. cit.: 382, 1980).

2. **Kalanchoe hirta** Harv. in Harv. & Sond., F.C. **2**: 379 (1862). — M. Wood, Handb. Fl. Natal: 46 (1907). — R.-Hamet in Bull. Herb. Boiss., Sér. 2, **8**: 36 (1908). — Berger in Engl. & Prantl, Nat. Pflanzenfam. ed. 2, **18a**: 406 (1930). — Riley, Fam. Flow. Pl. S. Afr.: 157 (1963). — Jacobs., Das Sukk. Lexikon: 252 (1970). — J. H. Ross, Fl. Natal: 179 (1972). Type from S. Africa (Cape Prov.).

A perennial (or biennial?) succulent, glandular-pubescent throughout; hairs of the indumentum up to 1 mm. long, capitate-glandular, slightly rigid, spreading, whitish or pale tawny, ± dense, rather dense on the inflorescence. Stem curved at the base, then erect, robust, terete, smooth, pale; lower internodes ± 1 cm. long, the uppermost up to 6 cm. Leaves not or slightly connate, not decurrent along the stem, attenuate into a petiole or cuneate into a subpetiolar base, the uppermost sessile; lamina up to 8·5(11) × 6·5(9) cm., ovate, rounded to subacute at the top, irregularly dentate, sinuate-dentate or repand-dentate at the margin, flat, succulent but not thick, nearly membranous on drying, dull green, pubescent-glandular on both sides; petiole up to 3·5 cm. long. Flowers erect in dense cymes, forming either a solitary terminal corymb or a panicle-like inflorescence up to 37 cm. long; branches of the inflorescence up to 24 cm. long; pedicels up to 5·75 mm. long, thick, rather broadened below the calyx. Calyx 3·5–5·5 mm. long, green in flower, yellowish to hyaline in fruit, membranous on drying; tube (0·5)1–1·5 mm. long, truncate and circumscissile at the base after anthesis; sepals 2–3 mm. broad at the base, deltate to ovate, acute (but neither attenuate nor cuspidate). Corolla 13·5–17 mm. long; tube rather swollen in the lower half to nearly spherical in fruit, contracted into a cylindrical-tubular portion above the carpels, sparsely pubescent, yellowish; corolla-lobes 4·5–7·25 × 2·5–3·25 mm., elliptic to obovate-oblong, acute, erect, connivent in fruit, orange. Stamen-filaments inserted above the middle of corolla-tube, those of lower stamens c. 1 mm. long; anthers ± 0·5 mm. long, included, the upper ones c. 2 mm. below the base of corolla-lobes. Follicles 5·5–7·5 mm. long (with the styles); styles 1·25–2 mm. long. Scales 2–2·75 mm. long, linear.

Zimbabwe. S: collected in Zimbabwe and grown at Division of Plant Industry, Pretoria, s.d., *P. Koch* 8666 (K; PRE).

Also in S. Africa (Cape Prov. and Natal). In dry places.

Similar to *K. lanceolata*, from which it differs in the shorter indumentum; the not quadrangular stem; the petiolate, relatively broader leaves which have dentate or sinuate-dentate margins and not or only slightly connate base; thicker pedicels; shorter, circumscissile calyx; neither attenuate nor cuspidate sepals; more swollen tube of corolla, etc.

3. **Kalanchoe crenata** (Andr.) Haw., Synops. Pl. Succ.: 109 (1812); in Philos. Mag. **6**: 303 (1829). — Trattin., Ausgew. Gartenpfl. **1**: 109, t. 59 (1821). — DC., Prodr. **3**: 395 (1828); Hist. Pl. Grasses: t. 176 (1832). — Steud., Nom. Bot. ed. 2, **1**: 252 (1840). — Harv. in Harv. & Sond., F.C. **2**: 379 (1862). — Britten in F.T.A. **2**: 394 (1871) pro parte. — Engl., Hochgebirgsfl. Trop. Afr.: 232 (1892) pro parte. — K. Schum. & Rümpler, Die Sukk.: 89 (1892). — Schwéinf. in Bull. Herb. Boiss. **4**, App. 2: 201 (1896) in adnot. — Bak. f. & al. in Journ. Linn. Soc., Bot. **37**: 151 (1905). — A. Chev., Expl. Bot. Afr. Occ. Fr. **1**: 254 (1920). — De Wild., Pl. Bequaert. **2**: 44 (1923). — Hutch. & Dalz., F.W.T.A. **1**, 1: 105, fig. 34 (1927). — Berger in Engl. & Prantl, Nat. Pflanzenfam. ed. 2, **18a**: 406 (1930). — A. Chev., Fl. Afr. Occ. Fr. **1**: 279 (1938). — Exell, Cat. Vasc. Pl. S. Tomé: 173 (1944); Suppl.: 19 (1956). — Robyns, Fl. Parc. Nat. Alb. **1**: 227 (1948). — Toussaint in F.C.B. **2**: 565 (1951) excl. syn. pro parte. — Keay in F.W.T.A. ed. 2, **1**, 1: 118, fig. 39 (1954). — Roberty, Petite Fl. Ouest-Afr.: 239 (1954). — Cufod. in Bull. Jard. Bot. Brux. **27**: 713 (1957). — Exell in Bull. Br. Mus., Bot. **3**: 99 (1963); op. cit. **4**: 340 (1973). — Agnew & Hanid, Fl. Upl. Kenya: 9 (1966). — Hulstaert, Notes Bot. Mongo: 55 (1966). — Cufod. in Österr. Bot. Zeit. **116**: 314 (1969). — Jacobs., Das Sukk. Lexikon: 251 (1970). — Raadts in Willdenowia, **8**: 126 (1977). — R. Fernandes in Bol. Soc. Brot., Sér. 2, **53**: 336 (1980). Type: a cultivated plant at Vere's Garden (London) (BM).

Verea crenata Andr. in Bot. Repos.: t. 21 (1798), "Vereia". — Willd. in L., Sp. Pl. ed. 4, **2**: 471 (1799). — Spreng., Syst. Veg. ed. 16, **2**: 260 (1825). — Dietr., Synops. Pl. **2**: 1328 (1840). Type as above.

Cotyledon crenata (Andr.) Vent., Jard. Malm. **1**: t. 49 (1804). — Sims in Curtis, Bot. Mag.: t. 1436 (1812). Type as above.

Cotyledon verea Jacq., Hort. Schoenbr. **4**: t. 435 (1804), *nom. illegit.*

Kalanchoe verea (Jacq.) Pers., Synops. Spec. Pl. **1**: 446 (1805), *nom. illegit.*

Cotyledon brasilica Vellozo, Fl. Flum.: 197 (1825) et **4**: t. 184 (1835). Type from Brasil.

Kalanchoe brasiliensis Cambess. in St.-Hill., Fl. Bras. Merid. **2**: 196 (1830). — Walp., Repert. **2**: 257 (1843). — C. B. Clarke in Hook., Fl. Br. India, **2**: 415 (1878). — Eichl. in Martius, Fl. Bras. **14**, 2: 382, t. 89 ii (1872). — Schönl. in Engl. & Prantl, Nat. Pflanzenf. **3**, 2a: 34 (1891). — Britton, Fl. Bermuda: 160, fig. 185 (1918). — Britton & Millspaugh, Bahama Fl.: 153 (1920). — Jacobs., Das Sukk. Lexikon: 250 (1970). Type from Brazil.

Kalanchoe afzeliana Britten, op. cit.: 393 (1871), *nom. illegit.* Type as for *Verea crenata.*

Kalanchoe coccinea Welw. ex Britten, op. cit.: 395 (1871). — Engl., Pflanzenw. Ost-Afr. C: 189 (1895) excl. specim. Mossamb. — Hiern, Cat. Afr. Pl. Welw. **1**, 1: 328 (err. 823) (1896). — De Wild., Miss. Laurent. **1**: 236 (1906). — Th. & H. Dur., Syll. Fl. Cong.: 193 (1909). — Hutch. & Dalz., loc. cit. (1927). — Exell in Journ. of Bot. **66**, Suppl. Polypet.: 161 (1928). Type from Angola.

Kalanchoe crenata var. *collina* Engl., loc. cit. (1895) pro parte.

Kalanchoe aegyptiaca sensu Hiern, loc. cit. quoad *Welwitsch* 2488 non (Lam.) DC. (1801).

Kalanchoe laciniata sensu R.-Hamet in Bull. Herb. Boiss., Sér. 2, **7**: 897–899 (1907) pro parte; op. cit. **8**: 17–19 (1908) pro parte. — Keay, tom. cit.: 117 (1954). — Exell, loc. cit. (1973) non (L.) DC. (1802), neque auct. fl. Afr.

Kalanchoe petitiana sensu Hutchs. & Dalz., loc. cit. (1927) non A. Rich. (1848).

Kalanchoe brasilica (Vellozo) Stellf. in Trib. Farm. Bras. **15**: 93 (1947). Type as for *Cotyledon brasilica.*

Kalanchoe integra sensu C. A. Backer, Fl. Males. Ser. 1, **4**: 202 (1953). — Chittend., Index Vols. 1–164 of Curtis, Bot. Mag.: 135 (1956). — Cufod. in Österr. Bot. Zeit. **116**: 312 (1969) non (Medic.) Kuntze (1891).

Kalanchoe crenata var. *crenata* — Cufod. in Bull. Jard. Bot. Brux. **27**: 713, fig. 68 (1957). — Jacobs., Das Sukk. Lexikon: 251 (1970). Type as for *Kalanchoe crenata.*

Kalanchoe crenata var. *verea* (Jacq.) Cufod., tom. cit.: 714, fig. 69 (1957). — Exell, loc. cit. (1973).

Kalanchoe crenata var. *coccinea* (Welw. ex Britten) Cufod., tom. cit.: 717, fig. 70 (1957). Type as for *Kalanchoe coccinea.*

Kalanchoe integra var. *integra* — Cufod., tom. cit.: 317 (1969).

Kalanchoe integra var. *verea* (Jacq.) Cufod., loc. cit. (1969).

Kalanchoe integra var. *crenato-rubra* Cufod., tom. cit.: 320 (1969). Type from Tanzania.

Kalanchoe integra var. *crenata* (Andr.) Cufod., loc. cit. (1969). Type as for *Kalanchoe crenata.*

A perennial succulent herb 0·3–2 m. high. Stem erect or ascending, fleshy, terete, up to 2 cm. in diameter at the base, usually simple, sometimes branched, glabrous or glabrescent towards the base, more or less pubescent-glandular above (hairs of the indumentum short, usually not longer than 0·5 mm., hyaline or pale tawny, spreading, thin, capitate-glandular) or also glabrous. Leaves decussate, horizontal to deflexed, petiolate, not very crowded below; lamina 4·3–25(30) × 1·5–12(20) cm., ovate or oblong-ovate to spathulate, usually with the breadth subequalling or less than half the length, the median passing gradually (in shape & size) into the bract-like upper ones, rounded at the top, irregularly doubly crenate to sometimes sublobed at the margin, sometimes edged with red, cuneate at the base and decurrent along the petiole, all glabrous or sometimes the upper ones sparsely pubescent-glandular, fleshy (but less thick than the majority of other species), flattish, concave, pale green changing to dark green or brownish on drying and then rather thin to membranous; petiole up to 4 cm. long, flattened and grooved above, broadened and ± embracing the stem at the base but not or slightly connate with the opposite one. Flowers in many-flowered cymes, grouped in corymbs forming terminal, usually large (up to 40 cm. or more long) panicles, sometimes only the terminal corymb present; branches of the panicle at ± 45° with the axis, ± pubescent-glandular or glabrous, the lower ones up to 35 cm. long, floriferous only in the terminal ¼–⅓; pedicels 2–7(10) mm. long, glabrous or ± glandular-pubescent; bracts linear, very attenuate and acute to nearly filiform. Calyx 2·4–10 mm. long; tube 0·1–1(1·5) mm. long; sepals 1–1·5 mm. broad at the base, lanceolate to linear-lanceolate, ± attenuate, very acute, rather scattered between them, green, sometimes lineolate with red, pubescent-glandular to glabrous, thin on drying. Corolla 11·5–22 mm. long; corolla-tube papery and somewhat rigid in fruit, white below, coloured ± like the lobes upwards, glandular-pubescent or glabrous; corolla-lobes 4·5–7·5 × (2·5)3·5–5 mm., oblong-lanceolate to elliptic, acute or subacute and with a somewhat long apiculum (up to 1 mm. long) at apex, sulphur-yellow to bright or deep yellow, or bright salmon to red, deep red, orange or brick. Anthers 0·5–0·7(1) mm. long, all included or the upper ones ± exserted. Follicles 6–8·5 mm., fusiform, attenuate; styles 0·75–2(2·5) mm. long. Scales 2·5–3·5 mm. long, linear. Seeds c. 0·75 mm. long, oblong, ribbed.

Sepals (3)3·5–10 mm. long; corolla up to 19(22) mm. long; anthers of the two verticils distinctly included; pedicels not or slightly dilated upwards   -   -   -   - subsp. *crenata*
Sepals 2·3–5(6) mm. long; corolla up to 15(16) mm. long; anthers of the upper stamens ± exserted; pedicels rather dilated upwards   -   -   -   -   - subsp. *nyassensis*

Subsp. **crenata**. — R. Fernandes in Bol. Soc. Brot., Sér. 2, **53**: 356 (1980)

Zambia. N: Mwenzo, c. 1760 m., viii.1938, *Champion* 460B (K). W: Solwezi Distr., just E. of R. Kabompo, 31.vii.1930, *Milne-Redhead* 806 (K). **Zimbabwe.** E: Melsetter, Bridal Veil Falls, from edge of falls, 11.x.1950, *Sturgeon & Panton* (BR; K; LISC; LMA; P; SRGH 30433). **Malawi.** N: Viphya, Chikangawa, c. 1920 m., 20.vii.1962, *Chapman* 1671 (K; LISC; SRGH).
Also in Egypt(?) (cultivated and naturalised), Uganda, Kenya, Tanzania, W. Africa (from Guinea to Angola) and S. Africa (Cape Prov.) and naturalised (?) in tropical America, India and Malaysia. In sunny places at the edge of forests, along roadsides, by streams etc.

A very polymorphic taxon where varieties based on the presence or absence of indumentum (on the stem, inflorescence, calyx and corolla) and on the colour of the corolla have been recognised. We consider the latter character of small importance, a similar variation in corolla colour existing in some other species, e.g. *K. lateritia* and *K. lanceolata*. Moreover, the colour of the corolla seems to change to some degree with age as was noted by various collectors. In the F.Z. area, plants with yellow (pale or deep), salmon, orange and brick corollas are found in the same places, e.g. in Vumba Mts. (Zimbabwe) and in Solwezi Distr. (Zambia). The density of the indumentum is also quite variable, some plants being somewhat pubescent-glandular, as the specimen *Bally in Champion* 460B (K) from Mwenzo (Zambia), whereas others are completely glabrous or have sparse hairs (all plants from Zimbabwe). However, hairiness, at least on the inflorescence, is a constant character in some regions, e.g. Zaire and Angola.
Most Zimbabwean specimens differ in several features from material of Cameroon, Zaire, Angola and E. Africa: the pedicels are somewhat broadened above and usually grooved (when dry); the calyx is slightly longer ((4)5–10 mm. long compared with 3–8 mm. in plants from other countries) and differs in shape, being truncate and finally circumscissile at the base rather than rounded and persistent; and the styles measure (1·5)2(2·5) mm. compared with 0·75–1·6 mm. Specimens from Tanzania were seen in which the calyx was truncate and circumscissile (1 specimen) and had longer styles (several specimens). Possibly the Zimbabwean plants could be treated as a separate variety or subspecies, which should include also some Tanzanian material.
The main features of subsp. *crenata*, common to plants throughout the distribution-range, are both the sepal shape (linear-lanceolate, attenuate) and anther position (all included, the upper (1)1·5–2(2·5) mm. below the mouth of the corolla tube). In specimens from the F.Z. area the filaments of the lower stamens are 0·75–1·75 mm. long, rarely almost absent, and those of the upper ones usually 1·5 mm. long. The collection *Chapman* 1671 has the apex of the anthers just appearing at the mouth of the corolla tube, and thus makes a transition to subsp. *nyassensis*. The differences between subsp. *crenata* and *K. lateritia* are discussed under the latter (No. 4). Some Zambian specimens from the Mbala (Abercorn) area, i.e. *Bullock* 3976 (BR; K), *Gamwell*

62 (BM), *Richards* 15289 (EA; K) have sparsely hairy leaves and a generally longer indumentum, and seem intermediate between subsp. *crenata* and *K. lateritia*. They may possibly represent hybrids between these taxa.

Subsp. **nyassensis** R. Fernandes in Bol. Soc. Brot., Sér. 2, **52**: 199 (1978); op. cit. **53**: 361 (1980). Type: Malawi, Mafinga Hills, 3 miles W. of Chisenga, alt. c. 2208 m., *Tyrer* 585 (BM, holotype; BR; SRGH).

This subspecies is distinguished from subsp. *crenata* by the stem prostrate at the base; by the calyx usually shorter (usually 2·5–5·5 mm., rarely 6·5 mm. long) and with less connate sepals (0·1–0·2(0·75) mm., not 0·5–1·5 mm. connate), by the smaller corolla, usually not longer than 15 mm., rarely 16 mm. long, and with shorter apiculum at the apex of the lobes; by the shorter filaments of the stamens (those of the upper stamens 0·3–0·75 mm. long, compared with usually 1·5 mm. long in subsp. *crenata*, those of the lower ones 0·2–0·3 mm. long compared with 0·75–1·75 mm. long in subsp. *crenata*); but mainly by the upper anthers which are usually completely exserted or sometimes ½ exserted.

Zambia. N: Makutus, 26.x.1972, *Fanshawe* F11 518 (K). **Malawi.** N: Mzimba Distr., S. Mzuzu, High Viphya, 10.ix.1956, *Jackson* 2048 (BR; K; LISC; SRGH).

Known only in Zambia and Malawi but perhaps existing also in Tanzania. Frequent in grassland or in crevices of rocks in evergreen forest or *Brachystegia* woodland.

As in Zimbabwean plants of subsp. *crenata*, in the subsp. *nyassensis* the pedicels are sulcate and broadened upwards, the calyx truncate at the base and circumscissile (on account of the very short calyx-tube, the sepals often fall independently of one another) and the styles (1·5)–2 mm. long, but the scales are a little longer (3–4(4·5) mm., while in subsp. *crenata* they are 2·5–3·5 mm. long). A variation in colour of the corolla from dark yellow to dark vermilion or brick-red is referred to by collectors, but pale or bright yellow is not indicated, orange-red, flame-red or salmon appear to be the most frequent colours. However, on drying, the corolla is always of a deep pinkish tone with darker veins.

All specimens, except *Hall-Martin* 1657 (PRE) which is completely glabrous, have the upper part of the stem, inflorescence, calyx and corolla shortly pubescent-glandular (hairs c. 0·25 mm. long).

By its clavate pedicels this taxon approaches *K. densiflora* Rolfe, but in the latter the pedicels are even more broadened above, the leaves relatively broader and with shorter petioles, the corolla not so long and with relatively broader lobes, the anthers broader and all included; besides this, *K. densiflora* is usually a glabrous plant whereas *K. crenata* subsp. *nyassensis* is usually pubescent-glandular.

4. **Kalanchoe lateritia** Engl. [in Abh. Königl. Akad. Wiss. Berl. **1894**: 19 (1894) *nom. nud.*]; Pflanzenw. Ost-Afr. **C**: 189 (1895). — R.E. Fr., Wiss. Ergebn. Schwed. Rhod.-Kongo-Exped. **1**: 59 (1916). — Bally in Journ. E. Afr. Uganda Nat. Hist. Soc. **15**: 13 (1940). — R.-Hamet in Bull. Jard. Bot. Brux. **19**: 437 (1949); in Bol. Soc. Brot., Sér. 2, **24**: 97–98 (1950). — Toussaint in F.C.B. **2**: 564 (1951). — R.-Hamet & Marnier-Lapostolle in Arch. Mus. Nat. Hist. Nat. Paris, Sér. 7, **8**: 78, t. 2 fig. I et K, t. 27 fig. 89–90, t. 28 fig. 91 (1964). — Binns, H.C.L.M.: 41 (1968). — Raadts in Willdenowia, **8**: 131 (1977). Syntypes from E. Africa (Lectotype *Holst* 2986 (B)).
  Kalanchoe coccinea var. *subsessilis* Britten in F.T.A. **2**: 395 (1871). Type: Malawi, Manganja Hills, *Meller* s.n. (K, lectotype).
  Kalanchoe caccinea sensu Engl., loc. cit. (1895) pro parte quoad specim. region. Mossamb. non Welw. ex Britten (1871).
  Kalanchoe crenata var. *collina* Engl., loc. cit. (1895) pro parte quoad specim. mossamb. *Braga* s.n. (COI).
  Kalanchoe cuisinii De Wild. & Th. Dur. in Bull. Soc. Roy. Bot. Belg. **38**, 2: 122 (1899); in Ann. Mus. Congo, Bot., Sér. 3, **1**: 82 (1901). — Th. & H. Dur., Syll. Fl. Cong.: 193 (1909). — R.-Hamet, loc. cit. (1949). Type from Zaire.
  Kalanchoe kirkii N.E. Br. in Gard. Chron., Ser. 3, **32**: 110 (1902). — Hook. f. in Curtis, Bot. Mag. **58**: t. 7871 (1902). — R.-Hamet & Marnier-Lapostolle, tom. cit.: t. 1 fig. F (1964). Type a cultivated plant at Kew Garden (K).
  Kalanchoe velutina sensu R.-Hamet in Bull. Herb. Boiss., Sér. 2, **8**: 36 (1908) pro parte quoad specim. Afr. Or. et Centr. — Berger in Engl. & Prantl, Nat. Pflanzenfam. ed. 2, **18a**: 406 (1930) pro parte. — Robyns. Fl. Parc Nat. Alb. **1**: 228 (1948). — Jacobsen, Das Sukk. Lexicon: 256 (1970) pro parte, non Welw. ex Britten (1871).
  Kalanchoe zimbabwensis Rendle in Journ. of Bot. **70**: 90 (1932). — Hutchs., Botanist in S. Afr.: 469 (1946). — J. Blake, Gard. E. Afr.: 181 (1950). — Jacobs., Das. Sukk. Lexikon: 257 (1970). Type: Zimbabwe, Chibopopo River, 11.viii.1928, *Rendle* 259 (BM, lectotype).
  Kalanchoe lateritia var. *zimbabwensis* (Rendle) Brenan in Mem. N.Y. Bot. Gard. **8**, 5: 435 (1954). Type as above.

Kalanchoe integra var. *subsessilis* (Britten) Cufod. in Österr. Bot. Zeit. **116**: 317 (1969). Type as for *Kalanchoe coccinea* var. *subsessilis*.

A succulent perennial, pubescent-glandular throughout, 19–150 cm. high (incl. the inflorescence); hairs of the indumentum up to 1(1·5) mm. long, spreading, thin, capitate-glandular, ± dense, tawny to dark-brown or rusty-red on drying. Stem erect, usually simple, sometimes forked into two ± long floriferous axes, 0·5–1·8 cm. in diameter at the base, terete, leafless at fruiting time; lower internodes (from below upwards) 0·3–7 cm. long, the median ones 6–15 cm. long, the two terminal ones usually very elongate, (12)18–29 cm. long. Leaves condensed towards the lower part of stem, there petiolate and spreading-decurved, the median and the upper ones usually rather scattered, distinctly smaller than the lower, sub-bract-like, shortly petiolate to subsessile, ascending; lamina of the lower leaves 3·5–16(21) × 3–10(12·5) cm., subcircular or broadly obovate, less than twice as long as broad, rounded or obtusely acuminate at the top, crenate or doubly crenate at margin, rounded or cuneate at the base, the others obovate-spathulate to linear-cuneate, nearly or quite entire, flat in all leaves and rather fleshy, not very thin and somewhat rigid, dark brown or dark green when dry; petiole 0·5–5 cm. long, channelled, rather broad. Flowers in ± dense cymes grouped either in a single ± dense small terminal corymb or in a larger looser one (up to 20 cm. in diameter) or sometimes in several corymbs disposed in thyrsoid or panicle-like inflorescences, very densely rusty-pubescent-glandular; pedicels 1–3·5(5) mm. long. Calyx (4)5–7(8) mm. long; tube (0·3)0·5–1(2) mm. long; sepals (1)1·5–2·5(3) mm. broad at the base, ovate-oblong to oblong-lanceolate, rarely ovate, ± attenuate, acute or subacute, green, lineolate with red. Corolla 13·5–17 mm. long; tube 8·5–14 mm. long, dilated at base, yellow-orange turning white-hyaline below and pale-pink upwards at fruiting; corolla-lobes 4·5–7(8·5) × 2–4·2 mm., ovate or elliptic, attenuate or abruptly contracted at the apex, long apiculate, usually orange-red or brilliant scarlet-orange to brick-red, rarely yellow. Filaments ± 2·5 mm. long, inserted above the middle of corolla-tube; anthers ± 0·75 mm. long, oblong, included. Follicles (4)–6–8(9) mm. long, attenuate into the 0·75–2 mm. long styles. Seeds c. 1 × 0·33 mm., oblong-linear, longitudinally slightly ribbed, curved at the top. Scales 1·75–4 mm. long, linear.

This description applies to var. *lateritia*. Var. *prostrata* Raadts, with prostrate stem and glabrous leaves and stem, and var. *pseudolateritia* Raadts, with a slightly longer corolla-tube, were described from Kenya and Tanzania.

Zimbabwe. W: Victoria, Kyle National Park Game Reserve, Chembira Hill, 22.v.1971, *Grosvenor* 528 (SRGH). C: Salisbury, Physic. Lab. Gard., 25.v.1943, *Sturgeon* (K; SRGH). E: Melsetter, Lavinas Rust Farm Hillside, 30.vii.1950, *Crook* M. 38 (K; LISC; SRGH). S: Zimbabwe, c. 1120 m., viii.1929, *Dame Alice & Misses Godman* 162 (BM). **Malawi.** S: 2 miles W. of Chiradzulu Mts. (cult. and flower. at Greendale) 21.vii.1958, *Leach* 7194 (K; PRE; SRGH); Likabula, Mulanje Mt., c. 800 m., vii.1958, *Chapman* H/734 (SRGH). **Mozambique.** N: between Muêda and Chomba, 25.ix.1948, *Pedro & Pedrógão* 5351 (LMA). Z: Alto Molucué, at km. 21 on the road to Alto Ligonha, c. 600 m., 29.x.1967, *Torre & Correia* 16276 (C; COI; LISC; MO). MS: Cheringoma, Inhaminga, 29.vii.1967, *Moura* 212 (COI; LMU). GI: 9 miles S. of Cheline on Maxixe-Mambone road, 5.x.1936, *Leach & Bayliss* 11841 (K; LISC; SRGH).

Also in Kenya, Tanzania, Rwanda and Zaire. In *Brachystegia* woodland, in grass, hollow of rocks etc.

All specimens seen except *Banda* 1061 (SRGH), from Blantyre, Ndiranda Forestry Plant. area, Malawi, and *Banda* 1158 (K; SRGH), cultivated in Chancellor College Biol. Bot. Gard. Malawi, have corollas more or less tinged with red which, according to collectors, with age or on drying turn to ± brick-red. However, the two above mentioned specimens have completely yellow corollas, brighter so on the lobes, not darkening on drying.

This species was originally considered as a variety of *K. crenata* (= *K. coccinea*). In fact, it is very similar to the latter mainly to the forms with an indumentum of glandular hairs on the stem and inflorescence, and corollas with identical variation in colour. Nevertheless, *K. lateritia* has some characters which permit one to distinguish it from all forms of *K. crenata*: hairs of the indumentum longer (up to 1–(1·5) mm., not only up to 0·5 mm. long as in *K. crenata*), denser and darker; leaves more condensed towards the base of the stem, the lower and median contrasting distinctly in form and shape with the 2–3 upper pairs which are not only more scattered but also smaller, nearly bract-like; in *K. lateritia* the leaves are hairy on both surfaces, while they are glabrous in nearly all forms of *K. crenata*; their lamina is relatively shorter (the breadth usually more than half of the length), thicker, turning darker in colour and more rigid on drying whereas in *K. crenata* the lamina is membranous or nearly so in the dry state; the inflorescence is usually denser, the pedicels shorter, the sepals relatively shorter and broader, more close (sinus narrower in *K. lateritia*), less attenuate, a little more connate at the base than in *K. crenata*. Thus we prefer to consider this taxon at specific level, following Engler, De Wildeman, N. E. Brown and many other authorities.

The specimen from Mozambique (Niassa, Cabo Delgado, Quissanga, between Mahate and Metuge, 1.x.1946, *Barbosa* 2344 (LISC; LMA)) has flowers like *K. fernandesii* (no. 5), but its leaves are petiolate.

5. **Kalanchoe fernandesii** R.-Hamet in Bol. Soc. Brot., Sér. 2, **24**: 112, t. 3 (1950); tom. cit.: 98–99 et 107–112 (1950). — Jacobs., Das Sukk. Lexikon: 252 (1970). — Raadts in Willdenowia, **8**: 134 (1977) in adnot. Type: Mozambique, Nampula, Rio Monapo, *Torre* 907 (COI, holotype; LISC).

Very similar to *K. lateritia* differing by the leaves which are similar throughout the stem, the lower ones not so condensed, the uppermost not so scattered, usually indistinctly petiolate or sessile and with not such thick laminas; by the internodes not longer than 7·5 cm.; by the longer pedicels (2·5–5 mm. long rather than 1–3·5 mm. long); by the usually longer calyx-tube (1·4–2 mm. long rather than only up to 1(2) mm. long); by the longer and broader sepals which are ovate or ovate lanceolate and very attenuate.

Mozambique. N: Nampula, Rio Monapo, 2.viii.1936, *Torre* 907 (COI; LISC).
Only the type known. In xerophytic forest near the river or in open places in forests.

This taxon is perhaps a variety or subspecies of *K. lateritia,* although it approaches *K. lanceolata* (No. 1) in several features: in the leaves which are sessile to indistinctly petiolate of relatively thin texture and little differentiated from base to apex of stem, and in the longer calyx tube and broader sepals. The possibility that *K. fernandesii* is a hybrid between *K. lateritia* and *L. lanceolata* must also be considered. However, the material is too scanty to allow any firm conclusion.

6. **Kalanchoe laciniata** (L.) DC., Hist. Pl. Grasses **2**: 100 (1802) quoad basion; Prodr. **3**: 395 (1828). — Pers., Synops. Sp. Pl. **1**: 446 (1805). — Haw., Synops. Pl. Succ.: 109 (1812); in Philos. Mag. **4**: 302 (1829). — Wight, Ic. Pl. Ind. Or.: t. 1158 (1846). — Hook. & Thomson in Journ. Linn. Soc., London, **2**: 91 (1858). — Britten in F.T.A. **2**: 392 (1871). — C. B. Clarke in Hook., Fl. Br. India, **2**: 415 (1878) pro parte. — Schönl. in Engl. & Prantl, Nat. Pflanzenfam. **3**: 2a: 35 (1890). — Engl., Hochgebirgsfl. Trop. Afr.: 232 (1892). — Hiern, Cat. Afr. Pl. Welw. **1**: 326 (1896). — R.-Hamet in Bull. Herb. Boiss., Sér. 2, **7**: 897, 899, 890 (1907) pro parte; op. cit. **8**: 17–18 (1908) quoad distr. geogr. pro parte. — Mildbraed, Wiss. Ergebn. Zweit. Deutsch. Z. Afr. Exped., Bot.: 221 (1914). — Schönl. in S. Afr. Journ. Sci. **17**, 2: 187 (1921). — Berger in Engl. & Prantl, Nat. Pflanzenfam. ed. 2, **18a**: 406 (1930) quoad syn. pro parte et distr. geogr. pro parte. — Wild, Guide Fl. Vict. Falls: 143 (1953). — Cufod. in Bull. Jard. Bot. Brux. **24**, Suppl.: 168 (1954). — Binns, H.C.L.M.: 41 (1968). — Friedr. in Prodr. Fl. SW. Afr. **52**: 38 (1968). — Jacobs., Das Sukk. Lexikon: 253 (1970) pro parte. — R. Fernandes in Bol. Soc. Brot., Sér. 2, **53**: 368 (1980). — G. E. Wickens in Kew Bul. **36**, 4: 672 (1982) non auct. plurim. fl. Asiat. et Malesian. Type a (cultivated?) specimen in Hortus Siccus Cliffortianus (BM).
   Cotyledon laciniata L., Sp. Pl. **1**: 430 (1753). — Murr., Syst. Veg. ed. **14**: 429 (1784). — Lam., Encycl. Méth., Bot. **2**: 142 (1786). — Willd. in L., Sp. Pl. ed. 4, **2**, 1: 758 (1799). Type as above.
   Verea laciniata (L.) Willd., Enum. Pl.: 433 (1809). — Spreng., Syst. Veg. ed. 12, **2**: 260 (1825). Type as above.
   Kalanchoe schweinfurthii Penzig in Estr. Atti Congr. Intern. Bot. Genova: 32 (1892). — Schweinf. in Bull. Herb. Boiss. **4**, App. 2: 199 (1896). — Engl., Pflanzenw. Afr. **1**, 1: 119, 121 et 122 (1910). — Cufod., tom. cit.: 169 (1954); in Webbia **19**, 2: 732 (1965). — Raadts in Willdenowia, **8**: 144 (1977). Type from Ethiopia.
   Kalanchoe rohlfsii Engl. in Ann. R. Ist. Bot. Roma, **9**: 252 (1902). — Cufod., loc. cit. (1954) tom. cit.: 736 (1965). — Agnew, Upl. Kenya Wild Fl.: 107 (1974). Lectotype from Ethiopia.

A succulent perennial (or also biennial?) up to 1·20 m. high (with the inflorescence). Stem erect, simple, terete, smooth, reddish, leafless at fruiting time, glabrous below, ± puberulous above, the hairs usually c. 0·5 mm. long, straight, spreading, capitate-glandular or with indistinct terminal gland, hyaline to pale tawny; lower internodes 0·3–4 cm. long, the upper ones up to 12 cm. long. Leaves glabrous, succulent, flat, membranous and pale to dark green on drying, petiolate, up to 23·5 cm. long (with the petiole); lamina usually divided (sometimes those of lower and the uppermost pairs simple, rarely all simple), 3-sect or -foliolate to pinnatifid or pinnate, the primary segments simple, bifid or 3-sect or pinnatisect, various combinations of foliar division occurring sometimes in the same specimen; foliar segments up to 14 c. long and 4 cm. broad, the terminal one the largest, ovate-lanceolate to linear, acute at the apex, petiolulate or attenuate towards the base, entire, dentate, or shallow to deeply lobed; main rachis up to 4 mm. broad, canaliculate; petiole up to 6(8) mm. long, flattish, canaliculate, enlarged into a clasping base but not connate with the opposite one. Flowers erect in ± dense cymes forming corymbs either terminal and solitary or grouped in large panicle-like inflorescences up to 40 × 30; branches puberulous-glandular, at ±45° with the axis, naked in the lower ¾–½, the lower ones up to 29 cm. long; internodes of panicle up to 15·5 cm. long; pedicels up to 10 mm. long, glandular-puberulous. Calyx (2·5)4–5·5(7) mm. long, sparsely pubescent-glandular, thin, pale green, sometimes lineolate with brownish-red; tube 1–1·5 mm. high; sepals lanceolate-triangular, ovate or oblong, acute, separated by broad sinuses. Corolla 12–14(16) mm. long, yellow to yellow-green or orange, sparsely puberulous on the outside; tube rounded below, then narrowing gradually to above the anthers, hyaline or yellowish, 4-gonous and with 4 longitudinal raised nerves at the angles in fruit; corolla-lobes 3·5–4·5 × 1·5–1·75 mm., oblong or ovate-lanceolate, apiculate. Filaments of the upper stamens c. 1 mm. long; anthers ± 0·5 mm. long, included, the upper ones c. 1·5 mm. below the base of corolla-lobes. Follicles (5)6·5–7·25 mm. long, fusiform, very attenuate and acute; styles 0·75–1·25(1·5) mm. long. Scales 2–2·25 × 0·25 mm., linear, acute.

Zimbabwe. N: Lomagundi, Whindale Farm, northern portion, c. 1230 m., 4.v.1969, *W. Jacobsen* 3909 (K; PRE). E: Umtali, Maranka Reserve, c. 864 m., veg. 27.ii.1953, *Chase* 4794 (BM; SRGH). S: Chibi, 1·5 km. S. of Chibi Admin. Centre, at forest of granite Whaleback, ± 900 m., 6.v.1970, *Biegel & Pope* 3271 (COI; SRGH). **Malawi.** N: Nyika Plateau, vi.1896, *Whyte* 243 (K). C: Kasungu Distr., by quarry 8 km. N. of Kasungu, 1060 m., 7.v.1970, *Brummitt* 10431 (K). S: Upper Shire, Chikala Hill, c. 9 mil. on Ntaja road, small river at foot of Chikala Hill with falls, c. 768 m., 14.v.1961, *Leach & Rutherford-Smith* 10843 (BM; K; PRE; SRGH). **Mozambique.** T: at km. 148 of railway, 17.v.1948, *Mendonça* 4301 (LISC; LMU).

Also in Ethiopia, Uganda, Kenya, Tanzania, Angola, Namibia and India. In stony places in shady situations, in sandy or humus-rich soil.

The specimen *Whyte* 243, from Malawi (Nyika Plateau), has no leaves, but the indumentum, the shape and size of calyx, corolla, fruits, etc., allow perhaps its attribution to *K. laciniata*.

K. laciniata is somewhat similar to *K. lanceolata* (no. 1), from which it differs in having usually divided leaves, shorter indumentum with less distinctly capitate-glandular hairs, a not 4-gonous stem, petiolate leaves, shorter calyx tube, narrower corolla lobes, more attenuate follicles, etc. *K. laciniata* also resembles *K. crenata* (No. 3) in its short indumentum, terete stem and petiolate leaves, but differs by having only indistinct capitate-glandular hairs, divided leaves, a shorter calyx, more connate at base and with not attenuate sepals, a smaller corolla with shorter and narrower lobes, etc.

7. **Kalanchoe lobata** R. Fernandes in Bol. Soc. Brot., Sér. 2, **52**: 200 (1978). Type: Zimbabwe, Umtali, *Plowes* 2176 (K, holotype; SRGH).

A succulent herb, probably perennial, up to 65 cm. high. Stem erect, rigid, woody below, terete, smooth, densely glandular-hairy (hairs up to 1 mm. long, spreading, hyaline to pale tawny). Leaves petiolate; lamina at least in the median leaves, irregularly incised-lobed to 3-lobed, cuneate at the base, sparsely hairy, flat, fleshy, membranous and pale green to yellowish on drying; petiole narrow, broadened at the base, not connate with the opposite one; nodes somewhat raised. Inflorescence a panicle formed by a dense terminal corymb up to 6 cm. in diameter flanked by lateral corymbs; pedicels 2·5–5·5 mm. long. Calyx 6–7·5 mm. long, rounded at the base, densely and shortly glandular-hairy; tube 2·5–3·5 mm. long; sepals 2·5–3·5 mm. broad at base, ovate-lanceolate, acute, not or shortly apiculate. Corolla 17·5–19 mm. long, orange or yellow; corolla-tube dilated at the base, very contracted just above the follicles, then narrowly cylindric-tubular; corolla-lobes 5–5·5 × 2·75–3 mm., rather apiculate. Anthers c. 0·7 mm. long, included. Follicles ± 7·5 mm. long; styles ± 0·5 mm. long.

Zimbabwe. N: Mrewa-Salisbury, 25.vi.1963, *Hall* 389/57 (B; M). E: Umtali, near Umtali town, c. 1280 m., v.1961, *Plowes* 2176 (K; SRGH).

Not known from elsewhere.

By the terete stem, petiolate and lobed leaves, this species approaches *K. laciniata*; by the indumentum, formed by more distinct glandular-headed hairs than in that species, length of calyx-tube, shape and size of sepals, shape and size of corolla-tube and lobes, it is very similar to *K. lanceolata*; in the length of indumentum-hairs, it is intermediate between both.

8. **Kalanchoe hametiorum** R.-Hamet in Bol. Soc. Brot., Sér. 2, **37**: 25, t. 3 (1963); op. cit. **43**: 201–204, t. 1 (1969). — Jacobs., Das Sukk. Lexikon: 252 (1970). Type: Mozambique, Nampula, *Torre* 1513 (COI, holotype; LISC).

A succulent perennial, 0·24–1·5 m. high (incl. the inflorescence), covered everywhere with a dense indumentum of very short, thin, simple, pluricelular, spreading, orange-brown to brownish hairs which are denser and capitate-glandular on the inflorescence and flowers. Stem 1–(2) from a woody (horizontal?) base, erect, straight, simple, terete, not nodose, 4–9 mm. in diam. below and 2–4 mm. in diameter under the lowest node of the inflorescence, usually almost leafless towards fruiting-time; lowermost internodes up to 2·5 cm. long, the next distal ones shortening up to 0·5 cm., then elongating up to (median ones) 7 cm., shortening again upwards (the infrafloral-one) to 2 cm. Leaves 4·6–11 × 0·4–1·5(2·7) cm., linear or oblong to narrowly elliptic, very attenuate and subacute or obtuse at the apex, attenuate below into a subpetiolar base, entire or the median and lower ones sinuate or remotely and shallowly crenate, ± fleshy, canaliculate above along the median part but flat at the base and subcylindrical towards the apex, sessile, subamplexicaul but not connate, not decurrent, usually all opposite or the uppermost approximate in pseudo-pairs or distinctly alternate. Flowers in few-flowered cymes grouped in ± dense corymb-like inflorescences 3·5–12 × 7–16 cm.; pedicels 3·5–11 mm. long, those of dichotomies up to 17 mm. Calyx fusiform-urceolate, 11–13 mm. long in flower, with the maximum diameter in the lower ⅓, up to 17·5 mm. long in fruit, subequalling or a little longer than the tube of the corolla, green in the lower ⅔, yellowish-green above, shortly hairy outside, capitate-glandular-hairy on the inside, the glands minute; tube 5·5–8 mm. long, rupturing irregularly from the sinus between the sepals downwards; sepals 4·8–9(10·5) × 2·3–3·5 mm., lanceolate to subovate, attenuate and acute, conduplicate or with the margins slightly inflexed, finally subcylindric-canaliculate and subulate at the apex, fleshy. Corolla 11–16 mm. long, up to 20 mm. at fruiting, shortly capitate-glandular-hairy on the outside; corolla-tube 8·85–11·25 mm. long, tubular-urceolate, obtusely quadrangular, with the maximal diameter below the middle, narrowing slightly upwards, yellowish-green; corolla-lobes 3·65–5·8 × 1·6–1·8 mm., oblong-lanceolate or subovate, attenuate and acute, erect to spreading-reflexed, glandular-capitate-hairy on both faces, yellowish to salmon or brownish-yellow. Stamens with the filaments inserted above the middle of the corolla-tube, the alternipetalous ± 0·6 mm. long, the oppositipetalous 1–1·75 mm. long; anthers c. 1 × 0·5 mm., ovate-oblong, those of the oppositipetalous stamens attaining the apex of the corolla-tube. Carpels 6–7·9 mm. long; styles 1·6–2·5 mm. long, up to 3·5 mm. long in fruit; stigmas included. Follicles c. 10·5 mm. long. Seeds c. 0·5 × 0·2 mm., subovoid. Scales 1·3–2 mm. long, linear or linear-subrectangular, several times longer than wide.

Mozambique. N: 74 km. E. of Nampula, c. 320 m., 21.v.1961, *Leach & Rutherford-Smith* 10959 (COI; K; PRE; SRGH). Z: Lugela, 60 km. from Mocuba at the cross with road to Milange, c. 200 m., 30.x.1967, *Torre & Correia* 15836 (C; COI; LISC; LMU).
Known only from Mozambique. In Rupideserta, on rocks.

9. **Kalanchoe velutina** Welw. ex Britten in F.T.A. **2**: 396 (1871). Type from Angola.

Subsp. **chimanimanensis** (R. Fernandes) R. Fernandes in Bol. Soc. Brot., Sér. 2, **53**: 430 (1980). Type: Zimbabwe, Chimanimani Mt., alt. c. 1728 m., *Leach* 9050 (K, holotype; SRGH).
Kalanchoe chimanimanensis R. Fernandes in Bol. Soc. Brot., Sér. 2, **52**: 202 (1978). Type as above.

A perennial succulent, up to c. 95 cm. tall, densely hispidulous throughout, the hairs subrigid, acute (not glandular-headed), tawny to ferrugineous. Stem erect, up to 8 mm. in diameter at base, subquadrangular and dark brown below, subterete and reddish brown to ferrugineous above, nearly leafless at fruiting-time; lower internodes 3–4 cm. long, thence decreasing successively up to the ¼–⅓ of the length of stem, there rather short (up to 0·7 cm. or less), finally the 3 uppermost ones elongating suddenly (the two terminal ones respectively 13·5–18 cm. and 17–20 cm. long); nodes somewhat raised. Leaves decussate, erect to spreading, petiolate;

lamina 1·5–4·5(6·5) × 0·7–2(2·3) cm., spathulate to obovate, rounded at the top, irregularly crenate-serrate except towards the base or the whole margin entire, cuneate below and decurrent into the petiole, flat, somewhat thick, rigid and dark brown or greenish-brown on drying; petiole up to 1·3 cm. long, neither amplexicaul nor connate with the opposite one. Flowers yellow to orange-yellow (brownish-red on drying), in dense cymes forming terminal dense corymbs 6–12 cm. in diameter; lower branches up to 6 cm. long, suberect; pedicels up to 6·5 mm. long, sulcate, slightly clavate towards the apex. Calyx 1·75–3 mm. long, fleshy, somewhat thick, rigid and reddish-brown on drying; tube up to 0·3 mm. high, sometimes 0; sepals ± 1·25 mm. broad at the base, triangular, contracted and then attenuate towards the acute apex, separated by open rounded sinuses, somewhat convex on the back. Corolla 15–16 mm. long, reddish-brown on drying; tube 11·5–12 mm. long, swollen-rounded at the base, thence distinctly 4-angled and attenuate up to the throat, densely hispid, mainly above; corolla-lobes 3–4 × 2·1 mm., elliptic, obtuse or rounded and shortly apiculate at the apex. Filaments of lower stamens c. 0·6 mm. long, inserted c. 8 mm. above the base of the corolla-tube; filaments of the upper stamens 2·2–2·5 mm. long, inserted c. 2 mm. below the base of the corolla-lobes; anthers 0·5–0·75 mm. long, the lower included, the upper ones protruding above the base of the corolla-lobes. Follicles 6·5–9 mm. long, very attenuate into a very slender apical portion; styles 2–2·5 mm. long. Scales 1·75–2 × 0·2–0·4 mm., linear or oblong, slightly emarginate.

Zimbabwe. E: Chimanimani Plateau, among rocks, c. 1792 m., 27.v.1959, *Leach* 9046 (SRGH).
Known only from Zimbabwe. Among rocks on slopes of mountains.

This subspecies differs from the type of *K. velutina,* an Angolan taxon, mainly in habit, the latter being a suffrutex, but also in the petiolate leaves, shorter pedicels and shorter corolla lobes (*K. velutina* subsp. *velutina* has ± sessile leaves, narrower than those in subsp. *chimanimanensis* and pedicels up to 12 mm. long).
K. citrina Schweinf. (in Bull. Herb. Boiss. **4**, App. 2: 199, 1896) has a similar non-glandular indumentum but differs from subsp. *chimanimanensis* in many characters, as follows:

Indumentum of tawny, subrigid, very short hairs; calyx 1·75–3 mm. long; calyx and corolla
    reddish-brown and rigid or subrigid on drying; pedicels up to 6·5 mm. long
                                        K. velutina subsp. *chimanimanensis*
Indumentum of whitish, softer and longer hairs; calyx 3–4 times longer; both calyx and corolla
    pale and soft on drying; pedicels 1–2 mm. long -   -   -   -   - *K. citrina*

10. **Kalanchoe latisepala** N.E. Br. in Kew Bull. **1908**: 435 (1908). — R.-Hamet in Bol. Soc. Brot., Sér. 2, **43**: 204–207, t. 2 et 3 (1969). — Raadts in Willdenowia, **8**: 149 (1977) in adnot. Type: a cultivated plant in Kew Gardens raised from seeds from Malawi (K, holotype).

A perennial semi-shrubby succulent 28–96 cm. high with a ± dense indumentum of short, spreading, glandular-headed hairs, except on the base of the stem and on the lower leaves which are glabrous. Stem erect, usually simple, woody and 1–1·5 cm. thick (sometimes nearly globose-conical) below, covered there with a smooth and papery bark, then terete, rugose and with the scars of the fallen leaves ± raised, finally subquadrangular and up to 0·5 cm. in diameter above, green or greenish-yellow to purple-brown mainly below; median internodes 0·5–3 cm. long, the two uppermost ones respectively 2–3 cm. and 4·5–5·5 cm. long. Leaves usually sessile or sometimes with a subpetiolar broad base, amplexicaul, erect-spreading, flat, fleshy, carinate beneath, canaliculate above, green, shining; lower leaves 11–12·5 × 7·2–9 cm., obovate-cuneate, rounded at the top, irregularly crenate or subcrenate-dentate at the margin, except along the base; median leaves 4·5–8·5 × 3·3–3·7 cm., oblong-cuneate; upper leaves obovate-oblong to oblong-spathulate, smaller, concave. Flowers in cymes forming a terminal, usually dense corymb-like inflorescence 9–18 cm. in diam.; pedicels 5–10 mm. long, erect; bracts elliptical-oblong, acute, apiculate. Calyx 8–16 mm. long, green, shining; calyx-tube 2·5–5 mm. long; sepals 7–13 × 4–6 mm., oblong, ovate-lanceolate or ovate, ± attenuate, acute, fleshy, shortly glandular-pubescent on both faces. Corolla-tube 30–34·5 mm. long, suburceolate-quadrangular in the lower 12 mm. and there c. 8 mm. in diameter, then attenuate into a narrow, quadrangular-tubular upper portion up to 3·5 mm. in diameter, straight, white or greenish-white, glandular-pubescent; corolla-lobes

12–15 × 6·5–9 mm., ovate or elliptical-ovate, contracted at the base, acute and apiculate at the apex, spreading or reflexed, white, shortly glandular-pubescent on the outside. Filaments of the lower stamens very short, inserted ± 25 mm. distant from the base of the corolla; filaments of the upper ones 1·5–2 mm. long, inserted at the base of the corolla-lobes; anthers yellow, the lower ones c. 2 mm. long, broadly ovate, included, with the base ± 6 mm. distant from the sinus between the lobes of the corolla, the upper ones c. 1 mm. long, oblong, exserted. Follicles 12–15 mm. long; styles 17·5–20 mm. long, exserted for c. 3–5 mm. Scales 3·5–7 mm. long, usually linear.

Malawi. C: Dedza, Kalichero Hill, vi.1961, *Chapman* 1368 (K; PRE; SRGH). S: Zomba, Domasi Valley, c. 1600 m., 14.iv.1959, *G. Jackson* 2322 (SRGH). **Mozambique.** N: Lichinga (Villa Cabral), Massangulo Mt., c. 1400 m., 25.ii.1964, *Torre & Paiva* 10823 (LISC; LMU). Known only from Malawi and Mozambique. On desert-rocks.

The specimen *Leach & Rutherford-Smith* 11047 (K; SRGH), collected in Mozambique (Niassa, Mt. Massangulo) has the leaves flabellate-lobate or 3-lobate at the top, sessile and thin, some of them nearly membranous on drying. The leaf-lobes are retuse or emarginate or sometimes again lobate. The leaves present circular pustules, possibly of a parasitic nature. In other characters the specimen is like *K. latisepala*. Besided this, from the above-mentioned plant, normal individuals of *K. latisepala* were raised at Greendale (*Leach & Rutherford-Smith* 11047A (COI; K; SRGH)), presenting unlobed leaves without pustules. This suggests that the parasitic attack was the cause of the abnormal leaves of the wild specimen.

14. **Kalanchoe dyeri** N.E. Br. in Gard. Chron., Ser. 3, **35**: 354 (Jun. 1904); in Curtis, Bot. Mag., Ser. 3, **60**: t. 7987 (Dec. 1904). — Anonym. in Gard. Chron., Ser. 3, **39**: 296, t. fac. p. 296 (1906). — R.-Hamet in Bull. Jard. Bot. Bruxelles **19**: 440 (1949) in adnot. — Cufod. in Webbia **19**, 2: 724–725 (1965) in adnot. — R.-Hamet in Bol. Soc. Brot., Sér. 2, **43**: 204–205 (1969). Type a cultivated specimen at Kew Gardens raised from seeds from Malawi (K, holotype).
    Kalanchoe quartiniana sensu R.-Hamet in Bull. Herb. Boiss., Sér. 2, **8**: 27–28 (1908). — Berger in Engl. & Prantl, Nat. Pflanzenfam. ed. 2, **18a**: 405 (1930). — Cufod. in Bull. Jard. Bot. Brux. **24**, Suppl.: 168 (1954). — Jacobs., Das Sukk. Lexikon: 255 (1970) quoad synon. et specim. Malaw. non A. Rich. (1847).

A completely glabrous, 60–75 cm. high, perennial succulent. Stem 1·8–2·5 cm. in diameter at the base, erect, stout, terete, smooth, glaucous. Leaves petiolate, spreading; lamina 4·8–19 × 3·1–112·5 cm., elliptic or, in the upper leaves, ovate, obtuse to nearly acute at the apex, rounded or cuneate at the base, irregularly crenate-dentate at the margin, fleshy, flat or with the sides ± inflexed, green above (with purplish midrib and main nerves), glaucous beneath; petiole 3·8–7·5 cm. long, 8–12·6 mm. broad and of approximately equal thickness, semiterete (convex beneath, flattened or canaliculate above), dilated at the base, semiamplexicaul; leaves of the uppermost pair 1·3–5 cm. long bract-like, oblong or spathulate, entire, sessile. Flowers large, upright, in 3–8-flowered cymes, forming corymbose inflorescences 22–30 × 15–22 cm., with suberect branches; pedicels 16–19 mm. long and 1·5–2 mm. in diameter, glaucous. Calyx-tube 1–1·75 mm. long; sepals 6–12 mm. long and 1–4 mm. broad at the base, oblong, attenuate, obtuse to acute, separated by rounded sinuses, fleshy. Corolla-tube 43–50 mm. long and 7–12·6 mm. in diameter at the subquadrangular base, thence attenuate, pale green; corolla-lobes 18–25 × 9·5–15 mm., lanceolate to elliptic, acute but not apiculate, pure white, spreading. Filaments of the lower stamens c. 1·6 mm. long, those of the upper ones c. 2 mm. long; anthers apiculate, yellow, those of the lower stamens 1·6–2 mm. long and c. 8 mm. below the sinuses between the corolla-lobes, those of the upper ones c. 1–1·25 mm. long, slightly exserted. Carpels 16–22 mm. long; styles 22–24 mm. long, filiform; stigmas included, c. 3 mm. below the base of corolla-lobes. Scales c. 10·5 mm. long, linear, bifid.

Malawi. N: Panda Peaks, c. 2041 m., ix.1902, *McClounie* 35 (K).
Known only from Malawi.

Perhaps not distinct from *K. quartiniana* A. Rich., but in this Ethiopian plant the flowers seem smaller (not more than 5 cm. in total length; the calyx is c. half as long, the lower anthers are only c. 1 mm. (not 1·6–2 mm.) long; the styles 13·5–17 mm. (not 22–24 mm.) long; and the scales are only c. 0·6 mm. (not c. 10 mm.) long; and the stem, leaves and pedicels are not glaucous. Nevertheless, more material from Malawi is needed to decide whether or not *K. dyeri* is distinct.

12. **Kalanchoe elizae** Berger in Monatsschr. Kakteenk. **13**: 69 (1903); in Engl. & Prantl, Nat. Pflanzenfam. ed. 2, **18a**: 405 (1930). — R.-Hamet in Bull. Herb. Boiss., Sér. 2, **8**: 39 (1908); in Bull. Soc. Bot. Fr. **57**: 22 (1910); in Rév. Gén. Bot. **28**: 81 (1916). — Jacobs., Das Sukk. Lexikon: 251 (1970). — Hilliard & B. L. Burtt in Notes Roy. Bot. Gard. Edinb. **31**, 1: 26 (1971). — R. Fernandes in Bol. Soc. Brot., Sér. 2, **53**: 362 (1980). Type: cultivated material from Tropical Africa.

Cotyledon insignis N.E. Br. in Curtis, Bot. Mag., Ser. 4, **1**: t. 8036 (1905). Type: a cultivated plant at Kew Gardens from seeds sent from Malawi by *J. Mahon* (K, holotype).

Kalanche insignis (N.E. Br.), N.E. Br., Handlist Dicot., Gnetaceae R. B. Gard. Kew, ed. 2: 109 (1931). Type as above.

Kalanchoe laurensii R.-Hamet in Rev. Gén. Bot. **28**: 81 (1916); in Bol. Soc. Brot., Sér. 2, **37**: 30–32, t. 5 (1963). Type: Mozambique, Namúli, Makua Country, *Last* s.n. (K, holotype).

A completely glabrous perennial succulent, up to 1·8 m. high (with the inflorescence). Stems 1 to few, 0·20–1·5 m. high, stout, up to 6·5 cm. in diameter at the base, erect, terete, red or pale green suffused with red, leafless at flowering time. Leaves decussate, 6·5–20·8 × 2·5–12 cm., broadly obovate or suborbicular to oblong, rounded and contracted at the base into a broad petiole, or subcuneate and attenuate into a subpetiolar flat base, the uppermost obovate-spathulate to spathulate and sessile, all rounded or obtuse at the top, subcrenate or undulate at the margin or entire, flat, fleshy (up to 4 mm. thick), green, slightly shining; midrib somewhat impressed above, slightly prominent beneath; petiole up to 1·5 × 1 cm. Flowers slightly zygomorphic, with the mouth directed upwards, grouped in loose to dense cymes forming a ± ample, loose, panicle-like inflorescence 9–28 × 12–40 cm., with strong axis and branches; branches decussate, divaricate, with flowers only in the upper ½–⅓; leaves of the base of inflorescence up to 6 × 3 cm.; pedicels 4–18 mm. long, rather thickened towards the apex; bracts oblong-linear to elliptic. Calyx 2·5–8(10) mm. long, campanulate, dark red, falling off with the corolla; tube 1·5–2·75 mm. long; sepals broadly triangular and broader than long, or lanceolate and then slightly longer than broad, acute, apiculate, separated by broad, retuse sinuses. Corolla 30–55 mm. long, dark yellow or orange to red; tube 21·5–37·5(45) mm. long, with the maximum diameter a little above the base, thence attenuate up to the middle and subcylindrical and slightly curved upwards, obscurely 4-angled; lobes 10–16 × 3·5–6(7·25) mm., lanceolate or ovate-oblong, asymmetrical towards the acute apex (rounded at one margin, nearly rectilinear at the other), rather apiculate, reflexed, yellow-orange or light red on the back and yellowish-green suffused with light red on the upper face. Filaments of the lower stamens inserted 11 mm. above the base of corolla-tube, those of the upper ones 13 mm. above the same; anthers 1·5–2 mm. long, oblong to ovate, dark red, much exserted, those of the upper stamens attaining the upper ¼ of the corolla-lobes, those of the lower ones attaining nearly the middle of corolla-lobes. Carpels 10·5–13·5(15) mm. long, attenuate into the styles; styles 15–25(30) mm. long, filiform, exserted, the stigmas attaining or exceeding the apex of corolla-lobes. Scales 1·6–6 mm., bilobed or truncate.

Malawi. S: Mulanje Mts., N. slopes of Great Ruo Gorge, 1500 m., 19.vi.1962, *E. A. Robinson* 5387 (K); also cultivated at Greendale, fl. 24.ii.1960, from plants collected on Mulanje Mt., *Leach* 9771 (B; BR; K; LISC; LMA; PRE; SRGH). **Mozambique.** N: Serra de Ribáuè, 12.x.1935, *Torre* 1122 (COI; LISC). Z: Lugela, 60 km. from Mocuba in the cross of road to Milange, c. 200 m., 30.x.1967, *Torre & Correia* 15836A (C; LISC; LMU).
Known only from Malawi and Mozambique. On desert-rocks.

A very interesting and well-characterised species whose large flowers present a slight zygomophism, with curved corolla-tube, asymmetrical corolla-limb and stamens with the filaments also curved.

K. laurensii, described from a wild specimen, as well as most of the specimens seen by us, has smaller flowers and relatively shorter and broader leaves than *K. insignis* which was described from cultivated plants. The differences are in all probability due to the cultivation of the latter.

In *Mendonça* 2170 (LISC), from Serra do Gúruè, Pico Namúli (loc. class. of *K. laurensii*), some flowers with normal 4-merous calyx and corolla have, however, 5 carpels.

13. **Kalanchoe humilis** Britten in F.T.A. **2**: 397 (1871). — R.-Hamet in Bull. Herb. Boiss., Sér. 2, **8**: 39 (1908); in Bull. Soc. Bot. Fr. **57**: 23 (1910). — Berger in Engl. & Prantl, Nat. Pflanzenfam. ed. 2, **18a**: 405 (1930). — Jacobs., Das Sukk. Lexikon: 252 (1970). — R. Fernandes in Bol. Soc. Brot., Sér. 2, **53**: 366 (1980). Type: Mozambique, Morrumbala, *Waller* s.n. (K, holotype).

Kalanchoe prasina N.E. Br. in Gard. Chron., Ser. 3, **35**: 211 (1904). Type a cultivated plant at Kew Gardens, raised from a Malawi stock (K, holotype).
Kalanchoe baumii sensu R.-Hamet in Bull. Herb. Boiss., Sér. 2, **7**: 895 (1907), quoad synon. *K. prasina.* — Schönl. in S. Afr. Journ. Sci. **17**, 2: 188 (1921) quoad aream geogr. (Malawi). Non Engl. & Gilg (1903).
Kalanchoe figueiredoi Croizat in Bull. Jard. Bot. Brux. **14**: 366, fig. 26 et t. 9 (1937). — Guillaumin in Rev. Hort., N. Sér., **28**: 306 (1943). — R.-Hamet & Marnier-Lapostolle in Arch. Mus. Nation. Hist. Nat. Paris, Sér. 7, **8**: 102, t. 35 fig. 122 et t. 36 fig. 123 (1964). — Jacobs., loc. cit. — Hilliard & Burtt in Notes Roy. Bot. Gard. Edinb. **32**, 3: 386 (1973). — Raadts in Willdenowia, **8**: 149 (1977). Type: A cultivated plant raised at Jard. Bot. Brux. from a stock sent from Mozambique, Metónia, by *Gomes e Sousa* (B; BKL; BR, holotype; EA; K; P).

A completely glabrous, perennial succulent up to 83 cm. tall (with the inflorescence). Stem 2·5–11 cm. long (up to 20 cm. (?) in cultivated plants) arising from a subhorizontal rhizomatous woody base up to 1 cm. thick, erect or oblique, simple or with few branches, purple or slightly glaucous; lower internodes ± short, the upper ones increasing up to 8·5 cm. Leaves (1·5)3–10(13) × (0·7)1·5–3·7(6) cm., obovate to spathulate, rounded or subtruncate at the top, all entire or the upper ones irregularly and obtusely crenate towards the apex, cuneate below and tapering into a broad petiolar base or subsessile, fleshy, flattish, rigid and subchartaceous on drying, pale green with a slight glaucous bloom on both sides, sometimes with purplish spots, decussate, sometimes crowded and subrosulate or, mainly the upper ones, ± scattered, the lower spreading, the upper suberect; nerves faintly conspicuous. Flowers small, erect or horizontal, in dichasiums grouped in a terminal, sometimes diffuse panicle, 12–38 cm. long, very ample, very floriferous and with flexuous axis in cultivated plants, with a glaucous bloom; peduncle (5)20–41 cm. long, straight or flexuous, terete, smooth, with 1–3 pairs of lanceolate, barren bracts; primary panicle-branches in 2–4 pairs, up to 11 cm. long, erect to erect-spreading, slender, subtended by bracts ± 3·5 mm. long, the lower ones repeatedly dichotomous; cymes 4–7 cm. long, unilateral, with 4–13 flowers; pedicels 2–10(15) mm. long, much longer than the minute bracteoles, thickened upwards, dull purple. Calyx 1–1·5(2) mm. long and 2 mm. in diameter, dull purple, falling off with the corolla; tube 0·25–0·5 mm. long; sepals broadly triangular, acute, glaucous. Corolla 5–7(8) mm. long; tube 4·5–5 mm. long, obsoletely 4-angled and scarcely dilated at the base, c. 4·5 mm. in diameter at the top and there distinctly 4-angled, greenish (lineolate with violet) to lilac; corolla-lobes 1·75–2(3) × 1–1·8 mm., oblong, obtuse, erect or slightly spreading, whitish or greenish, with the median part dull purple or with 5–7 longitudinal purplish veins. Filaments minutely papillose, inserted above the middle of corolla-tube, those of the lower stamens c. 0·5 mm. long, those of the upper ones 0·75–1·5 mm.; anthers 0·5 or 1 mm. long, suborbicular or oblong, apiculate, those of the lower stamens ± at the level of the mouth of corolla-tube, those of the upper ones exserted (attaining the middle of corolla-lobes). Follicles 4·5–5 mm. long; styles 0·5–1 mm. long. Scales 1·5–2 mm. long, linear, bifid.

Malawi. S: Mt. Mdima, 5.ii.1967, *Hilliard & Burtt* 4673 (E; also cultivated in R.B.G., Edinburgh).

Mozambique. N: Ribáuè, c. 960 m., 19.vii.1962, *Leach & Schelpe* 11418 (K; LISC; LMA; P; PRE; cult.); Serra de Ribáuè, Mepáluè, c. 800 m., 23.iii.1964, *Torre & Paiva* 11336 (C; LISC; LMU). Z: rocky island in Ruo R., at Zoa falls and rapids on Mozambique-Malawi border, 16.iii.1971, *Leach & Royle* 14814 (K; SRGH); Ile, Errego, c. 3 km. from M. Ile, c. 900 m., 3.iii.1966, *Torre & Correia* 14970 (LISC).
Known also from Tanzania. On rocks and in crevices.

Cultivated plants, as those described respectively by L. Croizat (as *K. figueiredoi*) and N. E. Brown (as *K. prasina*), are taller and more robust, with larger leaves, ampler inflorescences and slightly larger flowers than spontaneous ones. *K. humilis* was based on a small, slender specimen, but with its characteristic small purplish flowers it cannot be separated from the two preceding taxa.

14. **Kalanchoe luciae** R.-Hamet in Bull. Herb. Boiss., Sér. 2, **8**: 256 (1908); in Bull. Soc. Bot. Fr. **57**: 21 (1910). — Burtt Davy, F.P.F.T. **1**: 144 (1926). — Berger in Engl. & Prantl, Nat. Pflanzenfam. ed. 2, **18a**: 407 (1930). — R.-Hamet, Crass. Ic. Select. **4**: t. 78–80 (1960). — Letty, Wild Fl. Transv.: 152 (1962). — R.-Hamet & Marnier-Lapostolle in Arch. Mus. Nation. Hist. Nat. Paris, Sér. 7, **8**: 91, t. 33 fig. 113–114 (1964). — Jacobs., Das Sukk.

Lexikon: 253 (1970). — J. H. Ross, Fl. Natal: 179 (1972). — R. Fernandes in Bol. Soc. Brot., Sér. 2, **52**: 204 (1978). Type from S. Africa (Transvaal).

Kalanchoe aleurodes Stearn in Journ. of Bot. **69**: 164 (1931); in Gard. Chron., Ser. 3, **89**: 475, fig. 239 (1931). — Tutin in Gard. Chron., Ser. 3, **90**: 9 (1931). Type a cultivated plant in the Cambridge Botanical Garden, raised from Zimbabwean seeds, collected by *S. G. Arden* at Domboshawa (BM; CGE, holotype).

Kalanchoe albiflora Forbes in Bothalia, **4**: 37, t. pag. 39 (1941). Type from S. Afr. (Natal).

Kalanchoe thyrsiflora sensu Burtt Davy, op. cit.: 143 (1926) pro parte quoad specim. *Rogers* 24085. — Hutch., Botanist in S. Afr.: 465 (1946) non Harv. (1862).

Kalanchoe luciae subsp. *luciae* — Tölken in Journ. S. Afr. Bot. **44**, 1: 89 (1978).

A stout biennial succulent, 0·5–2·5 m. tall, completely glabrous but covered with a white mealy layer falling off with the age. Single flowering stem arising from the dense rosette of leaves formed in previous year, curved at base, then erect and straight, terete or sub-4-angled, up to 2 cm. in diam. at base, finally leafless below and marked there by the very close scars of fallen leaves, leafy upwards. Leaves decussate, sessile, semiamplexicaul, shortly connate and each decurrent in two raised lines down the stem, rounded or obtuse at the apex, entire, the oldest ones red at the margin, fleshy, flat; basal ones 6–18(23·5) × 5·5–15 cm., subcircular or obovate to oblong-spathulate, contracted into a very short and broad subpetiolar base (up to 3 cm. broad), very dense, spreading; median leaves 3·5–9 × 3–3·5 cm., oblong to spathulate or suborbicular, ± scattered, the uppermost ones smaller; lower internodes short (the shortest ones c. 2 mm. long), the median ones increasing to 16·5 cm. Flowers erect, usually in many-flowered, dense, shortly pedunculate, upright, scorpioid, opposite, axillary cymes, forming a narrow to somewhat broad, continuous or interrupted thyrsoid inflorescence up to 32·5(43) × 6·5(14) cm. or rarely a panicle; internodes of the inflorescence 2–8, longer than the cymes in the interrupted inflorescences, shorter to subequalling them in the continuous ones; leaves of the flowering axis bract-like, oblong, shorter than the cymes; pedicels up to 10 mm. long. Calyx-tube 0·75–1·5 mm. long; sepals 3–6·5 × 2·25–4 mm., oblong to lanceolate, obtuse to acute. Corolla 10–15(16) mm. long; tube 6–10 mm. long and ± 8 mm. in diameter at the middle, distinctly 4-angled, very contracted just below the limb, pale green, mealy; corolla-lobes 4·5–7 × 2·5–3 mm., oblong or ovate-triangular, obtuse to nearly acute, at first suberect, finally reflexed, pale yellow, pale pink, whitish or greenish, turning brownish-red when dry. Filaments 4 or 5 mm. long (at fruiting time), inserted ± at the ¾ of corolla-tube; anthers c. 1·5 × 1·5 mm., with divergent thecae, reddish, exserted. Follicles 7–11·5 mm. long; styles 2·75–3·5 mm. long, exserted. Scales 1–2 × 2–2·5 mm., usually broader than long, subquadrate-cuneate, entire, emarginate or 3-lobed.

Zimbabwe. N: Mazoe, ± 3 miles NE. of Mtoroshanga, iv.1965, *Leach & Bullock* 12823 (K; PRE; SRGH). C: Salisbury, Cleveland, 2.ix.1946, *Wild* 1203 (K; SRGH). E: Inyanga, ± 4 miles from Inyanga Village along Troutbeck road, c. 2016 m., 26.iv.1967, *Rushworth* 747 (SRGH). S: Lundi River kopjes, 30.vi.1930, *Hutchinson & Gillett* 3306 (BM; K; LISC; SRGH). **Mozambique. MS.**: Chimoio, 55 km. to Rotanda, in cross-road from Garuzo to Chicamba, c. 700 m., 25.xi.1965. *Torre & Correia* 13285 the stock origin of a plant cultivated at Sintra, Portugal [5.iv.1968, *Mendes* (LISC)]. Also cultivated in the garden of I.I.C.M. [Umbelúzi, 28.vii.1965, *Fidalgo de Carvalho* 773 (K; LISC)] brought from "Vila Pery region where it grows in *Parinari curatellifolia* and *Uapaca kirkiana* forest".

Known from Zimbabwe, Mozambique and S. Afr. (Natal and Transvaal). On granite slopes in woodland or savanna.

Specimens of *K. luciae* have sometimes been determined in herbaria as *K. thyrsiflora* Harv. This is a very similar species, especially in habit, but it differs from *K. luciae* in the following characters: calyx usually shorter relative to the corolla tube, and sepals slightly narrower and less connate; corolla larger with more swollen, less distinctly 4-angled tube, less contracted below the mouth; corolla lobes shorter relative to the tube, and broader than in *K. luciae*, filaments shorter and inserted higher in the corolla tube (just at or a little below the mouth); scales usually longer than broad, oblong; fruits larger.

The specimen *Rushworth & Mavi* 1827 (SRGH) from Zimbabwe (Wedza, SE. slopes of Wedza Mt., 64 km. S. of Marandellas, c. 1760 m., 21.v.1968) could be *K. thyrsiflora* or *K. crundallii* Verdoorn by the sepal shape, the stamen insertion position and size and shape of scales, but on the other hand the size of calyx, corolla-tube and follicles agrees with *K. luciae*; more material is needed from the same locality to decide whether *Rushworth & Mavi* 1827 is a small-flowered form of *K. thyrsiflora* or a hybrid between the latter (which so far is unknown from Zimbabwe) and *K. luciae* (relatively common in Zimbabwe). A similar plant, intermediate between the two species at least in scale size, was figured by R.-Hamet (loc. cit., 1960); it originated in the Transvaal (Lydenburg, *Crundall* (PRE)).

15. **Kalanchoe wildii** R.-Hamet ex R. Fernandes in Bol. Soc. Brot., Sér. 2, **52**: 204 (1978).
Type: Zimbabwe, Zimbabwe, *Ball* 16 (SAM).
Kalanchoe aleurodes sensu Suesseng. & Merxm. in Trans. Rhodes. Sci. Assoc. **43**: 15 (1951) non Stearn (1931).
Kalanchoe wildii R.-Hamet, Crass. Ic. Select. **2**: t. 27–30 (1956) *nom. nud.*

A completely glabrous, 30–70 cm. high biennial succulent, covered everywhere with a ± persistent chalky-white powdery clothing (very dense on the inflorescence). Stem arising singly from the rosette of leaves of the previous year, usually simple, rarely forked near the base, arched below, then erect, stout, 12–20 mm. in diameter at the base, there leafless and marked with the annular continuous scars of fallen leaves, terete below, sub-4-angled above; lower internodes 0·5–2·2 cm. long, median ones 2·3–7 cm. long, the uppermost ones 2·3–4·5 cm. long. Leaves decussate, denser towards the base, erect and ± applied to the stem or the lowermost ± spreading, usually longer than the internodes; lower and median leaves 2–10·5 × 1·7–4 cm., obovate to oblong-spathulate or subcircular, the largest narrowing into a base up to 2 cm. broad, the uppermost ones smaller, bract-like, all rounded at the top, entire, sessile, slightly connate, amplexicaul and decurrent in two raised lines down the stem. Flowers erect in ± dense opposite axillary cymes grouped in terminal interrupted or continuous spike-like inflorescences or in thyrsoid panicles, 7–26 × 6–8 cm.; sometimes smaller inflorescences, 4·5–10 × 4 cm., also present in the axils of the lower and median leaves; peduncles of the cymes upright, those of the lower ones up to 7 cm. long; pedicels up to 10 mm. long. Calyx 5–6·5 mm. long, rather fleshy; tube 1–1·5 mm. long; sepals 2·75–3·5 mm. broad at the base, ovate or broadly elliptic, obtuse or nearly rounded to subacute at the apex. Corolla 10–14 mm. long, urceolate, narrowing from the base of the tube to the apex of the lobes; corolla-tube 7–11 mm. long, 4–4·5 mm. in maximum diameter, 4-angled, not constricted below the lobes, orange-pink or yellow within; corolla-lobes 3–5·5 × 2·25–2·75 mm., oblong or subrectangular, rounded or truncate at the apex, not apiculate, inflexed at the margins, subkeeled, fleshy, rigid on drying, erect and connivent. Filaments 1 or 2 mm. long, inserted near the mouth of corolla-tube; anthers c. 1 × 1 mm., suborbicular, those of the lower stamens included or subexserted (below, attaining or slightly above the sinuses of corolla-lobes), those of the upper stamens distinctly exserted (reaching to ⅓ or ½ of corolla-lobes). Follicles ± 7·5 mm. long; styles c. 0·5 mm. long to almost absent; stigmas included or slightly exserted. Scales 1·5 × 2·5–2·75 mm. transversely subrectangular, rather broader than high, entire.

Zimbabwe. W: Matobo, Farm Besna Kobila, c. 1536 m., i.1957, *Miller* 4393 (SRGH). C: Makoni, c. 1568 m., vii.1917, *Eyles* 729 (BM; SRGH). S: Belingwe, ± 3 miles SE. of Mnene Mission, 10.vii.1966, *Leach & Bullock* 13318 (SRGH).
Known only from Zimbabwe. On rocks.

16. **Kalanchoe thyrsiflora** Harv. in Harv. & Sond., F.C. **2**: 380 (1862). Hook. f. in Curtis, Bot. Mag. **125**: t. 7678 (1899). — M. Wood & Evans, Natal Pl. **1**: 43, t. 52 (1899). — R.-Hamet in Bull. Herb. Boiss., Sér. 2, **7**: 894 (1907); in Bull. Soc. Bot. Fr. **57**: 19 (1910). — M. Wood, Handb. Fl. Natal: 46 (1907). — Burtt Davy, F.P.F.T. **1**: 143 (1926). — Pole Evans, Fl. Pl. S. Afr. **9**: 341 (1929). — Berger in Engl. & Prantl, Nat. Pflanzenfam. ed. 2, **18a**: 407, fig. 196 H–L (1930). — Hutch., Botanist in S. Afr.: 286 (1946). — R.-Hamet & Marnier-Lapostolle in Arch. Mus. Nation. Hist. Nat. Paris, Sér. 7, **8**: 92, t. 34 fig. 115–116 (1964). — Jacobs., Das Sukk. Lexikon: 256, t. 106 fig. 2 (1970). — Guillarmod, Fl. Lesotho: 180 (1971). — J. H. Ross. Fl. Natal: 179 (1972). Syntypes from S. Afr. (Cape Prov.).
Kalanchoe alternans sensu Eckl. & Zeyh., Enum. Pl. Afr. Austr. **3**: 305 (1837) non Pers. (1805).

A completely glabrous biennal 0·75–1·30 m. high, succulent ± pulverulent with white, sticky powder. Stem single, arising from the rosette of leaves of the previous year, simple, erect, stout, terete, up to 2·5 cm. in diameter at the base and ± 1·25 cm. in diameter below the inflorescence, smooth, leafless below at flowering time and marked there with the very close scars of fallen leaves. Leaves decussate, denser towards the base of the stem, obtuse or rounded at the top, entire, sometimes with pinkish margin, sessile, connate at the base and semi-amplexicaul, slightly decurrent, thick, fleshy, concave on the upper face, somewhat convex on the under one, the lower 7–17·5 × 7–12·5 cm., obovate or obovate-spathulate, the median and the upper ones successively smaller, oblong or spathulate. Flowers erect in ± dense axillary erect cymes forming a terminal thyrsoid panicle up to 30 cm. long and ± 8 cm. in diameter; pedicels 6·6–12 mm. long, somewhat thick. Calyx-tube 1–1·5 mm. high; sepals 3–5·5(7) × 2–2·7 mm., ovate to oblong-lanceolate, acute or nearly obtuse. Corolla-tube 11·5–15 mm. long and 6–7 mm. in diameter, suburceolate or ovoid-oblong, slightly contracted at the top, terete towards the base, distinctly 4-angular upwards, yellow or glaucous, pale and scarious-papyraceous on drying; corolla-lobes 3·25–5 × 3–4·5 mm., ovate to suborbicular, obtuse or rounded at the top, obsoletely apiculate, golden-yellow on the inner face, reflexed. Filaments of the upper (oppositipetalous) stamens 1·5–2 mm. long, inserted 1 mm. below the base of corolla-lobes, those of the lower ones (alternipetalous) c. 1 mm. long, inserted c. 1·5 below the sinuses between the corolla-lobes; anthers 1·3–1·4 mm. long, broadly ovate to subcircular, emarginate at the base, apiculate, those of the upper stamens completely exserted, the others included or their apices slightly above the sinuses of the corolla-lobes. Follicles (12·5)13·5–15 mm. long, oblong-lanceolate; styles (1·5)2·5–3 mm., partially exserted. Scales (1·75)2–3 × (1)1·25–2 mm., rectangular, emarginate. Seeds ± 0·3 mm., rounded at the top, longitudinally ribbed.

Botswana. SE: At Gaborone-Lobatse Road, c. 35 km. from Lobatse, 1200 m., 17.viii.1977, *O. J. Hansen* 3159 (C; GAB; K; PRE).

Also in Lesotho and S. Afr. (Natal, Transvaal, Orange Free State and E. Cape Prov.). In rocky ground with *Combretum apiculatum*, *Euclea undulata*, tree-*Aloes*, etc.

Similar to *K. luciae* and *K. wildii*. Differences with the former are given on p. 58. Relatively to *K. wildii*, *K. thyrsiflora* differs mainly by the larger flowers (corolla-tube 11·5–15 mm. long and 6–7 mm. in diameter while in *K. wildii* the corolla-tube is 7–11 mm. long and 4–4·5 mm. in diameter); by the relatively narrower sepals and broader, reflexed corolla-lobes (in *K. wildii* these are erect and connivent after anthesis); by the larger anthers (1·3–1·4 mm. long instead of c. 1 mm.) but relatively narrower, not so deeply emarginate at the base and distinctly apiculate (in *K. wildii* they are not apiculate or are obsoletely so); by the larger follicles ((12·5)13·5–15 mm. long instead of ± 7·5 mm. long); by the longer styles ((1·5)2·5–3 mm. long while in *K. wildii* they are only c. 0·5 mm. long or almost absent); and by the different scales (rectangular, longer than broad and emarginate; in *K. wildii* they are transversely subrectangular, broader than high and entire).

The Botswana specimen has both longer sepals and shorter styles than some examined specimens from other regions.

17. **Kalanchoe rotundifolia** (Haw.) Haw. in Philos. Mag. Journ. **66**: 31 (1825); in Philos. Mag. **6**: 304 (1829). — DC., Prodr. **3**: 395 (1828). — Eckl. & Zeyh., Enum. Pl. Afr. Austr. **3**: 305 (1837). — Harv. in Harv. & Sond., F.C., **2**: 379 (1862). — Balfour in Trans. Roy. Soc. Edinb. **31**: 90 (1888). — Schönl. in Engl. & Prantl, Nat. Pflanzenfam. **3**, 2a: 35 (1891). — Bak. f. in Journ. of Bot. **37**: 434 (1899). — M.-Wood & Evans, Natal Pl. **1**: 76, t. 94 (1899). — Schinz in Mém. Herb. Boiss. no. **10**: 38 (1900). — Engl. in Sitz. Königl. Preuss. Akad. Wiss. **52**: 878 (1906). — R.-Hamet in Bull. Herb. Boiss., Sér. 2, **7**: 895 (1907). — M.-Wood, Handb. Fl. Natal: 46 (1907). — Schönl. in S. Afr. Journ. Sci. **17**, 2: 188 (1921). — Burtt. Davy, F.P.F.T. **1**: 144 (1926). — Berger in Engl. & Prantl, Nat. Pflanzenfam. ed. 2, **18a**: 406 (1930). — Rendle in Journ. of Bot. **70**: 90 (1932). — Hutch., A Botan. in S. Afr.: 673 (1946). — Wilman, Check List Fl. Pl. Ferns Griqualand W.: 79 (1946). — Uhl in Amer. Journ. Bot. **35**: 397 (1948). — M. Hulme, Wild Fl. Natal: t. 25 fig. 1 (1954). — Popov in Journ. Linn. Soc., Bot. **55**: 718 (1957). — Mogg in Macnae & Kalk, Nat. Hist. Inhaca Island: 145 (1958) pro parte. — Letty, Wild Fl. Transv.: 149, t. 75 fig. 2 (1962). — Riley, Fam. Fl. Pl. S. Afr.: 157 (1963). — R.-Hamet & Marnier-Lapostolle, Arch. Mus. Nation. Hist. Nat. Paris, Sér. 7, **8**: 87, t. 32 fig. 108–109 (1964). — D. Edwards, Bot. Surv. S. Afr., Mem. **36**: 107, 125 et 264 (1967). — Friedr. in Prodr. Fl. SW. Afr. **52**: 38 (1968). — Gledhill, Eastern Cape Veld Fl.: 125, t. 28 fig. 2 (1969). — Jacobs., Das Sukk. Lexikon: 255 (1970). — Schijff & Schoonr. in Bothalia **10**, 3: 489 (1971). — Venter in Journ. S. Afr. Bot. **37**, 2: 106 (1971). — Raadts in Willdenowia, **8**: 120 (1977). — R. Fernandes in Bol. Soc. Brot., Sér. 2, **53**: 405 (1980). — Type a cultivated plant at Kew ⟶ (1822).
  Crassula rotundifolia Haw. in Philos. Mag. Journ. **64**: 188 (1824). — DC., Prodr. **3**: 384 (1828). — Harv. in Harv. & Sond., F.C. **2**: 365 (1862). — Schinz, loc. cit. Type as above.
  Verea rotundifolia (Haw.) Dietrich, Synops. Pl. **2**: 1328 (1840). Type as above.

A completely glabrous, 0·2–2 m. high, perennial (or sometimes annual) plant, suc-culent, sometimes with sterile small-leaved shoots at the base. Stems 1–(2), usually simple or sometimes with a branch near the base, rarely branched above, 2–5(8) mm. in diameter below, rather slender above, erect, terete, smooth, glaucous, often leafless after anthesis, sometimes several floriferous stems from a prostrate base of a main stem; nodes not or slightly raised. Leaves ± spreading to erect, at least the lower ones petiolate, condensate towards the base of stem, ± scattered upwards, caducous; lamina 1–5(8·5) × (0·5)0·7–2·5(5·5) cm., usually small, succulent but not very fleshy, rather thin to nearly translucent on drying, pale green to glaucous. Flowers erect, in usually loose cymes (sometimes reduced to 1–2 flowers), grouped in a corymb- or panicle-like or sometimes thyrsoid inflorescence, usually small (2–14 × 3·5–10 cm.), sometimes attaining 37(38) cm. in length; main branches of inflorescence up to 8 pairs, slender, erect, floriferous only at the extremity, the lower ones up to 27 cm. long; pedicels of dichotomies 2·5–8 mm. long. Calyx 1·5–2·5(3) mm. long; sepals 0·8–1 mm. broad at the base, lanceolate or triangular, acute, nearly free, rigid on drying, caducous at fruiting time. Corolla 10–12(13) mm. long; tube up to 4 mm. in diameter and ovoid to subglobose in the lower ⅔, contracted above the carpels into a narrower cylindrical-tubular portion twisted at fruiting-time, pale glaucous to green at the base, bright pink to scarlet above, turning pale pink to hyaline and bursting easily in fruit; lobes 2·5–4(5) × 1–1·25(1·5) mm., lanceolate to elliptic, subfalcate, very acute (but not apiculate), twisted at the apex, bright pink, orange, salmon to dull red or brick. Filaments 0·75–1·25 mm. long; anthers 0·3–0·4(0·5) mm., ovate, dark red, included, the upper ones just below the base of the corolla-limb. Follicles 4·25–6·75 mm. long, dark red; styles 0·3–0·5 mm. long.

1. Lower and median leaves peltate; petiole 1·7–4·7 cm. long - - - forma *peltata*
- Leaves not peltate; petiole not more than 1·2 cm. long - - - - - - 2
2. Leaves unlobed - - - - - - - - - - forma *rotundifolia*
- Leaves (at least the lower and median ones) ± deeply 3-lobed - - - forma *tripartita*

Forma **rotundifolia**

    Meristostylus brachycalyx Klotzsch in Peters, Reise Mossamb., Bot. **1**: 270 (1861). — Harms in Engl. & Prantl, op. cit.: 404 (1930), in adnot. Type: Mozambique, Inhambane, *Peters* s.n. (B).

    Kalanchoe integerrima Lange, Ind. Sem. Hort. Haun. (nom. emend.): 5 (1872); in Bot. Tidsskr., Ser. 3, **1**: 139, t. 5 (1878). Type a cultivated plant in Bot. Gart. Copenhagen.

    Kalanchoe luebbertiana Engl., Bot. Jahrb. **39**: 463 (1907). — R.-Hamet in Bull. Herb. Boiss., Sér. 2, **8**: 40 (1908). — Dinter in Fedde, Repert. **18**: 434 (1922). Type from Namibia.

    Kalanchoe holstii sensu Hutch., Botanist in S. Afr.: 469 (1946) non Engl. (1894).

    Kalanchoe guillauminii R.-Hamet in Bull. Mus. Nation. Hist. Nat., Sér. 2, **20**: 465 (1948). Type: a cultivated specimen at Jardin des Plantes, Paris.

    Kalanchoe rotundifolia var. *pseudo-leblanciae* R.-Hamet, Crass. Ic. Select. **3**: t. 59 (1958) *nom. nud.*; pro parte quoad inflorescent.

    Kalanchoe rotundifolia var. *strictifolia* R.-Hamet, op. cit. **4**: t. 70 (1960) *nom. nud.*

    Kalanchoe rotundifolia var. *guillauminii* (R.-Hamet) R.-Hamet, tom. cit.: t. 71 (1960); op. cit. **5**: t. 93 (1963).

    ?*Kalanchoe rotundifolia* var. *aequimagnisepala* R.-Hamet, tom. cit. **5**: t. 90–91 (1963) *nom. nud.*

    Kalanchoe rotundifolia var. *genuina* R.-Hamet, tom. cit.: t. 92 (1963). — J. H. Ross, Fl. Natal: 179 (1972) *nom. nud.*

    Kalanchoe decumbens Compton in Journ. S. Afr. Bot. **33**: 297 (1967). Type from Swaziland.

    Lower leaves circular, obovate to spathulate, rarely linear-spathulate, the median similar but narrower, the upper ones spathulate to lanceolate or linear, rounded, obtuse to subacute at the apex, usually entire or sometimes shallowly crenate towards the apex, attenuate or contracted at the base, at least the lower petiolate; petiole up to 1·2 cm. long, neither connate with the opposite one nor deccurent along the stem.

    Botswana. SE: Mochudi, 1914, *Rogers* 6446 (K). **Zimbabwe.** W: Matobo, Breadleaf (?) Farm, c. 1500 m., iii.1959, *Miller* 5813 (K; LISC; PRE; SRGH). C: 8 miles S. of Gweru, c. 1440 m., 2.v.1967, *Biegel* 2135 (K; LISC; PRE; SRGH). E: Odzi, c. 1120 m., 29.v.1939, *Eyles* 8617 (BR; K). S: Victoria, Kyle National Park Game Reserve, base of Mtunumushava Hill, n. Staff Houses, 21.v.1971, *Grosvenor* 513 (K; PRE; SRGH). **Mozambique.** Z: Quelimane, c. 32 m., 1908, *T. R. Sim* 20835 (PRE). MS: Maringua village, 6 miles S. of Sabi R., 23.vi.1950, *Chase* 2245 (BM; K; SRGH). GI: Gaza, from Guijá to Mabalane n. Barragem-village, left bank of Limpopo R., 3.vi.1959, *Barbosa & Lemos* 8583 (COI; K; LISC). M: Namaacha, below the waterfall, banks of Impamputo R., c. 1 km. from Lagoa dos Patos, 14.xi.1968, *A. Fernandes, R. Fernandes & M. Correia* 90 (COI; LMU); Inhaca Island Ponta Ponduini, 2.vi.1970, *Correia & Marques* 1958 (LMU).

Also in Socotra, Tanzania, S. Africa (Transvaal, Natal, E. Cape Prov.), Swaziland and Namibia. In woodland, open and secondary forests, savannas, open veld, etc., on sandy or rocky soils or on rocks, either in dry or wet habitats, sometimes in salt marshes.

The most common form of this very polymorphic species.

Forma **tripartita** R.-Hamet ex R. Fernandes in Bol. Soc. Brot., Sér. 2, **52**: 207 (1978). Type from S. Afr.(Natal).
    Kalanchoe rotundifolia var. *tripartita* R.-Hamet, Crass. Icon. Select. **2**: t. 33 (1956) *nom. nud.* — J. Ross, loc. cit. (1972) *nom. nud.*

Differs from the typical form by the lower and median leaves which are $\pm$ deeply 3-lobed, with the median lobe up to 3·3 × 1 cm., oblong, the lateral ones shorter, triangular or oblong, equal or subequal, ascending, all 3 lobes obtuse or rounded at the apex and entire. In outline, the leaves are rhomboidal, attenuate-cuneate at base, attaining 6–7 × 2·5–3·5 cm. The smaller leaves have relatively shorter lateral lobes which finally reduce sometimes to two subacute teeth; the upper leaves are unlobed, either rhomboidal with obtuse lateral angles or oblong.

Botswana. SE: 5 miles S. of Palapye Road, c. 1056 m., 14.iii.1961, *Leach* 10749 (K; LISC; M; PRE; SRGH, "plant originally collected by *Leach & Noel* in 1953, cultivated and flowering at Greendale"). **Zimbabwe.** W: Bulalima-Mangwe, about 10 km. N. of Marula on road to Mananda Dam, on a stream that runs into Mananda R., 20.iv.1972, *Grosvenor* 749 pro parte (quoad specimen. K et SRGH). **Mozambique.** M: between Massingir and Mapulanguene, n. the limits of Magude-Guijá circunscr., 8.v.1937, *M.F. Carvalho* 178 (LMA).
    Also in S. Africa (Natal and Transvaal).

Perhaps due to cultivation, the specimen *Leach* 10749 has flowers a little larger (predominantly 13 mm. long) than usual. Specimen *M.F. Carvalho* 621, from Mozambique, Magude, is intermediate between the type and forma *tripartita*.

Forma **peltata** R.-Hamet ex R. Fernandes in Bol. Soc. Brot., Sér. 2, **52**: 208 (1978). Type from Natal.
    Kalanchoe rotundifolia var. *peltata* R.-Hamet, Crass. Icon. Select. **4**: t. 72 (1960) *nom. nud.* — J. H. Ross, loc. cit. (1972) *nom. nud.*
    Kalanchoe neglecta Tölken in Journ. S. Afr. Bot. **44**, 1: 90 (1978). Type from S. Afr. (Natal).

This form differs from the type by its lower and median leaves which are broadly ovate, ovate-cordate or broadly oblong and peltate, not cuneate at the base but truncate or emarginate and bilobed (lobes broad, rounded). Their lamina is usually large (up to 8·5 × 5·5 cm.) and the petiole, longer (1·7–4·7 cm., not only up to 1·2 cm. long) and stronger than in the type, is attached at 0·5–1 cm. from the lamina base. The upper leaves, however, are not peltate and those subtending the inflorescence-branches are oblong-spathulate as in the type.

Mozambique. M: Bela Vista, between Salamanga and Futi R., 28.v.1963, *M.F. Carvalho* 618 (K).
    Also in S. Africa (Natal). Habitat as for the type.

We do not believe that plants with peltate leaves can be isolated as an independent species. In the flowers and other characters, forma *peltata* is like typical *K. rotundifolia*.

18. **Kalanchoe paniculata** Harv. in Harv. & Sond., F.C. **2**: 380 (1862). — R.-Hamet in Bull. Herb. Boiss., Sér. 2, **8**: 40 (1908); in Bull. Soc. Bot. Fr. **57**: 24 (1910). — Schönl. in S. Afr. Journ. Sci. **17**, 2: 188 (1921). — Burtt Davy, F.P.F.T. **1**: 144 (1926) pro parte. — Wilman, Check List Fl. Pl. Ferns Griqual. W.: 79 (1946). — Dyer in Fl. Pl. Afr.: t. 1007 (1947). — Letty, Wild Fl. Transv.: 152 (1962). — R.-Hamet & Marnier-Lapostolle in Arch. Mus. Nat. Hist. Nat., Paris, Sér. 7, **8**: 86, t. 31 fig. 104–106, t. 32 fig. 107 (1964). — J. H. Ross, Fl. Natal: 179 (1972). Type from S. Afr. (Orange Free State (Vetrivier)).
    Kalanchoe oblongifolia Harv., tom. cit.: 379 (1862). — R.-Hamet in Bull. Herb. Boiss., Sér. 2, **7**: 894 (1907); in Bull. Soc. Bot. Fr. **57**:: 19 (1910). — Jacobs., Das. Sukk. Lexikon: 254 (1970). Type from S. Africa (Cape Prov.).

A usually biennial completely glabrous succulent, up to 1·30 m. tall (incl. the inflorescence). Stem 12–18 cm. long in the first year, with numerous subrosulate leaves, elongating in the second year into the floriferous stem with $\pm$ scattered leaves, stout, subquadrangular at the base, rigid, erect, simple. Lower subrosulate leaves very large, up to 24·5 cm. long (including the flat, up to 5 cm. long subpetiolar

base) and 16 cm. broad, subcircular, broadly obovate or ovate-spathulate, rounded at
the apex, attenuate or contracted below; cauline leaves decussate, in 2–4 pairs,
smaller (the uppermost bract-like), oblong-ovate, sessile, decurrent, all entire, ±
conduplicate (the lamina folded upwards), arched-reflexed, rather fleshy, rigid-
coriaceous on drying. Flowers uniformly reddish-brown on drying, in dense, several
times trichotomous cymes, forming a flat-topped corymb-like terminal inflorescence
up to 38 cm. (or more) long and 40 cm. in diameter; main branches stout, straight,
suberect; pedicels 4–7 mm. long. Calyx green, rather shorter than the corolla; calyx-
tube 1–2 mm. long; sepals 1–3·5 × 1·5–2·1 mm. deltate, acute. Corolla 11·5–14 mm.
long; corolla-tube ± 11 mm. long, broader (c. 3 mm. in diam.) near the base,
distinctly 4-angled, the angles with a raised subrigid nerve at maturity prolonged
along the lobes, narrowing successively towards the base of the limb and contracted
there, yellow-green; corolla-lobes (2)2·5–3 × 1·5–2 mm., ovate, acute, apiculate,
yellow-green or yellow outside, yellow or orange-yellow on the inner surface, erect
but not connivent, finally reflexed or subreflexed. Filaments 2·1 or 2·9 mm. long;
anthers 0·75–1 mm. long, oblong, the lower ones below the sinus between the
corolla-lobes, the upper ones slightly exserted. Follicles 8–9 mm. long; styles 1–1·5
mm. long. Seeds c. 0·5–1 mm. long. Scales 2–2·5 mm. long, lanceolate or linear-
lanceolate, emarginate.

Zimbabwe. W: Matopos, c. 1470 to c. 1570 m., ii.1919, *Eyles* 1518 (BM; SRGH).
Mozambique. M: Matola, road Boane-Changalane, 12.v.1968, *Balsinhas* 1260 (COI;
LMA).
Also in S. Afr. (Natal, Transvaal and Orange Free State). In stony, dry places.

19. **Kalanchoe brachyloba** Welw. ex Britten in F.T.A. **2**: 392 (1871). — Engl.,
Hochgebirgsfl. Trop. Afr.: 232 (1892). — Hiern., Cat. Afr. Pl. Welw. **1**: 326 (1896). — R.-
Hamet in Bull. Herb. Boiss., Sér. 2, **7**: 896 (1907); in Bol. Soc. Brot., Sér. 2, **37**: 6–9
(1963). — Friedr. in Prodr. Fl. SW. Afr. **52**: 37 (1968). — Jacobs., Das Sukk. Lexikon:
250 (1970). — R. Fernandes in Bol. Soc. Brot., Sér. 2, **53**: 329 (1980). Type from Angola.
Kalanchoe multiflora Schinz in Verh. Bot. Ver. Brand. **30**: 172 (1888); in Bull. Herb.
Boiss. **5**, App. 3: 100 (1897). — N.E. Br. in Kew Bull. **1909**: 110 (1909). — Dinter in
Fedde, Repert. **18**: 434 (1922). — Hutch, Botanist in S. Afr.: 456 (1946). — R.-Hamet &
Marnier-Lapostolle in Arch. Mus. Nation. Hist. Nat. Paris, Sér. 7, **8**: 83, t. 30 fig.
100–101 (1964). Type: Botswana, Lake Ngami, *Schinz* 177 (K, isotype).
Kalanchoe paniculata sensu Bak. f. in Journ. of Bot. **37**: 434 (1899). — Engl. in Sitz.-
Ber. Königl. Preuss. Akad. Wiss. **52**: 878 (1906). — Burtt Davy, F.P.F.T. **1**: 144 (1926)
pro parte. — R.-Hamet, Crass. Ic. Select. **5**: t. 94–95 (1963) non Harv. (1862).
Kalanchoe baumii Engl. & Gilg in Warb., Kunene-Samb.-Exped. Baum: 242 (1903).
— R.-Hamet in Bull. Herb. Boiss. **7**: 895 (1907) pro parte; op. cit. **8**: 255 (1908); in Bull.
Soc. Bot. Fr. **57**: 21 (1910). Type from Angola.
Kalanchoe pyramidalis Schönl. in Rec. Albany Mus. **2**, 2: 154 (1907). — R.-Hamet in
Bull. Soc. Bot. Fr. **57**: 54 (1910). Type: a cultivated plant at Garden of Albany Museum
from seeds collected by *Schönland* in Botswana, Serowe (AMG, holotype; K).
Kalanchoe pruinosa Dinter in Fedde, Repert. **18**: 434 (1922); op. cit. **19**: 147 (1923),
descript. emend. Type from Namibia.
Kalanchoe praesindentis-vervoerdii R.-Hamet, Crass. Ic. Select. **5**: t. 80–83 (1963) *nom.
nud.*, pro parte quoad inflorest., fl. et fr. specim. *Holub* (Zimbabwe, Matabele Country,
Shashi R.(K)).

A completely glabrous, 0·60–2 m. tall perennial (or also biennial) succulent. Stem
from a subspherical or napiform tuber or woody rootstock, stout, terete up to 2·5 cm.
in diameter at the base, erect, simple or sometimes forked, green, leafless at flowering
time (at least below); lowermost internodes short (only up to 2 cm. long), the
following increasing successively up to 16 cm.; nodes somewhat thickened. Leaves
decussate, the lower the largest, (5·5)12–26(28) × (2·5)3–7(8) cm., subrosulate,
ovate-lanceolate or elliptic to oblong, spathulate or sometimes ovate, obtuse or
rounded at the apex, usually serrate-lobed or crenate or entire and undulate at the
margin, contracted or attenuate below into a subpetiolar base up to 12 mm. broad
and shortly sheathing, the median ones relatively narrower, lanceolate to oblong or
linear, also crenate or remotely dentate or entire, sessile, not connate, amplexicaul or
not, each ± decurrent in 2 raised lines, the upper shortening successively to the
uppermost bract-like ones, all fleshy and very succulent, glaucous or glaucous-green

when young and fresh, ± rigid or sometimes chartaceous on drying, ± caducous.
Flowers erect in ± dense cymes forming partial flattish corymbs, disposed
in a repeatedly dichotomous terminal flat-topped corymb-like inflorescence
(5)13–35(45) × (5)13–25 cm.; main branches of inflorescences in up to 8 pairs, the
lowermost up to 35 cm. long, straight, at ± 45° angles with the axis; pedicels of
dichotomies 6–11(15) mm. long. Calyx 1·5–4(5) mm. long, pale green; tube 0·75–1·5
mm. long; sepals 0·75–2·5(3·5) × 1·5–2·25 mm., lanceolate to shortly triangular,
subacute to obtuse, fleshy, rigid when dry. Corolla 13·5–16(17) mm. long, subconical
after anthesis, narrowing gradually from the roundish base to the apex of the
approximate lobes; corolla-tube (10)11–13·5(15) mm. long, and ± 4 mm. in
diameter near the base, obtusely 4-angled below, distinctly so above, not contracted
below the lobes after anthesis, yellow-green, turning reddish-brown scarious when
dry; corolla-lobes 2–5·25 × 2–4·25 mm., ovate or deltate-semiorbicular, acute,
slightly apiculate, lemon-yellow to orange-yellow, sometimes whitish- or red-edged,
erect and usually connivent, flat or twisted (together with the uppermost part of
tube) even after anthesis, only spreading in the afternoon. Filaments 0·5 or 1·5 mm.
long; anthers 0·5 or 0·75 mm., those of the lower stamens below or attaining the
sinuses between the corolla-lobes, those of the upper ones slightly above the base
of corolla-lobes. Follicles 8·5–12 mm. long; styles 0·5–1·5 mm. long. Scales
1·5–3·25 × 0·5 mm., linear-lanceolate, emarginate, hyaline or brownish-scarious.

Botswana. N: Kwebe Hills, 14.iv.1898, *Lugard* 224 (K). SE: 8 km. E. of Lothlekane,
24.iii.1965, *Wild & Drummond* 7258 (K; SRGH). **Zambia.** N: Luapula Distr., Mbesuma
Ranch, c. 1408 m., 24.iii.1961, *Astle* 873 (K; SRGH). W: Ndola, 10.vii.1954, *Fanshawe* 1370
(K; SRGH). C: Between Rufunsa and Lusaka, 1180 m., 26.iii.1955, *Exell, Mendonça & Wild*
1207 (BM; LISC; SRGH). S: Masabuka, 20.v.1961, *Fanshawe* 6595 (K). **Zimbabwe.** N:
Gokwe, Sasami River Test Herd, below Sasami River Gorge, 22.iv.1964, *Bingham* 1253
(LISC; SRGH). W: Wankie, Kazuma Range, 1000 m., v.1972, *Russell* 1956 (COI; SRGH). C:
Chilimanzi, 30 mil. S. of Umvuma, 15.iii.1958, *Leach* 7360 (SRGH). S: Gwanda, outcrop by
Bubye River n. Bubye Ranch homestead, c. 576 m., 3.v.1958, *Drummond* 5539 (BR; K; LISC;
SRGH). **Malawi.** N: Rumphi Distr., Njakwa Gorge, S. Rukuru R., c. 1120 m., 2.vii.1970,
Pawek 3526 (K). **Mozambique.** MS: Mossurize, between Mossurize R. and Massangena,
11.vi.1942, *Torre* 4302 (C; LISC; LMU; MO). M: Magude, between Motase and Mohine,
16.vii.1953, *Myre & Carvalho* 1696 (LMA).
    Also in Zaire, Angola, Namibia, Swaziland and S. Africa (Transvaal). In woodland, in
fissures of rocks, on sandy or stony ground.

    Similar to *K. paniculata* (no. 18) but distinguished by the leaves being relatively longer and
narrower, usually crenate, dentate or crenate-dentate; by the inflorescence not so dense with
less fastigiate ultimate branches; by the flowers slightly larger with a not so strongly marked 4-
angled corolla-tube and relatively shorter corolla-lobes which are usually connivent or twisted.

20. **Kalanchoe leblanciae** R.-Hamet in Fedde, Repert. **11**: 294 (1912). — J. H. Ross, Fl.
    Natal: 179 (1972). — R. Fernandes in Bol. Soc. Brot., Sér. 2, **53**: 396 (1980). Type:
    Moçambique, Delagoa Bay, *Junod* 443 (G, holotype; BR; P; Z).
      Kalanchoe mossambicana Resende & Sobrinho in Rev. Fac. Ci. Univ. Lisboa, Sér. 2, **2**:
    199 (1952) pro parte quoad *Mendonça* 1884A ((err. cit. 18841) (LISC)) et 1825 (LISC pro
    parte).
      Kalanchoe rotundifolia var. *pseudo-leblanciae* R.-Hamet, Crass. Ic. Select. **3**: t. 59 (1958)
    nom. nud., pro parte quoad folia specim. *Hornby* 503 (PRE).
      Kalanchoe paniculata sensu Mogg in Macnea & Kalk, Nat. Hist. Inhaca I., Moçamb.:
    145 (1958) pro parte, non. Harv. (1862).
      Kalanchoe rotundifolia sensu Mogg, loc. cit. pro parte, non (Haw.) Haw. (1825).

    A perennial (or biennial?) succulent herb, completely glabrous, 0·30–1·60 m. tall.
Stem erect, usually simple or sometimes 2-forked at the top (by abortion of the main
axis), up to 7 mm. thick and woody at base, slender upwards, ± distinctly 4-angled,
smooth, reddish and covered with a very tenuous whitish bloom, slightly dilatate at
nodes, leafless at flowering time at least below; lower internodes 0·5–1·7 cm. long, the
upper ones increasing up to 13 cm. Leaves up to 10 × 2(2·8) cm., oblong to oblong-
spathulate, rarely oblong-obovate, rounded or obtuse at the apex, crenate, sinuate-
crenate or irregularly serrate-lobed at the margin, sessile, but attenuate into a
broad, sometimes subpetiolar base, neither connate nor amplexicaul, fleshy,
flat, chartaceous and dark brown on drying, decussate, spreading to upright, the
uppermost bract-like and appressed to the stem, very caducous. Flowers erect in
dense dichotomous cymes forming usually a terminal corymb-like, reddish-brown

inflorescence 3–15 × 3–13 cm., or sometimes panicles up to 30 × 17 cm.; branches of the inflorescence opposite or nearly so, upright or erect-patent, the lower ones of the panicles up to 17·5 cm. long; internodes of panicles up to 14·5 cm. long; pedicels 3–4·5 mm. long. Calyx-tube up to 0·9 mm. long; calyx-lobes 1–1·25 mm. long, oblong-lanceolate to lanceolate or triangular, acute, appressed to the corolla-tube. Corolla 8–11·5 mm. long, yellow-green to yellow or salmon to bright red, reddish-brown throughout or with the lobes darker than the tube on drying; tube 4-angled, finally swollen in the lower half in fruit, contracted but not twisted above the follicles; corolla-lobes 2–3·1 × 1·75–2 mm., suboblong to ovate, rounded or obtuse and apiculate at the apex, flat or folded along the middle, usually not twisted, erect (but not connivent) or subspreading. Filaments ± 1·5 mm. long, inserted above the middle of corolla-tube; anthers 0·5–0·8 mm., oblong, orange or reddish, usually all included, the upper ones with the apices ± at corolla-throat level or sometimes half-exserted. Follicles 5·5–6 mm. long, attenuate; styles 0·75–1·25 mm. long, thick, reddish, included. Scales ± 1·5 × 0·4 mm., linear-lanceolate, subacute.

Mozambique. MS: Búzi, ix.1946, *Simão* 875 (LMA). GI: Inhambane, between Inharrime and Panda, 29.vii.1944, *Mendonça* 1884A (K; LISC, syntype of *K. mossambicana*). M: Maputo (Lourenço Marques), Bela Vista, on road from Timonganine to Freire, 8.viii.1957, *Barbosa & Lemos* 7792 (LMA; LISC); Inhaca Island, from sea level to 200 m., Hlanganyani Hill, 19.vii.1958, *Mogg* 28093 (K; PRE; SRGH).
Known also from S. Africa (Natal, Orange Free State?). In open forest and shrubby savanna near the coast, on sandy or sandy-clayey soil.

Similar to *K. brachyloba* (no. 19) but not so stout, with the stem ± 4-angled, smaller leaves, inflorescences with rather smaller flowers whose corolla-lobes are not connivent, etc.

21. **Kalanchoe sexangularis** N.E. Br. in Kew Bull. **1913**: 120 (1913). — Burtt Davy, F.P.F.T. **1**: 144 (1926) excl. specimen *Rogers* 2681. Type from S. Africa (sent from Transvaal and cultivated in Cambridge Bot. Gard.).
   Kalanchoe vatrinii R.-Hamet in Journ. of Bot. **54**, Suppl. 1: 9 (1916). — Berger in Engl. & Prantl, Nat. Pflanzenfam. ed. 2, **18a**: 406 (1930). — Jacobs., Das Sukk. Lexikon: 256 (1970). — R. Fernandes in Bol. Soc. Brot., Sér. 2, **53**: 420 (1980). Type: Zambia, Livingstone, *Rogers* 7444 (K, holotype).
   Kalanchoe mossambicana Resende in Bol. Soc. Portug. Ci. Nat. **17**: 184 (1949); in Portugaliae Acta Biol. (A), vol. **Goldsch.**: 731 (1950) *nom. nud.*
   Kalanchoe paniculata sensu Mogg, in Macnae & Kalk, Nat. Hist. Inhaca I., Moçamb.: 145 (1958) pro parte non Harv. (1862).
   Kalanchoe mossambicana Resende ex Resende & Sobrinho in Rev. Fac. Ci. Univ. Lisboa, Sér. 2, **2**, 2: 199, t. 1 (1952). — R.-Hamet & Marnier-Lapostolle in Arch. Mus. Nation. Hist. Nat. Paris, Sér. 7, **8**: 82, t. 29 fig. 98, t. 30 fig. 99 (1964). — Jacobs, op. cit.: 254. Type: Mozambique, Maputo, Goba, *Mendonça* 1825 (LISU, holotype; LISC pro parte).

Perennial, glabrous throughout. Stems 1 to few from a woody base, the floriferous ones 0·6–2 m. high, ± robust, indistinctly 4-angled or terete upwards, somewhat 4-angled towards the base, simple, erect, woody below, reddish, those of the sterile shoots distinctly 4-angled. Leaves opposite-decussate, petiolate or with a ± distinct petiole-base, spreading or deflexed, rather scattered at the middle of the stem, approximate towards the base (the shortest basal internodes ± 7 mm. long) as well those of the shoots, fleshy, coriaceous and rigid on drying; lamina 5–13 × 3–5 cm., oblong-obovate, broadly elliptic, obovate or oblong, rounded at the top, coarsely crenate or undulate-crenate at the margin except at the cuneate base, green to reddish; petiole up to 4(4·5) cm., not connate with the opposite one. Inflorescence up to 30 × 28 cm., panicle-like, not very dense, formed by subhemispherical partial corymbs; panicle-branches in 4–7 main pairs, the lower ones up to 13 cm. long, erect-patent or nearly divaricate, somewhat arched, at the axils of small bract-like leaves; pedicels 2–7 mm. long. Flowers green-yellow to bright-yellow. Calyx 2–3·25 mm. long, green, somewhat rigid and brownish on drying; sepals 1·5–2·25 mm. long and ± 1·25 mm. broad at the base, triangular, acute. Corolla 10–12(13) mm. long; corolla-tube 8–10 mm. long, broader below the middle, attenuate upwards, 4-angled (maxim. breadth of each face 3 mm.) neither contracted nor twisted above the follicles, pale pink hyaline at the lateral faces, salmon or reddish-brown along the raised nerves of the angles on drying; corolla-lobes 1·75–2·5(3) × 1·5–1·75 mm., ovate to subcircular, somewhat attenuate to or rounded at the apex, shortly apiculate, erect but not connivent, salmon or brownish-red with pale margin on drying.

Filaments of the upper stamens $\pm$ 1·5 mm. long, inserted 1–1·5 mm. below the base of corolla-lobes; anthers 0·4–0·5 mm., suborbicular, apiculate by a minute globose tubercule between the thecae, those of the upper stamens slightly exserted, just a little above of the base of corolla-lobes, those of the lower stamens included. Follicles 6·5–8·5 mm. long, attenuate into the styles; styles (1·75)2–2·5(3) mm. long, minutely papillose, with the stigmas at the level of the lower anthers. Scales 2·25–2·5 mm. long, linear.

Corolla 10–12(13) mm. long; corolla-tube 8–10 mm. long; follicles 6·5–8·5 mm. long; styles usually 2–2·5 mm. long; leaves usually with a distinct petiole up to 4(4·5) cm. long or a petiole-base - - - - - - - - - - - var. *sexangularis*
Corolla (13)14–16(17) mm. long; corolla-tube 10–13 mm. long, more attenuate; follicles 7–10 mm. long; styles 3–4 mm. long; leaves shortly petiolate - - - var. *intermedia*

Var. **sexangularis**

The above description applies to var. *sexangularis* only.

Zambia. S: Livingstone, N. bank of Zambesi, c. 960 m., Long. 25° 55′ E., Lat. 17° 54′ S., i.1914, *Rogers* 13122 (Z). **Zimbabwe.** E: Umtali, Commonage, c. 1152 m., 13.ix.1953, *Chase* 5159 (BM; K; LISC; SRGH). S: Victoria, Kyle National Park Game Reserve, Chembira Hill, 22.iv.1971, *Grosvenor* 527 (SRGH). **Mozambique.** GI: Gaza, Caniçado, 25 km. from Lagoa Nova to Chimai, 23.vii.1969, *Correia & Marques* 1040 (LMU). M: Namaacha, on Nat. Road no. 2, at km. 34, 5.vii.1967, *A. Marques* 2042 (LMU); Inhaca from Ponta Ponduini to Saco, in E. coast, 2.vi.1970, *Correia & Marques* 1574 (LMU).
Also in S. Africa (Transvaal). In rocky places.

Var. **intermedia** (R. Fernandes) R. Fernandes in Bol. Soc. Brot., Sér. 2, **55**: 99 (1981). Type: Zimbabwe, Salisbury, *Eyles* 8846 (K, holotype).
   Kalanchoe vatrinii var. *intermedia* R. Fernandes in Bol. Soc. Brot., Sér. 2, **53**: 422 (1980). Type as above.

Differs from the type by the floriferiferous stems subcylindric or indistinctly 4-angled; by the leaves usually smaller, sometimes roundish and entire, with a shorter petiole or a very short petiole-base, often red or reddish at least at the margin; by the corolla usually longer (14–16 mm.) with a longer tube more attenuate upwards and larger lobes (usually 3–3·5 × 2–3 mm. not 1·75–2·5 × 1·5–1·75 mm.); by the anthers slightly larger, the upper ones a little more exserted; and by the longer styles (3–4 mm. not 2–2·5(3) mm. long).

Zimbabwe. N: Mrewa, Chikukwe R., E. of the river, 8.ix.1962, *Leach* 11520 (K; LISC; SRGH, cultivated at Greendale). W: Matobo, Farm Besna Kobila, c. 1500 m., v.1955, *Miller* 2834 (K; LISC; SRGH). C: Makoni Kop., Rusape, vii.1916, *A. Hislop* s.n. (K). E: Umtali, Ziminya's Reserve, ix.1960, *Corner* s.n. (LISC; SRGH). S: Belingwe Native Reserve, $\pm$ 15 mil. N. of Sandawana Emerald Mine, 9.vii.1966, *Leach & Bullock* 13317 (PRE; SRGH). **Mozambique.** MS: between Vila Gouveia and Vandúzi, c. 800 m., 18.vii.1969, *Leach & Cannell* 14352 (LISC; SRGH).
   On stony and rocky slopes and summits among rocks.

3. **BRYOPHYLLUM** Salisb.

Bryophyllum Salisb., Parad. Lond.: t. 3 (1805).

Succulent herbs, undershrubs or shrubs, usually with erect stem. Leaves opposite, verticillate or alternate, simple and unlobed or lobed to pinnatifid or sometimes pinnate, usually petiolate, rarely sessile, flat or sometimes terete, $\pm$ fleshy-succulent. Inflorescence terminal, usually corymb-like, loose to dense, few- to many-flowered, rarely panicle-like. Flowers 4-merous, large, in $\pm$ long pedicels, usually pendulous, brightly coloured. Calyx shorter than the corolla; sepals 4, rarely free, usually $\pm$ connate below, sometimes into a $\pm$ long cylindric, campanulate or swollen-urceolate tube. Corolla gamopetalous; tube subcylindric or urceolate, $\pm$ distinctly 4-angled, straight or slightly curved; lobes ovate, triangular or semicircular, usually shorter than the corolla-tube, spreading or reflexed. Stamens 8, inserted at the base or below the middle of corolla-tube; filaments $\pm$ exserted. Carpels 4, $\pm$ connate below; styles longer than the ovaries, usually exserted. Follicles with numerous seeds. Seeds oblong or obovoid with rugulose tegument. Scales semicircular, quadrate or linear.

More than 30 species, one of them widespread throughout nearly all tropical regions. Some species are cultivated in gardens from which they have sometimes escaped becoming naturalised in many places.

1. Leaves terete, up to 13 × 0·55 cm., sessile, ternate to alternate -   -   -   1. *tubiflorum*
- Leaves not terete, flattish, petiolate, opposite, decussate -   -   -   -   -   -   2
2. Leaves never divided, ovate, not more than 8 × 3·9 cm.; calyx-tube 5–7·2 mm. long
        2. *miniatum*
- Leaves divided (pinnatisect or pinnate, sometimes reduced to the terminal leaflet); calyx-tube not less than 13 mm. long   -   -   -   -   -   -   -   -   -   3
3. Pedicels papillose; flowers often replaced by pseudobulbils; calyx-tube 13–16 mm. long; leaves pinnatisect, up to 15 × 5 cm.   -   -   -   -   -   -   3. *proliferum*
- Pedicels not papillose; pseudobulbils not present in the inflorescence; calyx-tube 21–31 mm. long; leaves imparipinnate, up to 20 × 12 cm.   -   -   -   -   -   4. *pinnatum*

1. **Bryophyllum tubiflorum** Harv. in Harv. & Sond., F.C. **2**: 380 (1862). — Schönl. in Engl. & Prantl, Nat. Pflanzenfam. **3**, 2a: 34 (1890). — Berger in Engl. & Prantl, op. cit., ed. 2, **18a** 411 (1930). — Guillaumin in Rev. Hort., N. Sér., **34**: 437, fig. 207 et 208 (1935). — J. Blake, Gard. E. Afr. ed. **3**: 66 et 177 (1950). Type: Mozambique, Delagoa Bay, *Forbes* s.n. (K).

    Kalanchoe delagoensis Eckl. & Zeyh., Enum. Pl. Afr. Austr. **3**: 305 (1837). — Walp., Repert. **2**: 256 (1843). — R.-Hamet in Bull. Herb. Boiss., Sér. 2, **8**: 39 (1908) *nom. nud.*

    Kalanchoe verticillata Scott Elliot in Ann. Bot. **5**: 354 (1891); in Journ. Linn. Soc., Bot. **29**: 14, t. 3 (1891). — R.-Hamet in Bull. Herb. Boiss., Sér. 2, **7**: 887 (1907). — Baron, Graham & Stewart in Trans. & Proc. Bot. Soc. Edinb. **30**, 2: 70 (1929). — É. François in Rev. Hort., N. Sér. **26**: 65, fig. 44 (1938). Type from Madagascar.

    Bryophyllum delagoense (Eckl. & Zeyh.) Schinz in Mém. Herb. Boiss. **10**: 38 (1900). — Druce in Rep. Bot. Exch. Cl. Br. Isles, **1916**: 611 (1917).

    Bryophyllum verticillatum (Scott Elliot) Berger, loc. cit. Type as for *Kalanchoe verticillata*.

    Kalanchoe tubiflora (Harv.) R.-Hamet in Bot. Centr. Beih. **29**, 2: 41 (1912). — R.-Hamet & P. de la Bâthie in Ann. Mus. Col. Marseille, Sér. 3, **2**: 125 (1914). — Stapf in Curtis, Bot. Mag. **155**: t. 9251 (1931). — Jahand. in Rev. Hort., N. Sér. **22**: 32, t. fac. p. 32 fig. 1–2 (1930). — F. H. Wright in Gard. Chron., Ser. 3, **97**: 77, fig. 32 et 34 (1935). — R.-Hamet & Marnier-Lapostolle in Arch. Mus. Nation. Hist. Nat. Paris, Sér. 7, **8**: 60, t. 22 fig. 70 et t. 23 fig. 71 (1964). — Jacobs., Das Sukk. Lexikon: 256, t. 103 fig. 3 (1970) *nom. illegit.*

A completely glabrous perennial succulent. Stem 0·20–1·20 m. high, erect, simple, terete, with sterile shoots at the base. Leaves ternate or alternate in the adult plant, opposite in the first nodes of the young shoots, ± caducous at flowering time, 1·5–13 cm. long and 2–6 mm. in diameter, terete, slightly canaliculate on the upper face, with 2–9 small apical teeth producing pseudbulbils in their axils, sessile, usually straight, erect to spreading, fleshy, greyish- or dull-green, spotted with dark green or red. Flowers in cymes forming a terminal, rather dense corymbose inflorescence up to 10–15 cm. in diameter; pedicels 6–20 mm. long. Calyx equalling ¼–⅓ of corolla, campanulate, green striped with red; tube 2·5–5 mm. long; lobes 5·8–7·7 × 3·7–5·7 mm., lanceolate, very acute. Corolla reddish-orange or reddish-ochre below, pale yellow suffused with red upwards; tube 22·5–35 mm. long, rather constricted just above the carpels, subfunnel-shaped-campanulate upwards; lobes 7–10·2 × 7–9·8 mm., obovate, obtuse or truncate, apiculate, spreading. Stamens attaining the top of corolla; filaments inserted below the middle of the corolla-tube; anthers 1–2 × 1–1·7 mm., broadly ovate. Carpels c. 6·15 mm. long, adnate for c. 1·65 mm.; styles 19·5 mm. long. Scales 0·7–1·6 × 0·85–1·4 mm., semicircular to trapeziform. Seeds 0·6 × 0·3 mm., obovoid.

Zimbabwe. W: Bubi Distr., Inyati Mission, 45 mil. N. of Bulawayo, cult. 19.v.1947, *R. W. J. Keay* (K; SRGH). C: Lalapanzi, iv.1955, *Ingle* (SRGH). **Mozambique.** M: Maputo (Lourenço Marques), gardens of CICA, 22.vi.1952, *Gomes Pedro* 3955 (LMA).

Native of Madagascar. Cultivated and occasionally naturalised. Can be a garden weed.

2. **Bryophyllum miniatum** (Hilsenb. & Boj. ex Tulasne) Berger in Engl. & Prantl, Nat. Pflanzenfam., ed. 2, **18a**: 412 (1930). Type from Madagascar.

    Kalanchoe miniata Hilsenb. & Boj. ex Tulasne in Ann. Sci. Nat. Paris, Bot., Sér. 4, **8**: 149 (1857). — Baillon in Bull. Soc. Linn. Paris, **1**: 469 (1885). — R.-Hamet in Bull. Herb. Boiss., Sér. 2, **8**: 20 (1908). — Turrill in Curtis, Bot. Mag. **173**: t. 378 (1962). — R.-Hamet & Marnier-Lapostolle in Arch. Mus. Nation. Hist. Nat. Paris, Sér. 7, **8**: 53, t. 2 fig. 2 et t. 18 fig. 52–53 (1964). — Jacobs., Das Sukk. Lexikon: 254 (1970). Type as above.

Kitchingia miniata (Hilsenb. & Boj. ex Tulasne) Bak. in Journ. of Bot. **20**: 109 (1882). Type as above.

A completely glabrous, perennial succulent plant. Stem 30–80 cm. high, ascending, rooting at the base. Leaves decussate, ± scattered, petiolate; lamina 2·5–8 × 1·2–4 cm., ovate, obtuse to subacute at the apex, crenate at the margin, rounded or subcordate at the base, flat, fleshy; petiole 1–3 cm. long. Flowers erect, in cymes forming a terminal, corymbose or paniculate inflorescence, 6·5–20 × 6–26 cm., sometimes some flowers abortive and replaced by pseudobulbils of minute, condensate leaves; pedicels 7–20 mm. long. Calyx-tube 4·5–7·2 mm. long, green-yellowish; lobes 4–8·4 × 5–8 mm., deltoid-semiorbicular, acute. Corolla-tube 22–31 mm. long, suburceolate, rather constricted below the middle, green-yellow in the part included in the calyx, bright red upwards; corolla-lobes 4–6 × 5–6·8 mm., semiorbicular, cuspidate, yellow or orange. Filaments attaining the base of corolla-lobes; anthers c. 1·5 mm. long. Carpels 8–11 mm. long, adnate for 1·75–2·25 mm.; styles 20–22 mm. long. Scales 0·9–1·4 × 0·6–1·6 mm., nearly as long as wide, subquadrate, emarginate.

Zimbabwe. C: Salisbury, cultivated and flowering at Greendale, 21.vii.1958, *Leach* 7216 (SRGH).
Native of Madagascar. Cultivated and sometimes escaping from gardens.

3. **Bryophyllum proliferum** Bowie ex Hook. in Curtis, Bot. Mag., Sér. 3, **15**: t. 5147 (1859). — Baillon in Bull. Soc. Linn. Paris, **1**: 468 (1885). — Schönl. in Engl. & Prantl, Nat. Pflanzenfam. **3**, 2a: 34 (1890). — Berger, op. cit. ed. 2, **18a**: 410, fig. 197 A–C (1930). — J. Blake, Gard. E. Afr. ed. 3: 177 (1950). Type from Madagascar.
    Kalanchoe prolifera (Bowie ex Hook.) R.-Hamet in Bull. Herb. Boiss., Sér. 2, **8**: 19 (1908). — C. A. Backer in Fl. Males., Ser. 1, **4**, 3: 200 (1953). — R.-Hamet & Marnier-Lapostolle in Arch. Mus. Nation. Hist. Nat. Paris, Sér. 7, **8**: 56, t. 20 fig. 60–62 (1964). — Jacobs., Das Sukk. Lexikon: 255, t. 105 fig. 4 (1970). Type as above.

A succulent perennial, up to c. 4 m. tall. Stem stout, nearly woody below, erect or ascending, simple, terete, glabrous, emitting sterile shoots at the base. Leaves decussate, scattered, impari-pinnatisect, rarely undivided, glabrous, petiolate; segments 7–15 × 1·5–5 cm., oblong, ovate-oblong or oblong-lanceolate, asymmetric and decurrent at the base, crenate or again pinnatisect, obtuse at the apex; petiole up to 16 (or more?) cm. long, rather broadened at the base, amplexicaul. Flowers pendulous, in dense cymes disposed in a terminal 40–80 × 20–40 cm. panicle, frequently ± numerous abortive flowers replaced by pseudobulbils of minute leaves; pedicels 8–15 mm. long, very slender, densely papillose (papillae short, obtuse). Calyx inflated-subcampanulate, bluntly 4-angled, green, shortly papillose; tube 13–16 mm. long; lobes 3·25–4 × 5·5–7 mm., semiorbicular, shortly acuminate-cuspidate. Corolla-tube 18–24 mm. long, constricted above the carpels, suburceolate upwards; corolla-lobes 2·75–3·3 × 3–4 mm., subovate, suddenly acuminate-cuspidate, spreading, yellow. Stamens inserted below the middle of the corolla-tube, distinctly exserted; anthers 2–2·6 × 1·3–1·45 mm., ovate. Carpels 7–8·2 mm. long, connate at the base for c. 2 mm.; styles 17–20 mm. long. Scales 1·3–1·6 × 2–2·4 mm., semicircular, emarginate.

Zimbabwe. C: Salisbury, Alexandra Park, Botanic Garden grounds, viii.1961, *Drummond* (K; LISC; SRGH).
Native of Madagascar. Cultivated and sometimes a garden escape.

4. **Bryophyllum pinnatum** (Lam.) Oken in Allg. Naturgesch. **3**, 3: 1966 (1841). — Baillon in Bull. Soc. Linn. Paris, **1**: 467 (1885). — Aschers. & Schweinf., Ill. Fl. Egypte: 79 (1887). — Kuntze, Rev. Gen. Pl. **1**: 228 (1891). — Engl., Pflanzenw. Ost-Afr. C: 188 (1895). — Muschl., Man. Fl. Egypt, **1**: 448 (1912). — Burtt Davy, F.P.F.T. **1**: 144 (1926). — Hutch. & Dalz., F.W.T.A. **1**, 1: 105 (1927). — Berger in Engl. & Prantl, Nat. Pflanzenfam., ed. 2, **18a**: 410 (1930). — Irvine, W. Afr. Bot.: 121 (1933). — Exell, Cat. Vasc. Pl. S. Tomé: 174 (1944); Suppl: 19 (1956). — F. W. Andr., Fl. Pl. Anglo-Egypt. Sudan: 75 (1950). — Keay in F.W.T.A. ed. 2, **1**, 1: 116 (1954). TAB. **7**. Type from Madagascar.
    Cotyledon pinnata Lam., Encycl. Méth., Bot. **2**: 141 (1786). Type as above.
    Cotyledon pinnata var. *β* Lam., loc. cit. (1786).
    Kalanchoe pinnata (Lam.) Pers., Synops. Sp. Pl. **1**: 446 (1805). — R.-Hamet in Bull. Herb. Boiss., Sér. 2, **8**: 21 (1908). — É. François in Rev. Hort., N. Sér. **26**: 64 et 66 (1938–1939). — Toussaint in F.C.B. **2**: 568 (1951). — C. A. Backer in Fl. Males., Ser. 1, **4**, 3: 199, fig. 1 (1953). — R.-Hamet & Marnier-Lapostolle in Arch. Mus. Nation. Hist. Nat.

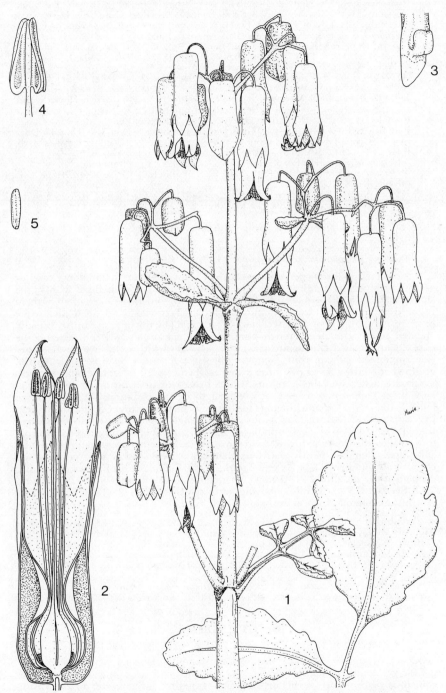

Tab. 7. BRYOPHYLLUM PINNATUM. 1, inflorescence and leaves (× ⅔), from *Williams* 716; 2, flower in longitudinal section (× 2); 3, hypogynous scale (× 4); 4, anther (× 6); 5, seed (× 20), 2–5 from *Brenan* 8740. From F.T.E.A.

Paris, Sér. 7, **8**: 55, t. 3 fig. R, t. 19 fig. 57–58 et t. 20 fig. 59 (1964). — Jacobsen, Das Sukk. Lexikon: 254 (1970). — M. Drar., Bot. Exp. Sudan 1938 in Publ. Cairo Univ. Herb. No. **3**: 41 (1970). — J. H. Ross, Fl. Natal: 179 (1972). Type as for *Cotyledon pinnata*.

Kalanchoe pinnata var. *floripendula* Pers., loc. cit. (1805).

Bryophyllum calycinum Salisb., Parad. Lond.: t. 3 (1805). — Sims in Curtis, Bot. Mag. **84**: t. 1409 (1811). — Haw., Synops. Pl. Succ. **3**: 110 (1812). — DC., Prodr. **3**: 396 (1828). — Walp., Repert. **2**: 257 (1843). — Tulasne in Ann. Sci. Nat., Paris, Bot., Sér. 4, **8**: 148 (1857). — Hook. & Thomson in Journ. Linn. Soc., Bot. **2**: 90 (1858). — Schweinf. & Aschers. in Schweinf., Beitr. Fl. Aethiop, Aufzähl.: **1**: 271 (1867). — Britten in F.T.A. **2**: 390 (1871). — Bak., Fl. Maurit. & Seychelles: 98 (1877). — C. B. Clarke in Hook., Fl. Br. India, **2**: 413 (1878). — Schönl. in Engl. & Prantl, op. cit. **3**, 2a: 34 (1890). — K. Schum. & Rümpler, Die Sukk.: 86 (1892). — De Wild., Miss. Laur.: 236 (1906); Ann. Mus. Cong., Bot., Sér. 5, **3**: 187 (1910). — M. Wood, Handb. Fl. Natal: 46 (1907). — Th. & H. Dur., Syll. Fl. Cong.: 193 (1909). — Broun & Massey, Fl. Sudan: 68 (1929). — J. Blake, Gard. E. Afr. ed. 3: 66 (1950). Type a cultivated plant (from a Calcutta Garden).

Cotyledon rhizophylla Roxb., Hort. Bengal.: 34 (1814). Type from India.

Cotyledon calycina (Salisb.) Roth, Nov. Pl. Sp.: 217 (1821). Type as for *Bryophyllum calycinum*.

Verea pinnata (Lam.) Spreng., Syst. Veg. **2**: 260 (1825). Type as for *Cotyledon pinnata*.

Bryophyllum germinans Blanco, Fl. Filip. **2**: 47, t. 147 (1878). Type from the Philippines?

Bryophyllum pinnatum var. *simplicifolium* Kuntze, loc. cit. (1891). Type from Madagascar.

Crassuvia floripendia Commers. ex Hiern, Cat. Afr. Pl. Welw. **1**: 326 (1896), *nom. illeg.* Type as for *C. pinnata* var. *β* Lam.

A completely glabrous perennial, succulent plant. Stem up to 2 m. long, erect or ascending, stout, terete, simple, spotted or striped with red, sometimes with sterile shoots at the base. Leaves decussate, imparipinnate, sometimes some or all of them reduced to the terminal leaflet; lamina 6–20 × 4–12 cm.; leaflets 3–5, circular, ovate or ovate-oblong to oblong-spathulate, the terminal one the largest, obtuse, broadly crenate, doubly crenate or crenate-dentate at the margin, usually contracted at the base or sometimes attenuate into a petiolule, flat, green with violet lines on the upper side; petiole 2–10 cm. long, semi-amplexicaul. Flowers pendulous, in cymes grouped in terminal, ample loose panicles; pedicels 10–25 mm. long, slender, divaricate, reflexed at the extremity. Calyx broadly cylindric-tubular, sunk at the base, green, sometimes striped with red or red-violet; calyx-tube 21–31 mm. long; lobes 7–10·5 × 7–11·25 mm., deltate to subsemiorbicular, acute. Corolla longer than the calyx; tube 25–40 mm. long, suburceolate-8-angled at the base, constricted above the carpels and then elongate-tubular-4-angled, green below, bright red or purple upwards; corolla-lobes 9–14 × 4·3–6·5 mm., deltate, abruptly acuminate, very acute, spreading. Stamens slightly exserted; filaments inserted below the middle of the corolla-tube; anthers 2·6–3 × 1·6–2·2 mm., ovate. Follicles 12–14 mm. long, ovoid, connate at the base for 2·2–3·5 mm., attenuate into the styles; styles 22·5–30 mm. long. Seeds c. 0·8 × 0·35 mm., obovoid, obtuse. Scales 1·8–2·6 × 1·4–1·8, subquadrate, obtuse or emarginate at the apex.

Zimbabwe. C: Salisbury, cult. & fl. at Greendale, 3.ix.1959, *Leach* 9349 (SRGH). E: Hondy (Honde?) Valley, Mpanga River, an escape, 24.viii.1949, *Chase* 1784 (BM; LISC; SRGH). **Malawi.** S: Mulanje, cult. in District Commissioners Garden, 560 m., 25.xii.1965, *Binns* 126 (SRGH). **Mozambique.** Z: Low Zambézia, Dombe (Chinde R.) Zambezi delta, 26.vi.1899, *Sarmento* 63 (COI). M: Marracuene, at km. 21 on road Maputo (Lourenço Marques) to Vila Luiza, 7.ix.1968, *Ferreira-Marques* 54 (COI).

Perhaps native to Madagascar. Introduced and naturalised in many parts of tropical and subtropical Africa as well as in Asia, Oceania, North America (Mexico), Central and South America.

4. COTYLEDON L.

Cotyledon L., Sp. Pl. **1**: 429 (1753); Gen. Pl. ed. 5: 196 (1954).

Succulent undershrubs or shrubs, completely glabrous or ± glandular-pubescent, tomentose or mealy. Leaves opposite and decussate, or verticillate, sometimes subrosulate, sessile or subpetiolate, fleshy, flat to terete, usually entire, persistent or caducous. Inflorescences terminal on a ± long peduncle, cymose, panicled or racemose. Flowers 5-merous, showy, mostly large, pedicelled and pendulous or sessile. Calyx shorter than corolla-tube. Corolla gamopetalous, deep red to yellow or greenish; tube subcylindrical, not or ± ventricose at the base, ± 5-

angled, usually narrowing to the throat; lobes spreading, reflexed or revolute, spirally twisted in bud. Stamens 10, inserted at the base of corolla-tube, usually exserted; filaments often hairy at the base; anthers ovate to oblong. Carpels 5, free, many-ovulate; styles usually longer than the ovaries, subulate; stigmas subcapitate. Follicles with numerous seeds.

A genus of c. 30 African species, occurring mainly in S. Africa.

Corolla glandular-pubescent, with the tube strongly swollen-5-gonous below, rather contracted above the carpels; calyx glandular-pubescent, with the lobes 6–10 mm. long; nectar-scales oblong, tubular above, dilatate and irregularly lobed at the margin
1. *barbeyi*
Corolla glabrous, not swollen below, subcylindrical, not or slightly contracted above the carpels; calyx glabrous, whitish-bloomed or mealy, with the lobes 3·5–4·5 mm. long; nectar-scales semi-circular, flat, slightly emarginate at the apex - - - 2. *oblonga*

1. **Cotyledon barbeyi** Schweinf. ex Bak. in Gard. Chron., Ser. 3, **13**: 624 (1893). — Engl. in Ann. R. Ist. Bot. Roma, **9**: 251 (1902). — Berger in Engl. & Prantl, Nat. Pflanzenfam. ed. 2, **18a**: 414 (1930). — Poellnitz in Fedde, Repert. **42**: 37 (1937). — Bally in Journ. E. Afr. Uganda Nat. Hist. Soc. **15**: 11 (1940). — Hutch., Botanist in S. Afr.: 456 (1946). — J. Blake, Gard. E. Afr. ed. 3: 179 (1950). — Cufod. in Bull. Jard. Bot. Brux. **24**, Suppl.: 165 (1954). — Jacobs., Das Sukk. Lexikon: 132, t. 38 fig. 1 (1970). Type: a cultivated plant in Kew Gardens from seeds collected by Schweinfurth in Yemen.
   Cotyledon barbeyi Schweinf. ex Penzig in Estr. Atti Congr. Intern. Bot. Genova: 341 (1892) *nom. nud.*
   Cotyledon barbey Schweinf. in Bull. Herb. Boiss. **4**, App. 2: 196 (1896). Syntypes from Ethiopia and Yemen.
   Cotyledon wickensii Schönl. in Rec. Albany Mus. **3**, 2: 141 (1915); in S. Afr. Journ. Sci. **17**, 2: 187 (1921). — Pole Evans, Fl. Pl. S. Afr. **4**: t. 154 (1924). — Burtt Davy, F.P.F.T. **1**: 143 (1926). — Poellnitz, tom. cit.: 39 (1937). — Letty, Wild Fl. Transv: 153 (1962). — Jacobs., Das Sukk. Lexikon: 136 (1970). — J. H. Ross, Fl. Natal: 179 (1972). Type from S. Afr. (Transvaal).
   ?Cotyledon wickensii var. *glandulosa* Poellnitz, tom. cit.: 40 (1937). — Jacobsen, loc. cit. Type from S. Afr. (Transvaal).
   Cotyledon wickensii var. *rhodesica* Schönl. ex Jacobs., loc. cit. *nom. non rite publ.* Type: Zimbabwe, Umtali, Odzani River Valley, *Teague* 169 (K; PRE, holotype?).

A succulent shrub up to 2(4) m. tall, forming clusters several feet across. Stem strong, ascending, branching from the base, leafless like the branches nearly up to the peduncles, glabrous. Leaves decussate, condensed to subrosulate at the end of sterile shoots and below the peduncles, obovate to oblong-obovate or subcircular, 2·5–10 × 3–6·5 cm., obtuse or rounded at the apex, entire and flat at the margin or the upper ones undulate, sessile or contracted to attenuate into a terete petiole-base, or lanceolate to spathulate, shortly acuminate and acute, apiculate, or sometimes oblong to narrowly linear up to 16 cm. long and 3 mm. broad, flat, concave or channelled above, ± convex beneath, sometimes semi-terete, fleshy, rigid on drying, glabrous and glaucous tinged pink or red or whitish, or shortly glandular-pubescent. Inflorescence cymose, up to 20 cm. in diameter, many-flowered; peduncle 19–40 cm. long, terete, sparsely glandular-pubescent just below the inflorescence, glabrous lower down; pedicels 1–3·7 cm. long, shortly and densely glandular-pubescent. Flowers pendulous. Calyx densely glandular-pubescent on the outside; tube 1–2 mm. long to absent; sepals 6–10 × 4–5 mm., ovate, ovate-oblong or triangular, acute or subacute. Corolla 2·5–3·7 cm. long, deep red, orange-red or coral-red (in F.Z. area), glandular-pubescent on the outside; tube c. 1·2–1·5 cm. long, rather swollen at the base and there 5-angled and subwinged at the angles, contracted above the sepals and thence subcylindrical; corolla-lobes c. 1·8–2 cm. long and ± 2 mm. broad, oblong-linear or linear-lanceolate, oblique at the apex, obtuse, apiculate, reflexed. Stamens exserted; filaments c. 2·2 cm. long, dilatate and hairy at the base, the hairs long, dense; anthers c. 1·5 mm. and suborbicular or 2·25–2·5 × 1·25 mm. and oblong (the two anther-types present in flowers of a single inflorescence, sometimes even in one flower), yellow. Follicles 1·6 cm. long, attenuate into the styles, glabrous; styles c. 9 mm. long, exserted; stigmas capitellate. Seeds c. 0·6 mm. long, oblong. Scales 2·5–3 mm. long, oblong, tubular above, dilatate and irregularly lobed or cut at the margin.

Zimbabwe. N: Urungwe, Karoi, c. 1280 m., 3.iii.1957, *Drewe* 41 (SRGH). W: Matopos Hills near Fort Master, vii.1954, *Plowes* 1784* (K; PRE; SRGH). C: Makoni, c. 1536 m., vi.1919, *Eyles* 1657 (BM; PRE; SRGH). E: Umtali, Clydesdale North, mountain summit, c. 1280 m., 15.vi.1958, *Chase* 6935 (EA; K; LISC; PRE; SRGH). S: Chingwarara Hill, Belingwe Tribal Trustland, E. of Belingwe/W. Nicholson road N. of Nuanetsi road, ± 1100 m., 3.vii.1968, *Müller* 777* (K; SRGH). **Mozambique.** GI: between Massingire and Mapulanguene, near the frontier between circumscr. Magude-Guijá, 8.v.1957, *M.F. Carvalho* 175 (PRE). M: Gaza, Magude, 1 km. after Mapulanguene to Macaene, 24.vi.1969, *Correia & Marques* 805 (LMU).

Also in Somalia, S. Arabia, E. Africa and S. Africa (Natal and Transvaal). In stony places, among rocks, in rock-crevices, etc.

Specimens marked with * have shortly pubescent-glandular leaves; the others have glabrous leaves.

The specimen *Plowes* from the Matopos Hills is numbered 1748 in SRGH but 1784 in both K and PRE.

Specimens of E. Africa seen by us are slightly different from those of Zimbabwe, Mozambique and S. Africa: the indumentum-hairs are shorter; the sepals are relatively broader and not so acute and sometimes subobtuse; the angles of the base of corolla-tube are less raised; the corolla-tube is relatively shorter; the styles slightly longer; the scales broader. Moreover, E. African plants apparently have yellow corollas whereas only ± red corollas have been found in the F.Z. area and in S. Africa. If the above differences are constant in the plants of E. Africa and not caused by the ageing of flowers then the southern representatives of the species might be included in a separate subspecies or variety.

2. **Cotyledon oblonga** Haw., Synops. Pl. Succ.: 106 (1812). — Eckl. & Zeyh., Enum. Pl. Afr. Austr. **3**: 306 (1837). TAB. **8**. Type: Plate 792 (K, lectotype).

Cotyledon undulata Haw., Suppl. Pl. Succ.: 20 (1819). — DC., Prodr. **3**: 396 (1828). — Eckl. & Zeyh., loc. cit. — Harv. in Harv. & Sond., F.C. **2**: 377 (1862). — Schönl. & Bak. f. in Journ. of Bot. **40**: 13 (1902). — Hemsley in Curtis, Bot. Mag., Ser. 3, **59**: t. 7931 (1903). — Poellnitz in Fedde, Repert. **42**: 35 (1937). Type from S. Afr. (Cape Prov.).

Cotyledon coruscans Haw., op. cit.: 21 (1819); in Philos. Mag. **1825**: 32 (1825). — Sims in Curtis, Bot. Mag. **52**: t. 2601 (1825). — Lodd., Bot. Cab.: t. 1030 (1825). — DC., Prodr. **3**: 396 (1828). — Eckl. & Zeyh., loc. cit. — Harv. in Harv. & Sond., F.C. **2**: 371 (1862). — Schönl. & Bak. f. tom. cit.: 17 (1902). — Schönl. in Rec. Albany Mus. **3**: 137 (1915). — Burtt Davy, F.P.F.T. **1**: 143, fig. 14C–D (1926). — Poellnitz, tom. cit.: 29 (1937). Type unknown.

Cotyledon canaliculata Haw., op. cit.: 22 (1819). Type unknown.

Cotyledon canalifolia Haw., tom. cit.: 327 (1825), *nom. illegit.* Type as for *C. coruscans.*

Cotyledon viridis Haw. in Philos. Mag. **1827**: 273 (1827). Type unknown.

Cotyledon crassifolia Haw., loc. cit. (1827). — Eckl. & Zeyh., loc. cit. Type from S. Africa (Cape Prov.).

Cotyledon cuneiformis Haw. in Philos. Mag. **1828**: 185 (1828). Type unknown.

Cotyledon orbiculata var. *oblonga* (Haw.) DC., Prodr. **3**: 396 (1828). — Tölken in Bothalia, **12**, 4: 619 (1979). Type as for *Cotyledon oblonga.*

Cotyledon virescens Schönl. & Bak. f., tom. cit.: 14 (1902). — Schönl., op. cit.: 136 (1915).

Cotyledon whiteae Schönl. & Bak. f., tom. cit.: 19 (1902), "whitei". Type from S. Africa (Cape Prov.).

Cotyledon galpinii Schönl. & Bak. f., tom. cit.: 16 (1902). Type from S. Africa (Cape Prov.).

Cotyledon orbiculata sensu Burtt Davy in Fl. Pl. S. Afr. **5**: t. 161 (1926); F.P.F.T. **1**: 143, fig. 14A, B (1926). — Guillarmot, Fl Lesotho: 180 (1971). Non L. (1753).

Cotyledon leucophylla C. A. Smith in Steyn, Toxic. Pl. S. Afr.: 224 (1934). — Dyer in Fl. Pl. S. Afr. **17**: t. 652 (1937). Type from S. Africa (Transvaal).

Cotyledon flavida Fourcade in Trans. R. Soc. S. Afr. **21**: 34 (1934). Type from S. Afr. (Cape Prov.).

Cotyledon decussata var. *flavida* (Fourcade) Poellnitz, tom. cit.: 31 (1937). Type as for *Cotyledon flavida.*

Cotyledon decussata var. *rubra* Poellnitz, loc. cit. Type from S. Africa (Cape Prov.).

Cotyledon macrantha var. *virescens* (Schönl. & Bak. f.) Poellnitz, tom. cit.: 33 (1937). Type from S. Afr. (Cape Prov.).

Cotyledon zuluensis Poellnitz, tom. cit.: 41 (1937). — Jacobs. Das Sukk. Lexikon: 136 (1970). — Venter in Journ. S. Afr. Bot. **37**, 2: 106 (1971). — J. H. Ross, Fl. Natal: 179 (1972). — Compton, Fl. Swaziland: 218 (1976). Type from S. Afr. (Natal).

Cotyledon obermeyerana Poellnitz, loc. cit. Type from S. Afr. (Cape Prov.).

Cotyledon rudatsii Poellnitz, tom. cit.: 42 (1937) *nom. provis.*

Cotyledon simulans Schönl. ex Poellnitz, op. cit. **48**: 111 (1940) *nom. tantum.*

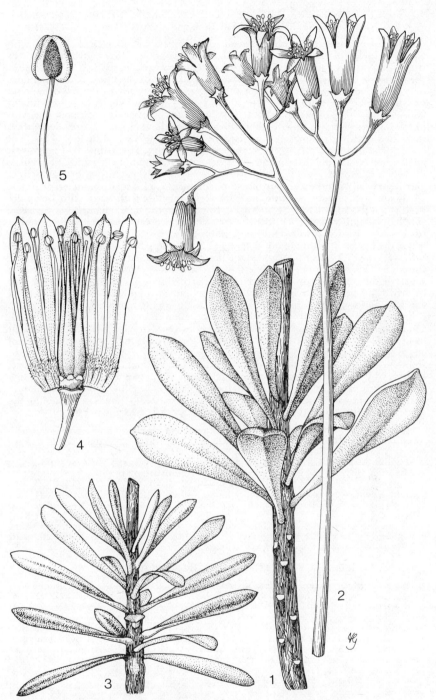

Tab. 8. COTYLEDON OBLONGA. 1, part of a branch with leaves (× ⅔); 2, inflorescence (× ⅔), 1–2 from *Balsinhas* 1267; 3, part of a branch with smaller leaves (× ⅔), from *Gomes Pedro* 3927; 4, flower opened to show stamens and pistils (× ⅘); 5, stamen (× 6), 4–5 from *Gomes Pedro* 3928.

Cotyledon simulans var. *spathulata* Schönl. ex Poellnitz, loc. cit.: (1940), *nom tantum*.
Cotyledon decussata sensu Guillarmod, Fl. Lesotho: 180 (1971) non Sims (1824).

A succulent undershrub, up to 1 m. tall. Stem and old branches thick, leafless, with a pale, wrinkled bark, marked above with the scars of fallen leaves, glabrous towards the base, pubescent above. Leaves of the current year condensed for 2–9·5 cm. below the peduncles, subopposite-decussate, usually all erect or sometimes the upper ones only so or sometimes all reflexed or the under ones only so, 3–10 × 1·7–7 cm. and obovate to obovate-acuminate or 6–10·5 × 1·8–2·5 cm. and obovate- to spathulate-cuneate, or oblong-linear up to 7 × 0·8 cm., rounded or truncate at the apex and with or without a short mucro, or shortly acuminate and acute, entire, sometimes margined with red, contracted or narrowing into a ± long subpetiole-base, not amplexicaul, fleshy, concave or channelled above, convex beneath or the narrowest ones subterete-channelled (but all ± flattish, ± rigid and not very thick on drying), densely and shortly glandular-pubescent (hairs whitish, patent, without a distinct glandular head). Inflorescences cymose, not dense, up to 15 cm. in diameter; peduncle 14–35 cm. long, terete, 5–7 mm. in diameter at the base without or with 1 or more bracts, reddish and densely glandular-pubescent at the base, sparsely so upwards, subglabrous and whitish below the inflorescence; pedicels 1·2–4·5 cm. long, rather thickened-obconical towards the apex, glabrous, whitish-bloomed or mealy like the calyx, the longest ones arched. Flowers ± 3·5 cm. long, usually pendulous, glabrous except inside for the corolla-tube at the insertion of filaments. Calyx 4–5 mm. long; tube 0·5–0·75 mm. long, truncate and circumscissile (with corolla-tube) below; sepals deltate, attenuate or contracted into a narrow, acute apex, separated at the base by broad and rounded sinuses. Corolla salmon, pinkish or orange-yellow; tube 1·5–2 cm. long and 6–10 mm. in diameter at the base, 6–8 mm. in diameter at the mouth, cylindrical or subcylindrical (neither angulate nor contracted above the carpels), sometimes narrowing slightly upwards; corolla-lobes 15–17 × 5–7 mm., oblong or lanceolate, acute, finally reflexed. Stamens exserted, but not attaining the apex of corolla-lobes; filaments 18–25 mm. long, dilatate below, inserted 4–5 mm. distant from the bottom of the corolla-tube, with a tuft of dense, long, white hairs at the base; anthers ovate, c. 2 mm. long or oblong, 3–4 mm. long (sometimes short and long anthers in the same inflorescence or in the same flower), yellow. Follicles c. 20 mm. long, narrowly fusiform; styles ± 15 mm. long, exserted; stigmas not attaining the anthers. Scales broader than long, semicircular, slightly emarginate, flat.

Mozambique. M: Sabié, Libombos Mts., M'ponduine, near the border (Namaacha) with Swaziland, 25.iv.1947, *Pedro & Pedrógão* 723 and 733 (LMA); Maputo, Matola, on road Boane-Changalane, 12.v.1968, *Balsinhas* 1267 (COI; LMA).
Also in S. Africa (Natal, Transvaal, Orange Free State and eastern Cape Prov.), Swaziland and Lesotho. On rocks and in rock crevices in exposed situations of slopes.

In the synonymy given above we have followed Tölken (in Bothalia **12**, 4: 619–620 (1979)). However, we have considered the taxon at specific level, not as a variety of *C. orbiculata* L. as Tölken does.
Our description was based on Mozambican specimens, which accord very well with the description of *C. zuluensis* Poellnitz, whose type was not seen by us. Nor have we seen the plate 792 (K), the designed lectotype of *C. oblonga* Haw.

Cotyledon glandulosa N.E. Br. in Kew Bull. **1913**: 300 (1913).

This taxon which, according to its author, is similar to *C. glutinosa* Schönl. (it is indentical with *C. papillaris* L. f., according to Tölken, in sched.), is doubtfully indigenous in Zambia. N. E. Brown says "Simpson-Hayward does not remember where he found it, but informs us that most of the plants were collected in Northern Rhodesia. The species to which it is most nearly allied is, however, a South African plant." As no other specimens have been collected since then in the F.Z. area, we exclude this taxon from our treatment of F.Z. Cotyledons. Schönland (in S. Afr. Journ. of Sci. **17**, 2: 187 (1921) cites the same *Simpson-Hayward* specimen for "Northern Rhodesia, s. loc.", based on N. E. Brown's reference.

95. VALERIANACEAE
By J. F. M. Cannon

Perennial or, less commonly, annual herbs, often with aromatic rhizomes; rarely slightly shrubby. Leaves opposite or occasionally basal; exstipulate, frequently pinnately divided, sometimes entire; bases often sheathing. Inflorescence a many-flowered compound dichasial cyme, thyrse or monochasium, sometimes sub-capitate; bracts and bracteoles usually present. Flowers hermaphrodite (occasionally unisexual and the plants then dioecious); normally 5-merous and irregular to nearly regular. Calyx obsolete or minute, sometimes expanded in fruit and lobed to form a pappus. Corolla tubular or funnel-shaped, often with a basal spur or sacoate; limb 3–5 fid, oblique or divided into 2 distinct lips. Stamens 1–4, alternating with the lobes of the corolla, epipetalous; anthers versatile, 2–4 lobed. Ovary inferior, with 3 locules only 1 of which is fertile; ovule solitary, anatropous; style, stigma subtruncate, entire or shortly 2–3 lobed. Fruit an achene. Seed pendulous; embryo straight with oblong cotyledons; endosperm sparse or lacking.

A family of some 13–15 genera and c. 400 species. Widely distributed throughout Eurasia, Africa and the Americas. Four genera (*Centranthus, Fedia, Valeriana* and *Valerianella*) occur in Africa, but the first two are confined to areas north of the Sahara. Kokwaro in F.T.E.A. reports that "Corn Salad" (*Valerianella locusta* (L.) Laterade is sometimes grown in agricultural experimental stations in East Africa, and *Centranthus ruber* has been grown as an ornamental in Zimbabwe.

VALERIANA L.

Valeriana L., Sp. Pl. **1**: 31 (1753); Gen. Pl., ed. 5: 19 (1754).

Perennial herbs, rarely climbers, occasionally subshrubs which are somewhat woody at the base; glabrous or sometimes sparsely pubescent with simple hairs; often with a characteristic pungent smell of "valerian" from the rizomes, especially when dry. Some species gynodioecious or polygamodioecious at anthesis. Inflorescence a dichasial cyme or thyrse, rarely subcapitate. Flowers hermaphrodite or unisexual. Calyx small and inrolled but usually developing in fruit into 5–15 plumose awns. Corolla imbricate, funneliform or campanulate, slightly saccate at the base, with (3–4) 5 lobes. Stamens usually 3, epipetalous, alternating with the corolla lobes, frequently exserted. Style with 3 short lobes or slightly emarginate. Fruit an achene, compressed, with 6 filiform ribs.

About 200 species, widely distributed in north temperate regions and in western South America. 3 species are known from Africa south of the Sahara. For a general account of this genus in tropical and South Africa see Meyer, F.G. in Journ. Linn. Soc. Bot. **55**: 766 (1958).

Valeriana capensis Thunb., Prod. Pl. Cap.: 7 (1784). — Sond. in Harv. & Sond., F.C. **3**: 40 (1865). — Meyer, F.G. in Journ. Linn. Soc. Bot. **55**: 766 (1958). — Chapman, Veg. Mlanje Mt.: 57 (1962). — Binns, H.C.L.M.: 103 (1968). — Kokwaro in F.T.E.A., Valerianaceae: 6 (1968). TAB. **9**. Type from South Africa.
Valeriana capensis var. *lanceolata* NE. Br. in Kew Bull. **1895**: 156 (1895). Syntypes from South Africa and Malawi, Mt. Mulanje, *Whyte* sn. (K. Holo; BM, Iso).

Erect perennial herbs of up to 1 m. at maturity, with fleshy rhizomes and fibrous roots; stolons occasionally present. Stems simple and unbranched below the inflorescence; generally glabrous except at the prominent nodes, but may rarely be quite densely covered with long, lax white hairs; internodes generally with well-developed ridges. Leaves mainly basal, but also present as 3–5 gradually reduced opposite pairs on the stem; most commonly pinnate, sometimes pinnatifid, more rarely subentire; terminal leaflets larger than the lateral ones, subrotund to linear-lanceolate, up to 13 cm. long; lateral leaflets ± irregularly dentate, becoming reduced in size towards the petiole. Basal leaves generally with long petioles of up to 18 cm. (the length probably much dependant on the density of surrounding

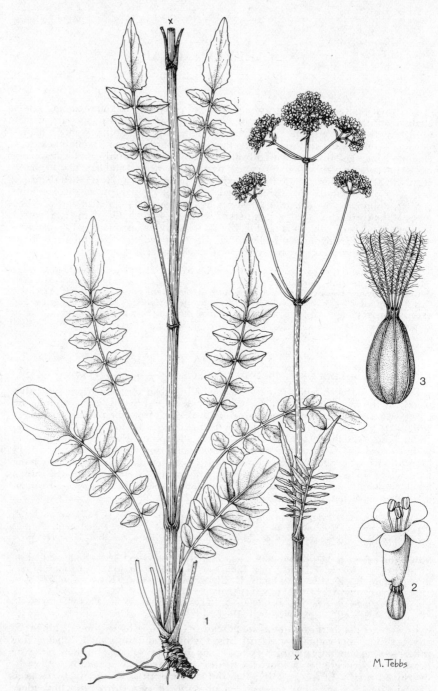

Tab. 9. VALERIANA CAPENSIS. 1, habit (× ½); 2, flower (× 5); 3, fruit with pappus (× 5), 1 from *Robson* 250, 2–3 from *Chapman* 416.

vegetation). Inflorescence a dichasial thyrse, initially subcapitate, but often expanding into a system of up to 30 × 10 cm. (even larger inflorescence have been reported from outside our area). Bracts linear to linear-lanceolate, 0·5–2 × 1–3 mm., sometimes with small lateral lobes. Inflorescence axis and branches generally glabrous with pubescent nodes, but may be sparsely hairy like the main stem. Flowers perfect or imperfect. Calyx inrolled at anthesis, limb segments forming a feathery pappus in fruit. Corolla infundibuliform, white sometimes flushed with pink and distinctly pink in bud; slightly irregular, up to 5 mm. long; throat often with long, dense white hairs, exterior glabrous. Stamens 3, distinctly exserted at anthesis, on long slender filaments. Style slender, with 3-lobed stigmatic surface at maturity, slightly exceeding the filaments. Achenes 3–5 × 1·5–3 mm., narrowly ovoid, strongly compressed, glabrous with 6 distinct filiform ribs, 3 dorsal, 1 ventral and 2 marginal.

Zambia. E: Nyika, fl. 29.xii.1962, *Fanshawe* 7291 (K). **Zimbabwe.** E: Umtali, Sheba Estate near Stapleford, fl. 25.x.1955 *Chase* 5884 (BM; K; LISC; SRGH). **Malawi.** N: Nyika Plateau, valley c. 2·5 miles SW. of Rest House, fl. 22.x.1958, *Robson* 250 (BM; K; LISC; PRE; SRGH). S: Mt. Mulanje, Tuchila Plateau, fl. 26.vii.1956, *Newman & Whitmore* 204 (BM). **Mozambique.** N: Námuli, fl. 1887, *Last* s.n. (K). MS: Manica, Zuira Mtns., fl. 4.ii.1965, *Torre & Pereira* 12650 (LISC).
From Kenya and Tanzania, through our area to South Africa. Marshy areas and streamsides in upland grasslands, 1800–2700 m.

The plants from Zomba and Mulanje Mountains (Malawi) are unusual in having the basal leaves more or less reduced to large, narrowly lanceolate, terminal lobes on long petioles. Some individuals are also exceptionally densely pubescent. They were recognised (along with some other specimens from South Africa) by N. E. Brown in 1825 as var. *lanceolata* and, seen by themselves are quite striking, but they are nevertheless linked by partial intermediates to the general range of variation of the species. However, while this suggests that the Mulanje populations have been isolated for a long period, the variety is probably not worthy of formal recognition, unless future experimental investigation suggests otherwise.

96. DIPSACACEAE
By Margaret J. and J. F. M. Cannon

Usually perennial herbs or subshrubs, less commonly annual or biennials. Leaves opposite, estipulate, entire to pinnatifid, to finely divided, sometimes noticeably heterophyllous, rarely connate at the base. Inflorescence usually capitate on a long peduncle, rarely verticillate or in lax panicles. Heads surrounded by well-developed involucres. Receptacle sometimes with short hairs, otherwise glabrous or with a paleaceous bract subtending each flower. Flowers sessile, hermaphrodite (rarely the outer female only); the ovary surrounded by an involucel of fused bracts, which are frequently extended to form a cupule. Calyx cupular, often with a toothed margin and long-spreading bristles. Corolla gamopetalous, somewhat zygomorphic or otherwise irregular, and with a 4-5 lobed limb. Stamens 2–4, inserted towards the top of the tube and alternating with the corolla lobes. Anthers frequently on rather long filaments and exserted at anthesis. Style filiform, simple, clavate or shortly 2-lobed. Ovary inferior with a solitary ovule. Fruit an achene within the persistent involucel and frequently with a conspicuous persistent calyx.

A family of 11 genera and some 350–400 species centred on the Mediterranean and Near East areas, and absent from the New World. Superficially very similar to the Compositae but readily distinguished on closer examination by the free, protruding anthers, and the fruit crowned by a persistent calyx of teeth or bristles. Represented in tropical Africa by four genera, where *Dipsacus* is confined to upland areas of East and North-east tropical Africa, *Pterocephalus* having a similar distribution but with an outlying species in Mozambique, while *Scabiosa* and *Cephalaria* are widely distributed throughout the area and extend into South Africa.

1. Calyx without persistent bristles - - - - - - - - - **1. Cephalaria**
   - Calyx with persistent bristles - - - - - - - - - - 2
2. Calyx bristles plumose - - - - - - - - - **2. Pterocephalus**
   - Calyx bristles not plumose - - - - - - - - - **3. Scabiosa**

1. CEPHALARIA Schrader

Cephalaria H. A. Schrader ex Roem. & Schult., Syst. Veg. **3**: 1, 43 (1818). — Szabo in Mat. Term. Közlem. **38**, 4: 1–248 (1940) *nom. conserv.*

Annual or perennial herbs or sub-shrubs. Leaves very variable, most species heterophyllous, entire or toothed or pinnatifid. Inflorescence of terminal, globose or ± cylindrical capitula with (1)2 or more rows of bracts. Involucral bracts scarious or pubescent, usually obtuse, oblanceolate, ovate or lanceolate. Receptacular bracts obtuse, acuminate, often cuspidate or pungent with a hard tip, scarious, pubescent or hairy. Flowers 4-partite, with a 4 angled, furrowed involucel, crowned with 4 angular hairy teeth, or with a membranous ± glabrous corona, entire, crenate or with 4 ± obtuse teeth. Calyx small, cupiliform ± lobed, glabrous or pilose. Corolla white, cream or yellowish, or mauve, of 4 lobes, those of the outer flowers often longer and more irregular than the inner. Stamens 4, style entire, filiform.

A genus of about 60 species from S. Europe, the near East, tropical and southern Africa.

1. Corona membranaceous, glabrous or minutely hairy on the edge, entire hairy, entire or with
   4 broad lobes -   -   -   -   -   -   -   -   -   -   -   -   -   -   -   - 2
 - Corona hairy, entire or forming 4 long teeth -   -   -   -   -   -   -   -   -   - 3
2. Plants with many leafy shoots at the nodes, receptacle bracts ovate-lanceolate - *2. goetzei*
 - Plants without leafy shoots at the nodes, receptacle bracts marked by cuspidate *1. pungens*
3. Receptacle bracts lanceolate or acuminate -   -   -   -   -   -   *3. katangensis*
 - Receptacle bracts markedly cuspidate -   -   -   -   -   -   -   *4. integrifolia*

A specimen at Kew — *Milne-Redhead* 864 — 10.viii.1930 from Mwinilunga District of Zambia (W) appears to be intermediate between *C. integrifolia* Napper and *C. retrosetosa* Engl. & Gilg from Angola. It differs from *C. retrosetosa* in having long attenuate to cuspidate receptacle bracts, and from *C. integrifolia* in its dense pubescence.

1. **Cephalaria pungens** Szabo in Engl., Bot. Jahrb. **57**: 642 (1922) and in Mat. Term.
   Közlem. **38**: 128 (1940). — Napper in Kew Bull. **21**: 468 (1968) and in F.T.E.A.,
   Dipsacaceae: 4 (1968). — Richards & Morony, Checklist Fl. Mbala Distr. 177 (1969). —
   Moriarty, Wild Flowers of Malawi: 112, t. 56. 4 (1975). — TAB. **10**. Type from Tanzania.
       C. attenuata sensu Binns, H.C.L.M.: 48 (1968) and Richards & Morony, loc. cit.: 177
   (1969) non (Linn. f.) Roem. & Schult.
       C. centaurioides sensu Hiern in F.T.A. **3**: 251 (1877) quoad spec. Tanzania & Malawi. —
   Binns, loc. cit.: 49 (1968).

Perennial herb with a woody rhizome up to 2 m., stems simple or branched, glabrous in the upper parts, hairy in the lower parts with simple and short hairs, and retrorse long hairs with many-celled glandular bases. Lower leaves elongate, lanceolate, usually entire, crenate or slightly toothed, occasionally with a few narrow lobes at base, rarely lyrate, with a long petiole, up to 3 times as long as the blade. Leaves hairy on margins and veins of the underside, hairs glandular with a many celled base. Upper stem leaves narrow linear, entire or with 1–2 pairs of entire, linear lobes at the base, hairy, scabrous or glabrous. Capitula globose, up to 25 mm. in diameter in flower. Involucral bracts blackish, ovate-ovate lanceolate, receptacular bracts ovate-lanceolate to oblanceolate, suddenly cuspidate, dark, hairy on the edges, hispid at the centre portion of the outer surface. Involucel furrowed, hairy in the lower parts and on the veins, corona membranaceous, entire, crenate or with 4 ± irregular broadly triangular lobes. Corolla white, cream or yellow, calyx villous. Mature fruit ± glabrous, slightly hairy above, secund, up to 6 mm. × 1·75 mm.

Zambia. N: Mbala (Abercorn), fl. iv.1954, *Nash* 60 (BM). W: Ndola, fr. 14.iii.1954 *Fanshawe* 964 (K; SRGH). C: Serenje Distr., Kundalila Falls, fl. & fr. 4.ii.1973, *Strid* 2828 (K). **Zimbabwe.** N: Miami, fl. iv.1926, *Rand* 88 (BM). C: Alveston, fl. & fr. 1.iv.1946, *Wild* 982 (SRGH). E: Odzani River Bridge, fl. 27.xi.1950, *Chase* 3225 (BM; COI; K; SRGH). **Malawi.** N: Rumphi Distr., Nyika Plateau, fl. 5.i.1959, *Richards* 10505 (K; SRGH). S: Zomba mountain road, fl. 14.xii.1936, *Lawrence* 209 (K). **Mozambique.** N: Lichinga (Vila Cabral), Lichinga Plateau, fl. iii.1934, *Torre* 22 (COI; LISC). MS: Manica, Rotanda, Tandara, fl. 19.xi.1965, *Torre & Correia* 13172 (LISC).

Throughout tropical Africa to Swaziland. In upland grassland, streamsides and swampy places. 1200–1800 m.

2. **Cephalaria goetzei** Engl. in Bot. Jahrb. **30**: 418 (1902). — Szabo in Magyar Bot. Lap. **24**:
   14 (1926); in Mat. Term. Közlem, **38**, 4: 130 (1940). — Napper in Kew Bull. **21**: 468
   (1968) & in F.T.E.A., Dipsacaceae: 4 (1968). Type from Tanzania.

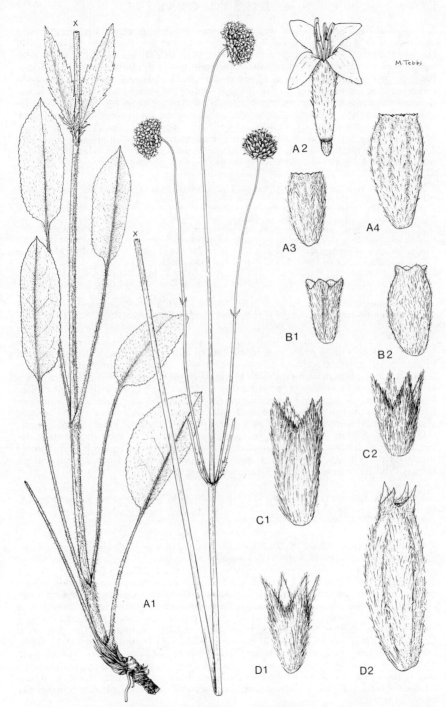

Tab. 10. A. — CEPHALARIA PUNGENS. A1, habit (×½); A2, flower (×4), both from *Chase* 4354; A3, involucel (×5); A4, fruit (×5), both from *Pawek* 6348. B. — CEPHALARIA GOETZII. B1, involucel (×5); B2, fruit (×5) both from *Robinson* 1355. C. — CEPHALARIA KATANGENSIS. C1, involucel (×5); C2, fruit (×5), both from *Richards* 9281. D. — CEPHALARIA INTEGRIFOLIA. D1, involucel (×5); D2, fruit (×5), both from *Tyrer* 517.

80   96. DIPSACACEAE

Perennial herb with a thick woody rhizome, producing stems with many leafy shoots up to 1·25 m. Stems furrowed and hollow, glabrous in the upper parts, slightly hairy in the lower parts with a few scabrid hairs on the leaf sheaths. Basal leaves very variable, narrowly elliptic or lanceolate in outline, entire or toothed or pinnatipartite, hairy on the margins and veins of the underside. Upper stem leaves glabrous or pubescent, sometimes hairy on the margins, sessile, pinnatifid with 2–4 lobes linear to linear-lanceolate in outline, the terminal lobe larger than the others. Capitula globose 20–25 mm. in diameter in flower. Involucral bracts ovate, short haired, dark coloured. Receptacular bracts longer, ovate-lanceolate, shortly hairy with dark tips. Involucel tube furrowed and hairy, corona membranaceous, glabrous or shortly ciliate on the edge, with 4 rather broad obtuse lobes. Calyx villous. Mature fruit 3–4 mm. × up to 1·5 mm., shortly hairy all over (1 spec. only examined).

Zambia. S: Mazabuka Distr., 22 miles N. of Choma, fl. & fr. 20.ii.1956, *Robinson* 1353 (SRGH). **Malawi.** C: Dedza Distr. Chongoni Forestry School, fl. 13.ii.1967, *Salubeni* 562 (K; LISC; SRGH). Also in Tanzania.
Swampy places in upland woodland; 1100 m.

3. **Cephalaria katangensis** Napper in Kew Bull. **21**: 464 (1968). — Richards & Morony in Checklist Fl. Mbala Distr. 177 (1969). Type from Zaire.
    C. attenuata var. *longifolia* De Wild. in Ann. Mus. Congo. Bot., Sér. 4, **1**: 164 (1903). Type from the Congo. Non *C. longifolia* E. Mey. in Zwei Pfl. Docum.: 128 & 152 (1843) nom. nud.
    C. humilis sensu Szabo in Mat. Term. Kozlem **38**: 124 (1940), pro parte. — De Wild; Contrib. Fl. Kat., Suppl. **1**: 95 (1927), non (Thunb.) Roem. & Schult.

Perennial herb with tough woody rhizome, plants (1)1·5–(2) m. Stems simple or branched, often hollow and with 6 pale coloured ridges, glabrous or slightly scabrid in the lower parts. Basal and lower stem leaves elongate, lanceolate or linear-lanceolate in outline, with long petioles, up to 3 times as long as blade. Leaves entire or toothed in the upper part, or sometimes with a few linear or linear-lanceolate lobes, hairy on the undersides and margins. Upper leaves narrowly linear, entire or with 1–2 pairs of entire linear lobes at the base, hairy, scabrous or glabrous. Capitula globose 20–(25) mm. in diameter in flower. Involucral bracts widely ovate-obtuse, scabrid with minute hairs on the margins, dark coloured at the tip. Receptacular bracts lanceolate to ovate-lanceolate, sub acuminate, with dark tips, pilose. Involucel furrowed, long hairy, corona of 4 angular acuminate teeth. Corolla white, calyx villous. Mature fruit up to 7·5 × 2 mm., with long bristly hairs.

Zambia. N: Luwingu Distr., fl. iv.1922, *Jeff* 33 (BM). W: Near Kasempa, Lufupa River, fl. 1921–22, *Foster* s.n. (BM; SRGH).
Known also from Zaire, Shaba Prov. In long grass, woodlands 1200–1275 m.

4. **Cephalaria integrifolia** Napper in Kew Bull. **21**: 463 (1968) and in F.T.E.A., Dipsacaceae: 16 (1968). Type from Tanzania.

A perennial rhizomatous herb, up to 1·5 m. Lower stem thick and woody, upper hollow, simple or sparingly branched. Lower parts of stems glabrous or with a few rigid hairs, or slightly scabrous, bridged, the upper stem with 4 ridges. Basal leaves elongate, linear lanceolate, long cuneate below, entire or toothed, glabrous or rarely slightly setose. Upper stem leaves entire or more rarely pinnati-partite, glabrous. Capitula globose, up to 30 mm. in diameter in flower. Involucral bracts blackish, ovate-lanceolate with short adpressed hairs. Receptacle bracts dark tipped, long acuminate or cuspidate with a long spinous tip, hairy except at the tip. Involucel ± furrowed, hairy, corona of 4 long teeth, densely hairy. Corolla white, calyx villous. Mature fruit up to 10 × 3·5 mm., slightly secund, hairy all over except on the prominent ridges.

Zambia. E: Lundazi, fl. 5.vii.1971, *Pawek* 4983 (K). **Malawi.** N: Nyika Plateau near Nganda Hill, fl. & fr. 7.ix.1962, *Tyrer* 910 (BM). **Mozambique.** Z: Serra do Gúruè, Chá Moçambique, source of R. Malema, fl. 4.i.1968, *Torre & Correia* 16877 (LISC).
Known also from Tanzania. Rough grassland 1550–1770 m.

2. PTEROCEPHALUS Adans.

Pterocephalus Adans., Fam. Pl. 2: 152 (1763).

Annual or perennial herbs or subshrubs. Capitula hemispherical or globular, with 1–2 rows of foiaceous receptacular bracts. Receptacle scales hairy or O. Involucel ± sulcate, 8 furrowed, toothed or with a ± membranous corona. Calyx short with 5–24 plumose setae of equal length. Corolla usually 5-fid or with 4–6-fid flowers in the same head, rarely entirely 4-fid. Marginal flowers often larger and with more unequal corolla lobes, inner flowers often smaller and more regular. Stigma oblique, entire or slightly 2 lobed.

A genus of about 25 species, found mainly in the Middle East, spreading to Southern Europe, India, China and Thailand, a few species occurring in the Canary Isles and North and East Africa.

Pterocephalus centennii M. J. Cannon in Bol. Soc. Brot., Sér. 2, 44: 243, tab. 1 (1981). TAB. 11. Types: Mozambique, Manica and Tsetsera Plateau, Torre & Pereira 12745 (LISC, holotype; BM; C; COI; LMU; MO; SRGH; WAG, isotypes).

Woody sub-shrub c. 2 m., with erect leafy shoots bearing several globular capitula. Stems hollow, branched; the lower parts with long, glandular based hairs and a short bristly pubescence; the upper part pubescent without long hairs. All leaves similar, 20–45 × 8–12 mm., opposite, connate, toothed, lanceolate or linear-lanceolate; with 1–2 pairs of linear-lanceolate or linear leaflets at the base, appearing almost whorled, subtending leafy shoots from about half way up the stem to the lowest peduncle; glabrous or slightly hispid especially on the underside of the clasping leaf bases and the small leaflets. Upper leaves shorter and more linear than the lower, but still with minute linear leaflets. Capitula spherical, c. 25 mm. in diameter. Involucral bracts greenish or purplish with green tips, broadly ovate, acute, c. 8 × 4 mm., pubescent. Receptacle bracts lanceolate, acute, purplish above, whitish and scarious below, the lower intermediate with the bracts of the involucre. Involucel furrowed, with 8 raised veins, densely hairy, with a corona of 4 broadly ovate teeth. Calyx small, patelliform, with 18–20 plumose bristles up to 7 mm. in young fruit, occasionally 1–2 in each calyx expanded and slightly laminate, dirty mauvish. Corolla whitish, c. 10 mm.; tube a little shorter than the calyx bristles, with 4 ± equal, spreading, broadly triangular lobes; the outer flowers scarcely differing from the inner. Stamens 4, stigma oblique, entire or slightly 2-lobed. Fully mature fruit unknown.

Mozambique. MS: Tsetsera Plateau, Chimoio (Vila Pery) fl. & fr. 6.ix.1965, Torre & Pereira 12745 (LISC, holotype; BM; C; COI; LMU; MO; SRGH; WAG, isotypes).
Known only from the type specimen. At edge of cloud forest dominated by Podocarpus milanjianus, 2000 m.

3. SCABIOSA L.

Scabiosa L., in Sp. Pl. 1: 98 (1753); Gen. Pl., ed. 5: 43 (1754).

Annual or perennial herbs, often heterophyllous. Leaves opposite, simple or deeply divided. Capitula hemispherical or cylindrical, often on long peduncles, involucral bracts herbaceous, in 1 to several rows. Receptacle bracts usually linear-lanceolate, shorter than the flowers, glabrous or pubescent. Involucel tube cylindrical, smooth or furrowed, 4–8 ribbed, expanded above into a membranous, many veined, erect or spreading corona. Calyx small, usually with 5 bristle-like teeth, spreading in fruit. Corolla of 5 unequal lobes, (rarely 4–6), and a short tube, usually longer in the marginal flowers. Stamens 4.

A genus of about 100 species, widely distributed in Europe and the temperate regions of Asia, spreading into North Africa, with a few species occurring in Africa south of the Sahara.

Capitula large (2–)2·5–3·5(–4) cm. in diameter, stems solid - - - - 1. S. columbaria
Capitula small 1·5–2 cm. in diameter, stems hollow - - - - 2. S. drakensbergensis

1. Scabiosa columbaria L., Sp. Pl. 1: 98 (1753). — Swynnerton in J. Linn. Soc. Bot. 40: 103 (1911). — Eyles in Trans. Roy. Soc. S. Afr. 5: 498 (1916). — Brenan Mem. N.Y. Bot. Gard. 8: 458 (1954). — Binns, H.C.L.M.: 49 (1968). — Napper in F.T.E.A., Dipsacaceae: 7 (1968). — Richards & Morony, Checklist Fl. Mbala Distr.: 177 (1969). TAB. 12. Type from Europe.

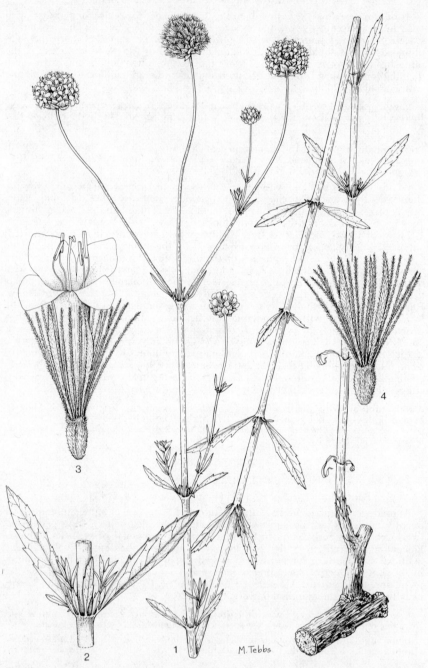

Tab. 11. PTEROCEPHALUS CENTENNII. 1, habit (×½); 2, stem node and leaves (×1); 3, flower (×5); 4, immature fruit (×5), all from *Torre & Pereira* 12745 (LISC).

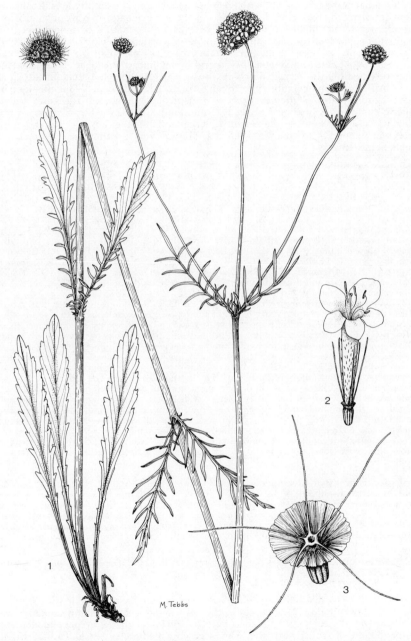

Tab. 12. SCABIOSA COLUMBARIA. 1, habit (×½); 2, flower (× 3); 3, fruit (× 3), all from *Saunders Davies* s.n.

Scabiosa austro-africana Heine in Mitt. Bot. Staatssamml. München **1**: 445 (1954). —
Napper in F.T.E.A., Dipsacaceae: 9 (1968). — Richards & Morony, Checklist Fl. Mbola
Distr.: 177 (1969). — Moriarty, Wild Flowers of Malawi: 112, t. 56. 5 (1975). Type from
S. Africa.

Perennial herb up to 1·5 m., with erect branched stems, heterophyllous. Leaves of
non-flowering rosettes and basal stem leaves ovate-lanceolate, obovate or lanceolate,
simple, lyrate or pinnatifid, upper stem leaves 1–2 pinnatifid or pinnatisect, the
segments lanceolate to linear, the terminal segment larger than the lateral, hispid
or pubescent. Flower heads on long peduncles, globular or hemispherical
(2–)2·5–3·5(–4) cm. in diameter. Involucral bracts linear lanceolate, acute, shorter
than or equalling the marginal flowers, pubescent. Receptacle bracts linear, acute
3–5 mm. long. Involucel tube cylindrical, 8-furrowed, pubescent, 2–3·5 mm. long in
fruit. Corona glabrous, membranaceous, conspicuously veined, 0·5–1·5 mm. Calyx
pilose, cupuliform, bearing 5 setae, 1–8 mm. long. Corolla mauve or white, or the
tube white and lip mauve, 5-lobed with 3 larger and 2 smaller lobes, larger and more
irregularly lobed in the marginal flowers. Mature fruit 2–4 mm. long, corona and
calyx persistent.

Botswana. SE: Kolobeng valley, 24 km. W. of Gaberones, fl. 24.x.1976, *Wollard* 249
(SRGH). **Zambia.** N: Abercorn, fl. 6.x.1952 *Nash* 188 (BM). W: Ndola Dist., fl. 21.x.1953,
Fanshawe 363 (K; LISC; SRGH). C: Mt. Makulu, fl. & fr. 9.xi.1956, *Angus* 1440 (COI; K;
LISC; PRE; SRGH). E: Nyika Plateau, fl. & fr. 23.xi.1955 *Lees* 53 (K). S: Mumbwa, fl. & fr.
vii.1912, *Macaulay* 732 (K). **Zimbabwe.** N: Mazoe, fl. xi.1906, *Eyles* 452 (BM; SRGH). W:
Matobo, fl. & fr. i.1954, *Miller* 2044 (PRE; SRGH). C: Hatfield, Salisbury, fr. 14.v.1934
Gilliland 133a (BM). E: Melsetter Dist., Chimanimani Mts., fl. & fr. 3.ii.1958, *Hall* 342 (BM).
S: Fort Victoria, fl. & fr. 14.xii.1947, *Newton & Juliasi* 56 (SRGH). **Malawi.** N: Nyika
Plateau, Lake Kaulime, fl. & fr. 24.x.1958, *Robson* 323 (BM; K; LISC; PRE; SRGH). C: Dedza
Mt., fl. 15.x.1937, *Longfield* 41 (BM). S: Tung Station, near Limbe, fr. 17.viii.1950, *Jackson*
110 (BM; K). **Mozambique.** MS: Garuso fl. iv.1935, *Gilliland* 1843 (BM). GI: Macia,
between Lagoa Pate and Bilene, fl. 26.iii.1954, *Barbosa & Balsinhas* 5477 (BM; LISC). LM:
Goba near R. Maiuana, fl. 2.xi.1960, *Balsinhas* 160 (BM; COI; LISC; PRE).

Widely distributed, from S. Africa, through East Africa to Ethiopia, the Near East, North
Africa and Europe. Upland grasslands, mountain slopes, open woodland etc. 1000–2000 m.

Scabiosa columbaria sens. lat. is the most widespread and variable species of the genus, there
being two groups of closely related taxa in Europe distinguished mainly by flower colour. In our
area there appear to be 2 main forms, a white flowered group, with rather globular heads and
long calyx setae (*S. austro-africana* Heine) and a mauve flowered group with ± hemispherical
heads and shorter calyx setae. However, white flowered plants with short calyx setae are not
infrequent, and mauve flowered plants with long setae also occur. The differences seen by
Napper in East Africa in leaf form do not appear to hold good in our area. We are greatly
indebted to B. L. Burtt for comments based on wide experience of the group in the field, and
concur with his opinion that *S. columbaria* has produced numerous local variants in Eastern and
Southern Africa. For the purposes of the present account it does not seem possible to produce
an adequate treatment of the variation in formal taxonomic terms. A complete review of the
whole group throughout its extensive range, preferably involving modern supplementary
taxonomic techniques, needs to be undertaken before a definitive classification can be
established.

Two specimens from Mozambique. (MS: R. Murorue, fl. 22.iv.1948, *Barbosa* 1519 (LISC)
and Mossurize, Espungabera, fl. 7.vi.1942, *Torre* 4256 (LISC)) are somewhat dissimilar from
the usual forms of *S. columbaria* in that the plants are extremely leafy, right to the top, the
terminal segment of the upper leaves being almost linear, with 2–3 small linear segments paired
at the base. The capitula are rather small, the flowers of the *Barbosa* specimen pinkish-lilac, the
other white. Further fieldwork is needed to establish the status of these interesting forms.

2. **Scabiosa drakensbergensis** B. L. Burtt in Notes Roy. Bot. Gard. Edin. **30**: 125 (1970).
   Type from S. Africa.
   Scabiosa africana auct. non L.

A robust perennial herb up to 1·25 m., stems densely pubescent with retrorse
hairs, the lower hollow. Leaves sessile, the lower lyrate-pinnatifid or pinnatisect, the
leaflets ovate-lanceolate to oblanceolate, the terminal ovate, crenate to dentate;
upper leaves with much narrower linear segments. Peduncles densely pubescent, up
to 25 cm., bearing small capitula 1·25–2 cm. in diameter. Involucral bracts linear,
pubescent, 7–10 mm., longer than the marginal flowers, becoming somewhat
reflexed in fruit. Receptacle bracts 3–5 mm., linear, pubescent. Involucel tube
cylindrical, 8-furrowed, pubescent, 2·5–3 mm. long in fruit. Corona glabrous,

membranaceous, c. 1 mm. Calyx setae 3·5–5 mm. long. Corolla pinkish or white. Mature fruit c. 3 mm. long, corona and calyx persistent.

Zimbabwe. E: Inyanga, fl. & fr. 27.i.1941, *Hopkins* 11687 (SRGH; K).

A plant from the same area, Nuzu plateau, fl. & fr. iii.1935, *Gilliland* 1615, (BM; K) appears to have affinities with *S. tysonii* L. Bolus, in its very leafy appearance, with broadly ovate terminal segments of the leaves right to the top of the stem, and small capitula. The specimen consists only of the top part of the plant, but in view of its hollow stems and robust nature it is best considered here, rather than to regard it as an extension of the range of *S. tysonii* from S. Africa.

98. GOODENIACEAE
By E. Launert

Herbs, small shrubs or rarely shrubs or small trees without latex, rarely spinescent. Leaves usually alternate, rarely opposite, sometimes all radical, entire, simple, exstipulate. Flowers sometimes solitary and axillary but more often arranged in a spike, raceme or panicle, hermaphrodite, protandrous. Calyx tubular, adnate to the ovary or very rarely free; limb consisting of 5 persistent lobes, rarely entire. Corolla sympetalous, 5-merous, bilabiate or very rarely 1-lipped; tube long, often villous inside. Lobes valvate, subequal. Stamens 5, alternating with the corolla-lobes and inserted at the base of the corolla, usually free but sometimes joined with the anthers; anthers 2-thecous, with the theca parallel and opening longitudinally. Disc sometimes present. Ovary mostly inferior, 1–2 (–4)-locular; ovules 1-several, erect or ascending. Style simple or rarely divided, emerging laterally from the split corolla-tube; stigma truncate or bilobed, with an apical indusium. Fruit a drupe, capsule or nutlike. Seeds with endosperm, small, flat, with a thin or thick and often crustaceous testa.

A family of 14 genera and over 300 species, predominantly in Australia but also in New Zealand, antarctic South America, tropical Asia, and Africa.

SCAEVOLA L.
Scaevola L., Mant. Pl. **2**: 145 (1771) *nom. conserv.*

Leaves alternate or rarely opposite, entire or with the margins dentate, sessile or petiolate, sometimes clasping the stem. Inflorescence axillary, cymose, usually pedunculate, rarely flowers solitary. Calyx much shorter than the corolla, usually fleshy; tube cylindrical to subglobose. Corolla 5-lobed, slightly fleshy; tube with a longitudinal split to the base, villous or pubescent inside; lobes subequal, spreading, with broad membranous margins. Stamens free; anthers with an apical projection. Ovary usually completely inferior, 2-locular, thick-walled, fleshy, style, semi-terete, simple or bilobed, thick; stigma surrounded by a ciliate cup-shaped membrane. Fruit a drupe, with a succulent or membranous exocarp and a woody or thin and cructaceous endocarp. Seeds discoid, with a parchment-like testa and a succulent endosperm.

A genus of more than 80 species; mostly indigenous to Australia, with a few in Asia, the Pacific Islands and Africa.

Scaevola plumieri (L.) Vahl, Symb. Bot. **2**: 36 (1791). — Dale & Greenway, Kenya Trees and Shrubs: 230 (1961). — Hepper in F.W.T.A., ed 2, **2**: 315 (1963). — Guillaumet in Fl. Mascareignes **110**: 2, fig. 1, 7–9 (1976). — F. G. Davies in F.T.E.A., Goodeniaceae: 3, fig. 10–12 (1978). TAB. **13**. Type based on an illustration of a plant from the Bahamas.
  Lobelia plumieri L. Sp. Pl. **2**: 929 (1753) pro parte. Type from Ceylon.
  Lobelia frutescens Mill., Gard. Dict., ed 8 (1768) *nom. superfl.* based on *L. plumieri* L.
  Scaevola lobelia Murr. in Syst. Veg., ed **13**: 178 (1774). — Hiern in F.T.A. **3**: 462 (1877) nom. superfl. based on *Lobelia plumieri* L.
  Scaevola thunbergii Eckl. & Zeyh., Enum. Pl. Afr. Austr.: 387 (1837). Type from S. Africa.

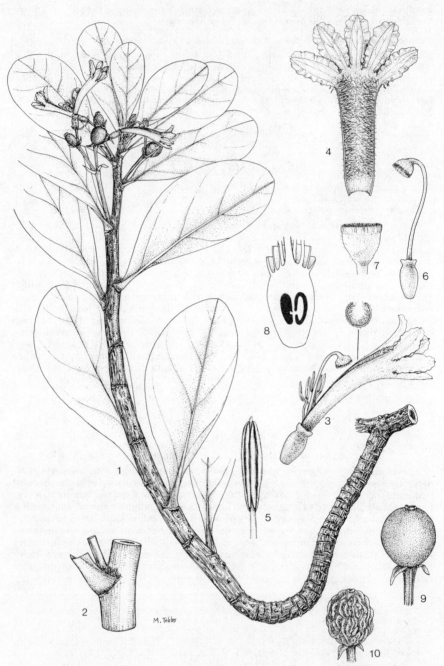

Tab. 13. SCAEVOLA PLUMIERI. 1, flowering branch (× ½), from *Gomes e Sousa* 1869; 2, leaf-axil
(× 2), from *Mogg* 26850; 3, flower (× 3); 4, corolla spread out (× 3); 5, stamen (× 6), 3–5
from *Brewer* 3527; 6, calyx and pistil (× 3), from *Mogg* 26850; 7, apex of style showing
indusium (× 3); 8, ovary in longitudinal section (× 3), 7–8 from *Brewer* 3527; 9, fresh fruit
(× 1); 10, dried fruit (× 1), 9–10 from *Mogg* 26850.

Evergreen small tree or shrub, 30–90(150) cm. tall, succulent, almost glabrous. Stems greenish or yellowish, covered with prominent oblique leaf-scars. Leaves in dense terminal clusters, very succulent, 5–9(11·5) cm. long, 2–5(7·5) cm. broad, obovate and tapering towards the base, sessile or shortly petiolate, with the apex rounded, entire, glabrous, yellow-green; lateral veins rather obscure (better visible in dried leaves). Petiole winged. Flowers sessile, arranged in short bracteate axillary cymes; bracts linear, glabrous or with hairs in the axils. Calyx very short, 1·75–2·5 mm. long, lobes hardly discernible. Corolla white or greenish, often yellowish inside, up to 23 mm. long; tube 9–13 mm. long, up to 3 mm. wide, pubescent outside, villous inside; lobes up to 11 mm. long, with a succulent green centre and whitish membranous crenulate margins. Ovary bilocular, with only one ovule developing (always?); style pubescent, with the apical indusium finely ciliate. Fruit black or bluish, ellipsoid to globose, 10–15 mm. wide, rather fleshy, wrinkled when dry. Seeds single, contained within the fruit.

Mozambique. N: Mafamede I., fl. 20.x.1965, *Gomes e Sousa* 4875, 4876 (COI; K). Z: 32 km. north of Quelimane, fl. 10.viii.1962, *Wild* 5866 (LISC; K; SRGH). MS: Beira, fl. 12.vii.1966, *Simon* 800 (K; LISC; SRGH). GI: Bazaruto I., fr. 21.x.1958, *Mogg* 28488 (K; LISC). M: Inhaca I., fl. & fr. 22.ix.1957, *Mogg* 31723 (K); Ponta do Ouro beach, east of Maputo, fl. & fr. 28.xii.1948, *Gomes e Sousa* 3927 (COI; K).

Widely distributed along sea shores in Africa (Somalia, Kenya, Tanzania to the Cape Prov. and in Western Africa from S. Tomé to Angola). Also in the Mascarenes, Ceylon, Florida and tropical S. America. On coastal sand dunes and recent sandstone formations but also known as a coloniser of lava flows (Réunion).

99. CAMPANULACEAE
By M. Thulin

Annual or perennial herbs, subshrubs, or rarely small shrubs, laticiferous. Leaves alternate, rarely opposite, simple, entire, dentate to incised or rarely variously lobed, exstipulate. Inflorescences generally cymose, panicle-, raceme-, spike- or head-like, or flowers solitary. Flowers ⚥, usually protandrous, regular, (3–) 5(–10)-merous, mostly with a bract and 2 bracteoles. Calyx ± adnate to the ovary; lobes usually free, persistent, valvate in bud. Petals connate to various degrees, sometimes almost free, valvate in bud. Stamens alternating with the corolla-lobes, free or rarely adnate to the corolla; anthers very rarely entirely or partly connate, introrse; filaments usually dilated at the base. Ovary ± inferior, rarely superior, 2–10-locular; ovules few-many, anatropous, on axile placentas; style 1, furnished with pollen-collecting hairs on and usually below the style-lobes. Fruit capsular, variously dehiscing by apical or lateral valves or pores, or ± baccate. Seeds 1-many, albuminous; embryo straight, terete.

About 35 genera and some 700 species, especially well represented in the Mediterranean region and S. Africa, but relatively sparsely developed in the tropics.

In keys and descriptions the style is often said to be hairy or hairy below. This does not include the pollen-collecting hairs that are always present on and usually somewhat below the style-lobes (however, they disappear by invagination after anthesis) but concerns the normal hairs often present further down on the style.

1. Corolla ± 5 cm. long, orange or red; flowers 6-merous; fruit baccate - **1. Canarina**
 – Corolla less than 3 cm. long, never orange or red; flowers (3–)5-merous; fruit capsular 2
2. Capsule with apical dehiscence; seeds without hair-like projections - - - 3
 – Capsule indehiscent, tardily opening by the irregular decomposition of the pericarp between the persistent, lateral nerves; seeds sometimes with hair-like projections; annuals
 3. Gunillaea
3. Capsule dehiscing by apical valves; annuals or perennials - - **2. Wahlenbergia**
 – Capsule dehiscing by an apical lid; perennial often mat-forming herbs **4. Craterocapsa**

1. CANARINA L.

Canarina L., Mant. Pl. Alt.: 148, 588 (1771) *nom. conserv.*

Glabrous terrestrial or epiphytic perennial herbs containing abundant white latex. Roots thickened, fleshy. Stems herbaceous, terete and hollow, di- or trichotomously branched from the leaf-axils; most leaf-axils also producing small, usually rudimentary, accessory shoots. Leaves opposite or ternate, petiolate. Flowers solitary in dichasial forks or terminal, large, pendent, (5–) 6 (–7)-merous throughout. Calyx-lobes entire or sometimes dentate, erect, spreading or reflexed. Corolla funnel-shaped or campanulate with short lobes. Filament-bases almost linear to broad, shield-like. Ovary inferior; style shorter than the corolla, markedly thickened towards the apex, with short style-lobes. Fruit baccate, with persistent calyx. Seeds numerous; testa finely pitted or striate.

Three species, two in tropical Africa, one in the Canary Is. The latter, *C. canariensis* (L.) Vatke, is a common greenhouse plant.

Canarina eminii Aschers. ex Schweinf. in Sitz.-Ber. Ges. Nat. Fr. Berl. **1892**: 173 (1892). — Engl., Pflanzenw. Ost-Afr. **C**: 400, t. 36 (1895). — B. L. Burtt in Curtis, Bot. Mag. **161**: t. 9531 fig. A–D (1938). — Robyns, Fl. Parc Nat. Alb. **2**: 402 (1947). — F. W. Andr., Fl. Pl. Anglo-Egypt. Sudan **3**: 70 (1956). — Hedberg et al. in Svensk Bot. Tidskr. **55**: 54, t. 4 fig. 6, 9, 13 (1961). — Binns, H.C.L.M.: 25 (1969). — Agnew, Upland Kenya Wild Fl.: 509, 510 (fig.) (1974). — Thulin in F.T.E.A., Campanulaceae: 4, fig. 1 (1976); in F.A.C., Campanulaceae: 3, fig. 1 (1977). TAB **14**. Type from Zaire.
    Canarina elegantissima T. C. E. Fries in Notizbl. Bot. Gart. Berl. **8**: 392, fig. 1 (1923). Type from Kenya.
    Canarina eminii var. *elgonensis* T. C. E. Fries, tom. cit.: 395, fig. 2 (1923). Type from Kenya.

Epiphytic or terrestrial usually glaucous herb. Root thick, often with a corky surface layer. Stems erect and scandent or pendent, up to several m. long, dichotomously branched, usually with a fine purplish mottling. Petioles less than half the length of the leaf-laminas, not coiled; leaf-laminas 2·5–10 × 1·5–9 cm., ± triangular to ovate, acute with cordate to truncate base, obtusely to sharply dentate to doubly dentate or doubly serrate. Hypanthium obconical, distinctly 6-ribbed, with the ribs projecting into the calyx-lobes. Calyx-lobes 1·8–3·8 × 0·5–1 cm., acute to acuminate, free, entire, erect or spreading. Corolla 4·3–7·6 cm. long, funnel-shaped, orange to orange-red with darker venation. Stamens with broad shield-like filament-bases; anthers 5·5–10 mm. long. Seeds 2·0–2·6 × 0·6–0·8 mm., elliptic-oblong, dark brown, finely striate.

Malawi. N: Misuku Hills, Willindi Forest, 12.i.1959, *Richards* 10624 (K; SRGH).
In eastern tropical Africa with its southern limit in Malawi. In upland or riverine forest, epiphytic or among rocks.

2. WAHLENBERGIA Schrader ex Roth

Wahlenbergia Schrader ex Roth, Nov. Pl. Sp.: 399 (1821) *nom. conserv.* — von Brehmer in Engl., Bot. Jahrb. **53**: 9–143 (1915). — Thulin in Symb. Bot. Ups. **21**, 1: 70 (1975).
    Lightfootia L'Hérit., Sertum Angl.: 4, t. 4 (1789) *nom. illegit.*
    Cervicina Delile, Fl. d'Egypte: 7, Atlas t. 5/2 (1813) *nom. rej.*
    Cephalostigma A.DC., Monogr. Camp.: 117 (1830).

Annual or perennial herbs, subshrubs or small shrubs. Leaves alternate or rarely opposite, mostly sessile, simple, entire, dentate or rarely lobed or incised. Inflorescences panicle-, raceme-, spike- or head-like, or flowers solitary. Calyx-lobes (3–) 5. Corolla ± deeply (3–) 5-lobed, or split almost to the base, ± pubescent inside near the base, rarely with long, slender hairs in the corolla-tube (but not in the F.Z. area). Stamens (3–) 5, free; filament-bases variously dilated or linear, usually ciliate. Ovary subinferior to rarely subsuperior, 2–5-locular; ovules many; style shorter or longer than the corolla, the upper part with pollen-collecting hairs, glabrous or hairy below, eglandular or with small glands present at or near the base of the 2–5 style-lobes. Capsule loculicidal, dehiscing by as many apical valves as there are loculi in the ovary. Seeds numerous, ± elliptic in outline; testa smooth or variously reticulate.

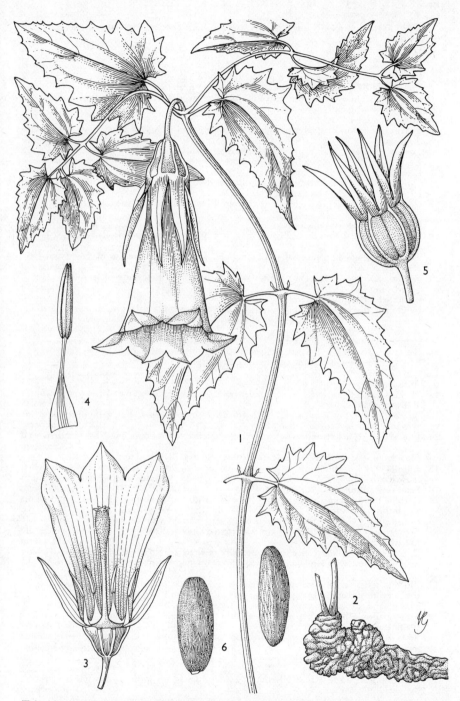

Tab. 14. CANARINA EMINII. 1, habit (× ⅔), from *Chojnacki* 8893 & *Norman* 223; 2, part of root
(× ⅔), from *Mooney* 7109; 3, flower with two stamens and calyx-lobes and three petals
removed (× ⅔); 4, stamen (× 2), 3–4, from *Hedberg* 158; 5, fruit (× ⅔); 6, seeds (× 12), 5–6,
from *Procter* 2298 & *Norman* 223. From F.T.E.A.

Some 200 species, mainly distributed in the southern hemisphere, especially abundant in S. Africa.

1. Leaves opposite- - - - - - - - - - - - 4. *madagascariensis*
 - Leaves alternate, except rarely for a few at the base - - - - - 2
2. Corolla-tube ⅓–½ the length of the corolla, at least 1 mm. long - - - 3
 - Corolla split almost to the base, the tube usually considerably less than 1 mm. long - 9
3. Style with tiny glands at the base of the lobes - - - - - - 4
 - Style eglandular - - - - - - - - - - - 6
4. Leaves few and inconspicuous, usually less than 7 mm. long and 2 mm. wide; gynoecium 2-merous - - - - - - - - - - - - - 2. *virgata*
 - Leaves ± numerous and larger; gynoecium 3-merous - - - - - 5
5. Leaves rosulate at the base of the stem; testa smooth - - - - 3. *androsacea*
 - Leaves not rosulate; testa reticulate - - - - - - - 1. *undulata*
6. Tiny annual, ± procumbent herb; corolla less than 3 mm. long - 8. *campanuloides*
 - Perennial, usually ± erect herbs; corolla at least 5 mm. long - - - - 7
7. Corolla-tube less than ½ the length of the corolla and shorter than the calyx-lobes
  6. *denticulata*
 - Corolla-tube at least ½ the length of the corolla and longer than the calyx-lobes - 8
8. Testa reticulate; leaves up to 30 mm. long, margin not incrassate; corolla 5–13 mm. long
  5. *capillacea*
 - Testa almost smooth; leaves up to 12 mm. long, margin incrassate; corolla 5–7·5 mm. long
  7. *banksiana*
9. Perennials; leaves ± erect and scattered, often not overlapping, lanceolate-subulate, up to 7 mm. long - - - - - - - - - - - 9. *subaphylla*
 - Perennials or annuals, if perennial leaves not as above - - - - 10
10. Perennials; leaves ± spreading, numerous and densely set, overlapping, linear to narrowly lanceolate, up to 15 mm. long; flowers subsessile in ± contracted inflorescences
  10. *huttonii*
 - Perennials or annuals, if perennial leaves and/or inflorescence not as above - 11
11. Seeds compressed, ± 2-faced; gynoecium 3-merous; hypanthium ± 10-nerved - 12
 - Seeds trigonous; gynoecium 2–3-merous; hypanthium 5–10-nerved - - - 17
12. Perennials - - - - - - - - - - - - 13
 - Annuals - - - - - - - - - - - - - 14
13. Flowers sessile in a strongly contracted inflorescence with a dominant dense terminal head; style glabrous below - - - - - - - - - - 12. *capitata*
 - Flowers in a lax or contracted, often spike-like inflorescence, but without a dominant terminal head; style usually hairy below - - - - - 11. *napiformis*
14. Flowers sessile in a strongly contracted inflorescence with a dominant dense terminal head
  12. *capitata*
 - Flowers in ± lax inflorescences, rarely subsessile in axillary and terminal clusters - 15
15. Corolla (3)3·5–7 mm. long; inflorescence lax or with flowers in axillary and terminal ± loose head-like clusters; flowers subsessile or on pedicels up to 20 mm. long; hypanthium hairy or glabrous - - - - - - - - - - 13. *pulchella*
 - Corolla 1·5–3·5 mm. long; inflorescence lax with pedicels up to 20 (–35) mm. long; hypanthium hirsute - - - - - - - - - - - 16
16. Corolla blue to white; inflorescence usually ± pyramidal with the main axis distinct nearly to the top; stem and pedicels hirsute with mixed hairs of different length; leaves lanceolate to narrowly ovate - - - - - - - - - - 14. *erecta*
 - Corolla yellow; inflorescence spreading, with main axis not particularly distinct; stem and pedicels hirsute with long hairs of almost uniform length; leaves narrowly ovate to ovate
  15. *flexuosa*
17. Flowers in dense, spherical or elongated terminal heads; upper leaves large, ± forming an involucrum surrounding the inflorescence; hypanthium 5-nerved - - - 18
 - Flowers in lax inflorescences without involucrum; hypanthium 5–10-nerved - - 19
18. Style glabrous below; hypanthium glabrous; corolla 5–7 mm. long; leaves undulate-dentate; annuals or perennials - - - - - - - 17. *collomioides*
 - Style with long hairs below; hypanthium hirsute; corolla ± 4 mm. long; leaves almost entire; annuals - - - - - - - - - - 18. *cephalodina*
19. Perennials - - - - - - - - - - - - 20
 - Annuals - - - - - - - - - - - - - 21
20. Leaves linear, up to 1·2–3·5(5) cm. long, less than 1 mm. wide; hypanthium 5-nerved; gynoecium 2-merous - - - - - - - - - - 20. *upembensis*
 - Leaves not as above; hypanthium ± 10-nerved; gynoecium 3-merous - 16. *abyssinica*
21. Gynoecium 3-merous; hypanthium hirsute - - - - - - 19. *hirsuta*
 - Gynoecium 2-merous; hypanthium glabrous or hirsute - - - - 22
22. Slender ± straggling and trailing herbs, glabrous, leaves up to 1·2 cm. long, ± 1 mm. wide
  23. *paludicola*
 - More or less erect herbs, ± hirsute at least below, rarely all glabrous; leaves larger - 23

23. Leaves usually ± obtuse, mucronulate, up to 1·8–5·7 cm. long, 0·5–1·6 cm. wide, markedly
   undulate-crenate at the margin; hypanthium 5-nerved; style subcapitate at the apex,
   densely hairy below - - - - - - - - - 22. *perrottetii*
- Leaves acute or subacute, up to 0·7–3 cm. long, 0·1–0·7 cm. wide, usually not undulate-
   crenate at the margin; hypanthium 5–10-nerved; style distinctly 2-lobed to subcapitate at
   the apex, glabrous or hairy below - - - - - - - 21. *ramosissima*

1. **Wahlenbergia undulata** (L.f.) A.DC., Monogr. Camp.: 148 (1830). — Sonder in Harv. &
   Sond., F.C. **3**: 579 (1865). — von Brehmer in Engl., Bot. Jahrb. **53**: 122 (1915). — Thulin
   in Symb. Bot. Ups. **21**, 1: 76, fig. 6A & B, 9A, 11A & B, 14A, N, U, 15B (1975); in
   F.T.E.A., Campanulaceae: 8, fig. 2A, N, U (1976); in Fl. Madag. 187. Campanulaceae: 5,
   t. 1 fig. 1–6 (1978). Type from S. Africa (Cape Prov.).
      Campanula undulata L.f., Suppl. Pl.: 142 (1781). Type as above.
      Wahlenbergia bojeri A.DC. in DC., Prodr. **7**: 435 (1839). Type from Madagascar.
      Wahlenbergia caledonica Sonder in Harv. & Sond., loc. cit. — von Brehmer in Engl.,
   loc. cit. — Brenan in Mem. N.Y. Bot. Gard. **8**: 492 (1954). — Binns, H.C.L.M.: 25 (1969).
   Type from S. Africa (Orange Free State).
      Wahlenbergia oatesii Rolfe in Oates, Matabeleland and the Victoria Falls, ed. 2, app. 5:
   402 (1889). Type: Zimbabwe, Matabeleland, *Oates* 4/78 (K, holotype).
      Wahlenbergia cyanea Engl. & Gilg in Warb., Kunene-Samb.-Exped. Baum: 395 (1903).
   Type from Angola.
      Wahlenbergia caledonica var. *cyanea* (Engl. & Gilg) von Brehmer in Engl., tom. cit.: 105
   (1915). Type as above.
      Wahlenbergia engleri von Brehmer in Engl., loc. cit. Syntypes: Zimbabwe, Umtali,
   Engler 3167 (B†); Salisbury at Norton, *Engler* 3026 (B†).
      Wahlenbergia dinteri (incl. varieties) von Brehmer in Engl., tom. cit.: 106 (1915). —
   Markgraf in Engl., Bot. Jahrb. **75**: 215 (1950). — Roessler in Prodr. Fl. SW. Afr. **136**: 8
   (1966). Syntypes from S. Africa and Namibia.
      Wahlenbergia scoparia von Brehmer in Engl., tom. cit.: 108 (1915). Syntypes from S.
   Africa and Namibia.

Perennial or annual erect herb, 20–90 cm. tall. Stems few to many, ± hirsute
towards the base or glabrous. Leaves 10–70 × 1·5–10 mm., alternate, scattered or
somewhat crowded towards the base, sessile, lanceolate to linear, acute to subacute
with cuneate to truncate base, ± hirsute or glabrous; margin incrassate, ± undulate-
crenate; midvein protruding beneath, lateral veins faint. Inflorescence lax; pedicels
up to 10–25 mm. long, glabrous; bracts ± ciliate to dentate or glabrous and entire.
Hypanthium obconical to hemispherical, ± 10-nerved, glabrous. Calyx-lobes 1·5–6
mm. long, ± ciliate to dentate or glabrous and entire. Corolla (5) 8–16 mm. long,
campanulate, blue, white, yellow or various intermediate colours, deeply 5-lobed,
puberulous inside near the base, glabrous outside; corolla-tube 1·5–5 mm. long.
Stamens with filament-bases broadly dilated to almost cross-shaped, ciliate; anthers
± 2·5–5 mm. long. Ovary 3-locular, subinferior; style somewhat shorter than
corolla, hairy or glabrous below; lobes 3 with 3 glands present between their bases.
Capsule 3–12 mm. long, 3-locular, ± 10-nerved; valves 3, 1–3 mm. long. Seeds
0·35–0·55 mm. long, ± elliptic in outline, ± compressed; testa reticulate.

Caprivi Strip. 11 km. S. of Katima Mulilo on road to Ngoma, 900 m., 22.xii.1958, *Killick &
Leistner* 3023 (SRGH). **Botswana.** SE: Farm Springfield, 3 km. S. of Lobatsi, 17.i.1960, *Leach
& Noel* 138 (SRGH). **Zambia.** B: Mongu, 6.i.1966, *Robinson* 6775 (BR; EA; K; LISC; M;
MO; SRGH). S: Choma, 1300 m., 22.xii.1958, *Robinson* 2952 (BR; EA; K; LISC; M; PRE;
SRGH). **Zimbabwe.** N: Mazoe, Iron Mask Hill, 1500 m., vi.1915, *Eyles* 613 (BM; K; SRGH).
W: Shangani Distr., Gwampa Forest Reserve, ii.1955, *Goldsmith* 80/55 (K; PRE; SRGH). C:
Salisbury Distr., between Avondale West and Mabelreign, 16.x.1955, *Drummond* 4908 (K;
LISC; PRE; SRGH). E: Melsetter, near Bridal Veil Falls, 17.v.1962, *Noel* 2438 (EA; K; LISC;
M; MO; PRE; SRGH). S: Belingwe East Reserve, 7.vii.1953, *Wild* 4124 (K; LISC; MO; PRE;
SRGII). **Malawi.** N: Nyika Plateau, Nyamkowa, 2030 m., 23.ii.1978, *Pawek* 13855 (K). S:
Zomba Plateau, 1450 m., 5.vi.1946, *Brass* 16260 (BR; K; MO; PRE; SRGH). **Mozambique.**
MS: Chimanimani Mts., Musapa Gap, 6.x.1950, *Munch* 334 (K; LISC; SRGH). GI: Gaza,
Xai-Xai, 10.xii.1940, *Torre* 2308 (LISC). M: Inhaca I., to 200 m., xi.1962, *Mogg* 30117 (K;
PRE; SRGH).
Also in Tanzania, Angola, Namibia and S. Africa, and in Madagascar. In various types of
grassland, often in rocky or seasonally moist places, also on roadsides, cultivated or waste
ground.

W. undulata, over its wide geographical and habitat range, shows much variation particularly
in flower size and colour, shape of leaves and fruits, and in pubescence. Usually it is perennial
but apparently annual forms occur in the coastal regions of Mozambique. Various local forms

may sometimes be distinguished. For example *Miller* 2167 and 2168 from the Western Prov. of Zimbabwe represent two forms with blue and yellow flowers respectively and with different leaf shape, which according to *Miller* grow together without intergradation. However, when the total material is considered all of these forms merge ± imperceptibly into each other which makes almost impossible any clear subdivision of the species. The perhaps most distinctive form within the F.Z. area occurs in Barotseland and adjacent parts of the Southern Prov. of Zambia and the Caprivi Strip. This is entirely glabrous with long, linear leaves and seems to prefer wetter habitats than what is normal for this species. It comes close to the type of *W. cyanea* from southern Angola which however is pubescent at the base.

2. **Wahlenbergia virgata** Engl., Pflanzenw. Ost-Afr. C: 400 (1895). — von Brehmer in Engl., Bot. Jahrb. **53**: 121 (1915). — R. E. Fries, Wiss. Ergebn. Schwed. Rhod.-Kongo-Exped. **1**: 315 (1916). — Brenan in Mem. N.Y. Bot. Gard. **8**: 492 (1954). — F. W. Andr., Fl. Pl. Anglo-Egypt. Sudan **3**: 72 (1956). — Binns, H.C.L.M.: 25 (1969). — Agnew, Upland Kenya Wild Fl.: 512 (fig.), 513 (1974). — Thulin in Symb. Bot. Ups. **21**, 1: 81, fig. 2A, 6C & D, 12 O, 16 (1975); in F.T.E.A., Campanulaceae: 9, fig. 3 (1976); in F.A.C., Campanulaceae: 11, fig. 3 (1977). TAB. **15**. Type: Malawi, Mt. Mulanje, *Whyte* (B, syntype †; K, lectotype; G, Z, isolectotypes).

    Wahlenbergia sparticula Chiov. in Ann. Bot. Roma **10**: 389 (1912). Type from Ethiopia.
    Wahlenbergia recurvata von Brehmer in Engl., op. cit. **51**: 232 (1914); op. cit. **53**: 119 (1915). Type from Tanzania.
    Wahlenbergia virgata var. *longisepala* von Brehmer in Engl., tom. cit.: 122 (1915). Type from Tanzania.
    Wahlenbergia virgata var. *tenuis* von Brehmer, loc. cit. Type from Tanzania.
    Wahlenbergia virgata var. *valida* von Brehmer, loc. cit. Type: Malawi, *Buchanan* 911 (B, syntype †; K, lectotype; BM, isolectotype).

Perennial ± erect herb, up to 70 cm. tall, from a ± woody taproot. Stems few, glabrous or ± hirsute at least towards the base, furrowed, usually with many long ± erect branches. Leaves 2–7(22) × 0·5–2(6) mm., alternate, few, often scale-like, widely scattered on the stem, lanceolate, acute, glabrous or ± hirsute; margin incrassate, sparsely denticulate. Inflorescence lax; pedicels 1–5 cm. long, glabrous or glabrescent. Hypanthium usually narrowly obconical, 10-nerved, glabrous. Calyx-lobes 1·5–4 mm. long, entire or almost so, rarely ± ciliate. Corolla 8–10(13) mm. long, white, ± bluish or yellowish, lobed to c. ⅔ of the length, puberulous inside near the base, glabrous outside; tube 2–5 mm. long. Stamens with filament-bases markedly dilated to almost cross-shaped, ciliate; anthers 2·5–4 mm. long. Ovary 2-locular, subinferior; style shorter than the corolla, thickened in the upper part, glabrous below; lobes 2, c. 1·5–2·5 mm. long with 2 or 4 glands present between and, if 4, also below their bases. Capsule 4–10(16) mm., long, 2-locular, usually narrowly obconical; valves 0·8–1·6 mm. long. Seeds 0·5–0·9 mm. long, narrowly oblong to elliptic in outline; testa reticulate.

Zambia. N: Mbala, far end of L. Chila, 7.x.1954, *Richards* 1981 (BR; K). E: Nyika Plateau, by main road 4 km. SW. of Rest House, 2150 m., 22.x.1958, *Robson* 242 (K; LISC; SRGH). **Zimbabwe.** W: Matopos, iii.1906, *Flanagan* 2974 (K). C: Marandellas, 1500 m., 22.ix.1946, *Rattray* 726 (K; SRGH). E: Mt. Nuza, 1750 m., 23.vi.1934, *Gilliland* 456 (BM; K; SRGH). **Malawi.** N: Nyika Plateau, near Nganda Hill, 2300 m., 6.vii.1962, *Tyrer* 815 (BM; BR; SRGH). C: Dedza Mt., 22.x.1956, *Banda* 286 (BM; K; SRGH). S: Mt. Mulanje, W. slopes, 1500 m., 24.vi.1946, *Brass* 16409 (BM; BR; K; MO; PRE; SRGH). **Mozambique.** N: between Unango and Metónia, 1896, *Johnson* (K). Z: Namuli Peaks, W. face, 1350 m., 26.vii.1962, *Leach & Schelpe* 11468 (K; SRGH). MS: Chimanimani Mts., between Skeleton Pass and the Plateau, 1700 m., 27.ix.1966, *Grosvenor* 213 (K; LISC; MO; PRE; SRGH).
    Also in Sudan, Ethiopia, Uganda, Kenya, Tanzania, Burundi, S. Africa (Transvaal, Natal), Lesotho and Swaziland. In upland grassland, often in patches of open soil, e.g. eroded places or roadsides.

3. **Wahlenbergia androsacea** A.DC., Monogr. Camp.: 150, t. 19A (1830). — Sonder in Harv. & Sond., F.C. **3**: 582 (1865). — von Brehmer in Engl., Bot. Jahrb. **53**: 136 (1915). — Markgraf in Engl., Bot. Jahrb. **75**: 211 (1950). — Roessler in Prodr. Fl. SW. Afr. **136**: 7 (1966). — Thulin in Symb. Bot. Ups. **21**, 1: 87, fig. 2B, 5B, 10A & B, 12D, 17A (1975). Type from S. Africa (Cape Prov.).
    Wahlenbergia arenaria A.DC. in DC., Prodr. **7**: 436 (1839). — von Brehmer in Engl., tom. cit.: 137 (1915). Type from S. Africa (Cape Prov.).
    Wahlenbergia inhambanensis Klotzsch in Peters, Reise Mossamb. Bot. **1**: 303 (1862). — Hemsley in F.T.A. **3**: 480 (1877). — Engl., Pflanzenw. Ost-Afr. C: 400 (1895). — von Brehmer in Engl., tom. cit.: 108 (1915). Type: Mozambique, Inhambane, *Peters* (B, holotype†).

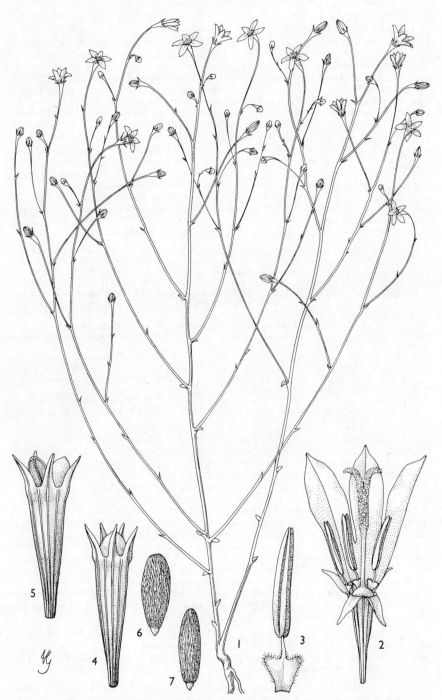

Tab. 15. WAHLENBERGIA VIRGATA. 1, habit ($\times\frac{2}{3}$); 2, flower with two stamens and petals removed ($\times 4$); 3, stamen ($\times 8$), 1–3, from *Thomas* 2143; 4, capsule ($\times 4$); 5, same, after dehiscence; 6, 7, seed, two views ($\times 24$), 4–7, from *Thulin* 326. From F.T.E.A.

Wahlenbergia gracilis sensu von Brehmer, tom. cit.: 112 (1915), pro parte quoad specim. *Bolus* 7839.

Annual erect herb, 10–55 cm. tall, from a white taproot. Stems few to many, glabrous or sometimes sparsely hirsute or puberulous towards the base. Leaves up to 25–130 × 6–18 mm., ± rosulate, flat, narrowly obovate to narrowly oblanceolate or spathulate, acute to subacute with attenuate base often narrowing into a distinct petiole, ciliate towards the base, ± hirsute on both sides or only above, or glabrous; margin not or very slightly incrassate, ± dentate, sometimes pinnately lobed at least towards the base; midvein prominent beneath, lateral veins faint. Inflorescence lax; pedicels up to 14–45 mm. long, glabrous; bracts and bracteoles usually ciliate at least towards the base. Hypanthium obovoid to hemispherical, c. 10-nerved, glabrous. Calyx-lobes 1–2·5 mm. long, narrowly triangular, entire, glabrous. Corolla 5–10 mm. long (forms with larger corollas, up to 22 mm. long, that probably belong here, occur in S. Africa and Namibia), pale blue or mauve, widely campanulate to almost rotate, deeply 5-lobed; tube 1–2 mm. long, puberulous inside near the base. Stamens with filament-bases broadly dilated to almost cross-shaped, shortly ciliate; anthers ± 2–2·5 mm. long. Ovary 3-locular, subinferior; style somewhat shorter than the corolla, hairy or glabrous below, 3-lobed, with 3 glands present below the pollen-collecting hairs. Capsule 3-locular, c. 10-nerved; valves 3, 0·8–1·2 mm. long. Seeds 0·4–0·6 mm. long, elliptic in outline, ± compressed; testa almost smooth.

Botswana. SE: on road from Francistown to Lobatsi, ix.1967, *Lambrecht* 305 (SRGH). **Zimbabwe.** N: Trelawney, 1200 m., 3.ix.1940, *Brain* 10985 (SRGH). C: Goromonzi Distr., Chindamora Native Res., 9.ix.1960, *Rutherford-Smith* 59 (K; LISC; PRE; SRGH). E: Inyanga Distr., Nyamaziwa R., 1650 m., 12.i.1951, *Chase* 3677 (BM; COI; K; LISC; SRGH). S: Victoria Distr., 1905, *Monro* 1206 (BM; Z). **Mozambique.** GI: Gaza, Xai-Xai, Inhamissa, 13.viii.1957, *Barbosa & Lemos* 7818 (COI; K; LISC). M: between Marracuene and Bobole, 14.ix.1961, *Lemos & Balsinhas* 192 (BM; COI; K; LISC).
Also in Namibia, S. Africa (Transvaal, Orange Free State, Natal and Cape Prov.), and Lesotho. On river banks, in grassland or sometimes ruderal, usually in sandy and damp places.

4. **Wahlenbergia madagascariensis** A.DC., Monogr. Camp.: 139 (1830). — Brenan in Mem. N.Y. Bot. Gard. **8**: 492 (1954). — Binns, H.C.L.M.: 25 (1969). — Thulin in Symb. Bot. Ups. **21**, 1: 106, fig. 4A, 7A & B (1975); in Fl. Madag. 187, Campanulaceae: 8, t. 1 fig. 7–11 (1978). Type from Madagascar.
Wahlenbergia hilsenbergii A.DC. in DC., Prodr. **7**: 429 (1839). Type from Madagascar.
Wahlenbergia oppositifolia A.DC., loc. cit. — von Brehmer in Engl., Bot. Jahrb. **53**: 134 (1915). Type from S. Africa (Cape Prov.).

Perennial ?short-lived herb with prostrate or decumbent stems, 10–50 cm. long, often rooting at the nodes, dichotomously branching, ± hirsute or rarely glabrescent. Leaves 8–30(50) × 2–8(17) mm., opposite, sessile to subpetiolate, elliptic to narrowly ovate, acute or subacute, slightly mucronate at the apex, cuneate at the base, ± hirsute to glabrescent; margin incrassate, ± undulate-crenate; midvein prominent beneath, lateral veins usually obscure. Inflorescence leafy with flowers in dichotomous forks or terminal, but usually becoming overtopped and then appearing lateral; pedicels 0·8–35 mm. long, usually ± hirsute. Hypanthium hemispherical, 10-nerved, hirsute or rarely glabrous. Calyx-lobes 1·6–3 mm. long, narrowly triangular, ± hirsute, denticulate, usually recurved in fruit. Corolla 2·4–6·4 mm. long, white, usually glabrous; tube 0·8–2·6 mm. long. Stamens with filament-bases slightly broadened, ciliate; anthers 0·8–1·8 mm. long. Ovary subinferior, 3-locular; style shorter than the corolla, eglandular, slightly thickened in the upper part, usually glabrous below; lobes 3, up to 1 mm. long. Capsule 3-locular with 3 thin and low valves. Seeds 0·5–0·6 mm. long, elliptic in outline, bluntly trigonous; testa reticulate, pale brown to almost black.

Zimbabwe. E: Umtali Distr., Banti Forest, 1830 m., 4.ii.1955, *E.M. & W.* 197 (BM; LISC; SRGH). S: Bikita Distr., S. base of Mt. Horsi, 1100 m., 9.v.1969, *Biegel* 3110A (K; PRE; SRGH). **Malawi.** S: Zomba Mt., Mulunguzi R., 1670 m., 9.iii.1955, *E.M. & W.* 766 (BM; LISC; SRGH). **Mozambique.** Z: Gúruè (Vila Junqueiro), 1600 m., 19.ix.1944, *Mendonça* 2116 (LISC).
Also in S. Africa (Transvaal, Natal, Cape Prov.), Swaziland and in Madagascar. In upland forest, forest edges or clearings, usually in shady or moist conditions.

5. **Wahlenbergia capillacea** (L.f.) A.DC., Monogr. Camp.: 156 (1830). — von Brehmer in Engl., Bot. Jahrb. **53**: 94 (1915). — Phillips in Ann. S. Afr. Mus. **16**, 1: 180 (1917). —

Thulin in Symb. Bot. Ups. **21**, 1: 111 (1975); in F.T.E.A., Campanulaceae: 15 (1976); in F.A.C., Campanulaceae: 16 (1977). Type from S. Africa (Cape Prov.).
  Campanula capillacea L.f., Suppl. Pl.: 139 (1781). Type as above.

Perennial erect herb, up to 50 cm. tall, from a thick taproot; stems several, glabrous or ± puberulous; old stems with persistent leaf-bases. Leaves 5–30 × 0·2–1·2(3) mm., alternate, sessile, usually many and densely set, often fascicled (in southern Africa), linear or rarely narrowly oblanceolate, margin not incrassate, flat or ± involute, entire or lower leaves sparsely dentate (in southern Africa rarely narrowly lobed or incised). Inflorescence lax; pedicels 3–25 mm. long, glabrous or puberulous, or rarely flowers subsessile. Hypanthium 10-nerved, glabrous or rarely ± puberulous. Calyx-lobes (1·2)1·5–4(6) mm. long, entire, glabrous or ± puberulous. Corolla 5–13 mm. long, blue, lobed to ⅓ or ¼ of the length; tube 3·5–9 mm. long, puberulous inside near the base, sometimes the inside ± densely set with long hairs (in southern Africa). Stamens with filament-bases narrowly dilated, ciliate; anthers 1·5–3·5 mm. long. Ovary 3-locular, semi-inferior; style shorter than the corolla, eglandular, not markedly thickened in the upper part, glabrous or hairy below; lobes 3, c. 0·8 mm. long. Capsule 3-locular, obconical or rarely hemispherical, 10-nerved, valves 2–3 mm. long. Seeds 0·4–0·9 mm. long, narrowly oblong to broadly elliptic in outline, ± wrinkled; testa ± distinctly reticulate.

Subsp. **tenuior** (Engl.) Thulin, tom. cit.: 113, fig. 1B, 7K & L, 9K, 12F, 14D & S, 19B (1975); op. cit.: 15, fig. 2D & S (1976); op. cit.: 16, fig. 2D & S (1977). Type from Tanzania.
  Wahlenbergia kilimandscharica Engl. in Abh. Königl. Preuss. Akad. Wiss. Berl. **1891**: 412 (1892). — von Brehmer in Engl., tom. cit.: 78, fig. 2F–J (1915). — Agnew, Upland Kenya Wild Fl.: 512 (fig.), 513 (1974). Syntypes from Tanzania.
  Wahlenbergia oliveri Schweinf. apud Engl. in Abh. Königl. Preuss. Akad. Wiss. Berl. **1891**: 412 (1892) *nom. nud.* Type from Tanzania.
  Wahlenbergia capillacea var. *tenuior* Engl., Bot. Jahrb. **30**: 418 (1901). — von Brehmer in Engl., tom. cit.: 94, fig. 2A–E (1915). Type as for *Wahlenbergia capillacea* subsp. *tenuior*.
  Wahlenbergia kilimandscharica var. *intermedia* von Brehmer, loc. cit. Type from Tanzania.
  Wahlenbergia aberdarica T. C. E. Fries in Notizbl. Bot. Gart. Berl. **8**: 395, fig. 5 (1923). Type from Kenya.

Leaves not fascicled. Inside of corolla without long hairs.

Zimbabwe. E: Chimanimani Mts., where Bundi leaves upper Bundi plain, 1600 m., 1.ii.1957, *Phipps* 284 (K; PRE; SRGH). **Malawi.** C: Lisasadzi, near Kasungu, 9.iv.1955, *Jackson* 1623 (EA, pro parte). S: Mulanje Mt., Chambe Plateau, 23.iii.1958, *Jackson* 2194 (K; SRGH). **Mozambique.** MS: Chimanimani Mts., Musapa Gap, 17.xi.1968, *Mavi* 691 (K; LISC; MO; SRGH).
Also in Kenya, Tanzania and Burundi. In upland grassland, often in rocky places.

Subsp. *capillacea* occurs in S. Africa (Natal, Cape Prov.), Swaziland and Lesotho.

6. **Wahlenbergia denticulata** (Burch.) A.DC., Monogr. Camp.: 152, t. 16 (1830). — Thulin in Symb. Bot. Ups. **21**, 1: 116, fig. 10G, 14H & T (1975); in F.T.E.A., Campanulaceae: 15, fig. 2H & T (1976); in F.A.C., Campanulaceae: 17, fig. 2H & T (1977). Type from S. Africa (Cape Prov.).
  Campanula denticulata Burch., Trav. Int. S. Afr. **1**: 538 (1822). Type as above.
  Lightfootia denticulata (Burch.) Sonder in Harv. & Sond., F. C. **3**: 559 (1865). — Phillips in Ann. S. Afr. Mus. **16**, 1: 178 (1917). — Markgraf in Engl., Bot. Jahrb. **75**: 209 (1950). — Adamson in Journ. S. Afr. Bot. **21**: 170 (1955). — Roessler in Prodr. Fl. SW. Afr. **136**: 2 (1966). — Agnew, Upland Kenya Wild Fl.: 511 (1974). Type as above.
  Lightfootia tenuifolia A.DC. in Ann. Sci. Nat., Bot. Sér. 5, **6**: 327 (1866). — Hemsley in F.T.A. **3**: 475 (1877). — Hiern, Cat. Afr. Pl. Welw. **1**: 629 (1898). Type from Angola.
  Wahlenbergia spinulosa Engl., Bot. Jahrb. **10**: 271 (1888) *nom. illegit.* Type from Namibia.
  Lightfootia goetzeana Engl., op. cit. **30**: 419 (1901). Type from Tanzania.
  Lightfootia laricifolia Engl. & Gilg in Warb., Kunene-Samb.-Exped. Baum: 397 (1903). Type from Angola.
  Lightfootia denticulata var. *podanthoides* Markgraf in Notizbl. Bot. Gart. Berl. **15**: 466 (1941); in Engl., loc. cit. Type from Namibia.

Perennial, rarely annual, often shrubby herb. Stems 0·1–1 m. long, usually many, ± decumbent and ascending, rarely erect, ± hirsute, puberulous or glabrescent; hairs mostly retrorse; old stems with persistent leaf-bases. Leaves 3–20(35) × 0·3–2·5 mm., alternate, sessile, ± spreading, linear to linear-lanceolate, acute to subacute, glabrous or pubescent; margin ± incrassate, usually markedly denticulate. Inflorescence lax or variously contracted; pedicels up to 20 mm. long, glabrous or shortly pubescent, or rarely flowers subsessile. Hypanthium often broadly obconical, 10-nerved, glabrous or pubescent. Calyx-lobes 1·5–5 mm. long, ± recurved and denticulate, glabrous or pubescent. Corolla 5–8(10) mm. long, blue to white or yellow, deeply split into linear-lanceolate lobes, puberulous inside near the base, glabrous or shortly pubescent outside; tube 1–3·2 mm. long. Stamens with filament-bases ± rhombic, ciliate; anthers 1·5–2·8 mm. long. Ovary subinferior to semi-superior, 3-locular; style usually slightly longer than the corolla, thickened in the upper part, eglandular, hairy below; lobes 3, c. 0·6–1 mm. long. Capsule 3-locular, 10-nerved. Seeds 0·4–0·6 mm. long, elliptic in outline, ± compressed or bluntly trigonous, usually very narrowly winged; testa almost smooth.

Zambia. N: 70 km. S. of Mbala (Abercorn), 1500 m., 18.vii.1930, *Hutchinson & Gillett* 3837 (BM; K). C: Chakwenga Headwaters, 100–129 km. E. of Lusaka, 25.viii.1963, *Robinson* 5619 (EA; K; SRGH). E: Jewa to Chipata, 28.viii.1929, *Burtt Davy* 20995 (K). S: Siamambo, 17.v.1961, *Fanshawe* 6574 (BR; K; LISC; SRGH). **Zimbabwe.** N: near Gokwe, 24.iv.1962, *Bingham* 240 (K; PRE; SRGH). W: near World's View, 14.iv.1955, *E.M. & W.* 1516 (BM; LISC; SRGH). C: 10 km. SE. of Gwelo, 1400 m., 25.v.1966, *Biegel* 1195 (K; SRGH). E: Inyanga Distr., near Southlands, Cumberland Block, 1700 m., 24.v.1954, *Chase* 5236 (BM; COI; K; LISC; SRGH). S: Belingwe Distr., 32 km. N. of West Nicholson, 17.iii.1964, *Wild* 6408 (BR; K; LISC; SRGH). **Malawi.** C: Lisasadzi near Kasungu, 9.iv.1955, *Jackson* 1623 (BR; EA, pro parte (see also sp. no. 5); K; LISC; PRE; SRGH).

Also in Kenya, Tanzania, Zaire, Angola, Namibia, S.Africa (Transvaal, Natal, Orange Free State, Cape Prov.) and Lesotho. Grassland or woodland, roadsides, old cultivations, usually in sandy or rocky places.

7. **Wahlenbergia banksiana** A.DC., Monogr. Camp.: 154 (1830). — von Brehmer in Engl., Bot. Jahrb. **53**: 83 (1915). — Thulin in Symb. Bot. Ups. **21**, 1: 119, fig. 8A & B, 20B (1975). Type from S. Africa.
 Wahlenbergia leucantha Engl. & Gilg in Warb., Kunene-Samb.-Exped. Baum: 396 (1903). Type from Angola.
 Wahlenbergia mashonica N.E. Br. in Kew Bull. **1906**: 165 (1906). — von Brehmer in Engl., tom. cit.: 78 (1915). — Type: Zimbabwe, between Salisbury and Headlands, *Cecil* 157 (K, holotype).
 Wahlenbergia okavangensis N.E. Br., op. cit. **1909**: 118 (1909). — von Brehmer in Engl., loc. cit. Type: Botswana, Ngamiland, Okavango Valley, *Lugard* 258 (K, holotype).
 Wahlenbergia saginoides S. Moore in Journ. Bot. Lond. **49**: 153 (1911). Type: Zimbabwe, Victoria, *Monro* 649 (BM, holotype; Z, isotype).
 Wahlenbergia rhodesiana S. Moore in Journ. Linn. Soc., Bot. **40**: 125 (1911). Type: Zimbabwe, northern Melsetter, *Swynnerton* 6225 (BM, holotype).
 Wahlenbergia multiflora Conrath in Kew Bull. **1914**: 134 (1914). Type from S. Africa (Transvaal).
 Wahlenbergia foliosa von Brehmer in Engl., tom. cit.: 84 (1915) *nom. illegit.* — Markgraf in Engl., Bot. Jahrb. **75**: 215 (1950). — Roessler in Prodr. Fl. SW. Afr. **136**: 8 (1966). Type from Namibia.
 Wahlenbergia gracillima S. Moore, op. cit. **56**: 9 (1918). Type from S. Africa (Transvaal).

Perennial, ± erect herb, 15–50 cm. tall. Stems many from a taproot, usually with persistent leaf-bases, ± hirsute at the base with usually retrorse hairs. Leaves up to 12 × 1·5 mm., alternate, sessile, erecto-patent, linear, acute with truncate base, ± hirsute or glabrous; margin incrassate, denticulate. Inflorescence lax; pedicels up to 10 mm. long, shortly pubescent to glabrous. Hypanthium obconical, 10-nerved, shortly pubescent to glabrous. Calyx-lobes 1·5–2·5 mm. long, ± straight, glabrous, sparsely denticulate, often involute. Corolla 5–7·5 mm. long, lobed to less than ½ the length, blue to white, puberulous inside near the base; tube 2·5–4 mm. long. Stamens with filament-bases ± rhombic, ciliate; anthers 1·5–2 mm. long. Ovary 3-locular, semi-inferior; style somewhat shorter than the corolla, eglandular, thickened in the upper 1·5–2 mm., hairy below; lobes 3, c. 0·6 mm. long. Capsule 3-locular, 10-nerved; valves 3, c. 1–1·5 mm. long. Seeds 0·3–0·4 mm. long, elliptic in outline, ± trigonous; testa almost smooth.

Caprivi Strip. Andara, shore and islands of Okavango R., *Merxmüller* 1981 (BR; M). **Botswana.** N: Ngamiland, Okavango Valley, 900 m., vi.1898, *Lugard* 258 (K). **Zambia.** B:

Mongu, 20.i.1966, *Robinson* 6819 (EA; K; M; MO; SRGH). C: Chakwenga Headwaters, 100–129 km. E. of Lusaka, 1.xii.1963, *Robinson* 5876 (K; M; SRGH). **Zimbabwe.** W: Matobo Distr., Farm Besna Kobila, 1450 m., xi.1956, *Miller* 3914 (K; PRE; SRGH). C: Salisbury Distr., Cleveland Dam, 1500 m., xii.1918, *Eyles* 1382 (BM; PRE; SRGH). E: Inyanga Distr., Pungwe R., 1800 m., 23.x.1946, *Wild* 1481 (K; SRGH). S: Fort Victoria Distr., Makaholi Experimental Station, 14.xii.1947, *Newton & Juliasi* 54 (SRGH). **Mozambique.** GI: Gaza, Bilene, 28.i.1982, *Pettersson* 2050 (UPS).

Also in Angola, Namibia and S. Africa (Transvaal, Natal). In sandy grassland, riversides, often in seasonally flooded areas.

The application of the name *W. banksiana* for this plant is not entirely certain. The type specimen, *Oldenburg* 813, is very poor and without locality. As far as S. Africa is concerned I have only seen specimens from Natal and the Transvaal, provinces which *Oldenburg* probably never visited. Nevertheless the type of *W. banksiana* agrees so well with the present plant that use of the name seems inevitable.

8. **Wahlenbergia campanuloides** (Delile) Vatke in Linnaea **38**: 700 (1874). — Hepper in F.W.T.A. ed. 2, **2**: 309 (1963). — Thulin in Symb. Bot. Ups. **21**, 1: 123 (1975). Type from Egypt.
  Cervicina campanuloides Delile, Fl. d'Egypte: 6, t. 5/2 (1813). Type as above.
  Wahlenbergia cervicina A.DC., Monogr. Campan.: 156 (1830) *nom. illegit.* — Hemsley in F.T.A. **3**: 479 (1877). — von Brehmer in Engl., Bot. Jahrb. **53**: 130 (1915). Type as above.
  Wahlenbergia humifusa Markgraf in Notizbl. Bot. Gart. Berl. **15**: 760 (1942); in Engl., Bot. Jahrb. **75**: 213 (1950). — Roessler in Prodr. Fl. SW. Afr. **136**: 8 (1966). Type from Namibia.

Annual herb, shortly hirsute; hairs often somewhat retrorse. Stems up to 3–13 cm. long, usually many, decumbent, ± dichotomously branched, from a taproot. Leaves up to 3·2–9 × 1·2–1·6 mm., alternate, linear to very narrowly elliptic or oblanceolate, acute, shortly hirsute mainly beneath; margin incrassate, sparsely denticulate; midvein protruding beneath. Inflorescence lax, leafy, ± monochasial; pedicels up to 3–7 mm. long, shortly hirsute. Hypanthium (4)8-nerved, shortly hirsute. Calyx-lobes (3)4(5), 1·0–3·6 mm. long usually ± unequal, one often rudimentary, ± oblanceolate, denticulate, shortly hirsute mainly inside. Corolla 2–2·6 mm. long, pale blue, shortly (3)4-lobed, shortly hirsute or glabrescent outside; tube 1·6–2 mm. long. Stamens (3)4; filament-bases linear, glabrous or ciliate; anthers 0·7–0·8 mm. long. Ovary subinferior, 2(–3)-locular; style shorter than the corolla, not thickened in the upper part, eglandular, glabrous below; lobes 2(3), c. 0·6 mm. long. Capsule almost spherical, 2(3)-locular, (4)8-nerved; valves short. Seeds 0·3–0·4 mm. long, elliptic in outline, ± irregularly angular, sometimes narrowly winged, almost smooth.

Zambia. B: Sesheke Distr., *Gairdner* 64 (K).
Also in Namibia, S. Africa (Cape Prov.) and scattered in northern Africa from Senegal to Egypt. In seasonally moist, sandy places.

9. **Wahlenbergia subaphylla** (Baker) Thulin in Symb. Bot. Ups. **21**, 1: 124 (1975); in F.T.E.A., Campanulaceae: 16 (1976); in F.A.C., Campanulaceae: 18 (1977); in Fl. Madag. 187, Campanulaceae: 10, t. 3 fig. 1–6 (1978). Type from Madagascar.
  Lightfootia subaphylla Baker in Journ. Linn. Soc., Bot. **20**: 193 (1883). Type as above.

Perennial erect herb, 20–50 cm. tall, from a pale rootstock. Stems few to many, usually sparsely branched, striate, glabrous or puberulous (usually at the base only), grey-green. Leaves 1·5–7(10) × 0·5–1 mm., alternate, sessile, ± erect, lanceolate-subulate; margin ± incrassate, entire or with a few teeth. Inflorescence ± lax or strongly contracted; pedicels up to 5(8) mm. long, glabrous or puberulous. Hypanthium obconical, ± 10-nerved, glabrous or puberulous. Calyx-lobes 1·2–2·8 mm. long, narrowly triangular, acute, glabrous, margin sometimes with a few reflexed teeth. Corolla 5·5–9 mm. long, white or bluish, deeply split into linear lobes, puberulous inside near the base, glabrous outside; tube 0·1–0·2 mm. long. Stamens with filament-bases ± angularly obovate, ciliate-pubescent; anthers 1·6–3·2 mm. long. Ovary 3-locular, semi-inferior; style c. as long as the corolla or shorter, eglandular, slightly thickened in the upper part, glabrous below; lobes 3, 1–1·2 mm. long. Capsule 3-locular; valves 3, 1·5–2·5 mm. long. Seeds 0·6–0·8 mm. long, elliptic in outline, obscurely trigonous, sometimes slightly winged; testa almost smooth.

Inflorescence ± lax and many-flowered; leaves flat with distinctly incrassate margins
                                                             subsp. *thesioides*
Inflorescence contracted with usually less than 10 flowers; leaves with slightly involute,
    scarcely incrassate margins   -   -   -   -   -   -   -   - subsp. *scoparia*

Subsp. **thesioides** Thulin, tom. cit.: 126, fig. 8E & F, 10E & F, 14E, 21 (1975); op. cit.: 17, fig.
    2E (1976); op. cit.: 18, fig. 2E (1977). Type from Tanzania.

Stem glabrous or puberulous only at the base, rarely the whole plant puberulous.
Leaves many, up to 7(10) mm. long, often overlapping, flat; margin incrassate.
Inflorescence ± lax, usually much branched and many-flowered. Seeds rather
bluntly and irregularly trigonous.

Malawi. N: Nyika Plateau, bridge below Chelinda Camp, 2200 m., 16.xi.1958, *Robson* 646
(K; SRGH).
Also in Tanzania and Zaire. In upland grassland, often appearing after burning.

Subsp. **scoparia** (Wild) Thulin, tom. cit.: 127 (1975). Type: Zimbabwe, Chimanimani Mts.,
    Mt. Peza, 15.x.1950, *Wild* 3608 (SRGH, holotype; K, PRE, isotypes).
    Lightfootia scoparia Wild in Kirkia **4**: 161 (1964). Type as above.

Stem glabrous or plant puberulous at least in the upper part. Leaves up to 6 mm.
long, sometimes overlapping, slightly involute; margin scarcely incrassate.
Inflorescence contracted with usually less than 10 flowers. Seeds irregularly
trigonous.

Zimbabwe. E: Umtali Distr., Vumba Mts., 30.x.1950, *Chase* 3104 (BM; K; LISC; SRGH).
Mozambique. MS: Chimanimani Mts., 7.vi.1949, *Munch* 182 (SRGH).
In upland grassland, often in rocky places.

10. **Wahlenbergia huttonii** (Sonder) Thulin in Symb. Bot. Ups. **21**, 1: 129, fig. 1C, 22B
    (1975); in F.T.E.A., Campanulaceae: 17 (1976). Type from S. Africa (Cape Prov.).
    Lightfootia huttonii Sonder in Harv. & Sond., F.C. **3**: 556 (1865). — Adamson in Journ.
    S. Afr. Bot. **21**: 207 (1955). Type as above.
    Lightfootia lycopodioides Mildbr. in Notizbl. Bot. Gart. Berl. **11**: 686 (1932) *nom.
    illegit.* Type from Tanzania.

Perennial erect herb or subshrub, 10–35(50) cm. tall, from a woody rootstock.
Stems several to many, puberulous, hirsute or glabrous, often unbranched; old stems
with persistent leaf-bases. Leaves 5–15 × 0·5–3 mm. (in S. Africa up to 25 × 8 mm.),
alternate, numerous and densely set, sessile, ± erect or spreading, linear to narrowly
lanceolate, acute, glabrous, ciliate or hirsute beneath; margin not or slightly
incrassate, entire. Flowers rather few in ± contracted often capitate inflorescences or
solitary, subsessile. Hypanthium obconical, obscurely 10-nerved, glabrous. Calyx-
lobes 2·5–4 mm. long, glabrous or ciliate. Corolla 6–10 mm. long, blue or mauve,
deeply split into linear lobes, puberulous inside near the base; tube less than 1 mm.
long. Stamens with filament-bases narrowly angularly obovate, ciliate; anthers 2–2·5
mm. long. Ovary 3-locular, semi-inferior; style c. as long as the corolla or somewhat
shorter, eglandular, glabrous below; lobes 3, c. 0·6–0·8 mm. long. Capsule 3-locular,
± 10-nerved; valves 3, c. 2 mm. long. Seeds c. 0·6 mm. long, elliptic in outline, ±
compressed; testa almost smooth.

Malawi. N: Rumphi Distr., Nyika south towards Mbzuinandi, 2300 m., 29.iii.1970, *Pawek*
3411 (K).
Also in Tanzania, S. Africa (Transvaal, Natal, Orange Free State, Cape Prov.) and
Swaziland. In upland grassland, often in rocky places.

11. **Wahlenbergia napiformis** (A.DC.) Thulin in Symb. Bot. Ups. **21**, 1: 133, fig. 5J, 8G &
    H, 10H & I, 12J, 14R, 22C (1975); in F.T.E.A., Campanulaceae: 18, fig. 2R (1976); in
    F.A.C., Campanulaceae: 19, fig. 2R (1977). Type from Angola.
    Lightfootia napiformis A.DC. in Ann. Sci. Nat., Bot. Sér. 5, **6**: 328 (1866). — Hemsley
    in F.T.A. **3**: 475 (1877). — Hiern, Cat. Afr. Pl. Welw. **1**: 629 (1898). Type as above.
    Lightfootia marginata A.DC., tom. cit.: 326 (1866). — Hemsley, tom. cit.: 474 (1877).
    — Hiern, loc. cit. — Engl. & Gilg in Warb., Kunene-Samb.-Exped. Baum: 396 (1903). —
    Lambinon & Duvigneaud in Bull. Soc. Roy. Bot. Belg. **93**: 45 (1961) pro parte. Type from
    Angola.
    Lightfootia glomerata Engl., Bot. Jahrb. **19**: Beibl. 47: 52 (1894); Pflanzenw. Ost-Afr. **C**:
    400 (1895). — Agnew, Upland Kenya Wild Fl.: 511 (1974). Type from Tanzania.

Lightfootia abyssinica Hochst. ex A. Rich. var. *glaberrima* Engl., loc. cit. Type from Sudan.
Lightfootia abyssinica var. *cinerea* Engl. & Gilg, op. cit.: 397 (1903). Type from Angola.
Lightfootia kagerensis S. Moore in Journ. Linn. Soc., Bot. **37**: 176 (1905). Type from Uganda.
Lightfootia campestris Engl. in Mildbr., Wiss. Ergebn. Zweit. Deutsch. Zentr.-Afr. Exped., **2**: 343 (1911). Type from Rwanda.
Lightfootia graminicola Scott in Kew Bull. **1915**: 45 (1915). Type from Angola.
Lightfootia abyssinica sensu R. E. Fries, Wiss. Ergebn. Schwed. Rhod.-Kongo-Exped. **1**: 315 (1916). — F. W. Andr., Fl. Pl. Anglo-Egypt. Sudan **3**: 71 (1956) non Hochst. ex A. Rich. (1850).
Lightfootia marginata var. *lucens* Lambinon in Bull. Soc. Roy. Bot. Belg. **93**: 46 (1961). Type: Zimbabwe, vicinity of Umvukwe Mts., near Darwendale, *Rodin* 4331 (K, holotype; S, SRGH, isotypes).

Perennial or ? biennial herb from a taproot. Stems up to 0·2–1 m. tall, usually few, erect, rarely ± decumbent or straggling, glabrous or hirsute or puberulous. Leaves up to 10–80 × 0·8–15 mm., alternate, linear to lanceolate or very narrowly elliptic, acute, glabrous or ± pubescent; margin incrassate, slightly revolute, ± undulate-dentate; midvein prominent beneath, lateral veins obscure or visible. Inflorescence leafy, ± spike-like, sometimes very dense or sometimes with flowers ± loosely clustered in the upper leaf axils; flowers subsessile or on ± hirsute or glabrous up to 5(23) mm. long pedicels. Hypanthium 10-nerved, ± hirsute or glabrous. Calyx-lobes 1·2–3·5 mm. long, ± hirsute or glabrous, usually with ± incrassate and sometimes shallowly denticulate margins. Corolla 3·5–7·5 mm. long, blue, purplish, yellowish or whitish, split almost to the base into linear lobes, puberulous inside near the base, ± hirsute or glabrous outside. Stamens with filament-bases ± broadly dilated, ciliate; anthers 1·6–2·8 mm. long. Ovary 3-locular, subinferior to semi-superior; style c. as long as the corolla, eglandular, hairy or rarely glabrous below; lobes 3, 0·4–0·8 mm. long. Capsule 3-locular, 2·8–5·5 mm. long, 10-nerved; valves 3, 1·2–3 mm. long. Seeds 0·4–0·7 mm. long, elliptic-oblong in outline, ± compressed and 2-faced, almost smooth.

Caprivi Strip. c. 80 km. from Singalamwe on road to Katima Mulilo, 900 m., 3.i.1959, *Killick & Leistner* 3290 (K; M; SRGH). **Botswana.** N: near Gwetshaa I., 22.iv.1973, *Smith* 534 (K; M; MO; PRE; SRGH). **Zambia.** B: 85 km. from Zambesi on Kabompo road, 26.iii.1961, *Drummond & Rutherford-Smith* 7374 (K; LISC; PRE; SRGH). N: Mbala Distr., road to Ningi Pans, 1500 m., 25.viii.1960, *Richards* 13163 (K; MO; SRGH). W: Luano, 20.v.1967, *Mutimushi* 1970 (K; SRGH). C: 10 km. E. of Chisamba, 19.v.1957, *Best* 128 (BR; K; SRGH). E: upper Loangwa R., Loangwene, iii.1897, *Nicholson* (K). S: Namwala Distr., Ngoma, 20.iii.1963, Mitchell 19/19 (BR; LISC; M; PRE; SRGH). **Zimbabwe.** N: Miami, 7.iii.1947, *Wild* 1790 (K; SRGH). W: Matobo Distr., Farm Besna Kobila, 1450 m., ii.1954, *Miller* 2204 (K; PRE; SRGH). C: Salisbury, 8.vi.1956, *Lennon* 22/56 (K; SRGH). E: Umtali Distr., Fern Valley, 1050 m., 21.ii.1954, *Chase* 5196 (BM; K; MO; PRE; SRGH). S: Victoria, 1909, *Monro* 940 (BM; Z). **Malawi.** N: Mzuzu, 5 km. SW. at Katoto, 1350 m., 4.iv.1973, *Pawek* 6520 (K; PRE). C: Lilongwe, Agricultural Research Station, 16.iv.1956, *Banda* 249 (BM; BR; LISC). S: L. Chilwa, near Fisheries Research Unit, 1.vii.1969, *Banda* 1117 (SRGH). **Mozambique.** N: Mandimba (Belém), 700 m., 24.ii.1964, *Torre & Paiva* 10744 (LISC). T: 14·2 km. from Furancungo to Vila Gamito, 10.vii.1949, *Barbosa & Carvalho* 3542 (K). MS: Rotanda, 13.iv.1948, *Barbosa* 1452 (LISC).
Also in Central African Republic, Sudan, Ethiopia, Uganda, Kenya, Tanzania, Republic of Congo, Zaire, Rwanda, Burundi, Angola and Namibia. In deciduous woodland or grassland, old cultivations, roadsides, usually on sandy or rocky soils.

W. napiformis is fairly uniform in Malawi, Zimbabwe, Mozambique, Botswana and in most of Zambia, with ± spike-like inflorescences and hairy styles. It is usually hirsute, but almost or entirely glabrous forms also occur (*Lightfootia marginata* var. *lucens*). Plants with the inflorescence ± branched and with at least some of the flowers distinctly pedicelled occur in many areas, particularly towards the west (*Lightfootia abyssinica* var. *cinerea*).
In N. and NW. Zambia there is a form with a ± lax and glabrous inflorescence, often glabrous style and with a habit approaching *W. abyssinica* with which it has usually been associated. Similar forms predominate in other areas, e.g. in Zaire, W. Tanzania, Uganda and W. Kenya. They intergrade with typical *W. napiformis* particularly in Tanzania, but also in Zambia. *W. napiformis* is best distinguished from *W. abyssinica* by having compressed not trigonous seeds.

12. **Wahlenbergia capitata** (Baker) Thulin in Symb. Bot. Ups. **21**, 1: 139, fig. 12G, 23 (1975); in F.T.E.A., Campanulaceae: 19, fig. 4 (1976); in F.A.C., Campanulaceae: 21, fig.

5 (1977). TAB. **16**. Type: Malawi, Nyika Plateau, *Whyte* (K, holotype; G, P, isotypes).
    Lightfootia capitata Baker in Kew Bull. **1898**: 158 (1898). Type as above.
    Lightfootia glomerata Engl. var. *subspicata* Engl., Pflanzenw. Ost.-Afr. **C**: 400 (1895).
Type: Malawi, *Buchanan* 40 (B, holotype †; K, lectotype; BM, G, isotypes).
    Lightfootia bequaertii De Wild. & Ledoux in Contr. Fl. Katanga, Suppl. **3**: 145 (1930).
Type from Zaire (Shaba).
    Lightfootia glomerata sensu Brenan in Mem. N.Y. Bot. Gard. **8**: 491 (1954). —
Lambinon & Duvigneaud in Bull. Soc. Roy. Bot. Belg. **93**: 47 (1961). — Binns,
H.C.L.M.: 25 (1969) non Engl. (1894).
    Lightfootia glomerata var. *capitata* (Baker) Lambinon in Bull. Soc. Roy. Bot. Belg. **93**:
47 (1961) pro parte excl. syn. *Lightfootia polycephala* Mildbr. Type as for *Wahlenbergia
capitata*.
    Lightfootia leptophylla sensu Binns, loc. cit. non C. H. Wright.
    Lightfootia elegans Gilli in Ann. Nat. Mus. Wien **77**: 56 (1973). Type from Tanzania.

Annual or usually perennial (although probably often short-lived) ± erect herb,
up to 1 m. tall, from a taproot. Stem with long branches from near the base or
unbranched, ± ribbed, hirsute at least below. Leaves up to 15–60(80) × 1·5–10(15)
mm., alternate, sessile, linear or narrowly lanceolate to lanceolate or elliptic, acute
with truncate to cuneate base, ± hirsute especially on veins beneath, or glabrous;
margin incrassate, dentate, often ± undulate; midvein prominent beneath, lateral
veins less so, but usually clearly visible. Inflorescence leafy, capitate, sometimes only
a terminal head present, but usually also a number of smaller lateral sessile or
pedunculate ones; flowers subsessile or rarely some flowers on pedicels up to 1·5 mm.
long. Hypanthium obconical, ± 10-nerved, ± hirsute or rarely glabrous. Calyx-
lobes 2·4–4 mm. long, ciliate-pubescent, rarely glabrous; margins incrassate. Corolla
5–6·5 mm. long, blue, white or mauve, split almost to the base into linear lobes, ±
hirsute outside on the midvein or glabrous. Stamens with filament-bases broadly
dilated, sometimes almost cross-shaped, ciliate-pubescent; anthers 1·2–2·4 mm.
long. Ovary 3-locular, semi-inferior; style c. as long as or somewhat longer than the
corolla, eglandular, slightly thickened in the upper part, glabrous below; lobes 3,
0·5–0·8 mm. long. Capsule 3-locular, ± 10-nerved; valves 3, 1·5–2·5 mm. long.
Seeds 0·5–0·8 mm. long, elliptic-oblong in outline, compressed; testa almost
smooth.

Zambia. N: Mbala Distr., Illuno Village, 1750 m., 9.vii.1970, *Sanane* 1245 (BR; K; SRGH).
W: Ndola, 13.v.1953, *Fanshawe* 5 (BR; K; SRGH). C: between Kabwe and Chiwefwe,
14.vii.1930, *Hutchinson & Gillett* 3668 (BM; K; SRGH). E: upper Loangwa R., x.1897,
Nicholson (K). S: between Mumbwa and Nangoma, 20.iii.1963, *van Rensburg* 1720 (K;
SRGH). **Zimbabwe.** N: Zwipani, Urungwe Trib. Tr. Land, 19.iv.1958, *Goodier* 565 (K;
PRE; SRGH). C: Selukwe Distr., Ferny Creek, 2.iv.1967, *Biegel* 2045 (K; SRGH). E: Umtali
Distr., Menine R., 1050 m., 20.iv.1954, *Chase* 5223 (BM; BR; BRLU; COI; K; LISC; M; MO;
P; SRGH). **Malawi.** N: Nyika Plateau, Nchena-chena Spur, 2000 m., 20.viii.1946, *Brass*
17348 (BM; BR; EA; K; MO; PRE; SRGH). C: Lilongwe, 1.iv.1955, *Jackson* 1556 (K;
SRGH). S: Mandala, 1100 m., 18.vi.1946, *Brass* 16354 (K; MO; PRE; SRGH).
Mozambique. N: 8 km. S. of Massangulo, 1050 m., 26.v.1961, *Leach & Rutherford-Smith*
11025 (K; LISC; SRGH).
    Also in Uganda, Tanzania, Zaire, Rwanda and Burundi. In deciduous woodland, bushland
and upland grassland, old cultivations, usually on sandy soils.

A notable deviating form of *W. capitata* is represented by *Mutimushi* 2188 (SRGH) from
Zambia, Petauke. This is a slender, almost glabrous plant with only a few hairs present near the
base and with long linear leaves. In all these features it agrees with the type of *Lightfootia elegans*
from S. Tanzania. As intermediates to normal *W. capitata* occur in the F.Z. area, however, I do
not think it worth recognising.

13. **Wahlenbergia pulchella** Thulin in Symb. Bot. Ups. **21**, 1: 147 (1975); in F.T.E.A.,
    Campanulaceae: 21 (1976); in F.A.C., Campanulaceae: 22 (1977). Type from Burundi.

Annual ± erect often ± spreading or straggling herb, 5–40 cm. tall. Stem with few
to many usually long branches, rarely unbranched, ± hirsute along its entire length
or glabrous in the upper part, rarely all glabrous. Leaves up to 7–40 × 1·5–10 mm.,
alternate, linear to ovate (basal leaves usually short and broad), acute to subacute,
with cuneate to truncate base, ± hirsute, especially beneath; margin incrassate, ±
dentate; midvein prominent beneath, lateral veins ± obscure. Inflorescence lax or
with flowers in ± dense axillary and terminal clusters; flowers subsessile or on
glabrous or hairy pedicels up to 20 mm. long. Hypanthium ± obconical, ± 10-

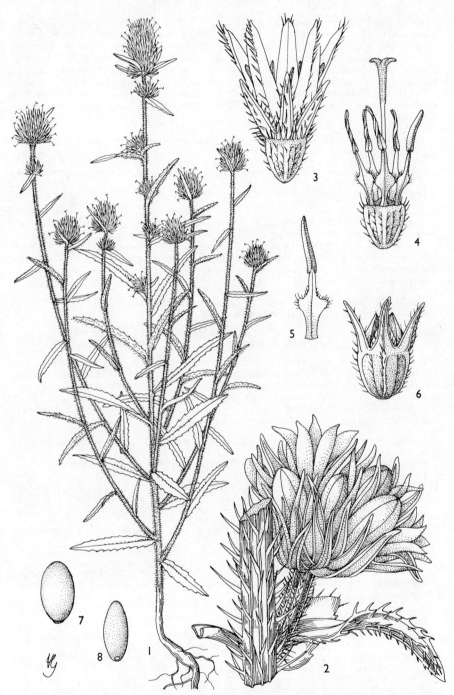

Tab. 16. WAHLENBERGIA CAPITATA. 1, habit (× ⅔); 2, detail of leaf and stem with young axillary inflorescence (× 6); 3, young flower (style not shown) (× 6); 4, flower with calyx-lobes and corolla removed (× 6); 5, stamen (× 8), 1–5, from *Milne-Redhead & Taylor* 10160; 6, capsule (× 6); 7, 8, seed, two views (× 30), 6–8, from *Thulin & Mhoro* 1174. From F.T.E.A.

nerved, $\pm$ hairy or glabrous. Calyx-lobes 0·8–3·2 mm. long, $\pm$ ciliate-pubescent or glabrous; margin incrassate. Corolla 3–6·5 mm. long (outside the F.Z. area 1·5–7 mm.), blue, white or yellow, split almost to the base into linear lobes, $\pm$ hairy or glabrous outside. Filament-bases $\pm$ broadly dilated, ciliate; anthers 0·5–1·6 mm. long. Ovary 3-locular, semi-inferior; style c. as long as or somewhat shorter than the corolla, eglandular, slightly thickened in the upper part, glabrous or rarely hairy below; lobes 3, 0·3–0·6 mm. long. Capsule 3-locular, $\pm$ 10-nerved; valves 3, 0·8–2·5 mm. long. Seeds 0·4–0·7 mm. long, elliptic-oblong in outline, compressed; testa almost smooth.

Stem $\pm$ hirsute below, but glabrous in the upper part; flowers with distinct pedicels 3–20 mm.
    long; corolla usually glabrous outside -   -   -   -   -   -   - subsp. *paradoxa*
Stem $\pm$ hirsute along its entire length; flowers subsessile (pedicels $\pm$ 0·5 mm. long) or
    occasionally pedicelled; corolla $\pm$ hairy outside -   -   -   -   - subsp. *mbalensis*

Subsp. **paradoxa** Thulin, tom. cit.: 153, fig. 26A (1975); op. cit.: 22 (1976). Type from Tanzania.

Erect herb, 10–35 cm. tall. Stem $\pm$ hirsute below, but glabrous or almost so in the upper part. Leaves up to 11–40 × 2–6(8) mm., lanceolate to linear. Inflorescence $\pm$ lax; pedicels 3–20 mm. long, glabrous. Hypanthium $\pm$ hirsute or glabrous. Calyx-lobes 1–2·4 mm. long. Corolla 3·5–6·4 mm. long, blue, glabrous or rarely hairy outside. Anthers 1–1·6 mm. long. Style glabrous or rarely hairy below. Seeds 0·4–0·7 mm. long.

Zambia. N: Shiwa Ngandu, 1500 m., 2.vi.1956, *Robinson* 1562 (K; SRGH).
Also in Tanzania. In sandy grassland or woodland, eroded places.

Subsp. **mbalensis** Thulin, tom. cit.: 150, fig. 26B (1975); op. cit.: 22 (1976). Type: Zambia, Mbala Distr., Kalambo Falls, *Richards* 23219 (K, holotype; BR, P, isotypes).

Erect herb, 10–40 cm. tall. Stem hirsute along its entire length. Leaves up to 14–32 × 5–10 mm., lanceolate to ovate or $\pm$ narrowly elliptic. Inflorescence with flowers in axillary and terminal clusters; pedicels c. 0·5 mm. long, but occasionally up to 7 mm. long, $\pm$ hirsute. Hypanthium $\pm$ hirsute. Calyx-lobes up to 2–3·2 mm. long. Corolla 3·5–5·5 mm. long, blue or white, $\pm$ hairy outside. Anthers (0·8)1·2–1·6 mm. long. Style glabrous below. Seeds 0·5–0·6 mm. long.

Zambia. N: Inono Valley close to Mpulungu road, 900 m., 27.iv.1952, *Richards* 1506 (K; W).
Also in Tanzania. In deciduous woodland or bushland, often on sandy or rocky soils.

14. **Wahlenbergia erecta** (Roth ex Roem. & Schult.) Tuyn in Fl. Males., ser. 1, **6**: 113, fig. 1 h–i (1960). — Thulin in Symb. Bot. Ups. **21**, 1: 155, fig. 5M, 27A & C, 29F & G (1975); in F.T.E.A., Campanulaceae: 23, fig. 5F & G (1976). TAB. **17**. Type from India.
    Dentella erecta Roth ex Roem. & Schult., Syst. Veg. **5**: 25 (1819). Type as above.
    Dentella perotifolia Willd. ex Roem. & Schult., loc. cit. *nom. invalid.*, not accepted by the author. Type from India.
    Wahlenbergia perotifolia Wight & Arn., Prodr. Fl. Penins. Ind. Or. **1**: 405 (1834) *nom. illegit. superfl.* Type as for *Wahlenbergia erecta.*
    Cephalostigma schimperi Hochst. ex A. Rich., Tent. Fl. Abyss. **2**: 2 (1850). — Clarke in Hook., Fl. Brit. Ind. **3**: 428 (1881). — Hiern, Cat. Afr. Pl. Welw. **1**: 628 (1898). Type from Ethiopia.
    Lightfootia arenaria A.DC. in Ann. Sci. Nat., Bot. Sér. 5, **6**: 329 (1866). — Hemsley in F.T.A. **3**: 476 (1877). Type from Angola.
    Cephalostigma erectum (Roth ex Roem. & Schult.) Vatke in Linnaea **38**: 699 (1874) quoad syn. pro parte. — Brenan in Mem. N.Y. Bot. Gard. **8**: 490 (1954) pro parte. — Binns, H.C.L.M.: 25 (1969). Type as for *Wahlenbergia erecta.*
    Cephalostigma hirsutum sensu Hemsley, tom. cit.: 472 (1877) non Edgew. (1851).
    Cephalostigma perotifolium (Wight & Arn.) Hutch. & Dalz. in F.W.T.A. **2**: 191 (1931) *nom. illegit.*, quoad specim. pro parte. Type as for *Wahlenbergia erecta.*
    Cephalostigma perrottetii sensu Hepper in F.W.T.A., ed. 2, **2**: 311 (1963) pro parte.
    Lightfootia perotifolia (Willd. ex Roem. & Schult.) E. Wimmer ex Agnew, Upland Kenya Wild Fl.: 511 (1974) *nom. invalid.* Type as for *Dentella perotifolia.*

Annual erect herb, 6–35 cm. tall, with main stem usually distinct nearly to the top of the plant, hirsute with mixed hairs of different length. Branches usually numerous, spreading. Leaves 8–27 × 3–8 mm., alternate, sessile, lanceolate to

narrowly ovate or ovate (the lowest only), acute, with cuneate to truncate base, ±
hirsute; margin incrassate, undulate-dentate; midvein prominent beneath, lateral
veins almost invisible. Inflorescence lax, ± pyramidal; pedicels 3–12(20) mm. long,
hirsute with mixed hairs of different length, rarely glabrous. Hypanthium
hemispherical, 10-nerved, hirsute. Calyx-lobes 1·5–3 mm. long, usually abruptly
broadened at the base, hirsute in lower part only. Corolla 1·5–2·5 mm. long, ± pale
blue or white, split almost to the base into linear-lanceolate lobes, ± hirsute outside.
Stamens with filament-bases usually narrowly triangular, sometimes slightly
broadened, glabrous or sparsely and shortly ciliate; anthers 0·4–0·7 mm. long. Ovary
3-locular, semi-inferior; style c. as long as the corolla, slightly thickened just below
the 3 short lobes, eglandular, glabrous below. Capsule 3-locular, 10-nerved; valves 3,
1·2–1·6 mm. long. Seeds 0·5–0·6 mm. long, elliptic-oblong in outline, compressed;
testa almost smooth.

Zambia. W: Ndola, 30.iii.1954, *Fanshawe* 1040 (K). C: 18 km. S. of Lusaka, Mt. Makulu
Research Station, 16.iv.1956, *Angus* 1231 (SRGH). E: Lundazi Distr., Lukusuzi National
Park, 800 m., 12.iv.1971, *Sayer* 1204 (SRGH). S: across Kaleya R., S. of Mapangazia,
3.iv.1969, *Anton-Smith* 568 (SRGH). **Zimbabwe.** N: Gokwe Distr., near Sasami R.,
3.iv.1964, *Bingham* 1246 (SRGH). C: Hartley Distr., Poole Farm, 1200 m., 7.iv.1954, *Wild*
4560 (K; LISC; MO; PRE; SRGH). E: Umtali Distr., S. boundary of Umtali Golf Course,
1100 m., 18.iv.1961, *Chase* 7463 (K; LISC; M; MO; P; S; SRGH). **Malawi.** N: Rumphi,
Chelinda R., 1100 m., 4.v.1974, *Pawek* 8617 (K; MO). C: Ntchisi Forest Reserve, 1590 m.,
26.iii.1970, *Brummitt* 9419 (K). S: Ntcheu Distr., Chirobwe, 1580 m., 18.iii.1955, *E.M. & W.*
1018 (BM; LISC; SRGH).
Also in Nigeria, Sudan, Ethiopia, Kenya, Tanzania, Angola and in India and Malaysia. In
grassland or deciduous woodland, waste places, cultivations, often on sandy soils.

15. **Wahlenbergia flexuosa** (Hook. f. & Thomson) Thulin in Symb. Bot. Ups. **21**, 1: 158, fig.
    2D, 5L, 27B, 29H & I (1975); in F.T.E.A., Campanulaceae: 24, fig. 5H & I (1976); in
    F.A.C., Campanulaceae: 26, fig 6H & I (1977). TAB. **17**. Type from India.
    Cephalostigma flexuosum Hook. f. & Thomson in Journ. Linn. Soc., Bot. **2**: 9 (1857). —
    Clarke in Hook., Fl. Brit. Ind. **3**: 428 (1881). Type as above.
    Cephalostigma erectum (Roth ex Roem. & Schult.) Vatke var. *luteum* Chiov. in Ann. Bot.
    Roma **9**: 79 (1911). Type from Ethiopia.
    Lightfootia hirsuta sensu Hepper in Kew Bull. **15**: 61 (1961) pro parte; in F.W.T.A., ed.
    2, **2**: 311 (1963) pro parte.

Annual erect herb 7–35 cm. tall, diffusely branched with the main stem not
particularly distinct, hirsute with long hairs of almost uniform length. Leaves
8–30(38) × 5–16(19) mm., alternate, sessile, narrowly ovate to ovate or elliptic,
acute with rounded to cuneate base, ± hirsute; margin incrassate, undulate-dentate;
midvein prominent beneath as are the lateral veins at least on larger leaves.
Inflorescence lax; pedicels 5–20(35) mm. long, hirsute with long hairs of almost
uniform length or glabrous. Hypanthium hemispherical, 10-nerved, hirsute. Calyx-
lobes 1·2–3·2 mm. long, often broadened at the base, ± hirsute. Corolla 1·5–3·5 mm.
long, yellow, split almost to the base into linear-lanceolate lobes, ± hirsute outside.
Stamens with filament-bases distinctly broadened, ciliate; anthers 0·6–1·6 mm. long.
Ovary 3-locular, semi-inferior; style c. as long as the corolla, eglandular, slightly
thickened below the 3 short lobes, glabrous below. Capsule 3-locular, 10-nerved;
valves 3, c. 2 mm. long. Seeds c. 0·6 mm. long, elliptic-oblong in outline,
compressed; testa almost smooth.

Malawi. N: Kondowe to Karonga, vii.1896, *Whyte* (G; K; Z). C: Dedza Distr., Chongoni
Forest Reserve, Chencherere Hill, 1675–1800 m., 23.iv.1970, *Brummitt* 10076 (K; SRGH).
Also in Nigeria, Cameroon, Central African Republic, Ethiopia, Uganda, Tanzania, Zaire,
Oman and India. Grassland or deciduous woodland, old cultivations.

16. **Wahlenbergia abyssinica** (Hochst. ex A. Rich.) Thulin in Symb. Bot. Ups. **21**, 1: 160,
    fig. 8I, 10J, 12A & B, 14G, P & Q, 22A (1975); in F.T.E.A., Campanulaceae: 24, fig. 2G,
    P & Q (1976); in F.A.C., Campanulaceae: 27, fig. 2G, P & Q (1977); in Fl. Madag. 187,
    Campanulaceae: 14, t. 4 fig. 1–6 (1978). Type from Ethiopia.
    Lightfootia abyssinica Hochst. ex A. Rich., Tent. Fl. Abyss. **2**: 1 (1850). — Hemsley in
    F.T.A. **3**: 474 (1877). — Brenan in Mem. N.Y. Bot. Gard. **8**: 491 (1954). — Adamson in
    Journ. S. Afr. Bot. **21**: 209 (1955). — Binns, H.C.L.M.: 25 (1969). — Agnew, Upland
    Kenya Wild Fl.: 511, 512 (fig.) (1974). — Williamson, Useful Pl. Malawi: 148 (1975).
    Type as above.
    Lightfootia madagascariensis A.DC., Monogr. Camp.: 116 (1830). Type from
    Madagascar.

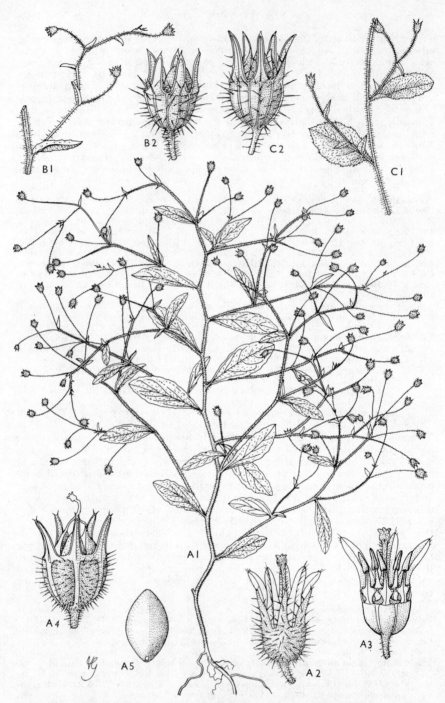

Tab. 17. WAHLENBERGIA HIRSUTA. A1, habit (×⅔), from *Renvoize* 2238; A2, flower (×8); A3, same, with calyx-lobes, two petals and pubescence of hypanthium removed (×8); A4, capsule (×6); A5, seed (×30), A2–A5 from *Polhill & Paulo* 2019. WAHLENBERGIA ERECTA. B1, flowering branch (×1); B2, capsule (×6), B1–B2, from *Burtt* 3606. WAHLENBERGIA FLEXUOSA. C1, flowering branch (×1); C2, capsule (×6), C1–C2 from *Letouzey* 7339. From F.T.E.A.

Lightfootia abyssinica var. *tenuis* Oliver in Journ. Linn. Soc., Bot. **21**: 401 (1885). Type from Kenya.
Lightfootia sodenii Engl., Bot. Jahrb. 19, Beibl. **47**: 52 (1894) as "*sodeni*". Syntypes from Tanzania (B†).
Lightfootia madagascariensis var. *glabra* Engl., Pflanzenw. Ost-Afr. **C**: 400 (1895). Type from Tanzania.
Lightfootia rupestris Engl., Bot. Jarhrb. **30**: 419 (1901). Type from Tanzania.
Lightfootia divaricata Engl., op. cit. **32**: 117 (1902) *nom. illegit.* Type from Ethiopia (B†).
Lightfootia grandifolia Engl., op. cit. **40**: 48 (1907). Syntypes from Tanzania (B†).
Lightfootia subulata Engl., loc. cit. *nom. illegit.* Type from Kenya (B†).
Lightfootia ellenbeckii Engl., tom. cit.: 49 (1907). Type from Ethiopia (B†).
Lightfootia elata Chiov., Racc. Bot. Miss. Cons. Kenya: 74 (1935). Type from Kenya.
Lightfootia arenicola Meikle in Kew Bull. **3**: 466 (1949). Type: Zambia, Mwinilunga Distr., source of Matonchi Dambo, *Milne-Redhead* 2964 (K, holotype; BM, BR, isotypes).
Lightfootia sp. A & *sp. B.* — Agnew, loc. cit.

Perennial or annual herb from a taproot, which is in old perennating specimens ± long and woody. Stems 12–90 cm. tall, few-many, erect or rarely ± decumbent, glabrous, hirsute or puberulous (hairs often ± curled). Leaves up to 8–75 × 1–12 mm., alternate, sessile, linear to lanceolate or very narrowly elliptic, acute, glabrous or ± pubescent; margin incrassate, usually slightly revolute, ± undulate-dentate; midvein protruding beneath, lateral veins obscure or prominent. Inflorescence lax, or sometimes ± contracted, rarely almost spicate; pedicels up to 2–15 mm. long, glabrous or rarely pubescent. Hypanthium (7)10-nerved, glabrous or rarely pubescent. Calyx-lobes 0·8–4 mm. long, usually glabrous, sometimes with markedly incrassate margins. Corolla 2–8 mm. long, blue, white, yellowish or sometimes reddish, split almost to the base into linear lobes, puberulous inside near the base, glabrous outside. Stamens with filament-bases ± broadly dilated, ciliate; anthers 0·7–2·5 mm. long. Ovary 3-locular, semi-inferior; style c. as long as the corolla, eglandular, glabrous or hairy below; lobes 3, 0·1–0·8 mm. long. Capsule 3-locular, 2·4–6 mm. long, (7)10-nerved; valves 3, 0·5–2 mm. long. Seeds 0·4–0·7 mm. long elliptic in outline, usually with acute ends, ± trigonous; testa almost smooth.

Zambia. B: 155 km. W. of Mankoya on road to Mongu, 8.xi.1959, *Drummond & Cookson* 6257 (K; PRE; SRGH). W: Chichele-Ndola, 4.viii.1953, *Fanshawe* 194 (BR; K; LISC; SRGH). E: upper Loangwa R., Chefumbazi, *Nicholson* (K). **Zimbabwe.** N: Umvukwe Mts., 1·5 km. N. of Mtoroshanga Pass, 19.xi.1961, *Leach* 11280 (K; LISC; PRE; SRGH). C: Salisbury, Avondale, 1500 m., 6.x.1955, *Drummond* 4889 (B; BR; K; LISC; PRE; S; SRGH). E: Inyanga Distr., Murambi Gardens near Earl Grey's Plot, 900 m., 1.viii.1954, *Chase* 5275 (BM; K; PRE; SRGH). **Malawi.** N: 8 km. E. of Mzuzu, 1300 m., 1.xi.1969, *Pawek* 2931 (K). C: Chintembwe, 1400 m., 9.ix.1946, *Brass* 17587 (K). S: Zomba, near Agriculture Research Services, 14.vii.1963, *Salubeni* 64 (SRGH). **Mozambique.** N: Malema (Entre Rios), 600 m., xi.1931, *Gomes e Sousa* 817 (K; S). Z: Mocuba, Namagoa, vi.1945, *Faulkner* 209 (BR; G; K; PRE; S; SRGH; UPS). T: between Furancungo and Angónia, 26.viii.1941, *Torre* 3345 (LISC). MS: Manga Distr., 6·5 km. N. of Maculi Beach, 8.ix.1962, *Noel* 2461 (K; LISC; SRGH). GI: Quissico, 28.ii.1955, *E.M. & W.* 708 (BM; LISC; SRGH). M: Delagoa Bay, 30 m., viii.1886, *MacOwan & Bolus* 1196 (G; P; UPS; W).
Also in Ethiopia, Somalia, Kenya, Tanzania, Zaire, Angola, S. Africa (Natal) and Madagascar. In woodland or grassland, forest clearings, old cultivations, road-sides, usually on sandy or rocky soils.

Only the perennial subsp. *abyssinica* occurs in the F.Z. area. Subsp. *parvipetala* Thulin is annual with the corolla less than 3 mm. long and is known from Kenya and Tanzania.

Subsp. *abyssinica* is very polymorphic although less so in the F.Z. area than in East Africa. The variation largely follows a ± geographical pattern. Thus the coastal forms in Mozambique ± agree with coastal forms in Tanzania, S. Africa (Natal) and Madagascar and have lax inflorescences, small leaves, sharply trigonous seeds and usually hairy styles. Further inland in Malawi, Mozambique and Zimbabwe as well as in Kenya and Tanzania, there is a trend towards larger leaves, more contracted inflorescences and glabrous styles. Finally the plants from Barotseland as well as from adjacent parts of Zaire and Angola are generally of a smaller size, have more bluntly trigonous seeds, small tough leaves, short corollas (3–3·5 mm. long) and usually hairy styles. However, as these forms are all ± vaguely defined and connected by intermediates I have preferred not to subdivide this taxon.

17. **Wahlenbergia collomioides** (A.DC.) Thulin in Symb. Bot. Ups. **21**, 1: 168 (1975); in F.T.E.A., Campanulaceae: 26 (1976); in F.A.C., Campanulaceae: 28 (1977). Type from Angola.

Lightfootia collomioides A.DC. in Ann. Sci. Nat., Bot. Sér. 5, **6**: 328 (1866). — Hemsley in F.T.A. **3**: 475 (1877). — Hiern, Cat. Afr. Pl. Welw. **1**: 630 (1898). — Engl. & Gilg in Warb., Kunene-Samb.-Exped. Baum: 396 (1903). — Lambinon & Duvign. in Bull. Soc. Roy. Bot. Belg. **93**: 45 (1961). Type as above.

Lightfootia leptophylla C. H. Wright in Hook., Ic. Pl. **27**: t. 2659 (1900). — Lambinon & Duvign., tom. cit.: 52 (1961). Type: Mozambique, Niassa, between Unango and L. Shire, *Johnson* 40 (K, holotype).

Lightfootia napiformis sensu De Wild., Pl. Bequaert. **2**: 124 (1923) non A.DC. (1866).

Lightfootia collomioides subsp. *katangensis* Lambinon in Bull. Soc. Roy. Bot. Belg. **93**: 49, fig. 3 (1961). Type from Zaire.

Annual or ?short-lived perennial herb, up to 1 m. tall. Stems erect, rarely unbranched, ± hirsute or glabrescent. Leaves 30–60 × 3–12(17) mm., alternate, sessile to subpetiolate, very narrowly elliptic to elliptic, acute with cuneate base, ± hirsute; margin incrassate, undulate-dentate; upper leaves of ± the same size as the lower ones, forming a sort of involucre. Inflorescences terminal, dense, capitate, round or somewhat elongated; flowers subsessile or sometimes with glabrous or hairy pedicels up to 4 mm. long. Hypanthium obconical or hemispherical, glabrous or almost so, 5-nerved. Calyx-lobes 2–4 mm. long, glabrous or ciliate. Corolla 5–7 mm. long, white or blue, split almost to the base into lanceolate lobes, glabrous outside, ± puberulous inside near the base; tube 0·1–0·4 mm. long. Filament-bases dilated, very shortly and densely ciliate; anthers 1·5–3 mm. long. Ovary 3-locular, semi-inferior, style longer than the corolla, thickened in the upper part, eglandular, glabrous below; lobes 3, 0·6–0·8 mm. long. Capsule 3-locular, 5-nerved, valves c. 1 mm. long. Seeds 0·5–0·7 mm. long, elliptic in outline with acute ends, trigonous; testa almost smooth.

Zambia. N: Mbala Distr., Ndundu, 1680 m., 2.v.1957, *Richards* 9514 (K; SRGH). W: Ndola, 10.v.1954, *Fanshawe* 1184 (BR; K; SRGH). S: Mumbwa, 1911, *Macaulay* 795 (K). **Malawi.** N: Likoma I., viii.1887, *Bellingham* (BM). **Mozambique.** N: 21 km. W. of Ribáuè, 600 m., 23.v.1961, *Leach & Rutherford-Smith* 10988 (K; LISC; SRGH). Z: Ile, 29.vi.1943, *Torre* 5545 (LISC).

Also in Central African Republic, Tanzania, Republic of Congo, Zaire and Angola. In deciduous woodland or on cultivated ground, usually on sandy or stony soils.

In Mozambique as well as in Tanzania *W. collomioides* is always an annual with often shortly pedicelled flowers. In Zambia and further towards the west ± perennial plants become increasingly common and the flowers are always subsessile in very dense heads.

18. **Wahlenbergia cephalodina** Thulin in Symb. Bot. Ups. **21**, 1: 171, fig. 50, 28A (1975). Type: Zambia, 21 km. WSW. of Kabompo on Zambesi road, *Drummond & Rutherford-Smith* 7276 (K, holotype; SRGH, isotype).

Annual erect herb, up to 25 cm. tall, hirsute. Leaves up to 40 × 10 mm., alternate, sessile to subpetiolate, narrowly elliptic to oblanceolate, acute to obtuse with attenuate base, sparsely hirsute; margin incrassate, almost entire, very shallowly denticulate; all leaves of c. the same size and shape, the upper ones forming a sort of involucre. Inflorescences terminal, dense, capitate; flowers sessile. Hypanthium obconical, 5-nerved, densely hirsute. Calyx-lobes c. 2·5 mm. long, hirsute. Corolla c. 4 mm. long, blue, split almost to the base into lanceolate lobes, glabrous outside, puberulous inside near the base; tube c. 0·5 mm. long. Stamens with filament-bases abruptly dilated, ciliate; anthers c. 1·6 mm. long. Ovary 3-locular, style longer than the corolla, thickened in the upper part, eglandular, with long hairs below; lobes 3, c. 0·2 mm. long. Capsule and seeds not seen.

Zambia. B: 21 km. WSW. of Kabompo on Zambesi road, 24.iii.1961, *Drummond & Rutherford-Smith* 7276 (K; SRGH).
Known only from Zambia. In woodland on Kalahari sand.

19. **Wahlenbergia hirsuta** (Edgew.) Tuyn in Fl. Males., ser. 1, **6**: 113 (1960). — Thulin in Symb. Bot. Ups. **21**, 1: 174, fig. 11N, 12H, 27D, 29A–E (1975); in F.T.E.A., Campanulaceae: 27, fig. 5A–E (1976); in F.A.C., Campanulaceae: 29, fig. 6A–E (1977); in Fl. Madag. 187, Campanulaceae: 16, t. 4 fig. 7–12 (1978). TAB. **17**. Type from India.

Cephalostigma hirsutum Edgew. in Trans. Linn. Soc. **20**: 81 (1846). — Clarke in Hook., Fl. Brit. Ind. **3**: 429 (1881). Type as above.

Cephalostigma erectum (Roth ex Roem. & Schult.) Vatke in Linnaea **38**: 699 (1874) quoad syn. pro parte. — Brenan in Mem. N.Y. Bot. Gard. **8**: 490 (1954) pro parte, excl. typ.

Cephalostigma erectum var. *coeruleum* Chiov. in Ann. Bot. Roma **9**: 79 (1911). Type from Ethiopia.
Cephalostigma perotifolium (Wight & Arn.) Hutch. & Dalz. in F.W.T.A. **2**: 191 (1931) *nom. illegit.*, pro parte excl. typ.
Lightfootia hirsuta (Edgew.) E. Wimmer ex Hepper in Kew Bull. **15**: 61 (1961) pro parte; in F.W.T.A., ed. 2, **2**: 311 (1963) pro parte. — Agnew, Upland Kenya Wild Fl.: 511 (1974). Type as for *Wahlenbergia hirsuta*.

Annual ± erect herb, 4–30 cm. tall. Stem usually much branched with widely spreading branches, ± hirsute. Leaves 10–55 × 4–25 mm., alternate, sessile to subpetiolate, narrowly oblanceolate to obovate or broadly elliptic to ovate, obtuse to subacute, rarely almost acuminate, with attenuate to cuneate base, ± hirsute; margin incrassate, undulate-dentate; midvein and lateral veins prominent beneath. Inflorescence lax, spreading; pedicels 5–20 mm. long, with very short but usually dense pubescence, often mixed with longer hairs. Hypanthium hemispherical, 5-nerved, ± hirsute. Calyx-lobes 1–3(5) mm. long. Corolla 1·5–4 mm. long, blue to pale blue or white, split almost to the base into lanceolate lobes, glabrous or hairy outside. Stamens with filament-bases ± rhombic, ciliate; anthers 0·5–1 mm. long. Ovary 3-locular, semi-inferior; style c. as long as or longer than the corolla, thickened in the upper part, eglandular, subcapitate or shortly 3-lobed at the apex, glabrous or hairy below. Capsule 3-locular, 5-nerved; valves 3, 1–2·5 mm. long. Seeds 0·4–0·6 mm. long, elliptic in outline with acute ends, trigonous; testa almost smooth, brown or sometimes almost black.

Zambia. N: Kasama Distr., Mungwi, 12.iv.1961, *Robinson* 4591 (EA; K; M; SRGH). W: Solwezi, 1350 m., 10.iv.1960, *Robinson* 3511 (K; SRGH). C: between Chilanga and Kafue, 8.iv.1963, *van Rensburg* 1885 (K; SRGH). S: Mapanza Mission, 1150 m., 29.iii.1953, *Robinson* 149 (K). **Zimbabwe.** N: Miami, iv.1926, *Rand* 80 (BM). C: Salisbury Distr., Rumani, 10.v.1948, *Wild* 2485 (SRGH). **Malawi.** N: Chitipa Distr., Misuku Hills, Kaseye Mission, 1250 m., 26.iv.1972, *Pawek* 5253 (SRGH). S: Limbe, 27.iii.1948, *Goodwin* 108 (BM).
Also in Senegal, Ghana, Nigeria, Cameroon, Central African Republic, Sudan, Ethiopia, Uganda, Kenya, Tanzania, Zaire, Rwanda, Burundi, Angola, Comoro Is., Madagascar, Yemen, India and Nepal. In grassland or woodland, roadsides, or on waste or cultivated ground.

20. **Wahlenbergia upembensis** Thulin in Sym. Bot. Ups. **21**, 1: 184, fig. 3A, 10K, 11O & P, 31B (1975); in F.A.C., Campanulaceae: 34 (1977). Type from Zaire.

Perennial erect herb, 20–50 cm. tall. Stems few to several, wiry, of purplish colour at least towards the base, usually branched in upper part only, glabrous or rarely puberulous at the base. Leaves up to (7)12–35(50) × 0·5–1 mm., alternate, linear, acute, glabrous, rarely sparsely pubescent; margin revolute with a few denticles visible on upper side of the leaf; midvein broad and prominent beneath, leaving a ± deep furrow on the upper side. Inflorescence lax; pedicels up to 30 mm. long, glabrous. Hypanthium obconical, 5(7)-nerved, glabrous. Calyx-lobes 0·8–1·6 mm. long, entire. Corolla (4)5–8 mm. long, white to yellowish or ?pale pink, split almost to the base into linear lobes. Stamens with filament-bases dilated, ciliate-pubescent; anthers 1·6–2·4 mm. long. Ovary 2-locular, semi-inferior; style c. as long as the corolla, thickened in the upper 1–2·5 mm., eglandular, glabrous below; lobes 2, broad and distinct. Capsule 2-locular, 5(–7)-nerved; valves 2, c. 1 mm. long. Seeds (0·5)0·7–0·8 mm. long, elliptic in outline with acute ends, trigonous; testa almost smooth.

Zambia. N: Shiwa Ngandu, 1500 m., 2.vi.1956, *Robinson* 1542 (K; SRGH).
Also in Zaire. In grassland or among rocks, often appearing after fires.

21. **Wahlenbergia ramosissima** (Hemsley) Thulin in Symb. Bot. Ups. **21**, 1: 187 (1975); in F.T.E.A., Campanulaceae: 30 (1976); in F.A.C., Campanulaceae: 35 (1977). Type from Cameroon.
Cephalostigma ramosissimum Hemsley in F.T.A. **3**: 472 (1877). — Hutch. & Dalz. in F.W.T.A. **2**: 191 (1931). Type as above.
Lightfootia ramosissima (Hemsley) E. Wimmer ex Hepper in Kew Bull. **15**: 61 (1961); in F.W.T.A. ed. 2, **2**: 311 (1963). Type as above.

Annual ± erect herb, 10–50 cm. tall. Stem ± hirsute or rarely glabrous. Leaves alternate, sessile, basal ones ± ovate, caducous; stem leaves 5–30(–50) × 1–5(9) mm., linear to elliptic, acute to subacute with attenuate to truncate base, ± hirsute or glabrous; margin incrassate, usually flat or slightly undulate, ± denticulate; midvein prominent beneath, lateral veins ± obscure. Inflorescence lax; pedicels (1)3–30 mm. long, glabrous or rarely hirsute. Hypanthium obconical or obovoid to hemispherical, 5–10-nerved, glabrous or ± hirsute. Calyx-lobes 0·5–2·5 mm. long, entire or sparsely denticulate, glabrous or ± hairy. Corolla 1·2–5 mm. long, white to blue or yellowish, split almost to the base into linear lobes, glabrous or hirsute outside. Filament-bases ± dilated, ciliate; anthers 0·3–1·6 mm. long, rounded to elongated. Ovary 2-locular, semi-inferior; style c. as long as or somewhat shorter than the corolla, eglandular, glabrous or hairy below, slightly thickened to subcapitate in the upper part; lobes 2, very short to distinct. Capsule 2-locular, 5–10-nerved; valves 2, 0·5–1·2 mm. long. Seeds 0·4–0·6 mm. long, elliptic in outline with acute ends, trigonous; testa almost smooth.

W. ramosissima is an aggregate of small, annual, probably autogamous forms with trigonous seeds and 2-merous gynoecia, very well represented in the F.Z. area. Most of these forms have rather narrow, often mutually exclusive ranges, but there are also forms with partly overlapping areas of distribution. I have chosen to classify the morphologically ± well recognizable forms within the complex as subspecies, but as the material of many forms is very limited the classification is rather preliminary.

1. Hypanthium 10-nerved   -   -   -   -   -   -   -   -   -   -   -   -   -   2
 - Hypanthium 5-nerved -   -   -   -   -   -   -   -   -   -   -   -   -   4
2. Corolla hirsute outside; stem usually branching from near the base; leaves narrowly elliptic
    or lanceolate   -   -   '-   -   -   -   -   -   -   -   b. subsp. *zambiensis*
 - Corolla glabrous outside; branching and leaves various   -   -   -   -   -   -   3
3. Stem usually branching from near the base; leaves narrowly elliptic to elliptic, up to 17 mm.
    long; corolla c. 3 mm. long   -   -   -   -   -   -   -   a. subsp. *ramosissima*
 - Stem branching mainly in upper part; leaves usually linear, up to 12–45 mm. long; corolla
    3·2–5·2 mm. long   -   -   -   -   -   -   -   -   -   d. subsp. *lateralis*
4. Plant glabrous or almost so; stem leaves linear, up to 1·5 mm. wide; corolla (2·5–)4 mm. long;
    style distinctly 2-lobed, glabrous below   -   -   -   -   -   g. subsp. *richardsiae*
 - Plant hairy at least at the base; stem leaves, corolla and style various   -   -   -   5
5. Style subcapitate at the apex, densely hairy below -   -   -   -   -   -   6
 - Style distinctly 2-lobed at the apex, glabrous or hairy below   -   -   -   -   7
6. Leaves linear, up to 2(–3) mm. wide   -   -   -   -   -   e. subsp. *centiflora*
 - Leaves lanceolate to ovate, up to 4–5 mm. wide   -   -   -   c. subsp. *subcapitata*
7. Hypanthium usually hirsute; corolla c. 1·2 mm. long   -   -   f. subsp. *oldenlandioides*
 - Hypanthium glabrous; corolla 2–3 mm. long -   -   -   -   a. subsp. *ramosissima*

a. Subsp. **ramosissima**. — Thulin in Symb. Bot. Ups. **21**, 1: 189, fig. 32A (1975).
    Wahlenbergia ramosissima subsp. *A*. — Thulin, tom. cit.: 190, fig. 32B, 33B (1975).

Stem 6–35 cm. tall, usually branching from near the base as well as above, ± hirsute at least below. Leaves 7–17 × 3–5(7) mm., narrowly elliptic to elliptic or lanceolate, acute to subacute, ± hirsute. Pedicels up to 10–30 mm. long, glabrous or almost so.' Hypanthium 5–10-nerved, glabrous or rarely sparsely hirsute. Calyx-lobes 0·7–1·2 mm. long. Corolla 2–3 mm. long, blue. Anthers c. 0·4–0·8 mm. long. Style thickened in the upper part, shortly but distinctly 2-lobed, glabrous or hairy below.

Malawi. N: Rumphi Distr., L. Kaulime, 2340 m., 16.v.1970, *Brummitt* 10782 (K; MAL; SRGH; UPS).
Also in Cameroon, Nigeria and Angola. In upland grassland, roadsides, often in sandy or rocky places.

At the time of my revision of this group (loc. cit.) I had seen material of typical subsp. *ramosissima* only from Nigeria and Cameroon. Some very similar specimens from Angola, differing mainly in having 10-nerved instead of 5-nerved hypanthia were placed in subsp. *A*. Two recent collections from Malawi (*Brummitt* 10782 and *Pawek* 5600), however, are intermediate in having 5–10-nerved hypanthia and I now include them as well as the Angola form in subsp. *ramosissima*. The range of this subspecies is thus much larger than was previously thought. All specimens seen from Malawi and Angola have glabrous styles.

b. Subsp. **zambiensis** Thulin in Symb. Bot. Ups. **21**, 1: 191, fig. 32C (1975); in F.A.C., Campanulaceae: 36 (1977). Type: Zambia, Saisi R., 32 km. from Mbala (Abercorn), *McCallum Webster* 608 (K, holotype).

Stem 10–28 cm. tall, usually branching from near the base as well as above, hirsute at least below. Leaves 5–23(30) × 2–7(9) mm., narrowly elliptic to lanceolate Pedicels up to 15 mm. long, glabrous or rarely hirsute. Hypanthium 10-nerved, hirsute. Calyx-lobes 1–2·5 mm. long. Corolla c. 3 mm. long, blue or white, outside generally hirsute. Anthers c. 0·8–1·2 mm. long. Style shortly but distinctly 2-lobed, glabrous or hairy below.

Zambia. N: Mpika Distr., Luitikila R., 1200 m., 6.iv.1961, *Richards* 14982 (K; SRGH), pro parte. W: Luanshya, 30.iii.1956, *Fanshawe* 2852 (K, pro parte; SRGH). S: Kafue National Park, Chunga, 19.v.1975, *Malaisse* 8613 (UPS).
Also in Zaire. In dry river banks, rocky hillsides, grassland or woodland.

The collection *Robinson* 6585 from NW. Zambia is provisionally placed here. It has hairy styles and narrower leaves than usual for the subspecies. More material may enable taxonomic distinction.

c. Subsp. **subcapitata** Thulin in Symb. Bot. Ups. **21**, 1: 192, fig. 32D (1975) pro parte; in F.T.E.A., Campanulaceae: 30 (1976). Type from Tanzania.

Stem up to 40 cm. tall, usually branching also from near the base, hirsute at least below. Leaves up to 17 × 4–5 mm., narrowly elliptic, acute or subacute, ± hirsute. Pedicels up to 30 mm. long, glabrous or almost so. Hypanthium 5-nerved, somewhat hirsute. Calyx-lobes up to 1·2 mm. long. Corolla up to 2·8 mm. long, blue with white centre, glabrous. Anthers c. 0·6 mm. long. Style subcapitate, very shortly lobed at the apex, densely hairy below.

Malawi. N: Nyika National Park, track to Chisanga Falls, 1800 m., 26.iv.1973, *Pawek* 6590 (K).
Also in Tanzania. In grassland or on disturbed ground.

Some collections from N. Malawi (*Pawek* 4819, 7211, 11374) differ from typical subsp. *subcapitata* in having glabrous hypanthia, up to 7 mm. wide leaves and a more spreading branching although without branches from the base. They may well represent another, still undescribed, subspecies.

d. Subsp. **lateralis** (von Brehmer) Thulin in Symb. Bot. Ups. **21**, 1: 193, fig. 32E (1975). Type from Angola.
    Lightfootia exilis A.DC. in Ann. Sci. Nat., Bot. Sér. 5, **6**: 330 (1866). — Hemsley in F.T.A. **3**: 476 (1877). — Hiern, Cat. Afr. Pl. Welw. **1**: 631 (1898). Type as for *Wahlenbergia ramosissima* subsp. *lateralis*.
    Lightfootia debilis A.DC., tom. cit.: 331 (1866). — Hemsley, loc. cit. — Hiern, loc. cit. Type from Angola.
    Wahlenbergia lateralis von Brehmer in Engl., Bot. Jahrb. **53**: 128 (1915). — Markgraf in Engl., Bot. Jahrb. **75**: 214 (1950). — Roessler in Prodr. Fl. SW. Afr. **136**: 8 (1966). Type as for *Wahlenbergia ramosissima* subsp. *lateralis*.
    Cephalostigma pyramidale Schinz in Viert. Naturf. Ges Zürich **60**: 421 (1915). Type from Namibia.

Stem 15–50 cm. tall, usually much branched but mainly in upper part, ± hirsute below or rarely glabrous. Leaves up to 12–45 × 1–3·5 mm., linear to narrowly lanceolate. Pedicels up to 10–25 mm. long, glabrous. Hypanthium 10-nerved, glabrous or occasionally with a few hairs. Calyx-lobes 0·6–2·0 mm. long. Corolla 3·2–5·2 mm. long, pale blue or blue with white centre. Anthers 0·8–1·2 mm. long. Style distinctly 2-lobed, hairy or glabrous below.

Botswana. N: 62 km. NE. of Maun towards Moremi, iv.1968, *Lambrecht* 513 (K; SRGH).
Zimbabwe. W: Wankie National Park, Bulawayo Pan, c. 6 km. NW. of Main Camp, 19.iv.1972, *Grosvenor* 721 (SRGH).
Also in Angola and Namibia. In grassland or woodland, usually in moist depressions on sandy ground.

e. Subsp. **centiflora** Thulin in Symb. Bot. Ups. **21**, 1: 194, fig. 32F, 34A (1975); in F.T.E.A., Campanulaceae: 31 (1976); in F.A.C., Campanulaceae: 36 (1977). Type: Zambia, old Mpulungu road between D'Hulmiti and Katula, *Richards* 5565 (K, holotype; BR, isotype).

Stem 10–30 cm. tall, usually much branched in upper part only, ± hirsute below. Leaves up to 12–30 × 1–2(3) mm., linear to narrowly lanceolate, basal leaves occasionally up to 5 mm. wide. Pedicels 3–23 mm. long, glabrous. Hypanthium 5-nerved, glabrous or occasionally hairy. Calyx-lobes 0·5–1 mm. long. Corolla 2–3·2 mm. long, blue to white. Anthers 0·4–0·6(0·8) mm. long. Style subcapitate, very shortly lobed at the apex, usually densely hairy below. Seeds c. 0·4 mm. long.

Zambia. N: Mbala Distr., Tasker's Deviation, Chilongowelo, 1500 m., 21.iv.1952, *Richards* 1474 (K; W). W: Mwinilunga Distr., Warnibobo R., 7.viii.1930, *Milne-Redhead* 857 (BR; K). **Malawi.** S: Mt. Mulanje, N. of Tuchila Hut, 1800 m., 22.vii.1956, *Newman & Whitmore* 137 (BM).

Also in Tanzania, Zaire and Angola. In woodland or grassland, usually in moist places on sandy or rocky ground.

The single specimen cited from Malawi is very poor and the record is in need of confirmation. *Robinson* 6872 from Zambia (Barotseland) has 10-nerved hypanthia and is ± intermediate to subsp. *lateralis*. Three collections from the Ndola area in Zambia (*Fanshawe* 2856, 9671 and *Wilberforce* 74) are similar to subsp. *centiflora* and subsp. *oldenlandioides* in habit, but are distinguished from both in having (7)10-nerved hypanthia. They also differ from subsp. *centiflora* by their distinctly 2-lobed, glabrous styles. This form can probably be regarded as a distinct subspecies, but more material is desired.

f. Subsp. **oldenlandioides** Thulin in Symb. Bot. Ups. **21**, 1: 196, fig. 32G, 33A (1975); in F.T.E.A., Campanulaceae: 31 (1976). Type from Tanzania.

Stem 4–10 cm. tall, much branched in upper part only, hirsute at the base. Leaves 5–15 × up to 1·5(2) mm., linear (basal leaves ovate), acute. Pedicels up to 15 mm. long, glabrous. Hypanthium 5-nerved, hirsute or glabrous. Calyx-lobes 0·7–0·9 mm. long. Corolla c. 1·2 mm. long, white. Anthers c. 0·3 mm. long, rounded. Style shortly but distinctly 2-lobed at the apex, glabrous below. Seeds c. 0·4 mm. long.

Zambia. W: 2 km. S. of Solwezi, 20.iii.1961, *Drummond & Rutherford-Smith* 7108 (BR; K; LISC; MO; P; SRGH; UPS). **Malawi.** N: Mzimba Distr., 53 km. SW. of Mzuzu, 1050 m., 18.iv.1974, *Pawek* 8359 (K).
Also in Tanzania. In *Brachystegia* woodland on sandy or stony ground.

g. Subsp. **richardsiae** Thulin in Symb. Bot. Ups. **21**, 1: 197, fig. 32H (1975). Type: Zambia, Mbala, road to Uningi Pans, *Richards* 18106 (K, holotype; BR, EA, MO, SRGH, isotypes).

Stem 7–22 cm. tall, branched mainly in upper part, ± erect, glabrous or almost so. Leaves 5–20 × up to 1·5 mm., linear, acute (basal leaves up to 2 mm. wide, ovate). Pedicels up to 20 mm. long, glabrous. Hypanthium 5-nerved, glabrous. Calyx-lobes 0·7–1 mm. long. Corolla 2·5–4 mm. long, blue or pale yellow. Anthers 0·8–1 mm. long. Style with 2 distinct lobes, c. 0·4 mm. long, glabrous below. Seeds c. 0·4 mm. long.

Zambia. N: Mbala Pans, 1500 m., 26.iii.1957, *Richards* 8875 (K; SRGH).
Known only from Zambia. In ± damp grassland on sandy soils.

22. **Wahlenbergia perrottetii** (A.DC.) Thulin in Symb. Bot. Ups. **21**, 1: 199, fig. 10M, 12K, 14F & O, 17C (1975); in F.T.E.A., Campanulaceae: 31, fig. 2F & O (1976); in F.A.C., Campanulaceae: 37, fig. 2 F & O (1977); in Fl. Madag. 187, Campanulaceae: 18, t. 5 f. 7–10 (1978). Type from Senegal.
    Cephalostigma perrottetii A.DC., Monogr. Camp.: 118 (1830). — Hemsley in F.T.A. **3**: 472 (1877). — De Wild., Pl. Bequaert. **4**: 425 (1928). — Hutch. & Dalz. in F.W.T.A. **2**: 191 (1931). — Hepper in F.W.T.A., ed. 2, **2**: 311 (1963) pro parte. Type as above.
    Cephalostigma prieurii A.D.C., loc. cit. — Hemsley, tom. cit.: 473 (1877), "*prieurei*". Type from Senegal.

Annual erect herb, 0·2–0·6(1) m. tall. Stem hirsute at the base. Leaves up to 18–57 × 5–16 mm., alternate, narrowly elliptic to lanceolate, ± obtuse and mucronulate, with cuneate base, hirsute at least on the lower leaves; margin incrassate, undulate-crenate; midvein and lateral veins prominent beneath. Inflorescence lax; pedicels up to 25 mm. long, glabrous or hirsute. Hypanthium obconical, 5-nerved, glabrous or hirsute. Calyx-lobes 0·7–1·6 mm. long. Corolla 1·8–3 mm. long, white to blue, split almost to the base into linear-lanceolate lobes, glabrous or with a few hairs outside. Filament-bases dilated, ciliate; anthers 0·6–0·9 mm. long. Ovary 2-locular, semi-inferior; style as long as or longer than the corolla, often blue, eglandular, hairy below, subcapitate with 2 very short lobes at the apex. Capsule 2-locular, 5-nerved, glabrous or hirsute; valves 2, c. 1 mm. long. Seeds 0·5–0·6 mm. long, elliptic in outline with acute ends, trigonous; testa almost smooth.

Malawi. N: Nkhata Bay Distr., Chintece Beach, 460 m., 23.v.1971, *Pawek* 4834 (K).
Also in western tropical Africa from Senegal to Angola, Uganda, Tanzania, Zaire, Burundi,

Madagascar, Comoro Is. and S. America. In grassland, woodland, waste places, or on cultivated ground.

W. perrottetii is close to *W. ramosissima*, and I treat it as a distinct species mainly because of its relative uniformity in a very large area of distribution including most of tropical Africa, Madagascar and S. America. Most of the subspecies of *W. ramosissima* instead have a very restricted distribution. The same applies for the next species, *W. paludicola*. Although very unlike *W. perrottetii*, *W. paludicola* is also close to some of the forms in the polymorphic *W. ramosissima*.

23. **Wahlenbergia paludicola** Thulin in Symb. Bot. Ups. **21**, 1: 202, fig. 35A (1975); in F.T.E.A., Campanulaceae: 32 (1976); in F.A.C., Campanulaceae: 39 (1977); in Fl. Madag. 187, Campanulaceae: 20, t. 5 fig. 11–16 (1978). Type: Zambia, Bangweulu, Kamindas, *Fries* 891 (UPS, lectotype).
     Lightfootia gracillima R. E. Fries, Wiss. Ergebn. Schwed. Rhod.-Kongo-Exped. **1**: 316 (1916). Type as for *Wahlenbergia paludicola*.
     Lightfootia abyssinica var. *tenuis* sensu Hepper in Kew Bull. **15**: 61 (1961); in F.W.T.A., ed. 2, **2**: 311 (1963).

Annual slender ± trailing or straggling herb, 10–50 cm. tall, glabrous. Stem usually branching only in upper part, furrowed. Leaves 2–12 × 0·2–1(1·6) mm., alternate, sessile, basal ones ± ovate, caducous, stem leaves linear to narrowly lanceolate, acute; margin incrassate with a few denticles; midvein prominent beneath, lateral veins obscure. Inflorescence lax; pedicels 6–32 mm. long. Hypanthium 5(–10)-nerved, glabrous. Calyx-lobes 0·6–1·4 mm. long. Corolla 2·4–3·5(5) mm. long, white or bluish, split almost to the base into lanceolate lobes. Filament-bases narrowly dilated, ± ciliate; anthers 0·35–0·8 mm. long. Ovary 2-locular, semi-inferior; style c. as long as or somewhat shorter than the corolla, eglandular, glabrous or ± hairy below, slightly thickened in the upper part, subcapitate and with 2 very short lobes at the apex. Capsule 2-locular, 5(–10)-nerved; valves 2, 0·6–1·4 mm. long. Seeds 0·45–0·6 mm. long, elliptic in outline with acute ends, trigonous; testa almost smooth.

Zambia. N: 72 km. on Kasama-Mbala road, 1500 m., 30.iii.1955, *Richards* 5215 (K; SRGH). W: Chingola, 1300 m., 16.iii.1960, *Robinson* 3400 (K; M; SRGH). C: 19 km. N. of Kabwe, 23.ix.1947, *Brenan & Greenway* 7942 (K). S: Mazabuka Distr., Choma, 1300 m., 14.iv.1958, *Robinson* 2841 (K; M; PRE; SRGH). **Zimbabwe.** C: Marandellas Distr., Digglefold Vlei, 1600 m., 18.iv.1952, *Wild* 3810 (K; PRE; SRGH). **Malawi.** N: Mzimba Distr., Mzuzu, 1375 m., 28.vi.1969, *Pawek* 2510 (K).
     Also in Cameroon, Uganda, Tanzania, Zaire, Angola and Madagascar. In boggy grassland, often on sandy soils.

3. GUNILLAEA Thulin
Gunillaea Thulin in Bot. Notiser **127**: 166 (1974).

Annual herbs. Leaves alternate, sessile, flat. Inflorescences ± leafy, monochasial. Flowers ± regular, protandrous. Calyx-lobes (3)4–5, accrescent. Corolla campanulate, (3)4–5-lobed. Stamens (3)4–5, free; filament-bases almost linear to broadly dilated, ciliate or glabrous. Ovary inferior, 2-locular; ovules numerous; style shorter than the corolla, 2-lobed, upper part with pollen-colleting hairs, lower part glabrous or hairy; 2 glands sometimes present at the base of the style-lobes. Capsule thin-walled, indehiscent, tardily opening by the irregular decomposition of the pericarp between the persistent lateral nerves. Seeds numerous; testa sulcate, sometimes with hair-like projections.

Two species in tropical Africa, one extending to Madagascar.

Corolla less than 5 mm. long, shorter or longer than the calyx-lobes; style without glands
     - - - - - - - - - - - - - - - - - 1. *emirnensis*
Corolla 5–12 mm. long, longer than the calyx-lobes; style with 2 tiny glands at the base of the style-lobes - - - - - - - - - - - - - - - 2. *rhodesica*

1. **Gunillaea emirnensis** (A.DC.) Thulin in Bot. Notiser **127**: 166, fig. 1, 3F–K, 10C & F (1974); in F.T.E.A., Campanulaceae: 34, fig. 6 (1976); in F.A.C., Campanulaceae: 41, fig. 7 (1977); in Fl. Madag. 187, Campanulaceae: 22, t. 6 (1978). TAB. **18**. Type from Madagascar.
     Wahlenbergia emirnensis A.DC. in DC., Prodr. **7**: 432 (1839). Type as above.

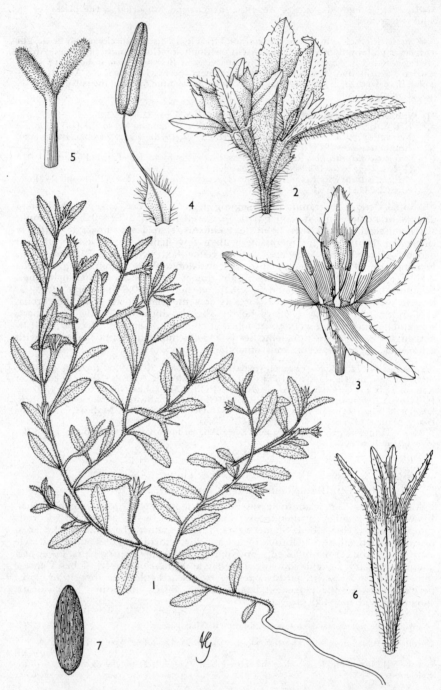

Tab. 18. GUNILLAEA EMIRNENSIS 1, habit (× ⅔); 2, flower and leaf with sympodial shoot in the leaf-axil (× 4); 3, flower with two petals removed (× 6); 4, stamen (× 18); 5, style (× 12); 6, capsule (× 3); 7, seed (× 54). All from *Robson* 667. From F.T.E.A.

Wahlenbergia huillana A.DC. in Ann. Sci. Nat., Bot. Sér. 5, **6**: 333 (1866). — Hemsley in F.T.A. **3**: 479 (1877). Type from Angola.
Wahlenbergia huillana var. *pusilla* A.DC., loc. cit. Type from Angola.
Wahlenbergia rutenbergiana Vatke in Abh. Naturw. Ver. Bremen **9**: 123 (1885). Type from Madagascar.
Cervicina huillana (A.DC.) Hiern, Cat. Afr. Pl. Welw. **1**: 631 (1898). Type as for *Wahlenbergia huillana*.

Annual ascending or decumbent, rarely erect herb, 4–40 cm. tall, usually much branched from the base. Stem ± hirsute, rarely glabrescent. Leaves up to 8–40 × 2–10 mm., narrowly elliptic to elliptic or oblanceolate, acute to almost rounded with attenuate base, ± hirsute or glabrous; margin slightly incrassate, ± undulate-crenate; midvein protruding beneath, lateral veins obscure. Inflorescence not well demarcated, lax. Flowers sessile or shortly pedicellate; pedicel elongating in fruit up to 10 mm. Hypanthium narrowly obconical, 5–10-nerved, glabrous or hirsute. Calyx-lobes 3–5, 1·5–5(10) mm. long, often of varying length, narrowly triangular to oblanceolate or narrowly oblong, glabrous or ± hirsute, sparsely denticulate. Corolla 2·4–4(5) mm. long, white or blue; lobes 3–4(5), united c. halfway. Stamens 3–4(5), 2–3·2 mm. long; filament-bases narrowly triangular to broadly angular-obovate, glabrous or ciliate; anthers 0·6–1·6 mm. long. Style glabrous or rarely hairy below, eglandular; lobes 2, 0·5–1·6 mm. long, often slightly unequal in length; pollen-collecting hairs present on style-lobes only. Capsules up to 10 mm. long, obovoid to narrowly obconical or cylindrical, often slightly curved upwards, prominently 5–10 nerved, glabrous or hirsute. Seeds 0·5–0·8 mm. long, elliptic in outline, often ± reniform, slightly compressed; testa sulcate, rarely with hair-like projections.

Zambia. B: Kaoma Distr., near Luena R., 20.xi.1959, *Drummond & Cookson* 6642 (SRGH). N: 50 km. SE. of Kasama, Chambeshi Flats, 23.i.1961, *Robinson* 4308 (EA; K; M; SRGH). C: Luangwa Valley, Mtuwe, 4.xii.1965, *Astle* 4162 (SRGH). E: Lundazi R. above dam, 1200 m., 19.xi.1958, *Robinson & Fanshawe* 667 (BM; BR; K; LISC; SRGH). S: Mapanza, 1050 m., 8.xi.1953, *Robinson* 361 (BR; K). **Zimbabwe.** N: Urungwe Distr., Sanyati R., near junction of Fulechi R., 750 m., 11.x.1957, *Phipps* 752 (BM; BR; EA; K; LISC; P; SRGH). C: Salisbury, 1500 m., xi.1919, *Eyles* 1944 (K; SRGH; Z). **Malawi.** N: Rumphi, 16.xii.1964, *Robinson* 6287 (M; SRGH).
Also in Tanzania, Angola, Zaire and Madagascar. In temporarily wet places such as river banks and marsh edges.

2. **Gunillaea rhodesica** (Adamson) Thulin in Bot. Notiser **127**: 168, fig. 2, 3A–E, 8C, 10A, B, E & H, 11A, C & E (1974); in F.A.C., Campanulaceae: 42 (1977). Type: Zimbabwe, Victoria Falls, *Flanagan* 3162 (BOL, holotype; PRE, isotype).
Prismatocarpus rhodesicus Adamson in Journ. S. Afr. Bot. **17**: 123 (1951). — Roessler in Prodr. Fl. SW. Afr. **136**: 4 (1966). Type as above.

Annual usually erect herb, up to 60 cm. tall, often much-branched from the base. Stems ± hirsute, at least below. Leaves up to 12–50 × 2–6 mm., narrowly elliptic to linear or narrowly oblanceolate, attenuate, ± acute, ± hirsute; margin incrassate, ± undulate-crenate with small projecting teeth. Inflorescence usually well demarcated, lax. Flowers sessile or shortly pedicellate; pedicel elongating in fruit up to 25 mm. Hypanthium narrowly obconical, c. 10-nerved, glabrous or glabrescent. Calyx-lobes (4)5, 3–6(9) mm. long, narrowly triangular to narrowly oblanceolate, sometimes revolute, glabrous or ciliate, usually sparsely denticulate. Corolla 5–12 mm. long, blue or white, puberulous inside towards the base; lobes (4)5, united up to c. halfway. Stamens (4)5, 4–5 mm. long; filament-bases dilated, almost cross-shaped, ciliate; anthers 2–2·5 mm. long. Style hairy below; style-lobes c. 0·8 mm. long, with 2 tiny glands at the base; pollen-collecting hairs present on style-lobes and immediately below them. Capsules up to 2 cm. long, cylindrical or narrowly obconical, often slightly curved upwards, prominently c. 10-nerved and with the persistent style-base forming a hard cone on the top. Seeds 0·5–0·7 mm. long, narrowly elliptic to elliptic; testa sulcate, usually with hair-like projections.

Caprivi Strip. Kakumba I., 900 m., 17.i.1959, *Killick & Leistner* 3416 (K; M; SRGH). **Botswana.** N: Kasane, floodplain of Chobe R., 900 m., 20.i.1972, *Biegel & Russell* 3692 (SRGH). **Zambia.** B: Mongu, 1.xii.1962, *Robinson* 5506 (K; M; SRGH). N: Mansa, 13.xi.1964, *Mutimushi* 1061 (SRGH). W: L. Ishiku-Ndola, 18.x.1953, *Fanshawe* 430 (BR; K). S: Namwala Distr., Kafue R. at Kalala, 7.xii.1963, *Mitchell* 24/11 (BR; K; LISC; SRGH). **Zimbabwe.** N: Sebungwe Distr., Zambesi R. near Binga, 6.xi.1958, *Phipps* 1363 (K; LISC;

SRGH). W: Victoria Falls, xii.1904, *Allen* 118 (K; SRGH). S: Ndanga Distr., Lundi R., Chipinda Pools, 300 m., 16.x.1960, *Goodier* 12 (SRGH). **Mozambique.** MS: Gorongosa, Chitengo, margin of Pungue R., 23.x.1965, *Balsinhas* 1002 (COI).

Also in Angola, Zaire and Namibia. In seasonally wet places such as river banks and lake shores, on sand or mud.

4. CRATEROCAPSA Hilliard & Burtt

Craterocapsa Hilliard & Burtt in Notes Roy. Bot. Gard. Edin. **32**: 314 (1973).

Perennial prostrate herbs. Leaves alternate or sometimes opposite, sessile or almost so. Flowers single or few together, terminal or appearing lateral, sessile or shortly pedicelled. Calyx-lobes 5, united at the base. Corolla funnel-shaped, 5-lobed. Stamens 5, free; filament-bases dilated, ciliate. Ovary inferior, (2–)3-locular with very delicate septa; ovules numerous; style thickened at the base or encircled by a fleshy disk, (2–)3-lobed at the apex, upper part with pollen-collecting hairs. Capsule dehiscing by an apical lid, unilocular at maturity. Seeds numerous, ellipsoid, ± trigonuos.

A genus of 4 species, distributed in eastern S. Africa, Lesotho and Swaziland, 1 species extending also to Zimbabwe.

Craterocapsa tarsodes Hilliard & Burtt in Notes Roy. Bot. Gard. Edin. **32**: 324, fig. 5A (1973). TAB. **19**. Type from S. Africa (Natal).
       Wahlenbergia montana A.DC. var. *glabrata* Sonder in Harv. & Sond., F.C. **3**: 573 (1865). — von Brehmer in Engl., Bot. Jahrb. **53**: 131 (1915). Type from S. Africa (Drakensberg).

Perennial mat-forming herb with prostrate branching stems radiating from a thick taproot. Stems glabrous or hairy, leafy and finally rough by persisting leaf-bases. Leaves (5)10–25(40) × (2)3–6(12) mm., mostly rosulate at the tips of the branches, lanceolate, elliptic or oblanceolate, acute at the apex, tapering to a broad flat petiole-like base; margin incrassate, sparsely denticulate, ± undulate, thinner and usually ciliate towards the base, otherwise leaf generally glabrous, occasionally sparsely pubescent. Flowers sessile in the leaf-rosettes, opening one at a time. Hypanthium 10-nerved, glabrous. Calyx-lobes 3–8 mm. long, united for 1–2 mm. at the base, lanceolate, acute, with incrassate, ciliate margins. Corolla 12–22 mm. long, funnel-shaped, blue to white, outside puberulous or glabrous, inside pubescent at least below; tube 6–12 mm. long. Filament-bases narrowly dilated, ciliate; anthers 2–3 mm. long. Ovary 3-locular. Style shorter than the corolla, eglandular, hairy below, at the base broadened to a shallow cone, thickened towards the apex, shortly 3-lobed. Capsule 2–3 mm. long, subspherical, 10-nerved, crowned by the hardened calyx-lobes, dehiscing by an apical lid formed by the accrescent style-base, or sometimes by a ring around the base of the persistent style. Seeds 0·7–0·8 mm. long, trigonous, reticulate, red-brown.

Zimbabwe. E: Mt. Inyangani, near summit, 2550 m., 6.xii.1959, *Wild* 4906 (K; SRGH).
Also in S. Africa (Cape Prov., Natal, Orange Free State, Transvaal), Swaziland and Lesotho. In stony grassland on bare soil or on broken rock sheets.

100. SPHENOCLEACEAE
By M. Thulin

Annual glabrous helophytic herbs. Leaves alternate, simple, entire, exstipulate. Inflorescences terminal, dense, spicate. Flowers bisexual, regular, subtended by a bract and 2 bracteoles. Calyx-tube adnate to the ovary; lobes 5, quincuncial, shortly connate, persistent. Corolla campanulate-urceolate, epigynous, caducous; lobes 5, imbricate. Stamens 5, inserted on the corolla-tube, alternating with the petals; filaments very short; anthers rounded, 2-celled, dehiscing longitudinally. Ovary semi-inferior, 2-locular, with numerous anatropous ovules on large stipitate axile placentas; style short, glabrous, with a capitate slightly 2-lobed stigma. Fruit a

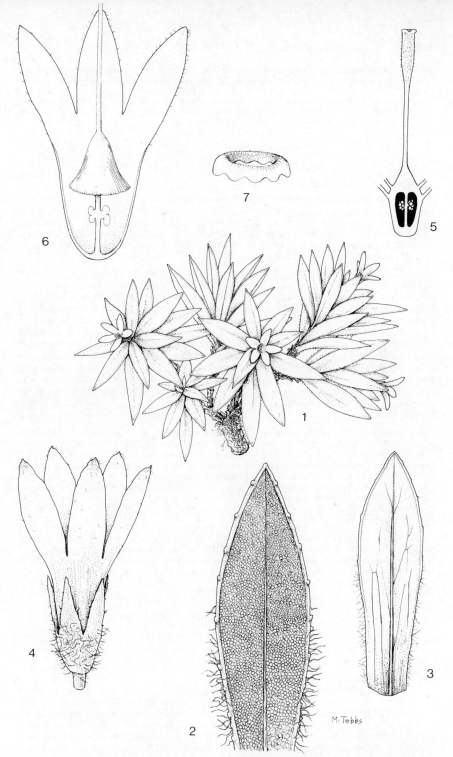

Tab. 19. CRATEROCAPSA TARSODES. 1, habit (× 1); 2, leaf, superior surface (× 4); 3, leaf, inferior surface (× 2); 4, flower (× 6), 1–4 from *Dawson* 21; 5, gynoecium in longitudinal section (× 6); 6, fruit after dehiscence in longitudinal section (the plug replaced intact) (× 6); 7, ring of tissue from apex of ovary (× 6), 5–7 after *Hilliard & Burtt*, tom. cit.: 315.

circumscissile capsule, obovate-obconic, 2-locular, membranous. Seeds numerous, minute; testa irregularly plicate-costate; endosperm scanty or absent.

A single almost pantropical genus. Sometimes placed in the *Campanulaceae*, but distinguished by the quincuncial-imbricate perianth-lobes, the glabrous style, the circumscissile capsule and the apparent absence of latex canals.

SPHENOCLEA Gaertn.

Sphenoclea Gaertn., Fruct. **1**: 113, t. 24/5 (1788).

Characters as for the family.

Genus of 2 species; one, *S. dalzielii* N.E. Br., is confined to West Africa.

Sphenoclea zeylanica Gaertn., Fruct. **1**: 113, t. 24/5 (1788). — Hemsley in F.T.A. **3**: 481 (1877). — Hepper in F.W.T.A., ed. 2, **2**: 307 (1963). — Roessler in Prodr. Fl. SW. Afr. **138** (1966). — Airy Shaw in F.T.E.A., Sphenocleaceae: 1, fig. 1 (1968). — Thulin in F.A.C., Sphenocleaceae: 2 (1973); in Fl. Madag. **187** bis., Sphenocleaceae: 26 (1978). — Monod in Adansonia, sér. 2, **20**: 147 (1980). TAB. **20**. Type from Ceylon.

Herb, up to 1·5 m. tall. Stem hollow, often much branched. Leaf-blade lanceolate to narrowly ovate or elliptic, up to 12 × 5 cm., attenuate at the base, acute or subacute at the apex; petiole up to 1·5 cm. long. Spikes cylindrical-conical, up to 12 cm. long but usually much shorter, and c. 1 cm. in diameter, with only a few flowers open at a time; peduncle up to 8 cm. long; bracts and bracteoles oblanceolate-spathulate, the tips arched over the flowers except at anthesis. Flowers sessile, attached longitudinally to the rhachis by a linear base. Calyx-lobes broadly triangular, rounded at apex, 1–1·5 mm. long. Corolla white to greenish white or pink, 2–3 mm. long; lobes ovate-triangular, united about half-way. Filaments slightly dilated at base; anthers c. 0·5 mm. long. Ovary obovoid, 1·5–2 mm. long. Capsule 4–5 mm. in diameter, dehiscing below the calyx-lobes which fall with the lid, leaving the scarious base persistent on the rhachis. Seeds yellowish brown, oblong, c. 0·5 mm. long.

Botswana. SE: Eastern Bamangwato Territory, *Holub* s.n. (K). **Zambia.** W: Mafumbwe R., 25.vi.1953, *Fanshawe* 127 (K; SRGH). C: Luangwa valley, Lubi R., Mtuwe, 600 m., 3.iii.1967, *Astle* 5040 (K; SRGH). E: Luangwa R., near Kamunshya, 27.x.1966, *Lawton* 1467 (K; SRGH). S: Masonsa dambo, 9·5 km. N. of Mapanza, 1060 m., 11.iv.1954, *Robinson* 681 (K; SRGH). **Zimbabwe.** N: Binga Distr., Mwenda Research Station, 12.iv.1966, *Jarman* A (8) (K; SRGH). W: Wankie National Park, 5 km. ENE. of Nehimba, 62 km. WSW. of Main Camp, 6.iii.1969, *Rushworth* 1648 (K; SRGH). S: Nuanetsi R., 11 km. downstream from Malipate, 29.iv.1961, *Drummond & Rutherford-Smith* 7585 (COI; K; LISC; SRGH). **Malawi.** S: Chikwawa Distr., Lengwe Game Reserve, 100 m., 8.iii.1970, *Brummitt & Hall-Martin* 8961 (K; SRGH). **Mozambique.** T: Boroma, ii.1892, *Menyhart* 605 (K). MS: Gorongosa National Park, SE. Urema Plains, iii.1972, *Tinley* 2454 (SRGH). M: Namaacha, R. Movene, 29.iv.1967, *Marques* 2003 (BM; COI; LISC; PRE; SRGH).

Widespread in tropical Africa and also in tropical Asia and America (probably introduced). In wet places, such as in and near pools, swamps, streamsides and irrigation channels.

101. LOBELIACEAE

By M. Thulin

Annual or perennial herbs, subshrubs, shrubs or trees, laticiferous. Leaves alternate or more rarely opposite or verticillate, simple, entire to dentate or incised, more rarely lobed or dissected, exstipulate. Inflorescences generally racemose, or flowers solitary in leaf-axils; pedicels mostly with 2 bracteoles. Flowers bisexual or rarely unisexual, usually protandrous, zygomorphic, 5-merous, often resupinate. Calyx regular or somewhat 2-lipped; lobes valvate in bud. Petals usually united into a tube, which is often ± split on the back (or on the lower side in non-resupinate flowers), rarely 2 or all petals free or almost so, 2-lipped or rarely 1-lipped or subregular, valvate in bud. Stamens alternating with petals, free or ± adnate to the corolla; filaments linear, rarely dilated at the base, ± connate, rarely free; anthers

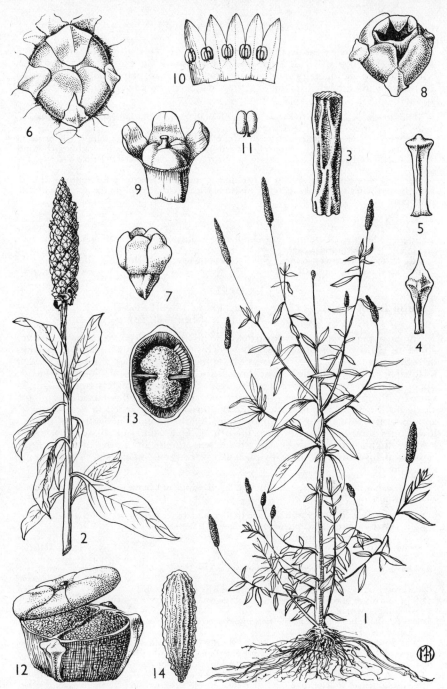

Tab. 20. SPHENOCLEA ZEYLANICA. 1, habit (× ½); 2, part of flowering branch (× ⅔), 1, 2 from
Milne-Redhead & Taylor 7463; rhachis of inflorescence, showing scars left by fallen
capsules (× 4); 4, bract (× 4); 5, bracteole (× 4); 6, flower-bud, apical view (× 4), 3–6 from
Jones in FHI 18808; 7, bud (from another plant) beginning to open, side view (× 4); 8,
flower, showing opening corolla, oblique view (× 4); 9, gynoecium and calyx (with two
sepals removed) showing cuneate base (× 4); 10, corolla opened out (× 4); 11, stamen
(× 8), 7–11 from *Deighton* 132a; 12, fruit partly dehisced (× 4); 13, transverse section of
fruit (× 4), 12, 13 from *Jones* in FHI 18808. From F.T.E.A.

united into a tube or rarely free (*Cyphia*), introrse. Ovary ± inferior, rarely ± superior, (1)2-locular; ovules few–many, anatropous, on axile placentas, style 1, with pollen-collecting hairs at or near the apex. Fruit capsular, usually dehiscing by 2(4) apical valves, rarely by a lid, or ± baccate. Seeds 1-many, albuminous, embryo straight.

About 30 genera and some 1000 species, mainly in the tropics and subtropics, most numerous in the New World.

The aberrant genus *Cyphia* is treated here for convenience. Along with some small, probably only distantly related New World genera it is usually considered a distinct subfamily, either within Lobeliaceae or Campanulaceae sensu lato, but could also well be regarded a family of its own.

The flowers in almost all species of *Lobelia* and in some *Monopsis* are resupinate. The 3-lobed lower lip in such flowers therefore morphologically corresponds to the upper lip of non-resupinate ones.

1. Anthers united into a tube - - - - - - - - - - - - **2**
– Anthers free - - - - - - - - - - - - - **3. Cyphia**
2. Stigma-lobes short and broad, surrounded by a ring of pollen-collecting hairs; bracteoles minute, ± linear, rarely absent - - - - - - - - - **1. Lobelia**
– Stigma-lobes linear with a ring of pollen-collecting hairs well below them; bracteoles (in the F.Z. area) resembling small leaves - - - - - - - **2. Monopsis**

1. LOBELIA L.

Lobelia L., Sp. Pl. **2**: 929 (1753); Gen. Pl., ed. 5: 401 (1754). — E. Wimmer in Engl., Pflanzenr. **IV**. 276 b: 408 (1953).

Annual or perennial herbs, subshrubs or small trees with a palm-like habit. Leaves alternate, entire to incised or lobed. Flowers bisexual or rarely unisexual in racemes or solitary in leaf-axils, nearly always resupinate. Calyx-lobes 5, often somewhat unequal. Corolla split to the base or almost so on the back only, or sometimes the two upper petals are almost free to the base or all the petals ultimately free, usually 2-lipped with upper lip 2-lobed, lower 3-lobed, more rarely all petals are forming a single lip. Stamens 5, free from the corolla or slightly attached to it at the base; anthers united into a tube, the two lower or all of them penicillate at the tip, the two lower ones somewhat shorter than the others. Ovary inferior to subsuperior, 2-locular; stigma-lobes 2, short and broad, surrounded by a ring of hairs. Fruit capsular dehiscing by 2, rarely 4, apical valves, or rarely indehiscent; seeds usually numerous.

A large genus of some 300 species, present on all continents, but most species in the tropics and subtropics.

1. Stout erect plants, 1–9 m. tall when in flower, sometimes woody below; leaves densely set and spirally arranged on upper part of vegetative stem - - - - - **2**
– Much smaller, erect to prostrate herbs; leaves not as above - - - - - **4**
2. Inflorescence branched - - - - - - - - - 1. *stricklandiae*
– Inflorescence an unbranched raceme - - - - - - - - - / **3**
3. Leaves narrowly oblanceolate to oblong-obovate, 25–90 × 4–14 cm., serrate or dentate
2. *giberroa*
– Leaves usually linear, 15–40 × 2–4(6) cm., entire or almost so - - 3. *mildbraedii*
4. The two lower anthers only penicillate at the tip - - - - - - **5**
– All anthers penicillate at the tip - - - - - - - - - - **24**
5. All corolla-lobes subequal, the two upper free or almost so; lower lip with a crescent-shaped gland inside at the base - - - - - - - - 31. *angolensis*
– The two upper corolla-lobes smaller and shorter than the others, connate with lower lip for most of their length; no such gland present at inside of lower lip - - - **6**
6. Capsular valves about as long as or usually longer than the rest of the capsule; calyx-lobes very narrowly triangular; stem not or scarcely winged - - - - 6. *trullifolia*
– Capsular valves shorter than the rest of the capsule, or if occasionally about as long, other characters not as above - - - - - - - - - - **7**
7. Prostrate or decumbent herb; leaves suborbicular to subreniform, incised-dentate; pedicels ebracteolate - - - - - - - - - - - 7. *lobata*
– Erect or ascending, scarcely prostrate herbs; leaves not as above; pedicels minutely bracteolate, usually at the base - - - - - - - - - **8**
8. Calyx-lobes completely reflexed after anthesis, ciliate; lower leaves broadly ovate, pubescent on both sides; bracts equalling or shorter than pedicels - - 21. *quarreana*
– Calyx-lobes erect or spreading, if ± reflexed leaves and bracts not as above - - **9**
9. Upper corolla-lobes linear, not broadened at the base; pedicels 2–8(10) mm. long; leaves

linear-oblong to spathulate; upper 3 anthers each with a tuft of hairs near the apex on the
back; capsule cylindrical-obconical, c. 7–12 mm. long - - - - - 22. *anceps*
– Characters not combined as above - - - - - - - - - 10
10. Leaves up to 13 mm. long and 6 mm. wide; calyx-lobes denticulate; stem not or scarcely
winged - - - - - - - - - - - - - - 11
– Leaves usually much larger; other characters not combined as above - - - 13
11. Annual; leaves up to 3–5 × 1–1·5 mm.; raceme one-sided - - - 11. *uliginosa*
– Rhizomatous perennials; leaves 5–13 × 1–6 mm. - - - - - - 12
12. Upper leaves markedly narrower than the lower ones; calyx-lobes 2·5–6 mm. long;
bracteoles ± 1 mm. long - - - - - - - - - 9. *welwitschii*
– All leaves of similar shape; calyx-lobes 1·2–2·5 mm. long; bracteoles 0·2–0·5 mm. long
10. *livingstoniana*
13. Stem terete to ± triangular, not or narrowly winged; corolla 7–15 mm. long - - 14
– Stem distinctly triangular, at least after drying broadly winged, or if only narrowly winged
the corolla less than 6 mm. long - - - - - - - - - 16
14. Stem terete to triangular, often ribbed but not winged; corolla 7–13 mm. long- 4. *erinus*
– Stem narrowly winged; corolla 9–15 mm. long - - - - - - 15
15. Calyx-lobes 3–5·5 mm. long, ciliate or pubescent; leaves up to 25–70 mm. long; corolla
9–13 mm. long - - - - - - - - - - - 8. *flaccida*
– Calyx-lobes 2–3·2 mm. long, glabrous or sparsely ciliate; leaves up to 28 mm. long; corolla
13–15 mm. long - - - - - - - - - - - 5. *kirkii*
16. Leaves 7–16 × 2–5 mm.; calyx-lobes erect, entire, not ciliate; pedicels 4–7 mm. long
14. *chireensis*
– Leaves usually much larger, or if almost as small, other characters not combined as above
17
17. Calyx-lobes denticulate with one or two pairs of teeth at the margin - - - 18
– Calyx-lobes entire or occasionally with a single tooth at the margin - - - 19
18. Pedicels longer than bracts; leaves up to 25–55 × 10–35 mm. - - - - 12. *sapinii*
– Pedicels equalling or shorter than bracts; leaves up to 7–20 × 6–12 mm. - 17. *inconspicua*
19. Capsule narrowly obconical to subcylindrical, 6–13 mm. long, valves excluded; corolla
4·5–6 mm. long, white with purplish or greenish markings in throat - - 15. *molleri*
– Capsule subglobose to narrowly obconical, shorter; corolla not as above - - - 20
20. Corolla 4–11 mm. long, if less than 6 mm. always with calyx-lobes recurved in fruit; usually
decumbent to ascending herb to 60 cm. long - - - - - 13. *fervens*
– Corolla 3–5·5 mm. long; calyx-lobes erect or spreading in fruit; usually erect herbs to 35
cm. high - - - - - - - - - - - - - 21
21. Calyx-lobes narrowly triangular, 0·6–1 mm. wide at the base, ciliate - 20. *intercedens*
– Calyx-lobes subulate, more narrow at the base, glabrous or ciliate - - - 22
22. Pedicels up to 5–8 mm. long; capsule excluding valves ± 2·5 mm. long; calyx-lobes ciliate
19. *lasiocalycina*
– At least some pedicels longer; capsule excluding valves 2·5–5·5 mm. long; calyx-lobes
usually glabrous - - - - - - - - - - - - 23
23. Stem and leaves glabrous; upper bracts lanceolate to ovate, longer to shorter than pedicels
16. *heyneana*
– Stem and leaves ± sparsely pubescent with long hairs; upper bracts linear to lanceolate,
equalling or shorter than pedicels - - - - - - - 18. *adnexa*
24. Flowers solitary, scattered in leaf-axils, subsessile or on up to 5 mm. long pedicels; capsule
4-valved; plant prostrate - - - - - - - - - 30. *thermalis*
– Flowers in racemes, pedicelled; capsule 2-valved; usually erect or ascending herbs - 25
25. Leaves sessile or almost so - - - - - - - - - - 26
– At least lower leaves distinctly petiolate - - - - - - - - 29
26. Corolla 8–13(14) mm. long - - - - - - - - - 23. *goetzei*
– Corolla 16–24 mm. long - - - - - - - - - - - 27
27. Leaves up to 10–16 mm. long - - - - - - - - - 24. *ovina*
– Leaves longer - - - - - - - - - - - - - 28
28. Leaves narrowly elliptic to narrowly ovate, 7–18 mm. wide, coarsely dentate or crenate
25. *blantyrensis*
– Leaves ± linear, up to 5 mm. wide, very sparsely denticulate or entire - 26. *caerulea*
29. Stem, leaves, pedicels and calyx densely spreading-pilose - - - 27. *cobaltica*
– Stem, leaves, pedicels and calyx glabrous or ± pubescent - - - - 30
30. At least lower leaves cordate, very coarsely crenate or dentate with few teeth
28. *pteropoda*
– Leaves cuneate to truncate or sometimes subcordate, coarsely dentate or serrate with ±
numerous teeth - - - - - - - - - - - 29. *baumannii*

1. **Lobelia stricklandiae** Gilliland in Journ. Bot., Lond., **73**: 248 (1935), as *"stricklandae"*.
— E. Wimmer in Engl., Pflanzenr. **IV**. 276 b: 677, fig. 101 i (1953) & 276 c: 886 (1968). —
Mabberley in Kew Bull., **29**: 562 (1974). Type: Zimbabwe, Penhalonga, *Gilliland 798*
(BM, holotype; K).
Lobelia stricklandiae f. *uncinata* E. Wimmer, loc. cit. (1968). Type from Tanzania.

Plant 2–6 m. tall in flower, ± erect, suckering from the base. Stem 4–8 cm. thick at base, terete, hollow, pubescent above. Rosette-leaves of unflowered plant sessile, ± narrowly oblanceolate, up to 50 × 10 cm., acute or acuminate at the apex, attenuate at the base, weakly pubescent on both sides, more densely pubescent on veins beneath; margin serrate to doubly serrate or denticulate; venation protruding beneath. Inflorescence up to 3–4 m. long, lax, branched with a terminal many-flowered raceme and usually many axillary racemes on leafy peduncles below; pedicels 8–25 mm. long, spreading, pubescent; upper bracts linear to lanceolate and 10–20 mm. long, lower bracts ovate and up to 30–40 mm. long; bracteoles usually in upper part of the pedicel, linear, 1·5–4 mm. long. Hypanthium obovoid, 10-nerved, ± pubescent. Calyx-lobes narrowly triangular, flat, 10–16 mm. long, weakly pubescent on both sides, entire. Corolla 25–35 mm. long, mauve or pinkish purple, split to the base on the back, ± papillose-pubescent outside and pubescent inside on the bottom of the tube; two lateral petals linear, united with middle petals below, but often becoming free to the base; three middle petals soon splitting up at the apex into three ± 6 mm. long lobes. Filaments linear, somewhat dilated at the base, connate almost to the base and forming a firm tube, papillose mainly towards the base, free from corolla. Anther-tube 6–7 mm. long, all anthers glabrous at the apex, but sparsely pilose on the back. Ovary subinferior. Capsule obovoid to subspherical, 10–15 mm. long, 10-nerved, weakly pubescent or glabrescent with two short valves. Seeds ± broadly elliptic-oblong in outline, compressed, narrowly winged along one margin, 0·6–0·7 mm. long, finely striate, pale brown.

Zimbabwe. E: 10 km. SW. Melsetter, Belmont Forest, 16.ix.1950, *Crook* M 128 (K; SRGH; W). **Malawi.** C: Dedza Distr., on the Lake View road below Chirobwe, 30.v.1961, *Chapman* 1337 (K; SRGH). **Mozambique.** MS: Gorongosa Mts., road to Mt. Gogogo, 1300 m., 25.ix.1943, *Torre* 5936 (LISC). Also in Tanzania and NE. Transvaal. Forest margins or glades, often on stream banks.

Very closely related to *Lobelia xongorolana* E. Wimmer in Angola, and could probably as well be treated as a subspecies of this.

2. **Lobelia giberroa** Hemsl. in F.T.A. **3**: 465 (1877). — Engl. in Abh. Königl. Preuss. Akad. Wiss. Berlin **1891**: 409 (1892). — Bak. f. in J. Linn. Soc., Bot. **38**: t. 17/3 (1908). — Engl. in Mildbr., Deutsch. Zentr.-Afr.-Exped., **2**: 344 (1911). — R.E.Fr., Wiss. Ergebn. Schwed. Rhod.-Kongo-Exped. **1**: 316 (1916). — R.E. & T.C.E. Fr. in Svensk Bot. Tidskr. **16**: 397, fig. 3a–d (1922). — Hauman in Mém. Inst. Roy. Col. Belg. 8°, **2**: 8, 28 (1934). — E. A. Bruce in Kew Bull. **1934**: 79 (1934). — Robyns, Fl. Parc Nat. Alb. **2**: 411 (1947). — E. Wimmer in Engl., Pflanzenr. **IV**. 276 b: 670 (1953). — F. W. Andr., Fl. Pl. Anglo-Egypt. Sudan, **3**: 76 (1956). — E. Wimmer, op. cit. **IV**. 276 c: 885 (1968). — Mabberley in Kew Bull., **29**: 564, fig. 1E, 2B, 3, 5A–C, 6B, 8 (1974). — Agnew, Upland Kenya Wild Fl.: 515–516 (fig.) (1974). TAB. **21**. Type from Ethiopia.

Tupa schimperi Hochst. ex A. Rich., Tent. Fl. Abyss. **2**: 10, t. 63 (1850). Type as above.
Lobelia volkensii Engl., Bot. Jahrb. **19**, Beibl. 47: 49 (1894). — R.E. & T.C.E. Fr., tom. cit.: 399, fig. 3 h (1922). — E. A. Bruce, loc. cit. (1934). Type from Tanzania.
Lobelia volkensii var. *ulugurensis* Engl. in Notizbl. Bot. Gart. Berl. **1**: 106 (1895). Syntypes from Tanzania.
Lobelia squarrosa Bak. in Kew Bull. **1898**: 157 (1898). — R.E. & T.C.E. Fr., tom. cit.: 416 (1922). — Hauman, tom. cit.: 30 (1934). Type: Malawi, Masuku Plateau, *Whyte* 306 (K, holotype).
Lobelia usafuensis Engl., Bot. Jahrb. **30**: 420 (1901). — R.E. & T.C.E. Fr., tom. cit.: 399, fig. 3f–g (1922). — E. A. Bruce, tom. cit.: 78 (1934). Type from Tanzania.
Lobelia ulugurensis (Engl.) R.E. & T.C.E. Fr., tom. cit.: 398 (1922). Type as for *Lobelia volkensii* var. *ulugurensis*.
Lobelia giberroa var. *volkensii* (Engl.) Hauman, tom. cit.: 8 (1934). Type as for *Lobelia volkensii*.
Lobelia giberroa var. *ulugurensis* (Engl.) Hauman, tom. cit.: 9 (1934). — Robyns, tom. cit.: 412 (1947). Types as for *Lobelia volkensii* var. *ulugurensis*.
Lobelia giberroa var. *longibracteata* Hauman, tom. cit.: 10 (1934). — E. Wimmer, tom. cit.: 672 (1953). Syntypes from Zaire and Rwanda.
Lobelia giberroa var. *usafuensis* (Engl.) Hauman, tom. cit.: 7, 11 (1934). — E. Wimmer, loc. cit. (1953) & 276 c: 886 (1968). Type as for *Lobelia usafuensis* Engl.
Lobelia intermedia Hauman, tom. cit.: 31, fig. 4B (1934). — E. A. Bruce, tom. cit.: 274 (1934). Type from Zaire.
Lobelia giberroa var. *usafuensis* (Engl.) Hauman, tom. cit.: 7, 11 (1934). — E. Wimmer, in tom. cit.: 672 (1953) & 276 c: 885 (1968). Type from Tanzania.
Lobelia giberroa var. *intermedia* (Hauman) Robyns, tom. cit.: 413 (1947). — E. Wimmer, tom. cit.: 885 (1968). Type as for *Lobelia intermedia*.

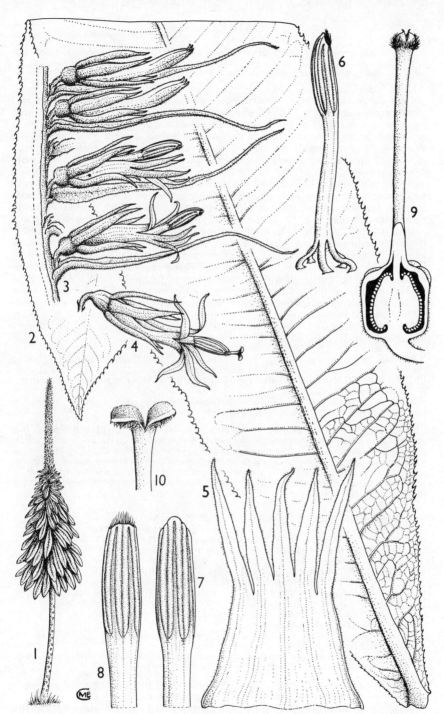

Tab. 21. LOBELIA GIBERROA. 1, habit, from photograph *Pawek* 10046; 2, leaf (×⅔), from
Hedberg 1953; 3, portion of sectioned inflorescence (×1), from *Bally* 7986; 4, older flower
to show extruded style (×1), from *Hedberg* 1953; 5, corolla spread out (×2); 6, stamens
with anther-tube (side view) (×2); 7, anther-tube from above (×3); 8, anther-tube from
below (×3); 9, style and ovary of a young flower (×3), 5–9 from *Bally* 7986; 10, style of an
older flower (×4), from *Hedberg* 1953. From F.T.E.A.

Lobelia giberroa var. *mionandra* E. Wimmer in Annal. Naturhist. Mus. Wien **56**: 368 (1948); in tom. cit.: 673 (1953). Type from Zaire.

Lobelia giberroa subsp. *squarrosa* (Bak.) Mabberley, tom. cit.: 566 (1974). Type as for *Lobelia squarrosa*.

Plant 2–9 m. tall in flower, suckering from the base. Stem erect, 6–12 cm. thick and woody at the base, terete, hollow, ± pubescent above. Rosette-leaves of unflowered plant sessile, narrowly oblanceolate to oblong-ovate, 25–90 × 4–15 cm., acute to acuminate at the apex, attenuate at the base, pubescent beneath especially on the veins or glabrous, weakly pubescent or glabrous above; margin doubly serrate, dentate or denticulate; venation protruding beneath with costae in ± 30–40 pairs arising at an angle of 60–90° to the midrib. Inflorescence 1–3 m. long, dense, cylindrical, unbranched; pedicels 4–8 mm. long, pubescent, with two minute linear bracteoles; bracts linear to narrowly lanceolate, 25–75 × 2–4 mm. Hypanthium obovoid, ± truncate at the base, 10-nerved, pubescent. Calyx-lobes narrowly triangular to linear, flat, 10–28 mm. long, entire, ± pubescent especially on the outside. Corolla 25–35 mm. long, greenish or greenish white often tinged with blue or purple, split to the base on the back, ± pubescent on the outside at least at the apex, glabrous inside; two lateral petals linear, ± united with middle petals below; three middle petals united to form a lip with linear lobes. Filaments linear, connate for most of their length, ± papillose at the margins, free from corolla. Anther-tube (6)9–15 mm. long, the two lower anthers apically barbate, otherwise glabrous or occasionally pubescent on connectives. Ovary subinferior. Capsule ± 10 mm. long, 10-nerved, pubescent, crowned by the hardening calyx-lobes and with two short valves. Seeds elliptic-oblong in outline, compressed, narrowly winged, ± 0·8 mm. long, finely striate, pale brown.

Zambia. N: Isoka Distr., Mafingi Mts., 20.xi.1952, *Angus* 801 (BM; FHO; K). E: Nyika Plateau, 2·5 km. SW. of Rest House, 2150 m., 30.x.1958, *Robson & Angus* 478 (BM; K; LISC; SRGH). **Malawi.** N: Mzimba Distr., Viphya Plateau, 13.vi.1960, *Chapman* 733 (BM; FHO; K; SRGH).

Also in Sudan, Ethiopia, Uganda, Kenya, Tanzania, Zaire, Rwanda and Burundi. In upland, often secondary, forest, at forest edges, stream sides or swamp edges.

Lobelia giberroa is the most widespread of the African Giant Lobelias and many forms of this variable plant have been given specific or varietal rank in the past. In the last revision of the group (Mabberley, op. cit.) only two subspecies were recognised within *L. giberroa*: one northern including the type of the species with bracts longer than the flowers, and one southern, subsp. *squarrosa*, with bracts shorter than the flowers and comprising all material from the F.Z. area. As numerous intermediate forms are found, however, particularly in Tanzania, I have chosen here to treat it as a single variable taxon.

3. **Lobelia mildbraedii** Engl. in Mildbr., Deutsch. Zentr.-Afr.-Exp. **2**: 344 (1911). — R.E. & T.C.E. Fr. in Svensk Bot. Tidskr. **16**: 405 (1922). — Hauman in Mém. Inst. Roy. Col. Belg. 8°, **2**: 31 (1934). — E. A. Bruce in Kew Bull. **1934**: 73, fig. 5/2 (1934). — Robyns, Fl. Parc. Nat. Alb. **2**: 410 (1947). — E. Wimmer in Engl., Pflanzenr. **IV**. 276 b: 668, fig. 101f (1953). — Brenan in Mem. N.Y. Bot. Gard. **8**: 490 (1954). — E. Wimmer, op. cit. IV. 276 c: 884 (1968). — Binns, H.C.L.M.: **25** (1969). — Mabberley in Kew Bull. **29**: 571 (1974). — Moriarty, Wild Flowers of Malawi: **73**, t. 37, fig. 7 (1975). Type from Rwanda.

Lobelia utshungwensis R.E. & T.C.E. Fr., loc. cit., fig. 6e–f. — E. A. Bruce, loc. cit. — E. Wimmer, tom. cit.: 669 (1953) & op. cit. 276 c: 885 (1968). Type from Tanzania.

Lobelia suavibracteata Hauman, tom. cit.: 33, fig. 4A (1934). — E. Wimmer, op. cit. IV. 276 b: 669 (1953). Type from Zaire.

Lobelia mildbraedii var. *robynsii* E. Wimmer in Annal. Naturhist. Mus. Wien **56**: 368 (1948); in tom. cit.: 669 (1953). Type from Rwanda.

Lobelia mildbraedii f. *acutifolia* E. Wimmer, tom. cit.: 885 (1968) *nom. invalid.* Orig. colls. from Uganda and Zaire.

Plant 1–3·5 m. tall in flower, suckering from the base. Stem ± erect, to 8 cm. thick at base, terete, hollow above, glabrous to pubescent. Leaves sessile, linear to narrowly lanceolate, 15–40 × 2–4(6) cm., acute to rounded and mucronate at the apex, rounded to auriculate at the base, glabrous or puberulous and often shining above, ± puberulous beneath; margin entire or at least in young leaves denticulate by the projecting hydathodes; venation protruding beneath with numerous costae arising at an angle of c. 45° to the midrib distal to the phyllodic leaf-base with ± parallel venation. Inflorescence up to 1 m. long, dense, cylindrical, unbranched; pedicels 5–6(8) mm. long, pubescent, with two minute bracteoles; bracts narrowly oblong to linear, (30)40–60 × 3–8 mm., acute or rounded and mucronate at the apex, ± pubescent. Hypanthium obovoid, truncate at the base, 10-nerved, pubescent. Calyx-lobes narrowly triangular, flat, 8–15(20) mm. long, entire, pubescent on both sides. Corolla 30–35 mm. long, greenish, lilac or pale blue, split to the base on the back, pubescent outside; two lateral petals linear, ± united with middle petals below; three middle petals united to form a lip with linear lobes. Filaments linear, connate for most of their length, ± papillose towards the base, free from corolla. Anther-tube 9–11 mm. long, the two lower anthers apically barbate, otherwise glabrous or almost so. Ovary subinferior. Capsule subspherical, ± 7 mm. long, 10-nerved, pubescent, with two valves, 2–3 mm. long. Seeds elliptic-oblong in outline, compressed, winged on one side, 1·2–1·5 mm. long, finely striate, pale brown.

Malawi. N: Nyika Plateau, 1·6 km. N. of rest house at Chelinda, 2150 m., 1.ix.1962, *Tyrer* 730 (BM; SRGH).

Also in Uganda, Tanzania, Zaire, Rwanda and Burundi. In upland swamps. Also recorded from NE. Zambia by Fanshawe (Checkl. Woody Pl. Zambia) but no specimen seen.

4. **Lobelia erinus** L., Sp. Pl.: 932 (1753) pro parte. — Sonder in Harv. & Sond., F.C. **3**: 544 (1865) pro parte. — E. Wimmer in Annal. Naturhist. Mus. Wien **56**: 349 (1948); in Engl., Pflanzenr. **IV**. 276 b: 519, fig. 87 e–g (1953). Type from S. Africa (Cape Prov.).

Lobelia bellidiflora L. f., Suppl. Pl.: 396 (1781), as *"bellidifolia"* by Thunb. and later authors. Type from S. Africa (Cape Prov.).

Rapuntium flexuosum Presl, Prodr. Monogr. Lobel.: 16 (1836) *nom. illegit.* Type from S. Africa (Natal), not seen.

Lobelia natalensis A.DC. in DC., Prodr. **7**: 369 (1839). — Sonder, tom. cit.: 545 (1865) pro parte. Type as for *Rapuntium flexuosum*.

Lobelia senegalensis A.DC., tom. cit.: 372 (1839). — Hemsl. in F.T.A. **3**: 469 (1877). — E. Wimmer, tom. cit.: 551, fig. 90 d (1953) pro parte; in F.W.T.A., ed.2, **2**: 313 (1963); in Engl., Pflanzenr. **IV**. 276 c: 869 (1968) pro parte. — Binns, H.C.L.M.: 25 (1969). Type from Senegal.

Lobelia lavendulacea Klotzsch in Peters, Reise Mossamb., Bot. **1**: 302 (1862). — E. Wimmer, tom. cit.: 551 (1953). Type from Zanzibar.

Lobelia kohautiana Vatke in Linnaea **38**: 719 (1874) *nom. illegit.* Type as for *Lobelia senegalensis*.

Lobelia nuda Hemsl., loc. cit. — Hiern, Cat. Afr. Pl. Welw. **1**: 628 (1898). — R.E. Fr., Wiss. Ergebn. Schwed. Rhod.-Kongo-Exped. **1**: 317 (1916). — E. Wimmer, tom. cit.: 464, fig. 78 b (1953). — Roessler in Prodr. Fl. SW. Afr. **137**: 3 (1966). — E. Wimmer, op. cit. **IV**. 276 c: 857 (1968). — Binns, H.C.L.M.: 25 (1969). Type: Zambia, Batoka Country, *Kirk* s.n. (K, holotype).

Lobelia pubescens Dryand. var. *simplex* O. Ktze. in Jahrb. Bot. Gart. Berlin **4**: 267 (1886). Type from Namibia.

Lobelia benguellensis Hiern, tom. cit.: 626 (1898). — E. Wimmer, op. cit. **IV**. 276 b: 545 (1953) & 276 c: 868 (1968). Type from Angola.

Lobelia chilawana Schinz in Mem. Herb. Boiss. **10**: 70 (1900). Type: Mozambique, Delagoa Bay, *Junod* 398 (G, holotype).

Lobelia rosulata S. Moore in Journ. Bot. **41**: 402 (1903). — Markgraf in Engl., Bot. Jahrb. **75**: 217 (1950). Type from S. Africa (Transvaal).

Lobelia jugosa S. Moore in Journ. Linn. Soc., Bot. **40**: 125 (1911). Type: Zimbabwe, between Lusitu R. and Nyahodi R., *Swynnerton* 2037 (BM, holotype).

Lobelia transvaalensis Schltr. in Engl., Bot. Jahrb. **57**: 622 (1922). Type from S. Africa (Transvaal).

Lobelia trierarchi R. Good in Journ. Bot., Lond. **62**: 139 (1924). Type from Sudan.

Lobelia turgida E. Wimmer in Feddes Repert. Spec. Nov. **38**: 82 (1935). — Markgraf in Engl., Bot. Jahrb. **75**: 218 (1950). Type: Mozambique, *Dawe* 349 (BR, holotype; K).

Lobelia nuda f. *hirtella* E. Wimmer in Annal. Naturhist. Mus. Wien **56**: 343 (1948); in tom. cit.: 465 (1953). Type from Namibia.

Lobelia filiformis Lam. var. *natalensis* (A.DC.) E. Wimmer in Annal. Naturhist. Mus. Wien **56**: 354 (1948); in Engl., Pflanzenr. **IV**. 276 b: 541, fig. 89 c (1953). — Roessler, loc. cit. Type as for *Lobelia natalensis*.

Lobelia filiformis var. *natalensis* f. *albiflora* E. Wimmer in Annal. Naturhist. Mus. Wien **56**: 355 (1948); in Engl., Pflanzenr. **IV**. 276 b: 542 (1953). Syntypes from S. Africa (Natal) and Tanzania.

Lobelia filiformis var. *natalensis* f. *multipilis* E. Wimmer in Annal. Naturhist. Mus. Wien **56**: 355 (1948); in Engl., Pflanzenr. **IV**. 276 b: 542 (1953). Syntypes from Angola, Zambia: Chirihutu, *R.E. Fries* 287 (B; UPS; W) and Mozambique: Beira, *Braga* 81 (B).

Lobelia filiformis var. *natalensis* f. *muzandazora* E. Wimmer in Annal. Naturhist. Mus. Wien **56**: 355 (1948); in Engl., Pflanzenr. **IV**. 276 b: 542 (1953). Syntypes from Zimbabwe: St. Trias Hill Mission, *Mundy* 3176 (K), and without precise locality, *Hislop* 59 (K).

Lobelia nuda var. *rosulata* (S. Moore) E. Wimmer in Kew Bull. **7**: 138 (1952); in Engl., Pflanzenr. **IV**. 276 b: 775 (1953) & 276 c: 857 (1968). Type as for *Lobelia rosulata*.

Lobelia altimontis E. Wimmer in loc. cit. (1952); in Engl., Pflanzenr. **IV**. 276 b: 780 (1953) & 276 c: 866 (1968). Syntypes from Zimbabwe: Marandellas, *Walters* 2364 (K, syntype; SRGH), Pork Pie, *Chase* 1424 (BM; K; SRGH; W) and Pungwe, *Hopkins* in GHS 9396 (SRGH; W, syntype).

Lobelia polyodon E. Wimmer in tom. cit.: 140 (1952); in Engl., Pflanzenr. **IV**. 276 b: 778 (1953) & 276 c: 864 (1968). Syntypes from Zimbabwe: Inyanga Distr., Chipungu stream, *Chase* 1850 (SRGH; W, syntype) and Mozambique: Chimanimani Mts., *Wild* 2902 (K; SRGH; W) and *Munch* 69 (SRGH; W, syntype).

Lobelia wildii E. Wimmer in tom. cit.: 141 (1952); in Engl., Pflanzenr. **IV**. 276 b: 781 (1953) & 276 c: 866 (1968). Type: Zimbabwe, Chimanimani Mts., Mt. Peza, *Wild* 3619 (K; SRGH; W).

Lobelia wildii var. *arcana* E. Wimmer in tom. cit.: 141 (1952); in Engl., Pflanzenr. **IV**. 276 b: 781 (1953). Type: Zimbabwe, Marandellas Cave, *Wild* 3289 (SRGH; W, holotype).

Lobelia raridentata E. Wimmer in Engl., Pflanzenr. **IV**. 276 b: 538 (1953). Type: Zimbabwe, Vumba Mts., *Peter* 30616 (B, holotype; W).

Lobelia keilhackii E. Wimmer in Engl., Pflanzenr. **IV**. 276 b: 543 (1953). Type: Zimbabwe, Belingwe, *Keilhack* s.n. (B, holotype).

Lobelia nuzana E. Wimmer in Engl., Pflanzenr. **IV**. 276 b: 780 (1953). Type: Zimbabwe, Mt. Nuza, *Gilliland* 313 (BM; J, holotype; SRGH).

Lobelia senegalensis var. *turgida* (E. Wimmer) E. Wimmer in Engl., Pflanzenr. **IV**. 276 b: 552 (1953) & 276 c: 869 (1968). Type as for *Lobelia turgida*.

Lobelia filiformis var. *filiformis* sensu E. Wimmer in Engl., Pflanzenr. **IV**. 276 b: 541 (1953) pro parte. — Binns, H.C.L.M.: 25 (1969) pro parte.

Lobelia melsetteria E. Wimmer in Engl., Pflanzenr. **IV**. 276 c: 866 (1968) *nom. invalid.* Orig. colls.: Zimbabwe, Melsetter Distr., *Crook* M15 (PRE); Zambia, Kansuaja, *Stohr* 83 (PRE).

Lobelia lydenburgensis sensu E. Wimmer in Engl., Pflanzenr. **IV**. 276 c: 856 (1968) pro parte.

Annual or sometimes perennial, decumbent to erect herb, 5–70 cm. long or tall. Stem terete to ± triangular or ribbed, glabrous to pubescent at least below with rather rigid to very fine and soft hairs. Leaves up to 15–75(100) × (1)4–20 mm., sparsely serrulate to dentate, crenate or almost pinnatifid, the upper linear to narrowly elliptic, the lower sometimes ± rosulate, oblanceolate to spathulate, acute to obtuse at the apex, narrowing below into a petiole-like base or an up to 15 mm. long petiole, glabrous to pubescent on both surfaces, or ciliate at the leaf-bases only. Flowers in lax racemes; pedicels up to 5–45 mm. long, glabrous or occasionally pubescent; bracts similar to upper stem leaves, much shorter to equalling the pedicels; bracteoles c. 0·4–1·2 mm. long, linear, at or near the base of the pedicel, occasionally absent. Hypanthium narrowly obconical to obovoid, 8–10-nerved, glabrous or occasionally pubescent. Calyx-lobes narrowly triangular to subulate, erect òr somewhat spreading, 1·2–5 mm. long, entire, glabrous or occasionally pubescent or ciliate. Corolla 7–13 mm. long, blue or rarely mauve or white, with two bumps in the mouth of the tube, split to 0·6–2·8(3·2) mm. from the base on the back, short pubescent or. papillose on the inside of the tube, glabrous or occasionally pubescent on the outside. Stamens 3–7 mm. long; filaments linear, puberulous and attached to the corolla-tube at the base; anther-tube 1·2–2·4 mm. long, ± pubescent on the back of the upper thecae, or occasionally glabrous, the two lower anthers each with a tuft of hairs and a hyaline appendage at the apex. Ovary subinferior. Capsule 8–10-nerved with 2 valves ± 1 mm. long, inferior part of capsule 3–8 mm. long. Seeds elliptic to broadly elliptic in outline, somewhat compressed, 0·3–0·4 mm. long, very finely striate, brown.

Caprivi Strip. Mpola, 24 km. on Katima Mulilo-Ngoma road, 910 m., 5.i.1959, *Killick & Leistner* 3297 (PRE; SRGH). **Botswana.** N: Shaile Camp on Linyanti R., 28.x.1972, *Pope, Biegel & Russell* 896 (COI; K; SRGH). SE: Gaberones, ix.1967, *Lambrecht* 297 (K; SRGH). **Zambia.** B: Mongu Lealui, 10 km. E. of Mongu, 24.x.1965, *Robinson* 6689 (B; EA; K; MO; SRGH). N: Mbala Distr., Kasesha Village, 1520 m., 15.v.1969, *Sananẹ* 709 (K). W: Kitwe, 30.vii.1963, *Mutimushi* 346 (K; SRGH). C: Chakwenga Headwaters, 100–129 km. E. of Lusaka, 25.viii.1963, *Robinson* 5634 (K; SRGH). **Zimbabwe.** N: Noro Mine, 13 km. N. of Mtoroshanga, 17.iv.1964, *Wild* 6531 (K; LISC). W: Zambesi R., near Victoria Falls, m., vii.1908, *F. A. Rogers* 5134 (K). C: Hatcliffe South, 14 km. NNE. of Salisbury, 1540 m., 26.ix.1955, *Drummond* 4879 (B; EA; K; SRGH; W). E: Inyanga Distr., Hill Fort, Rhodes Estate, 1830 m., 19.iv.1953, *Chase* 4912 (BM; COI; K; MO; SRGH). S: Chibi Distr., near

Madzivire dip, 6 km. N. of Lundi R. bridge, 3.v.1962, *Drummond* 7879 (K; SRGH). **Malawi.**
N: Viphya Plateau, 95 km. SW. of Mzuzu, 1890 m., 22.xi.1975, *Pawek* 10371 (K; MO; SRGH).
C: Dedza Distr., Chongoni Forest boundary, Kadzera dambo, 4.i.1968, *Salubeni* 928 (K;
SRGH). **Mozambique.** N: Lichinga (Vila Cabral), 13.vi.1934, *Torre* 166 (LISC). MS:
Chimoio (Vila Pery), Choa Mts., 1300 m., 25.v.1971, *Torre & Correia* 18660 (LISC). T: 2 km.
N. of Mlangeni near Ntcheu-Dedza road, 1385 m., 25.v.1970, *Brummitt* 11104 (K; LISC). GI:
Bazaruto I., 5 m., 28.x.1958, *Mogg* 28686 (LISC). M: Marracuene, Incomati valley, 2.x.1957,
Barbosa & Lemos 7918 (COI; K; LISC).

Widespread in tropical and southern Africa. In swamps, river banks, sandy ground, upland
grassland, also on disturbed or cultivated ground.

L. erinus in the broad sense taken here is an exceedingly variable species, particularly in
eastern Zimbabwe, where two or three forms may occur in the same area as on the Chimanimani
Mts. Some of the more characteristic forms are here briefly described. In many parts of the FZ-
area there is an annual, erect, rather tall, subglabrous or sparsely pubescent form with narrow,
non-rosulate leaves and an angular stem. This agrees with the type of *L. senegalensis* and is the
predominant form in the northern part of the range of the species, but very similar plants also
occur in the Cape area. It is usually found at lower altitudes in ± wet conditions. At the coast of
Mozambique and Tanzania there is a similar form with ± erect pedicels less than 10 mm. long
(*L. lavandulacea, L. chilawana*). Another form in Zambia, Zimbabwe as well as in Angola,
Namibia, SE. Botswana and Transvaal is ± pubescent with ± rosulate leaves and a ± terete
stem (e.g. *L. rosulata, L. transvaalensis, L. jugosa*). Also in this case very similar forms occur at
the Cape. In the uplands of eastern Zimbabwe and western Mozambique there are, for
example, distinctly perennial, often densely pubescent and ± decumbent forms (*L. altimontis*),
and annuals with ± petiolate, deeply dentate to pinnatifid leaves (*L. polyodon*). All these forms,
however, ± completely intergrade, and intermediates are legion. Under these circumstances I
have not been able to make any satisfactory subdivision of the species.

L. erinus has in tropical Africa very often been called *L. filiformis* or *L. filiformis* var.
natalensis. *L. filiformis*, however, in my opinion is a closely related species endemic to the
Mascarene Islands.

5. **Lobelia kirkii** R.E. Fr., Wiss. Ergebn. Schwed. Rhod. – Kongo-Exped. **1**: 317 (1916). — E.
Wimmer in Engl., Pflanzenr. **IV**. 276 b: 518 (1953) & 276 c: 864 (1968). Syntypes from
Zimbabwe: Victoria Falls, *R.E. Fries* 50 (UPS) & 59 (UPS, syntype; W).
Lobelia natalensis sensu Hemsl. in F.T.A. **3**: 469 (1877).
Lobelia kirkii var. *microphylla* Schltr. in Engl., Bot. Jahrb. **57**: 618 (1922). — E.
Wimmer, op. cit. **IV**. 276 b: 519 (1953) & 276 c: 864 (1968). Syntypes from Zimbabwe:
Victoria Falls, *Engler* 2981 (B) & 2984a (B).

Annual or short-lived perennial, decumbent to ascending, usually much-
branched herb, 8–40 cm. long. Stem sulcate and narrowly winged, glabrous (in the
F.Z. area). Leaves 10–28 × 2–9 mm., crenate to crenate-serrate, the upper narrowly
elliptic, the lower narrowly obovate or spathulate, usually ± obtuse at the apex,
narrowing below into a petiole-like base, glabrous except for the often ciliate leaf-
bases. Flowers in lax, few-flowered racemes; pedicels 8–15(20) mm. long, glabrous;
bracts similar to upper stem leaves, shorter than the pedicels; bracteoles c. 0·2–0·4
mm. long, at the base of the pedicel, occasionally absent. Hypanthium obconical,
8–10-nerved, glabrous. Calyx-lobes narrowly triangular, ± spreading, 2–3·2 mm.
long, entire, glabrous or sparsely ciliate. Corolla 13–15 mm. long, blue, with two
bumps in the mouth of the tube, split to 1–1·5 mm. from the base on the back,
sparsely papillose on the inside of the tube, glabrous outside. Stamens (6)8–9 mm.
long; filaments linear, puberulous and attached to the corolla-tube at the base;
anther-tube 2–2·4 mm. long, glabrous or shortly pubescent on the back of the upper
thecae, the two lower anthers each with a tuft of hairs and a hyaline appendage at the
apex. Ovary subinferior. Capsule 8–10-nerved with 2 valves ± 1 mm. long, inferior
part of capsule ± 4 mm. long. Seeds oblong-elliptic in outline, 0·4 mm. long, very
finely striate, pale brown.

Zambia. S: Victoria Falls, 26.vii.1961, *Angus* 3040 (K; SRGH). **Zimbabwe.** W: Victoria
Falls, 880 m., 17.ix.1954, *Greenway* 8806 (K; SRGH). S: Zimbabwe Ruins, 1.vii.1930,
Hutchinson & Gillett 3333 (K, pro parte).
Also in Angola. At the Victoria Falls growing on rocks and moist ground in the spray zone of
the Falls. The specimen from the Zimbabwe Ruins is the only one in the F.Z. area which is not
from the Victoria Falls and the record needs confirmation.

L. kirkii is an attractive plant, closely related to *L. erinus*. It is distinguished mainly by its
narrowly winged stem, usually shorter pedicels and calyx-lobes, longer stamens and smaller
leaves.

6. **Lobelia trullifolia** Hemsl. in F.T.A. **3**: 466 (1877). — E. Wimmer, in Engl., Pflanzenr. **IV**. 276 b: 490, fig. 80 g 1 (1953). — Brenan in Mem. N.Y. Bot. Gard. **8**: 490 (1954). — E. Wimmer, op. cit. **IV**. 276 c: 861 (1968). — Binns, H.C.L.M.: 25 (1969). Type: Malawi, Manganja range, Mt. Chiradzulu, *Meller* s.n. (K, holotype).

Lobelia melleri Hemsl., op. cit.: 468 (1877). — E. Wimmer, tom. cit.: 529, fig. 88 h (1953) & **IV**. 276 c: 865 (1968), excl. var. *grossidens* E. Wimmer. — Binns, H.C.L.M.: 25 (1969). Type: Malawi, Dakanamaio Island, *Meller* s.n. (K, holotype).

Lobelia nyassae Engl., Pflanzenw. Ost.-Afr. **C**: 401 (1895). Type: Malawi, *Buchanan* 479 (B, holotype; K).

Lobelia buchananii Bak. in Kew Bull. **1898**: 156 (1898). — E. Wimmer, tom. cit.: 491 (1953). Syntypes from Malawi: *Buchanan* 316 (B; BM; E; K, syntype; W) and Mt. Zomba, *Whyte* s.n. (K).

Lobelia intertexta Bak., tom. cit.: 157 (1898). — Hook. f. in Bot. Mag.: t. 7615 (1898). — E. Wimmer, tom. cit.: 529, fig. 88 g (1953). — Brenan in Mem. N.Y. Bot. Gard. **8**: 490 (1954). — E. Wimmer, op. cit. **IV**. 276 c: 865 (1968). — Binns, H.C.L.M.: 25 (1969). — Moriarty, Wild Flowers of Malawi: 793, t. 37 fig. 2 (1975). Type: Malawi, Nyika Plateau, *Whyte* s.n. (B; E; K, holotype).

Lobelia nyikensis Bak., loc. cit. — E. Wimmer, tom. cit.: 529, fig. 88 f (1953). — Binns, H.C.L.M.: 25 (1969). Type: Malawi, Nyika Plateau, *Whyte* s.n. (B; K, holotype).

Lobelia wentzeliana Engl., Bot. Jahrb. **30**: 420 (1901). — E. Wimmer, tom. cit.: 499, fig. 82 e (1953). Type from Tanzania.

Lobelia saliensis E. Wimmer in Notizbl. Bot. Gart. Berl. **12**: 107 (1934). Type from Tanzania.

Lobelia usambarensis Engl. var. *hispidella* E. Wimmer, loc. cit.: 868 (1968). Type from Tanzania.

Lobelia brassiana E. Wimmer in Kew Bull. **7**: 139 (1952); tom. cit.: 777 (1953). — Brenan in Mem. N.Y. Bot. Gard. **8**: 490 (1954). — Binns, H.C.L.M.: 25 (1969). Type: Malawi, Zomba, *Brass* 16040 (K, holotype).

Lobelia melleri f. *pilosula* E. Wimmer, op. cit. **IV**. 276 b: 529 (1953). Type: Malawi, *Buchanan* 120 (B; BM; K, holotype).

Lobelia melleri var. *pulchra* E. Wimmer, loc. cit. Type: Malawi, *Buchanan* 1104 (B; BM; E; G; K; W).

Lobelia intertexta f. *arida* E. Wimmer, tom. cit.: 530 (1953). Type: Malawi, Mt. Mulanje, *Adamson* s.n. (K, holotype).

Lobelia intertexta var. *diversifolia* E. Wimmer, loc. cit. — Binns, H.C.L.M.: 25 (1969). Type: Malawi, Mt. Mulanje, *Forbes* 96 (EA, holotype).

Lobelia trullifolia var. *saliensis* (E. Wimmer) E. Wimmer, tom. cit.: 490 (1953). Type as for *Lobelia saliensis*.

Lobelia trullifolia var. *saliensis* f. *glabricalycina* E. Wimmer, op. cit. **IV**. 276 c: 861 (1968). Type: Malawi, Mt. Mulanje, *Goodier* 267 (K, holotype).

Lobelia melleri var. *trichella* E. Wimmer, tom. cit.: 865 (1968). Type: Malawi, Mt. Mulanje, *Forbes* 157 (EA, holotype).

Lobelia zombaënsis E. Wimmer, tom. cit.: 866 (1968). Type: Malawi, Zomba Plateau, *Benson* 1054 (PRE, holotype; W).

Lobelia usambarensis var. *calantha* E. Wimmer, tom. cit.: 868 (1968). Type: Malawi, Nyika Plateau, *Jackson* 871 (holotype said to be in PRE, but despite a thorough search it has not been found).

Annual or short-lived perennial, erect to decumbent herb, 4–60 cm. tall, often purplish tinged towards the base. Stem usually much branched, very narrowly winged or ribbed, sparsely to densely pubescent or occasionally glabrous. Leaves up to 5–40 × 4–30 mm., ovate to subreniform or rarely narrowly elliptic, narrowing upwards, obtuse to subacute at the apex, cuneate to cordate or rarely attenuate at the base, with a petiole up to 1·5–10(15) mm. long, ± crenate to pinnatifid, densely pubescent to glabrous on both sides or only the upper side pubescent. Flowers in lax leafy racemes or rarely scattered in the upper leaf-axils; pedicels (10)20–45 mm. long, filiform, glabrous or pubescent; upper bracts linear to elliptic or rarely ovate, usually much shorter than the pedicels; bracteoles 0·1–0·8 mm. long, at or near the base of the pedicel, rarely absent. Hypanthium broadly obconical, usually very short, glabrous or pubescent. Calyx-lobes very narrowly triangular, ± erect or spreading, (1·5)2–5(7) mm. long, entire, ciliate, ciliate-pubescent or rarely glabrous. Corolla 3·2–16 mm. long, blue to pale pink, split to 0·6–2 mm. from the base on the back, short pubescent to sparsely papillose inside, glabrous or pubescent on the lobes outside. Filaments linear, at least the two lower pubescent below, rarely all subglabrous, all attached to the corolla-tube at the base. Anther-tube 0·6–1·6(2) mm. long, short pubescent at least on the back, or glabrous; the two lower anthers each with a tuft of hairs and a hyaline appendage at the apex. Ovary subsuperior to semisuperior. Capsule ± obovoid, glabrous or pubescent with prominent valves

1·5–3·2 mm. long, about as long as or usually longer than the inferior part of the capsule. Seeds elliptic to broadly elliptic in outline, ± compressed, 0·3–0·5 mm. long, very finely striate, brown.

1. Corolla 7–16 mm. long; leaves various - - - - - - - - - - 2
– Corolla 3–6·5 mm. long; leaves broadly ovate to subreniform, usually with cordate base 3
2. Annual or perennial up to 60 cm. tall; leaves ± crenate - - - - subsp. *trullifolia*
– Annual up to 15 cm. tall; leaves pinnatifid - - - - - subsp. *pinnatifida*
3. Stem and leaves pubescent with hairs up to 0·5–1·2 mm. long; corolla 3–5 mm. long; annual
   subsp. *minor*
– Stem and leaves pubescent with hairs up to 0·3(–0·4) mm. long; corolla 4–6·5 mm. long;
   annual or perennial - - - - - - - - - - subsp. *rhodesica*

Subsp. **trullifolia**

Annual or short-lived perennial, erect to decumbent herb, 4–60 cm. tall, sparsely to densely pubescent with hairs up to (0·2)0·4–1·5 mm. long, or glabrous. Leaves up to 8–40 × 4–30 mm., ovate to subreniform or rarely narrowly elliptic, cuneate to truncate or rarely cordate or attenuate at the base, ± crenate; petiole up to 2–10(15) mm. long. Corolla (7)9–14(16) mm. long, short pubescent inside, glabrous or pubescent on the lobes outside. Anther-tube 1–1·6(2) mm. long. Seeds 0·4–0·5 mm. long.

Zambia. E: Nyika Plateau by Zambian Rest House, 23.iv.1971, *Anton-Smith* in GHS 214848 (SRGH). **Malawi.** N: Rumphi Distr., Nyika road E. of Kaulime, vi.1960, *Chapman* 765 (BM; SRGH). C: Ntchisi Distr., Ntchisi Mt., 1600 m., 26.vii.1946, *Brass* 16972 (K; SRGH). S: Mt. Mulanje, Luchenya Plateau, near Mission Cottage, 2150 m., 8.v.1963, *Wild* 6135 (K; LISC; SRGH). **Mozambique.** Z: Gúruè, Gúruè Mts., 1400 m., 18.ix.1944, *Mendonça* 2093 (LISC). MS: W. face of Mt. Gorongosa, 1220–1300 m., 10.vii.1969, *Leach & Cannell* 14284 (K; LISC; SRGH).
Also in Ethiopia, Kenya and Tanzania. Grassland, forest margins, stream sides, road sides, often in rocky places or on bare ground.

Subsp. *trullifolia* is very variable in habit, pubescence, leaf-shape, flower-size etc., but as transitional forms are always numerous it seems best not to give taxonomic recognition to any of its variants. More or less narrow-leaved forms are found on Mt. Mulanje in Malawi and on Mt. Gorongosa in Mozambique, and on Gúruè Mts. in Mozambique there is, along with normal plants, also a conspicuous form with the corolla up to 16 mm. long (e.g. *Mendonça* 2176).

Subsp. **pinnatifida** Thulin, subsp. nov.*
Type: Zambia, 130 km. SSW. of Mpika, Muso Hills, *Kornaś* 1810 (K, holotype).

Annual erect herb up to 15 cm. tall, pubescent with hairs up to 0·5 mm. long. Leaves up to 7 × 5·5 mm., ± ovate with cuneate to truncate base, pinnatifid; petiole up to 1·5 mm. long. Corolla blue, 8–10 mm. long, glabrous outside. Anther-tube 1–1·2 mm. long. Seeds 0·4 mm. long.

Zambia. N: Serenje-Mpika road, 1200 m., 5.iv.1961, *Richards* 15002 (EA; K; SRGH).
Not known from elsewhere. Rocky hillsides on moist ground.

Subsp. **minor** Thulin, subsp. nov.†
Type: Zambia, Mbala Distr., rocky hill above Ndundu, *Richards* 11070 (K, holotype).
Lobelia milneana E. Wimmer in Engl., Pflanzenr. **IV**. 276 c: 879 (1968) *nom. invalid.*, quoad specim. pro parte. Orig. colls. from Tanzania.

Annual erect or ascending herb up to 15 cm. tall, pubescent with hairs up to 0·5–1·2 mm. long. Leaves up to 5–13 × 5–14 mm., broadly ovate to subreniform, lower leaves with truncate to cordate base, ± crenate; petiole up to 3–6 mm. long. Corolla pale mauve to pale blue, 3·2–5 mm. long, sparsely puberulous inside, glabrous outside. Anther-tube 0·6–0·8 mm. long, short pubescent to subglabrous on the back. Seeds 0·3–0·35 mm. long.

Zambia. N: 6·5 km. NE. of Kasama, 17.iii.1957, *Savory* 161 (K; SRGH). **Malawi.** N: Rumphi Distr., N. Viphya, Uzumara rain forest, 1750–2050 m., 7.viii.1972, *Pawek* 5621 (K).
Also in Tanzania. Grassland, rocky outcrops or in rock crevices, usually in shade.

* Differt a *L. trullifolia* subsp. *trullifolia* foliis pinnatifidis, usque 7 × 5·5 mm., petiolo usque 1·5 mm. longo.

† Differt a *L. trullifolia* subsp. *rhodesica* habitu semper annuo, corolla 3–5 mm. longa, caulibus et foliis pubescentibus pilis usque 0·5–1·2 mm. longis.

Subsp. **rhodesica** (R.E. Fr.) Thulin, comb. nov. Type: Zambia, Bwana Mkubwa, *R.E. Fries* 449 (B; BR; K; UPS, holotype; W).

  Lobelia rhodesica R.E. Fr., Wiss. Ergebn. Schwed. Rhod.-Kongo-Exped. **1**: 316, t. 20/1–2 (1916). — E. Wimmer in Engl., Pflanzenr. **IV**. 276 b: 489 (1953). Type as above.

Annual or short-lived perennial, erect to decumbent herb, 5–20 mm. tall, ± densely puberulous with hairs up to 0·3(–0·4) mm. long. Leaves up to 5–12 mm. long, 5–12 mm. wide, broadly ovate to subcircular or subreniform, lower leaves truncate to subcordate at the base, ± incised-crenate; petiole up to 13–22 mm. long. Corolla blue or pale blue, 4–6·5 mm. long, sparsely puberulous inside, short pubescent on the lobes outside. Anther-tube 0·6–1·2 mm. long, short pubescent on the back. Seeds 0·3–0·4 mm. long.

  Zambia. N: Mpika Distr., Serenje-Mpika road, 1200 m., 5.iv.1961, *Richards* 14992 (K). C: Mkushi, 27.iii.1961, *Angus* 2524 (FHO). W: Chati, 8.iii.1964, *Fanshawe* 8395 (K; SRGH). Not known from elsewhere. Rock crevices, sandy places.

7. **Lobelia lobata** E. Wimmer in Engl., Pflanzenr. **IV**. 276 c: 864, fig. 24 (1968). Type: Zimbabwe, Matopo Hills, *Eyles* 1040 (BM; PRE, holotype; SRGH).

  Lobelia trullifolia sensu Gibbs in Journ. Linn. Soc., Bot. **37**: 451 (1906) non Hemsl.

  Lobelia rhodesica sensu E. Wimmer, tom. cit.: 860 (1968) non R.E. Fr.

Annual or short-lived perennial, prostrate or decumbent, up to 6–30 cm. long herb, often forming tangling masses. Stems sparsely to much branched, angular, not winged, sparsely to densely short pubescent, often rooting at lower nodes. Leaves petiolate, up to 5–15 × 5–15 mm., suborbicular to subreniform or upper leaves rarely elliptic, usually acute at the apex, truncate or rarely cuneate at the base, incised-dentate, short pubescent on both sides or rarely subglabrous; venation prominent beneath; petiole up to 6–12 mm. long, pubescent. Flowers solitary, terminal or axillary; pedicels 8–25 mm. long, filiform, glabrous or short pubescent, ebracteolate. Hypanthium obovoid to obconical, 8–10-nerved, glabrous or short pubescent. Calyx-lobes very narrowly triangular or subulate, erect or spreading, 1·6–2·4 mm. long, entire, glabrous or short pubescent. Corolla 8–10 mm. long, blue or pale blue, split to 2–2·5 mm. from the base on the back, puberulous inside towards the base. Filaments linear, all somewhat pubescent below, attached to the corolla-tube at the base. Anther-tube ± 1·2 mm. long, glabrous or short pubescent towards the apex on the back; the two lower anthers each with a tuft of hairs and a hyaline appendage at the apex. Ovary semi-inferior. Capsule obovoid to obconical, 8–10-nerved, glabrous or short pubescent, with 2 valves ± 1 mm. long, inferior part of capsule 2–2·8 mm. long. Seeds broadly elliptic in outline, compressed, ± 0·4 mm. long, finely striate, brown.

  Zimbabwe. W: Matobo Distr., Besna Kobila Farm, 1490 m., iv.1962, *Miller* 8230 (K; LISC; SRGH).
  Not known from elsewhere. Growing under overhanging rocks in moist and shady places.

8. **Lobelia flaccida** (Presl) A.DC. in DC., Prodr. **7**: 360 (1839). — E. Wimmer in Engl., Pflanzenr. **IV**. 276 b: 508, fig. 84 a (1953). Type from S. Africa (Cape Prov.).

  Rapuntium flaccidum Presl, Prodr. Monogr. Lobel.: 13 (1836). Type as above.

  Lobelia krebsiana A.DC., tom. cit.: 385 (1839). Type from S. Africa (Cape Prov.).

  Lobelia erinus sensu Sonder in Harv. & Sond., F.C. **3**: 544 (1865) pro parte.

  Lobelia filiformis Lam. var. *krebsiana* (A.DC.) E. Wimmer, tom. cit.: 542, fig. 84 c (1953). Type as for *Lobelia krebsiana*.

  Lobelia filiformis var. *krebsiana* f. *rusticana* E. Wimmer, tom. cit.: 542 (1953). Syntypes from S. Africa.

Annual or short-lived perennial, decumbent to erect herb, 5–60 cm. tall. Stem angular, narrowly winged, glabrous or pubescent on the wings with rather rigid hairs. Leaves (5)20–70 × 2–10(17) mm., crenate to serrate or subpinnatifid, the upper linear to lanceolate, the lower linear to spathulate or suborbicular, acute to obtuse at the apex, sessile or narrowing below into a petiole-like base or an up to 7 mm. long petiole, glabrous or lower leaves ± pubescent. Flowers in lax racemes; pedicels 5–50 mm. long, glabrous or papillose to short pubescent; bracts similar to upper stem leaves, shorter to longer than the pedicels; bracteoles c. 0·4–1·6 mm. long, linear, at or near the base of the pedicel. Hypanthium narrowly obconical to obovoid, 8–10-nerved, glabrous or papillose to short pubescent. Calyx-lobes narrowly triangular, erect or spreading, 1·2–6 mm. long, with one to three pairs of teeth at the margin or entire, glabrous or ciliate to pubescent. Corolla 8–15 mm. long, blue, mauve or white, with two bumps in the mouth of the tube, split to 0·3–1·5 mm. from the base on the back, papillose on the inside of the tube, glabrous or pubescent on the outside. Filaments linear, puberulous at the base or ± along their entire length, attached to the corolla-tube at the base. Anther-tube (1·2)2–2·4 mm. long, ± pubescent on the back of the upper thecae, or glabrous, the two lower anthers each with a tuft of hairs and a hyaline appendage at the apex. Ovary subinferior. Capsule 8–10-nerved with 2 valves 1–1·5 mm. long, inferior part of capsule 3–7 mm. long. Seeds elliptic to broadly elliptic in outline, somewhat compressed, 0·35–0·5 mm. long, very finely striate, brown.

Subsp. **mossiana** (R. Good) Thulin, comb. nov. Type from S. Africa (Transvaal).
    Lobelia mossiana R. Good in Journ. Bot., Lond. **62**: 49 (1924). — E. Wimmer in Engl., Pflanzenr. **IV**. 276 b: 551 (1953) & 276 c: 868 (1968) pro parte. Type as for *Lobelia flaccida* subsp. *mossiana*.
    Lobelia senegalensis A.DC. var. *subaspera* E. Wimmer, tom. cit.: 552 (1953) & **IV**. 276 c: 869 (1968) pro parte. Syntypes from Transvaal, Zimbabwe: Chirinda, *Swynnerton* 513 (B), Umtali, *Engler* 3205 (B) & Odzani R., *Teague* 559 (K) and Mozambique: *Johnson* 116 (K).

Leaves up to 25–70 × 2–10 mm., serrate to sparsely serrulate, linear to lanceolate, the basal ones linear to obovate. Pedicels 5–14(20) mm. long, papillose or glabrous; bracts usually equalling or longer than pedicels. Calyx-lobes 3–5·5 mm. long, entire or very occasionally toothed, ciliate to pubescent. Corolla usually white, pale mauve or ± pale blue, 9–13 mm. long.

    Zimbabwe. E: Umtali Golf Course, 1100 m., 5.xii.1950, *Chase* 3232 (BM; COI; SRGH). S: Belingwe Distr., Mt. Buhwa, 1000 m., 1.v.1973, *Biegel, Pope & Simon* 4250 (MO; SRGH). **Mozambique**. MS: Serra Mocuta, 1250 m., 3.vi.1971, *Biegel & Pope* 3525 (LISC; SRGH). M: Libombos, near Namaacha, Mt. M'Ponduine, 800 m., 22.ii.1955, *Exell, Mendonça & Wild* 506 (BM; LISC; SRGH).
    Also in Transvaal, Natal and Swaziland. Woodland, grassland, stream sides and on disturbed or cultivated ground.

    In all material from Zimbabwe and Manica e Sofala the pedicels and calyces are papillose, while in material from S. Mozambique as well as from parts of Natal the pedicels are glabrous and the calyx-lobes only are ciliate. Such specimens may be difficult to distinguish from *L. erinus*. The narrow wings along the stem in *L. flaccida* subsp. *mossiana* is the most reliable character.
    Subsp. *flaccida* is distributed from Transvaal to E. Cape.

9. **Lobelia welwitschii** Engl. & Diels in Engl., Bot. Jahrb. **26**: 116 (1898). — R.E. Fr., Wiss. Ergebn. Schwed. Rhod.-Kongo-Exped. **1**: 317 (1916). — E. Wimmer in Engl., Pflanzenr. **IV**. 276 b: 544 (1953) & 276 c: 867 (1968). TAB. **22**. Type from Angola.
    Lobelia welwitschii var. *albiflora* Hiern, Cat. Afr. Pl. Welw. **1**: 627 (1898). — E. Wimmer, loc. cit. (1953). Type from Angola.
    Lobelia stolzii Schltr. in Engl., Bot. Jahrb. **57**: 621 (1922). Type from Tanzania.

Perennial, erect or ascending, rhizomatous herb, glabrous or very occasionally pubescent, 10–50 cm. tall. Stems not or sparsely branched, angular, not or very narrowly winged. Leaves up to 12 × 1–6 mm., sparsely denticulate; upper leaves narrowly lanceolate, acute, ± erect; lower leaves lanceolate to elliptic or broadly ovate, sometimes obtuse, ± spreading; epidermis above with large cells (0·05–0·1 mm. long) particularly towards the apex and along margins. Flowers in lax racemes; pedicels up to 30 mm. long; bracts similar to upper stem leaves; bracteoles ± 1 mm. long, linear, on lower half of the pedicel. Hypanthium obconical to obovoid, c. 10-nerved. Calyx-lobes narrow, often subulate, erect, 2·5–6 mm. long, with 1–2 pairs of teeth at the margin. Corolla 6·5–10·5 mm. long, blue or rarely white, with two bumps in the mouth of the tube, split to c. 0·4 mm. from the base on the back, short pubescent inside on the bottom of the tube. Filaments linear, pubescent at the base, connate for most of their length, attached to the corolla-tube at the base. Anther-tube 1·5–2 mm. long, shortly pubescent on the back of the upper thecae or glabrous, the two lower anthers each with a tuft of hairs and a hyaline appendage at the apex. Ovary subinferior. Capsule c. 10-nerved with 2 valves ± 1 mm. long, inferior part of capsule 4–7 mm. long. Seeds elliptic-oblong in outline, somewhat compressed, 0·4–0·5 mm. long, almost smooth, brown.

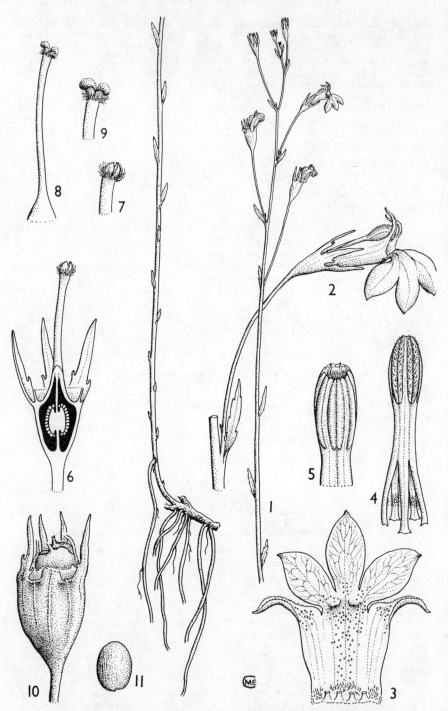

Tab. 22. LOBELIA WELWITSCHII. 1, habit (× 1), from *Semsei* 2482; 2, flower (× 5); 3, corolla spread out (× 6); 4, stamens with anther-tube (× 12); 5, anther-tube from below (× 16), 6, longitudinal section of ovary from young flower (× 10); 7, style-apex from young flower (× 16); 8, style and stigma from older flower (× 10); 9, style-apex from older flower (× 16), 2–9 from *Milne-Redhead* 2654; 10, capsule (× 6); 11, seed (× 40), 10–11 from *Semsei* 2482. From F.T.E.A.

Zambia. N: 61 km. W. of Isoka, 1400 m., 12.viii.1966, *Gillett* 17419 (BR; EA; K; LISC; SRGH). W: Mwinilunga Distr., Kalenda dambo, 8.x.1937, *Milne-Redhead* 2654 (B; BM; BR; K). C: Chakwenga Headwaters, 100–129 km. E. of Lusaka, 27.iii.1965, *Robinson* 6474 (B; K; SRGH). E: Lunkwakwa, 12.x.1967, *Mutimushi* 2140 (K). **Zimbabwe.** C: Marandellas, 20.iii.1948, *Corby* 41 (SRGH). E: Reserve road N. of Inyanga village, 1820 m., 28.i.1958, *Norman* R 34 (K). S: Zimbabwe, Chilopepo R., 11.viii.1929, *Rendle* 314 (BM). **Malawi.** N: Nyika Plateau, Lake Kaulime, 2150 m., 15.xi.1958, *Robson* 631 (BM; LISC; SRGH). C: 5 km. NE. of Mchinji on Kasungu road, 1215 m., 27.iv.1970, *Brummitt* 10204 (K; LISC; MAL; SRGH). **Mozambique.** N: Maniamba, 1100 m., 2.iii.1964, *Torre & Paiva* 10943 (LISC).
Also in Cameroun, Ethiopia, Uganda, Kenya, Tanzania, Zaire and Angola. Bogs, swampy ground, lake edges and stream sides.

10. **Lobelia livingstoniana** R.E. Fr., Wiss. Ergebn. Schwed. Rhod.-Kongo-Exped. **1**: 317, t. 20/3, 4 (1916). — E. Wimmer in Engl., Pflanzenr. **IV**. 276 b: 544 (1953). Type: Zambia, Victoria Falls, Livingstone I., *R.E. Fries* 181 (B; UPS, holotype; W).
    Lobelia borleana E. Wimmer, op. cit. **IV**. 276 c: 867, fig. 26a (1968). Type: Zambia, Barotse Valley, Sefula, *Borle* 238 (PRE, holotype).

Perennial, erect or ascending, rhizomatous herb, 7–25 cm. tall. Stems not or sparsely branched, angular, narrowly winged, glabrous. Leaves 5–13 × 2–4 mm., denticulate, all leaves of similar shape, lanceolate to narrowly ovate or elliptic, acute, rather spreading; epidermis with small cells on both surfaces. Flowers in lax racemes; pedicels up to 22 mm. long, glabrous or papillose; at least lower bracts similar to stem leaves; bracteoles linear, 0·2–0·5 mm. long, at or near base of pedicel. Hypanthium obconical to obovoid, c. 10-nerved. Calyx-lobes narrowly triangular, erect or spreading, 1·2–2·5 mm. long, with 1–2 pairs of teeth at the margin. Corolla 7–9·5 mm. long, blue, with two distinct bumps in the mouth of the tube, split to c. 0·4 mm. from the base on the back, papillose inside on the bottom of the tube. Filaments linear, the 2 lower pubescent at the base, connate for most of their length, attached to the corolla-tube at the base. Anther-tube ± 1·5 mm. long, shortly pubescent on the back of the upper thecae, the two lower anthers each with a tuft of hairs and a hyaline appendage at the apex. Ovary subinferior. Capsule c. 10-nerved with 2 valves ± 1 mm. long, inferior part of capsule 2–4 mm. long. Seeds elliptic-oblong in outline, compressed, 0·4–0·5 mm. long, almost smooth, brown.

Zambia. B: Sefula, 910 m., viii.1921, *Borle* 238 (PRE). W: Mwinilunga Distr., Luakera Falls, 15.vi.1938, *Price* M4 (K). S: Victoria Falls, Livingstone I., 910 m., 27.viii.1911, *Rogers* 7415 (BM).
Also in Angola. Swamps, river sides, apparently in very wet situations.

Very closely allied to *Lobelia welwitschii* but distinguished by its leaves being of a more uniform width along the entire length of the stem, by shorter calyx-lobes, shorter bracteoles, somewhat more broadly winged stems and by the cells of the leaf epidermis being small and of a uniform size on both sides of the leaf.

11. **Lobelia uliginosa** E. Wimmer in Engl., Pflanzenr. **IV**. 276 c: 867, fig. 25a (1968). Type from Tanzania.

Annual erect glabrous herb, 15–40 cm. tall. Stem not or sparsely branched, angular, very narrowly winged, purplish below. Leaves inconspicuous, up to 3–5 × 1–1·5 mm., sparsely denticulate; upper leaves lanceolate, acute, erect; lower leaves often ovate or rounded, sometimes obtuse, spreading or reflexed. Flowers in lax, up to 15-flowered, one-sided racemes; pedicels up to 10(–16) mm. long; bracts similar to upper stem leaves; bracteoles minute, at the base of the pedicel. Hypanthium obovoid to subglobose, c. 10-nerved, glabrous or almost so. Calyx-lobes narrowly triangular to subulate, erect, 1–3 mm. long, with 1–2 pairs of teeth near the base. Corolla 6–9 mm. long, blue, with two bumps in the mouth of the tube, split to c. 0·4 mm. from the base on the back, sparsely papillose inside on the bottom of the tube. Filaments linear, pubescent at the base, connate for most of their length, attached to corolla-tube at the very base only. Anther-tube 1·2–1·8 mm. long, shortly pubescent on the back of the upper thecae, the two lower anthers each with a tuft of hairs and a hyaline appendage at the apex. Ovary subinferior. Capsule c. 10-nerved, glabrous or almost so with 2 valves ± 1 mm. long, inferior part of capsule 2–3 mm. long. Seeds elliptic-oblong in outline, compressed, ± 0·4 mm. long, almost smooth, brown.

Zambia. N: Kawambwa, 1340 m., 21.vi.1957, *Robinson* 2341 (K; SRGH).
Also in Tanzania. Bogs, swampy ground, wet sand or on wet rocks.

12. **Lobelia sapinii** De Wild. in Bull. Jard. Bot. Brux. **3**: 261 (1911). — E. Wimmer in Engl.,
Pflanzenr. **IV**. 276 b: 511 (1953); F.W.T.A., ed. 2, **2**: 313 (1963); in Engl. op. cit. 276 c: 863
(1968). Type from Zaire.
  Lobelia ledermannii Schltr. in Engl., Bot. Jahrb. **57**: 619 (1922). Syntypes from
Cameroon.
  Lobelia lelyana E. Wimmer, op. cit. **IV**. 276 b: 465, fig. 78c (1953); F.W.T.A., ed. 2, **2**:
313 (1963). Type from Nigeria.

Annual erect herb, 8–40 cm. tall. Stem sparsely to much branched, triangular with
narrowly winged edges, sparsely pubescent. Leaves up to 25–55 mm. long, 10–35
mm. wide, ovate, acute or obtuse at the apex, abruptly narrowing below into a
petiole-like base, serrate to doubly dentate or crenate, sparsely pubescent on both
sides; venation prominent beneath. Flowers in lax terminal and axillary racemes;
pedicels 3–12 mm. long, glabrous or pubescent; upper bracts 1·5–3·5 mm. long,
much shorter than pedicels; bracteoles minute, ± ovate, at the base of the pedicel.
Hypanthium narrowly obconical, 8-nerved, usually glabrous. Calyx-lobes very
narrowly triangular, ± erect, 2–2·5(3·5) mm. long, with a pair of teeth at the base,
glabrous or ciliate. Corolla 3·5–4·5 mm. long, blue, or white with blue markings, split
practically to the base on the back, papillose inside on the bottom of the tube.
Filaments linear, at least the two lower pubescent below, attached to the corolla-tube
at the base. Anther-tube 0·6–1 mm. long, ± pilose on the back of the upper thecae;
the two lower anthers each with a tuft of hairs and a hyaline appendage at the apex;
the upper anthers ± papillose at the apex. Ovary subinferior. Capsule narrowly
obconical, 8-nerved, with 2 valves ± 0·8 mm. long, inferior part of capsule 5–7 mm.
long. Seeds lenticular, 0·4–0·5 mm. long, finely striate, brown.

Malawi. N: between Kondowe and Karonga, 610–1820 m., vii.1896, *Whyte* s.n. (K).
  Also in Sierra Leone, Ivory Coast, Ghana, Togo, Nigeria, Cameroun, Zaire and Tanzania.
Grassland and woodland, usually in rocky places.

13. **Lobelia fervens** Thunb., Fl. Cap. **2**: 46 (1818). — Sonder in F.C. **3**: 548 (1865). — Hemsl.
in F.T.A. **3**: 468 (1877). Type perhaps from Madagascar, but erroneously said to be from
the Cape.
  Lobelia madagascariensis Roem. & Schultes, Syst. Veg. **5**: 67 (1819). Type from
Madagascar.
  Lobelia pterocaulon Klotzsch in Peters, Reise Mossamb. Bot. **1**: 299 (1862). Type from
the Comoro Is.
  Lobelia subulata Klotzsch, tom. cit.: 300 (1862). Type from Zanzibar.
  Lobelia asperulata Klotzsch, loc. cit. (1862). Type from Zanzibar.
  Lobelia humilis Klotzsch, tom. cit.: 301 (1862). Type: Mozambique, Cabaceira, *Peters*
s.n. (B, holotype †).
  Lobelia petersiana Klotzsch, tom. cit.: 302 (1862). Type: Mozambique, Cabaceira,
Peters s.n. (B, holotype).
  Lobelia fervens var. *asperulata* (Klotzsch) Sond., loc. cit. Type as for *Lobelia asperulata*.
  Lobelia trialata Hamilt. ex D. Don var. *grandiflora* Chiov., Result. Scient. Miss.
Stefanini-Paoli Somalia Ital.: 108 (1916). Type from Somalia.
  Lobelia mearnsii De Wild., Pl. Bequaert. **1**: 293 (1922) as *"mearnsi"*. — E. Wimmer in
Engl., Pflanzenr. **IV**. 276 b: 469 (1953). Type from Kenya.
  Lobelia odontoptera Schltr. in Engl., Bot. Jahrb. **57**: 620 (1922). — E. Wimmer, tom.
cit.: 467 (1953). Type from Tanzania.
  Lobelia anceps sensu E. Wimmer in Notizbl. Bot. Gart. Berl. **12**: 103 (1934); in Engl.,
Pflanzenr. **IV**. 276 b: 477 (1953) pro parte; in Fl. Madag. **186**. Lobéliacées: 7, fig. 1/5–7
(1953); in Pflanzenr. **IV**. 276 c: 859 (1968) pro parte. — Badré & Cadet in Adansonia, **11**:
673 (1972). — Badré in Fl. Mascareignes **111**. Campanulacées: 5, fig. 3: 8–13 (1976) non
L.f.
  Lobelia anceps L.f. var. *asperulata* (Klotzsch) E. Wimmer in Annal. Naturhist. Mus.
Wien **56**: 346 (1948); tom. cit.: 479 (1953); in Fl. Madag. **186**. Lobéliacées: 7 (1953). —
Badré & Cadet, tom. cit.: 674 (1972). — Badré, tom. cit.: 6 (1976). Type as for *Lobelia
asperulata*.
  Lobelia odontoptera var. *depilis* E. Wimmer, tom. cit.: 468 (1953). Type from Tanzania.

Annual or short-lived perennial, erect, ascending or decumbent herb, 10–60 cm.
tall. Stems often rooting at lower nodes, sparsely branched, triangular and broadly
winged, usually glabrous. Leaves up to 20–50 × 3–20 mm., linear or narrowly elliptic
to ovate, narrowing upwards, acute to obtuse at the apex, attenuate to cuneate below,
often with an up to 5 mm. long petiole-like base, serrate-crenate, glabrous or rarely
pubescent; at least the midvein prominent beneath. Flowers in long lax terminal and
axillary leafy racemes; pedicels up to 5–13 mm. long, glabrous, papillose or
pubescent; upper bracts linear to narrowly elliptic, longer than or rarely equalling

pedicels; bracteoles 0·3–0·8(2) mm. long, at the base of the pedicel. Hypanthium narrowly obconical to subspherical, c. 8-nerved, glabrous or rarely pubescent. Calyx-lobes narrowly triangular to subulate, ± erect, spreading, or recurved in fruit, 1·2–3·6(5) mm. long, entire or rarely with a single tooth at the margin, glabrous or rarely ciliate. Corolla 4–11 mm. long, blue with darker markings, split to 0·1–1 mm. from the base on the back, glabrous outside, papillose or short pubescent inside on the bottom of the tube. Filaments linear, at least the two lower pubescent below, attached to the corolla-tube at the base. Anther-tube 0·8–2 mm. long, short pubescent on the back at least near the apex or rarely glabrous; the two lower anthers each with a tuft of hairs and a hyaline appendage at the apex. Ovary subinferior. Capsule narrowly obconical to subspherical, c. 8-nerved, glabrous or rarely pubescent, with 2 valves 0·6–1·2 mm. long, inferior part of capsule 2·4–6 mm. long. Seeds elliptic in outline, compressed, 0·35–0·45 mm. long, ± distinctly reticulate to finely striate, brown.

Calyx-lobes erect or ± spreading in fruit; corolla 6–11 mm. long, split to 0·1–0·3(0·6) mm. from the base on the back; inferior part of capsule 2·8–6 mm. long; seeds finely and ± distinctly reticulate - - - - - - - - - - - - subsp. *fervens*
Calyx-lobes recurved in fruit; corolla 4–8 mm. long, split to (0·3) 0·4–0·6(1) mm. from the base on the back; inferior part of capsule 2·4–3·2 mm. long; seeds finely striate
subsp. *recurvata*

Subsp. **fervens**

Zimbabwe. E: Melsetter Distr., Ngorima T.T.L. (East), bank of Haroni R., 400 m., 25.xi.1967, *Simon & Ngoni* 1303 (K; LISC; SRGH). **Mozambique.** N: Angoche, Missão Malatane, 7.xi.1936, *Torre* 963 (COI; LISC). Z: Maganja da Costa, 15 m., 15.xi.1966, *Torre & Correia* 14659 (LISC). T: M'tamba, *Stocks* 93 (K), MS: Cheringoma, 26.v.1942, *Torre* 4213 (LISC). M: Matola, 10 m., 10.xii.1897, *Schlechter* 11698 (B; BM; G; K; W).
Also in Ethiopia, Somalia, Kenya, Tanzania, Comoro Is., Madagascar, Réunion and in Brazil (introduced). Grassland, forest margins, road sides, stream sides or on coastal sand, often in damp places.

Subsp. **recurvata** (E. Wimmer) Thulin, comb. nov. Type from Tanzania.
Lobelia anceps L.f. var. *recurvata* E. Wimmer in Notizbl. Bot. Gart. Berl. **12**: 103 (1934); in Engl., Pflanzenr. **IV**. 276 b: 479 (1953). Type as for *Lobelia fervens* subsp. *recurvata*.
Lobelia fervens sensu Hiern, Cat. Afr. Pl. Welw. **1**: 627 (1898).

Zambia. N: Mweru-wa-ntipa swamp, Kasongole, 940 m., 3.viii.1962, *Tyrer* 231 (BM). **Zimbabwe.** S: Nuanetsi Distr., Malangwe R., SW. Mateke Hills, 610 m., 6.v.1958, *Drummond* 5567 (K; PRE; SRGH). **Mozambique.** Z: Mocuba, Namagoa, 60–120 m., vii.1945, *Faulkner* 238 (BR; K; PRE; SRGH).
Also in Uganda, Kenya, Tanzania, Zaire, Burundi and Angola. On moist ground in grassland or woodland.

The two subspecies are easily distinguished within the F.Z. area, but intermediates occur in Tanzania.
Two specimens from N. Malawi (*Pawek* 170 and 4072) as well as a single collection from S. Tanzania cannot be satisfactorily placed in any of the two subspecies recognised here. In their broad bracts, often shorter than the pedicels, ± reflexed calyx-lobes, and comparatively large seeds (0·5 mm.) they come very close to *L. dichotoma* Miq. (1856), a species known from India, Ceylon and Java. In Fl. Malesiana *L. dichotoma* is regarded as a synonym of *L. heyneana*, an opinion with which I cannot agree. Instead it is closely related to *L. fervens*.

14. **Lobelia chireensis** A Rich., Tent. Fl. Abyss. **2**: 6 (1850). — Hemsl. in F.T.A. **3**: 468 (1877). — E. Wimmer in Engl., Pflanzenr. **IV**. 276 b: 467 (1953). Type from Ethiopia.
Lobelia victorialis E. Wimmer, tom. cit.: 776 (1953). Type: Zimbabwe/Zambia, Victoria Falls, Herb. *Moss* 18672 (J; W, holotype).

Annual, ± erect herb, 8–30 cm. tall. Stem not or sparsely branched, triangular with broadly winged eges, glabrous or with a few short and stout hairs. Leaves 7–16 × 2–5 mm., subcircular to elliptic at the base of the plant, gradually narrowing upwards to the linear-oblong bracts, obtuse to acute, denticulate, glabrous or with a few short stout hairs mainly on midvein beneath. Flowers in lax leafy racemes; pedicels 4–7 mm. long, glabrous, papillose or with a few short hairs; bracteoles linear, 0·4–0·5 mm. long, at or near base of pedicel. Hypanthium subglobose, 8–10-nerved, glabrous or with short retrorse hairs. Calyx-lobes narrowly triangular, erect, 1·4–2·8 mm. long, entire, glabrous or with a few short hairs on the midvein outside. Corolla 3·5–4·5 mm. long, blue with darker markings and two bumps in the mouth of the tube, split practically to the base on the back, pubescent inside on the bottom of the tube. Filaments linear, the 2 lower pubescent at the base, connate for most of their length, attached to the corolla-tube at the base. Anther-tube $\pm$ 0·8 mm. long, sparsely and shortly pubescent on the back of the upper thecae, the two lower anthers each with a tuft of hairs and a hyaline appendage at the apex. Ovary semi-inferior. Capsule subglobose, 8–10-nerved with 2 valves $\pm$ 1 mm. long, inferior part of capsule 2–2·5 mm. long. Seeds elliptic-oblong in outline, somewhat compressed, 0·4–0·45 mm. long, almost smooth, brown.

Zambia. S: Mazabuka-Kaleya, 15.iv.1963, *van Rensburg* 1921 (K; SRGH). **Zimbabwe.** W: Victoria Falls, S. bank in front of Main Falls, 30.viii.1947, *Greenway & Brenan* 8029 (K). Also in Ethiopia, Burundi and Tanzania. Wet and muddy places in grassland.

15. **Lobelia molleri** Henriq. in Bol. Soc. Brot. **10**: 137 (1893). — E. Wimmer in Engler, Pflanzenr. **IV**. 276 b: 472 (1953); F.W.T.A., ed. 2, **2**: 313 (1963); op. cit. **IV**. 276 c: 858 (1968). Type from S. Tomé.
    Lobelia thomensis Engl. & Diels in Engl., Bot. Jahrb. **26**: 114 (1898). Type from S. Tomé.
    Lobelia butaguensis De Wild. in Rev. Zool. Afr. **8**, Suppl. Bot.: 25 (1920). Types from Zaire.
    Lobelia dealbata E. Wimmer in Notizbl. Bot. Gart. Berl. **12**: 104 (1934); in Engl., Pflanzenr. **IV**. 276 b: 543, fig. 89d (1953) & 276 c: 867 (1968). Type from Tanzania.
    Lobelia molleri var. *butaguensis* (De Wild.) E. Wimmer ex Robyns, Fl. Parc Nat. Alb. **2**: 409 (1947). Type as for *Lobelia butaguensis*.
    Lobelia molleri f. *latifolia* E. Wimmer in Kew Bull. **7**: 138 (1952); in tom. cit.: 766 (1953) & **IV**. 276 c: 858 (1968). Type: Zimbabwe, Chirinda, *Wild* 2142 (K, holotype; SRGH).

Annual or short-lived perennial, decumbent, 10–75 cm. long. Stems usually many and much branched, often rooting at lower nodes, triangular with narrowly winged edges, glabrous or sparsely pubescent. Leaves up to 20–70 × 7–20(30) mm., narrowly elliptic or lanceolate to ovate, gradually narrowing upwards on the stem, acute or obtuse at the apex, gradually or abruptly narrowing below into a petiole-like base, serrate or crenate, sparsely pubescent on both sides or only above, ciliate at the base; venation prominent beneath. Flowers in lax leafy racemes: pedicels 10–20 mm. long, shorter than or equalling the bracts, glabrous, with minute bracteoles at the base. Hypanthium narrowly obconical, 8–10-nerved, glabrous. Calyx-lobes narrowly triangular to almost filiform, erect or somewhat spreading, 2·5–4(5) mm. long, entire, ciliate. Corolla 4·5–6 mm. long, white with purplish or greenish markings in throat, split practically to the base on the back, shortly pubescent inside on the bottom of the tube. Filaments linear, all somewhat pubescent below, connate for most of their length, attached to the corolla-tube at the very base. Anther-tube 1·2–1·6 mm. long, glabrous on the back, the two lower anthers each with a tuft of hairs and a hyaline appendage at the apex. Ovary subinferior. Capsule narrowly obconical to subcylindrical, 8–10-nerved, with 2 valves $\pm$ 1 mm. long, inferior part of capsule 6–13 mm. long. Seeds broadly elliptic in outline, somewhat compressed, 0·45–0·6 mm. long, striate, brown.

Zimbabwe. E: Inyanga Distr., descent to Pungwe, 1430 m., 10.vi.1957, *Goodier & Phipps* 103 (BR; K; SRGH). **Malawi.** N: 8 km. E. of Mzuzu, Roseveare's, 1220 m., 20.vii.1973, *Pawek* 7215 (BR; K; MO; SRGH). Also in S. Tomé, Fernando Po, Cameroon, Sudan, Uganda, Tanzania, Zaire and Burundi. Upland forest, forest margins, stream sides, usually in moist and shady places.

16. **Lobelia heyneana** Roem. & Schultes, Syst. Veg. **5**: 50 (1819), as "*heyniana*". — E. Wimmer in Ann. Naturhist. Mus. Wien **56**: 344 (1948); in Engl., Pflanzenr. **IV**. 276 b: 473 (1953), excl. var. *intercedens* and var. *inconspicua*. — Moeliono in Fl. Males., ser. 1. **6**: 129 (1960) pro parte. Type from India.
    Lobelia trialata Hamilt. ex D. Don, Prodr. Fl. Nepal.: 157 (1825). — Clarke in Fl. Br. Ind. **3**: 425 (1881). Type from Nepal.
    Lobelia laurentia sensu A. Rich., Tent. Fl. Abyss. **2**: 7 (1850) pro parte, non L.
    Lobelia umbrosa Hochst. ex Hemsl. in F.T.A. **3**: 468 (1877). Type from Ethiopia.
    Lobelia trialata var. *umbrosa* (Hochst. ex Hemsl.) Chiov., Result. Scient. Miss. Stefanini-Paoli Somalia Ital.: 109 (1916). Type as above.
    Lobelia trialata var. *asiatica* Chiov., loc. cit. (1916) *nom. illegit.* Type as for *Lobelia trialata*.
    Lobelia heyneana var. *viridissima* E. Wimmer, op. cit. **IV**. 276 c: 858 (1968). Type: Zambia, Mapanza Mission, *Robinson* 107 (K, holotype).

Annual erect or ascending herb, 5–25 cm. tall. Stem sparsely to much branched, triangular and broadly winged, glabrous or rarely pubescent (not in the FZ-area). Leaves up to 13–35 × 8–20 mm., ovate to suborbicular, gradually narrowing upwards, acute or obtuse at the apex, gradually or abruptly narrowing below into a petiole-like base, usually rather bluntly and shallowly dentate-crenate, glabrous or rarely pubescent on both sides (not in the FZ-area); venation prominent beneath. Flowers in lax terminal and axillary leafy racemes; pedicels up to 10–20 mm. long, glabrous or papillose; upper bracts lanceolate to ovate or elliptic, shorter to longer than pedicels; bracteoles 0·2–0·8 mm. long, at or near the base of the pedicel. Hypanthium obovoid to obconical, c. 8-nerved, glabrous. Calyx-lobes subulate, erect or somewhat spreading, 1·2–3·2 mm. long, entire or rarely with a single tooth at the margin, glabrous or ciliate. Corolla 3·5–5·2 mm. long, pink, pale mauve or pale blue, usually with darker markings, split to 0·4–0·5 mm. from the base on the back, pubescent inside on the bottom of the tube. Filaments linear, the two lower pubescent below, attached to the corolla-tube at the base. Anther-tube 0·6–1 mm. long, pubescent at least towards the apex on the back or glabrous; the two lower anthers each with a tuft of hairs and a hyaline appendage at the apex. Ovary subinferior. Capsule obovoid to narrowly obconical, c. 8-nerved, with 2 valves ± 1 mm. long, inferior part of capsule 2·5–5 mm. long. Seeds broadly elliptic in outline, compressed, 0·35–0·45 mm. long, finely striate, brown.

Zambia. N: 42 km. from Mporokoso on the Sumbu road, 7.iv.1970, *Anton-Smith* 674 (SRGH). E: 13 km. W. of Kachalola, 400 m., 17.iii.1959, *Robson* 1751 (K). S: Mapanza Mission, 1060 m., 28.ii.1953, *Robinson* 107 (K).
Also in Ethiopia, Tanzania, Zaire, Oman and from India east to China. Grassland or woodland at road sides, on termite hills, or in cultivations; usually in shady or damp places.

17. **Lobelia inconspicua** A. Rich., Tent. Fl. Abyss. **2**: 8 (1850). — Hemsl. in F.T.A. **3**: 467 (1877). Type from Ethiopia.
    Lobelia maranguensis Engl., Pflanzenw. Ost-Afr. **C**: 401 (1895). — E. Wimmer in Engl., Pflanzenr. **IV**. 276 b: 498, fig. 84 d (1953) & 276 c: 862 (1968). Type from Tanzania.
    Lobelia ilysanthoides Schltr. in Engl., Bot. Jahrb. **57**: 617 (1922). Lectotype from Cameroon.
    Lobelia heyneana Roem. & Schultes var. *inconspicua* (A. Rich.) E. Wimmer, tom. cit.: 475, fig. 79 c (1953); in F.W.T.A., ed. 2, **2**: 313 (1963); in Engler., Pflanzenr. **IV**. 276 c: 858 (1968). Type as for *Lobelia inconspicua*.

Annual erect or ascending herb, 3–15(35) cm. tall. Stem sparsely to much branched, triangular and rather broadly winged, sparsely to densely hirsute. Leaves up to 7–20 × 6–12 mm., broadly ovate to elliptic, gradually narrowing upwards, usually acute at the apex, gradually or abruptly narrowing below into a petiole-like base, serrate or dentate with usually acute teeth, at least lower leaves sparsely to densely pubescent on both sides; venation prominent beneath. Flowers in lax terminal and axillary leafy racemes; pedicels up to 5–10(13) mm. long, glabrous, finely papillose or pubescent; upper bracts lanceolate to ovate or elliptic, equalling or longer than pedicels; bracteoles c. 0·2 mm. long, at the base of the pedicel. Hypanthium obovoid to obconical, c. 8-nerved, glabrous or rarely pubescent. Calyx-lobes subulate, erect or somewhat spreading, 1·6–3·2 mm. long, with one or two pairs of teeth or rarely only a single tooth at the margin, glabrous or ± pubescent. Corolla 2·8–4 mm. long, pale blue or white, usually with darker markings, split to 0·1–0·2 mm. from the base on the back, glabrous or with a few hairs on the lobes outside, pubescent inside on the bottom of the tube. Filaments linear, at least the two lower pubescent below, attached to the corolla-tube at the base. Anther-tube 0·6–0·8 mm. long, glabrous or with a few short hairs on the back; the two lower anthers each with a tuft of hairs and a hyaline appendage at the apex. Ovary subinferior. Capsule obovoid to narrowly obconical, c. 8-nerved, with 2 valves ± 1 mm. long, inferior part of capsule 3–5 mm. long. Seeds broadly elliptic in outline, compressed, 0·35–0·45 mm. long, finely striate, brown.

Malawi. N: Nkhata Bay Distr., 8 km. E. of Mzuzu, Roseveare's, 1220 m., 20.vii.1973, *Pawek* 7217 (K; SRGH).
Also in Nigeria, Cameroon, Ethiopia, Uganda, Kenya, Tanzania, Zaire, Burundi and in India. Upland grassland and woodland, roadsides, old cultivations, often on rocky or gravelly ground.

L. inconspicua has often been confused with the closely allied *L. heyneana* but is pubescent and has denticulate calyx-lobes, shorter pedicels, and often smaller leaves with usually more acute teeth. Also the corolla is split practically to the base dorsally in *L. inconspicua* while in *L. heyneana* the split ends about 0·5 mm. from the base. In herbarium material, however, this can usually only be seen after dissection of a young flower. *L. inconspicua* is also generally found at higher altitudes than *L. heyneana*.

18. **Lobelia adnexa** E. Wimmer in Ann. Naturhist. Mus. Wien **56**: 352 (1948); in Engl., Pflanzenr. **IV**. 276 c: 511, fig. 88 n (1953). Type: N. Malawi, *Whyte* s.n. (B, holotype; K).

Annual erect herb, 6–35 cm. tall. Stem branching from near the base or unbranched, triangular and rather broadly winged, sparsely pubescent with long hairs or glabrous. Leaves up to 11–30 × 6–14 mm., elliptic to broadly ovate, gradually narrowing upwards, acute or subacute at the apex, gradually or abruptly narrowing below into a petiole-like base, serrate-dentate, sparsely pubescent on both sides or glabrous at least beneath; venation prominent beneath. Flowers in lax terminal and axillary leafy racemes; pedicels up to 9–15 mm. long, usually upcurved near the apex in fruit, usually finely papillose; upper bracts linear or lanceolate, equalling or shorter than pedicels; bracteoles 0·3–0·4 mm. long, linear, at or near the base of the pedicel. Hypanthium obovoid to narrowly obconical, 8–10-nerved, glabrous or occasionally sparsely pubescent. Calyx-lobes subulate, erect or spreading, 1·6–3·2 mm. long, entire, glabrous. Corolla 3–4·5 mm. long, blue to pale violet, split to 0·5–0·8 mm. from the base on the back, pubescent with ± retrorse hairs inside on the bottom of the tube. Filaments linear, the two lower pubescent below, attached to the corolla-tube at the base. Anther-tube 0·6–0·8 mm. long, pubescent or glabrous on the back; the two lower anthers each with a tuft of hairs and a hyaline appendage at the apex. Ovary subinferior. Capsule obovoid to narrowly obconical, 8–10-nerved with 2 valves ± 1 mm. long, inferior part of capsule 2·5–5·5 mm. long. Seeds elliptic in outline, somewhat compressed, ± 0·4 mm. long, very finely striate, brown.

Malawi. N: Rumphi Distr., Livingstonia escarpment, 1030 m., 24.iv.1969, *Pawek* 2326 (K). **Zimbabwe.** N: Darwin Distr., Umsengedzi R., 14.v.1955, *Watmough* 125 (SRGH).

Also in Sierra Leone, Cameroon, Nigeria and S. Tanzania. Shady or rocky places, roadside banks.

19. **Lobelia lasiocalycina** E. Wimmer in Fedde, Repert. Spec. Nov. **38**: 81 (1935); in Engl., Pflanzenr. **IV**. 276 b: 473 (1953). Type from Zaire.
   Lobelia ingrata E. Wimmer, op. cit. **IV**. 276 c: 859 (1968). Type: Zambia, Chilongowelo, 19.iv.1952, *Richards* 1468 (K, isotype; the holotype was said to be in PRE, but despite a thorough search it has not been found).

Annual erect herb, 7–20 cm. tall. Stem sparsely to much branched, triangular and rather broadly winged, ± pubescent with long hairs. Leaves up to 10–30 × 7–18 mm., broadly ovate, gradually narrowing upwards, obtuse to acute at the apex, abruptly narrowing below into a petiole-like base, serrate-dentate, ± pubescent on both sides; venation prominent beneath. Flowers in lax terminal and axillary leafy racemes; pedicels up to 5–8 mm. long, usually upcurved near the apex in fruit, finely papillose and sometimes with a few hairs as well; upper bracts linear or lanceolate, equalling or shorter than pedicels; bracteoles ± 0·4 mm. long, linear, at or near the base of the pedicel. Hypanthium obovoid, 8-nerved, pubescent. Calyx-lobes subulate, erect, 1·8–2·5 mm. long, entire, ciliate. Corolla 3·5–4·5 mm. long, white or pale mauve, split to c. 0·6 mm. from the base on the back, short pubescent inside on the bottom of the tube, glabrous or with a few hairs on the lobes outside. Filaments linear, the two lower pubescent below, attached to the corolla-tube at the base. Anther-tube 0·6–0·8 mm. long, pubescent on the back; the two lower anthers each with a tuft of hairs and a hyaline appendage at the apex. Ovary subinferior. Capsule obovoid, 8-nerved with 2 valves ± 1 mm. long, inferior part of capsule ± 2·5 mm. long. Seeds elliptic in outline, somewhat compressed, ± 0·4 mm. long, very finely striate, brown.

Zambia. N: Kalambo Falls, 900 m., 16.iv.1966, *Richards* 21442 (K) in part. **Malawi.** N: Chitipa Distr., Kaseye Mission, 1250 m., 26.iv.1972, *Pawek* 5257 (K; MO; SRGH).

Also in Zaire. Gravelly or stony places, also as a weed.

20. **Lobelia intercedens** (E. Wimmer) Thulin, comb. nov. Type: Malawi, Fort Hill, *Whyte* s.n. (K, holotype).
   Lobelia heyneana Roem. & Schultes var. *intercedens* E. Wimmer in Engl., Pflanzenr. **IV**. 276 b: 475 (1953). Type as above.

Annual erect herb, 6–25 cm. tall. Stem sparsely to much branched, triangular and rather broadly winged, sparsely pubescent or subglabrous below. Leaves up to 15–35 × 10–17 mm., broadly ovate, gradually narrowing upwards, acute to obtuse at the apex, gradually or abruptly narrowing below into a petiole-like base, serrate-dentate, sparsely pubescent on both sides or glabrous; venation prominent beneath. Flowers in lax terminal and axillary leafy racemes; pedicels up to 6–12 mm. long, finely papillose; upper bracts linear or lanceolate, equalling or shorter than pedicels; bracteoles 0·4–0·5 mm. long, linear, inserted at the base, middle or top of the pedicel. Hypanthium obconical, 8–10-nerved, glabrous. Calyx-lobes narrowly triangular, ± spreading, 1·6–4 mm. long, 0·6–1 mm. wide at the base, entire, ciliate. Corolla 4·5–5·5 mm. long, blue or pale blue, split to 1–1·4 mm. from the base on the back, papillose inside on the bottom of the tube. Filaments linear, all pubescent below, attached to the corolla-tube at the base. Anther-tube 0·6–1 mm. long, sparsely pubescent on the back; the two lower anthers each with a tuft of hairs and a hyaline appendage at the apex. Ovary semi-inferior. Capsule obconical, sometimes narrowly so, 8–10-nerved with 2 valves 1·2–2 mm. long, inferior part of capsule 3–4·5 mm. long. Seeds elliptic in outline, somewhat compressed, 0·4–0·5 mm. long, very finely striate, brown.

Zambia. W: Ndola, 18.iv.1966, *Fanshawe* 9675 (K; SRGH). **Zimbabwe.** N: Darwin Distr., near Mutepatepa, 14.iii.1963, *Wild* 6079 (K; SRGH). **Malawi.** N: Chitipa Distr., Misuku Hills, Mughesse, 1580 m., 8.vii.1973, *Pawek* 7093 (K) in part.

Not known from elsewhere. Woodland or dry evergreen forest, at road sides or on termite hills, usually in shade.

21. **Lobelia quarreana** E. Wimmer in Fedde, Repert. Spec. Nov. **38**: 80 (1935); in Engl., Pflanzenr. **IV**. 276 b: 473 (1953). Type from Zaire.

Annual erect herb, 5–40 cm. tall. Stem sparsely to much branched, triangular and rather broadly winged, glabrous or sparsely pubescent with long hairs below. Leaves up to 15–40 × 10–24 mm., broadly ovate, gradually narrowing upwards, usually acute at the apex, gradually or abruptly narrowing below into a petiole-like base, serrate-dentate, sparsely pubescent on both sides; venation prominent beneath. Flowers in lax terminal and axillary leafy racemes; pedicels up to 6–20 mm. long, finely papillose; upper bracts linear or lanceolate, equalling or shorter than pedicels; bracteoles 0·5–1 mm. long, linear, at the base or near the middle of the pedicel. Hypanthium obconical, 8–10-nerved, glabrous. Calyx-lobes narrowly triangular, completely reflexed after the anthesis, 1·6–5 mm. long, c. 1 mm. wide at the base, entire, ciliate. Corolla 4·5–6·5 mm. long, pale blue, pale mauve or white with darker markings, split to c. 1 mm. from the base on the back, glabrous or papillose inside on the bottom of the tube. Filaments linear, all glabrous or the two lower pubescent below, attached to the corolla-tube at the base. Anther-tube 0·6–1 mm. long, glabrous or with a few short hairs on the back; the two lower anthers each with a tuft of hairs and a hyaline appendage at the apex. Ovary semi-inferior. Capsule obconical, 8–10-nerved, with 2 valves 1·5–3 mm. long, inferior part of capsule 1·5–4 mm. long. Seeds elliptic in outline, somewhat compressed, ± 0·5 mm. long, very finely striate, brown.

Zambia. N: Mbala Distr., Kloof Dhul'miti, 1490 m., 14.v.1952, *Richards* 1712 (K; W). S: Mumbwa, *Macaulay* 805 & 830 (K).

Also in Zaire. Woodland or bushland at road sides, on termite hills, or in old cultivations.

22. **Lobelia anceps** L.f., Suppl. Pl.: 395 (1781). — Thunb., Prodr. Fl. Cap. **1**: 40 (1794). — A.DC. in DC., Prodr. **7**: 375 (1839). — Sonder in Harv. & Sond., F.C. **3**: 547 (1865). Type from S. Africa.

Lobelia alata Labill., Nov. Holl. Pl. **1**: 51, t. 52 (1804). — E. Wimmer in Notizbl. Bot. Gart. Berl. **12**: 103 (1934); in Ann. Naturhist. Mus. Wien **56**: 343 (1948); in Engl., Pflanzenr. **IV**. 276 b: 469, fig. 79 a (1953). Type from N. Australia.

Lobelia repens Thunb., tom. cit.: 41 (1794). Type from S. Africa.

Perennial decumbent to erect herb up to 60 cm. tall; stems usually glabrous, winged, often rooting at lower nodes. Leaves somewhat fleshy, usually glabrous; lower ones linear-oblong to spathulate, often petiolate and ± obtuse at the apex, up to 80 × 21 mm., sparsely dentate-serrate or denticulate; upper ones narrower, sessile, often acute. Racemes often lax; pedicels 2–8(10) mm. long, puberulous, shorter or longer than the bracts; bracteoles ± 1 mm. long, at base of pedicel.

Hypanthium narrowly obconical, c. 10-nerved, puberulous or glabrous. Calyx-lobes narrowly triangular, erect, 1–3 mm. long, glabrous or puberulous. Corolla 6–15(18) mm. long, blue or violet, split to ± 0·5 mm. from the base on the back, short pubescent on the inside of the tube near the base, glabrous or sparsely and minutely hairy on the outside; upper lobes linear, not broadened at the base. Stamens 4–11 mm. long, attached to the corolla at the base; filaments linear, connate for most of their length, sparsely pubescent mainly at the base; anther-tube 1–2·4 mm. long, the two lower anthers with a short hyaline appendage and a tuft of hairs at the apex, the three upper anthers usually with a tuft of hairs on the back near the apex. Capsule cylindrical-obconical, up to 12 mm. long, puberulous or glabrous, 10-nerved; valves ± 1·5 mm. long. Seeds broadly elliptic to subcircular in outline, somewhat compressed, ± 0·4 mm. long, finely striate.

Mozambique. M: Bela Vista, between Zitundo and Ponta do Ouro, 3.xii.1968, *Balsinhas* 1411 (LISC, part).
Coastal areas of SE. Africa, Australia, New Zealand and Chile. In boggy grassland, on moist rocks and river banks.

This widespread species was well-known as *L. anceps* for more than 150 years, until Wimmer (1934, 1948, 1953) started to use this name instead for the plant previously known as *L. fervens* Thunb. As this change was based on an obvious mistake in the typification of *L. anceps* (Thulin, in press) there are all reasons to retain this name in its earlier sense.
Wimmer (1953) recognised 5 varieties within this very variable species (as *L. alata*). The single collection known from the F.Z.-area is an erect, robust, large-flowered plant nearest corresponding with var. *ottoniana* (Presl.) E. Wimmer.

23. **Lobelia goetzei** Diels in Engl., Bot. Jahrb. **28**: 501 (1900). — E. Wimmer in Engl., Pflanzenr. **IV**. 276 b: 596 (1953) & 276 c: 876 (1968). Type from Tanzania.
    Lobelia graciliflora E. Wimmer in Notizbl. Bot. Gart. Berl. **12**: 105 (1934); loc. cit. (1953). Type from Tanzania.
    Lobelia chamaedryfolia (Presl) A.DC. var. *confinis* E. Wimmer in Kew Bull. **7**: 142 (1952); tom. cit.: 783 (1953). Type: Zimbabwe, Inyanga, Stapleford, *Hopkins* in GHS 12481 (K; MO; SRGH; W, holotype).
    Lobelia holstii Engl. var. *subhirsuta* E. Wimmer in Kew Bull. **7**: 142 (1952); loc. cit. (1953) & loc. cit. (1968) as "*subhirta*". Type: Zimbabwe, Inyanga Downs, *Sturgeon* in GHS 16954 (SRGH; W, holotype).
    Lobelia chamaedryfolia sensu E. Wimmer, tom. cit.: 595 (1953) quoad specim. ex Zimbabwe.
    Lobelia holstii var. *holstii* sensu E. Wimmer, loc. cit. (1968) quoad specim. ex Zimbabwe.

Annual or perennial ± erect or ascending herb from a taproot. Stems few to many, 15–70 cm. tall, ribbed, ± densely hirsute with ± spreading hairs at least below. Leaves oblanceolate to obovate or elliptic, up to 15–55 × 7–22 mm., acute to obtuse at the apex, attenuate into a petiole-like base or cuneate at the base, hirsute on both sides but with ± appressed hairs above, crenate-serrate; venation distinct beneath. Flowers in lax, 3–10-flowered, pedunculate racemes; pedicels 3–12 mm. long, erect in fruit, strigose; bracts lanceolate to narrowly ovate, 2–5·5 mm. long, entire; bracteoles usually at or near the base of the pedicel, but sometimes higher up, linear-lanceolate, 1·5–3·2 mm. long. Hypanthium broadly obconical to hemispherical, somewhat oblique, ± 8-nerved, ± strigose. Calyx-lobes narrowly triangular, 2–4 mm. long, entire, glabrous to shortly ciliate, somewhat gibbous at the base. Corolla 8–13(14) mm. long, pink, mauve, purple or rarely white, with two longitudinal yellowish crests in the mouth of the tube, split to the base on the back, pubescent outside on the lobes, rarely glabrous, papillose inside near the mouth of the tube; lower lip 5–6 mm. long with obovate-oblong lobes, ± 1·5–2·5 mm. wide; upper lobes lanceolate, ± recurved, 2–2·5 mm. long. Filaments linear, connate above, attached to corolla-tube at the base, densely pubescent on the inside in the upper part, otherwise glabrous. Anther-tube 1·2–2 mm. long, all anthers densely hairy at the apex, otherwise glabrous. Ovary semi-inferior. Capsule ± 8-nerved, ± strigose, with 2 valves 2–3·5 mm. long. Seeds ± broadly elliptic in outline, trigonous, ± 1 mm. long, very finely reticulate, brown to almost black.

Zimbabwe. C: Makoni Distr., 49 km. on Rusape-Inyanga road, 1800 m., 14.xii.1975, *Best* 1235 (SRGH). E: Melsetter, Ngorima T.T.L., confluence of Haroni R. and Lusitu R., 23.xi.1967, *Ngoni* 48 (BR; K; LISC; MO; SRGH). **Malawi.** N: North Viphya, Uzumara rain forest, 2050 m., 7.viii.1972, *Pawek* 5612 (K. MO). **Mozambique.** Z: Gurué Mts., W. of

Namuli Mts., near the source of Malema R., 1700 m., 5.i.1968, *Torre & Correia* 16936 (LISC). MS: Gorongosa Mts., Nhandore Mt., 1840 m., *Torre & Pereira* 12414 (LISC).

Also in S. Tanzania and Transvaal. In upland grassland, on rocky hillsides or at forest margins, often on disturbed ground.

The name *L. chamaedryfolia* has usually been used for this plant in the F.Z. area. In my opinion *L. chamaedryfolia* is a distinct but closely related species apparently confined to Natal. It differs mainly by its decumbent habit, often very broadly obovate leaves, blue flowers, and by its pedicels being reflexed in fruit. In a single specimen from Zimbabwe, *Noel* 2015, the flowers are also said to be blue, but this needs confirmation. Also *L. holstii* Engl., a widespread species in tropical E. Africa, is closely related to *L. goetzei* and differs mainly in its larger, mostly 14–18 mm. long corolla with broader lobes of the lower lip and by being subglabrous or sparsely appressed-pubescent on stems and leaves.

24. **Lobelia ovina** E. Wimmer in Notizbl. Bot. Gart. Berl. **12**: 106 (1934); in Engl., Pflanzenr. **IV**. 276 b: 581 (1953) & 276 c: 871 (1968). — Binns, H.C.L.M.: 25 (1969). Types from Tanzania.

Perennial erect herb from a woody rootstock. Stems few to many, 15–80 cm. tall, rather coarse, ribbed, pubescent with short spreading or ± appressed hairs. Leaves sessile, linear to ovate, up to 10–16 × 1·2–9 mm., acute to obtuse at the apex, truncate at the base, shortly often appressed pubescent on both sides, mainly at the margins and along the midvein beneath; margin entire or denticulate; venation, at least the midvein, distinct beneath. Flowers in ± lax, 4–7(15)-flowered, pedunculate racemes; pedicels 5–18(30) mm. long, densely strigulose to white-pubescent; bracts lanceolate, 3–6(10) mm. long, entire; bracteoles usually at or near the base of the pedicel, linear, 1–3(5) mm. long. Hypanthium broadly obconical, ± oblique, ± 8-nerved, strigulose to densely white-pubescent. Calyx-lobes narrowly triangular, 3–7 mm. long, entire, glabrous to sparsely strigulose or pubescent, somewhat gibbous at the base. Corolla 16–23 mm. long, pink or mauve with two longitudinal crests in the mouth of the tube, split to the base on the back, pubescent outside, papillose inside, at least near mouth of the tube; lower lobes obovate-oblong, 7–13 mm. long; upper lobes lanceolate, recurved, 5–7 mm. long. Filaments linear, ± pilose in upper part, connate above, attached to corolla-tube at the base. Anther-tube 2·4–3 mm. long, all anthers densely hairy at the apex and ± pilose on the connectives outside. Ovary semi-inferior to semi-superior. Capsule ± 8-nerved, strigulose to densely white-pubescent, with 2 valves 3–6 mm. long. Seeds ± broadly elliptic in outline, trigonous, 0·8–1 mm. long, very finely reticulate, brown.

Zambia. E: Nyika, 13.i.1974, *Fanshawe* 12173 (K). **Malawi.** N: Nyika Plateau, near Chelinda Camp, 2200 m., 28.x.1958, *Robson* 453 (K; LISC; SRGH).

Restricted to S. Tanzania and the Nyika Plateau in E. Zambia and N. Malawi. Upland grassland, often appearing after burning.

25. **Lobelia blantyrensis** E. Wimmer in Annal. Naturhist. Mus. Wien **56**: 363 (1948). — Brenan in Mem. N.Y. Bot. Gard. **8**: 489 (1954). — E. Wimmer in Pflanzenr. **IV**. 276 b: 595 (1953) & 276 c: 876 (1968). — Binns, H.C.L.M.: 25 (1969). — Moriarty, Wild Flowers of Malawi: 73, t. 37 fig. 3 (1975). Syntypes from Malawi: Blantyre, *Buchanan* 119 (B, syntype; BM; W, fragment) and Mt. Mulanje, *McClounie* 26 (B).

Annual or perennial erect or ascending herb from a taproot. Stems few to several, up to 90 cm. tall, rather coarse, ribbed, pubescent with spreading or ± appressed up to 1 mm. long hairs. Leaves narrowly elliptic to lanceolate or narrowly ovate, up to 30(50) × 12(18) mm., acute to obtuse and mucronulate at the apex, attenuate into a petiole-like base or cuneate at the base, with spreading or ± appressed pubescence on both sides; margin coarsely dentate or crenate; venation distinct below. Flowers in lax, (2)4–8-flowered, pedunculate racemes; pedicels 4–8(16) mm. long, strigose; bracts linear-lanceolate, up to 9 mm. long, entire, strigose; bracteoles usually at the base of the pedicel, linear, 1·5–3 mm. long, strigose. Hypanthium broadly obconical to hemispherical, ± 8-nerved, strigose. Calyx-lobes narrowly triangular, 3–6 mm. long, entire, strigose, somewhat gibbous at the base. Corolla 17–23 mm. long, deep blue or violet with two whitish longitudinal crests in the mouth of the tube, split to the base on the back, sparsely strigose outside on the tube and along main veins of the lobes, short pubescent inside near mouth of the tube; lower lobes obovate-oblong, 9–14 mm. long; upper lobes lanceolate, recurved, 4–5 mm. long. Filaments linear, glabrous, connate above, attached to corolla-tube at the base. Anther-tube 2–2·5 mm. long, all anthers densely hairy at the apex, otherwise glabrous except for some long slender hairs on the connectives outside. Ovary semi-inferior. Capsule ± 8-nerved, strigose, with 2 valves 2–3 mm. long. Seeds broadly elliptic in outline, trigonous, ± 0·8 mm. long, very finely reticulate, brown.

Malawi. S: Mt. Mulanje, Tuchila Plateau, 1980 m., 21.vii.1956, *Newman & Whitmore* 91 (BM; BR; COI; SRGH). **Mozambique.** Z: Gúruè Mts., 1500 m., 24.ix.1944, *Mendonça* 2273 (LISC).
Restricted to S. Malawi and an adjacent area in Mozambique. Rocky grassland, forest edges or along forest paths.

26. **Lobelia caerulea** Sims in Curtis's Bot. Mag. **53**: t. 2701 (1826). — E. Wimmer in Engl., Pflanzenr. **IV**. 276 b: 593 (1953) & 276 c: 875 (1968). Type from S. Africa.
    Lobelia coronopifolia L. var. *caerulea* (Sims) Sonder in Harv. & Sond. F.C. **3**: 543 (1865). Type as above.

Annual or perennial ± erect or ascending herb from a taproot. Stems few to many, up to 70 cm. tall, ribbed, ± densely short-pubescent below. Leaves linear to narrowly oblanceolate, up to 25–50 × 1·5–5 mm., acute to obtuse at the apex, tapering into a petiole-like base, glabrous to densely short-pubescent; margin sparsely dentate to crenate, or entire. Flowers in ± lax (2)3–5-flowered, pedunculate racemes; pedicels up to 17 mm. long, erect in fruit, glabrous or almost so; bracts linear-lanceolate, 3–5 mm. long, entire, ciliate; bracteoles in the upper part of the pedicel, ± 1 mm. long, ciliate. Hypanthium very short, ± oblique, ± 8-nerved, ± strigose. Calyx-lobes narrowly triangular, 2–4 mm. long, entire, ciliate, somewhat gibbous at the base. Corolla 19–24 mm. long, deep blue or violet with two longitudinal yellowish crests in the mouth of the tube, split to the base on the back, strigulose outside on the lobes, papillose inside near the mouth of the tube; lower lip about half the length of the corolla with obovate-oblong lobes; upper lobes lanceolate, ± recurved, ± 5 mm. long. Filaments linear, connate above, attached to corolla-tube at the base, densely pubescent inside near the top, otherwise glabrous. Anther-tube ± 2·5 mm. long, all anthers densely hairy at the apex and the 3 upper anthers often with some long slender hairs on the connectives, otherwise glabrous. Ovary semi-superior. Capsule ± 8-nerved, ± strigose, with 2 valves 5–6·5 mm. long. Seeds ± broadly elliptic in outline, trigonous with 2 flat sides and one convex side, ± 1 mm. long, very finely reticulate, dark brown.

Var. **macularis** (Presl) E. Wimmer in Pflanzenr. **IV**. 276 b: 593 (1953) & 276 c: 875 (1968). Type from S. Africa (Natal).
    Rapuntium maculare Presl, Prodr. Monogr. Lobel.: 20 (1836). Type as above.
    Rapuntium maculare var. *procerum* Presl, op. cit.: 21 (1836). Type from S. Africa.
    Lobelia coronopifolia L. var. *macularis* (Presl) Sonder, tom cit.: 543 (1865). Type as for *L. caerulea* var. *macularis.*
    Lobelia macularis (Presl) A. DC. in DC., Prodr. **7**: 364 (1839). Types as for *L. caerulea* var. *macularis.*
    Lobelia caerulea var. *macularis* f. *procera* (Presl) E. Wimmer, tom. cit.: 594 (1953) & loc. cit. (1968). Type as for *Rapuntium maculare* var. *procerum.*

Leaves glabrous or ciliate near the base and sometimes also pubescent on the midvein beneath.

Mozambique. GI: Bilene, near the S. Martinho beach, 5.vii.1944, *Torre* 6679 (LISC). Also in Natal and the Cape Province. On coastal dunes or in littoral grassland.

Lobelia caerulea var. *caerulea* has the leaves short-pubescent all over and occurs in coastal areas of the SE. Cape Province. Wimmer's classification has been followed here, but the distinction between *L. caerulea* and some other South African species, notably *L. tomentosa* L.f., is very weak and in need of further study.

27. **Lobelia cobaltica** S. Moore in Journ. Linn. Soc., Bot. **40**: 124 (1911). — E. Wimmer in Engl., Pflanzenr. **IV**. 276 b: 583 (1953) & 276 c: 872 (1968). Type: Zimbabwe, Chimanimani Mts., *Swynnerton* 2036 (BM, holotype).
    Lobelia longipilosa E. Wimmer in Kew Bull. **7**: 142 (1952); tom. cit.: 782 (1953). Type: Zimbabwe, Chimanimani Mts., *Wild* 2869 (K; SRGH; W).

Annual (?) decumbent herb from a taproot. Stems few to several, up to 60 cm. tall, rather robust, ribbed, usually densely pilose with spreading hairs much varying in size, the largest 1·5 mm. long. Leaves with the lamina deltoid to very broadly ovate, up to 20 × 25 mm., acute to obtuse and mucronulate at the apex, truncate to subcordate or upper leaves sometimes cuneate at the base, pilose on both sides, in all but the uppermost leaves narrowing into an up to 10(–20) mm. long, winged petiole; margin coarsely dentate or crenate; venation distinct beneath. Flowers in lax, 2–5-flowered, pedunculate racemes; pedicels 4–8 mm. long, pilose; bracts linear-lanceolate, up to 5 mm. long, entire, pilose; bracteoles near the base of the pedicel, linear, 1–2·5 mm. long, pilose. Hypanthium broadly obconical, ± oblique, ± 8-nerved, pilose. Calyx-lobes very slender, up to 7 mm. long, entire, pilose, somewhat gibbous at the base. Corolla 17–23 mm. long, deep blue, with two whitish longitudinal crests in the mouth of the tube, split to the base on the back, sparsely pubescent outside on the tube and along the midveins of the lobes, papillose inside near mouth of the tube; lower lobes obovate-oblong, 9–13 mm. long; upper lobes lanceolate, recurved, ± 4 mm. long. Filaments linear, glabrous, connate above, attached to corolla-tube at the base. Anther-tube ± 2·5 mm. long, all anthers densely hairy at the apex, otherwise glabrous or sparsely pubescent on the connectives outside. Ovary semi-inferior. Capsule ± 8-nerved, pilose, with 2 valves 1·5–3 mm. long. Seeds broadly elliptic in outline, trigonous, ± 0·8 mm. long, very finely reticulate, pale to dark brown.

Zimbabwe. E: Chimanimani Mts., W. of Mawenje Peak, "Stonehenge", 1830 m., 8.v.1958, *Chase* 6905 (COI; K; LISC; S). Mozambique. MS: Chimanimani Mts., 6·5 km. E. of the Saddle, 11.v.1965, *Whellan* 2235 (K; SRGH).
Known only from the Chimanimani Mts. Rocky places, in rock crevices, gulleys, usually in shady and sheltered conditions.

28. **Lobelia pteropoda** (Presl) A. DC. in DC., Prodr. **7**: 364 (1839). — E. Wimmer in Engl., Pflanzenr. **IV**. 276 b: 585 (1953). Type from S. Africa (Cape Prov.).
Rapuntium pteropodum Presl., Prodr. Monogr. Lobel.: 21 (1836). Type as above.

Annual or perennial (?) ± erect or decumbent herb up to 80 cm. tall. Stems rather slender, ribbed, glabrous or sparsely pubescent. Leaves with the lamina triangular to very broadly ovate, up to 50 × 40 mm., acute to obtuse and mucronulate at the apex, truncate to cordate at the base, ± pubescent on both sides; all leaves with a winged ± pubescent petiole up to 10–60 mm. long; leaf-margin very coarsely dentate or crenate with mucronulate teeth; venation distinct beneath. Flowers in lax, up to 6-flowered, pedunculate racemes; pedicels 3–7 mm. long, pubescent; bracts linear, up to 7 mm. long, entire, ± pubescent; bracteoles at or near the base of the pedicel, linear, up to 1 mm. long, often broadened at the tip, ± pubescent. Hypanthium narrowly to broadly obconical, ± 8-nerved, glabrous or ± pubescent. Calyx-lobes narrowly triangular, 3–4·5 mm. long, entire, ± pubescent, somewhat gibbous at the base. Corolla 16–23 mm. long, pale blue to mauvish or white, with two longitudinal crests in the mouth of the tube, split to the base on the back, ± pubescent on the outside of the lobes, puberulous inside on the bottom of the tube. Filaments linear, glabrous, connate above. Anther-tube 2–2·5 mm. long, all anthers densely hairy at the apex, otherwise glabrous or with a few straight hairs on the connectives outside. Ovary semisuperior to subinferior. Capsule ± 8-nerved, glabrous or ± pubescent, with 2 valves ± 3 mm. long. Seeds ± broadly elliptic in outline, somewhat compressed, ± 0·7 mm. long, almost smooth, dark brown.

Mozambique. MS: Mocuta Mts., 1200 m., 1.vi.1971, *Biegel* 3511 (LISC; SRGH).
Also in Natal, Orange Free State and Cape Province. The above collection is the only one known from the F.Z.-area and was collected among *Aframomum* and grass in open *Brachystegia* woodland on E.-facing slope.

Some very closely related species, *Lobelia vanreenensis* (Kuntze) K. Schum. and *L. preslii* A. DC. occur in SE. South Africa and may well be found to be conspecific with *L. pteropoda* which name, however, has priority.

29. **Lobelia baumannii** Engl., Bot. Jahrb. **19**, Beibl. 47: 51 (1894). — E. Wimmer in Engl., Pflanzenr. **IV**. 276 b: 586 (1953) & 276 c: 872 (1968), excl. specim. ex Nigeria. Binns, H.C.L.M.: 25 (1969). — Agnew, Upland Kenya Wild Fl.: 515 (1974). Types from Tanzania.
Lobelia baumannii var. *cuneata* E. Wimmer in Notizbl. Bot. Gart. Berl. **12**: 104 (1934); in tom. cit.: 587 (1953). Type from Tanzania.
Lobelia baumannii var. *cuneata* f. *alba* E. Wimmer, loc. cit. (1934); loc. cit. (1953). Type from Tanzania.

Perennial procumbent or straggling, rarely erect herb up to 70 cm. long, often rooting at lower nodes. Stems ribbed or narrowly winged, ± pubescent. Leaves with the lamina narrowly to very broadly ovate or triangular, up to 25–80 × 15–50 mm., acute to acuminate at the apex, cuneate or truncate to subcordate at the base, ± pubescent on both sides but mainly on the nerves; petiole up to 15–30 mm. long; leaf margin coarsely dentate or serrate; venation prominent, particularly beneath. Flowers in lax, 2–6(12)-flowered, shortly pedunculate racemes; pedicels up to 5–12 mm. long, ± pubescent; bracts linear, 2–7·5 mm. long, sometimes broadest above the middle; bracteoles usually near the base of the pedicel, linear, up to 1·5 mm. long, often broadened at the tip. Hypanthium narrowly to broadly obconical, 8–10-nerved, sparsely appressed pubescent in the lower part to spreading pubescent all over. Calyx-lobes narrowly triangular or linear, sometimes broadest above the middle, often ± recurved, 2·5–7·5 mm. long, ciliate and often with a few hairs also on the midvein, somewhat gibbous at the base. Corolla 13–30 mm. long, white, mauve, or blue with darker markings in throat, with two longitudinal crests in the mouth of the tube, split to the base on the back, glabrous or ± pubescent on the lobes outside, puberulous inside on the bottom of the tube. Filaments linear, pubescent below except sometimes for the upper one, puberulous inside at the top, connate or almost free above, attached to corolla-tube at the base. Anther-tube 2–3 mm. long, all anthers penicillate at the apex, otherwise glabrous or sparsely pubescent on the connectives outside. Ovary subinferior to semi-inferior. Capsule 8–10-nerved, ± pubescent, with 2 valves 2·5–4 mm. long, inferior part of capsule 4–10 mm. long. Seeds elliptic-oblong to broadly elliptic in outline, somewhat compressed, 0·9–1·2 mm. long, distinctly reticulate, pale brown.

Zambia. W: Mwinilunga, Kalene Hill, 18.v.1969, *Mutimushi* 3205 (SRGH). **Zimbabwe.** E: foot of E. slopes of Inyangani Mt., 900 m., 9.vii.1961, *Wild* 5511 (K; LISC; SRGH). **Malawi.** N: 8 km. E. of Mzuzu at Roseveare's, 1200 m., 2.ix.1975, *Pawek* 10072 (K; SRGH). S: Mulanje Mt., Great Ruo Gorge, 770 m., 18.vi.1962, *Richards* 16795 (K; SRGH). **Mozambique.** Z: Gúruè, 26.vi.1943, *Torre* 5592 (LISC).
Also in Kenya, Tanzania, E. Zaire and Burundi. Forest floor or forest margins, often on stream banks in shade.
The record of *L. baumannii* in F.W.T.A., ed. 2, **2**: 315 (1963) is based on a wrongly named specimen of *L. hartlaubii* Buchen.

30. **Lobelia thermalis** Thunb., Prodr. Fl. Cap. **1**: 40 (1794). — Hiern, Cat. Afr. Pl. Welw. **1**: 625 (1898). — Gibbs in Journ. Linn. Soc., Bot. **37**: 451 (1906). — E. Wimmer in Engl., Pflanzenr. **IV**. 276 b: 567, fig. 93 a (1953). — Roessler in Prodr. Fl. SW. Afr. **137**: 4 (1966). — E. Wimmer, op. cit. **IV**. 276 c: 870 (1968). Type from S. Africa (Cape Prov.).
   Lobelia leptocarpa Griessel. in Linnaea **5**: 419 (1830). Type from S. Africa.
   Lobelia mundtiana Cham. in Linnaea **8**: 215 (1833). Type from S. Africa (Cape Prov.).
   Parastranthus thermalis (Thunb.) A. DC. in DC., Prodr. **7**: 354 (1839). — Sond. in Harv. & Sond., F.C. **3**: 537 (1865). Type as for *Lobelia thermalis*.

Perennial prostrate herb. Stems radiating from a taproot, 4–50 cm. long, rooting at the nodes, ribbed or sulcate, densely hirsute to almost glabrous. Leaves usually numerous and rather densely set, up to 7–17 × 3–8 mm., narrowly to broadly elliptic, acute at the apex, sessile or narrowing at the base into a short petiole, glabrous, ciliate at the base, or occasionally hairy all over; margin incrassate, dentate-serrate. Flowers solitary, irregularly scattered in leaf-axils, subsessile or on pedicels up to 5 mm. long, pubescent; bracteoles linear, 0·8–1·6 mm. long. Hypanthium narrowly obconical, indistinctly nerved, finely pubescent to hirsute. Calyx-lobes narrowly triangular, erect, 4–5 mm. long, shortly united at the base, entire, ciliate or ± hairy. Corolla 14–18 mm. long, blue or pale blue with yellow markings, sometimes white, with two crests in the mouth of the tube, split to the base or almost so on the back, papillose on the inside of the tube, short pubescent on the outside usually with longer hairs on the midvein of the lobes. Stamens 10–14 mm. long, almost free from the corolla; filaments connate for most of their length, pubescent at the base; anther-tube 2·9–3·2 mm. long, all anthers penicillate at the tip, otherwise glabrous or almost so. Capsule cylindrical-obconical up to 17 mm. long, ± hairy, indistinctly nerved; pericarp rather tough in texture, finally tending to split longitudinally into 5–10, linear segments; apical valves 4, narrow, 2–3 mm. long. Seeds broadly elliptic to suborbicular in outline, somewhat compressed, 0·35–0·4 mm. long, finely striate, brown.

Botswana. SE: Derdepoort, 910 m., 29.xi.1954, *Codd* 8856 (K; PRE; SRGH). **Zimbabwe.** W: Matobo, 1.i.1949, *West* 2836 (SRGH). C: Gwelo Distr., between Fletcher High School and

Gwelo Teachers College, 1400 m., 6.ii.1968, *Biegel* 2536 (B; BR; K; SRGH). S: Victoria Distr., Kyle National Park, Mkonda Dam, 20.xii.1973, *Basera* in GHS 246822 (SRGH).

Also in Angola, Namibia, S. Africa and Lesotho. On moist ground such as river banks and edges of swamps and pools.

An isolated species deviating from other *Lobelia* species particularly in its 4-valved, often longitudinally splitting capsule. It seems also to be the only *Lobelia* without regularly resupinate flowers. For this reason it was included in *Parastranthus* (now *Monopsis*) by A. De Candolle (loc. cit.) and Sonder (loc. cit.). The style, however, is that of a *Lobelia*.

31. **Lobelia angolensis** Engl. & Diels in Engl., Bot. Jahrb. **26**: 114 (1898). — E. Wimmer in Engl., Pflanzenr. **IV**. 276 b: 604, fig. 95 d (1953). TAB. **23**. Syntypes from Angola.
   Metzleria depressa A. DC. in DC., Prodr. **7**: 350 (1839) *nom. illegit.*, non (L.f.) Presl. Type from S. Africa (Cape Prov.).
   Metzleria dregeana Sonder in Harv. & Sond., F.C. **3**: 533 (1865), non *Lobelia dregeana* (Presl) A. DC. Type as above.
   Lobelia lythroides Diels in Engl., Bot. Jahrb. **26**: 113 (1898). Syntypes from S. Africa (Transvaal).
   Lobelia dekindtiana Engl., Bot. Jahrb. **32**: 147 (1902). Type from Angola.
   Lobelia minutidentata Engl. & Gilg in Warb., Kunene-Samb.-Exped. Baum: 397 (1903). — Gibbs in Journ. Linn. Soc., Bot. **37**: 452 (1906). — E. Wimmer, tom. cit.: 603 (1953). Type from Angola.
   Lobelia sonderi A. Zahlbr. in Annal. Hofmus. Wien **18**: 404 (1903). Type as for *Metzleria depressa* A. DC.
   Lobelia seineri Schltr. in Engl., Bot. Jahrb. **57**: 616 (1922). Types from Namibia.
   Lobelia tsotsorogensis Bremek. & Oberm. in Ann. Trans. Mus. **16**: 437 (1935). Type: Botswana, Tsotsoroga Pan, *van Son* in HTM 28973 (PRE, holotype).
   Lobelia depressa sensu E. Wimmer in Notizbl. Bot. Gart. Berl. **15**: 633 (1941). — Markgraf in Engl., Bot. Jahrb. **75**: 216 (1950). — E. Wimmer, tom. cit.: 604, fig. 95 b (1953). — Roessler in Prodr. Fl. SW. Afr. **137**: 2 (1966). — Binns, H.C.L.M.: 25 (1969) non L.f. (1781).
   Lobelia depressa L.f. var. *dregeana* (Sonder) E. Wimmer in Notizbl. Bot. Gart. Berl. **15**: 633 (1941). — Markgraf, tom. cit.: 217 (1950). — E. Wimmer, tom. cit.: 605, fig. 95 b (1953). Type as for *Metzleria depressa* A. DC.
   Lobelia depressa var. *seineri* (Schltr.) E. Wimmer, loc. cit. (1941). — Markgraf, loc. cit. — E. Wimmer, tom. cit.: 605 (1953). Type as for *Lobelia seineri*.
   Lobelia depressa var. *linearifolia* E. Wimmer in Engl., Pflanzenr. **IV**. 276 c: 879 (1968). Type from Tanzania.

Annual or short-lived perennial herb, usually rhizomatous, prostrate, ascending or erect, glabrous. Stems usually numerous, 2–20 cm. long, somewhat sulcate. Leaves usually numerous, rather fleshy, up to 3–15 × 1–6 mm., linear to elliptic, ovate or spathulate, acute or obtuse at the apex, sessile or narrowing at the base into a short petiole, entire or rarely minutely denticulate. Flowers in upper leaf-axils; pedicels 2–20 mm. long, shorter to longer than the leaves; bracteoles linear, 0·5–1 mm. long. Hypanthium broadly obconical, 10-nerved, oblique, somewhat drawn out in the upper part on the ventral side, making the calyx bilabiate. Calyx-lobes triangular, erect, 1–2 mm. long, entire. Corolla 2·5–5 mm. long, white to blue, pink or purple, split to the base on the back, glabrous; lobes subequal, the upper two free or almost so, the lower three usually ± connate and with a crescent-shaped gland at the base. Stamens 2–4 mm. long, free from the corolla; filaments connate for most of their length; anther-tube dark, 1·2–2 mm. long, glabrous or very shortly pubescent on the back; the two lower anthers each with a hyaline appendage c. 0·4 mm. long and a tuft of minute hairs at the tip. Ovary semi-inferior. Capsule broadly obconical, 10-nerved; valves ± 1·2 mm. long. Seeds elliptic, often somewhat twisted, ± 0·3 mm. long, finely striate.

Caprivi Strip. E. of Cuando R., 940 m., x.1945, *Curson* 1127 (PRE). **Botswana.** N: Toten, just NE. of Lake Ngami, 18.iii.1965, *Wild & Drummond* 7139 (BR; K; LISC; SRGH). SE: 6 km. S. of Gaberones, 970 m., 18.viii.1974, *Woollard* 66 (SRGH). **Zambia.** B: near Senanga, 1020 m., 4.viii.1952, *Codd* 7400 (BM; SRGH). **Zimbabwe.** N: Gokwe Distr., Sengwa Research Station, 10.viii.1975, *Guy* 2359 (SRGH). W: Wankie National Park, Mtoa Pans, 38 km. WNW. of Main Camp, 1020 m., 20.i.1970, *Rushworth* 2394 (SRGH). C: Salisbury Distr., Groombridge vlei between Teviotdale Rd. and Waller Ave., 23.x.1969, *Pope* 186 (MO; SRGH). **Malawi.** C: Dedza Distr., Chongoni Forest Reserve, 21.ix.1967, *Salubeni* 840 (K; LISC; PRE; SRGH). **Mozambique.** GI: Guija area, along Limpopo R., vii.1915, *Gazaland Exped.* in HTM 15754 (PRE).

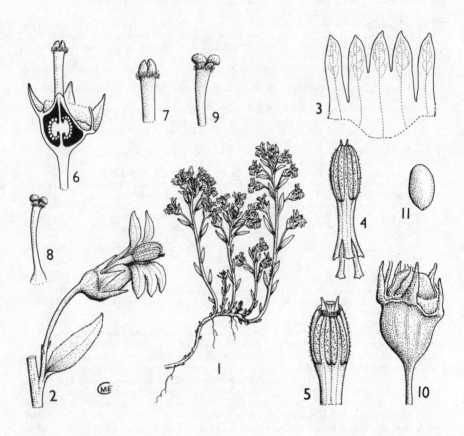

Tab. 23. LOBELIA ANGOLENSIS. 1, habit (× 1); 2, flower (× 6); 3, corolla spread out (× 8); 4, stamens with anther-tube (× 10); 5, anther-tube from below (× 12); 6, longitudinal section of ovary from young flower (× 10); 7, style-apex from young flower (× 14); 8, style and stigma from older flower (× 10); 9, style-apex from older flower (× 14); 10, capsule (× 8); 11, seed (× 40); all from *Robinson* 5438. From F.T.E.A.

Also in Kenya, Tanzania, Angola, Namibia and S. Africa. On moist ground such as river banks and edges of swamps and pools.

Belongs to *Lobelia* subgen. *Metzleria* (Presl) E. Wimmer, a small group of mainly S. African species characterised by subequal corolla-lobes, slightly bilabiate calyx, distinct appendages to the tips of the two lower anthers, and presence of a crescent-shaped gland at the base of the 3 lower corolla-lobes. The group has often been considered a distinct genus, *Metzleria* Presl. The delimitation of species within the group is not satisfactory at present.

L. angolensis has usually been called *L. depressa* L.f., a name which has never been typified. Indisputable type material recently found in S-LINN proved to be *Monopsis simplex* (L.) E. Wimmer = *Lobelia simplex* L. (1771), to which *L. depressa* therefore is a synonym. If *Metzleria* is treated as a genus of its own, *M. dregeana* Sonder will be the correct name for *L. angolensis*.

2. MONOPSIS Salisb.

Monopsis Salisb. in Trans. Hort. Soc. Lond. **2**: 37 (1817). — E. Wimmer in Engl., Pflanzenr. **IV**. 276 b: 698 (1953).

Parastranthus G. Don, General Syst. **3**: 716 (1834).

Dobrowskya Presl, Prodr. Monogr. Lobel.: 8 (1836).

Annual or perennial herbs. Leaves alternate, opposite or verticillate. Flowers resupinate or not, in racemes or solitary in leaf axils, pedicellate or sessile; bracteoles conspicuous resembling small leaves (in all tropical African species), or absent. Calyx 5-lobed. Corolla split to the base on the back (or on the lower side in non-resupinate flowers), 2-lipped to subregular, the lower lip (or the upper one in non-resupinate flowers) 3-lobed. Stamens 5; filaments connate in upper part or almost free from each other, usually attached to the corolla-tube at the base; anthers united into a tube, all of them penicillate at the tip, the two lateral ones somewhat longer than the others and with shorter hairs at the tip. Ovary inferior or almost so, 2-locular; stigma-lobes linear, becoming revolute and with a ring of hairs present on the style below them. Fruit dehiscing by 2 apical valves or ± indehiscent; seeds numerous, usually reticulate.

Genus of some 20 species, mainly S. African, with 3 species extending into tropical Africa.

1. Leaves opposite; flowers not resupinate - - - - - - - 1. *stellarioides*
- Leaves alternate; flowers resupinate - - - - - - - - - - 2
2. Leaves linear; corolla 10 mm. or more long - - - - - - -2. *decipiens*
- Leaves ± elliptic; corolla 4–5 mm. long- - - - - - - - 3. *zeyheri*

1. **Monopsis stellarioides** (Presl) Urb. in Eichl., Jahrb. Königl. Bot. Gart. Berl. **1**: 275 (1881). — E. Wimmer in Engl., Pflanzenr. **IV**. 276 b: 703, fig. 106/d (1953). Type from S. Africa.

Dobrowskya stellarioides Presl, Prodr. Monogr. Lobel.: 10 (1836). — Sond. in Harv. & Sond., F.C. **3**: 550 (1865). Type as above.

Parastranthus stellarioides (Presl) Vatke in Linnaea **38**: 717 (1874), pro parte. Type as above.

Lobelia stellarioides (Presl) Benth. & Hook. f. ex Hemsl. in F.T.A. **3**: 470 (1877), pro parte. Type as above.

Annual or perennial decumbent herb, often stoloniferous. Stems 5–60 cm. long, usually rooting at the lower nodes, ribbed, with harsh retrorse pubescence. Leaves opposite, subsessile or shortly petiolate, linear to elliptic, up to 10–35 × 2–9 mm., acute or obtuse and mucronulate at the apex, rounded to attenuate at the base, almost glabrous or usually with fine but harsh pubescence on both surfaces, usually retrorse beneath and at margin, but with forwardly directed hairs above; margin incrassate, slightly revolute, crenate or serrate; midvein prominent beneath. Flowers not resupinate, solitary, axillary, scattered on the stem; pedicels 8–35 mm. long, pubescent with retrorse hairs; bracteoles at the base of the pedicel, 6–16 mm. long, resembling small leaves. Hypanthium obconical or obovoid, ± 10-nerved, pubescent. Calyx-lobes 2–6 mm. long, acute, entire, ± pubescent at the margins and on the outside, often reflexed in fruit. Corolla 6·5–11·5 mm. long, dirty yellow, dull pink, brownish purple or violet (but always ± violet when dried), with two bumps in the mouth of the tube, split on the lower side almost to the base, two-lipped with the upper lip 3-lobed and the lower lip consisting of two spathulate lobes united with the tube for 0·5–2·5 mm; short pubescent outside, often with longer hairs on the midveins of the lobes, glabrous or puberulous inside. Filaments free from corolla,

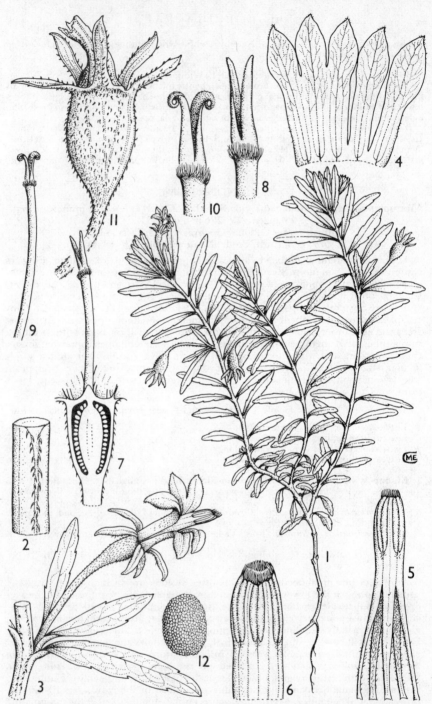

Tab. 24. MONOPSIS STELLARIOIDES subsp. SCHIMPERIANA. 1, habit (× 1); 2, portion of stem
showing retrorse pubescence (× 10); 3, flower with bract and bracteoles (× 4); 4, corolla
spread out (× 6); 5, stamens with anther-tube from below (× 10); 6, anther-tube from above
(× 16); 7, style and ovary from young flower (× 8); 8, stigmatic lobes (× 16), 1–8 from
Davis 56; 9, style from older flower (× 8); 10, stigmatic lobes (× 16), 9–10 from *Dawkins*
743; 11, fruit (× 6); 12, seed (× 20), 11–12 from *Schlieben* 4312. From F.T.E.A.

linear, connate above, all pubescent at least on the inside above. Anther-tube ± 1·6 mm. long, all anthers densely penicillate at the tip, otherwise glabrous. Stigma-lobes ± 0·6 mm. long. Capsule ± 10-nerved, obconical to oblong-obovoid, up to 10 mm. long, pubescent, dehiscing by valves up to 1·5 mm. long, or ± indehiscent. Seeds broadly elliptic or subcircular in outline, compressed, ± 0·6 mm. long, distinctly and regularly reticulate, dark brown.

Subsp. **schimperiana** (Urb.) Thulin in Bot. Notiser **132**: 131, fig. 1A, D (1979). TAB. **24**. Type from Ethiopia.
   Monopsis schimperiana Urb. in Eichl., loc. cit. (1881). Type as above.
   Lobelia violaceo-aurantiaca De Wild. in Rev. Zool. Afr. **8**, Suppl. Bot.: 28 (1920). Type from Zaire.
   Lobelia mokuluensis De Wild., tom. cit.: 26 (1920). Type from Zaire.
   Monopsis brevicalyx T.C.E. Fr. in Notizbl. Bot. Gart. Berl. **8**: 411 (1923). — E. Wimmer in Engl., Pflanzenr. **IV**. 276 b: 705 (1953) & 276 c: 890 (1968). Type from Kenya.
   Monopsis schimperiana var. *brevifolia* Chiov., Racc. Bot. Miss. Consol. Kenya: 73 (1935). Type from Kenya.
   Monopsis stellarioides f. *violaceo-aurantiaca* (De Wild.) E. Wimmer, tom. cit.: 704 (1953). Type as for *Lobelia violaceo-aurantiaca.*
   Monopsis stellarioides var. *schimperiana* (Urb.) E. Wimmer, loc. cit. — Hedb. in Symb. Bot. Upsal. **15**: 195, 334 (1957). — E. Wimmer in F.W.T.A., ed. 2, **2**: 315 (1963); loc. cit. (1968). Type as for *Monopsis stellarioides* subsp. *schimperiana.*
   Monopsis stellarioides sensu E. Wimmer in Fl. Madag. **186**. Lobéliacées: 28, fig. 8/4–8 (1953). — Binns, H.C.L.M.: 25 (1969) pro parte. — Agnew, Upl. Kenya Wild Fl.: 517, 519 (fig.) (1974). — Moriarty, Wild Flowers of Malawi: 73, t. 37 fig. 6 (1975).

Corolla usually dirty yellow, dull pink or brownish purple, puberulous inside, with the two lower lobes united with the upper lip for 0·5–1·0(1·5) mm. Capsule often oblong-obovoid and indehiscent, only rarely apical valves are formed.

Zambia. N: Lumangwe, 14.xi.1957, *Fanshawe* 4002 (K). **Malawi.** S: Zomba Plateau, Chagwa Dam, 11.i.1971, *Moriarty* 437 (MAL, seen in photo).
Also in Ethiopia, Fernando Poo, Cameroon, Uganda, Kenya, Tanzania, Zaire, Rwanda, Burundi and Comoro Is. In upland grassland and moor, forest margins, and rocky places.

2. **Monopsis decipiens** (Sonder) Thulin in Bot. Notiser **132**: 134, fig. 1B, E, 2 (1979). Type from Transvaal.
   Lobelia decipiens Sonder in Harv. & Sond., F.C. **3**: 540 (1865). — Rolfe in Oates, Matabeleland and the Victoria Falls, ed. 2, app. 5: 402 (1889). — Gibbs in Journ. Linn. Soc., Bot. **37**: 451 (1906). — E. Wimmer in Engl., Pflanzenr. **IV**. 276 b: 588 (1953). — Plowes & Drummond, Wild Flowers of Rhodesia: 137, t. 178 (1976). Type as above.
   Lobelia breynii Lam. var. *bragae* Engl., Pflanzenw. Ost-Afr. **C**: 402 (1895). Type: Mozambique, Beira (locality probably erroneous), *Braga* 135 (B, holotype).
   Lobelia dobrowskyoides Diels in Engl., Bot. Jahrb. **26**: 117 (1898). Type from Transvaal.
   Monopsis scabra sensu E. Wimmer, op. cit. 276 c: 891 (1968) quoad specim. ex Zimbabwe.

Annual or perennial erect or ascending herb. Stems 5–45 cm. long, usually sparsely branched, ribbed, ± pubescent with upturned hairs. Leaves alternate, fairly numerous and densely set, ± erecto-patent, linear, up to 8–15(25) × 0·8–1·5(2·5) mm., acute at the apex, harshly strigulose mainly along margins and on midrib beneath; margin incrassate, entire or rarely with one pair of denticles; midvein prominent beneath. Flowers resupinate, solitary, axillary or apparently terminal, confined to the upper part of the plant; pedicels 15–55 mm. long, pubescent with forwardly directed hairs; bracteoles resembling small leaves, sometimes opposite. Hypanthium hemispherical or broadly obconical, ± 10-nerved, appressed pubescent. Calyx-lobes 2–5 mm. long, acute, entire, strigulose on the outside, spreading. Corolla 10–18 mm. long, blue or blue and purple with yellow markings on lower lip, sometimes white or almost so, split on the upper side to ± 1 mm. from the base, 2-lipped with the lower lip broadly 3-lobed and the upper lip consisting of two reflexed lobes united with the tube for most of their length; appressed pubescent outside, sparsely pubescent near the base inside. Filaments almost free from corolla, linear, connate above, all pubescent on the inside above. Anther-tube 1·5–2·5 mm. long, all anthers densely penicillate at the tip, otherwise glabrous. Capsule ± 10-nerved, subglobose to ovoid, up to 6·5 mm. long, dehiscing by valves up to 2·5 mm. long. Seeds broadly elliptic in outline, compressed, 0·5–0·6 mm. long, reticulate, brown.

Zimbabwe. N: 8 km. on the Mtoko-Tete road, 910 m., viii.1956, *Davies* 2095 (SRGH). W: Matobo Distr., Farm Besna Kobila, 1460 m., xii.1960, *Miller* 7552 (LISC; SRGH). C: Salisbury Distr., between Wellesley and Darwendale, 25.ix.1955, *Drummond* 4876 (K; SRGH; W). E: Melsetter Distr., Charter Forest Estates, 11.iii.1966, *Plowes* 2766 (COI; K; LISC; MO; SRGH). S: Victoria, *Monro* 583 (BM; MO; SRGH). **Mozambique.** MS: Manica, Rotanda, Tandara, 1700 m., 19.xi.1965, *Torre & Correia* 13175 (LISC). M: Namaacha, Mt. M'Ponduine, 780 m., 25.vii.1980, *Pettersson* 2019 (UPS).

Also in S. Africa (Transvaal, Natal, Orange Free State and E. Cape Province), Swaziland and Lesotho. Wimmer (1953) also cited a specimen from Botswana without precise locality, *Holub* s.n. (PR not seen). In grassland, often in seasonally wet places.

3. **Monopsis zeyheri** (Sonder) Thulin in Bot. Notiser **132**: 135, fig. 1C (1979). Type from Transvaal.
    Lobelia zeyheri Sonder in Harv. & Sond., F.C. **3**: 539 (1865). — Markgraf in Engl., Bot. Jahrb. **75**: 209 (1950). — E. Wimmer in Engl., Pflanzenr. **IV**. 276 b: 581 (1953). — Roessler in Prodr. Fl. SW. Afr. **137**: 4 (1966). — E. Wimmer, op. cit. **IV**. 276 c: 871 (1968). — Binns, H.C.L.M.: 25 (1969). Type as above.
    Lobelia fonticola Engl. & Gilg in Warb., Kunene-Samb.-Exped. Baum: 398 (1903). — Gibbs in Journ. Linn. Soc., Bot. **37**: 452 (1906). Type from Angola.
    Cephalostigma nanellum R. E. Fr., Wiss. Ergebn. Schwed. Rhod.-Kongo-Exped. **1**: 315 (1916) as "*nanella*". Type: Zimbabwe, Victoria Falls, *R.E. Fries* 193 (UPS, holotype).

Slender annual erect or decumbent herb, often branched from the base. Stems 2–15(28) cm. tall, terete, somewhat ribbed, densely pubescent with spreading hairs. Leaves sessile, elliptic to narrowly ovate, up to 14 × 7 mm., acute at the apex, rounded or cuneate at the base, pubescent on both surfaces; margin coarsely dentate-serrate. Flowers resupinate, often cleistogamous, in lax racemes with leaf-like bracts diminishing in size upwards; pedicels up to 30 mm. long, pubescent; bracteoles at base of pedicel, resembling small leaves, 1–5 mm. long. Hypanthium broadly obconical to subglobose, ± 10-nerved, pubescent. Calyx-lobes 0·5–1·6 mm. long, entire, pubescent. Corolla 4–5 mm. long, white, pale blue or pink with two yellowish bumps in the mouth of the tube, split on the back to ± 0·6 mm. from the base, pubescent outside and inside on bottom of tube, 2-lipped with the lower lip 3-lobed and the upper lip consisting of 2 shorter lobes united with the tube for most of their length. Filaments ± attached to the corolla-tube at the base, linear, scarcely connate above, all sparsely pubescent on the inside near the tip, the two anterior ones also ciliate at the base. Anther-tube ± 0·6 mm. long, all anthers penicillate at the tip, otherwise glabrous. Capsule ± 10-nerved, 2–3·5 mm. long, pubescent; valves ± 1 mm. long. Seeds broadly elliptic in outline, strongly compressed, ± 0·35 mm. long, striate-verruculate, brown.

Zambia. B: 15 km. E. of Mongu, 12.ix.1962, *Robinson* 5465 (K; SRGH). C: Chakwenga Headwaters, 100–129 km. E. of Lusaka, 25.viii.1963, *Robinson* 5635 (EA; K; LISC; SRGH). E: Lunkwakwa/Fort Jameson, 21.ix.1966, *Mutimushi* 1446 (K; SRGH). S: Choma, 1310 m., 24.vi.1958, *Robinson* 2874 (EA; K; M; PRE; SRGH). **Zimbabwe.** N: Lomagundi Distr., Mangula, 1170 m., 14.ix.1969, *Jacobsen* 3944 (PRE). W: Matobo Distr., Farm Besna Kobila, 1430 m., ix.1956, *Miller* 3633 (K; SRGH). C: Gwelo Distr., Mlezu school farm, 1230 m., 10.vi.1976, *Biegel* 5311 A (SRGH). E: Umtali Distr., Chimedza R., about 3 km. from its junction with Tsungwesi R., 990 m., 8.ix.1955, *Drummond* 4867 (K; PRE; SRGH; W). S: Nuanetsi Distr., Malangwe R., SW. Mateke Hills, 620 m., 6.v.1958, *Drummond* 5592 (K; LISC; PRE; SRGH). **Malawi.** C: Dedza Distr., Chongoni Forest Reserve, 20.vii.1967, *Salubeni* 776 (SRGH).

Also in Kenya, Tanzania, Angola, Namibia and S. Africa (Transvaal). On seasonally wet ground, stream banks or shallow soils over rocks, often in sandy places.

3. CYPHIA Berg.

Cyphia Berg., Descr. Pl. Cap.: 172 (1767). — E. Wimmer in Engl., Pflanzenr. **IV**. 276 c: 935 (1968). — Thulin in Bot. Notiser **131**: 455–471 (1978).

Perennial herbs from a subterranean, subglobose or elongated root-tuber, laticiferous; stems herbaceous, erect or twining. Leaves alternate, sessile or petiolate. Flowers in racemes, zygomorphic. Calyx 5-lobed. Corolla either split to the base into 2 lips with the upper one 3-lobed and the lower consisting of 2 free petals, or all petals united into a tube with a 5-lobed limb. Stamens 5; filaments free or united towards the base; anthers free or apically loosely united, glabrous or ± pubescent on the back of the connective. Ovary subinferior to subsuperior, 2-locular; ovules many; style shorter than the stamens, obliquely clavate, crowned by hairs, in the upper part with a stigmatic cavity communicating with the air through a small, lateral aperture. Capsule 2-locular, dehiscing loculicidally by 2 apical, usually bifid valves. Seeds numerous.

Genus of some 60 species, entirely confined to Africa, especially abundant in South Africa.

1. Corolla tubular with a 5-lobed limb - - - - - - - - - 2
- Corolla split to the base into 2 lips, the upper 3-lobed and the lower consisting of 2 free petals - - - - - - - - - - - - - - 5
2. Plant erect; calyx-lobes about half as long as corolla; corolla 8–12 mm. long - - 3
- Plant twining, rarely suberect; calyx-lobes usually much less than half as long as corolla; corolla 8–21 mm. long - - - - - - - - - - - 4
3. Inflorescence dense, not interrupted; corolla c. 8 mm. long with the lobes papillose on the inside - - - - - - - - - - - -1. *nyikensis*
- Inflorescence fairly lax, interrupted; corolla 10–12 mm. long with the lobes hairy on the inside - - - - - - - - - - - 2. *mafingensis*
4. Corolla glabrous outside; inflorescence rhachis practically glabrous; seeds not winged, irregularly angular, 0·8–1 mm. long - - - - - - - 3. *stenopetala*
- Corolla crisped-pubescent outside; inflorescence rhachis crisped-pubescent at least towards the apex; seeds broadly winged, flat, c. 2·4–3·2 mm. long - - 4. *mazoensis*
5. Plant twining; corolla 3·5–4 mm. long - - - - - - - 8. *reducta*
- Plant erect or twining; corolla at least 5 mm. long - - - - - - 6
6. Hairs on the back of the connectives up to c. 0·1–0·5 mm. long or absent; seeds broadly winged, flat - - - - - - - - - - - - - 7
- Hairs on the back of the connectives up to c. 0·8–1·5 mm. long; seeds (not known in *C. richardsiae*) not or scarcely winged, coarsely reticulate - - - - - 10
7. Stems usually several from the base, ± erect; inflorescence rhachis practically glabrous; corolla 6·5–7·5 mm. long, papillose mainly along the sutures outside - - 7. *decora*
- Stem single, erect or twining; inflorescence rhachis pubescent at least towards the apex; corolla short pubescent outside - - - - - - - - - 8
8. Plant ± erect; corolla 5–6·5 mm. long - - - - - - - - 6. *alba*
- Plant ± twining; corolla 8–18 mm. long - - - - - - - 9
9. Inflorescence rhachis with crisped appressed hairs; stamens 3·5–6 mm. long with the hairs on the back of the connectives 0·1–0·2 mm. long or absent; inflorescence usually unbranched - - - - - - - - - - - 4. *mazoensis*
- Inflorescence rhachis with spreading, not crisped hairs; stamens 6–8 mm. long with the hairs on the back of the connectives 0·4–0·5 mm. long; inflorescence usually branched 5. *brummittii*
10. Stamens 2·5–3·2 mm. long; upper petals not or scarcely saccate at the base - - 11
- Stamens 4–8 mm. long; upper petals usually markedly saccate at the base - - 12
11. Plant erect or somewhat twisting, 20–50 cm. long; inflorescence fairly dense, up to c. 15 cm. long - - - - - - - - - - - - 9. *richardsiae*
- Plant twining, usually much longer; inflorescence lax, up to 100 cm. long 10. *brachyandra*
12. Plant usually erect; inflorescence dense to rather lax with a sulcate axis; anthers with a tuft of c. 0·5 mm. long hairs at the tip by which the anthers are firmly united apically at least in young flowers - - - - - - - - - - - 11. *erecta*
- Plant usually twining; inflorescence usually very lax with a ribbed axis; anthers with a tuft of 0·2–0·5 mm. long hairs at the tip, not or very loosely united apically - 12. *lasiandra*

1. **Cyphia nyikensis** Thulin in Bot. Notiser **131**: 461, fig. 4 (1978). Type: Malawi, Nyika Plateau, *Robinson* 4515 (SRGH, holotype).

Erect herb, 15–23 cm. tall; stem unbranched, ribbed, papillose at the base, short pubescent in the inflorescence region, otherwise glabrous. Leaves subsessile to shortly petiolate, crowded towards the base of the stem, 20–45 × 7–18 mm., usually with 1–2 smaller leaves further up on the stem, lanceolate to elliptic, acute at the apex, cuneate at the base, glabrous; margin serrate-crenate, slightly revolute; veins prominent beneath, papillose. Raceme dense, up to 3 cm. long, long-pedunculate; pedicels up to 2 mm. long, short pubescent; bracts up to 10 mm. long, dentate; bracteoles up to 5 mm. long, lanceolate-acuminate. Hypanthium broadly obconical, 10-nerved, short pubescent. Calyx-lobes narrowly triangular, erect, 3·2–5·6 mm. long, glabrous, with 1–2 pairs of teeth at the margin. Corolla c. 8 mm. long, pale pink, tubular with 5 subequal lobes c. 4 mm. long, at least the 2 lower petals with slits along their sutures at the base, tube short pubescent on the outside and on the inside towards the base, lobes papillose on both sides. Stamens c. 3·2 mm. long; filaments linear, free, slightly dilated at the base, sparsely pubescent; anthers narrowly elliptic-oblong, c. 1·5 mm. long, shortly pubescent on the back and with a short tuft of hairs at the tip. Style c. 1·8 mm. long. Capsule and seeds not known.

Malawi. N: Nyika Plateau, 2400 m., 14.iii.1961, *Robinson* 4515 (SRGH).
Not known elsewhere. In shallow soil over rock.

2. **Cyphia mafingensis** Thulin, sp. nov.* Type: Malawi, Mafinga Mts., *Brummitt, Polhill &*
Banda 16260 (K, holotype).

Erect herb, c. 25 cm. tall; stem unbranched, ribbed, papillose at the base, shortly
pubescent in the inflorescence region, otherwise glabrous. Leaves subsessile,
crowded towards the base of the stem, 20–70 mm. × 8–37 mm., elliptic to ovate, acute
at the apex, cuneate at the base, glabrous; margin serrate, slightly revolute; veins
prominent beneath. Inflorescence pedunculate, fairly lax, at least 10 cm. long,
interrupted, with the lower flowers fascicled on abbreviated branches; pedicels c. 2
mm. long, shortly pubescent; bracts 5–10 mm. long, serrate to denticulate;
bracteoles c. 1·5–3 mm. long, linear. Hypanthium broadly obconical, 10-nerved,
shortly pubescent. Calyx-lobes linear oblong, erect, 4–6 mm. long, glabrous, with 1–2
pairs of teeth at the margin. Corolla 10–12 mm. long, very pale lilac, tubular with 5
subequal c. 5 mm. long lobes, apparently early splitting along the ventral suture,
sparsely papillose on the outside, hairy within. Stamens c. 4 mm. long; filaments
linear, free, pubescent above, dilated and ciliate at the base; anthers narrowly
elliptic-oblong, c. 2 mm. long, dorsally shortly pubescent and with a short tuft of
hairs at the apex. Style c. 3·5 mm. long. Capsule and seeds not known.

Malawi. N: Mafinga Mts., north end of main ridge, 2300 m., 2.iii.1982, *Brummitt, Polhill &*
Banda 16260 (K).
Not known elsewhere. On rocky outcrop in grassland.

3. **Cyphia stenopetala** Diels in Engl., Bot. Jahrb. **26**: 112 (1898). — E. Wimmer in Engl.,
Pflanzenr. **IV**. 276 c: 945, fig. 42 A (1968). — Thulin in Bot. Notiser **131**: 461 (1978). Type
from Transvaal.
   Cyphia bechuanensis Bremek. & Oberm. in Ann. Transvaal Mus. **16**: 437 (1935). Type:
Botswana, Metsimaklaba R., near Gaberones, *van Son* in HTM 28788 (BM; PRE,
holotype).
   Cyphia stenopetala var. *johannesburgensis* E. Wimmer, tom. cit.: 946 (1968). Type from
Transvaal.

Twining to suberect herb, 25–90 cm. tall from a tuber c. 2–4 cm. in diameter; stem
not or sparsely branched, ribbed, practically glabrous in all its length. Leaves sessile
or subpetiolate, up to 35–70 × 1–10 mm., linear to linear-lanceolate, acute at the
apex, attenuate at the base, glabrous; margin serrulate, ± revolute; midvein
prominent beneath. Raceme rather lax, up to 20 cm. long; pedicels 1–4 mm. long,
very finely crisped pubescent to glabrous; bracts up to 5–35 mm. long; bracteoles
1·5–2·5 mm. long, lanceolate-acuminate. Hypanthium broadly obconical, 10-
nerved, very finely crisped pubescent to glabrous. Calyx-lobes narrowly triangular,
erect to ± reflexed, 1·5–3·6 mm. long, glabrous, with 2–3 pairs of denticles at the
margin. Corolla 11–21 mm. long, pink or purple, tubular with 5 subequal lobes, the 2
lower petals sometimes finally becoming ± free; all petals linear, glabrous outside,
short pubescent inside along the sutures towards the base. Stamens c. 4·5–5·5 mm.
long; filaments linear, dilated and connate at the base, pubescent; anthers narrowly
elliptic-oblong, c. 1·5 mm. long, glabrous or with a short tuft of hairs at the tip only.
Ovary semi-inferior; style c. 4 mm. long. Capsule subglobose to ovoid, inferior part
c. 3 mm. long with 10 distinct nerves connected by weak transversal nerves; valves c.
3 mm. long. Seeds oblong, irregularly angular, ± compressed, 0·8–1 mm. long,
reticulate, brown.

Botswana. SE: Metsimaklaba R., near Gaberones, 15.iii.1930, *van Son* in HTM 28788
(BM; PRE).
Also in Transvaal and NE. Cape Province. In grassland or rocky places.

4. **Cyphia mazoensis** S. Moore in Journ. Bot., Lond. **45**: 46 (1907). — E. Wimmer in Engl.,
Pflanzenr. **IV**. 276 c: 949, fig. 42B (1968). — Binns, H.C.L.M.: 25 (1969) pro parte. —
Thulin in Bot. Notiser **131**: 461 (1978). Type: Zimbabwe, Mazoe, *Eyles* 231 (BM,
holotype; SRGH).

* A *C. nyikensi* inflorescentia laxa interrupta, corolla 10–12 mm. longa lobis intus
pubescentibus differt.

Cyphia rivularis E. Wimmer in Kew Bull. **7**: 145 (1952); in tom. cit.: 942, fig. 37A (1968). Type: Zimbabwe, Inyanga Distr., Inyangombi R. above Nyamziwa Falls, *Chase* 590 (BM; SRGH; W, holotype).

Cyphia mazoensis var. *stellaris* E. Wimmer in Kew Bull. **7**: 144 (1952). Syntypes from Zimbabwe: Umtali Distr., Stapleford, *Hopkins* in GHS 12477 (SRGH; W, syntype), Cecil Heights Commonage, *Chase* 2139 (BM; LISC; SRGH; W, syntype) and Inyanga Distr., Pungwe Valley, *Chase* 894 (BM; LISC; SRGH; W, syntype).

? *Cyphia gamopetala* Duvign. & Denaeyer in Bull. Soc. Roy. Bot. Belg. **96**: 133 (1963). Type from Zaire (lost).

Cyphia rogersii S. Moore subsp. *rogersii* sensu E. Wimmer, tom. cit.: 947, fig. 39 (1968) pro parte.

Cyphia subscandens E. Wimmer, tom. cit.: 949, fig. 40B (1968), *nom. invalid.* Orig. coll.: Zambia, Sunzu Mt., *Bullock* 2184 (K; W).

Cyphia mazoensis f. *angustior* E. Wimmer, tom. cit.: 950 (1968), *nom. invalid.* Orig. colls. from Zimbabwe: Umtali Distr., Banti Forest, *Exell, Medonca & Wild* 210 (BM; SRGH) and Melsetter Distr., Iona Farm, *Chase* 4860 (SRGH).

Twining and climbing, rarely suberect herb from 20 cm. up to 1 m. or more long, from a tuber c. 3–5 × 1·5–2 cm.; stem not or sparsely branched, ribbed, glabrous below, ± densely, very finely crisped pubescent at least in the upper part of the inflorescence region. Leaves petiolate, up to 35–140 × (2)10–30(45) mm., linear to ovate, acute to acuminate at the apex, attenuate to cuneate at the base, glabrous; margin serrate or serrate-crenate, flat or ± revolute; venation prominent beneath; petiole up to 5–15 mm. long. Raceme twining, lax, up to 50 cm. long, sometimes with ± short and few-flowered branches below, or with 2–3-flowered branches almost to the top of the inflorescence; pedicels 1·5–18 mm. long, very shortly crisped pubescent; bracts up to 5–20 mm. long; bracteoles 1–3·5 mm. long, sometimes denticulate. Hypanthium broadly obconical, 10-nerved, shortly crisped pubescent at least towards the base. Calyx-lobes narrowly triangular, erect to ± reflexed, 1·5–8 mm. long, glabrous, with 1–8 pairs of teeth at the margin, basal teeth sometimes very long and resembling the calyx-lobes. Corolla 8–18 mm. long, white, greenish yellow, pink, purple or red, tubular with 5 subequal lobes, or at least the 2 lower petals sometimes becoming ± free; all petals linear, short pubescent outside at least towards the base, pubescent inside, at least towards the base. Stamens 3·5–6 mm. long; filaments linear, free, pubescent; anthers narrowly elliptic-oblong, 1·2–2 mm. long, with a short tuft of hairs at the apex and ± sparsely pubescent with hairs 0·1–0·2 mm. long or rarely glabrous on the back of the connective. Ovary semisuperior; style 2·5–5 mm. long. Capsule subglobose, inferior part 2–5 mm. long, membranous, with 10 distinct nerves connected by weak transversal nerves; valves 4–9 mm. long. Seeds almost round in outline, flat, 2·4–3·2 × 2–2·8 mm., broadly winged, brown.

Zambia. N: Chilongowelo, Plain of Death, 1460 m., 19.v.1955, *Richards* 4159A (BR; K; LISC). C: Lusaka, 5.ii.1965, *Fanshawe* 9177 (BR; K; SRGH). S: Mazabuka Distr., Mabwingombe Hills, 5.ii.1960, *White* 6826 (K). **Zimbabwe.** N: vicinity of Umvukwe Mts., near Toroshanga pass, 24–27.iv.1948, *Rodin* 4426 (K; MO; PRE; SRGH). C: Wedza Mt., 19.ii.1963, *Wild* 5997 (K; LISC; MO; SRGH). E: Umtali Distr., Inyamatshira Mt., 1520 m., 2.iv.1955, *Chase* 5538 (BM; BR; COI; K; LISC; SRGH). S: Belingwe Distr., Sikanajena Kop, 1100 m., 4.v.1973, *Biegel, Pope & Simon* 4295 (K; LISC; MO; SRGH). **Malawi.** C: Kongwe Mt., near Dowa, 1525 m., 18.ii.1959, *Robson* 1655 (K; LISC; SRGH). **Mozambique.** MS: Chimanimani Mts., near St. George's Cave between The Saddle and Poacher's Cave, 1520 m., 12.iv.1967, *Grosvenor* 394 (K; LISC; SRGH).

Probably also in Shaba. In woodland and grassland, usually on rocky ground.

Specimens from E. Zimbabwe with branched inflorescences, where the flowers are 2–5 together in the leaf-axils have been called *Cyphia rivularis* or *C. mazoensis* var. *stellaris*. There are too many intermediates for this distinction to be maintained, at least at the specific level.

C. mazoensis is very closely related to *C. rogersii* S. Moore (1918) in Transvaal, which differs mainly by being glabrous on the outside of the corolla and by always being glabrous on the back of the connectives. They might perhaps be treated as subspecies.

5. **Cyphia brummittii** Thulin, sp. nov.* Type: Malawi, Mt. Mulanje, path from Tuchila Hut to head of Ruo Basin, *Brummitt* 9649 (K, holotype; MAL).

---

* Differt a *C. mazoensi* inflorescentia plerumque ramosa pubescenti pilis patentibus, corolla bilabiata petalis duobus inferis liberis, staminibus 6–8 mm. longis, antheris dorsaliter pubescentibus pilis 0·4–0·5 mm. longis et seminibus 4–4·5 × 3 mm.

Twining and climbing herb up to 1 m. or more long; stem sparsely branched, ribbed, glabrous below, finely pubescent with spreading hairs at least in the upper part of the inflorescence region. Leaves petiolate, up to 75 × 30(45) mm., narrowly lanceolate to narrowly ovate, lower leaves sometimes hastate, acute at the apex, cuneate to subtruncate at the base, glabrous; margin serrate or serrate-crenate, flat; venation prominent beneath; petiole up to 20 mm. long. Inflorescence twining, lax, up to 80 cm. long, usually branched, except for in the uppermost part, with groups of 2–several flowers from the leaf-axils; pedicels 2–5 mm. long, shortly pubescent with spreading hairs; bracteoles 1–2 mm. long. Hypanthium broadly obconical, 10-nerved, shortly pubescent with spreading hairs at least towards the base. Calyx-lobes linear, erect, becoming spreading or somewhat reflexed in fruit, 5–9 mm. long, with 1–3 pairs of teeth at the margin. Corolla 9–14 mm. long, white, split to the base into 2 lips, the upper 3-lobed but finally disintegrating, saccate at the base, the lower consisting of 2 free petals; all petals linear, sparsely or densely pubescent outside with short spreading hairs, pubescent inside with longer hairs. Stamens 6–8 mm. long; filaments linear, free, dilated at the base, pubescent; anthers elliptic-oblong, 1·5–2·4 mm. long, pubescent with hairs 0·4–0·5 mm. long on the back of the connective but glabrous at the apex. Ovary semisuperior; style 4–6·5 mm. long. Capsule ovoid-globose, inferior part 3–4 mm. long, membranous, with 10 distinct nerves connected by weak transversal nerves; valves 7–8 mm. long. Seeds broadly elliptic in outline, flat, 4–4·5 × 3 mm., broadly winged, brown.

Malawi. S: Mt. Mulanje, Chambe Plateau, path to Litchenya, 1950 m., 12.ii.1979, *Blackmore, Brummitt & Banda* 350 (K; MAL).
Only known from Mt. Mulanje. In upland grassland and on shrubby hillsides.

6. **Cyphia alba** N. E. Br. in Kew Bull. **1906**: 165 (1906). — E. Wimmer in Engl., Pflanzenr. **IV**. 276 c: 973, fig. 56A (1968). — Thulin in Bot. Notiser **131**: 462 (1978). Type: Zimbabwe, Manika, N. of Umtali, *Cecil* 163 (K, holotype).
   Cyphia alba f. *purpurea* E. Wimmer, loc. cit. Type: Zimbabwe, Inyanga Distr., Mt. Inyangani, summit ridge, *Whellan & Davies* 987 (K, holotype; SRGH).

Erect or somewhat twining herb, 20–70 cm. tall. Stem single, unbranched or sometimes sparsely branched, ribbed, glabrous below, shortly crisped pubescent in the inflorescence region. Leaves sessile, up to 25–90 × 1·5–6 mm., linear to linear-lanceolate, often complicate, acute at the apex, attenuate at the base, thickened and sparsely serrulate at the margin, glabrous; midvein very prominent beneath. Raceme rather dense, sometimes with a short branch below, 3–10(18) cm. long, up to 30-flowered; peduncle up to 14 cm. long; pedicels 1·5–3·2 mm. long, shortly crisped pubescent; bracts linear-lanceolate, 1·5–12 mm. long, often denticulate; bracteoles 0·6–3·2 mm. long. Hypanthium cup-shaped, 10-nerved, shortly crisped pubescent to subglabrous. Calyx-lobes narrowly triangular, erect or sometimes reflexed, 1–2·8 mm. long, with 1–3 pairs of teeth at the margin, glabrous. Corolla 5–6·5 mm. long, white to purple, split to the base into 2 lips, the upper 3-lobed but finally disintegrating, the lower divided into 2 free petals; all petals oblanceolate-spathulate, ± densely short pubescent outside, pubescent inside at least in a zone near the base, but sometimes sparsely pubescent almost to the tip. Stamens 2–3 mm. long; filaments linear, free, somewhat dilated at the base, pubescent; anthers elliptic-oblong, 0·9–1·2 mm. long, short pubescent (hairs 0·1–0·2 mm. long) at the tip and usually also on the back of the connective. Ovary semi-inferior; style 1·2–2 mm. long. Capsule subglobose, with 10 distinct nerves connected by some weak transversal nerves; valves 2–2·5 mm. long, inferior part of capsule 2–3 mm. long. Seeds broadly elliptic in outline, flat, broadly winged, c. 2 × 1·5 mm.

Zimbabwe. E: Chimanimani Mts., summit of Mt. Peza, 2175 m., 30.xii.1957, *Goodier* 497 (SRGH).
Not known from elsewhere. In upland grassland.

7. **Cyphia decora** Thulin in Bot. Notiser **131**: 462, fig. 1, 5 (1978). Type: Malawi, Mt. Mulanje, Chambe Plateau, *Jackson* 2167 (K, holotype; SRGH).

Erect or ascending sometimes somewhat twining herb, 7–45 cm. tall from a root-tuber c. 2 cm. thick. Stems usually several from the base, ribbed, entirely glabrous or sparsely papillose in the inflorescence region only. Leaves sessile or shortly petiolate, ± crowded towards the base, up to 15–45 × 3–5 mm., linear to narrowly elliptic, acute at the apex, attenuate at the base, thickened and serrulate at the margin, glabrous; midvein very prominent beneath. Raceme rather dense, sometimes with a few short branches below, 3–12 cm. long, up to 12-flowered; peduncle up to 18 cm. long; pedicels 1–2·5 mm. long, papillose to subglabrous; bracts linear-lanceolate, up to 5 mm. long, often denticulate; bracteoles ± 1·5 mm. long. Hypanthium cup-shaped, 10-nerved, sparsely papillose to subglabrous. Calyx-lobes narrowly triangular, ± erect, 1·5–2·2 mm. long, with 1–2 pairs of teeth at the margin, glabrous. Corolla 6·5–7·5 mm. long, pink to purple, split to the base into 2 lips, the upper 3-lobed but finally disintegrating, the lower divided into 2 free petals; all petals oblanceolate-spathulate, papillose mainly along the sutures outside and in a zone near the base inside. Stamens 2·5–3·2 mm. long; filaments linear, free, somewhat dilated at the base, pubescent; anthers elliptic-oblong, 1·2–1·4 mm. long, densely short pubescent (hairs c. 0·2–0·3 mm. long) on the back of the connective but usually glabrous at the very tip. Ovary semi-inferior; style c. 2 mm. long. Capsule ovoid to subglobose, with 10 distinct nerves connected by weak transversal nerves; valves 2–3 mm. long, as long as or longer than inferior part of capsule. Seeds broadly elliptic in outline, flat, broadly winged, 2–2·4 × 1·2–1·8 mm.

Malawi. S: Mt. Mulanje, Chambe basin, 1820 m., 23.i.1967, *Hilliard & Burtt* 4591 (K; SRGH).
Only known from Mt. Mulanje. Upland grassland, often in rock cracks.

8. **Cyphia reducta** E. Wimmer in Kew Bull. **7**: 144 (1952); in Engl., Pflanzenr. **IV**. 276 c: 1001, fig. 72C (1968). — Thulin in Bot. Notiser **131**: 464, fig. 2A (1978). Type: Zimbabwe, Domboshawa, *Wild* 1662 (K; SRGH; W, lectotype).

Twining herb. Stems almost filiform, branching, ribbed, entirely glabrous or sparsely papillose in the inflorescence region only. Leaves shortly petiolate, up to 25–50 × 1·5–4 mm., linear, acute at the apex, attenuate at the base, thickened and sparsely serrulate at the margin, glabrous; midvein very prominent beneath. Racemes twining, lax, several-flowered; pedicels 1·5–5 mm. long, finely pubescent or papillose; bracts linear, up to 4 mm. long, often denticulate; bracteoles 0·4–1 mm. long. Hypanthium cup-shaped, 10-nerved, glabrous or almost so. Calyx-lobes narrowly triangular, ± erect, 0·5–1·2 mm. long, entire or with a pair of teeth at the base, glabrous. Corolla 3·5–4 mm. long, white, sometimes tinged with pink, split to the base into 2 lips, the upper 3-lobed but finally disintegrating, the lower divided into 2 free petals; all petals narrowly oblanceolate, glabrous outside, short pubescent inside in a zone near the base. Stamens 1·5–1·8 mm. long; filaments linear, free, somewhat dilated at the base, pubescent; anthers elliptic-oblong, 0·6–0·8 mm. long, short pubescent at the tip and usually also on the back of the connective. Ovary semi-inferior; style c. 1·2 mm. long. Capsule subglobose, with 10 distinct nerves connected by some weak transversal nerves, with 2 valves 1·4–2·4 mm. long, inferior part of capsule 1·4–2·4 mm. long. Seeds broadly elliptic, flat, broadly winged, 1·4–1·8 × 1–1·2 mm.

Zimbabwe. C: Goromonzi Distr., Domboshawa, Chindamora T.T.L., Ngomakurira, 1520 m., 14.iii.1965, *Loveridge* 1357 (SRGH). **Mozambique.** N: Lizombe, 110 km. NE. of Lichinga, 17.ii.1982, *Pettersson* 2117 (UPS).
Not known from elsewhere. Twining among grasses.

9. **Cyphia richardsiae** E. Wimmer in Engl., Pflanzenr. **IV**. 276 c: 973, fig. 57 (1968). — Thulin in Bot. Notiser **131**: 465 (1978). Type from Tanzania.

Erect or sometimes somewhat twisting herb, 20–50 cm. tall. Stem unbranched, ribbed, sparsely and finely pubescent in the inflorescence region. Leaves sessile, up to 20–75 × 2–8 mm., linear to linear-lanceolate, acute at the apex, attenuate at the base, revolute, thickened and serrulate at the margin, glabrous; midvein very prominent beneath. Raceme rather dense, sometimes with a short branch below, 4–16 cm. long, c. 10–30-flowered; pedicels 1·5–5·5 mm. long, shortly crisped pubescent; bracts linear, 5–15(30) mm. long, denticulate; bracteoles 1–1·5 mm. long. Hypanthium broadly obconical, 10-nerved, sparsely pubescent at least towards the base. Calyx-lobes narrowly triangular, ± erect or spreading, 1·2–3·5 mm. long, with 1–3 pairs of teeth at the margin, glabrous. Corolla 5–9·5 mm. long, pale pink, mauve, lavender or white, split to the base into 2 lips, the upper 3-lobed but finally disintegrating, the lower divided into 2 free petals; all petals narrowly oblanceolate to linear, glabrous outside, pubescent inside at the base only to ± pubescent in their entire length. Stamens 2·5–3 mm. long; filaments linear, abruptly dilated, ciliate and obscurely connate at the very base; anthers elliptic-oblong, 1·2–1·5 mm. long, pubescent with up to 0·8–1 mm. long, rather straight to curled hairs on the upper ¾ of the connective and with shorter hairs towards its base, also a small tuft of hairs c. 0·2 mm. long at the tip. Ovary semi-inferior; style c. 1·2–1·6 mm. long. Capsule and seeds not known.

Malawi. N: Chitipa Distr., Nyika Plateau, Nganda Peak, 2600 m., 16.iv.1975, *Pawek* 9282 (SRGH).
Also in Tanzania and Zaire (Shaba). In upland grassland.

The material from Malawi and adjacent parts of S. Tanzania differs from the type race in W. Tanzania in that the petals are longer (6·5–9·5 mm. versus 5–6·5 mm.), the inflorescence is somewhat denser and the leaves are more numerous. Possibly the plants from N. Malawi and S. Tanzania will eventually prove to be a distinct taxon, but the present evidence is too weak and more material is needed.

10. **Cyphia brachyandra** Thulin in Bot. Notiser **131**: 466, fig. 2B, 6 (1978). Type from Tanzania.

Twining slender herb, up to 2 m. long, from a subglobose or elongated root-tuber, up to 3 cm. long. Stem not or sparsely branched, ribbed, glabrous or very sparsely and finely pubescent in the inflorescence region. Leaves sessile or shortly petiolate, up to 35–85 × 2–30 mm., linear to ovate, acute at the apex, attenuate to cuneate at the base, serrate or crenate at the margin, glabrous; venation prominent beneath; petiole up to 12 mm. long, often curved. Raceme lax, up to 100 cm. long and c. 20–30-flowered; pedicels up to 3–13 mm. long, shortly crisped pubescent; bracts ± leaf-like, diminishing in size upwards; bracteoles 1–2 mm. long. Hypanthium broadly obconical, 10-nerved, glabrous or sparsely pubescent at least towards the base. Calyx-lobes narrowly triangular, ± erect or spreading, becoming reflexed in fruit, 2·4–4 mm. long, with 1–3 pairs of teeth at the margin, glabrous or sparsely ciliate. Corolla 9–12 mm. long, pale pink, pale purple or mauve, split to the base into 2 lips, the upper 3-lobed but finally disintegrating, the lower divided into 2 free petals; all petals linear, glabrous outside, pubescent inside at the base. Stamens c. 3 mm. long; filaments linear, abruptly dilated, ciliate and obscurely connate at the very base; anthers elliptic-oblong, 1·4–1·6 mm. long, pubescent with up to 0·8–1 mm. long, rather straight hairs on the upper ¾ of the connective and with shorter hairs towards its base, also a small tuft of hairs c. 0·2 mm. long at the tip. Ovary semi-inferior; style 1·5–2 mm. long. Capsule ovoid, with 10 distinct nerves connected by weak transversal nerves; inferior part cup-shaped, 2·4–4 mm. long; valves 2·8–3·2 mm. long. Seeds ± broadly elliptic in outline, ± compressed and irregularly angular, c. 1·2 mm. long, coarsely reticulate, brown.

Malawi. N: Rumphi Distr., Nyika Plateau, Chelinda bridge, 2280 m., 29.iii.1970, *Pawek* 3423 (K).
Also in S. Tanzania. At forest margins, in scrub or rocky places.

11. **Cyphia erecta** De Wild. in Ann. Mus. Congo Belge, Bot., sér. 4, **1**: 162, t. 38/5–8 (1903). — Milne-Redhead in Kew Bull. **5**: 378 (1951). — E. Wimmer in Engl., Pflanzenr. **IV**. 276 c: 974, fig. 59 (1968). — Thulin in Bot. Notiser **131**: 467 (1978). TAB. 25. Type from Zaire (Shaba).
  Cyphia scandens De Wild., tom. cit.: 163 (1903). — E. Wimmer, tom. cit.: 974, fig. 74C (1968). Type from Zaire (Shaba).
  Cyphia erecta f. *minor* De Wild. in Ann. Soc. Sci. Brux. **38**, Mém.: 444 (1914). — E. Wimmer in tom. cit.: 975 (1968). Type from Zaire (Shaba).
  Cyphia regularis E. Wimmer in Kew Bull. **7**: 143 (1952); tom. cit.: 975, fig. 48C (1968). Type: Zambia, upper Loangwa River, *Nicholson* s.n. (K, holotype).
  Cyphia erecta var. *ufipana* E. Wimmer, loc. cit., *nom. invalid.* Orig. colls. from Tanzania and Malawi: Kota-Kota, *Benson* 693 (PRE, not seen).
  Cyphia rhodesiaca E. Wimmer, tom. cit.: 975, fig. 58C (1968). Type: Zambia, Mwinilunga Distr., Kalanda Dambo, *Milne-Redhead* 3593 (BM; BR, holotype; K).

Erect or rarely somewhat twining herb, 8–80 cm. tall, from a subglobose tuber c. 2 cm. in diameter; stem not or sparsely branched, ribbed below, sulcate in the inflorescence region, glabrous or sparsely pubescent above, rarely all plant pubescent. Leaves sessile or shortly petiolate, up to 30–110 × 3–25 mm., linear to

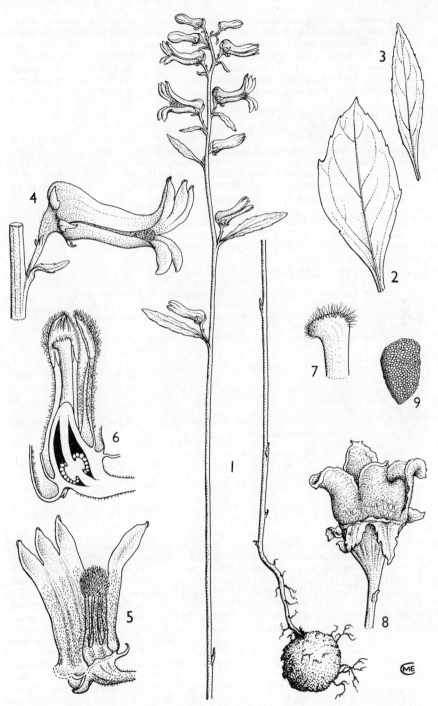

Tab. 25. CYPHIA ERECTA. 1, habit (×1), from *Milne-Redhead* 3953; 2, 3, leaves (×1), from *Fanshawe* 695; 4, flower (×4); 5, flower with some sepals and one petal removed (×4); 6, longitudinal section of ovary (×6); 7, style-apex (×12), 4–7 from *Milne-Redhead* 3953; 8, capsule (×4); 9, seed (×10), 8–9 from *Fanshawe* 695. From F.T.E.A.

broadly elliptic, acute or subacute at the apex, attenuate to cuneate at the base, serrate or serrate-crenate, glabrous or rarely short pubescent, mainly above; venation prominent beneath. Raceme dense or lax, up to 25 cm. long, sometimes branched; pedicels 1–6 mm. long, very finely pubescent or papillose; bracts leaf-like, diminishing in size upwards; bracteoles 1–3(7) mm. long, filiform to lanceolate, sometimes denticulate. Hypanthium broadly obconical, 10-nerved, shortly pubescent to papillose or glabrous. Calyx-lobes ± narrowly triangular, usually reflexed, 1·5–4·5 mm. long, with 1–3 pairs of teeth at the margin, glabrous. Corolla 9–16(20) mm. long, pale blue to mauve or purple, ± saccate at the base, split to the base into 2 lips, the upper 3-lobed but finally ± disintegrating, the lower divided into 2 free petals; all petals linear, glabrous or very rarely pubescent outside, white pubescent inside on the lower half or up to c. ⅔ of the length. Stamens 4–7 mm. long; filaments linear, dilated and connate at the base, ± pubescent; anthers elliptic-oblong, 1·6–3·2 mm. long, lanate with curled hairs c. 1·5 mm. long on the back of the connective, and at the tip with a tuft of hairs, ± 0·5 mm. long, by which the anthers are apically united at least in young flowers. Ovary semi-superior; style 2·5–5·5 mm. long. Capsule ovoid to subglobose, inferior part 2·5–4 mm. long, with 10 distinct nerves with ± transversal connections, membranous parts almost or entirely lacking; valves 3·5–6 mm. long. Seeds ± broadly elliptic in outline, ± compressed or irregularly angular, c. 1–1·5 mm. long, coarsely reticulate, brown.

Zambia. N: Mporokoso Distr., Kabwe Plain, Mweru-Wantipa, 1000 m., 14.xii.1960, *Richards* 13698 (K; SRGH). W: Ndola, 20.i.1954, *Fanshawe* 695 (BR; K; SRGH). C: Chakwenga Headwaters, 100–129 km. E. of Lusaka, 10.i.1964, *Robinson* 6187 (K; SRGH). E: 27 km. S. of Chitipa, 5 km. into Zambia, facing Mafinga Hills, 1070 m., 27.xii.1975, *Pawek* 6156 (K; SRGH). **Malawi.** N: Mzimba Distr., Viphya Plateau, 60 km. SW. of Mzuzu, 1670 m., 3.ii.1974, *Pawek* 8034 (SRGH). C: Mchine (Fort Manning), near District Offices, 1250 m., 7.i.1959, *Robson* 1070 (BM; K; LISC; SRGH).
Also in Tanzania and Zaire (Shaba). In grassland, woodland and thickets.

12. **Cyphia lasiandra** Diels in Engl., Bot. Jahrb. **26**: 111 (1898). — E. Wimmer in Engl., Pflanzenr. **IV**. 276 c: 991, fig. 66D (1968). — Thulin in Bot. Notiser **131**: 469 (1978). Type from Angola.
   Cyphia nyasica Bak. in Kew Bull. **1898**: 157 (1898). — E. Wimmer, tom. cit.: 1010, fig. 77 (1968). — Binns, H.C.L.M.: 25 (1969). Type: Malawi, between Kondowe and Karonga, *Whyte* s.n. (K, holotype).
   ? *Cyphia antunesii* Engl., Bot. Jahrb. **32**: 147 (1902). — E. Wimmer, tom. cit.: 983 (1968). Type from Angola (B†).
   Cyphia cacondensis Good in Journ. Bot., Lond. **65**, Suppl. 2: 68 (1927). — E. Wimmer, tom. cit.: 991, fig. 66B (1968). Type from Angola.
   Cyphia peteriana E. Wimmer in Kew Bull. **7**: 147 (1952); tom. cit.: 1010, fig. 78, 79E (1968). Type from Tanzania.
   Cyphia zernyana E. Wimmer, tom. cit.: 1006, fig. 74D (1968). Type from Tanzania.
   Cyphia exelliana E. Wimmer, tom. cit.: 1010, fig. 76E (1968). Type: Malawi, Dedza Mt., *Exell, Mendonça & Wild* 1080 (BM; LISC; SRGH, holotype).
   Cyphia floribunda E. Wimmer, tom. cit.: 1013, fig. 67C, 83 (1968). Type from Tanzania.

Twining and climbing herb, well-developed specimens forming tangling masses, usually more than 1 m. and up to several m. long, from a subglobose tuber, c. 2–3 cm. in diameter; stems sparsely branched, ribbed, glabrous or sparsely puberulous in the inflorescence region only. Leaves petiolate, up to 30–100 × (3)6–40(55) mm., linear to ovate, acute to acuminate or rarely ± obtuse at the apex, attenuate to truncate or subcordate at the base, serrate, glabrous or rarely short pubescent above; venation prominent beneath; petiole 1·5–25 mm. long, often curved. Raceme twining, usually very lax, often more than 50 cm. long, sometimes with ± short and few-flowered branches below; pedicels 2–10(20) mm. long, very finely crisped pubescent to almost glabrous; bracts leaf-like, diminishing in size upwards; bracteoles 0·8–5 mm. long, filiform to lanceolate-acuminate, sometimes denticulate. Hypanthium broadly obconical,. 10-nerved, shortly pubescent to glabrous. Calyx-lobes narrowly triangular, erect to reflexed, 1·5–4 mm. or up to 10 mm. long in fruit, with 1–5 pairs of teeth at the margin, glabrous. Corolla 9–17 mm. long, pale pink to purple or almost white, ± saccate at the base, split to the base into 2 lips, the upper 3-lobed but finally ± disintegrating, the lower divided into 2 free petals; all petals linear, glabrous or sparsely pubescent outside, white pubescent inside, at least on the lower half. Stamens 4–8 mm. long; filaments linear, dilated and connate at the base, ±

pubescent; anthers elliptic-oblong, 2–3·2 mm. long, lanate with curled hairs c. 1·2–1·5 mm. long on the back of the connective, and at the tip with a tuft of hairs, 0·2–0·5 mm. long. Ovary semi-inferior; style 2·5–5·5 mm. long. Capsule ovoid to subglobose, inferior part 3–5·5 mm. long, membranous, with 10 distinct nerves connected by weak transversal nerves; valves 2·5–5 mm. long. Seeds ± broadly elliptic in outline, ± compressed or irregularly angular, c. 0·8–1·6 mm. long, coarsely reticulate, brown.

Zambia. W: Solwezi Distr., 27 km. E. of Lunga R. on Kasempa-Kitwe road, 1260 m., 25.i.1975, *Brummitt, Chisumpa & Polhill* 14156 (BR; K; NDO; SRGH; UPS). C: Serenje Distr., Kundalila Falls, 1450 m., 4.ii.1973, *Strid* 2844 (C; MO). **Malawi.** N: Mzuzu, Marymount, 1370 m., 6.v.1969, *Pawek* 2373 (K). C: Dedza Mt., 9.iv.1968, *Jeke* 168 (SRGH). S: Mulanje Mt., Chambe Plateau, 23.iii.1958, *Jackson* 2170 (K; SRGH). **Mozambique.** N: Maniamba, Jéci Mts., Mt. Chicungulo, 1400 m., 2.iii.1964, *Torre & Paiva* 10978 (LISC).
   Also in Tanzania, Zaire (Shaba) and Angola. In grassland, woodland and thickets.

The specimens from Mulanje Mt. have narrower leaves and longer pedicels than the other material from the F.Z. area. They are matched in these characters by specimens from Tanzania and Zaire.

102. ERICACEAE
By R. Ross

Small trees, shrubs or sub-shrubs, never herbs. Leaves simple, alternate, opposite or whorled, exstipulate, usually evergreen. Flowers bisexual, actinomorphic or slightly zygomorphic, in the axils of the leaves, in terminal umbels, or on simple racemes which may be terminal or axillary or both. Calyx and corolla normally 4-merous or 5-merous, occasionally 3-merous. Sepals free or fused, normally equal but in a few genera one sepal larger than the others. Corolla normally gamopetalous, 3–5-lobed, lobes contorted or imbricate. Stamens normally twice as many as the petals, occasionally equal in number to them and alternating with them, or, more rarely, intermediate in number, inserted on a hypogynous or epigynous disc; filaments normally free, occasionally adnate to the corolla at their base; anthers innate or versatile, opening by terminal or lateral pores, often with appendages. Carpels normally equal in number to the petals, occasionally one less, or all but one aborted, united to form a normally 3–5-locular, occasionally 1-locular, ovary, which is normally superior but is inferior in some genera; seeds numerous in most genera, 1 or 2 per loculus in a few, placentation axile or basal. Fruit a capsule, usually loculicidal, a drupe or a berry.

A cosmopolitan family of some 80 genera, but not represented in Australia or New Zealand, mainly temperate or at higher altitudes in the sub-tropics and tropics. Of the six genera occurring in the Flora Zambesiaca region, two, *Agauria* and *Vaccinium*, belong to the subfamily Vaccinioideae Endl. as delimited by Stevens (in Bot. J. Linn. Soc. Lond. **64**: 31–44. 1971), and these have alternate leaves of moderate size without reflexed margins. The remaining four genera, to which 18 of the 21 species found in the region belong, are members of the subfamily Ericoideae, and have very small leaves usually arranged in whorls and with margins so strongly reflexed that they meet, or almost meet, along the midrib on the abaxial side of the leaf. In the descriptions of the genera and species of this subfamily, the line of the sharp angle where the leaf blade is reflexed is termed the apparent margin of the leaf.

1. Leaves 20 mm. or more long, more than 5 mm. broad; calyx and corolla pentamerous   2
 - Leaves less than 15 mm. long, less than 4·0 mm. broad; calyx and corolla tetramerous or trimerous in some flowers -   -   -   -   -   -   -   -   -   -   3
2. Ovary superior; fruit a cartilaginous to woody capsule -   -   -   - **1. Agauria**
 - Ovary inferior; fruit a succulent berry -   -   -   -   -   - **2. Vaccinium**
3. Calyx segments all equal; bract, and often bracteoles, present on pedicel -   -   4
 - One calyx segment longer than the other three; no bracts or bracteoles -   -   - 6
4. Stamens 4* -   -   -   -   -   -   -   -   -   -   -   -   -   - 5
 - Stamens more than 4* -   -   -   -   -   -   -   -   -   - **3. Erica**

* Some specimens of *Erica pleiotricha* have flowers with 4, 5, 6, 7, or 8 stamens on the same plant.

5. Hairs with side branches to their tips present on the branchlets; corolla tubular to narrowly
   campanulate, not expanded above- - - - - - - - - - **3. Erica**
 - No hairs with side branches to their tips present on the branchlets; corolla crateriform or
   tubular below and expanded above - - - - - - - - **4. Blaeria**
6. Stamens 4 - - - - - - - - - - - - - 7
 - Stamens 6–8 - - - - - - - - - - - **6. Philippia**
7. Leaves in whorls of 4; anthers with appendages; stigma infundibuliform - **5. Ericinella**
 - Leaves in whorls of 3; anthers without appendages; stigma disciform - **6. Philippia**

1. AGAURIA (DC.) Hook. f.

Leucothoe sect. **Agauria** DC., Prodr. **7**: 602 (1839).
Agauria (DC.) Hook. f. in Benth. & Hook., Gen. Pl. **2**: 586 (1876).

Small trees. Leaves sub-opposite to alternate, subsessile to petiolate, evergreen,
lanceolate to oblong to ovate or obovate. Inflorescence consisting of axillary and
terminal racemes. Pedicels short, 1-flowered, subtended by caducous bracts and
each bearing a pair of caducous bracteoles. Calyx and corolla actinomorphic and
hypogynous. Calyx segments 5, fused below, persistent in fruit. Corolla obconical,
lobes 5, small, imbricate, not persistent in fruit. Stamens 10, included, adhering
slightly to the corolla at the base and shed with it, filaments geniculate, anthers
dehiscing by terminal pores. Ovary superior, 5-celled, ovules few, placentation
basal; style 1, persistent until capsule dehisces. Capsule 5-valved, loculicidal,
without a central column after dehiscence.

A monospecific genus on high ground throughout tropical Africa, and in Madagascar and the
Mascarene islands.

Agauria salicifolia (Comm. ex Lam.) Hook. f. ex Oliv., F.T.A. **3**: 483 (1877). — Sleumer in
     Bot. Jahrb. **69**: 381 (1938). — Brenan, T.T.C.L.: 189 (1949); in Mem. N.Y. Bot. Gard. **8**:
     492 (1954). — Dale & Greenway, Kenya Trees & Shrubs: 178 (1961). TAB. **26**. Type
     from Réunion.
     Andromeda salicifolia Comm. ex Lam., Encycl. Méth., Bot. **1**: 159 (1783) "*salicisfolia*".
     — Hook. in Curtis's Bot. Mag. **60**: t. 3286 (1833). Type as above.
     Andromeda buxifolia Comm. ex Lam., loc. cit. (1783). Type from Réunion.
     Andromeda pyrifolia Pers., Syn. Pl. **1**: 481 (1805). Type from Réunion.
     Agarista salicifolia (Comm. ex Lam.) Don, Gen. Syst. **3**: 837 (1834). Type as for
     Agauria salicifolia.
     Agarista buxifolia (Comm. ex Lam.) Don, loc. cit. (1834). Type as for *Andromeda
     buxifolia*.
     Agarista pyrifolia (Pers.) Don, tom. cit.: 838 (1834). Type as for *Andromeda pyrifolia*.
     Leucothoe salicifolia (Comm. ex Lam.) DC., Prodr. **7**: 602 (1839). Type as for *Agauria
     salicifolia*.
     Leucothoe buxifolia (Comm. ex Lam.) DC., tom. cit.: 603 (1839). Type as for
     Andromeda buxifolia.
     Leucothoe salicifolia var. *pyrifolia* (Pers.) DC., tom. cit.: 603 (1839). Type as for
     Andromeda pyrifolia.
     Leucothoe littoralis DC., tom. cit.: 602 (1839). Type from Madagascar.
     Leucothoe bojeri DC., tom. cit.: 603 (1839). Type from Madagascar.
     Agauria salicifolia var. *pyrifolia* (Pers.) Oliv., loc. cit. (1877). Type as for *Andromeda
     pyrifolia*.
     Agauria polyphylla Baker in J. Linn. Soc. Lond., Bot. **20**: 194 (1883). Type from
     Madagascar.
     Agauria nummularifolia Baker in J. Linn. Soc. Lond., Bot. **25**: 332 (1890). Type from
     Madagascar.
     Agauria callibotrys Cordemoy, Fl. Ile de Réunion: 438 (1895). Type from Réunion.
     Agauria goetzei Engl. in Bot. Jahrb. **30**: 369 (1901). Type from Tanzania.
     Agauria buxifolia (Comm. ex Lam.) Perrier de la Bâthie in Cat. Pl. Madag., Eric. &
     Vaccin.: 7 (1934). Type as for *Andromeda buxifolia*.
     Agauria littoralis (DC.) Perrier de la Bâthie, loc. cit. (1934). Type as for *Leucothoe
     littoralis*.
     Agauria bojeri (DC.) Perrier de la Bâthie, loc. cit. (1934). Type as for *Leucothoe bojeri*.

Small tree to about 12 m. tall. Branchlets glabrous or pubescent with sparse to
moderately dense short simple hairs, sometimes accompanied by a few longer and
stouter glandular hairs, these hairs occasionally dense. Leaves subcoriaceous, entire,
in the Flora Zambesiaca area narrowly lanceolate to oblong, acuminate, glabrous
except on the midrib on the ventral side of the leaf, 20–120 mm. long, 8–35 mm.
broad; petioles 5–10 mm. long, pubescent or glabrous. Peduncles 15–30-flowered,

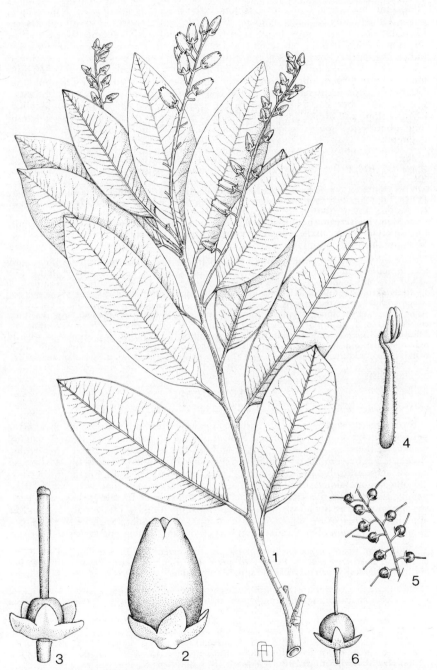

Tab. 26. AGAURIA SALICIFOLIA. 1, flowering branch (×⅔); 2, flower (×4); 3, calyx and ovary (×4); 4, stamen (×8); 5, fruiting branch (×⅔); 6, capsule with calyx and style (×2), all from *Robinson* 5706.

100–150 mm. long; pedicels 2–6 mm. long; peduncles and pedicels normally pubescent, occasionally glabrous. Bracts at the base of the pedicel, 3–5 mm. long, pubescent, broadly to narrowly ovate. Bracteoles inserted from ¼ to ¾ of the pedicel from its base, 1·0–1·5 mm. long, narrow, pubescent. Calyx segments 5, fused for about ⅓ of their length, triangular, c. 2·5 mm. long, shortly and rather sparsely pubescent on the outer surface, margins densely ciliate. Corolla pale green tinged red, to yellow, to creamy white, obconical to slightly urceolate, glabrous, 8–10 mm. long, 4–5 mm. broad, lobes 5, small, imbricate, triangular, obtuse, 0·7–0·8 mm. long, 1·0 mm. broad. Stamens 10, ½–¾ of the length of the corolla; filaments pubescent, geniculate; anthers c. 0·8 mm. long, thecae deeply divided. Ovary globose, 5-celled, c. 2 mm. long and broad, sparsely and shortly pubescent; style 1, glabrous, 6–8 mm. long, about equalling the corolla; stigma small, capitate. Ripe capsule 5–7 mm. in diameter.

Zambia. N: Mpika Distr., 48 km. S. of Shiwa Ngandu on the Mpika road, fr. 28.xi.1952, *Angus* 857 (BM; FHO). W: Mwinilunga, fr. 14.viii.1938, *Holmes* 1182 (K). C: Serenje Distr., Kundalila Falls, fl. 13.x.1963, *Robinson* 5706 (K). **Malawi.** N: Nyika Plateau, Lake Kaulime, 2200 m., fl. 23.x.1958, *Robson* 278 (BM; K; LISC; SRGH). S: Zomba Mt., Chingwe's Hole, 1830 m., fl. 9.iii.1955, *Exell, Mendonça & Wild* 758 (BM; LISC; SRGH). **Mozambique.** Z: Gùruè, Gùruè Mts., 1400 m., fr. 19.ix.1944, *Mendonça* 2118 (LISC).

On high ground throughout tropical Africa, also in Madagascar and the Mascarene islands. In riparian forest and on the margins of evergreen forest, occasionally scattered in grassland, 1375–2300 m.

Sleumer (loc. cit.) recognized a large number of infra-specific taxa within *A. salicifolia*, most of them confined to Madagascar or the Mascarene islands. These he distinguished on the basis of leaf size and shape and on the presence or absence of glandular hairs. He regarded all specimens from continental Africa as belonging to the same variety as the type of the species, which comes from Madagascar, calling this variety *A. salicifolia* var. *pyrifolia* (Pers.) Oliv. Although some specimens from the Flora Zambesiaca region have leaves so large that they correspond to *A. salicifolia* var. *megaphylla* Sleumer, a variety said by him to be confined to Madagascar, there is no discontinuity in the variation of the continental specimens as regards either leaf size and shape or pubescence and no question of more than one infraspecific taxon being represented there, whatever may be the case on Madagascar and the Mascarene islands.

2. VACCINIUM L.

Vaccinium L., Sp. Pl. **1**: 349 (1753); Gen. Pl., ed. 5: 166 (1754).

Small trees, shrubs or sub-shrubs. Leaves alternate, subsessile or shortly petiolate, evergreen or deciduous. Flowers borne on axillary or terminal, usually bracteate, racemes, or solitary in the axils of the leaves, 4-merous or 5-merous. Bracts, when present, usually small and scale-like but sometimes leaf-like. Calyx and corolla actinomorphic and epigynous. Calyx lobes usually small, persistent in fruit. Corolla gamopetalous, rotate to urceolate-globose, from very shallowly divided to so deeply divided that the lobes are almost free. Stamens 8–10, the anthers with each theca prolonged at the apex into a tubular horn opening by a terminal pore, anther appendages present or absent. Ovary inferior, 4–5-locular, ovules few (2–15 per loculus); style single. Fruit a succulent berry; corolla, stamens and style shed in fruit.

200–250 spp. mainly in the north temperate zone but extending on the mountains to the East Indies and in America south to Peru. Two species in Africa south of the Sahara with a further one in Madagascar which may also occur in Mozambique.

Corolla lobes about ¼ of the total length of the corolla, about 1 mm. long - - 1. *exul*
Corolla lobes about ½ of the total length of the corolla, 2·5–4 mm. long - 2. *madagascariense*

1. **Vaccinium exul** Bolus in Hook., Ic. Pl. **20**: t. 1941 (1890). — N. E. Brown in Thiselton-Dyer, Fl. Cap. **4** (1): 1 (1909). — Hutchinson, Botanist in Southern Afr.: 351 (1946). TAB. **27**. Type from South Africa (Transvaal).
    Vaccinium africanum Britten in Trans. Linn. Soc. Lond., Ser. 2, Bot. **4**: 23 (1894). Type: Malawi, Mt. Mulanje, *Whyte* s.n. (BM, holotype; K).
    Vaccinium exul var. *africanum* (Britten) Brenan in Mem. N.Y. Bot. Gard. **8**: 496 (1954). Type as for *Vaccinium africanum*.

Shrubs to 5 m. tall but usually less, branched from the base. Branchlets glabrous, minutely puberulous, or occasionally sparsely pubescent. Leaves alternate, narrowly ovate, acuminate, glandular serrate, 25–65 mm. long, 10–25 mm. broad, glabrous or

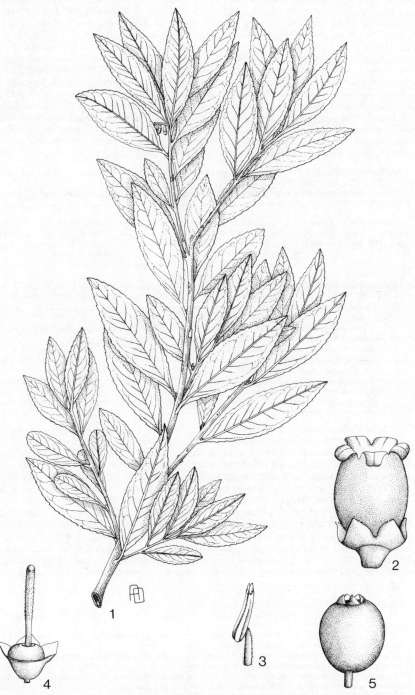

Tab. 27. VACCINIUM EXUL. 1, flowering branch (× ⅔); 2, flower (× 4); 3, stamen (× 4); 4, ovary with calyx (× 4); all from *Brass* 16816; 5, fruit (× 2), from *Brass* 16526.

sometimes with hairs on the basal part of the midrib on the ventral surface of the leaf; petioles glabrous or sometimes sparsely pubescent, 2–5 mm. long. Inflorescence consisting of axillary racemes, peduncles glabrous, 15–20-flowered, up to 80 mm. long. Flowers subtended by caducous leaf-like bracts, to 15 mm. long but usually considerably smaller. Pedicels glabrous, 3–7 mm. long in flower, extending to 8–13 mm. in fruit, and bearing two opposite bracteoles near the base; bracteoles leaf-like to narrow and membranous, 2–3 mm. long. Sepals 5, free, inserted at the top of the ovary, triangular, acute, glabrous, c. 1·2 mm. long, c. 1·5 mm. broad. Corolla 5-lobed, white or pink, obconical to narrowly urceolate, 5–6·5 mm. long, 3–4·5 mm. broad, lobes imbricate, obtuse, c. 1 mm. long, sinuses acute. Stamens 8–10, included; filaments pubescent above, 2 mm. long; anthers versatile, puberulous, 3·0–3·5 mm. long, opening by pores at the apex of two long horns, without appendages. Ovary glabrous. Style included, 4–5·5 mm. long, not swollen above, shed in fruit. Fruit a red berry, 5–8 mm. in diameter.

Malawi. S: Mulanje Distr., Mt. Mulanje, Litchenya Plateau, 2125 m., fr. 8.v.1963, *Wild* 6158 (BM; K; LISC; SRGH).
Also in South Africa (Transvaal). Forest edges and montane scrub on rocky ground, 1825–2450 m.

The presence of hairs on the young branchlets, the petioles and the basal part of the midrib in some specimens from Malawi indicates that Hutchinson's (loc. cit.) opinion that the specimens from Mt. Mulanje were not separable from those from the Transvaal is to be preferred to Brenan's (loc. cit.) recognition of separate varieties for the populations in the two areas.

2. **Vaccinium madagascariense** (Thouars ex Poir.) Sleumer in Engl., Bot. Jahrb. **71**: 424 (1941). Type from Madagascar.
　　Cavinium madagascariense Thouars ex Poir., Dict. Sci. Nat. **7**: 310 (1817). Type as above.
　　Vaccinium emirnense Hook., Ic. Pl. **2**: t. 131 (1837). — Perrier de la Bâthie in Cat. Pl. Madag., Eric. & Vaccin.: 13 (1934). Type from Madagascar.
　　Vaccinium secundiflorum Hook., tom. cit.: t. 134 (1837). Type from Madagascar.
　　Agapetes madagascariense (Thouars ex Poir.) Dunal in DC., Prodr. **7**: 555 (1839). Type as for *Vaccinium madagascariense*.
　　Vaccinium fasciculatum Bojer ex Dunal, tom. cit.: 570 (1839). Type from Madagascar.
　　Vaccinium laevigatum Bojer ex Dunal, tom. cit.: 571 (1839). Type from Madagascar.
　　Vaccinium forbesii Hook., Ic. Pl. **4**: t. 345 (1841). Type: Mozambique, Moçambique, *Forbes* s.n. (K, holotype).

A small shrub with glabrous branchlets. Leaves alternate, broad ovate, acute, cuneate at the base, minutely glandular-dentate, glabrous, 30–45 mm. long, 20–27 mm. broad; petioles 3·0–3·5 mm. long, glabrous. Inflorescence consisting of axillary racemes; peduncles glabrous, 3–5-flowered. Flowers subtended by small scale-like bracts c. 2 mm. long and 2 mm. broad, with acute apices and ciliate margins. Pedicels 3–3·5 mm. long in flower, glabrous, bearing two opposite bracteoles resembling the bracts at ⅔–¾ of their length from the base. Sepals 5, free, inserted at the top of the ovary, broadly triangular, glabrous, c. 0·5 mm. long, 1·5 mm. broad. Corolla 5-lobed, yellow to greenish white, obconical, c. 8 mm. long, 4–6 mm. broad, lobes imbricate, obtuse, 2·5–4 mm. long, sinuses acute. Stamens 8–10, about equalling or slightly exceeding the corolla; filaments densely pubescent, c. 6·5 mm. long; anthers versatile, c. 3 mm. long, opening by pores at the tips of two horns, glabrous, without appendages. Ovary glabrous, somewhat conical in flower. Style exserted, 8·5–10 mm. long, not swollen above, shed in fruit. Fruit a red berry.

Mozambique. N: Moçambique, *Forbes* s.n. (K).
Also in Madagascar. No information is available on its habitat in Mozambique; in Madagascar in sclerophyllous and ericoid scrub between 800 m. and 2500 m.

According to Perrier de la Bâthie (loc. cit.) there is only one variable species of *Vaccinium* in Madagascar. One specimen only of this species from the Flora Zambesiaca region is known. This is the type of *Vaccinium forbesii* at Kew, which is from Herb. Hooker and is labelled simply "Mozambique. Forbes" written on the sheet. Forbes collected in the neighbourhood of Moçambique between 5.ii.1823 and 7.ii.1823, and also in Madagascar between 29.xii.1822 and 3.i.1823. He died in Mozambique on 16.viii.1823 before the expedition of which he was a member returned to England. One cannot therefore rule out the possibility that the locality on this sheet is the result of a confusion and that the specimen actually came from Madagascar, especially as Forbes did not number his specimens.

3. ERICA L.

Erica L., Sp. Pl. **1**: 352 (1753); Gen. Pl., ed. 5: 167 (1754).

Trees, shrubs, or sub-shrubs, usually much branched with flexuously ascending stems. Branchlets slender. Leaves normally in whorls of 3–6 (3–4 in the F.Z. area), or occasionally spirally inserted, very small, the margins usually revolute and meeting in the centre line on the under side of the leaf to form a sulcus. Inflorescence normally consisting of capitate clusters of up to 20 flowers on the tips of the branchlets, but sometimes spicate with the flowers in the axils of leaves on branches whose vegetative growth may continue after flowering. Pedicels short and 1-flowered, normally bearing 1 bract and 2 opposite bracteoles, but sometimes either the bract or the bracteoles, very rarely both, missing. Calyx and corolla actinomorphic and hypogynous. Calyx segments 4, free or fused to a greater or lesser extent. Corolla globular to tubular, 4-lobed, the lobes usually appreciably shorter than the tube, in the F.Z. area rather to very small, always less than 1 cm. Stamens normally 8, rarely fewer (4–7) or more (12); filaments free, inserted on a hypogynous disc, often geniculate below the anthers; anthers often with appendages, dehiscing by lateral pores. Ovary 4 (8) loculate, placentation axile. Style 1, more or less expanded at the apex. Fruit a cartilaginous loculicidal capsule contained within the persistent calyx and corolla. Seeds several per loculus.

A genus of c. 600 species, mostly in the south-western part of Cape province, South Africa, but extending on the eastern side of Africa to Ethiopia, mostly on high ground, and around the Mediterranean region to Macaronesia and northwards in western Europe to Scandinavia.

In addition to the species described below, *Erica arborea* L., with hairs on the branchlets with side branches to the tip, leaves in whorls of 3, glabrous leaves, bracts and bracteoles, and included anthers is planted in Zimbabwe.

1. Leaves always in whorls of 3 - - - - - - - - - - - 2
 – Leaves in whorls of 4, sometimes (in no. 3) in whorls of 3 on some but not all branchlets 6
2. Anthers included in the corolla tube - - - - - - - - - 3
 – Anthers equalling or exceeding the tips of the corolla lobes - - - - 4
3. Anthers with appendages, corolla 2–2·5 mm., ovoid urceolate, leaf surface glabrous
 6. *austronyassana*
 – Anthers without appendages, corolla 1·5–2·0 mm., campanulate subglobose, leaf surface pubescent - - - - - - - - - - - - - 9. *natalitia*
4. Leaf surface without a pubescence of short simple hairs - - - - 1. *pleiotricha*
 – Leaf surface pubescent with short simple hairs - - - - - - 5
5. Corolla 3 mm. long or more, longer than broad and more or less constricted at the mouth
 2. *wildii*
 – Corolla 2 mm. long or less, no longer than broad, not constricted at the mouth - 7. *woodii*
6. Anthers included in the corolla tube - - - - - - - - 7
 – Anthers equalling or exceeding the tips of the corolla lobes - - - - 8
7. Branchlets and leaves glabrous - - - - - - - - 5. *whyteana*
 – Branchlets and leaves pilose - - - - - - - - 8. *johnstoniana*
8. Anthers without appendages - - - - - - - - 4. *milanjiana*
 – Anthers with appendages - - - - - - - - - - 9
9. Bracteoles 0·5 mm. or less, membranous; corolla not exceeding 3 mm. - 3. *lanceolifera*
 – Bracteoles 2–3 mm., not membranous; corolla 4 mm. or more - - - 8. *johnstoniana*

1. **Erica pleiotricha** S. Moore in J. Linn. Soc. Lond., Bot. **40**: 127 (1911). — Alm & Fries in Ark. Bot. **21A** (7): 5 (1927). — Brenan in Kirkia **4**: 146 (1964). — R. Ross in Bol. Soc. Brot., Sér. 2, **53**: 139 (1981). Type: Zimbabwe, Chimanimani Mts., 2100 m., 1906, *Swynnerton* 648a (BM, holotype; K).

A much branched sub-shrub 0·3–1·0 m. tall. Branchlets slender, the finest 0·2–0·3 mm. in diameter, without infrafoliar ridges, densely pubescent with very short simple hairs, and many longer hairs, some of them c. 0·15 mm. long and with side branches to the tip, others which may be either eglandular, 0·4–0·5 mm. long, with side branches on the lower part and often forked above, or glandular, 0·15–0·4 mm. long, with side branches on the lower part of the longer ones. Leaves in whorls of 3, patent to ascending, narrowly ovate to ovate, acute, 1·0–3·5 mm. long, 0·3–1·5 mm. wide, the margins reflexed from very slightly to so strongly that they meet along the mid line at the back of the leaf, glabrous, or with a single hair at the tip c. 0·2 mm. long with simple, glandular or occasionally forked tips, or with a row of hairs 0·1–0·3 mm. long along the apparent margin of the leaf or, where the margins of the leaf are scarcely reflexed, along a curved ridge about equidistant between mid-rib and leaf margin, these hairs with simple, glandular or forked tips; petioles 0·2–0·5 mm. long, glabrous. Flowers in clusters of 6–12 at the tips of the numerous side branches which become shorter and more closely spaced towards the apex of the stem. Pedicels 1·5–4 mm. long, pubescent with short simple hairs and longer hairs that may be glandular and unbranched or eglandular with side branches below and simple or forked tips. Bract varying in position from basal to near the calyx, leaf-like, 1·5–2·5 mm. long; bracteoles also varying in position and distance above the bract from near the base of the pedicel to just below the calyx, from 1·5 mm. long and leaf-like above to 0·5 mm. long, membranous throughout and very narrow; bracts and bracteoles with ciliate margins, the cilia glandular or eglandular with simple or forked tips. Calyx segments united at the base only, triangular to linear with acute to obtuse tips, 0·8–1·5 mm. long, with ciliate margins, the cilia 0·1–0·2 mm. long, glandular, forked or simple at the tip, sometimes with side branches at the base. Corolla urceolate to tubular-campanulate, pale purple to pink to white, glabrous, 1·5–4·5 mm. long, 1·0–3·0 mm. broad, lobes rounded, sinuses acute, c. 0·5 mm. deep. Stamens 4–8, equalling the corolla to exceeding it by c. 2 mm.; filaments flattened, slightly geniculate; anthers normally with aristate appendages 0·2–0·3 mm. long, these sometimes fused to a greater or lesser extent to the filaments, but rarely anthers muticous. Ovary globose, pubescent, style 2·5–4·5 mm., exceeding the stamens by c. 1 mm., usually glabrous, occasionally pubescent below, stigma capitate.

Branchlets with glandular hairs- - - - - - - - - var. *pleiotricha*
Branchlets with all hairs eglandular - - - - - - - - var. *blaeriodes*

Var. **pleiotricha.**—R. Ross in Bol. Soc. Brot., Sér. 2, **53**: 139 (1981).

Branchlets with many glandular hairs and few hairs with side branches to the tip. Stamens 8–7, never exceeding the corolla by more than the length of the anthers. Ovary densely pubescent with some hairs forked.

Zimbabwe. E: Melsetter Distr., Chimanimani Mts. near summit of Point 71, 2300 m., fl. 25.ix.1966, *Simon* 827 (K; LISC; PRE; SRGH). **Mozambique.** MS: Manica, Chimanimani Mts. Messurussero, fl. 11.vii.1949, *Pedro & Pedrogão* 7448 (LMA; PRE).
Known only from Zimbabwe and Mozambique. In damp places among rocks near the summits of mountains, 1800–2400 m.

Var. **blaeriodes** (H. Wild) R. Ross in Bol. Soc. Brot., Sér. 2, **53**: 140 (1981). Type: Mozambique, Manica, Chimanimani Mts., 2000 m., fl. 6.vi.1949, *Munch* 206 (K; LISC; SRGH, holotype).
     Erica thryptomenoides S. Moore in J. Linn. Soc. Lond., Bot. **40**: 126 (1911). — Alm & Fries in Ark. Bot. **21A**: (7): 8 (1927). Type: Zimbabwe, Chimanimani Mts., 2100 m., fl. 26.ix.1906, *Swynnerton* 647 (BM, holotype; K).
     Erica eylesii L. Bolus in Ann. Bolus Herb. **3**: 174, pl. 9D (1924). — Type: Zimbabwe, Melsetter, 1525 m., fl. x.1920, *Eyles* 2755 (K, holotype).
     Erica eylesii Alm & Fries in Ark. Bot. **21A** (7): 6 (1927), non *E. eylesii* L. Bolus. Type: as for *E. eylesii* L. Bolus.
     Erica eylesii var. *blaeriodes* H. Wild in Bothalia **6**: 429, fig. 5 (1954). Type: as for *E. pleiotricha* var. *blaeriodes*.

Branchlets with no glandular hairs and many hairs with side branches to their apex. Stamens 4–8, sometimes exceeding the corolla by more than the length of the anther. Ovary sparsely to moderately pubescent with short simple hairs only.

Zimbabwe. E: Melsetter Distr., south-west ridge of Musapa Mt., 2000 m., fl. 9.x.1950, *Chase* 2951 (BM; COI; K; LISC; SRGH). **Mozambique.** MS: Chimanimani Mts. summit of Nhamadima, fl. 18.iv.1960, *Goodier & Phipps* 365 (K; LISC; SRGH).
Known only from Zimbabwe and Mozambique. Amongst rocks, usually in rather damp places, 900–2300m.

2. **Erica wildii** Brenan in Kirkia **4**: 146 (1964). TAB. **28**. Type: Zimbabwe, Chimanimani Mts., 1600 m., fl. 16.x.1950, *Wild* 3643 (K, holotype; LISC; SRGH).

A branched prostrate to trailing sub-shrub not reaching more than c. 25 cm. tall. Branchlets slender, the finest 0·15–0·25 mm. diameter, infrafoliar ridges present but poorly developed, pubescent with short simple hairs less than 0·1 mm. long and sometimes also much sparser longer hairs, usually with side branches near the base, the tips glandular, or forked, or simple. Leaves in whorls of 3, ascending to appressed, narrowly ovate, acute, 0·8–3·5 mm. long, 0·5–1·0 mm. broad, the reflexed margins nearly or quite meeting along the mid line at the back of the leaf, upper surface of the leaf, including the reflexed part, densely pubescent with short simple

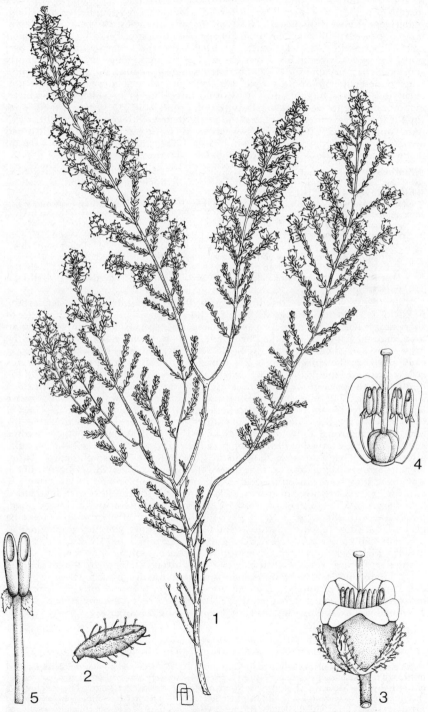

Tab. 28. ERICA WILDII. 1, flowering branch (× ⅔); 2, underside of leaf (× 8); 3, flower (× 6); 4, vertical section of flower (× 6); 5, stamen (× 8), all from *Grosvenor* 192.

hairs c. 0·05 mm. long, and a few longer hairs, 0·2–0·3 mm. long, on the apparent margin of the leaf with glandular, or forked, or simple tips; petioles 0·3 mm. long, puberulent. Flowers in clusters of 3–6 on the side branches. Pedicels 1·3–3·0 mm. long, pubescent with short simple hairs and sometimes scattered longer glandular hairs. Bract and bracteoles inserted high on the pedicel, narrow, linear, obtuse, purplish, pubescence as on leaves, bract 1·2–1·5 mm. long, bracteoles slightly shorter. Sepals fused at the base only, narrow, linear, obtuse, purplish, c. 1·5 mm. long, pubescent with short simple hairs, margins ciliate, cilia glandular or not, eglandular ones sometimes forked. Corolla urceolate to urceolate-campanulate, glabrous, mauve to deep pink to pale pink, 3–5·5 mm. long, 2·0–3·5 mm. broad, lobes rounded, sinuses acute, about 0·5 mm. deep. Stamens 8; filaments straight, 2·5–5 mm. long; anthers 1·0 mm., semi-exserted, very shortly pubescent, with broad aristate pubescent appendages c. 0·4 mm. long. Ovary globose, glabrous, c. 1·0 mm. diameter, style glabrous, exceeding corolla by c. 1·5 mm., stigma capitate.

Zimbabwe. E: Melsetter Distr., Chimanimani Mts, 1500 m., fl. 24.ix.1966, *Grosvenor* 182 (K; LISC; PRE; SRGH). **Mozambique.** MS: Upper Bundi River, 1675 m., fl. xii.1964, *Whellan* 2212 (BM; K; PRE; SRGH).

Known only from Zimbabwe and Mozambique. In upland grassland and savannah and amongst rocks, 1050–2400 m.

3. **Erica lanceolifera** S. Moore in J. Linn. Soc. Lond., Bot. **40**: 126 (1911).— Alm & Fries in Ark. Bot. **21A** (7): 8 (1927). — Brenan in Kirkia **4**: 146 1964. Type: Zimbabwe, Chimanimani Mts., 2100 m., fl. 26.ix.1906, *Swynnerton* 1288 (BM, holotype; K).

Erica gazensis Wild in Bothalia **6**: 428, fig. 4 (1954). Type: Zimbabwe, Melsetter District, Chimanimani National Park, 1850 m., fl. 8.vii.1950, *Thompson* 16 (BM; SRGH, holotype).

A sub-shrub, often with trailing branches, 20–60 cm. tall. Branchlets slender, the finest 0·2–0·3 mm. in diameter, without infrafoliar ridges, with a dense pubescence of short simple hairs c. 0·1 mm. long and a sparser pubescence of long hairs to 1·5 mm. long, the tips glandular or forked or simple, and often with side branches near the base. Leaves in whorls of 4, ascending, occasionally 3 on some branchlets, narrowly ovate to ovate, the apices obtuse, 2–5 mm. long, 0·6–1·3 mm. broad, the margins reflexed to a varying degree from only slightly to so strongly that they meet on the mid line at the back of the leaf, whole upper surface usually pubescent with short simple hairs c. 0·05 mm. long but this pubescence sometimes absent, and sparse longer hairs, to 0·5 mm., on the apparent margin of the leaf and on the abaxial part of the dorsal surface, these longer hairs glandular, or forked, or simple; petioles 0·4–0·7 mm. long, glabrous. Flowers in heads of 4–12 on side branches which become shorter and more crowded near the apex of the main stem. Pedicels 2–4 mm. long, pubescent with hairs similar to those on the stem. Bract narrow, spathulate, 1–2 mm. long, insertion varying from near the base of the pedicel to about three quarters of the way up it, pubescence as on the leaves. Bracteoles membranous, very narrow, 0·5 mm. long to much reduced or absent, ciliate with forked or simple or glandular hairs. Calyx segments free almost to the base, triangular ovate, acute, c. 1 mm. long, pubescent with short simple hairs over the surface and ciliate margins, the cilia glandular or some to most eglandular and forked or simple. Corolla campanulate to urceolate-campanulate, purple to deep pink, 2·0–2·5 mm. long, 1·7–2·2 mm. broad, usually glabrous, occasionally minutely pubescent along the margin, lobes triangular, obtuse, sinuses obtuse, c. 0·8 mm. deep. Stamens 8 (?–6), filaments 1·0–2·0 mm., anthers 0·8–1·0 mm., their tips equalling the corolla to exceeding it by c. 0·5 mm., with minutely pubescent appendages c. 0·4 mm. long, aristate, and broadened and irregularly toothed above. Ovary globose, pubescent; style glabrous, 2–3 mm. long, exceeding the corolla by 1–2 mm., stigma capitate.

Zimbabwe. E: Melsetter Distr., Nyamhunza River, Martin Forest Reserve, 1300 m., fl. 7.x.1950, *Chase* 2976 (BM; COI; K; LISC; SRGH).

Possibly also in Tanzania.* In montane grassland and woodland, 1050–2400 m.

* A consignment of plants collected in Tanzania by *C. F. M. Swynnerton* was received at BM in 1923. These specimens were not numbered by him but were assigned serial numbers from 1 onwards by the Museum. The data accompanying them was very scant. They included a specimen of *Erica lanceolifera* [1040] from 10,000 ft. (3000 m.) on Mt. Kilimanjaro and one of *Erica johnstonii* [2015] with no details. As these two species have never been collected by anyone else in Tanzania and both occur not infrequently in Swynnerton's main collecting area in Rhodesia, E, some doubt attaches to these records.

Swynnerton 1288, the type of *E. lanceolifera* S. Moore, has occasional glandular hairs on the bracts and calyx but none elsewhere, whilst on *Thompson* 16, the type of *E. gazensis* Wild, all the longer hairs on leaves, pedicels, bracts, bracteoles and calyx are glandular, as are all on the branchlets except for a very few with forked tips. However *Wild* 6272 has about equal numbers of glandular and forked eglandular hairs on branchlets, leaves, bracts, bracteoles and calyx, long hairs being absent from its pedicels; other specimens show a wide range in the proportions of these types of hair. In all other respects these specimens are very similar and they clearly represent a single species and not two that can be distinguished by the presence or absence of glandular hairs.

4. **Erica milanjiana** Bolus in Trans. S. Afr. Phil. Soc. **16**: 141 (1905). — Alm & Fries in Ark. Bot. **21A** (7): 7 (1927). — Brenan in Mem. N.Y. Bot. Gard. **8**: 493 (1954). Type: Malawi, Mulanje District, Mulanje Plateau, fl. iii.1896 *McClounie* s.n. (K, holotype).

A straggling, much branched sub-shrub, the stems up to 80 cm. long. Branchlets slender, the finest 0·2–0·3 mm. diameter, without infrafoliar ridges, pubescent with very short simple hairs c. 0·05 mm. long and longer simple glandular hairs. Leaves in whorls of 4, patent, narrowly ovate, acute, 1·5–6 mm. long, 0·4–2·0 mm. wide, the margins normally slightly revolute but sometimes more strongly reflexed, dorsal surface with a pubescence, often rather sparse, of very short simple hairs, distant glandular cilia along the apparent margin of the leaf and at the apex; petioles c. 0·5 mm. long, pubescent. Flowers in groups of 4–8 at the tips of the branchlets, rather sparse. Pedicels 4–6 mm. long, glandular pubescent. Bract and bracteoles leaf-like, bract 1·2–1·5 mm. long, inserted at ⅓–½ of the pedicel from its base, bracteoles 0·8–1·0 mm. long, inserted 1·0–1·5 mm. below the calyx. Calyx segments fused at the base only, narrowly triangular, acute, margins reflexed in the upper part, pubescent with short simple hairs and occasional glandular hairs, c. 1·5 mm. long. Corolla urceolate-campanulate, pink to white, glabrous, 2·5–3·0 mm. long, 2·0–2·5 mm. broad, lobes obtuse to rounded, sinuses broadly acute, c. 0·5 mm. deep. Stamens 7–8, exserted, filaments c. 3·0 mm. long, anthers c. 1·0 mm. long, without appendages. Ovary globose, pubescent, c. 0·8 mm. diam.; style glabrous, 3–5 mm. long, stigma capitate, small.

Malawi. S: Mulanje Distr., Mt. Mulanje, base of west peak, 2200 m., fl. 8.vi.1962, *Robinson* 5312 (K; LISC; SRGH).
Known only from Mt. Mulanje, Malawi. In shady places under rocks, 1950–2400 m.

5. **Erica whyteana** Britten in Trans. Linn. Soc. Lond., Ser. 2, Bot. **4**: 24, pl. 5 figs. 7–12 (1894). — S. Moore in Smithson. Misc. Coll. **68** (5): 10 (1917). — Alm & Fries in Ark. Bot. **21A** (7): 9 (1927). — Weimarck in Bot. Notis. **1940**: 53 (1940). — Brenan in Mem. N.Y. Bot. Gard. **8**: 493 (1954). — R. Ross in Bol. Soc. Brot., Sér. 2. **53**; 140 (1981). FRONTISP. Type: Malawi, Mulanje Distr., Mt. Mulanje, *Whyte* 59 (BM, holotype).
    Erica princeana Engl. in Bot. Jahrb. **43**: 363 (1909). — T.C.E. Fries in Notizbl. Bot. Gart. Mus. Berl. **8**: 689 (1924). — Alm & Fries in Ark. Bot. **21A** (7): 11 (1927). — Brenan, T.T.C.L.: 193 (1949). Type from Tanzania.
    Erica swynnertonii S. Moore in J. Linn. Soc. Lond., Bot. **40**: 128 (1911); in Smithson. Misc. Coll. **68** (5): 10 (1917). — Alm & Fries in Ark. Bot. **21A** (7): 12 (1927). — Plowes & Drummond, Wild Flowers of Rhodesia: pl. 120 (1976). Syntypes: Zimbabwe, near Melsetter, 1800 m., fl. 23.ix.1906, *Swynnerton* 648 (BM); Hills between Lusito & Nyaludi Rs., 1500 m., fl. iv.1907, *Swynnerton* 1063 (BM; K; SRGH); Chimanimani Mts., 1800 m., fl. 22.ix.1906, *Swynnerton* 1064 (BM; K; SRGH); Chimanimani Mts., 2100 m., fl. 26.ix.1906, *Swynnerton* 1065 (BM).
    Erica keniensis S. Moore in Smithson. Misc. Coll. **68** (5): 10 (1917). Type from Kenya.
    Erica whyteana subsp. *princeana* (Engl.) Hedberg in Symb. Bot. Upsal. **15** (1): 142, 297 (1957). Type as for *E. princeana*.

Erect or trailing, sparingly branched sub-shrub 10–90 cm. tall. Branchlets rather stout for the genus, 1·2–1·5 mm. diameter, occasionally more slender to 0·6 mm. diameter, glabrous, with strongly developed infra-foliar ridges. Leaves spirally arranged in 8 ranks or sub-verticillate in whorls of 4, occasionally 3 in the slenderest specimens, patent to appressed, very narrowly ovate, apices acute and shortly aristate, normally sulcate below but occasionally margins revolute and not meeting in the mid line at the back of the leaf, (3·5) 6·0–12·0 mm. long, (0·6) 0·8–1·5 (2·0) mm. broad, glabrous, usually with glandular teeth on the apparent margin more distant or absent towards the base; petioles glabrous, c. 1·5 mm. long. Flowers in the axils of the leaves on the upper 2–10 cm. of the stems. Pedicels 2·5–15 mm. long, glabrous. Bract leaf-like, 2·5–5 mm. long, inserted on the basal part of the pedicel, but sometimes absent. Bracteoles narrow, linear, acute, 0·7–2·5 mm. long, with very shortly stalked glands along the margin. Calyx segments free almost to the base, deep magenta to pink, broadly ovate, acute, 1·5–3·0 mm. long, usually (in specimens from the F.Z. area) reflexed, with minute glandular teeth along the margin. Corolla urceolate, deep magenta to pink to white, glabrous, 3·5–5·5 mm. long, 2·5–3·7 mm. broad, lobes obtusely pointed to rounded, sinuses acute, c. 0·8 mm. deep. Stamens 8, included, filaments c. 2 mm. long, anthers c. 0·8 mm. long, with aristate appendages c. 0·6 mm. long. Ovary globose, glabrous, c. 1 mm. in diameter; style stout, glabrous, 2·0–3·0 mm. long, about equalling the corolla, scarcely widened at the stigma.

Zimbabwe. E: Inyanga District, Mare River below Rhodes Hotel grounds, 1700 m., fl. 18.iv.1953, *Chase* 4901 (BM; COI; LISC; PRE; SRGH). **Malawi.** S: Mulanje Distr., Mt. Mulanje, Litchenya Plateau, 2140 m., fl. 27.vi.1946, *Brass* 16479 (BM; FHO; K; PRE; SRGH). **Mozambique.** MS: Chimanimani Mts., 1700 m., fl. 6.vii.1949, *Pedro & Pedrogão* 7183 (LMA).

Also in Tanzania and Kenya. In damp places, mostly beside rivers and streams in montane savannah and grassland, 1100–2500 m.

The specimens from Tanzania and Kenya, in which the calyx segments are normally not reflexed, may represent a separate subspecies, *Erica whyteana* subsp. *princeana* (Engl.) Hedb.

6. **Erica austronyassana** Alm & Fries in Ark. Bot. **21A** (7): 18 (1927). Type: Malawi, Mulanje Distr., Mt. Mulanje, *Adamson* 329 (K, holotype).

A much branched sub-shrub, the branches to 30 cm. long but forming a cushion no more than 15 cm. tall. Branchlets slender, the finest 0·2–0·3 mm. diameter, without infrafoliar ridges, minutely pubescent with simple hairs c. 0·02 mm. long. Leaves in whorls of 3, ascending to appressed, narrowly elliptic, obtuse, sulcate below, 2·0–3·0 mm. long, 0·3–0·5 mm. wide, the proximal part of the apparent margin with evanescent minute ciliolae, sessile or very shortly stalked glands at the apex and widely spaced along the apparent margin; petioles c. 0·5 mm. long, minutely ciliolate. Flowers in groups of 3, very occasionally 6–9, at the tips of the branchlets. Pedicels glabrous, c. 1·5 mm. long. Bract and bracteoles inserted on the upper half of the pedicel, bracteoles only very slightly above bract, narrowly ovate, acute, ciliolate, bract c. 0·4 mm. long, bracteoles c. 0·6 mm. long. Calyx segments united only at the base, narrowly ovate, acute, ciliolate, leaf-like at the tip, c. 1·3 mm. long. Corolla narrowly urceolate, pink to mauve, glabrous 2·0–2·5 mm. long, 1·5–2·0 mm. broad, lobes rounded, sinuses acute, c. 0·4 mm. deep. Stamens 8, included, filaments geniculate, c. 0·7 mm. long, anthers broad, c. 0·4 mm. long, with aristate appendages with fimbriate margins c. 0·3 mm. long. Ovary depressed globose, ridged, c. 0·5 mm. in diameter, minutely pubescent; style included, glabrous, c. 1·0 mm. long, stigma capitate.

Malawi. S: Mulanje Distr., Mt. Mulanje, Main Peak–Sapitwa, 3000 m., fl. 11.v.1963, *Wild* 6194 (BM; K; LISC; SRGH).

Known only from Mt. Mulanje, Malawi. Montane grassland, 2000–3000 m.

7. **Erica woodii** Bolus in J. Bot., Lond. **32**: 237 (1894). — Guthrie & Bolus in Thiselton-Dyer, Fl. Capensis **4** (1): 214 (1905). — J. M. Wood in Trans. S. Afr. Phil. Soc. **18**: 188 (1908). — Baker & Oliver, Ericas in Southern Africa: 120, pl. 115 (1967). — Trauseld, Wild Flowers of the Natal Drakensberg: 138 (1969). — Guillarmod, Fl. Lesotho: 221 (1971). — J. H. Ross, Fl. Natal: 269 (1972). — Gibson, Wild Flowers of Natal (Coastal Region): pl. 76 fig. 4 (1975). Type from South Africa (Natal).
    Erica rhodesiaca Alm & Fries in Ark. Bot. **21A** (7): 19 (1927). — Weimarck in Bot. Notis. **1940**: 54 (1940). Type: Zimbabwe, Inyanga Distr., 1912–13, *Mundy* s.n. (K, holotype).
    Erica woodii var. *rhodesiaca* (Alm & Fries) Dulfer in Ann. Naturhist. Mus. Wien **67**: 85 (1963). Type as for *E. rhodesiaca*.

Erect branched sub-shrub 20–60 cm. tall. Branchlets ascending, narrow, the finest 0·2–0·3 mm. in diameter, without infrafoliar ridges, pubescent with very short simple hairs c. 0·05 mm. long and sparser hairs up to 0·7 mm. long with many side branches on the lower part and the tips forked or glandular. Leaves in whorls of 3, patent to ascending, very narrowly ovate, obtuse, sulcate below, 1·0–4·0 mm. long, 0·4–1·2 mm. wide, with short simple hairs over the whole surface and distant cilia on the apparent margin with side branches on their basal parts and forked or glandular tips; petioles c. 0·3 mm. long, pubescent. Flowers in clusters of 6–9 at the tips of the side branchlets and in the axils of the upper whorls of leaves on the main branches, forming a narrow paniculate inflorescence up to 20 cm. long. Pedicels pubescent with short simple hairs, 0·5–1·0 mm. long. Flowers at the tips of the branchlets with a basal narrow spathulate bract 1·0–1·5 mm. long, flowers in axils of leaves on main branches without a bract. Bracteoles opposite, inserted on the upper half of the pedicel, membranous, up to 0·5 mm. long but sometimes much smaller or absent; pubescence of bracts and bracteoles the same as that of leaves but the marginal cilia closer. Calyx segments united at the base only, narrowly ovate, obtuse, c. 0·8 mm. long, pubescence the same as that of the leaves. Corolla broadly campanulate to crateriform, mauve, pink or white, glabrous except for the minutely ciliate margins, 1·0–1·8 mm. long, 1·0–2·5 mm. broad, lobes cuneate, obtuse, sinuses broadly acute, c. 0·5 mm. deep. Stamens 6–8, not quite equalling the corolla lobes but exceeding the sinuses, filaments c. 1·0 mm. long, anthers broad, c. 0·7 mm. long, with aristate appendages c. 0·3 mm. long. Ovary depressed globose, pubescent; style glabrous, exceeding the corolla, c. 1·3 mm. long, stigma capitate to peltate, 0·25–0·4 mm. in diameter.

Zimbabwe. N: Mt. Binga, 2300–2435 m., fl. 29.i.1966, *Pereira, Sarmento & Marques* 3 (LMU). E: Inyanga, 1950 m., fl. 25.ii.1951, *Chase* 3691 (BM; COI; LISC; PRE; SRGH).
Also in S. Africa from eastern Cape Province to Natal, and in Lesotho and Swaziland. In scrub and grassland, often by streams, 1700–2435 m.

Most specimens of this species are apparently eglandular, although detailed examination often reveals that a few leaves bear one or two glandular hairs. Other specimens have a considerable number of glandular hairs on stem, leaves, bracts, bracteoles and calyx, but there are always some with forked tips.
Dulfer (loc. cit.) does not give his reasons for maintaining *E. woodii* var. *rhodesiaca* as a separate variety within the species and I can see no significant differences between its type and that of *E. woodii.*

8. **Erica johnstoniana** Britten in Trans. Linn. Soc. Lond., Ser. 2, Bot. **4**: 23, pl. 5 figs. 1–6 (1894). — Alm & Fries in Ark. Bot. **21A** (7): 20 (1927). — Weimarck in Bot. Notis. **1940**: 54 (1940). — Brenan in Mem. N.Y. Bot. Gard. **8**: 493 (1954). Type: Malawi, Mulanje Distr., Mulanje Plateau, 1800–2000 m., fl. xi.1891, *Whyte* 4 (BM, holotype; K).

A much branched shrub 0·2–2·0 m. tall. Branchlets stout, the finest 0·4–1·0 mm. diameter, without infrafoliar ridges, densely pubescent with very short simple hairs, somewhat sparser hairs with side branches to the tip and stout white hairs 0·5–1·5 mm. long with side branches near the base. Leaves in whorls of 4, patent to ascending, narrowly ovate to ovate, acute, margins reflexed to a greater or lesser extent but not meeting along the mid line on the under side of the leaf, with stout simple hairs, sometimes with a few side branches at the base, 0·5–1·5 mm. long, along the apparent margin of the leaf and on the reflexed part of the upper surface, usually glabrous otherwise but sometimes with short simple hairs among the larger ones, and more rarely over the whole upper surface, 3·5–8·5 mm. long, 1·0–4·0 mm. broad; petioles 0·5–1·0 mm. long, reddish, glabrous or occasionally with 1 or 2 stout hairs. Flowers in clusters of 8–16, occasionally only 4, at the tips of side branchlets. Pedicels 2–3 mm. long, pubescent with long stout white hairs with side branches at the base and sometimes also shorter hairs with side branches to the tip. Bract and bracteoles similar, subverticillate, on the upper part of the pedicel, linear, acute, 2–3 mm. long, with stout white hairs with side branches at the base along the margins and at the apex, usually otherwise glabrous but sometimes with short simple hairs also. Calyx segments free or slightly united at the base, narrowly triangular or linear with acute tips, 3·0–4·5 mm. long, hairs as on the bracts, usually equalling the corolla but occasionally not much over ½ its length. Corolla ovoid, pink, occasionally with purple tips or very occasionally white with purple tips, minutely pubescent, 4·0–7·0 mm. long, 3·0–4·0 mm. wide, lobes rounded with fimbriate or ciliate margins, sinuses acute, c. 0·5 mm. deep. Stamens 8, normally included in the corolla tube but occasionally the anthers semi-exserted; filaments flattened, geniculate, 3·5–5·5 mm. long; anthers c. 0·8 mm. long, with rather large membranous appendages c. 0·6 mm. long with fimbriate margins, appendages usually semi-elliptical but sometimes with a downward projecting arista. Ovary globose, ridged, pubescent with short hairs; style 2·5–6·0 mm. long, usually equalling or slightly exceeding the corolla, occasionally projecting by 2–3 mm.; stigma small, capitate.

Zimbabwe. N: Mt. Binga, 2050–2200 m., bud 29.i.1966, *Pereira, Sarmento & Marques* 19 (LMU). E: Melsetter Distr., Red Star Farm c. 30 km. south-east of Melsetter, fl. 27.v.1955, *Crook* 566 (K; LISC; LMU; PRE; SRGH). **Malawi.** S: Mulanje Distr., Mt. Mulanje, Litchenya Plateau, 1870 m., fl. 27.vi.1946, *Brass* 16455 (BM; FHO; K; PRE; SRGH). **Mozambique.** MS: Tsetsera, 1980 m., fl. 8.ii.1955, *Exell, Mendonça & Wild* 301 (BM; LISC; SRGH).

Possibly also in Tanzania.* In grassland and at forest borders, sometimes by streams or amongst rocks, 1500–2550 m.

9. **Erica natalitia** Bolus in J. Linn. Soc. Lond., Bot. **24**: 187 (1887). — Guthrie & Bolus in Thiselton-Dyer, Fl. Capensis **4** (1): 307 (1905). — J. M. Wood in Trans. S. Afr. Phil. Soc. **18**: 188 (1908). — Letty, Wild Flowers of the Transvaal: 237, pl. 118 fig. 1 (1962). — J. H. Ross, Fl. Natal: 269 (1972). Type from South Africa (Natal).

A much branched shrub 1–3 m. tall. Branchlets slender, the finest 0·25–0·3 mm. in diameter, without infrafoliar ridges, with a dense pubescence of short simple hairs c. 0·05 mm. long and sparse slightly longer glandular hairs. Leaves in whorls of 3, patent to ascending, very narrowly ovate, obtuse, sulcate below, i.e. margins reflexed and meeting in the mid line at the back of the leaf, 2·0–3·5 mm. long, 0·5–0·8 mm. broad, rather sparsely pubescent with short simple hairs; petioles 0·5–0·6 mm. long, pubescent. Flowers in clusters of 3–9 at the tips of the side branches that become shorter and more crowded towards the tips of the stems, forming paniculate inflorescences up to 20 cm. long and 8 cm. wide at the base. Pedicels pubescent with short simple hairs, 1·0–1·5 mm. long. Bract inserted at about ⅓ of the pedicel from the base, appressed to the pedicel, membranous, very narrowly ovate, pubescent, c. 0·8 mm. long. Bracteoles inserted at about ⅔ of the pedicel from the base, appressed to the pedicel, membranous, narrow linear, pubescent, c. 0·5 mm. long. Calyx minutely pubescent, c. 0·8 mm., segments fused for almost their whole length, tips leaf-like, obtuse. Corolla campanulate to sub-globose, purple, minutely pubescent, 1·5–2·0 mm. long, 1·3–1·7 mm. broad, lobes obtuse, sinuses acute, c. 0·6 mm. deep. Stamens 8, the tips of the anthers about equalling the corolla, filaments narrow, c. 0·8 mm. long, anthers broad, c. 0·5 mm. long, without appendages. Ovary globose, glabrous; style exserted, glabrous, c. 1 mm. long; stigma peltate, 0·8–1·0 mm. in diameter.

Mozambique. GI: Manhiça, Chiau, fl. 20.vi.1950, *Gomes e Sousa* 4001 (COI; K; LMA). Also in Natal. On the edges of marshes in forest and savannah.

4. BLAERIA L.

Blaeria L., Sp. Pl. **1**: 112 (1753); Gen. Pl., ed. 5: 51 (1754).

Shrubs or subshrubs to about 1 m. tall, much branched with slender and flexous stems often with many dwarfed side branches in the axils of their leaves. Leaves in whorls of 3 or 4 (always 3 in the F.Z. area), very small, the margins reflexed. Inflorescence consisting of clusters of flowers at the tips of the branchlets. Pedicels short and 1-flowered, normally bearing one bract and two opposite bracteoles, but the bracteoles sometimes absent. Calyx and corolla hypogynous, tetramerous. Calyx actinomorphic, segments free to the base. Corolla crateriform to campanulate to slightly urceolate, or tubular below and campanulate to crateriform above, actinomorphic or when a tubular portion is present this is often somewhat bent. Stamens 4; filaments free, inserted on a hypogynous disc; anthers often with appendages, dehiscing by lateral pores. Ovary 4-loculate, placentation axile. Style 1, somewhat expanded at the apex. Fruit a cartilaginous loculicidal capsule contained within the persistent calyx and corolla. Seeds several per loculus.

A genus of c. 20 species, perhaps fewer, from South Africa and on high ground throughout tropical Africa.

Leaves patent (except on dwarfed side branches), straight or recurved, occasionally glabrous, usually with very short hairs over the whole dorsal surface and longer hairs on the apparent margins and sometimes also on the abaxial surface, these hairs widely spaced and usually without side branches on the basal part, but hairs with side branches near the base and bifurcate tips occasionally present   -   -   -   -   -   -   -   1. *kingaensis*

Leaves ascending, incurved, with very short hairs over the whole dorsal surface and longer hairs closely spaced on the apparent margins and the abaxial surface, these hairs with side branches on the basal part and tips forked into 3 or 4   -   -   -   -   -   2. *filago*

* See footnote to *Erica lanceolifera*, p. 166.

1. **Blaeria kingaensis** Engl. in Bot. Jahrb. **30**: 370 (1901). — Alm & Fries in Acta Hort. Berg.
`8 (8); 247 (1924). — Brenan, T.T.C.L.: 191 (1949). — R. Ross in Bol. Soc. Brot., Sér. 2,
53: 123 (1981). TAB. **29**. Type from Tanzania.

  Blaeria spicata var. *patula* Engl. in Phys. Abhandl. K. Akad. Wiss. Berlin **1891** (2): 325
(1892). Type: Malawi, Shire Highlands, near Blantyre, *Last* s.n. (K, holotype).

  Blaeria subverticillata Engl. in Bot. Jahrb. **30**: 372 (1901). Type from Tanzania.

  Blaeria patula (Engl.) Engl. in Bot. Jahrb. **43**: 364 (1909). — Alm & Fries, tom. cit.: 260
(1924). — Brenan, T.T.C.L.: 192 (1949); in Mem. N.Y. Bot. Gdn. **8**: 495 (1954). Type as
for *B. spicata* var. *patula*.

  Blaeria tenuifolia Engl. in Bot. Jahrb. **43**: 365 (1909). Type: Malawi, Masuku Plateau,
2100–2300 m., fl. vii. 1896, *Whyte* 276 (K, holotype).

  ? *Blaeria kiwuensis* Engl. in Bot. Jahrb. **43**: 364 (1909). — Alm & Fries, tom. cit.: 260
(1924) certe pro parte, quoad syn. *B. tenuifolia* Engl. et specim. leg. Stolz, Whyte et
Turner. — Brenan, T.T.C.L.: 192 (1949); in Mem. N.Y. Bot. Gard. **8**: 495 (1954). Type
from Zaire.

  Blaeria stolzii Alm & Fries, tom. cit.: 246 (1924). — Brenan, T.T.C.L.: 191 (1949).
Type from Tanzania.

  Blaeria friesii Weimarck in Bot. Notiser. **1940**: 60 (1940). Type: Zimbabwe, "ad radices
montis Inyangani", c. 2000 m., fl. 14.ii.1931, *Norlindh & Weimarck* 5027 (BM; K; LD,
holotype).

  Blaeria patula var. *minima* Brenan in Mem. N.Y. Bot. Gard. **8**: 496 (1954). Type:
Malawi, Nyika Plateau, 2200 m., fl. 17.viii.1946, *Brass* 17291 (K, holotype).

Small shrubs or shrublets to 1 m. tall, branched from the base, with more or less
dwarfed branchlets arising from the axils of the leaves. Branchlets without infrafoliar
ridges, pubescent with very short simple hairs and longer hairs, mostly with side
branches on their basal part and with tips either glandular or eglandular and usually
bifurcate, glandular hairs up to 0·9 mm. long, eglandular hairs to 2·0 mm. long, the
proportions of the two types varying from all of one to all of the other; occasionally
some branchlets of a plant glabrous. Leaves in whorls of 3, patent on the main stems,
somewhat ascending on the dwarfed side shoots, straight or recurved, margins
reflexed but normally not quite meeting in the centre line on the abaxial side of the
leaf, 1·5–5·0 mm. long, 0·2–1·0 mm. broad, normally pubescent with short simple
hairs over the whole dorsal surface of the leaf and widely spaced longer hairs, up to
0·8 mm., at the apex, on the apparent margins and on the abaxial part of the dorsal
surface of the leaf, these longer hairs either glandular or eglandular and usually
bifurcate, eglandular hairs occasionally with a few side branches at the base; short
simple hairs sometimes few and caducous, sometimes absent, and occasionally
longer hairs much reduced in number or leaf entirely glabrous with sessile glands
along the apparent margin; petioles c. 0·5 mm. long, minutely pubescent or glabrous.
Inflorescence consisting of clusters of 3–15 flowers at the tips of dwarfed branchlets
on the upper part of the stem. Pedicels 0·5–1·5 mm. long, minutely pubescent or
glabrous, sometimes also with sparse longer glandular hairs. Bracts leaf-like, 1·0–2·0
mm. long, inserted from very near the base of the pedicel to close under the calyx,
pubescence as on leaves. Bracteoles only present when bract inserted on lower half of
pedicel, when present very narrow, linear, 0·5–0·7 mm. long, with marginal hairs
that may be either glandular or eglandular and bifurcate and usually also with very
short simple hairs. Calyx segments 4, free to the base, narrowly oblong to narrowly
triangular, 0·9–1·5 mm. long, pubescent with marginal hairs that may be either
glandular or eglandular and bifurcate, and usually also with short simple hairs on the
outer surface, but these absent when there are no such hairs on the leaves. Corolla 4-
lobed, pink to pale lavender to purple, 1·0–3·0 mm. long, 1·0–1·5 mm. broad, the
shortest crateriform, the longer ones tubular below and crateriform to campanulate
above, the crateriform or campanulate part 1·0–1·5 mm. long, the tubular portion,
when c. 1 mm. long or more, usually somewhat curved, corolla lobes rounded,
sinuses acute, c. 0·5 mm. deep. Stamens 4, about equalling the corolla, sometimes the
anthers protruding; filaments narrow, somewhat expanded below the anthers.
Anthers 0·4–0·8 mm. long, deeply divided, thecae elliptical, opening by lateral pores,
with appendages sometimes absent, usually present and varying in length from
minute to 0·4 mm. long. Ovary globose, pubescent, with dense white hairs on the
upper part and sparser hairs below. Style exserted, 1·5–3·0 mm. long, stigma small,
capitate.

Zambia. N: Mafinga, on summit, fr. 26.viii.1958, *Lawton* 466 (K). **Zimbabwe.** E:
Inyangani Distr., Nyangani Farm, fl. 21.iii.1949, *Chase* 1301 (BM; COI; K; LISC; SRGH).

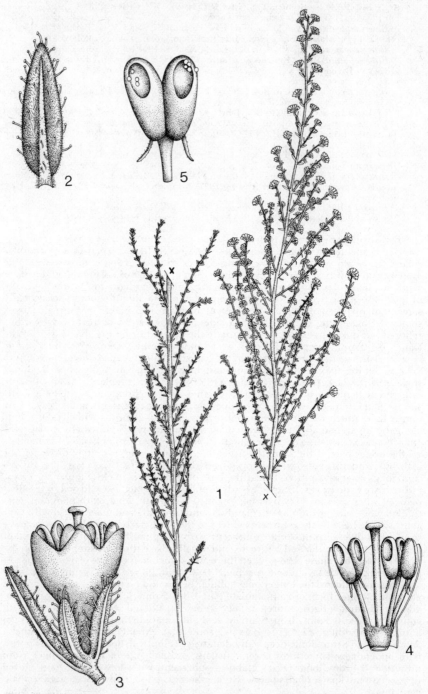

Tab. 29. BLARIA KINGAENSIS. 1, flowering branch (×⅔); 2, underside of leaf (×8); 3, flower
(×8); 4, vertical section of flower (×8); 5, stamen (×14), all from *West* 4805.

Malawi. N: Nyika, Chelinda, 2150 m., fl. 3.v.1963, *Chapman* 1995 (BM; K; SRGH). S: Mulanje Distr., Mt. Mulanje, Litchenya Plateau, 2000 m., fr. 27.vi.1946, *Brass* 16471(BM; FHO; K; SRGH). **Mozambique.** Z: Gúruè, N. of Namuli, fl. 12.viii.1949, *Andrada* 1861 (COI; LISC). MS: Gorongosa, Gogoga Peak, 1750 m., fl. 6.vii.1955, *A.C.L. Schelpe* 492 (BM; LISC; SRGH).

Also in south-western Tanzania, possibly extending further north to Ruanda and to northern Tanzania (see Ross, loc. cit.). In montane grassland or scrub, often by the side of streams, and scattered amongst rocks, 1600–2550 m.

2. **Blaeria filago** Alm & Fries in Notizbl. Bot. Gart. Berl. **8**: 691 (1924); in Acta Hort. Berg. **8** (8): 253 (1924). — Hedberg in Symb. Bot. Upsal. **15** (1): 147, 306 (1957). Type from Kenya.

 Blaeria afromontana Alm & Fries in Notizbl. Bot. Gart. Berl. **8**: 692 (1924). Type from Kenya.
 Blaeria viscosa Alm & Fries in Notizbl. Bot. Gart. Berl. **8**: 692 (1924); in Acta Hort. Berg. **8** (8): 254 (1924). Type from Kenya.
 Blaeria elgonensis Alm & Fries in Notizbl. Bot. Gart. Berl. **8**: 693 (1924). Type from Kenya.
 Blaeria filago var. *afromontanā* (Alm & Fries) Alm & Fries in Acta Hort. Berg. 8 (8): 253 (1924). Type as for *B. afromontana*.
 Blaeria viscosa var. *elgonensis* (Alm & Fries) Alm & Fries in Acta Hort. Berg. 8 (8): 254 (1924). Type as for *B. elgonensis*.
 Blaeria saxicola Alm & Fries in Acta Hort. Berg. **8** (8): 252 (1924). — Brenan, T.T.C.L.: 191 (1949). Type from Tanzania.
 Blaeria filago subsp. *saxicola* (Alm & Fries) Hedberg, tom. cit.: 148 (1957). Type as for *B. saxicola*.

Shrublet to 50 cm. tall, branched below, branches slender, flexuous, ascending, with a few dwarfed branchlets on the floriferous stems only. Branches pubescent with short simple hairs and longer hairs to 1·0 mm. long, with side branches on their basal parts and bifurcate tips. Leaves in whorls of 3, ascending, somewhat incurved, margins reflexed, c. 2·5 mm. long, 1·0 mm. broad, pubescent, the whole dorsal surface with very short simple hairs, stouter hairs, 0·5–0·7 mm. long, closely spaced along the apparent margin and between this and the true margin, most with side branches near the base and tips forked into 3 or 4 branches; petioles c. 0·5 mm. long, minutely pubescent. Inflorescence consisting of clusters of 3–9 flowers at the tips of branches and dwarfed branchlets. Pedicels 0·7–1·0 mm. long, pubescent with short simple hairs. Bracts leaf-like, c. 2 mm. long, inserted c. ⅓ up the pedicel, with pubescence as on the leaves. Bracteoles inserted just above the bracts, c. 1 mm. long, pubescence as on leaves. Calyx segments narrowly triangular, c. 1·5 mm. long, externally pubescent with short simple hairs and with close-packed stout marginal cilia 0·5–0·7 mm. long. Corolla 4-lobed, purplish pink, c. 2·0 mm. long, 1·5 mm. broad, tubular below, campanulate above. Stamens 4, about equalling the corolla; filaments c. 1·5 mm. long; anthers c. 0·6 mm. long, thecae elliptical, with short appendages c. 0·1 mm. long. Ovary globose, pubescent with dense white hairs on the upper part and sparser hairs below. Style exserted, c. 2 mm. long, stigma small capitate.

 Malawi. N: Nyika Plateau, 2350 m., fr. 19.viii.1946, *Brass* 17340a (SRGH).
 Tanzania and Kenya at high altitudes on the mountains. In open grassland, 2350 m.

The only specimen of this species from the Flora Zambesiaca region is a branch mounted on the sheet of *Brass* 17340 in SRGH. The remaining material on this sheet and all the material of *Brass* 17340 in BM and K is *B. kingaensis*. The description given above is based on this single specimen and does not take into account the variation in the specimens from the main area of distribution. The corolla can be up to 4 mm. long, the forked hairs may be replaced in part or entirely by glandular hairs, and the anther appendages can be absent to 0·5 mm. long.

5. **ERICINELLA** Klotzsch

Ericinella Klotzsch in Linnaea **12**: 222 (1838)

Much branched shrubs or sub-shrubs with slender branchlets. Leaves in whorls of 3 or 4, very small, acicular, the outer margin reflexed and meeting in the mid line on the under side of the leaf. Flowers small, in capitate clusters of 3–16 at the tips of the branchlets. Pedicels without bracts. Calyx and corolla tetramerous, hypogynous and persistent in fruit. Calyx with one sepal longer than the other 3, leaf-like, free and inserted immediately below the others. Corolla actinomorphic, campanulate to tubular. Stamens 4; anthers with appendages. Style slender; stigma narrowly to broadly infundibuliform. Ovary globose, quadrilocular, capsule loculicidal.

Species 3, two in Cape Province, South Africa, one in tropical Africa.

Ericinella microdonta (C. H. Wright) Alm & Fries in Acta Hort. Berg. **8**: 262 (1924); in K. Svenska Vetensk.-Akad. Handl., Ser. 3, **4** (4): 46 (1927). — Brenan in Mem. N.Y. Bot. Gard. **8**: 494 (1954). TAB. **30**. Syntypes: Malawi, Mulanje Distr., Mt. Mulanje, 1800 m., *McClounie* 55, 75, 95 (K).
    Blaeria microdonta C. H. Wright in Kew Bull. **1897**: 272 (1897). Type as above.
    Ericinella shinniae S. Moore in J. Bot., Lond. **54**: 287 (1916). Type: Malawi, Mulanje Distr., *Shinn* s.n. (BM.)
    Ericinella brassii Brenan in Mem. N.Y. Bot. Gard. **8**: 494 (1954). Type: Malawi, Mulanje Distr., Mt. Mulanje, Litchenya Plateau, 1870 m., fl. 27.vi.1946, *Brass* 16454 (BM; K, holotype; SRGH).
    Ericinella microdonta var. *craspedotricha* Brenan, tom. cit.: 495 (1954). Type: Malawi, Mulanje Distr., Mt. Mulanje, southwest ridge, 2400 m., fl. 28.vi.1946, *Brass* 16497 (K, holotype; SRGH).

Shrub 0·75–3·0 m. tall. Branchlets moderately stout, the finest 0·4–1·0 mm. in diameter, with a sparse to dense pubescence of hairs c. 0·2 mm. long with short dense side branches to the tip. Leaves in whorls of 4, patent to appressed, linear to lanceolate, acute, 1·5–5·0 mm. long, 0·7–1·0 mm. broad, with sessile glands along the apparent margin normally alternating with white conical hairs 0·1–0·2 mm. long, but these sometimes absent or present only on the proximal part of some leaves; petioles 0·3–0·8 mm. long, glabrous, sometimes with sessile glands on the margin. Flowers in clusters of 4–16 at the tips of the branchlets. Pedicels glabrous, 1·5–4·0 mm. long. Calyx 4-partite, the margins of the sepals ciliate with hairs similar to those on the leaves but shorter and closer, or with sessile glands in specimens with few or no hairs on the leaves, the 3 equal sepals free, oblong to triangular with obtuse apices, 0·7–1·0 mm. long, the longer sepal leaf-like, 1·0–1·8 mm. long. Corolla white, narrowly infundibuliform to tubular and expanded above, 2·0–3·0 mm. long, 1·0–1·5 mm. wide at the apex, persistent, the lower part expanded by the swelling capsule so that the corolla becomes cyathiform in fruit. Stamens 4, included; filaments flattened, geniculate above, 1·3–2·0 mm. long; anthers broad, deeply cleft, opening by elliptical pores in the upper part of the theca, c. 0·7 × 0·5 mm., with aristate puberulous appendages c. 0·3 mm. long. Ovary globose, c. 1·0 mm. in diameter, glabrous; style slender, exserted, 2·0–3·2 mm. long; stigma narrowly to broadly infundibuliform.

Malawi. N: South Nyika Mts., 1200–2100 m., fl. vii.1896, *Whyte* s.n. (K). S: Mulanje Distr., Mt. Mulanje, between Litchenya Forestry Hut and Simpson's Peak, 1800 m., fl. 30.iii.1960, *Phipps* 2787 (K; LISC; SRGH). **Mozambique.** Z: Milange, Serra do Chiperone, 1800 m., fl. 3.ii.1972, *Correia & Marques* 2514 (LMU).
    Also in south-west Tanzania.

Examination of a long series of specimens shows that there is no basis for separating *Ericinella brassii* from *E. microdonta* nor for recognizing the var. *craspedotricha* as distinct within that species. The types differ in density of indumentum on the branchlets, size of leaf and presence or absence of hairs on their apparent margins, corolla size, and stigma diameter, but in nothing else. The only one of these characters in which there is any indication of discontinuity in variation is leaf size; specimens with fully developed leaves between 2·0 and 3·0 mm. in length do not seem to occur. The specimens with short leaves, but not only these, have a dense indumentum, short corollas and wide stigmas, but there is a continuous range of variation in these characters among the specimens as a whole and the difference in leaf length by itself seems a quite inadequate basis for any taxonomic separation.

6. PHILIPPIA Klotzsch

Philippia Klotzsch in Linnaea **9**: 354 (1834). — Alm & Fries in K. Svenska Vetensk.-Akad. Handl., Ser. 3, **4** (4): 9 (1927).

Much branched trees or shrubs with flexuously ascending stems. Branchlets slender. Leaves in whorls of 3 or 4, very small, the outer margins reflexed, normally to such an extent that they meet along the mid-line of the under side of the leaf, forming a sulcus; often glandular at the tip and along the apparent margin. Flowers small, in capitate clusters of up to about 12 flowers, but usually fewer, at the tips of the branchlets. Pedicels without bracts. Calyx and corolla hypogynous, 4-partite, occasionally 3-partite in some flowers, regularly 3-partite in some species not in the

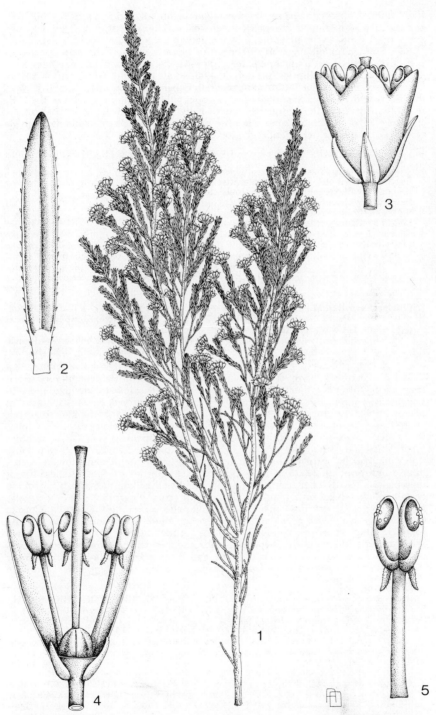

Tab. 30. ERICINELLA MICRODONTA. 1, flowering branch ($\times \frac{2}{3}$); 2, underside of leaf ($\times 14$); 3, flower ($\times 14$); 4, vertical section of flower ($\times 16$); 5, stamen ($\times 20$), all from *Hilliard* & *Burtt* 4505.

F.Z. region. Calyx with one sepal longer than the others, and this sepal usually free and leaf-like at the tip, other sepals connate to a variable extent, sometimes leaf-like at the tip. Corolla actinomorphic, campanulate to urceolate, not deeply cleft, lobes obtuse. Stamens in one species 4 and opposite the calyx lobes, in the other species (5) 6–8 (10), anthers without appendages, in some species connate, in others free, opening by narrowly elliptical pores on the outer side of the upper part of each lobe. Ovary 3–4-locular, placentation axile. Style 1, expanded at the apex normally into a disc bearing the stigmatic surfaces on its upper side, but in one species in the F.Z. region expansion at the tip of the style broadly conical and the areas bearing the stigmatic surfaces somewhat depressed. Fruit a cartilaginous loculicidal capsule contained within the persistent calyx and corolla.

A genus of c. 40 species from tropical and S. Africa, Madagascar and the Mascarene Islands.

1. Leaves in whorls of 3, anthers free after anthesis - - - - - - - 2
 - Leaves in whorls of 4, anthers connate after anthesis - - - - - - 5
2. Branchlets without hairs with side branches to the tip - - - - - - 3
 - Branchlets with hairs with side branches to the tip - - - - - - 4
3. Style more than 1 mm. long, exserted from the corolla - - - - - 1. *simii*
 - Style 0·5 mm. long or less, equalling the corolla - - - - - - 2. *evansii*
4. Stamens 6 in most or all flowers, ovary pubescent - - - - - 3. *mannii*
 - Stamens 4, ovary glabrous - - - - - - - - 4. *nyassana*
5. Style 0·3 mm. or less, equal sepals connate to ⅔ their length or more, stamens usually 8
                                                               5. *benguelensis*
 - Style 0·5 mm. or more, equal sepals connate to ½ their length or less, stamens usually 6
                                                               6. *hexandra*

1. **Philippia simii** S. Moore in J. Linn. Soc. Lond., Bot. **40**: 128 (1911). — R. Ross in Bull. Jard. Bot. Brux. **27**: 749 (1957). TAB. **31**. Type: Mozambique, Zambésia, Bajone, Maganja da Costa, *Sim* in Herb. *Bolus* 5688 (BM, lectotype).
   Philippia friesii Weimarck in Bot. Notis. **1940**: 58 (1940). Type: Zimbabwe, Inyanga, "ad dejectum fluminis Pungwe", *Fries, Norlindh & Weimarck* 3843 (BM; K; LD, holotype; PRE).

A shrub or tree 0·2–5 m. tall. Branchlets with a dense pubescence of very short, simple hairs, often too short to be visible with a hand lens; infrafoliar ridges present. Leaves in whorls of 3, ascending and often slightly curved outwards, linear to narrowly ovate, obtuse, with glands and minute hairs on the apparent margins, otherwise glabrous, 1·0–5·0 mm. long, 0·35–0·8 mm. wide; petioles c. 0·5 mm. long, very shortly ciliate on the margins. Flowers in clusters of normally only 3, sometimes 6, at the tips of the branchlets. Pedicels glabrous, expanded above, 1·5–2·5 mm. long. Calyx and corolla normally 4-partite, occasionally 3-partite in some flowers. Calyx infundibuliform, ciliate at the margins, otherwise glabrous, all the lobes leaf-like at the tips; the 3 equal lobes triangular, acute, usually exceeding the sinus between the corolla lobes, 1·3–2·2 mm. long, fused for about ½ their length; the 4th lobe usually slightly longer, sometimes considerably longer and exceeding the corolla. Corolla campanulate, yellowish green suffused crimson, 1·6–2·4 mm. long, lobes broadly rounded, sinuses narrowly rounded. Stamens 8, equalling the corolla, becoming free shortly after anthesis. Ovary 4-locular, glabrous; style with a broadly conical distal expansion somewhat depressed at the apex, shortly exserted in flower, more strongly exserted beyond the persistent corolla in fruit, 1·2–1·6 mm. long, 0·8–1·1 mm. in diameter at the apex.

Zimbabwe. E: Inyanga, 1550 m., fl. 24.i.1951, *Chase* 3570 (BM; COI; LISC; SRGH). **Mozambique.** N: Niassa, Ribáuè, 1600 m., fr. 28.i.1964, *Torre & Paiva* 10314 (LISC). Z: Pebane, fr. 11.ix.1964, *Gomes e Sousa* 4828 (K; LISC; LMA; LMU; PRE; SRGH). MS: Beira, Chinizina, fl. 23.iv.1957, *Gomes e Sousa* 4372 (BM; COI; FHO; LMA; PRE).
Also in the northern Transvaal. In open woodland and savannah or scattered in montane grassland, occuring on damp sands near the coast and inland often on rocky slopes beside streams or rivers, 0–2600 m.

2. **Philippia evansii** N. E. Brown in Thiselton-Dyer, Fl. Capensis **4** (1): 316 (1905). — Alm & Fries in K. Svenska Vetensk.-Akad. Handl., Ser. 3, **4** (4): 39 (1927). — Trauseld, Wild Flowers Natal Drakensberg: 140–141 (1969). — J. H. Ross, Fl. Natal: 269 (1972). Type from Natal.

A shrub or tree 1–10 m. tall. Branchlets very fine, densely pubescent with simple hairs up to 0·15 mm. long, occasionally glandular, without infrafoliar ridges. Leaves

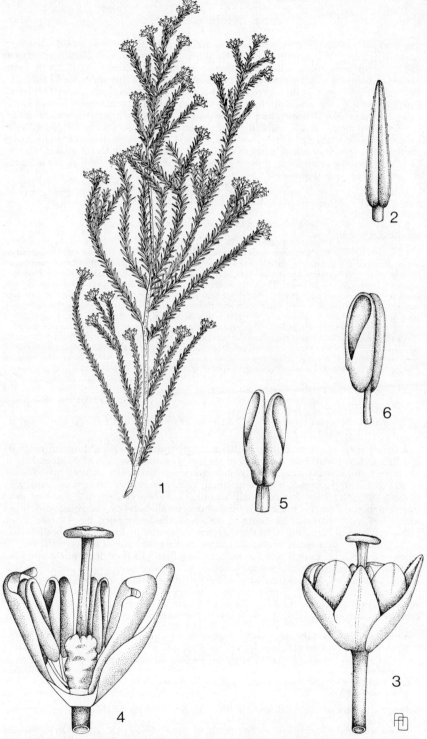

Tab. 31. PHILIPPIA SIMII. 1, flowering branch (×⅔); 2, underside of leaf (×6); 3, flower (×10); 4, vertical section of flower (×16); 5, stamen, front view (×20); 6, stamen, lateral view (×20), all from *Grosvenor* 330.

in whorls of 3, ascending to appressed, straight or slightly curved inwards, linear to narrowly ovate, obtuse, 0·7–4 mm. long, 0·3–0·8 mm. wide with glandular teeth at the apparent margin lost with age, otherwise glabrous; petioles 0·1–0·2 mm. long, glabrous. Flowers small for the genus, in clusters of 3–6 at the tips of the branchlets. Pedicels 1·0–1·3 mm. long, normally glabrous, occasionally with a very few hairs. Calyx spreading, not reaching the corolla sinuses, with minute glandular cilia on the margins, otherwise glabrous, lobes all leaf-like at the tip; the 3 equal lobes triangular, 0·4–0·6 mm. long, fused for about ½ their length; the 4th lobe little longer. Corolla urceolate, pink, glabrous, 1·0–1·2 mm. long, 1·0–1·6 mm. wide, lobes obtuse, sinuses shallow, obtuse. Stamens 8, anthers free, not quite equalling the corolla. Ovary 4-locular, glabrous; style stout, 0·3–0·5 mm. long, equalling the corolla, the terminal disc 0·6–0·7 mm. in diameter.

Zimbabwe. E: Umtali, Engwa, 1980 m., fl. 2.ii.1955, *Exell, Mendonça & Wild* 118 (BM; LISC; SRGH). **Mozambique.** MS: Manica, Serra Zuira, Tsetsera Plateau, 2050 m., fr. 4.xi.1965, *Torre & Pereira* 12639 (LISC).
Also on the Drakensberg mountains in Natal. At the edge of forest, or forming colonies in grassland, 1500–2600 m.

This species has not previously been recorded from either Zimbabwe or Mozambique although it is quite distinct from any other occurring in the F.Z. area and there are a number of specimens of it in herbaria; these have for the most part been identified erroneously as *Philippia mannii* (Hook. f.) Alm & Fries. The only differences that I can detect between the specimens from the F.Z. area and those from Natal are that in the former the glands along the apparent margin of the leaf are less well developed and the corolla is less deeply divided; the sinuses in specimens from the F.Z. area are only 0·2–0·3 mm. deep as against about 0·5 mm. in specimens from Natal. These differences seem inadequate to justify taxonomic separation at any level.
Alm & Fries include *Philippia evansii* in their key under the heading "calyx basi connatus", contrasted with "calyx fere ad medium vel altius connatus", but in the Natal specimens, as well as those from the F.Z. area, the three equal sepals are fused to the middle; as in all species of *Philippia*, there is less fusion of the larger sepal to those on either side of it.

3. **Philippia mannii** (Hook.f.) Alm & Fries in K. Svenska Vetensk.-Akad. Handl., Ser. 3, **4** (4): 37 (1927). — R. Ross in Hepper, F.W.T.A., ed. 2, **2**: 2 (1963); in Bol. Soc. Brot., Sér. 2, **53**: 142 (1981). Type from Fernando Po.
  Ericinella mannii Hook f. in J. Proc. Linn. Soc. Lond., Bot. **6**: 16 (1862). — Oliver, F.T.A. **3**: 484 (1877). Type as above.

A shrub or tree 1–10 m. tall. Branchlets densely pubescent with short simple hairs and longer hairs with side branches to the tip. Leaves in whorls of 3, ascending to appressed, not curved outwards, linear to narrowly lanceolate, acute, 2·0–6·0 mm. long, 0·6–1·25 mm. broad, with very shortly ciliate margins when young, otherwise glabrous; petioles 0·4–0·7 mm. long, glabrous. Flowers in clusters of 3–6 at the tips of the branchlets. Pedicels pubescent with short simple hairs or glabrous, sometimes with a few hairs with side branches to the tip on the lower part, 1·3–4·0 mm. long. Calyx infundibuliform, ciliate at the margins, otherwise glabrous, the lobes leaf-like at the tip; the 3 equal lobes acute, not quite reaching the sinuses between the corolla lobes, 0·8–1·5 mm. long, fused for rather more than ½ of their length; the 4th lobe slightly to considerably exceeding the other 3. Corolla campanulate, white to pale green suffused red, 1·2–2·0 mm. long, lobes broadly rounded, sinuses shallow, acute. Stamens usually 6 but 5, 7 or 8 in some flowers of some specimens, anthers free, equalling the corolla. Ovary 3-locular or 4-locular, pubescent; style exserted, 1·0–2·0 mm. long, the terminal disc 0·6–1·0 mm. in diameter.

A species with one subspecies (subsp. *mannii*) in south-eastern Nigeria and Fernando Po and two in East Africa and the F.Z. area.

Finest branchlets 0·3 mm. or less in diameter, without infrafoliar ridges; ovary 3-locular in most or all flowers - - - - - - - - - subsp. *pallidiflora*
Finest branchlets not less than 0·5 mm. in diameter, with infrafoliar ridges; ovary 4-locular in most or all flowers - - - - - - - - - subsp. *usambarensis*

Subsp. **pallidiflora** (Engler) R. Ross in Bol. Soc. Brot., Sér. 2, **53**: 143 (1981). Type from Tanzania.
  Philippia pallidiflora Engler, Bot. Jahrb. **43**: 370 (1909). — Alm & Fries in K. Svenska Vetensk.-Akad. Handl., Ser. 3, **4** (4): 40 (1927). — Brenan, T.T.C.L.: 194 (1949). — R. Ross in Bull. Jard. Bot. Brux. **27**: 751 (1957), quoad subsp. *pallidiflora*. — White, F.F.N.R.: 315 (1962). Type as above.
  Philippia uhehensis Engler, loc. cit. Type from Tanzania.

Finest branchlets 0·3 mm. or less in diameter, without infrafoliar ridges. Leaves not exceeding 4·0 mm. long, 1 mm. broad. Pedicels up to 2 mm. long, glabrous or with, at the base, a few hairs with side branches to the tip. Ovary 3-locular in most or all flowers.

Zambia. N: Mpika, fr. 2.ii.1955, *Fanshawe* 1934 (BR; K; SRGH). **Zimbabwe.** E: Manicaland, fr. ix.1934, *Gilliland* 712 (BM; FHO; K; SRGH). S: Belingwe, Mt. Buhwa, saddle on NW. slopes, 1150 m., fl. & fr. 28.iv.1973, *Pope* 973 (K; SRGH). **Malawi.** N: Nyika Plateau, 2400 m., fl. 28.x.1958, *Robson* 425 (BM; K; LISC; PRE; SRGH). **Mozambique.** MS: Báruè, Choa Mts., 1000 m., fl. 9.xii.1965, *Torre & Correia* 13434 (LISC).
Also in south western Tanzania, south eastern Zaire and eastern Angola. In *Brachystegia-Uapaca* woodland and montane scrub and grassland, 1000–2400 m.

Subsp. **usambarensis** (Alm & Fries) R. Ross in Bol. Soc. Brot., Sér. 2, **53**: 144 (1981). Type from Tanzania.
   Philippia usambarensis Alm & Fries in K. Svenska Vetensk.-Akad. Handl., Ser. 3, **4**(4); 35 (1927). — Brenan, T.T.C.L.: 194 (1949). Type as above.
   Philippia pallidiflora subsp. *usambarensis* (Alm & Fries) R. Ross in Bull. Jard. Bot. Brux. **27**: 752 (1957). — Dale & Greenway, Kenya Trees & Shrubs: 180 (1961). Type as above.

Finest branchlets 0·5 mm. or more in diameter, with prominent broad infrafoliar ridges. Leaves up to 6 mm. long, 1·25 mm. broad. Pedicels pubescent with short simple hairs, up to 4 mm. long. Ovary 4-locular in most or all flowers.

Mozambique. Z: Gúruè, Gúruè Mts., E. of Namuli Peak, 1800 m., fl. 8.ix.1967. *Torre & Correia* 16001 (LISC).
Also in southern Kenya and Tanzania.

4. **Philippia nyassana** Alm & Fries in K. Svenska Vetensk.-Akad. Handl., Ser. 3, **4** (4): 33 (1927). — Brenan in Mem. N.Y. Bot. Gard. **8**: 493 (1954). Type: Malawi, Mulanje Plateau, *McClounie* s.n. (K, holotype).

A shrub or tree 0·5–6 m. tall. Branchlets densely pubescent with short simple hairs and longer hairs with side branches to the tip, the slenderest branchlets c. 0·2 mm. diameter; no infrafoliar ridges. Leaves in whorls of 3, appressed and curved inwards, narrowly ovate, acute, with shortly ciliate margins, otherwise glabrous, 1·0–2·0 mm. long, 0·4–0·7 mm. broad; petioles c. 0·1 mm. long, glabrous. Flowers in clusters of 3–9 at the tips of the branchlets. Pedicels glabrous, 1·0–1·5 mm. long. Calyx bowl-shaped, ciliate at the margins, otherwise glabrous, the lobes leaf-like at the tip; the 3 equal lobes narrow, acute, not quite reaching the sinuses between the corolla lobes, 0·6–0·8 mm. long, fused for ¼ or less of their length; the 4th lobe slightly longer, reaching to or just exceeding the sinuses between the corolla lobes. Corolla campanulate, green suffused crimson, 1·2–1·5 mm. long, lobes broadly rounded, sinuses moderately deep and acute. Stamens 4, very occasionally 5, anthers free, slightly exceeding the corolla. Ovary 4-locular, glabrous; style exserted, 1·0–1·2 mm. long, the terminal disc 0·5–0·8 mm. in diameter.

Malawi. S: Mt. Mulanje, Luchenya Plateau, 2125 m., fl. 8.v.1963, *Wild* 6138 (K; LISC; PRE; SRGH).
Only known from Mt. Mulanje. On edges of montane forest and scattered in neighbouring grassland, 1800–2150 m.

This species was reported from the Inyanga district of Zimbabwe by Weimarck (Bot. Notiser **1940**: 58, 1940) on the basis of two gatherings, *Fries, Norlindh & Weimarck* 3070 and 3602. However, the specimen of no. 3070 in BM is *Philippia mannii* subsp. *pallidiflora*; the capsules have dehisced but are 3-locular and pubescent, and there are 6 stamens in one of the very few flowers in which these are still present. The BM specimen of no. 3602 has no flowers but its leaves are in whorls of 4 and there are no hairs with side branches to the tip on the stem. It is either *P. benguelensis* (Welw. ex Engl.) Britten or *P. hexandra* S. Moore, more probably the latter.

5. **Philippia benguelensis** (Welw. ex Engl.) Britten in Trans. Linn. Soc. Lond., Bot., Ser. 2, **4**: 24 (1894) "*benguellensis*". — Alm & Fries in K. Svenska Vetensk.-Akad. Handl., Ser. 3, **4** (4): 20 (1927). — Weimarck in Bot. Notis. **1940**: 54 (1940), excl. var. *intermedia*. — Robyns, F.P.N.A. **2**: 21 (1947). — Brenan, T.T.C.L.: 193 (1949). — Eggeling & Dale, Indig. Trees Uganda, ed. 2: 111 (1951). — Brenan in Mem. N.Y. Bot. Gard. **8**: 493 (1954). — Dale & Greenway, Kenya Trees & Shrubs: 179 (1961). — White, F.F.N.R.; 315 (1962). — R. Ross in Bol. Soc. Brot., Sér. 2, **53**: 144 (1981). Type from Angola.

Salaxis benguelensis Welw. ex Engl. in Abh. Preuss. Akad. Wiss. **1891**: 328 (1892). Type as above.
Philippia milanjiensis Britten & Rendle in Trans. Linn. Soc. Lond., Bot., Ser. 2, **4**: 24 (1894). Type: Malawi, Mulanje, *Whyte* s.n. (BM, holotype; K).
Philippia holstii Engl., Pflanzenw. Ost-Afr. **C**: 302 (1895). Type from Tanzania.
Philippia stuhlmannii Engl., loc. cit. Type from Zaire.
Philippia congoensis S. Moore in Journ. Bot., Lond. **57**: 212 (1919). Type from Zaire.
Philippia kundelungensis S. Moore., loc. cit. Type from Zaire.

A shrub or tree 1–6 m. tall. Branchlets with poorly developed infrafoliar ridges, with a dense pubescence of short simple hairs and, usually, sparser longer hairs that are often glandular. Leaves in whorls of 4, appressed to spreading, straight or slightly curved inwards, narrowly ovate, acute to obtuse, pubescent at least on the upper surface but the hairs sometimes evanescent, 1·5–4·5 mm. long, 0·4–0·7 mm. broad, but leaves on coppice shoots may be up to 10 mm. long and 3·5 mm. broad with the margins scarcely reflexed; petioles 0·5–0·7 mm. long, pubescent. Flowers in clusters of 4–12 at the tips of the branchlets. Pedicels 1–2 mm. long, pubescent with simple hairs less than 0·05 mm. long. Calyx and corolla occasionally 5-partite, usually 4-partite. Calyx bowl-shaped, shortly pubescent, the margin glandular or fimbriate, occasionally very shortly ciliate; the 3 equal lobes broadly triangular, about equalling the sinuses of the corolla, leaf-like tips scarcely developed, 0·9–1·2 mm. long, fused for ⅔ of their length, the sinuses rounded; the 4th lobe somewhat longer, almost equalling to slightly exceeding the corolla, leaf-like at the tip. Corolla bowl-shaped, pubescent without, 1·1–1·8 mm. long, 1·5–2·5 mm. wide, lobes broadly rounded, sinuses shallow, rounded. Stamens 8 in all flowers of most specimens, but from 5 to 9 in some flowers of some specimens, slightly exceeding the corolla, anthers remaining fused after dehiscence. Ovary 4-locular, pubescent; style short, 0·1–0·3 mm., exceeded by the stamens until the capsule begins to ripen, terminal disc 0·8–1·5 mm. diameter.

Leaves with sparse erect hairs or glabrous -   -   -   -   -   -   - var. *benguelensis*
Leaves with moderately dense appressed hairs on both surfaces   -   -   - var. *albescens*

Var. **benguelensis**

Leaves with sparse erect hairs, often confined to the upper surface and evanescent. Glandular hairs on stem and glands, often evanescent, on apparent margin and often at tip of leaf. Corolla pale green to green suffused crimson.

Zambia. N: Mpika, fl. 2.ii.1955, *Fanshawe* 1946 (FHO; K; SRGH). W: Mwinilunga, Dobeka Bridge, 1375 m., fl. 21.xii.1969, *Simon & Williamson* 1922 (K; SRGH). C: Serenje, Kundalila Falls, fl. 13.x.1963, *Robinson* 5707 (K; SRGH). E: Nyika Plateau, Kangampande Mt., 2150 m., fl. 6.v.1952, *White* 2740 (BM; FHO; K). **Zimbabwe.** N: Mazoe, fl. 9.ii.1947, *Wild* 1639 (K; SRGH). C: Makoni, 1600 m., fr. iv.1906, *Eyles* 336 (BM). E: Inyanga, in proclivitate montium, 1800 m., fl. 4.xi.1930, *Fries, Norlindh & Weimarck* 2570 (BM; LD; PRE; SRGH). **Malawi.** N: Nyika Plateau, 2300 m., fl. 16.viii.1946, *Brass* 17251 (BM; K; PRE; SRGH). C: Chongoni Mt., 2000 m., fl. 3.ii.1959, *Robson* 1437 (BM; K; LISC; SRGH). S: Mt. Mulanje, Sombani Plateau, 2450 m., fl. 13.v.1963, *Wild* 6236 (BM; K; LISC; SRGH). **Mozambique.** N: Niassa, Lichinga (Vila Cabral), Massangula Mts., 1450 m., fr. 25.ii.1964, *Torre & Paiva* 10787 (LISC). T: Macanga, between Casula and Furancungo, fr. 9.vii.1949, *Andrada* 1729 (LISC; LMA; PRE).
Also from Uganda, Tanzania, eastern Zaire and Angola. In open places within and on the edges of *Brachystegia* woodland and montane forest, in secondary scrub, and scattered or forming thickets in montane grassland.

Var. **albescens** R. Ross in Bull. Jard. Bot. Brux. **27**: 754 (1957).

Leaves with moderately dense white appressed hairs on both surfaces. No glands on stem or leaves. Corolla white.

Zambia. W: Mwinilunga Distr., Dobeka Dambo, 45 km. W. of Mwinilunga, 1350 m. fl. 22.i.1975, *Brummitt, Chisumpa & Polhill* 13986 (K). **Malawi.** N: Nyika Plateau, Chowo Rock, 2200 m., fl. 10.i.1974, *Pawek* 7942 (SRGH). **Mozambique.** Z: Gúruè, nr. Namuli Peak, 1500 m., fl. 9.ix.1943, *Torre* 5138 (LISC).
Also from south-western Tanzania. Amongst rocks and by a stream, 1500–2200 m.

6. **Philippia hexandra** S. Moore in J. Linn. Soc. Lond., Bot. **40**: 129 (1911). — Alm & Fries in K. Svenska Vetensk.-Akad. Handl., ser. 3, **4** (4): 22 (1927). — Weimarck in Bot. Notis. **1940**: 55 (1940). — R. Ross in Bol. Soc. Brot. Sér. 2, **53**: 144 (1981). Type: Zimbabwe, Melsetter, *Swynnerton* 1147 (BM, lectotype).

Philippia norlindhii Weimarck, tom. cit.: 56 (1940). Type: Zimbabwe. Inyanga, "prope montem Inyangani", *Fries, Norlindh & Weimarck* 3641 (BM; K; LD, holotype; PRE).
Philippia benguelensis var. *intermedia* Weimarck, tom. cit.: 55 (1940). Type: Zimbabwe, Inyanga, *Fries, Norlindh & Weimarck* 2621 (BM; LD, holotype; PRE).

A shrub or tree 0·5–7 m. tall. Branchlets with a dense pubescence of short simple hairs and, usually, sparser longer hairs 0·3–1·0 mm. long, these longer hairs sometimes glandular and sometimes with short branches near the base. Leaves in whorls of 4, appressed to spreading, straight or slightly curved inwards, linear to ovate, acute, 1·0–2·5 mm. long, 0·6–0·8 mm. broad; glabrous or pubescent with erect hairs, with glandular teeth on the apparent margin or sparse glandular hairs on the apparent margin and the lower surface, sometimes a glandular hair at the apex; petiole up to 0·5 mm. long, pubescent with very short hairs. Flowers in clusters of 4–16 at the tips of the branchlets. Pedicels 1·0–1·5 mm. long, pubescent, the hairs erect and c. 0·1 mm. long, sometimes very sparse. Calyx infundibuliform, normally pubescent with erect hairs c. 0·1 mm. long on the outer surface and along the margins, sometimes glabrous on the outer surface with either ciliate or glandular margins, all the lobes leaf-like at the tip; the 3 equal lobes ovate, acute 0·8–1·4 mm. long, reaching to about the sinuses of the corolla, fused to less than ½ their length, the sinuses acute; the 4th lobe equalling to appreciably exceeding the corolla. Corolla infundibuliform, pale green suffused red, normally glabrous, occasionally with a few hairs on the outer surface, shallowly divided, 1·0–2·0 mm. long, 0·9–1·6 mm. broad, lobes obtuse, sinuses acute. Stamens 6, occasionally 7 or 8 in some or all flowers, equalling the corolla, anthers remaining fused after dehiscence. Ovary 4-locular, pubescent, sometimes very sparsely so; style 0·6–1·0 mm. long, glabrous or sparsely pubescent at the base or throughout, equalling or slightly exceeding the anthers, the terminal disc 0·6–1·0 mm. in diameter.

Zambia. W: Mwinilunga, Dobeka Bridge, fl. 7.xii.1937, *Milne-Redhead* 3532 (BM; K; PRE). **Zimbabwe.** C: Wedza Mt., 1625 m., fl. 27.ii.1964, *Wild* 6341 (BM; K; LISC; PRE; SRGH). E: Inyanga, Pungwe River Bank, 1850 m., fl. 8.viii.1950, *Chase* 2877 (BM; COI; LISC; SRGH). S: Belingwe, Mt. Buhwa, E. slopes, 1400 m., fr. 9.vii.1968, *Müller* 788 (SRGH). **Mozambique.** T: Macanga, between Casula and Furancungo, fl. 9.vii.1949, *Barbosa & Carvalho* 3524A (LMA). MS: Manica, Mt. Vumba, 1200 m., fl. 25.viii.1962, *Gomes e Sousa* 4785 (K; LMA; PRE).
Not known outside the F.Z. area. On the edge of and in open places within *Brachystegia* woodland and montane forest, often forming dense thickets, in rocky places, on the banks of streams, and scattered in montane grassland, 1200–2600m.

103. PLUMBAGINACEAE
By A. R. Vickery

Perennial, rarely annual, herbs or small shrubs. Leaves exstipulate, alternate or in basal rosettes. Inflorescence various, often cymose. Bracts scarious. Flowers 5-merous, bisexual, actinomorphic. Petals free, slightly joined at base, or united to form long basal tube. Stamens inserted at base of corolla, antipetalous. Ovary superior, 1-celled with 1 anatropous ovule. Styles 5, or 1 with 5 stigma-lobes. Fruit a dry 1-seeded capsule, often enclosed in the persistent calyx, indehiscent, or operculate, or dehiscing irregularly. Seeds with abundant, scanty, or absent, mealy endosperm.

A family of about 17 genera, cosmopolitan, most frequent on sea coasts or saline inland habitats.
Species of *Ceratostigma*, small shrubs with bright blue flowers in compact heads, native of China and the Himalayas, are occasionally cultivated as garden ornamentals in Zimbabwe.

1. PLUMBAGO L.
Plumbago L., Sp. Pl. **1**: 151 (1753): G. Pl., ed. 5: 75 (1754).

Perennial herbs or shrubs. Leaves alternate, simple. Flowers in alternate 1-flowered spikelets grouped in elongated terminal spikes. Calyx tubular, 5-ribbed, scarious. Corolla with narrow tube. Stamens free. Ovary 1-locular; style 1 with 5 stigma lobes. Fruit a capsule, membranous, enclosed in persistent calyx, dehiscing in a complete ring near base and often splitting into 5 valves from below.

A genus of about 20 species occurring in tropical and warm temperate areas.

1. Peduncles densely covered with short white hairs; corolla pale blue -   -   3. *auriculata*
- Peduncles without short white hairs; corolla bright deep blue or white   -   -   -   2
2. Corolla lobes white, sometimes with slight blue flush, drying pale orange; leaves rarely auriculate - - - - - - - - - - - - - - -   1. *zeylanica*
- Corolla lobes bright blue, drying bluish-purple; leaves with prominent auricles
2. *amplexicaulis*

1. **Plumbago zeylanica** L., Sp. Pl. **1**: 151 (1753). — Oliv., F.T.A. **3**: 486 (1877). — Wright in Fl. Cap. **4**, 1: 425 (1906). — Fries, Ergebn. Schwed. Rhod.-Kongo Exped. **1**: 254 (1916). — Watt & Breyer-Brandwijk, Med. & Pois. Pl. S. & E. Afr., ed. 2: 850 (1962). — Dyer in Fl. S. Afr. **26**: 17 (1963). — Pohn., Roessler & Schreiber in Merxm., Prodr. Fl. SW. Afr. **105**: 4 (1967). — Wilmot-Dear, F.T.E.A. Plumbaginaceae: 6 (1976). Lectotype: Hort. Cliff.: 53 (BM).

Creeping herb, scandent or semi-scandent shrub, often much branched, to 2·5 m. tall. Stems striate, wiry, usually woody. Leaf-blades ovate, ovate-lanceolate, elliptic or oblong, rarely obovate, 2·5–13 cm. long, 1–6 cm. wide, acute, acuminate or mucronate, base cuneate, glabrous, often with white waxy dots on lower surface. Petiole 2–12 mm. long, amplexicaul at base, occasionally auriculate. Flowers sweetly scented. Peduncles bearing prominent sessile glands. Bracts ovate, ovate-lanceolate or lanceolate, 2–8 mm. long, 1·5–2 mm. wide. Calyx 10–12 mm. long, prominently ribbed, bearing abundant stiff-stalked glands. Calyx teeth up to 1·5 mm. long. Corolla white, sometimes with blue flush at base of lobes, drying pale orange; tube 17–26 mm. long, lobes obovate, 6–10 mm. long, 3–5 mm. broad, acute, with shortly excurrent central nerve.

Botswana. N: Savuti Channel, st. 26.x.1972, *Biegel, Pope & Russell* 4070 (SRGH). SE: Mahalapye, Experimental Station, fr. 11.iv.1962, *Yalala* 341 (SRGH). **Zambia.** B: Masese, fr. 15.vi.1960, *Fanshawe* 5736 (K). N: Mpika, south of South Reserve, 600 m., fl. 29.iii.1970, *Astle* 58228 (K; SRGH). W: Ndola, fl. & fr., 22.vii.1954, *Fanshawe* 1395 (K; LISC; SRGH). C: Chakwenga headwaters, 100–130 km. east of Lusaka, fr. 27.iii.1963, *Robinson* 6502 (K; SRGH). E: Petauke, Beit Bridge, Luangwa River, fl. & fr. 18.iv.1952, *White* 2403 (BM). S: between Livingstone and Choma, fl. 10.vii.1930, *Hutchinson & Gillett* 3513 (BM; K; LISC; SRGH). **Zimbabwe.** W: Wankie, Victoria Falls, near bridge, 880 m., fl. 7.iv.1976, *Simpathu* 11 (SRGH). C: Rusape, Dunedin, fl. & fr. 8.ii.1944, *Dayle* 11705 (SRGH). E: Umtali, slope of Mt. Sheni, "The Grove", 1150 m., fl. 5.iii.1950, *Chase* 1989 (BM; K; ISC; SRGH). S: Danga, Sabi River, Chitsas village, fl. & fr. 14.vi.1950, *Chase* 2445 (BM; SRGH). **Malawi.** N: Nkhata Bay, 760 m., fl. & fr. 23.vii.1972. *Pawek* 5532 (K; SRGH). S: Ntcheu Distr., Sharpe Vale, fl. & fr. 19.vii.1958, *Jackson* 2237 (K; SRGH). **Mozambique.** Z: Lugela-Mocuba area, Namagoa Estate, fl. & fr. ix., *Faulkner* 48 (K; SRGH). T: between Marueira and Songo, 3–4 km. from Songo, fl. 24.iii.1972, *Macedo* 5077 (K; SRGH). MS: Manica, near Mavita, fl. & fr. 26.iv.1948, *Barbosa* 1575 (LISC). GI: Gaza, fr. 1917–18, *Junod* 363 (G; LISC). M: c. 4·8 km. north of Maputo (Lourenco Marques), fl. & fr. 29.ii.1949, *Rodin* 4150 (K).

Widespread in tropics and subtropics. Forest, deciduous woodland, scrub and grassland, often by rivers, often on termite mounds.

2. **Plumbago amplexicaulis** Oliv. in Journ. Linn. Soc., Bot. **15**: 96 (1876); in F.T.A. **3**: 487 (1877). — Engl., Pflanzenw. Ost-Afr. **C**: 304 (1895). — Brenan, T.T.C.L.: 452 (1949). — White, F.F.N.R.: 318 (1962). — Wilmot-Dear, F.T.E.A., Plumbaginaceae: 9 (1976). TAB. **32**. Type from Tanzania.

Erect herb or subshrub, much branched, to 90 cm. high. Stems glabrous, striate. Leaves sessile or with short winged petioles, amplexicaul with large auricles. Lower leaf-blades obovate or oblong, 5–24 cm. long, 2–9 cm. wide, upper smaller, often narrowly oblong or lanceolate. Peduncles bearing dense short-stalked glands. Bracts ovate-lanceolate or lanceolate, glandular. Calyx 8–10 mm. long, weakly to prominently ribbed, bearing numerous stalked glands, stalks often curved; calyx teeth up to 1·5 mm. long. Corolla-tube mauve or reddish, 15–27 mm. long; corolla lobes bright deep blue, drying bluish-purple, 8–10 mm. long, 6 mm. broad, acute with shortly excurrent central nerve.

Zambia. N: Mbale (Abercorn) Distr., Kawimbe, 1680, fl. 29.i.1957, *Richards* 8006 (K; SRGH). W: Ndola, road from Kitwe-Ndola road to South Downs, 1260 m., fl. 11.xii.1960,

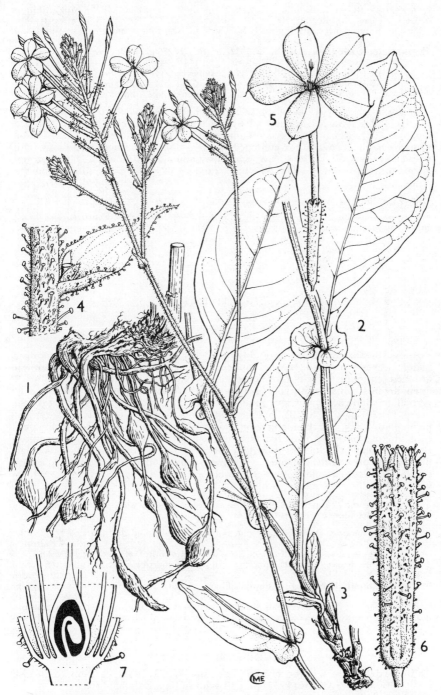

Tab. 32. PLUMBAGO AMPLEXICAULIS. 1, rootstock (× ⅔), from *Polhill & Paulo* 1577; 2, leaf (× ⅔), from *Richards* 7071; 3, flowering shoot (× ⅔); 4, detail of inflorescence-axis and bracts (× 8); 5, flower (× 2); 6, calyx (× 6); 7, longitudinal section of ovary (× 20), 3–7 from *Richards* 11585. From F.T.E.A.

Linley 30 (K; SRGH). C: Katissima Hills, fl. 21.xii. 1907, *Kassner* 2182 (BM). **Mozambique.** N: Cabo Delgado, Montepuez, 500 m., fl. 23.xii.1963, *Torre & Paiva* 9655 (LISC; SRGH). Woodland, scrub or grassland, usually associated with termite mounds. Also in Tanzania.

3. **Plumbago auriculata** Lam., Encycl. Méth., Bot. **2**: 270 (1786). — Merr., Fl. Man.: 361 (1912). — van Steenis in Fl. Males. **1**, 4 (2): 111 (1949). — Dyer in Fl. S. Afr. **26**: 19 (1963). — Wilmot-Dear, F.T.E.A., Plumbaginaceae: 5 (1976). Type from East Indies.
   Plumbago capensis Thunb., Prodr. Pl. Cap.: 33 (1794). — Wright in Fl. Cap. **4**, 1: 424 (1906). — Phillips in Fl. Pl. S. Afr.: 6, t. 222 (1926). Type from South Africa.

Shrub, often semi-scandent, up to 2 m. high. Stems striate, woody. Leaves with short petioles, obovate or elliptic, up to 5 cm. long, 2–2·5 cm. wide, obtuse, mucronate, glabrous, often lepidote on lower surface. Peduncle with dense short white hairs. Bracts ovate or ovate-lanceolate, 5–7 mm. long. Calyx 10–13 mm. long, ribbed, bearing short white hairs, and fairly abundant stalked, mostly recurved, glands on upper half. Corolla pale blue, tube up to 30 mm. long, lobes obovate, obtuse, 10–14 mm. long.

Mozambique. M: Movene, in quaries, fl. 14.iii.1958, *Barbosa & Lemos* 8266 (LISC).

Only one possibly wild specimen seen from our area. Native of South Africa; widely cultivated as an ornamental in tropical, subtropical and temperate regions, including F.Z.-area.

104. PRIMULACEAE
By F. K. Kupicha

Annual or perennial herbs or rarely shrubs; stems erect or prostrate and rooting at nodes. Leaves exstipulate, basal or cauline, alternate, opposite or verticillate, simple or lobed, entire or dentate. Flowers actinomorphic or very rarely zygomorphic, ♀, often heterostylous, solitary or in racemose, spicate, paniculate, umbellate or verticillate inflorescences. Calyx gamosepalous, free or rarely partially adnate to the ovary, (4)5(9)-partite, usually persistent. Corolla gamopetalous, rotate to campanulate with a very short to long tube and 4–9-lobed limb, rarely absent. Stamens equalling corolla-lobes, usually adnate to the corolla, oppositipetalous, sometimes alternating with staminodes. Ovary superior or rarely semi-inferior, unilocular; placentation free-central; ovules 2–3 or more, usually many. Fruit a capsule with valvate or circumscissile dehiscence, rarely indehiscent. Seeds few to numerous, often angular.

A family of some 22 genera and 500 species with almost world-wide distribution but found chiefly in north temperate regions.

1. Ovary semi-inferior; staminodes present; pedicels with bract inserted at or near the middle, thus geniculate - - - - - - - - - - - - - **4. Samolus**
   - Ovary superior; staminodes absent; pedicels with bract (or leaf) at base, not geniculate   2
2. Plants pilose; leaves lobate-dentate; flowers borne in short extra-axillary 1–few-flowered racemes - - - - - - - - - - - - - **1. Ardisiandra**
   - Plants glabrous; leaves entire; flower solitary, axillary, usually arranged in racemes or spikes - - - - - - - - - - - - - - - 3
3. Capsule valvate; robust perennials - - - - - - - **2. Lysimachia**
   - Capsule circumscissile or rarely indehiscent; small, relatively slender annual and perennial herbs - - - - - - - - - - - - - **3. Anagallis**

1. ARDISIANDRA Hook. f.

Ardisiandra Hook. f. in Journ. Linn. Soc., Bot. **7**: 205 (1864). — Weim. in Svensk Bot. Tidskr. **30**: 36 (1936). — P. Taylor in Kew Bull. **13**: 146 (1958).

Pilose creeping herbs with leafy stems. Leaves petiolate, alternate, orbicular, coarsely lobed and toothed. Racemes few-flowered, extra-axillary, short-peduncled; flowers pedicellate and bracteate, inconspicuous. Calyx segments 5, free or somewhat accrescent, persistent. Corolla campanulate, with 5 lobes ⅓–½ the length of the corolla. Stamens 5, oppositipetalous; filaments short, connate below into a continuous ring fused at the base to the corolla-tube; anthers apiculate. Ovary multiovulate. Capsule globose, dehiscent by 5 apical valves or indehiscent, many-seeded. Seeds angular, blackish, papillose.

A genus of 3 species endemic to the mountains of tropical Africa.

Fruit indehiscent, the pericarp uniformly translucent; inflorescence axes relatively long, with peduncles 0–8 mm. and pedicels 15–20 mm. long; corolla 6–8 mm. long; calyx about half as long as corolla at anthesis - - - - - - - - - - 1. *wettsteinii*
Fruit dehiscent, with pericarp translucent below, creamy-white opaque above, opening at apex by 5 triangular cartilaginous teeth; inflorescence axes relatively short, with peduncles 0–3 mm. and pedicels 5–8 mm. long; corolla 3–5 mm. long; calyx ± equalling corolla at anthesis - - - - - - - - - - - 2. *sibthorpioides*

1. **Ardisiandra wettsteinii** R. Wagner in Anzeiger Akad. Wiss. Wien **69**: 185 (1932). — P. Taylor in Kew Bull. **13**: 147, fig. 4, 1–11 (1958); in F.T.E.A., Primulaceae: 4, fig. 1, 1–11 (1958). — Binns, H.C.L.M.: 89 (1968). — Boutique in Fl. Afr. Centr., Primulaceae: 3, fig. 1c (1971). TAB. **33**. Type from Tanzania.
  Ardisiandra orientalis Weim. in Svensk Bot. Tidskr. **30**: 41, fig. 1a, 2 (1936). Syntypes from Kenya and Tanzania.
  Ardisiandra stolzii Weim., tom. cit.: 44, fig. 1e, 2 (1936). Type from Tanzania.

Creeping perennial herb. All vegetative parts pilose; hairs long, weak, multicellular and uniseriate, the cross-walls between the cells conspicuously reddish. Stems up to 1 m. long or more, reddish, partly prostrate and rooting at and between nodes, partly ascending to erect and very leafy. Petioles 1–3(4) cm. long, slender. Leaf-lamina 1–3(4) cm. in diameter, broadly ovate or orbicular, cordate at base, crenately 5–9-lobed, the lobes coarsely toothed. Racemes 1–4(6)-flowered, flowers drooping; peduncles up to 8 mm. long, pedicels 15–20 mm. long, each with a subtending lanceolate bract c. 3 mm. long at or near the base. Calyx-segments c. ½ as long as the corolla, ovate, tapering at apex, the margins overlapping at the base, pilose, ciliate. Corolla 6–8 mm. long, campanulate, white to pale pink or lilac, glabrous or sparsely pilose, divided to the middle into oblong-cuspidate lobes. Stamens c. 3 mm. long; anthers c. 1·5 mm. long. Ovary hemispherical, densely and stiffly pilose. Capsule c. 4 mm. in diameter, indehiscent, transparent-walled. Seeds 0·8–1 mm. long, numerous, ± tetrahedral, blackish, granular.

Zambia. E: Nyika, fl. & fr. 30.xii.1962, *Fanshawe* 7324 (K). **Zimbabwe**. E: Chimanimani Mts., Mt. Peza, 1980 m., 15.x.1950, *Wild* 3627 (K; LISC; SRGH). **Malawi**. N: Nyika Plateau, Lake Kauline, fl. & fr. 23.x.1958, *Robson & Angus* 276 (K). S: Zomba Distr., Zomba Plateau, 1720 m., fr. 21.viii.1972, *Brummitt* 12938 (K). **Mozambique**. MS: Manica, Serra Zuira, Tsetsera Plateau, c. 2000 m., 10.x.1965, *Torre & Correia* 12846 (K; LISC; LMU; SRGH).
  Also known from Uganda, Rwanda, Burundi, Kenya, Tanzania and Zaire. In deep shade on humus-rich substrates, in evergreen forest.

2. **Ardisiandra sibthorpioides** Hook. f. in Journ. Linn. Soc., Bot. **7**: 205, t. 1 (1864). — Weim. in Svensk Bot. Tidskr. **30**: 39, fig. 1b, 2 (1936). — P. Taylor in Kew Bull. **13**: 147, fig. 4, 12–13 (1958); in F.T.E.A., Primulaceae: 3, fig. 1, 12–13 (1958). — Boutique in Fl. Afr. Centr., Primulaceae: 2, fig. 1a & b (1971). TAB. **33**. Type from Fernando Po.
  Ardisiandra engleri Weim. in tom. cit.: 40, fig. 1d, 2 (1936). Type from Tanzania.

Creeping perennial herb. All vegetative parts pilose; hairs as in *A. wettsteinii*. Stems up to 1 m. long or more, reddish, partly prostrate and rooting at and between nodes, partly ascending to erect and very leafy. Petioles (0·5)1·5–3(4) cm. long, slender. Leaf-lamina up to 4 cm. in diameter, broadly ovate or orbicular, cordate at base, crenately 5–9-lobed, the lobes coarsely toothed. Racemes 1–3(5)-flowered, flowers drooping; peduncles 0–3 mm. long; pedicels 5–8 mm. long, each with a subtending lanceolate bract c. 2 mm. long. Calyx-segments ± equalling corolla, ovate-lanceolate, the margins overlapping at the base, pilose, ciliate. Corolla 3–5 mm. long, campanulate, white, glabrous, deeply divided into rounded-triangular lobes. Stamens c. 1·7 mm. long; anthers c. 1·2 mm. long. Ovary hemispherical, pilose. Capsule c. 3 mm. in diameter, globose, transparent-walled at the base, opaque, creamy-white and cartilaginous at the apex, here dehiscing into 5 recurving teeth. Seeds 0·8–1 mm. long, numerous, ± tetrahedral, blackish, granular.

Malawi. S: Mulanje Distr., Mulanje Mt., foot of Great Ruo Gorge, 1060 m., fl. & fr. 18.iii.1970, *Brummitt & Banda* 9212 (K; LISC; MAL; SRGH).

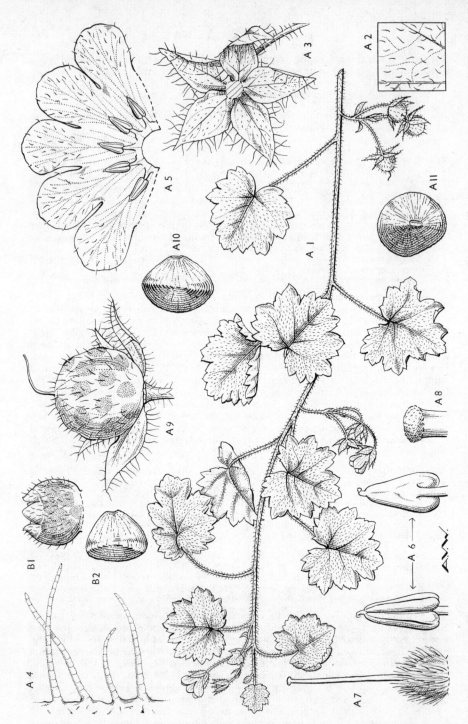

Tab. 33. ARDISIANDRA WETTSTEINII. A1, part of plant (× 1); A2, surface of leaf (× 4); A3, calyx
(× 5); A4, cilia on margin of calyx-lobes (× 30); A5, corolla, opened out (× 5); A6, anther,
dorsal and ventral view (× 15); A7, pistil (× 10); A8, stigma (× 40); A9, fruit (× 5); A10,
seed, lateral view (× 20); A11, seed, basal view (× 20), A1–A11 from *Drummond & Hemsley*
4284. ARDISIANDRA SIBTHORPIOIDES. B1, fruit (× 5); B2, seed, lateral view (× 20), B1–B2
from *Purseglove* 3746. From F.T.E.A.

Distributed from Fernando Po and Cameroon through Zaire to Uganda, Kenya, Sudan, Ethiopia and Tanzania. In evergreen forest in shade under boulders.

The specimen cited is the only collection known from the F.Z. area, and I am grateful to Dr. Brummitt for bringing it to my attention.

2. LYSIMACHIA L.

Lysimachia L., Sp. Pl. **1**: 147 (1753).

Perennial herbs or very rarely shrubs, prostrate or erect. Leaves alternate, opposite or verticillate, entire. Flowers axillary or in spike-like racemes or panicles. Calyx more or less deeply 5–6-lobed. Corolla rotate or campanulate, white or yellow or rarely (as in F.Z. species) pink to purple, 5–6-lobed, the segments contorted, entire or dentate. Stamens equalling petals, epipetalous and oppositipetalous, often alternating with staminodes. Ovary superior, ovoid or globose; style slender, with terminal stigma. Capsule usually dehiscing at apex by 5 valves, few–many-seeded. Seeds various, usually angular.

A widespread genus of over 100 species with its main centre of distribution in E. Asia. The tropical African members comprise a small group of closely related species.

Lysimachia ruhmeriana Vatke in Linnaea **40**: 204 (1876). — Oliver in F.T.A. **3**: 489 (1877). — Knuth in Engl., Pflanzenr. **IV**, 237: 292 (1905). — P. Taylor in Kew Bull. **13**: 142 (1958); in F.T.E.A., Primulaceae: 5, fig. 2 (1958). — Dyer in Fl. Southern Afr. **26**: 13 (1963). — Binns, H.C.L.M.: 89 (1968). — Bizzarri in Webbia **24**: 640, fig. 1 (1970). — Boutique in Fl. Afr. Centr., Primulaceae: 5, t. 1 (1971). TAB. **34**. Type from Ethiopia.
  Lysimachia parviflora Baker in Journ. Linn. Soc., Bot. **20**: 196 (1883). — Knuth, loc. cit. Syntypes from Madagascar.
  Lysimachia africana Engl., Pflanzenw. Ost-Afr. **C**: 304 (1895). — Bizzarri, tom. cit.: 644, fig. 3 (1970). Type from Tanzania.
  Lysimachia saganeitensis Knuth, loc. cit. — Bizzarri, tom. cit.: 642, fig. 2 (1970). Type from Ethiopia.
  Lysimachia woodii Knuth, loc. cit. Type from S. Africa (Natal).

Robust perennial herb up to 1 m. tall. Stems erect or decumbent, terete, reddish, leafy. Leaves up to 6 × 1·5 cm. (in F.Z. area), broadly to narrowly elliptic or lanceolate, ± epetiolate, often shortly acuminate at the apex, cuneate, obtuse or cordate at the base; lamina surfaces and margin almost always dotted with irregular blackish glands. Racemes up to 30 cm. long, spicate, lax or congested, terminal and axillary. Pedicels up to 5 mm. long (in fruit), each subtended by a bract 2–3 mm. long. Flowers covered with short-stalked glands on all parts, especially on inner surface of calyx and corolla and on stamens. Calyx divided almost to the base; lobes c. 2·5 mm. long, oblong-elliptic. Corolla c. 3·8 mm. long, campanulate, white, pink or mauve; lobes equalling or slightly longer than calyx, elliptic. Stamens c. 2 mm. long, inserted in a ring below the corolla lobes; connective produced above the anther into a fleshy apiculus. Ovary c. 1 mm. long, globose, with style c. 1 mm. long. Capsule 3–4(5) mm. in diameter, globose, dehiscing by 5 valves, these remaining erect, not recurving. Seeds 0·8–1 mm. long, ± tetrahedral, blackish, smooth or minutely granular.

Zambia. E: Nyika, fl. 30.xii.1962, *Fanshawe* 7335 (K; SRGH). **Zimbabwe.** E: Inyanga Distr., 11 km. beyond Pungwe at edge of Matendirere R., 1680 m., fl. & fr. 26.i.1951, *Chase* 3559 (BM; K). **Malawi.** N: Chitipa Distr., Misuku Hills, 1520 m., fl. & fr. 12.i.1959, *Robinson* 3177 (K; PRE; SRGH). C: Dedza Distr., near Bembeke Parish, fr. 15.xi.1967, *Jeke* 130 (MAL). **Mozambique.** MS: Tsetsera, 2130 m., fl. & fr. 4.iii.1954, *Wild* 4487 (K; LISC; SRGH).
  Distributed throughout tropical Africa from Ethiopia southwards to S. Africa, and in Cameroon and Madagascar. On damp soil in swamps, in woods and by rivers.

I agree with P. Taylor (op. cit.) that *L. ruhmeriana* should be regarded as a widespread, polymorphic species, not subdivided into 5 species as advocated by Knuth and Bizzarri. The type of *L. ruhmeriana* has semi-amplexicaul leaves, a congested inflorescence and relatively large (5 mm. long) capsules and so is rather different in appearance from the representatives in the F.Z. area, which tend to have leaves tapering at the base, lax spikes and smaller capsules. In Bizzarri's classification the latter all belong to *L. africana*. However, these characters appear to be uncorrelated, varying independently throughout Africa.

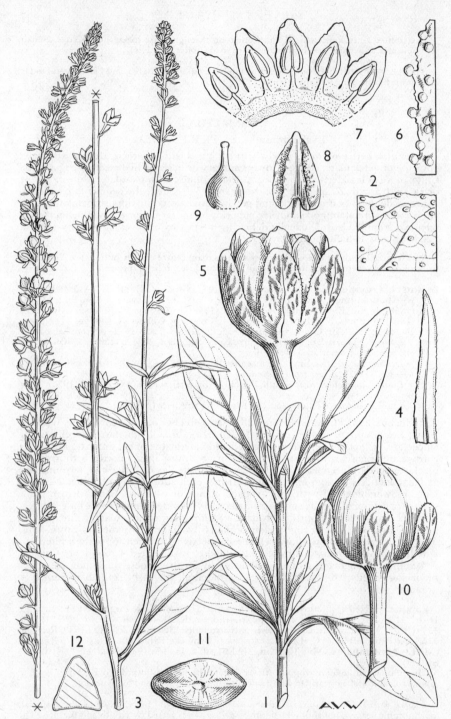

Tab. 34. LYSIMACHIA RUHMERIANA. 1, leaves and part of lower stem (× 1); 2, surface of leaf (× 5);
3, part of inflorescence (× 1); 4, bract (× 8); 5, flower (× 8); 6, margin of calyx-lobe (× 80);
7, corolla opened out (× 10); 8, anther (× 20); 9, pistil (× 10); 10, fruit (× 8); 11, seed, basal
view (× 30); 12, vertical section (inverted) across the seed through the hilum (× 30), all
from *Conrads* 5407. From F.T.E.A.

3. ANAGALLIS L.

Anagallis L., Sp. Pl. **1**: 148 (1753). — P. Taylor in Kew Bull. **10**: 321 (1956); op. cit. **13**: 133 (1958).

Glabrous annual and perennial herbs; stems erect or prostrate and rooting at the nodes. Leaves sessile or subpetiolate, alternate or opposite, elliptic to suborbicular or rarely capillary, entire. Flowers small, 5(6)-merous, pedicellate or rarely sessile, solitary in leaf axils or in terminal racemes. Calyx gamosepalous, divided almost to the base; segments often with hyaline margins. Corolla subcampanulate; lobes shortly connate at the base, usually white or pale pink, rarely (in F.Z. area) bright blue or red. Stamens 5, oppositipetalous and epipetalous, adnate to the base of the corolla; filaments free or connate into a short tube, often bearded with multicellular hairs; anthers basifixed, introrse, ± square to oblong. Ovary globose, 5-carpelled, unilocular, with terminal simple terete style and capitate stigma; placentation free-central. Fruit globose or obovoid, circumscissile or indehiscent, 1- or many-seeded. Seeds ellipsoid, 3-angled, dark brown, papillose.

A genus of c. 30 species centred in tropical Africa but extending to Europe, especially the Mediterranean area, India, Malaysia and Australia, and found also in N. and S. America.

1. Plants of dry habitats; leaves opposite; flowers bright blue or red; petals ± as broad as long, many nerved, fringed with short-stalked glands (Subgen. *Anagallis*)   -   - 1. *arvensis*
- Plants of wet habitats; leaves alternate or less often opposite or subopposite; flowers white or pale pink; petals usually much longer than broad, 3–5-nerved, not fringed with glands - - - - - - - - - - - - - - - 2
2. Plants prostrate, rooting at nodes; flowering shoots always leafy at apex, the flowers solitary and axillary (Subgen. *Jiresekia*) - - - - - - - - - - 3
- Plants erect or ascending, not rooting at nodes, sometimes aquatic; inflorescence terminal, a simple or branched raceme (Subgen. *Centunculus*) - - - - - 5
3. Capsule indehiscent, containing 1 seed; calyx-lobes c. 1 mm. long -   -   2. *oligantha*
- Capsule circumscissile, many-seeded; calyx-lobes 2 mm. long or more - - - 4
4. Filaments glabrous, not connate above point of insertion on corolla; leaves broadly ovate to elliptic - - - - - - - - - - - - - 3. *serpens*
- Filaments glandular-hairy below, connate to a variable extent above point of insertion on corolla; leaves narrowly to broadly obovate - - - - - - 4. *gracilipes*
5. *Leaves capillary or subulate to narrowly obovate, at least 3 times and usually more than 10 times as long as broad - - - - - - - - - - - 6
- *Leaves lanceolate to ovate or elliptic to spathulate, less than 3 times and usually less than twice as long as broad - - - - - - - - - - - - 9
6. Submerged aquatic, with capillary leaves 2–5 cm. long - - - - - 5. *kochii*
- Plants not aquatic, although sometimes growing on seasonally flooded ground; leaves not capillary, 1 cm. long or less - - - - - - - - - - - - 7
7. Plants (4) 15–25 cm. tall; leaves subulate to narrowly obovate, 5–10 mm. long; corolla-lobes 4–4·5 mm. long; anthers 0·3–0·5 mm. long - - - - - - 6. *elegantula*
- Plants 4–6(10) cm. tall; leaves subulate, up to 7 mm. long; corolla-lobes c. 3 mm. long; anthers 0·2–0·25 mm. long - - - - - - - - - - 8
8. Flowers pink; filaments glabrous - - - - - - - 8. *rhodesiaca*
- Flowers white; filaments bearded - - - - - - - 9. *acuminata*
9. Plants usually decumbent, stems up to 40 cm. long; corolla 4–5 mm. long; anthers 0·4–0·7 mm. long - - - - - - - - - - - 7. *tenuicaulis*
- Plants usually erect, stems up to 20 cm. tall; corolla less than 4 mm. long; anthers 0·25 mm. long or less - - - - - - - - - - - - - 10
10. Filaments adnate to the corolla for ⅓–⅙ of their length, bearded in the lower half

10. *barbata*
- Filaments adnate to the corolla for ¼–½ their length, glabrous - - - - 11
11. Leaves sessile, elliptic, 2–4 mm. long; seeds 0·2–0·3 mm. long - - 11. *pumila*
- Leaves petiolate, broadly spathulate, 5–13 mm. long; seeds 0·4–0·65 mm. long

12. *djalonis*

The species of *Anagallis* which occur in the F.Z. area include representatives of all three subgenera recognised by P. Taylor (in Kew Bull. **10**: 322, 1955), although no. 1, *A. arvensis*, is introduced. The members of Subgen. *Jiresekia* (Schmidt) P. Taylor (nos. 2–4) are creeping plants restricted to high altitudes. The majority of F.Z. species of *Anagallis* belong to Subgen. *Centunculus* (L.) P. Taylor, and these form a group of closely related species distinguished by small but apparently quite constant features. Many of these species are known from comparatively few and scattered localities, and more collections are needed to determine their actual distributions.

*This refers to the broadest leaves present on the plant.

1. **Anagallis arvensis** L., Sp. Pl. **1**: 148 (1753). — Knuth in Engl., Pflanzenr. **IV**, 237: 322 (1905). — P. Taylor in Kew Bull. **10**: 329 (1956); op. cit. **13**: 134 (1958); in F.T.E.A., Primulaceae: 10 (1958). Dyer in Fl. Southern Afr. **26**: 14 (1963). — Bizzarri in Webbia **24**: 652, fig. 5 (1970). Described from Europe.

Tufted annual herb with ascending, branching stems up to 50 cm. long. Leaves 6–20 mm. long, opposite and decussate or rarely whorled, sessile, elliptic, ovate or obovate, acute at apex. Flowers borne in axils of all but lowermost leaves; pedicels at first short and erect, becoming long and recurved in fruit. Calyx-segments c. 4–5 mm. long, lanceolate. Petals ± equalling calyx, broadly obovate, connate at the very base, fringed with short-stalked glands, deep blue or red. Stamens c. ⅔ as long as petals, adnate to base of corolla; filaments free, blue or red, bearded; anthers c. 0·6 mm. long, oblong. Capsule 3–5 mm. in diameter, circumscissile, many-seeded; seeds c, 1·2 mm. long.

Zimbabwe. C: Salisbury Distr., Meyrick Park, fl. & fr. iv.1963, *Grosvenor* 24 (SRGH). E: Umtali Distr., La Rochelle, Imbeza Valley, 1220 m., fr. 19.x.1953, *Chase* 5145 (BM; K; SRGH).
Native in the Mediterranean area and W. Europe; introduced as a weed throughout temperate regions, and found at higher altitudes in the tropics.

2. **Anagallis oligantha** P. Taylor in Kew Bull. **13**: 137, fig. 2, 9–12 (1958). — Binns, H.C.L.M.: 89 (1968). Type: Malawi, Mt. Mulanje, L. Ruo Plateau, 1770 m., 16.viii.1956, *Newman & Whitmore* 454 (BM, holotype).
   Anagallis sp. — P. Taylor, op. cit. **10**: 349 (1956).

Creeping perennial herb. Stems prostrate, rooting at nodes, simple or sparingly branched. Leaves 2·5–5 mm. long, alternate, suborbicular, acute at apex, with petiole 0·5–1 mm. long. Flowers lateral, shortly pedicelled. Calyx-lobes c. 1 mm. long, ovate-acuminate. Corolla white or pale pink; tube 0·6–0·8 mm. long, lobes 2–3 mm. long, elliptic, obtuse or subacute at apex. Stamens c. ⅔ as long as corolla, adnate to the corolla for c. 0·4 mm.; filaments connate into a tube c. 0·4 mm. high, glabrous; anthers c. 0·25 mm. long. Ovary c. 0·4 mm. in diameter with filiform style c. 1·3 mm. long; ovules 2–3. Fruit (immature) c. 1 mm. in diameter, 1-seeded, indehiscent. Mature seeds unknown.

Malawi. S: Mulanje Distr., base of West Peak, Mt. Mulanje, 2010 m., st. 11.vi.1962, *Richards* 16641 (K).
Known only from the Mt. Mulanje area. On peat, near streams.

3. **Anagallis serpens** Hochst. ex DC., Prodr. **8**: 668 (1844). — Knuth in Engl., Pflanzenr. **IV**, 237: 326 (1905). — P. Taylor in Kew Bull. **10**: 332 (1956). — Bizzarri in Webbia **24**: 661, fig. 6 (1970). Type from Ethiopia.
Subsp. **serpens**
   Lysimachia quartiniana A. Rich., Tent. Fl. Abyss. **2**: 16 (1851). — Oliver in F.T.A. **3**: 489 (1877). — Knuth, loc. cit. Type from Ethiopia.

Creeping perennial (or annual) herb. Stems prostrate, rooting at nodes, sparingly branched. Leaves 6–9 mm. long, opposite or alternate, broadly ovate to elliptic, usually acute at the apex, tapering abruptly at the base into a petiole up to 3 mm. long. Flowers lateral, with pedicels 3–23 mm. long, these erect at first but often recurving in fruit. Calyx-lobes 2–2·5 mm. long, broadly lanceolate. Corolla white or pale pink; lobes 5–6·5 mm. long, elliptic, obtuse. Stamens ½–⅔ as long as petals, the filaments glabrous, slender above but widening towards the base, not connate into a tube, adnate to the corolla for c. 0·5 mm.; anthers c. 1 mm. long. Capsule c. 2·5 mm. in diameter, globose, circumscissile, many-seeded. Mature seeds unknown.

Zimbabwe. E: Inyanga, Gairezi Ranch on border, 10 km. N. of Troutbeck, 1830 m., fl. 16.ix.1956, *Robinson* 1946 (K; PRE; SRGH).
Also found in Ethiopia and the Sudan. In bogs, on stream banks, in shady kloofs.

Note: *A. serpens* subsp. *meyeri-johannis* (Engl.) P. Taylor, a very variable taxon which differs from the type subspecies in having alternate leaves which are usually proportionately narrower, occurs in Uganda, Kenya and Tanzania; subsp. *serpens* thus has an unusual, disjunct distribution (see P. Taylor, loc. cit. 1956).

4. **Anagallis gracilipes** P. Taylor in Kew Bull. **13**: 135, fig. 1, 1–5 (1958). Type: Zimbabwe, Umtali Distr., Engwa, 1980 m., 8.ii.1955, *E. M. & W.* 304 (BM, holotype; SRGH).

Creeping perennial herb. Stems prostrate, rooting at the nodes, sparingly branched. Leaves 4–8 mm. long, alternate, narrowly to broadly obovate, subsessile to shortly petiolate. Flowers lateral, on pedicels 4–25 mm. long. Calyx 2·5–3·5 mm. long, lobes narrowly lanceolate. Corolla pink; lobes 5–7 mm. long, elliptic. Stamens c. ⅔ as long as petals, shortly adnate to corolla; filaments connate into a tube of very variable length, hairy at the base. Capsule c. 3 mm. in diameter, spherical, many-seeded, circumscissile. Seeds 0·9–1 mm. long.

Zimbabwe. E: Inyanga Distr., Mtendere bed, fl. 28.xii.1965, *West* 7124 (K; SRGH). **Mozambique.** MS: Sofala Prov., Gorongosa Mt., Gogogo Summit Area, fl. i.1972, *Tinley* 2322 (SRGH).
Known only from the F.Z. area. On very wet soil or in shallow water.

5. **Anagallis kochii** Hess in Ber. Schweiz. Bot. Ges. **63**: 213, t. 2 & 3 (1953). — P. Taylor in Kew Bull. **10**: 340 (1956). — Boutique in Fl. Afr. Centr., Primulaceae: 14 (1971). TAB. **35**. Type from Angola.
   Anagallis hurneri Hess, op. cit. **67**: 80, fig. 1 (1957). — P. Taylor, op. cit. **13**: 139 (1958). Type from Zaire.

Aquatic ?annual. Stems up to 45 cm. or more, fleshy, simple, the lower leafy part submerged, the inflorescence held erect above the water surface. Leaves 2–5 cm. long capillary; insertion very irregular, leaves sometimes in groups of 2 or 3. Inflorescence a terminal raceme. Pedicels up to 3·5 cm. long, each subtended by a subulate bract; bracts c. 14 mm. long at base of raceme, decreasing to c. 2 mm. at apex. Calyx-lobes c. 2 mm. long, lanceolate. Corolla lobes 4–5 mm. long, white marked with fine mauve lines, elliptic, obtuse at apex. Stamens ½–⅔ as long as petals, filaments adnate to corolla at base, otherwise free, conspicuously bearded in the lower half; anthers c. 0·7 mm. long. Capsule c. 3 mm. long, globose, many-seeded, circumscissile. Mature seeds 0·5 mm. long.

Zambia. W: Mwinilunga Distr., Kalenda Dambo, Matonchi, 1400 m., fl. & fr. 16.iv.1960, *Robinson* 3585 (K). C: Mkushi Distr., Matuku R. source, 1310 m., fl. & fr. 1.iv.1963, *Vesey-FitzGerald* 4068 (K).
Known also from Angola and Zaire. In flooded pans or dambos.

When *A. kochii* and *A. hurneri* were first described, the taxon was known only from these two type collections which, though identical in vegetative characters, differed considerably in flower size and petal shape: the type of *A. kochii* has corolla up to 15 mm. in diameter, with obtuse lobes, that of *A. hurneri* has corolla c. 5 mm. in diameter, with acute lobes. Since then, however, many additional collections have widened the known range of *A. kochii* and they show that it is a very variable species including specimens of the "*A. hurneri*" extreme. The above description of *A. kochii* is based on Zambian specimens.

6. **Anagallis elegantula** P. Taylor in Kew Bull. **10**: 341 (1956). — Boutique in Fl. Afr. Centr., Primulaceae: 13 (1971). Type from Angola.
   Anagallis pulchella Welw. ex Schinz in Bull. Herb. Boiss. **2**: 221 (1894). — Hiern, Cat. Afr. Pl. Welw. **3**: 635 (1898). — Knuth in Engl., Pflanzenr. **IV**, 237: 333 (1905), non Salisb. (1796). Syntypes from Angola.

Erect perennial herb. Stems 6–22(30) cm. tall, simple or very sparsely branched, sometimes pink below. Leaves 7–10 mm. long, alternate or subopposite, varying from subulate to narrowly obovate with mucronate apex, sessile or very shortly petiolate. Flowers on pedicels up to 8 mm. long, arranged in simple or branched terminal racemes; inflorescence open and extending for much of height of plant or forming a dense pyramid at the stem apex. Calyx-lobes c. 3·5 mm. long, lanceolate-acuminate. Corolla white or pink; lobes 4–4·5 mm. long, narrowly oblong with tridentate apex. Stamens c. ⅔ as long as petals, the filaments white or crimson, slender above, expanded below, here very densely bearded, not connate, adnate to the corolla for c. 0·7 mm.; anthers 0·3–0·5 mm. long, greenish. Capsule c. 1·5 mm. in diameter, circumscissile, many-seeded. Seeds 0·4–0·5 mm. long.

Zambia. B: Kale, Lwingishi Dambo, 1190 m., 1.i.1965, *Symoens* 11326 (K). N: 72 km. S. of Mbala on Kasama road, fl. & fr. 30.iii.1955, *E. M. & W.* 1343 (BM; SRGH). W: Mwinilunga Distr., Sinkabolo Swamp, 1200 m., fl. 20.xi.1962, *Richards* 17425 (K; PRE; SRGH). C: Mkushi Distr., Fiwila, Mkushi Dambo, 1220 m., fl. & fr. 6.i.1958, *Robinson* 2651 (K; SRGH).
Known also from Angola and Zaire. Among grass in boggy places.

See note under 7. *A. tenuicaulis*.

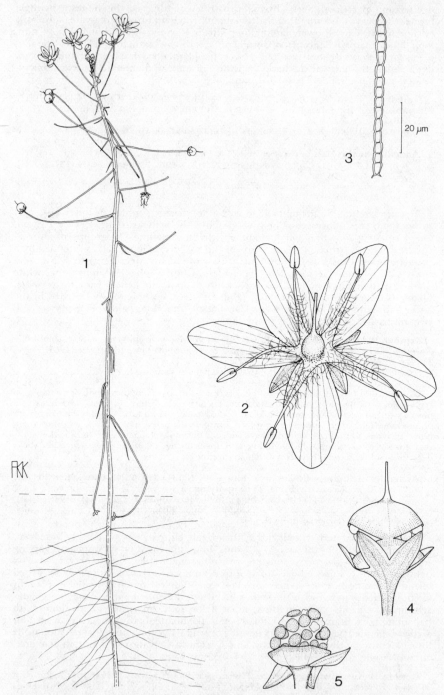

Tab. 35. ANAGALLIS KOCHII. 1, habit, lower ⅓ of plant submerged (×⅔); 2, flower (×6); 3, multicellular hair from filament; 4, dehiscing capsule (×6); 5, fruit after dehiscence showing seeds embedded in the free-central placenta (×6). All from *Vesey-FitzGerald* 4068.

7. **Anagallis tenuicaulis** Baker in Journ. Bot., Lond. **20**: 172 (1883). — P. Taylor in Kew Bull. **10**: 342, fig. 5, 14 & 15 (1956); op. cit. **13**: 140 (1958); in F.T.E.A., Primulaceae: 16 (1958). — Dyer in Fl. Southern Afr. **26**: 15 (1963). — Bizzarri in Webbia **24**: 667, fig. 8 (1970). — Boutique in Fl. Afr. Centr., Primulaceae: 10 (1971). TAB. **36**. Type from Madagascar.

Anagallis hanningtoniana Baker in Kew Bull. **1901**: 127 (1901). — S. Moore in Journ. Linn. Soc., Bot. **40**: 130 (1911). Syntypes from Tanzania and Zambia: Mbala Distr., Fwambo, *Carson* 45 & 72 (K, syntypes).

Weakly erect or straggling perennial herb. Stems up to 35 cm. long or more, branched or rarely simple, with narrowly winged ribs. Leaves 5–10 mm. long, opposite or alternate, broadly ovate to ovate, the apex acute or mucronate, subsessile or shortly petiolate. Flowers in short or long racemes or occasionally solitary and axillary, on pedicels 6–17 mm. long. Calyx-lobes 2·5–4 mm. long, lanceolate-acuminate. Corolla white; lobes 3·5–5 mm. long, narrowly elliptic with apex truncate, mucronate or tridentate. Stamens c. ⅔ as long as petals, the filaments slender above, expanded below, here densely bearded, briefly connate, adnate to the corolla for c. 0·5 mm., lower part of corolla tube sometimes with conspicuous red granules; anthers 0·4–0·7 mm. long, oblong. Capsule c. 1·5 mm. in diameter, many-seeded, circumscissile. Seeds 0·5–0·65 mm. long.

Zambia. N: Mbala Distr., Kali Dambo, 1580 m., fl. & fr. 2.i.1952, *Richards* 526 (K). **Zimbabwe.** E: Mutare Distr., Vumba, Burma Valley, Bomponi, 1220 m., fl. 4.xii.1961, *Wild & Chase* 5541 (BM; COI; K; SRGH). **Malawi.** N: Luwawa Dam, Viphya, 120 km. S. of Mzuzu, 1620 m., fl. 8.ii.1971, *Pawek* 4402 (K; MAL).

Found also in Ethiopia, Zaire, Cameroon, Angola, Kenya, Uganda, Tanzania, S. Africa (Natal) and Madagascar. In bogs and swamps.

A. tenuicaulis can be confused with *A. elegantula* (no. 6), in which the occasional specimen has basal leaves that are broadly obovate and less than 3 times as long as broad. The two species are similar in floral details, although *A. tenuicaulis* has slightly larger anthers and seeds. Their habit is different, however: in *A. tenuicaulis* the stems are long and trailing and usually branched, with rather widely spaced nodes; the inflorescence is lax and long, and the flowers long-pedicelled; in *A. elegantula* the stems are more stiffly erect, unbranched, with closer nodes, and the leaves often borne erect and appressed to the stem; the inflorescence is often in the form of a dense apical cylinder or pyramid, the flowers with relatively short pedicels.

8. **Anagallis rhodesiaca** R. E. Fries, Wiss. Ergebn. Schwed. Rhod.-Kongo-Exped. **1**: 253 (1916). — P. Taylor in Kew Bull. **10**: 341 (1956); op cit. **13**: 140 (1958); in F.T.E.A., Primulaceae: 15 (1958). Type: Zambia, Kali between Mansa (Fort Rosebery) and Bangweulu, 17.ix.1911, *Fries* 634 (K; UPS, holotype).

Erect annual herb 6–20 cm. tall. Stems reddish, simple or branched above. Leaves up to 7 mm. long, alternate, filiform to subulate. Flowers in terminal racemes, on pedicels up to 10 mm. long. Calyx-lobes 2–2·5 mm. long, lanceolate. Corolla c. 3 mm. long, deep pink, the lobes oblong-elliptic with rounded or mucronate apex. Stamens c. ½ as long as petals; filaments slender above, expanded below, not connate, adnate to corolla for 0·5 mm., glabrous; lower part of corolla-tube usually with conspicuous red granules; anthers 0·2 mm. long. Fruit c. 1·5 mm. in diameter, many-seeded, circumscissile. Seed 0·3–0·4 mm. long.

Zambia. N: Kali, between Mansa (Fort Rosebery) and Bangweulu, fl. 17.ix.1911, *Fries* 634 (K; UPS).

Known also from southernmost Tanzania. In seasonally flooded places.

Closely related to no. 9, *A. acuminata*.

9. **Anagallis acuminata** Welw. ex Schinz in Bull. Herb. Boiss. **2**: 221 (1894). — Hiern, Cat. Afr. Pl. Welw. **3**: 636 (1898). — Knuth in Engl., Pflanzenr. **IV**, 237: 333 (1905). — P. Taylor in Kew Bull. **10**: 341 (1956); op. cit. **13**: 140 (1958); in F.T.E.A., Primulaceae: 15 (1958). — Boutique in Fl. Afr. Centr., Primulaceae: 14 (1971). Type from Angola.

Erect annual herb 3–9 cm. tall. Stems reddish, simple or branched. Leaves up to 7 cm. long, opposite below, alternate above, subulate. Flowers present at most nodes, on pedicels 4–8 mm. long. Calyx-lobes 2·2–2·5 mm. long, lanceolate. Corolla white or pale cream; tube c. 0·5 mm. long, lobes c. 2·5 mm. long, elliptic, the apex acuminate or tridentate. Stamens c. ⅔ as long as the petals; filaments slender above, expanded below, not connate, adnate to corolla for 0·5 mm., bearded in lower half; lower part of corolla-tube usually with conspicuous red granules; anthers 0·2–0·25

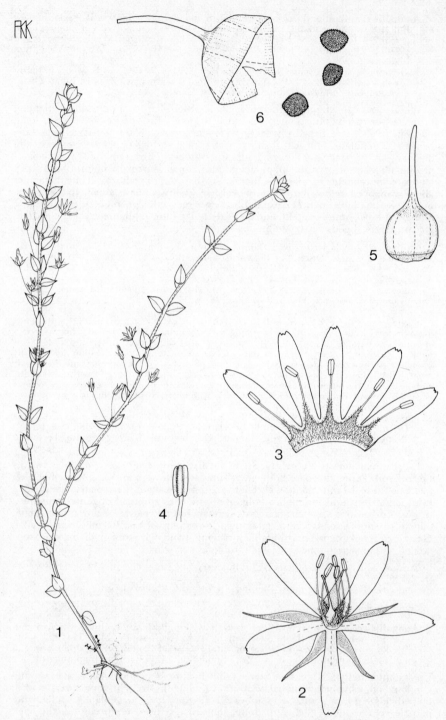

Tab. 36. ANAGALLISTENUICAULIS. 1, habit (× ⅔) *Pawek* 4402; 2, flower (× 6); 3, corolla opened out (× 6); 4, anther (× 12); 5, fruit (× 12); 6, separated cap of fruit with 3 seeds (× 12), 2–6 from *Bullock* 2116.

mm. long. Fruit c. 1·5 mm. in diameter, many-seeded, circumscissile. Seeds c. 0·3 mm. long.

Zambia. N: near Kasama, fl. & fr. 31.iii.1955, *E. M. & W.* 1385 (BM; SRGH). W: Mwinilunga Distr., fl. & fr. 16.v.1969, *Mutimushi* 3396 (K; SRGH). **Zimbabwe.** C: Salisbury Distr., Chindamora Reserve, Ngomokurira, 1680 m., fl. & fr. 25.iii.1952, *Wild* 3784 (K; SRGH).

Also known from Angola, Zaire and Tanzania. On wet peaty soil, sometimes partly submerged.

10. **Anagallis barbata** (P. Taylor) Kupicha, stat. nov. Type from Tanzania.
   Anagallis pumila Swartz var. *barbata* P. Taylor in Kew Bull. **10**: 345, fig. 5, 1–3 (1956); op. cit. **13**: 141 (1958); in F.T.E.A., Primulaceae: 17 (1958). — Binns, H.C.L.M.: 89 (1968). — Boutique in Fl. Afr. Centr., Primulaceae: 12 (1971). Type as above.

Erect or ascending annual herb 4–20 cm. tall. Stems reddish, simple or branched. Leaves 3–9 mm. long, alternate or subopposite, lanceolate to ovate, sessile. Flowers borne at most nodes or the upper ones, on pedicels 3–5 mm. long. Calyx-lobes 1·5–2·5 mm. long, ovate-acuminate. Corolla white or pale pink, slightly exceeding calyx, the lobes oblong-elliptic, acute or minutely 3-toothed at apex. Stamens $\frac{3}{4}$–$\frac{5}{6}$ as long as petals, adnate to the corolla for $\frac{1}{8}$–$\frac{1}{6}$ their length; filaments shortly connate at the base, free parts slender, bearded; lower part of the corolla-tube with conspicuous red granules; anthers c. 0·25 mm. long. Fruit c. 2 mm. in diameter, many-seeded, circumscissile. Seeds 0·2–0·3 mm. long.

Zambia. N: Mbala Distr., N'Kali Dambo, 1680 m., 8.v.1957, *Richards* 9605 (K). W: Solwezi, 1350 m., fl. 9.iv.1960, *Robinson* 3465 (K; SRGH). C: Kabwe, 1220 m., fl. & fr., *Rogers* 8159 (K). S: 35 km. N. of Choma, 1190 m., fl. & fr. 17.v.1954, *Robinson* 786 (K). **Zimbabwe.** N: Miami, iv.1926, *Rand* 66 (BM). W: Matobo, Farm Besna Kobila, 1460 m., fr. v.1959, *Miller* 5924 (K). C: Salisbury, 1490 m., fl. & fr. 27.iv.1948, *Greatrex* in SRGH 20042 (K; SRGH). **Malawi.** S: Ntcheu Distr., Lower Kirk Range, Chipusiri, 1460 m., fl. 17.iii.1955, *E. M. & W.* 933 (BM; SRGH). **Mozambique.** Z: Quelimane, fl. & fr. viii.1887, *Scott* (K). M: Namaacha, near Canada-Dry factory, fr. 17.v.1971, *Marques* 2276 (LMU).

Distributed more or less throughout tropical Africa. On waterlogged substrates of various kinds: on edge of pool in laterite pan, on sandy stream bank, on heavy black clay.

Note: *A. barbata* is very similar to *A. acuminata*; the latter, however, has narrower leaves and filaments which are relatively shorter and not connate at the base.

11. **Anagallis pumila** Swartz, Prodr. Veg. Ind. Occ. **1**: 40 (1788). — P. Taylor in Kew Bull. **10**: 342, fig. 5, 4–6 (1956); op. cit. **13**: 140 (1958); in F.T.E.A., Primulaceae: 16 (1958) excl. vars. — Dyer in Fl. Southern Afr. **26**: 15 (1963). — Friedrich-Holzhammer in Prodr. Fl. SW. Afr. **104**: 1 (1967). Type from Jamaica.

For extensive synonymy see P. Taylor, loc. cit. (1956).

Erect annual herb 2–10 cm. tall. Stems reddish, simple or branched. Leaves 2–4 mm. long, alternate, elliptic, acute at apex, sessile. Flowers occurring at most nodes, on pedicels 3–6 mm. long. Calyx-lobes 1–2 mm. long, lanceolate-acuminate. Corolla pale pink, very slightly longer than calyx; lobes elliptic, acute at apex. Stamens c. $\frac{2}{3}$ as long as petals, adnate to the corolla for $\frac{1}{4}$–$\frac{1}{3}$ their length; filaments shortly connate at the base, free parts slender, glabrous; lower part of corolla-tube with conspicuous red granules; anthers 0·2 mm. long. Fruit c. 1·5 mm. in diameter, globose, many-seeded, circumscissile. Seeds 0·2–0·3 mm. in diameter.

Zambia. S: Mazabuka Distr., between Mazabuka and Kaleya, fl. & fr. 15.iv.1963, *van Rensburg* 1911 (SRGH). **Zimbabwe.** W: Matobo Distr., Farm Besna Kobila, 1460 m., fr. v.1955, *Miller* 2860 (K; SRGH). C: Salisbury Distr., Makabusi R., 1370 m., fl. & fr. 24.iv.1948, *Wild* 2513 (K; SRGH). S: Bikita Distr., Dafana confluence with R. Turgwe, 1050 m., fl. 5.v.1969, *Biegel* 3012 (K; PRE; SRGH).

Widely distributed throughout the tropics. In F.Z. area found on shallow waterlogged soil over granite.

12. **Anagallis djalonis** A. Chev. in Journ. Bot., Paris, Sér. 2, **2**: 115 (1909). — P. Taylor in Kew Bull. **13**: 141 (1958); in F.T.E.A., Primulaceae: 17 (1958). — Boutique in Fl. Afr. Centr., Primulaceae: 12 (1971). Syntypes from Guinea.
   Anagallis pumila Swartz var. *djalonis* (A. Chev.) P. Taylor, op. cit. **10**: 346, fig. 5, 7–9 (1956). Syntypes as above.

Erect tufted annual herb 3–9 cm. tall. Stem simple or branched from the base. Leaves 5–13 mm. long, alternate, broadly spathulate, the apex acute, the base tapering abruptly into a short slender petiole. Flowers occurring at most nodes, on pedicels 6–9 mm. long. Calyx-lobes 2–2·5 mm. long, lanceolate-acuminate. Corolla white or pale pink, equalling or slightly shorter than calyx; lobes elliptic, acute at apex. Stamens c. ⅔ the length of the petals, adnate to the corolla for ⅓–½ their length; filaments shortly connate at the base, free parts either slender or flattened, glabrous; lower part of corolla-tube with conspicuous red granules; anthers 0·15 mm. long. Fruit c. 2·5 mm. in diameter, globose, many-seeded, circumscissile. Seeds (0·4) 0·45–0·65 mm. long.

Zambia. N: Kasama Distr., Chishimba Falls, fl. & fr. 31.iii.1955, *E. M. & W.* 1360 (BM; SRGH). W: Solwezi Distr., 8 km. E. of Solwezi, fl. & fr. 17.iii.1961, *Drummond & Rutherford-Smith* 6980 (K; PRE; SRGH). S: Kafue Gorge, 980 m., fr. 14.iv.1956, *Robinson* 1476 (K; SRGH). **Malawi.** N: Rumphi Distr., Manchewe Falls, top of Livingstonia Escarpment, 1160 m., fl. & fr. 14.iv.1971, *Pawek* 4912 (K; MAL).
Tropical Africa from Guinea to Kenya, Cameroon, Zaire and Angola. On damp shady ground beneath trees in *Brachystegia* woodland and riverine forest.

P. Taylor (see references) found that in *A. djalonis* the free parts of the filament are flattened and expanded, whereas in *A. pumila* they are slender. My survey has shown, however, that although the extreme forms belong to each taxon as reported, they are rather infrequent and most specimens of both species have somewhat expanded filaments. On the other hand, differences in leaf-shape and seed size appear to be very reliable diagnostic characters in this context.

4. SAMOLUS L.
Samolus L., Sp. Pl. 1: 171 (1753).

Annual and perennial herbs; stems erect or rarely procumbent, sometimes with a basal leaf-rosette. Leaves alternate, linear, oblong or spathulate, entire. Flowers in racemes or corymbs; bracts inserted at the base or near the middle of pedicels. Calyx partly adnate to ovary, the limb 5-lobed, persistent. Corolla subcampanulate, the limb 5-lobed. Stamens borne on the corolla-tube, oppositipetalous, alternating with small staminodes. Ovary globose; style short, stigma obtuse or capitellate; ovules numerous; placentation free-central. Capsule ovoid or globose, dehiscing by 5 valves. Seeds numerous, angular.

A genus of c. 9 species, one cosmopolitan, the rest confined to S. America, Australia or S. Africa.

Samolus valerandi L., Sp. Pl. 1: 171 (1753). — Knuth in Engl., Pflanzenr. IV, 237: 337 (1905). — P. Taylor in F.T.E.A., Primulaceae: 19, fig. 5 (1958). — Dyer in Fl. Southern Afr. 26: 10 (1963). — Friedrich-Holzhammer in Prodr. Fl. SW. Afr. 104: 2 (1967). — Bizzarri in Webbia 24: 687, fig. 13 (1970). — Boutique in Fl. Afr. Centr., Primulaceae: 8, fig. 2 (1971). TAB. 37. Described from England, Sweden and Africa.

Glabrous annual herb up to 50 cm. tall. Stems erect, simple or branched, leafy, arising from a basal rosette. Basal leaves up to 10 × 3 cm., fleshy, spathulate, tapering at the base into a petiole. Lowest stem leaves 5 cm. long or less, spathulate to suborbicular with a· short petiole; upper leaves decreasing in size, becoming bractiform, elliptic, sessile. Racemes many-flowered; pedicels with bract c. 1 mm. long inserted at or above the middle. Calyx cup-shaped, with tube c. 1·5 mm. long and triangular lobes c. ½ as long as tube. Corolla c. 2 mm. long, white, subcampanulate, with broadly spathulate lobes. Stamens 5, c. 1 mm. long, inserted at the base of the corolla-tube; staminodes 5, subulate, inserted in the corolla-tube between the lobes. Style c. 0·5 mm. long. Capsule c. 3 mm. long, globose; valves strongly reflexed after dehiscence. Seeds 0·3–0·6 mm. long, ± tetrahedral, minutely granular, dark brown.

Botswana. N: without locality, fl. iii.1967, *Guy* 55/67 (K; LISC; PRE). **Zambia.** N: Kasongola, Katanga border, 925 m., fl. & fr. 3.viii.1962, *Tyrer* 234 (BM; SRGH). **Zimbabwe.** W: Umzingwane Distr., 10 km. from Essexvale on Bulawayo road, fl. & fr. 4.ii.1974, *Mavi* 1534 (SRGH). C: Salisbury Distr., Mazoe road, 1220 m., fl. & fr. 7.xi.1931, *Brain* 7608 (SRGH). E: Chipinga Distr., Sabi Valley, c. 610 m., fr. x.1966, *Goldsmith* 94/66 (K; LISC; PRE; SRGH). S: Gwanda Distr., c. 14 km. W. of Fort Tuli, fl. & fr. 13.v.1959, *Drummond* 6133 (COI; K; PRE; SRGH). **Mozambique.** M: Inhaca I., Letivi freshwater swamp, fl. & fr. 19.vii.1958, *Mogg* 28086 (K; LMA; LMU; SRGH).
An almost cosmopolitan species. Growing at water level on stream banks, or in swamps.

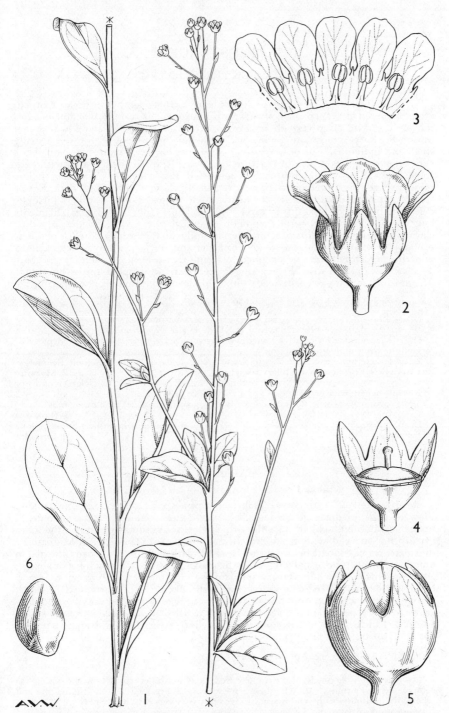

Tab. 37. SAMOLUS VALERANDI. 1, part of plant (× 1); 2, flower (× 12); 3, corolla opened out, with attached stamens and staminodes (× 12); 4, pistil, with two calyx-lobes removed (× 12); 5, fruit (× 12); 6, seed (× 40), from *Napper* 749. From F.T.E.A.

105. MYRSINACEAE
By F. K. Kupicha

Trees, shrubs and lianes. Leaves exstipulate, alternate, simple, entire or toothed, often clustered at branch ends, always with schizogenous resiniferous dots or lines but these varying from very obvious to obscure. Inflorescences lateral or terminal (not in F.Z. area), racemose, paniculate, umbellate or fasciculate. Flowers usually small and individually inconspicuous, 5- or 4-merous, actinomorphic, ⚥ or plants very often dioecious. Calyx of free or connate sepals, often ciliate, often with dark spots, aestivation valvate, imbricate or contorted. Petals usually white or pink, less frequently purple or yellow, free or more usually connate, often with dark dots or stripes, often papillose. Stamens as many as and opposite petals; anthers introrse, dehiscing by longitudinal slits or rarely by apical pores (not in F.Z. area); filaments long or short, adnate to corolla, sometimes almost free. Ovary globose, ovoid or clavate, superior or (in *Maesa*) semi-inferior, unilocular; number of carpels obscure, probably 3–4; style long and slender or short or rarely absent; stigma punctiform, capitate, discoid or lobed; placenta free-central, bearing few to many ovules in one or more rows. Fruit an indehiscent berry or drupe, 1-seeded except for *Maesa* which is many-seeded. Seeds with copious, often ruminate, endosperm.

A pantropical family of c. 30 genera and 900 species.

1. Ovary semi-inferior; fruit crowned with persistent calyx-remains; seeds many, crowded on the surface of the free-central placenta, polyhedral, with smooth endosperm - **1. Maesa**
   - Ovary superior; fruit with calyx sometimes persisting at the base; seed solitary, depressed-globose, with ruminate endosperm* - - - - - - - - - 2
2. Leaves numerous, crowded, not confined to branch ends, 6–12(20) mm. long, the margin minutely serrate; flowers in axillary fascicles - - - - - **2. Myrsine**
   - Leaves not as above but fewer, often confined to branch ends, usually much larger, entire or coarsely dentate; flowers in fascicles or inflorescence pedunculate - - - - 3
3. Trees; peduncles absent, flowers in dense fascicles; anthers sessile, adnate to corolla
   3. Rapanea
   - Lianes or shrubs; peduncles present, flowers in racemes or umbels; anthers borne on long filaments - - - - - - - - - - - - **4. Embelia**

1. MAESA Forssk.

Maesa Forssk., Fl. Aegypt.-Arab.: 66 (1775).

Shrubs, often sub-sarmentose, glabrous or pubescent. Leaves entire or dentate. Flowers usually unisexual (⚥ in F.Z. area), 5- or 4-merous, subtended by 2 bracteoles, borne in axillary racemes or panicles. Calyx adnate to ovary, the lobes usually not marked with resiniferous lines, persistent. Corolla campanulate or urceolate; petals connate below, usually marked with dark lines. Stamens inserted in mouth of corolla tube, usually with long filaments, these very rarely short or absent; anthers dehiscing by slits. Ovary semi-inferior or inferior, with short style and broad, subdiscoid, often lobed stigma; ovules usually numerous, arranged in several rows on the placenta, rarely few and uniseriate. Fruit dry or fleshy, ± globose, crowned with persistent sepals and style, usually many-seeded. Seeds small, embedded in the placenta, irregularly polyhedral, typically ± turbinate with the apex flat, the testa not pitted; endosperm smooth.

A genus c. 75 species occurring in the Old World tropics.

* During the development of the solitary seed, in all genera except *Maesa*, the placenta is compressed into a thin sheath lying between endocarp and testa. The testa itself is thin and follows the contours of the endosperm, to which it is closely attached. Species of *Myrsine*, *Rapanea* and *Embelia* in the F.Z. area all have seeds with ruminate endosperm, and, correlated with this, a pitted testa. The pits are filled with yellow crystalline material of placental origin, so that the seed appears to be smoothly globose, having a brownish testa with conspicuous yellow dots.

Maesa lanceolata Forssk., Fl. Aegypt.-Arab.: cvi & 66 (1775). — A.DC. in DC., Prodr. **8**: 78 (1844). — Baker in F.T.A. **3**: 492 (1877). — S. Moore in Journ. Linn. Soc., Bot. **40**: 130 (1911). — Gilg & Schellenb. in Mildbr., Wiss. Ergebn. Deutsch. Zentr.-Afr. Exped. 1907–1908, **2**: 512 (1913). — Eyles in Trans. Roy. Soc. S. Afr. **5**: 436 (1916). — Steedman, Trees etc. S. Rhod.: 60 (1933). — Burtt Davy, N.C.L.: 55 (1936). — Pardy in Rhod. Agric. Journ. **53**: 963 (1956). — F. White, F.F.N.R.: 316, fig. 57 (1962). — Hepper in F.W.T.A. ed. 2, **2**: 33, fig. 207 (1963). — Dyer in Fl. Southern Afr. **26**: 2 (1963). — Palmer & Pitman, Trees of Southern Afr. **3**: 1725, photos. pp. 1724 & 1725, fig. p. 1726 (1972). — R. B. Drumm. in Kirkia **10**: 265 (1975). — P. Halliday in F.T.E.A., Myrsinaceae: in print. TAB. **38**. Type from Yemen.

Baeobotrys lanceolata (Forssk.) Vahl, Symb. Bot. **1**: 19, t. 6 (1790). Type as above.

Maesa rufescens A.DC., tom. cit.: 81 (1844). — Gilg & Schellenb., loc. cit. Type from S. Africa (Natal).

Maesa angolensis Gilg in Notizbl. Bot. Gart. Berl. **1**: 72 (1895). Type from Angola.

Maesa mildbraedii Gilg & Schellenb. in Engl., Bot. Jahrb. **48**: 512 (1912); in Mildbr., tom. cit.: 512, t. 68 (1913). Syntypes from Zaire and Uganda.

Maesa sp. — Eyles, loc. cit.

An evergreen shrub or small, much-branched tree 2–10 m. high. Bark rough, grey or reddish-brown, the twigs often with rather prominent lenticels. Leaves 6–17 cm. long, varying from lanceolate (length 3 times breadth) to broadly elliptic (length twice breadth or less); apex acute to acuminate, base cuneate; petiole 1·5–3 cm. long; margin very shallowly to distinctly serrate, sometimes (in some narrow-leaved specimens) slightly revolute; lamina glabrous or pubescent with coarse, stiff hairs especially on nerves; midrib and lateral nerves conspicuous or inconspicuous when dry. Flowers pale yellow or whitish, scented, ☿, pentamerous, borne in axillary panicles usually shorter than leaves; panicle branches usually rufous-pubescent; pedicels 1–2 mm. long, subtended by a subulate bract c. 0·6 mm. long; bracteoles 2, c. 1 mm. long, ovate, opposite, inserted at base of flower. Calyx 1–1·6 mm. long, the lobes 0·8–1·2 mm. long, triangular to ovate, glabrous or ciliate. Corolla 1·5–1·8 mm. long, campanulate, the lobes ± equalling tube, orbicular, imbricate, undulate-margined, with glandular dots or lines. Stamens inserted in the corolla tube; anthers 0·4–0·7 mm. long, broadly oblong; filaments very short. Ovary semi-inferior, shallowly conical at apex, with fleshy style c. 0·6 mm. long; stigma truncate, somewhat lobed. Fruit 3–4 mm. in diameter, globose, crowned with the persistent calyx lobes, greenish to pale pink. Seeds numerous, blackish, angular, granular, embedded in the placenta.

Zambia. N: Kawambwa, fl. 23.viii.1957, *Fanshawe* 3558 (K). W: Mwinilunga Distr., S. of Dobeka Bridge, fl. 17.xi.1937, *Milne-Redhead* 3282 (BM; K; LISC; PRE; SRGH). C: river 27 km. NE. of Chiwefwe, fr. 14.vii.1930, *Hutchinson & Gillett* 3697 (BM; COI; LISC; SRGH). E: Chipata (Ft. Jameson), Lunkwakwa, 1150 m., fr. 23.iii.1955, *E. M. & W.* 1133 (BM; LISC; SRGH). **Zimbabwe.** N: Mazoe Distr., 1310 m., fl. iv.1906, *Eyles* 303 (BM). C: Mermaid's Pool E. of Salisbury, fl. 29.viii.1931, *Gilliland* 9 (BM). E: Umtali Distr., Vumba Mts. Hotel, 1650 m., fl. 27.ii.1956, *Chase* 5984 (BM; K; PRE; SRGH). S: 6·5 km. E. of Zimbabwe, fl. 1.vii.1930, *Hutchinson & Gillett* 3360 (K). **Malawi.** N: Mafinga Hills, 3 km. WSW. of Chisenga, 1740 m., fl. & fr. 25.viii.1962, *Tyrer* 572 (BM; SRGH). C: Dedza Mt., fr. 24.x.1956, *Banda* 325 (BM; FHO; MAL). S: Mt. Mulanje, below Chisongole, 1070 m., fl. 25.viii.1956, *Newman & Whitmore* 593 (BM). **Mozambique.** N: Lichinga, serra de Massangulo, c. 1600 m., fl. 5.iii.1964, *Torre & Paiva* 11035 (BR; LISC; LUA; M; WAG). Z: 8 km. W. of Gúruè, fr. immat. 1.vii.1942, *Hornby* 2702 (K; PRE). T: between Furancungo & Vila Gamito, fl. 10.vii.1949, *Barbosa & Carvalho* 3540 (K; LMA). MS: Báruè, serra de Choa, 17 km. from Catandica (Vila Gouveia), c. 1500 m., fl. 13.xii.1965, *Torre & Correia* 13612 (BR; K; LISC; M).

Widespread in tropical and southern Africa, extending into Arabia. In riverine forest in savanna, kloof forest and forest margins.

Maesa lanceolata shows much variation throughout its range. Within the F.Z. area there is some correlation between variation in leaf shape and presence or absence of indumentum, so that leaves tend to be either lanceolate and glabrous or broadly elliptic and pubescent on the lower surface. This correlation is very marked in Zimbabwe and moderately so in Malawi and Zambia, but breaks down in Mozambique, where most specimens have broad glabrous leaves. Plants with broad pubescent leaves have often been named "*Maesa rufescens*".

Swynnerton made extensive notes on the two Zimbabwean forms (on herbarium specimens at BM). He observed that the narrow-leaved form was a relatively tall, straight tree of high-altitude forest (at 1830–2130 m.), while the broad-leaved form was a smaller, wide-branched, fire-resistant shrub of lower-altitude open woodland and grass jungle. Having assembled a

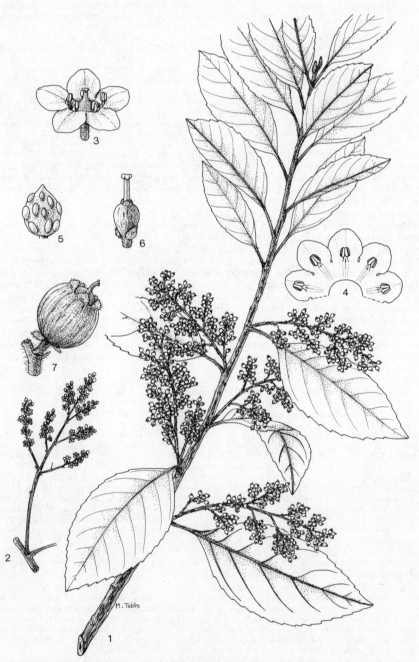

Tab. 38. MAESA LANCEOLATA. 1, fruiting branch (× ½), from *Banda* 325; 2, inflorescence (× ½);
3, flower (×6), 2–3 from *Borges* 168; 4, corolla spread out (×6), from *Banda* 325; 5,
placenta (×14); 6, young fruit (×6); 7, mature fruit (×6), 5–7 from *Forbes* 242.

large amount of herbarium material, I cannot confirm this distinction; there appears to be no correlation between leaf-shape and habit or habitat.

Specimens from Angola can be divided into two forms paralleling the Zimbabwean ones. The broad-leaved form is apparently the same as that from the F.Z. area, while the narrow-leaved form (*Maesa angolensis*), distinctive in having leaves with inconspicuous venation and revolute margins, merges gradually with the narrow-leaved form found in western Zambia.

The type of *M. rufescens*, from Natal, is said to have leaves c. 3 times as long as broad, and so is perhaps not identical with the pubescent-leaved form in the F.Z. area.

Within the rest of Africa north of the F.Z. area, *M. lanceolata* cannot readily be divided into different forms, although it is still variable. There is no indication that the Arabian type (which has fairly narrow, glabrous leaves) is more closely related to the narrow-leaved F.Z. form than to the F.Z. broad-leaved form. Since the variation in leaf-shape and hairiness does not appear to be correlated with any other character, and there is a mosaic of variation in these characters throughout the species' range, it seems best to treat *M. lanceolata* as a single species without infraspecific division.

One specimen from Malawi, S: Zomba Plateau, 1620 m., 1.vi.1957, *Boughey* 1820 (SRGH), appears to fall outside the usual range of variation of *M. lanceolata*: it is of a "suffrutescent herb" up to 1·3 m. high, growing in grassland, with obovate, obtuse leaves up to 4·5 × 2·75 cm., crenate at the margin. The leaf shape and general habit is suggestive of the S. African *M. alnifolia* Harvey, but *Boughey* 1820 has the much-branched many-flowered panicles characteristic of *M. lanceolata*, while *M. alnifolia* has few-flowered racemes. Perhaps the unusual features of this Malawi specimen are due to stunting by grazing.

2. MYRSINE L.

Myrsine L., Sp. Pl. **1**: 196 (1753).

Dioecious trees and shrubs with leaves very often serrate or crenate. Flowers 4–5-merous, borne in few-flowered axillary fascicles. Sepals free or shortly connate at the base, marked with resin-dots, the margins usually ciliate. Petals almost free or connate to the middle, marked with resin-dots, usually ciliate at the margin. Stamens included in or exceeding corolla; filaments present, inserted at base of corolla, often on a glandular ring; anthers dehiscing by slits. Ovary glabrous, ovoid or ellipsoid; style well developed; stigma large, discoid, often with incised margins; placenta with few ovules in 1 row. Fruit a globose 1-seeded fleshy drupe with crustaceous endocarp. Seed subglobose, with ruminate endosperm.

A genus of c. 5 species distributed from Africa to E. Asia.

Authors who have dealt with Myrsinaceae on a regional basis have sometimes sunk *Rapanea* into *Myrsine* (e.g. Hosaka in Occ. Pap. Bernice P. Bishop Mus., Honolulu **16**: 25–79, 1940 (Hawaii) and Stearn in Bull. Brit. Mus. Nat. Hist., Bot. **4**: 177, 1969 (Jamaica)), sometimes kept the two genera separate (e.g. A. C. Smith in Journ. Arnold Arb. **54**: 278, 1973 (Fijian region)). Since *Myrsine* and *Rapanea* are sparsely represented in Africa and are easily distinguished here, they are traditionally treated as separate. As Walker has stated (in Bot. Mag. Tokyo **67**: 248–249, 1954), the present classification is probably artificial, but it cannot be improved without a detailed world-wide study.

Myrsine africana L., Sp. Pl. **1**: 196 (1753). — Baker in F.T.A. **3**: 493 (1877). — Mez in Engl., Pflanzenr. **IV**, 236: 340, fig. 58 (1902). — Harvey ex Wright in Harv. & Sond., F.C. **4**: 434 (1906). — Gilg. & Schellenb. in Mildbr., Wiss. Ergebn. Deutsch. Zentr.-Afr. Exped. 1907–1908, **2**: 515 (1913). — S. Moore in Journ. Linn. Soc., Bot. **40**: 130 (1911). — Eyles in Trans. Roy. Soc. S. Afr. **5**: 436 (1916). — Marloth, Fl. S. Afr. **3**: 28, t. 8 (1932). — Burtt Davy & Hoyle, N.C.L.: 55 (1936). — F. White, F.F.N.R.: 316 (1962). — Dyer in Fl. Southern Afr. **26**: 5 (1963). — Killick in Fl. Pl. Afr. **40**: t. 1564 (1969). — R. B. Drumm. in Kirkia **10**: 265 (1975). — Taton in Bull. Jard. Bot. Nat. Belg. **46**: 449, fig. 1 (1976). — P. Halliday in F.T.E.A., Myrsinaceae: in print. TAB. **39**. Type from S. Africa.

Much branched, functionally dioecious perennial varying from a suffrutex 30 cm. high to a large shrub up to 3·5 m. tall. Bark smooth to finely rugose, brownish-grey. Branches ascending, rusty- or brownish-tomentose especially at apices. Leaves very numerous, crowded, 6–12 mm. long, varying from ± orbicular to elliptic or obovate, acute or obtuse and apiculate at apex, subsessile or tapering at base to a petiole c. 1 mm. long, the proximal part of the midrib on adaxial side tomentose, otherwise lamina glabrous, coriaceous, inconspicuously nerved apart from the prominent abaxial midrib, the margin slightly revolute and sparsely serrulate (teeth usually inconspicuous due to the curved margin). Flowers tetramerous, small, borne in several-flowered axillary fascicles, shortly pedicellate. Calyx c. ½ as long as corolla,

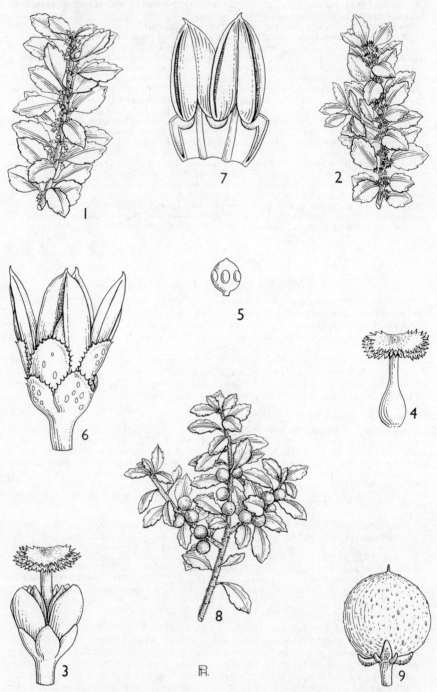

Tab. 39. MYRSINE AFRICANA. 1, habit (×1), from *Tweedie* 1994; 2, habit (×1), from *Williams* 12333; 3, ♀ flower (×12); 4, pistil (×12); 5, placenta (×18), 3–5, from *Tweedie* 1994; 6, male flower (×12); 7, stamens (×12), 6–7, from *Williams* 12333; 8, habit (×1); 9, fruit (×6), 8–9, from *Oteke* 66. From F.T.E.A.

divided almost to the base, the lobes ovate, ciliate, conspicuously glandular-
punctate. Corolla gamopetalous, with rounded-triangular ciliate lobes ± as long as
the tube, greenish, cream-coloured or pink, conspicuously punctate. ♂ flower:
anthers 1·3–2 mm. long, long-exserted from corolla (flower c. 2–2·5 mm. long),
oblong-triangular, apiculate, red; ovary vestigial, clavate. ♀ flower: anthers as in ♂
but shorter than and concealed by corolla (flower c. 2 mm. long), functionless; ovary
c. 1·2 mm. long, clavate, with prominent much-lobed capitate stigma. Fruit a
globose 1-seeded drupe, at first dry, greenish and c. 2–3 mm. in diameter, later
developing a purplish fleshy pericarp and swelling to 8 mm. in diameter.

Zambia. N: Isoka Distr., Mafingi Range above Chisenga, c. 2130 m., fl. buds 22.xi.1952,
Angus 829 (BM; FHO). W: Mwinilunga Distr., near Tshikundula stream by crossing with
Mwinilunga-Kabompo road, fl. buds 4.x.1952, *Angus* 591 (BM; FHO). **Zimbabwe.** N: Mazoe
Distr., Barwick Estate, S. of Mtoroshanga, fl. buds 24.vii.1971, *Orpen* GHS 213236 (PRE;
SRGH). W: Matobo Distr., mouth of Bambata Cave, Matopos National Park, st. 16.iv.1971,
Biegel 3504 (SRGH). C: Wedza Distr., Wedza Mt., 1620 m., ♂ fl. 4.xi.1971, *Biegel* 3635 (K;
LISC; PRE; SRGH). E: Inyanga Distr., Pungwe, ♂ fl. 8.viii.1950, *Chase* 2863 (BM; PRE;
SRGH). S: Belingwe Distr., Bukwa Mt., c. 1660 m., fr. 29.iv.1973, *Pope* 985 (SRGH).
Malawi. N: Nyika Plateau, near L. Kaulime, 2230 m., ♀ fl. 3.ix.1962, *Tyrer* 789 (BM; COI;
SRGH). C: Dedza Distr., Dedza Mt., 2160 m., fr. 12.xi.1977, *Brummitt* 15077 (MAL). S: Mt.
Mulanje, Tuchila Plateau, 1830 m., ♀ fl. & fr. 24.viii.1956, *Newman & Whitmore* 172 (BM;
COI; SRGH). **Mozambique.** N: Serra Ribaué, c. 1320 m., ♀ fl. 21.viii.1968, *Macedo &
Macuacua* 3500 (LMA). Z: Gúruè Distr., N. of Namúli, ♀ fl. & fr. 12.viii.1949, *Andrada* 1849
(COI; LISC). MS: serra Zuira, c. 1800 m., fl. 6.xi.1965, *Torre & Pereira* 12733 (COI; EA;
LISC; P; WAG).

Occurring in Africa from the Cape northwards throughout E. Africa to Ethiopia and Sudan,
and in Angola and Zaire; also in the Azores, Socotra, and across Asia to Central China. Near
water in *Brachystegia-Julbernardia* woodland, in clearings and at margins of montane forest.

The fruits of *M. africana* are often attacked by a fungus, referred to by Mez (op. cit.: 10) as
Capnodium fructicolum Pat. (syn. *Coryneliospora fructicola* (Pat.) Fitzpatrick); drupes so
infected are covered with hard black tubercles.

SPECIMINA INCERTAE SEDIS

Three specimens of *Myrsine* from the Eastern province of Zimbabwe may possibly belong to
the S. African species *M. pillansii* Adamson (in Journ. S. Afr. Bot. **7**: 204, 1941). They are:
Inyanga Distr., Pungwe Circles, st. 28.ix.1962, *West* 4183 (K; SRGH); same locality, Pungwe
Source, ♀ fl. 20.x.1946, *Wild* 1429 (K; SRGH); Inyanga, ♀ fl. 5.x.1946, *Mackintosh* 1/46
(FHO). The differences between this species and *M. africana* may be summarised as follows:

Suffrutex or shrub 0·3–3·5 m. tall; leaves 6–20 mm. long, with small sharp marginal serrations
M. africana

Shrub or small tree 2·5–10 m. tall; leaves 20–60 mm. long, entire or with a few blunt teeth
M. pillansii

The three Zimbabwean specimens cited above are intermediate between the two species in
vegetative characters. *M. pillansii* is problematical, in that although many specimens have been
collected from the Cape, Natal and the Transvaal, only ♀ flowers are known so far (see Dyer &
Killick in Bothalia **10**: 368–369, 1971).

3. RAPANEA Aubl.

Rapanea Aubl., Hist. Pl. Guiane Fr. **1**: 121, t. 46 (1775).

Trees and shrubs, glabrous or pubescent. Leaves entire or very rarely dentate,
often lepidote. Flowers borne in fascicles on very short lateral branches, 4–5(6–7)-
merous, ♀ or more usually unisexual, species often dioecious. Sepals free or shortly
connate, often ciliate at the margin and marked with resin-dots or lines. Petals almost
free or connate up to ⅔ their length, often papillose at the margin, often marked with
dots or lines. Stamens inserted in the throat of the corolla; filaments absent, the
anthers sessile, dehiscing by slits. Ovary globose or ellipsoid; style short or absent;
stigma variable in shape, elongated, scutate or capitate, often lobed or mor-
chelliform; placenta with few ovules in 1 row. Fruit dry or fleshy, 1-seeded, with
hard endocarp. Seed subglobose with a basal cavity at point of attachment, the
endosperm ruminate (in F.Z. species) or smooth.

A pantropical genus with c. 100 species.

Rapanea melanophloeos (L.) Mez in Engl., Pflanzenr. **IV**, 236: 375 (1902). — Burtt Davy,
N.C.L.: 55 (1936). — Pardy in Rhod. Agric. Journ. **53**: 520 (1956). — Dyer in Fl.
Southern Afr. **26**: 8, fig. 2, 1 (1963). — Palmer & Pitman, Trees of Southern Afr. **3**: 1729,
photo. pp. 1605 & 1728 (1972). — R. B. Drumm. in Kirkia **10**: 265 (1975). — P. Halliday
in F.T.E.A., Myrsinaceae: in print. TAB. **40**. Type from S. Africa (Cape Prov.).
    Sideroxylon melanophloeos L., Mant. Pl. Alt.: 48 (1767) excl. syn. Type as above.
    Myrsine melanophloeos (L.) R. Br., Prodr. Fl. Nov. Holl.: 533 (1810). — Harvey ex
Wright in Harv. & Sond., F.C. **4**: 434 (1906). — Baker in F.T.A. **3**: 494 (1877). Type as
above.
    Myrsine rhododendroides Gilg in Engl., Bot. Jahrb. **19**, Beibl. 47: 44 (1894); in Engl.,
Pflanzenw. Ost.-Afr. **C**: 303 (1895). Syntypes from Tanzania.
    Myrsine neurophylla Gilg in Engl., op. cit.: 45 (1894); in Engl., loc. cit. (1895). Syntypes
from Cameroon and Uganda.
    Myrsine runssorica Gilg in Engl., loc. cit. (1895). Type from Uganda.
    Rapanea runssorica (Gilg) Mez in Engl., tom. cit.: 373 (1902). Type as above.
    Rapanea ulugurensis Mez, tom. cit.: 374 (1902). Type from Tanzania.
    Rapanea rhododendroides (Gilg) Mez, tom. cit.: 374 (1902). — Mildbr., Wiss. Ergebn.
Deutsch. Zentr.-Afr. Exped. 1907–1908, **2**: 517 (1913). Syntypes as for *Myrsine
rhododendroides*.
    Rapanea neurophylla (Gilg) Mez, loc. cit. (1902). — Mildbr., loc. cit. — Hepper in
F.W.T.A. ed. 2, **2**: 31 (1963). Syntypes as for *Myrsine neurophylla*.
    Rapanea umbratilis S. Moore in Journ. Linn. Soc., Bot. **40**: 130 (1911). Type:
Zimbabwe, Melsetter, 1830 m., 10.x.1908, *Swynnerton* 6163 (K, holotype).
    Rapanea pellucido-striata Gilg & Schellenb. in Engl., Bot. Jahrb. **48**: 523 (1912). —
Mildbr., op. cit.: 515, t. 69 fig. A–F (1913). Syntypes from the Uganda/Zaire border.
    Rapanea pulchra Gilg & Schellenb., tom. cit.: 524 (1912). — Mildbr., loc. cit.: t. 69 fig.
G–M (1913). Syntypes from Zaire and Tanzania.
    Rapanea usambarensis Gilg & Schellenb., loc. cit. Syntypes from Tanzania.
    Rapanea schliebenii Mildbr. in Notizbl. Bot. Gart. Berl. **12**: 87 (1934). Type from
Tanzania.
    ?Rapanea gracilior Mildbr., tom. cit. 88 (1934). Syntypes from Tanzania.
    Rapanea sp. 1. — F. White, F.F.N.R.: 318 (1962).

    An evergreen tree 3–18 m. tall, with longitudinally furrowed, brownish-grey
to blackish bark. Leaves coriaceous, glabrous, clustered at branch ends; lamina
4–9(15) × 1–3·5(5) cm., elliptic to oblanceolate, entire, acute or rounded at apex,
tapering or acute at base; surfaces usually dotted with minute circular scales, often
also marked with resin-canals in the form of dots or short to long lines radiating from
the midrib; nervature prominent or inconspicuous; lamina dark green above, paler
beneath; petiole 5–10 mm. long, often reddish. Flowers fasciculate on short shoots in
axils of current and recently fallen leaves, whitish or greenish yellow, pentamerous, ⚥
or sometimes with functionless stamens or ovary, each subtended by a rust-coloured
triangular bract c. 1 mm. long; pedicels 1–1·5 mm. long, stout. Calyx cup-shaped;
sepals c. 1 mm. long, almost free, minutely papillose-ciliate. Corolla ellipsoid in bud,
petals almost free, c. 2·5 mm. long, oblong-acute, dotted with conspicuous resin-
ducts, densely and minutely papillose-ciliate at margins (in bud neighbouring petals
adhering by papillae). Anthers sessile, dorsifixed to corolla lobes, c. 1·5 mm. long,
introrse. Gynoecium c. 1·5 mm. long; ovary globose, tapering at apex to a short style;
stigma thick, conical. Fruit a globose purple 1-seeded drupe 3–5 mm. in diameter,
borne on previous year's wood.

    Zambia. N: Mpika, fr. 4.ii.1955, *Fanshawe* 1961 (FHO; K). W: Mwinilunga Distr., source
of R. Matonchi, fr. 16.ii.1938, *Milne-Redhead* 4593 (BM; K; PRE). **Zimbabwe.** N: Sipolilo
Distr., Great Dyke, Mpingi Pass, 1520 m., fr. 17.v.1962, *Wild* 5780 (SRGH). C: Makoni
Distr., 1695 m., st. 16.vi.1957, *Chase* 6521 (K; SRGH). E: Inyanga Distr., Rhodes Inyanga
Estate, fl. 8.xii.1950, *English* 2/50 (K; LISC; SRGH). **Malawi.** N: Viphya, Luwawa, c. 1680
m., fl. 17.xii.1961, *Chapman* 1508 (MAL; SRGH). C: Dedza Distr., Dedza Mt., fr. 28.x.1965,
Banda 707 (K; MAL; SRGH). S: Mulanje Distr., Litchenya Plateau, fl. immat. 13.vii.1946,
Brass 16815 (BM; K; PRE; SRGH). **Mozambique.** N: Ribáuè, serra Mepáluè, fl. 5.xii.1968,
Torre & Correia 16370 (LISC). Z: Gúruè (Vila Junqueiro), near source of R. Malèma, c. 1700
m., fl. 4.i.1968, *Torre & Correia* 16924 (LISC). MS: Manica, Mavita, Serra Mocuta, c. 1588
m., fl. 13.xii.1965, *Pereira & Marques* 1079 (COI; LISC; LMU; PRE; SRGH). M: Namaacha,
Mt. Ponduini, fl. 14.vii.1969, *Correia & Marques* 910 (BM; COI; K; LMA; LMU; SRGH).
    Widespread in tropical and southern Africa. In relict moist evergreen montane forest and
kloof forest.

    Rapanea melanophloeos is a very variable species, as the extensive synonymy implies. The
characters used to differentiate between the many "species" were: leaf shape, petiole length,

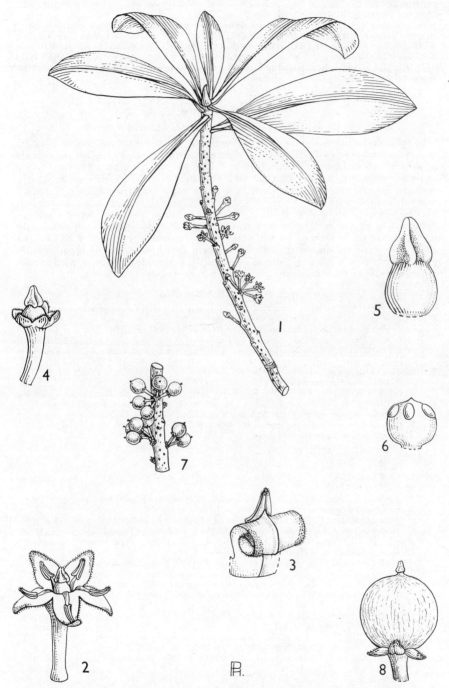

Tab. 40. RAPANEA MELANOPHLOEOS. 1, habit (× ⅔); 2, flower (× 4); 3, petal and anther (× 8); 4, calyx and pistil (× 4); 5, pistil (× 8); 6, placenta (× 12), 1–6 from *Thulin & Tidigs* 247; 7, fruits (× ⅔); 8, fruit (× 4), 7–8, from *Tweedie* 1805. From F.T.E.A.

occurrence of resin-lines on leaf-lamina, length of corolla tube, etc. These features do not, however, appear to be taxonomically significant.

Several specimens, from scattered localities, have been seen with fruits narrowly pear-shaped, instead of spherical; these include *Fanshawe* 1961 (FHO; K) from Zambia (N), *McGregor* 41/48 (FHO; SRGH), *Williams* 9572 (K; SRGH) and *Wild* 3524 (K; LISC) from Zimbabwe (E) and *Torre & Correia* 13593 (LISC) from Mozambique (MS). The sheet *Wild* 3524 (LISC) has both spherical and elongated fruits on a single branch, which suggests that the abnormal fruit shape is caused by a parasite.

4. EMBELIA Burm. f.

Embelia Burm. f., Fl. Ind.: 62, t. 23 (1768), *nom. conserv.*

Shrubs, very often scandent, rarely arborescent. Leaves entire or crenate. Flowers unisexual (and species dioecious), 4–5-merous, borne in axillary racemes or umbels or in terminal or lateral panicles (not in F.Z. area). Sepals almost free or shortly connate, imbricate, usually marked with resiniferous dots. Petals free or shortly connate, elliptic, obovate or oblong, the inner surface and sometimes the outer margins densely papillose. Stamens longer or shorter than corolla and inserted at varying levels, rarely almost free; filaments usually well developed; anthers dehiscing by slits or very rarely by pores (not in F.Z. area), usually with a dark spot on the abaxial side. Ovary of ♀ flower subglobose or ovoid, often pilose, with a long or short style; stigma disciform, entire or rarely lobed; placenta with few ovules arranged in 1 row. Fruit a ± globose 1-seeded drupe. Seed subglobose, with a small to large basal cavity; endosperm ruminate (in F.Z. species) or smooth.

A genus of c. 75 species occurring in the Old World tropics.

1. Leaves sparsely pilose on lower surface; inflorescences in axils of current leaves
                                                                                3. *upembensis*
– Leaves glabrous; inflorescences on short shoots in axils of fallen leaves -   -   - 2
2. Leaves entire; fruits with thin crustaceous endocarp   -   -   -   - 1. *schimperi*
– Leaves coarsely crenate in distal half; fruits with thick woody endocarp   - 2. *xylocarpa*

1. **Embelia schimperi** Vatke in Linnaea **40**: 206 (1876). — Mez in Engl., Pflanzenr. **IV**, 236: 329 (1902). — Hepper in F.W.T.A. ed. 2, **2**: 32 (1936). — Brenan in Mem. N.Y. Bot. Gard. **8**: 498 (1954). — F. White, F.F.N.R.: 316 (1962). — R. B. Drumm. in Kirkia **10**: 265 (1975). — P. Halliday in F.T.E.A., Myrsinaceae: in print. Type from Ethiopia.
    Embelia abyssinica Baker in F.T.A. **3**: 497 (1877). Type from Ethiopia.
    Embelia guineensis Baker, tom. cit.: 496 (1877). — Mez, tom. cit.: 331 (1902). — Hepper, loc. cit. Type from Guinea.
    Embelia kilimandscharica Gilg in Engl., Bot. Jahrb. **19**, Beibl. 47: 45 (1894). — Mez, tom. cit.: 326 (1902). Type from Tanzania.
    Pattara pellucida Hiern, Cat. Afr. Pl. Welw. **3**: 639 (1898). Type from Angola.
    Embelia pellucida (Hiern) K. Schum. in Just's Bot. Jahresb. **26**: 390 (1900). Type as above.
    ?*Embelia mujenja* Gilg in Engl, op. cit. **28**: 446 (1900). — Mez, tom. cit.: 330 (1902). Type from Tanzania.
    Embelia nyassana Gilg in Engl., op. cit. **30**: 96 (1901). — Mez, tom. cit.: 329 (1902). — S. Moore in Journ. Linn. Soc., Bot. **40**: 130 (1911). — Eyles in Trans. Roy. Soc. S. Afr. **5**: 436 (1916). — Burtt Davy & Hoyle, N.C.L.: 55 (1936). Type from "Nyassaland" (Malawi), *Buchanan* 42 (BM, isotype).
    ?*Embelia retusa* Gilg in Engl., tom. cit.: 95 (1901). — Mez, tom. cit.: 330 (1902). Syntypes from Zaire.
    Embelia gilgii Mez, loc. cit. Syntypes from Togo and Sierra Leone.
    ?*Embelia bambuseti* Gilg & Schellenb. in Engl., Bot. Jahrb. **48**: 520 (1912). Type from Tanzania.
    ?*Embelia tibatiensis* Gilg & Schellenb., tom. cit.: 521 (1912). Syntypes from Cameroon.
    Embelia tessmannii Gilg & Schellenb., tom. cit.: 522 (1912). Type from Equatorial Guinea.
    ?*Embelia dasyantha* Gilg & Schellenb., loc. cit. Type from Cameroon.
    Embelia batesii S. Moore in Journ. Bot. **63**: 147 (1925). Type from Cameroon.
    Embelia spp. — Eyles, loc. cit.
    Embelia sp. A. — Hepper in F.W.T.A. ed. 2, **2**: 32 (1963).
    Embelia sp. — Dyer in Fl. Southern Afr. **26**: 5 (1963).

Scandent shrub or liane climbing to 6 m. by means of hard persistent lateral short shoots, with long trailing branches up to 5 cm. or more in diameter; young stems glabrous; bark greyish or blackish, that of twigs longitudinally furrowed, with

prominent lenticels. Plants dioecious or more rarely flowers ♀. Leaves
2·5–10 × 1·5–6 cm., elliptic, obovate or suborbicular, apiculate, acute, obtuse or
emarginate at apex, almost truncate to tapering at base, entire, clustered at ends of
branches; petiole 0·5–1 cm. long; lamina glossy, pale to dark green, thinly fleshy to
coriaceous, inconspicuously to prominently nerved, with or without obvious black
resin-dots (these punctate or shortly linear), glabrous. Racemes 2·5–3 cm. long,
15–35-flowered, cylindrical, borne on lateral short shoots on old wood proximal to
current year's leaves, densely pubescent with short gland-tipped hairs; pedicels 3–6
mm. long, each subtended by a triangular bract c. 1 mm. long, with triangular to
ovate lobes c. 2 × length of tube, glandular-pubescent, often with a few black resin-
dots on abaxial side; petals free, 2·6–3·2 mm. long, oblong to narrowly ovate, white,
cream-coloured, greenish or yellowish, sparsely to densely glandular-pubescent
especially on margin and inner surface, often with black resin-dots on abaxial side. ♂
flower: stamens functional; anthers 1–1·2 mm. long, oblong, yellow, with black dot
on outer side near point of attachment; filament adnate to petal for a variable distance
(up to middle of petal), elongating during anthesis so that the anther is eventually
well exserted from flower; gynoecium 0·8–1·2 mm. long, rudimentary. ♀ flower:
stamens as in ♂ flower but anthers triangular, functionless and filaments scarcely
elongating; gynoecium 3–3·2 mm. long, clavate, the stigma conspicuously lobed. ♀
flower with stamens as in ♂ flower and ovary as in ♀ flower. Fruit c. 5 mm. in
diameter, globose or compressed-globose, 1-seeded, greenish at first becoming
scarlet when ripe. Seed globose with basal cavity.

Caprivi Strip. Mpila I., 910 m., fr. 15.i.1959, *Killick & Leistner* 3387 (K). **Zambia.** B:
Mongu Distr., Luanginga R., c. 6 km. NW. of Sandaula Pontoon, fl. & fr. 17.xi.1959,
Drummond & Cookson 6587 (K; LISC; PRE; SRGH). N: Mbala (Abercorn), near L. Chila,
1680 m., fl. 11.xii.1954, *Siame* 530 (K). W: Ndola, fr. immat. 18.xii.1954, *Fanshawe* 1730 (K;
SRGH). C: Roma Township, Lusaka, fl. buds 17.xii.1963, *Angus* 3815 (K). **Zimbabwe.** E:
Inyanga Distr., Eastern Highlands Tea Estate, fl. & fr. immat. 17.xi.1960, *Wild* 5277 (K; PRE;
SRGH). **Malawi.** N: Mzimba Distr., 1·5 km. W. of Lupaso School near Mzuzu, 1370 m., fr.
31.i.1971, *Pawek* 4351 (K; MAL). C: Dedza Distr., Chongoni Forest near Nchinje stream, fr.
19.xii.1957, *Adlard* 270 (MAL; SRGH). S: Thondwe (Ntondwe), fl. 21.x.1905, *Cameron* 159
(K). **Mozambique.** N: Massangulo, st. 15.v.1948, *Pedro & Pedrogão* 3529 (LMA). Z: Gúruè
(Vila Junqueiro), st. 7.vii.1942, *Hornby* 4556 (K; PRE). T: Angonia, between Cólubuè and Vila
Coutinho, c. 1500 m., fr. 7.iii.1964, *Correia* 175 (LISC; MO). MS: Báruè, 17 km. from
Catandica (Vila Gouveia), c. 1500 m., fr. immat. 13.xii.1965, *Torre & Correia* 13624 (LISC).
  Widespread in tropical Africa. On river banks, in ravine forests and at forest margins, and in
woodland on termitaria.

  This very widespread tropical African species is variable in leaf characters, and the many
synonyms were based on particular leaf-forms, these often being paralleled in distant regions.
The type specimen of *E. schimperi* (Ethiopia) has oblong-elliptic, rather coriaceous leaves
rounded-acute at both ends, with prominent reticulate venation and apparently no resin-dots.
The type specimens of *E. guineensis* (Sierra Leone), *E. pellucida* (Angola) and *E. nyassana*
(Malawi) have obovate leaves of thin chartaceous texture (when dry), rounded at apex, tapering
at base, with venation very inconspicuous or impressed and then the lamina puckered along
each nerve; there are numerous conspicuous resin-dots.
  In F.W.T.A., Hepper recognised these two extreme forms and a third, *species A,* with the
leaf-shape and resin-dots of "*E. guineense*" but the texture and strong venation of "*E.
schimperi*".
  Throughout tropical Africa, there is some degree of correlation between leaf-shape, leaf-
texture and geography, in that the elliptic, coriaceous, strongly veined leaf is predominant in
Ethiopia and countries of the F.T.E.A. area while the obovate, chartaceous, weak-veined,
punctate leaf is the usual form in the F.Z. area. In Angola, Zaire and W. tropical Africa, both
forms are equally represented.
  Despite this partial geographical separation, I am not convinced that if the specimens of each
extreme form were grouped together they would represent two genuine genetically distinct
taxa, and in any case many intermediates would remain. More work needs to be done to
discover whether the differences are taxonomically significant or merely environmental.
  Brenan (in Mem. N.Y. Bot. Gard. **8**: 490, 1954) took a similar view, though on a more limited
scale, when he proposed that *E. schimperi* was the correct name for *E. abyssinica, E.
kilimandscharica, E. nyassana* and (probably) *E. mujenja* and *E. pellucida.*
  A few specimens from the extreme W. of the F.Z. area have rather small leaves, probably
reflecting the relatively dry environment in these localities. They include *Killick & Leistner*
3387 (K) from the Caprivi Strip, which Dyer in Fl. Southern Afr. **26**: 5 (1963) mentions as
possibly belonging to *E. ruminata,* and *Codd* 7316 (BM; K) from Zambia (B) which P. Halliday
in Kew Bull. **32**: 294 (1978) cites under *E. xylocarpa. Drummond & Cookson* 6587 (K; LISC;
PRE; SRGH) from the same area is a similar form.

A single gathering from Zimbabwe (E: Inyanga Distr., Stapleford, Chisanza R., 1220 m., fr. 4.iv.1962, *Wild* 5696 (FHO; PRE)) is very unusual in having dentate leaves.

E. ruminata (E. Mey. ex. A. DC.) Mez, from S. Africa, is closely related to *E. schimperi*. It differs in having generally much smaller leaves (1–7 × 0·7–3 cm.), rather contorted branches with much closer nodes than in *E. schimperi,* and shorter, fewer-flowered racemes (up to 1 cm. long with up to 10 flowers).

2. **Embelia xylocarpa** P. Halliday in Kew Bull. **32**: 293, fig. 1 (1978); in F.T.E.A., Myrsinaceae: in print. TAB. **41**. Type from Tanzania.

Shrub or small tree 1·5–6 m. tall, with somewhat sarmentose branches and distinctive smooth pale grey to whitish bark on younger twigs. Plants dioecious. Leaves 3–7·5 × 2–4 cm., obovate, rounded or occasionally emarginate at apex, acute at base, serrate in distal half to rarely subentire, clustered at ends of lateral short shoots which remain hard and spiny after leaf-fall; petiole 3–14 mm. long; lamina membranous or subcoriaceous, with midrib impressed above and prominent below, resin-dots very conspicuous to obscure, glabrous. Flowers subsessile to shortly pedicelled, in dense fascicles or short racemes subtended by minute bracts, on old wood, mainly on the short shoots proximal to current leaves. Calyx 1·75–2 mm. long, cup-shaped, with resiniferous dots; lobes as long as tube, ovate or triangular, entire or somewhat dentate. Corolla 3·5–4·5 mm. long, black-dotted, petals free, narrowly oblong, greenish to yellowish or reddish. Stamens inserted on corolla at varying levels between $\frac{1}{4}$ and $\frac{2}{3}$ from base; anthers c. 0·75 mm. long, in ♂ flowers fertile and borne on long filaments, in ♀ flowers functionless and borne on short filaments, sometimes almost sessile. Gynoecium in ♂ flower rudimentary, in ♀ flower c. 5 mm. long, with globose ovary, slender style and 2-lobed stigma. Fruit up to 1·2 cm. in diameter, depressed-globose, without persistent style, (?) with thin fleshy mesocarp (not apparent in herbarium) and woody endocarp 1–2·5 mm. thick.

Mozambique. N: Namaita, between Nampula and Murrupula, c. 350 m., fr. 26.iii.1964, *Torre & Paiva* 11390 (LD; LISC; M; WAG). Z: Gúruè (Vila Junqueiro), c. 3 km. from Lioma, c. 700 m., fr. immat. & photogr. 10.xi.1967, *Torre & Correia* 16051 (COI; EA; LISC; MO). MS: Gorongosa National Park, near Serração de Urema, fr. 14.xii.1965, *Macedo* 1705 (LMA). Also known from Tanzania and Zaire. In *Brachystegia* woodland.

See note concerning the collection *Codd* 7316 under *E. schimperi.*

3. **Embelia upembensis** Taton in Bull. Jard. Bot. Nat. Belg. **47**: 197 (1977). Type from Zaire.

Shrub or small tree 1–5 m. high. Indumentum of multicellular simple and branched eglandular crispate hairs and shorter gland-tipped hairs (scarcely visible with a hand-lens). Young stems, petioles and inflorescences sparsely to densely whitish- or pale brownish-crispate-pilose; bark of older twigs pinkish-grey. Leaves chartaceous, not clustered at branch ends; lamina 1·8–6(8·4) × 1·3–2·2(3·7) cm., obovate, narrowly obovate or suborbicular; apex obtuse and often apiculate, or subacute, base acute or tapering; petiole 5–6(8) mm. long; upper leaf surface glabrous or sparsely glandular-hairy, with inconspicuous nervature, lower surface crispate-pilose especially on the prominent midrib and lateral nerves; margin entire or finely and unevenly dentate in the distal half. Plants dioecious; flowers pale green, pentamerous, borne in axillary 3–11(13)-flowered umbels or subumbellate racemes; peduncles 5–10 mm. long, pedicels 1·5–3 mm. long. Calyx cup-shaped, sepals 1·5 mm. long, almost free, triangular. Petals almost free, c. 3 mm. long, narrowly oblong-acute, densely papillose on adaxial surface. ♂ flowers: anthers c. 0·5 mm. long, elliptic or ovate, red; filaments 2·5–3 mm. long, attached to base of corolla; gynoecium rudimentary. ♀ flower: anthers triangular, functionless; filaments 1–1·5 mm. long; gynoecium with subglobose ovary 1 mm. long and slender, shortly 2-lobed style c. 2 mm. long. Fruit c. 4 mm. in diameter, purple, subglobose, faintly longitudinally ribbed when dry, subtended by the persistent reflexed sepals. Seed globose with deep basal cavity.

Zambia. N: Mbesuma Ranch, 1190 m., fr. 31.vii.1961, *Astle* 835 (SRGH). Known only from northern Zambia and Zaire (Shaba Prov.). In *Brachystegia* woodland, on lake shores and river terraces, often on termitaria.

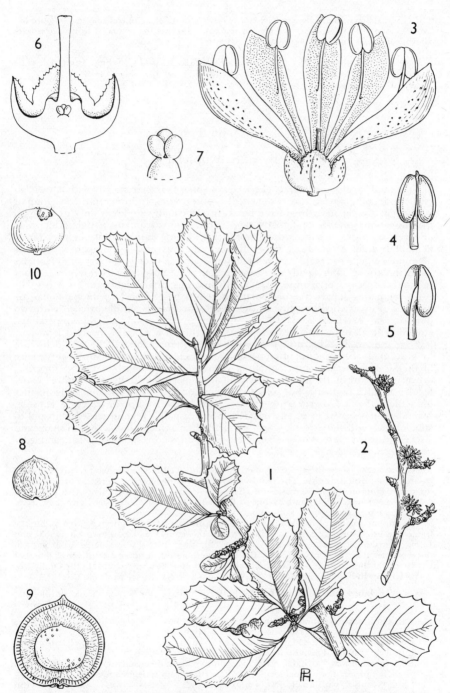

Tab. 41. EMBELIA XYLOCARPA. 1, habit (× ⅔), from *Milne-Redhead & Taylor* 10642; 2, habit (flowers)(× ⅔); 3, ♂ flower (× 8); 4, apiculate anther (× 12); 5, obtuse anther (× 12); 6, calyx and ovary in longitudinal section (× 16); 7, placenta, 2–7 from *Muir/Hay* 108; 8, fruit (× 1); 9, fruit in longitudinal section (× 2); 10, seed (enclosed in membrane) (× 2), 8–10, from *Milne-Redhead & Taylor* 10642. From F.T.E.A.

Closely related to the Ugandan *E. nilotica* Oliver, but differing in leaf shape: *E. nilotica* has broadly elliptic leaves usually rounded at the base, whereas the leaves of *E. upembensis* are obovate and tapering towards the base.

Note: The record of the W. African species *Embelia welwitschii* K. Schum. from Malawi by Burtt Davy & Hoyle, N.C.L.: 55 (1936), is erroneous.

106. SAPOTACEAE
By F. K. Kupicha

Trees and shrubs, very rarely (1 species) a liane, characteristically with latex in all parts; branching pattern often repeatedly subterminal. Indumentum consisting of T-shaped, Y-shaped or medifixed hairs (this is usually apparent only under high magnification; see Tab. 44, fig. 3). Leaves stipulate or exstipulate, petiolate, alternate or rarely opposite or subopposite (not in F.Z. area), simple, entire or very rarely dentate (not in F.Z. area), very often grouped at ends of branches. Flowers actinomorphic, ♀ or rarely ♀ by reduction of stamens, usually solitary or in fascicles in leaf axils or on older wood (occasionally plants cauliflorous), very rarely in racemes (not in F.Z. area). Calyx uniseriate, (4) 5 (6)-merous, or biseriate with $2 + 2$, $3 + 3$ or $4 + 4$ segments, sepals free or shortly united at base. Corolla gamopetalous, the number of lobes equalling that of calyx segments; petals simple or each with two dorsi-lateral petaloid appendages. Stamens oppositipetalous and epipetalous, equalling petals in number or more numerous (not in F.Z. area), inserted at various levels in the corolla-tube; anthers extrorse or less often introrse, 2-thecous, dehiscing longitudinally. Staminodes often present, always inserted on the corolla at the base of the sinus between lobes, equalling corolla-lobes in number. Ovary superior, syncarpous, conical or suborbicular, usually densely hairy, usually with as many locules as calyx segments but sometimes more or fewer; placentation axile; locules uniovulate. Fruit a berry with sticky, often edible, pulp, or rarely a capsule, several–many-seeded or often 1-seeded. Seeds usually ± compressed-ellipsoid to subglobose, with shiny testa; scar basal, lateral or basilateral, varying in shape and size; endosperm copious or ± absent and cotyledons correspondingly thin and leafy or swollen and fleshy.

The Sapotaceae comprise about 500–600 species almost all of which inhabit relatively humid environments in the tropics. The family itself is well-defined and coherent, but its generic limits are notoriously problematic; thus two recent world-wide treatments (Aubréville in Adansonia, Mém. 1 (1965) and Baehni in Boissiera **11** (1965)), both based on many years' research, are substantially in disagreement.

The present account of the family in south-east tropical Africa has been written within the framework established by A. Meeuse, the author of the family for Fl. Southern Afr. (1963), and J. H. Hemsley, its author for F.T.E.A. (1968); their two systems agree well, and all discrepancies have been discussed by the latter author. For a fuller discussion of different treatments of the Sapotaceae, and a helpful description of flower structure in the family, see J. H. Hemsley in F.T.E.A., Sapot.: 1–2 (1968).

N.B. The dimensions given for fruit are for fresh material; the berries shrink considerably when dried.

1. Calyx of 5 sepals arranged spirally or in 1 whorl -   -   -   -   -   -   2
 - Calyx of 2 distinct whorls of sepals, either $3 + 3$ or $4 + 4$ -   -   -   -   - 11
2. Corolla of 5 deeply 3-lobed petals united into a basal tube   -   -   **9. Inhambanella**
 - Corolla of 5 unlobed petals united into a tube of varying length   -   -   -   - 3
3. Leaf venation pattern characteristic: secondary nerves inconspicuous, surface of lamina divided into small polygonal areoles by the tertiary vein reticulation; seeds subspherical, faintly 5-ribbed, with small suborbicular basal scar   -   -   -   - **8. Sideroxylon**
 - Leaf venation not as above, secondary nerves easily distinguished and lamina not obviously marked into small isodiametric polygonal areoles; seed ± compressed-ellipsoid, with long lateral scar   -   -   -   -   -   -   -   -   -   -   -   -   -   4
4. Fruits 1–5-seeded; seeds containing abundant endosperm, cotyledons thin, leafy
                                                           1. Chrysophyllum
 - Fruits 1-seeded; seeds without endosperm, cotyledons thick and fleshy -   -   - 5

5. Lateral nerves of leaf closely spaced, straight and parallel, diverging at a wide angle from the midrib; lower leaf surface silvery or coppery sericeous with a dense indumentum of appressed hairs - - - - - - - - - **5. Bequaertiodendron**
– Leaves not as above, lateral nerves widely spaced and curving from midrib to margin, lamina not sericeous below - - - - - - - - - - - 6
6. Stipules present on youngest leaves - - - - - - - - - 7
– Stipules absent, even on younger leaves - - - - - - - - 9
7. Flowers solitary, sessile; free lobes of petal shorter than tube; persistent stipules forming a dense tuft at branch ends - - - - - - - - **7. Vincentella** p.p.
– Flowers densely fasciculate, pedicellate; free lobes of petals longer than tube; stipules soon caducous - - - - - - - - - - - - 8
8. Staminodes long and conspicuous, corolla-lobes completely reflexed at anthesis, revealing the ferrugineous-tomentose ovary - - - - - **7. Vincentella** p.p.
– Staminodes very small or absent; corolla-lobes erect at anthesis - **6. Pachystela**
9. Leaves with conspicuous characteristic venation pattern: main lateral nerves spanned by straight transverse tertiary veins - - - - - - - - - 10
– Leaves not as above, tertiary vein reticulation diffuse and faint - **2. Afrosersalisia**
10. Flowers pedicellate; staminodes present; anthers extrorse - - - **3. Aningeria**
– Flowers sessile; staminodes absent; anthers introrse - - - **4. Malacantha**
11. Calyx with 3 + 3 sepals; corolla of 6 deeply 3-lobed petals joined into a tube at the base; staminodes glabrous, standing away from the ovary; anthers relatively small (c. 1·5 mm. long)- - - - - - - - - - - - **12. Manilkara**
– Calyx with 4 + 4 sepals; corolla of 8 deeply 3-lobed petals joined into a tube at the base; staminodes densely pilose abaxially and on margins and loosely coherent into a cone round the ovary and style; anthers relatively large (2·5 mm. long or more) - - - 12
12. Leaves clustered at branch ends; seeds with large scar covering up to ½ surface area; endosperm absent - - - - - - - - - **11. Vitellariopsis**
– Leaves not clustered at branch ends; seeds with small basal scar, containing abundant endosperm - - - - - - - - - - **10. Mimusops**

1. CHRYSOPHYLLUM L.

Chrysophyllum L., Sp. Pl. **1**: 192 (1753); Gen. Pl., ed. 5: 88 (1754).

Gambeya Pierre, Notes Bot. Sapot.: 61 (1891).
Donella Pierre ex Baillon, Hist. Pl. **11**: 294 (1891).
Austrogambeya Aubrév. & Pellegrin in Adansonia **1**: 7 (1961).

Trees or shrubs, sometimes scandent. Leaves coriaceous or chartaceous, not or sometimes clustered at branch ends; lamina sometimes with lateral nerves very numerous, close, straight and parallel. Stipules absent. Flowers usually ☿ (flowers with reduced anthers occur sporadically in some species), 5-merous, borne in few- to many-flowered fascicles in axils of current leaves or on older wood. Sepals ± free to the base. Corolla cylindric, urceolate or campanulate; lobes entire, equalling or shorter than tube. Stamens not exserted; filaments inserted at level of base of lobes or in tube; anthers extrorse. Alternipetalous staminodes usually absent. Ovary conical, densely pilose, tapering into a short thick style, usually 5-locular; ovules solitary, with lateral or basi-lateral attachment. Fruit a subglobose to pear-shaped berry, often with lobes corresponding with number of seeds; pericarp fleshy or subcoriaceous. Seeds (1) 2–5, with long narrow lateral scar; testa hard, smooth and shiny; endosperm copious; cotyledons thin and leafy.

A genus centred in tropical America, with c. 40 species in the Old World tropics.

Aubréville considers that *Chrysophyllum* is an exclusively American genus, and that African representatives should be classified in *Gambeya, Donella, Austrogambeya* and *Gambeyobotrys* Aubrév.; see Adansonia **12**: 187 (1972).
   Chrysophyllum cainito L., native to tropical America, is known from a single collection from Malawi, S: Mandala, Scott's Mission, fl. 16.xii.1959, *Willan* 4 (K). This species, the type of the genus, is easily recognised by its leaves which are elliptic, with closely spaced lateral nerves, acute or shortly acuminate at apex, glossy above and strikingly ferrugineous-sericeous beneath. It is prized for its delicious fruit known as the Caimito or Star apple.
   C. albidum G. Don is recorded from Malawi by Burtt Davy & Hoyle (N.C.L.: 70, 1936), but this species does not occur in S.E. tropical Africa further south than Uganda and Kenya.

1. Leaves chartaceous; lateral nerves numerous, closely spaced (less than 2 mm. apart), parallel, joining in a prominent marginal vein; lower leaf surface glabrous, except for the usually brownish-pubescent midrib - - - - - - - - - 2

- Leaves coriaceous; lateral nerves not as above, well over 2 mm. apart and not joining in a prominent marginal vein; lower leaf surface with dense indumentum at least in young leaves
3

2. Tree up to 35 m. tall; petioles 5–8 mm. long; fruit subglobose, slightly 3–5-ribbed; filaments pilose - - - - - - - - - - - - - - 1. *viridifolium*
- Small shrub, or climber attaining up to 20 m.; petioles 2–4 mm. long; fruit ovoid, contracted into a narrow beak at apex, 3–4-ribbed; filaments glabrous - - - 2. *welwitschii*

3. Tree 5–45 mm. tall; leaves densely silvery- or ferrugineous-sericeous on lower surface; anthers without hairs at apex - - - - - - - - 3. *gorungosanum*
- Shrub or small tree 2–5(10) m. tall; lower leaf surface ferrugineous or whitish crispate-lanate; anthers with a small tuft of hairs at apex - - - - - 4. *bangweolense*

1. **Chrysophyllum viridifolium** Wood & Franks in Wood, Natal Pl. **6**: t. 569 (1911). — Gerstner in Journ. S. Afr. Bot. **12**: 48, fig. 3 (1946). — A. Meeuse in Bothalia **7**: 328 (1960); in Fl. Southern Afr. **26**: 35 (1963). — J. H. Hemsley in F.T.E.A., Sapotaceae: 16 (1968). — Palmer & Pitman, Trees of Southern Afr. **3**: 1739, with fig. (1972). — R. B. Drumm. in Kirkia **10**: 266 (1975). Type from S. Africa (Natal).
    Chrysophyllum welwitschii sensu Gomes e Sousa, Essen. Fl. Inhambane: 19 (1943); Dendrol. Moçamb. **2**: 626 (1967), non Engl.
    Donella viridifolium (Wood & Franks) Aubrév. & Pellegrin in Notul. Syst. **16**: 248 (1960). Type as for *Chrysophyllum viridifolium*.

Evergreen tree up to 35 m. tall; bole long and clear, sometimes deeply fluted at the base; bark fairly smooth, pale grey, shallowly longitudinally fissured; slash white, with red underbark. Twigs much branched in the horizontal plane, somewhat zig-zag, leafy throughout. Apical buds and young leaves ferrugineous-pubescent, soon glabrescent; bark of twigs grey, finely longitudinally wrinkled. Leaves chartaceous; lamina 4–9·5 × 1·5–5 cm., oblong or narrowly ovate-oblong, contracting abruptly at apex into a blunt-tipped mucro or acumen, cuspidate at base; margin finely undulate. Petiole 5–8 mm. long, brownish-pubescent. Upper leaf surface glossy, glabrous; midrib impressed, lateral nerves numerous, straight, parallel, joined by a continuous marginal vein. Lower surface mat, glabrous except for the very prominent brownish-pubescent midrib. Flowers white, in axillary fascicles; pedicels 2–5 mm. long, slender; buds c. 2 mm. long. Calyx c. 1·6 mm. long, cup-shaped; sepals strongly imbricate, outer surfaces appressed-pubescent, margins ciliate. Corolla c. 1·8 mm. long, campanulate; tube c. 0·8 mm. long, lobes ovate, ciliate. Stamens inserted near base of tube; filaments pilose with long white hairs; anthers c. 0·6 mm. long. Gynoecium c. 1·6 mm. long. Fruit up to 3·5 cm. in diameter, subglobose, slightly 3–5-ribbed, short-stalked, glabrous, green at first, becoming yellow and edible when ripe. Seeds 3–5, up to 1·8 cm. long.

Zimbabwe. E: Chipinga Distr., Chirinda Forest, c. 1100 m., fl. i.1962, *Goldsmith* 9/62 (BM; COI; FHO; K; LISC; PRE; SRGH). **Mozambique.** MS: E. slopes of Mt. Zembe, c. 850 m., st. 20.vii.1970, *Müller & Gordon* 1365 (K; LISC; PRE; SRGH). GI: Homoine, 150 m., st. x.1955, *Gomes e Sousa* 1678 (COI; K; PRE).
. Occurring also in Kenya and S. Africa (Natal) and Swaziland. In mixed evergreen woodland.

Closely related to *C. pruniforme* Engl., which is widespread in lowland rainforest in tropical Africa from Sierra Leone to Uganda; this species differs from *C. viridifolium* chiefly in the shape of its leaves and fruit.

2. **Chrysophyllum welwitschii** Engl., Bot. Jahrb. **12**: 521 (1890); Mon. Afr. Pflanzen. **8**: 41, t. 13 fig. A (1904). — Hiern, Cat. Afr. Pl. Welw. **1**: 641 (1898). — A. Meeuse in Bothalia **7**: 329 (1960). — F. White, F.F.N.R.: 321 (1962). — Heine in F.W.T.A. ed. 2, **2**: 27 (1963). — Fanshawe, Check-list Woody Pl. Zamb.: 10 (1973). Type from Angola.
    Sideroxylon sp. — Ficalho, Pl. Ut. Afr. Port.: 211 (1884).
    Micropholis angolensis Pierre, Notes Bot. Sapot.: 41 (1891). Type from Angola.
    Chrysophyllum klainei Pierre ex Engl., Mon. Afr. Pflanzen. **8**: 42, t. 14 fig. B (1904). Syntypes from Gabon.
    Chrysophyllum ealaense De Wild., Pl. Bequaert. **4**: 128 (1926). Type from Zaire.

Evergreen climber up to 20 m. high with very slender trunk, or shrub 3–5 m. tall; twigs long, leafy, often only sparingly branched. Apical buds, young stems and petioles appressed brownish-pubescent; older branches glabrescent, with smooth brown bark. Leaves chartaceous; lamina 4–11 × 1·5–4·5 cm., oblong, oblong-obovate or oblong-elliptic, apex cuspidate-acuminate with narrowly oblong tip 5–7 mm. long, base rounded or acute; petiole 2–4 mm. long. Lamina glabrous on both surfaces except for midrib of lower surface, this very prominent and sometimes brownish-pubescent. Lateral nerves numerous, closely spaced, parallel, joined into a continuous prominent marginal vein. Flowers white to greenish-yellow, in few- to many-flowered fascicles in leaf axils; pedicels 1–3 mm. long; buds c. 1 mm. long, globose. Calyx c. 1·5 mm. long, of 5 orbicular, cucullate, strongly imbricate sepals sparsely pilose on dorsal surface. Corolla c. 2 mm. long, urceolate, the tube ± equalling lobes, these elliptic, ciliate. Stamens inserted at base of corolla-tube; anthers c. 0·5 mm. long; filaments glabrous. Gynoecium c. 1 mm. long. Fruits up to 4·5 × 3 cm., ovoid, tapering at apex into a narrow beak, 3–4-seeded, 3–4-lobed (at least when dry), greenish. Seeds c. 1·7 cm. long, testa mottled yellowish-brown.

Zambia. W: Mwinilunga Distr., Muzera R., 16 km. W. of Kakoma, st. 29.ix.1952, *White* 3406 (FHO; K).

W. tropical Africa, Zaire and Angola. This species is found only in the extreme NW. part of the F.Z. area, here occurring in gallery forest. It is the only scandent member of the family.

3. **Chrysophyllum gorungosanum** Engl., Mon. Afr. Pflanzen. **8**: 44, t. 15 fig. B (1904). — Burtt Davy & Hoyle, N.C.L.: 70 (1936). — Brenan in Mem. N.Y. Bot. Gard. **8**: 498 (1954). — Topham, N.C.L. ed. 2: 94 (1958). — A. Meeuse in Bothalia **7**: 329, fig. 3 (1960). — Heine in F.W.T.A. ed. 2, **2**: 28 (1963). — J. H. Hemsley in F.T.E.A., Sapotaceae: 10, fig. 1 (1968). — Fanshawe, Check-list Woody Pl. Zambia: 10 (1973). — R. B. Drumm. in Kirkia **10**: 266 (1975). TAB. **42**. Type: Mozambique, Gorongosa Mts., *Carvalho* (B, holotype†; COI, isotype).

Chrysophyllum fulvum S. Moore in Journ. Linn. Soc., Bot. **40**: 131 (1911). — Eyles in Trans. Roy. Soc. S. Afr. **5**: 437 (1916). — Steedman, Trees etc. S. Rhod.: 62 (1933). — Burtt Davy & Hoyle, loc. cit. Type: Zimbabwe, Chipinga Distr., Chirinda Forest, *Swynnerton* 19 (BM, holotype; K, isotype).

Chrysophyllum boivinianum sensu F. White, F.F.N.R.: 320 (1962) non (Pierre) Baehni.

Regularly branched evergreen tree 5–45 m. tall, the larger specimens with fluted bole. Young branchlets densely ferrugineous appressed-tomentose, older twigs glabrous, with grey, roughish, longitudinally fissured bark. Leaves coriaceous, clustered at ends of branches. Lamina 5–20 × 1·5–5·5 cm., oblanceolate to narrowly obovate or less often elliptic, the apex usually cuspidate-acuminate or sometimes obtuse, the base narrowly acute. Petiole 8–20 mm. long. Upper leaf surface dark green, appressed-pilose to glabrescent, with inconspicuous impressed midrib and lateral nerves and a fine raised reticulation. Lower surface densely silvery- or ferrugineous-sericeous, with conspicuous prominent midrib and lateral nerves, the latter 16–21 on each side, diverging from the midrib at c. 60°, almost straight except near the margin, here becoming faint and curving towards apex. Flowers subsessile, borne in axils of current leaves or on slightly older wood, in dense fascicles or sometimes solitary. Calyx c. 3·2 mm. long, cup-shaped; lobes ovate, strongly imbricate, appressed-pubescent on both sides, the outer ones a little larger and more leathery. Corolla c. 3·6 mm. long, white; lobes ± equalling tube, ovate, slightly auricled at base, ciliate. Stamens inserted near base of corolla-tube; filaments c. 1·6 mm. long, glabrous; anthers c. 1·2 mm. long. Small fleshy staminodes sometimes present. Gynoecium c. 2·2 mm. long. Fruit up to 4 × 3 cm., globose to ellipsoid, sometimes somewhat apiculate, with dense reddish-brown pubescence often rubbing away in patches, 4–5-seeded. Seeds up to 2·8 × 1·2 cm., dark brown or blackish.

Zambia. N: Isoka Distr., Mafinga Mts., 6 km. W. of Chisenga Rest House, 1980 m., st. 21.xi.1952, *White* 3734 (K). **Zimbabwe.** E: Chirinda Forest, fl. 2.i.1947, *Chase* 427 (BM; K; LISC; SRGH). **Malawi.** N: Chitipa Distr., Misuku Hills, Mughese Forest Reserve, c. 1850 m., st. 15.ix.1970, *Müller* 1660 (K; MAL; SRGH). C: Ntchisi Mt., 1650 m., fr. 31.vii.1946, *Brass* 17067 (K; SRGH). S: Cholo [Thyolo] Mt., 1430 m., st. 3.iv.1963, *Chapman* 1841 (MAL; SRGH). **Mozambique.** N: Ribáuè, serra Mepáluè, c. 1400 m., fl. 5.xii.1967, *Torre & Correia* 16372 (LISC). Z: Gúruè, near source of R. Malema, c. 1750 m., fl. & fr. 4.i.1968, *Torre & Correia* 16909 (BR; COI; EA; LISC; MO). MS: Garuso Forest, st. iv.1935, *Gilliland* 1810 (BM; K).

Cameroon to Angola and Zaire, Uganda, Kenya and Tanzania. Common in primary rain forest.

4. **Chrysophyllum bangweolense** R. E. Fries, Wiss. Ergebn. Schwed. Rhod.-Kongo-Exped. **1**: 254, fig. 29 (1916). — F. White, F.F.N.R.: 320 (1962). — J. H. Hemsley in F.T.E.A., Sapot.: 14 (1968). — Fanshawe, Check-list Woody Pl. Zamb.: 10 (1973). Syntypes: Zambia, near L. Bangweulu, *Fries* 909 & 909a (UPS).

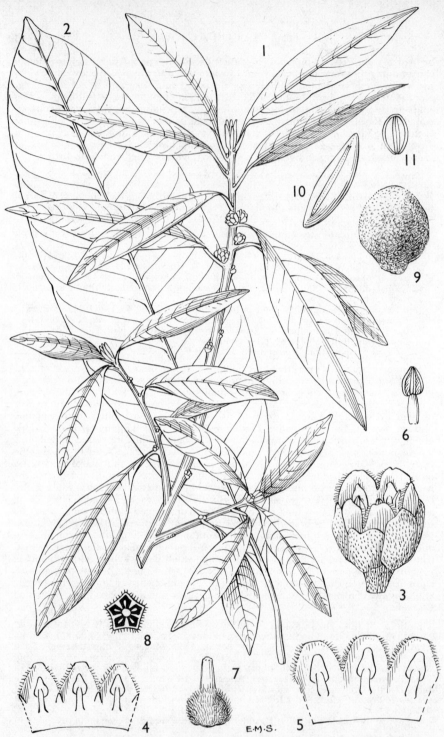

Tab. 42. CHRYSOPHYLLUM GORUNGOSANUM. 1, flowering branch (× ⅔), from *Hockliffe* in F.D. 1390; 2, sapling leaf (× ⅔), from *Drummond & Hemsley* 4364; 3, flower (× 6), from *Eggeling* 1510 in F.D. 1440; 4, section of corolla with stamens and small staminodes (× 6), from *Battiscombe* 560; 5, section of corolla lacking staminodes (× 6); 6, stamen (× 6); 7, ovary (× 6); 8, diagrammatic transverse section of ovary, 5–8 from *Eggeling* 1510 in F.D. 1440; 9, fruit (× ⅔); 10, seed (× 1); 11, transverse section of seed (× 1), 9–11, from *Drummond & Hemsley* 4365. From F.T.E.A.

Chrysophyllum cacondense Greves in Journ. Bot., Lond. **65**, Suppl. 2: 72 (description begins) (1927); op. cit. **67**, Suppl. 2: 73 (description ends) (1929). Type from Angola.
Austrogambeya bangweolensis (R. E. Fries) Aubrév. & Pellegrin in Adansonia **1**: 7, t. 1 (1961). Syntypes as for *Chrysophyllum bangweolense*.

Shrub or small tree 2–5(10) m. tall with rough reddish-brown bark. Twigs much branched, often in one plane, not or slightly zig-zag, leafy throughout, the apices densely ferrugineous-tomentose; older twigs with rough grey scaly bark. Leaves thick, leathery; lamina 4·5–9·5(12) × 2·25–3·5(5) cm., elliptic, oblong-elliptic or narrowly obovate, the apex contracted into a short rounded mucro, the base acute; petiole 2–4 mm. long; leaves at first rusty- or whitish-crispate-lanate, glabrescent on both sides but especially so above. Upper surface glossy, with impressed or level midrib and conspicuous, curved, not strictly parallel lateral nerves. Lower surface mat, with prominent midrib. Flowers in axillary fascicles; pedicels 1–3 mm. long; buds 2–2·5 mm. long. Calyx c. 2·4 mm. long, cup-shaped; lobes suborbicular, strongly imbricate, outer surface sparsely appressed-pubescent. Corolla c. 2·8 mm. long, white, campanulate; tube c. 1 mm. long, lobes suborbicular, ciliate. Stamens attached near base of tube; filaments flattened, glabrous; anthers c. 0·6 mm. long, with short tuft of hairs at apex. Gynoecium c. 1·8 mm. long. Fruit up to 4·5 cm. in diameter, depressed-subglobose, short-stalked, pale yellowish green. Seeds up to 2·1 cm. long, brown.

Zambia. B: near Mongu, 1045 m., fr. vi.1933, *Trapnell* 1293 (K). N: Luapula Distr., L. Mweru, fl. 27.v.1961, *Astle* 713 (K; SRGH). W: Mwinilunga Distr., near mile post 54 on Angola road, fl. 26.ix.1952, *Holmes* 924 (FHO; K; PRE; SRGH).
Known also from Zaire, Angola and Tanzania. In woodland and evergreen thicket.

According to Fanshawe, loc. cit., this species also occurs in the Central and Eastern provinces of Zambia.
Irregular flowers with one or more rudimentary stamens or with a stamen in an alternipetalous, staminodial position are sometimes found.
The comments of A. Meeuse (in Bothalia **7**: 329, 1960) about the affinities and geographical distribution of *C. bangweolense* are erroneous.

2. AFROSERSALISIA A. Chev.

Afrosersalisia A. Chev. in Rev. Bot. Appl. Agric. Trop. **23**: 292 (1943). — J. H. Hemsley in Kew Bull. **20**: 478 (1966).
Sersalisia sensu auct. mult. non R. Br. (1810).
Pouteria sensu Baehni in Candollea **7**: 489 (1938) pro parte et sensu A. Meeuse in Bothalia **7**: 341 (1960) pro parte, non Aubl. (1775).
Amorphospermum sensu Baehni in Boissiera **11**: 102 (1965) pro parte, non F. Muell. (1870).

Trees and shrubs. Stipules absent. Leaves moderately coriaceous, glabrous; lateral nerves fairly distant, curving towards apex. Flowers fascicled in axils of current and fallen leaves, pedicellate. Calyx of 5 sepals, united up to ½ their length. Corolla lobes equalling or longer than tube. Stamens inserted at mouth of tube; filaments short, anthers erect, extrorse. Staminodes present, very small, ± oblong, dentate at apex. Ovary ovoid or narrowly conical, pilose with short style. Fruit a fleshy red 1-seeded berry with calyx persisting at base. Seed ellipsoid, with wide scar extending the length of the seed; endosperm absent, cotyledons thick and fleshy.

A tropical and subtropical African genus of c. 5 species.

Leaves rounded to obtuse at apex; lamina surfaces ± smooth, not finely and evenly rugose; lateral nerves impressed above, prominent beneath - - - - 1. *cerasifera*
Leaves long-acuminate or cuspidate-acuminate at apex; both surfaces of lamina finely and evenly rugose, with impressed lateral nerves - - - - - - 2. *kassneri*

1. **Afrosersalisia cerasifera** (Welw.) Aubrév. in Bull. Soc. Bot. Fr. **104**: 281 (1957). — F. White, F.F.N.R.: 320 (1962). — Heine in F.W.T.A. ed. 2, **2**: 30 (1963). — J. H. Hemsley in Kew Bull. **20**: 482 (1966); in F.T.E.A., Sapotaceae: 42, fig. 8 (1968). — Fanshawe, Check-list Woody Pl. Zambia: 11 (1973). — R. B. Drumm. in Kirkia **10**: 266 (1975). TAB. **43**. Syntypes from Angola.
Sapota cerasifera Welw., Apont. in Ann. Cons. Ultramar., Parte Naõ Off., Sér. 1, **1858**: 585 (1859). Syntypes as above.
Chrysophyllum disaco Hiern, Cat. Afr. Pl. Welw. **3**: 642 (1898). Syntypes from Angola.

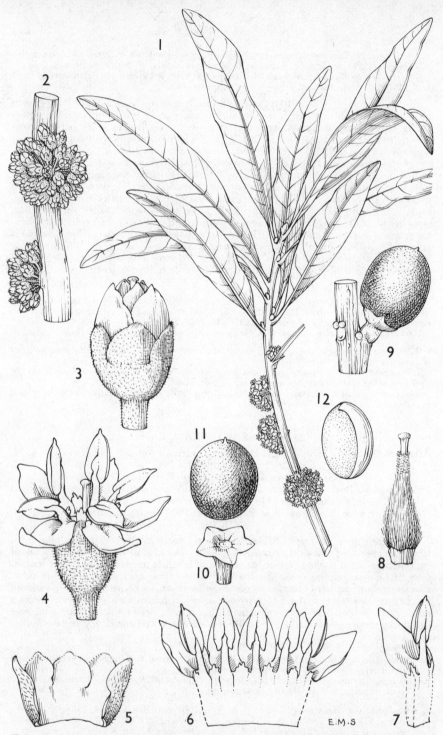

Tab. 43. AFROSERSALISIA CERASIFERA. 1, flowering branchlet (× ½), from *Semsei* 1406; 2, clusters of young flowers (× 1), from *Semsei* 885; 3, young flower (× 6); 4, flower (× 6); 5, dissected calyx (× 6); 6, corolla dissected to show stamens and small staminodes (× 6); 7, corolla segment and stamen (× 6); 8, ovary (× 6), 3–8, from *Semsei* 1406; 9, fruit, attached (× 1), from *Greenway* 1004; 10, fruiting pedicel and cupular calyx (× 1); 11, fruit (× 1); 12, seed (× 1), 10–12 from *Drummond & Hemsley* 3154. From F.T.E.A.

Chrysophyllum cerasiferum (Welw.) Hiern, op. cit.: 643 (1898). Syntypes as for *Afrosersalisia cerasifera*.

Sersalisia disaco (Hiern) Engl., Mon. Afr. Pflanzen. **8**: 30 (1904). — Syntypes as for *Chrysophyllum disaco*.

Sersalisia cerasifera (Welw.) Engl., loc. cit. Syntypes as for *Afrosersalisia cerasifera*.

Pouteria cerasifera (Welw.) A. Meeuse in Bothalia **7**: 341 (1960). Syntypes as above.

Pouteria disaco (Hiern) A. Meeuse, loc. cit. Syntypes as for *Chrysophyllum disaco*.

For fuller synonymy, see J. H. Hemsley, tom. cit.: 478–483 (1966).

Evergreen tree 7–40 m. tall; trunk straight, often unbranched for a considerable distance, with rough, pale to dark grey, scaly or longitudinally fissured bark; slash pale brown or pink, turning orange; lower part of trunk fluted and buttressed. Young shoots ± glabrous. Leaf-lamina 3–21 × 1·2–5·5 cm., elliptic to narrowly elliptic-obovate, the apex rounded, the base acuminate; petiole 1–4 cm. long. Upper leaf surface smooth, glabrous, somewhat glossy, with faintly impressed midrib and inconspicuous lateral nerves 7–14 on each side; lower surface glabrous, mat, with fairly prominent main nerves and a faintly visible reticulation; leaves often drying brownish. Flowers densely fasciculate in axils of current and fallen leaves; pedicels up to 7 mm. long. Calyx c. 2·5 mm. long, comprising 5 hard ovate sepals fused to half their length and forming a firm cylinder constricted at the apex. Corolla yellowish-cream to greenish, the lower half forming a narrow tube within the calyx, expanding above into 5 patent to reflexed free lobes c. 2·5 mm. long. Anthers c. 1·5 mm. long, apiculate. Staminodes c. 0·5 mm. long, inconspicuous. Ovary c. 4 mm. long, narrowly conical. Fruit solitary, up to 2·5 × 2 cm., a red ovoid to subglobose berry with persistent style, persistent calyx forming a woody cupule at base, and thick woody stalk 3–10 mm. long. Seed up to 2 cm. long; scar occupying ⅔ or more of surface area.

Zambia. N: S. of L. Young, 1520 m., fr. 23.vii.1938, *Greenway & Trapnell* 5481 (K). W: Mwinilunga Distr., Zambesi R. 6 km. N. of Kalene Hill, fl. buds 23.ix.1952, *Angus* 531 (BM; COI; FHO; K; PRE). **Malawi.** N: Nkhata Bay Distr., Chombe Estate, 30 m., fr. 21.vi.1960, *Adlard* 379 (FHO; MAL; SRGH). **Mozambique.** Z: Gúruè, near crossing of Licungo R. on Lioma to Namarroi road, fr. 15.viii.1949, *Andrada* 1876 (COI; LISC). MS: Beira, N. of Manga, st. 7.viii.1971, *Masterson* 1493 (LISC; SRGH).

Throughout most of tropical Africa. Riverine forest, lowland mixed woodland and evergreen forest.

2. **Afrosersalisia kassneri** (Engl.) J. H. Hemsley in Kew Bull. **20**: 483 (1966); in F.T.E.A., Sapotaceae: 44 (1968). Type from Kenya.

Sersalisia kassneri Engl., Mon. Afr. Pflanzen. **8**: 31 (1904) (genus queried by original author). Type as above.

Pouteria kassneri (Engl.) Baehni in Candollea **9**: 280 (1942) (genus again queried). Type as above.

Tulestea kassneri (Engl.) Aubrév. in Adansonia **12**: 191, t. 1 (1972). Type as above.

Small tree or shrub 2–3 m. high; bark brown, minutely fissured, slash pale pink. Branching repeatedly subterminal with leaves confined to branch ends. Young stems and petioles ferrugineous appressed-pubescent, older twigs glabrous with finely fissured brownish-grey bark. Leaves 6–12 × 2–3·3 cm., narrowly oblong-elliptic or elliptic-obovate, the apex long-acuminate or cuspidate-acuminate with rounded tip, the base very long-acuminate; margin finely undulate; both sides of leaf with midrib raised, lateral nerves very faint and surface finely rugose. Flowers (probably) fasciculate, borne in leaf axils and on older wood; pedicels up to 1·5 mm. long. Calyx c. 1·5 mm. long, cup-shaped, the lobes nearly free, broadly ovate, puberulous. Corolla whitish, up to 2·5 mm. long with tube c. 1 mm. long; lobes ovate. Staminodes very small, ± oblong, dentate at apex. Ovary c. 1 mm. long, ovoid. Fruit solitary, a fleshy red 1-seeded berry with flower parts persisting at base. Seed 12 × 7 × 3 mm., compressed-ellipsoid, glossy brown, with duller elliptic scar c. 5 mm. wide extending the length of the seed. Cotyledons large, plano-convex, endosperm 0.

Zimbabwe. E: Melsetter Distr., Makurupini-Haroni Forest, st. 22.iv.1973, *Mavi* 1437 (K; SRGH). **Mozambique.** MS: Haroni-Makurupini Forest, 400 m., fr. 4.xii.1964, *Wild, Goldsmith & Müller* 6645 (K; SRGH).

Known also from Kenya and Tanzania, and reported from Gabon (by Aubréville, loc. cit.). Medium-altitude moist evergreen forest.

The fruit and seed of this rare species, which are described here in some detail, were unknown until their appearance in the recent collection of *Wild et al.* from Mozambique.

3. ANINGERIA Aubrév. & Pellegrin

Aningeria Aubrév. & Pellegrin in Bull. Soc. Bot. Fr. **81**: 795 (1935).
Pouteria sensu Baehni in Candollea **7**: 418 (1938) pro parte et sensu A. Meeuse in
Bothalia **7**: 341 (1960) pro parte, non Aubl. (1775).
Rhamnoluma sensu Baehni in Boissiera **11**: 115 (1965) pro parte non Baill. (1891).

Tall trees. Leaves exstipulate, chartaceous or coriaceous, pellucid-punctate
(obscurely so in *A. adolfi-friedericii*), densely pubescent to subglabrous; lateral
nerves many-paired, looped distally and running parallel to leaf-margin or meeting it
at a very acute angle; tertiary veins conspicuous, spanning lateral nerves at right
angles. Flowers (4) 5 (6)-merous, pedicellate, fasciculate in axils of current or fallen
leaves. Sepals ± free to base. Corolla with tube longer than lobes. Stamens inserted
near top of tube, anthers extrorse, included. Staminodes present, small, ± subulate
or rarely petaloid. Ovary subglobose, densely pilose, with long cylindrical style
slightly expanded towards the apex. Fruit a subglobose to narrowly ellipsoid berry,
sometimes beaked. Seed solitary, ± ellipsoid; testa shining brown; scar covering up
to half surface area of seed; endosperm absent, cotyledons fleshy.

A tropical African genus of c. 4 species.

Lower leaf surface with principal nerves ferrugineous-pilose; petioles 1·5–2 cm. long; calyx c. 6
mm. long; fruit up to 4 cm. long - 　 - 　 - 　 - 　 1. *adolfi-friedericii* subsp. *australis*
Lower leaf surfaces completely glabrous or principal nerves whitish lanate-puberulous;
petioles up to 1 cm. long; calyx c. 4 mm. long; fruit up to 2 cm. long 　 - 　 - 2. *altissima*

1. **Aningeria adolfi-friedericii** (Engl.) Robyns & Gilbert in Robyns & Tournay, Fl. Parc
Nat. Alb. **2**: 43 (1947). Syntypes from Rwanda and Zaire.
　　Sideroxylon adolfi-friederici Engl. in Mildbr., Wiss. Ergebn. Deutsche Zentr.-Afr.
Exp. 1907–1908, **2**: 519 (1913). Syntypes as above.

Subsp. **australis** J. H. Hemsley in Kew Bull. **15**: 282 (1961); in F.T.E.A., Sapotaceae: 31
(1968). Type from Tanzania.
　　Aningeria adolfi-friedericii sensu F. White, F.F.N.R.: 320 (1962), non (Engl.) Robyns
& Gilbert (1947) sensu stricto.

Evergreen tree up to 50 m. tall or more, with strongly buttressed bole and mottled,
grey, shallowly longitudinally fissured bark; slash pale pinkish-brown with lighter
striations. Young stem ferrugineous-pubescent, older twigs tardily glabrescent, with
smooth bark. Leaves coriaceous to thinly membranous, very minutely and obscurely
pellucid-punctate, not confined to branch ends. Lamina 12–19 × 4–6 cm., obovate-
elliptic to narrowly oblong-obovate, the apex cuspidate-acuminate, the base
cuspidate to acute; petioles 1·5–2 cm. long. Lateral nerves 15–20 on each side. Upper
leaf-surface glabrous, glossy, with impressed nerves; lower surface with prominent
ferrugineous-pilose nerves, glabrous between. Pedicels up to 1 cm. long, densely
pubescent. Calyx up to 6 mm. long, sepals elliptic. Corolla up to 9 mm. long, with
long tube and short rounded lobes, pale yellowish. Anthers c. 1·5 mm. long. Style up
to 6·5 mm. long. Fruit up to 4 cm. long, narrowly ellipsoid, beaked, puberulous. Seed
up to 3 cm. long.

Zambia. E: Lundazi Distr., Kangampande Mt., Nyika Plateau, 2130 m., st. 7.v.1952, *White*
2764 (FHO; K). **Zimbabwe.** E: Inyanga Distr., Mtarazi Falls, c. 1220 m., st. 6.xii.1966,
Müller 584 (K; PRE; SRGH). **Malawi.** N: Misuku Hills, Mughesse Forest Reserve, c. 1850
m., st. 15.ix.1970, *Müller* 1663 (K; MAL; SRGH). C: Nkhota-Kota Distr., Ntchisi, st.
23.iii.1963, *Chapman* 1826 (SRGH).
　　Subsp. *australis* is known also from Tanzania. *Aningeria adolfi-friedericii* extends from
eastern Zaire to SW. Ethiopia and southwards to Kenya and Tanzania; it is characteristically
found in upland rainforest. According to Fanshawe, Check-list Woody Pl. Zamb.: 4 (1973), the
species also occurs in the Northern province of Zambia, but I have seen no material from that
region.

The five subspecies recognised by Hemsley, loc. cit., are distinguished mainly by differences
in leaf-size, shape and indumentum.

2. **Aningeria altissima** (A. Chev.) Aubrév. & Pellegrin in Bull. Soc. Bot. Fr. **81**: 796 (1935).
— Verdc. in Kew Bull. **11**: 453 (1957). — J. H. Hemsley in Kew Bull. **15**: 277 (1961); in
F.T.E.A., Sapotaceae: 27 (1968). — Heine in F.W.T.A. ed. 2, **2**: 24 (1963). — Fanshawe,
Check-list Woody Pl. Zambia: 4 (1973). TAB. **44**. Syntypes from Guinea.
　　Hormogyne altissima A. Chev. in Mém. Soc. Bot. Fr. **8**: 265 (1917). Syntypes as above.

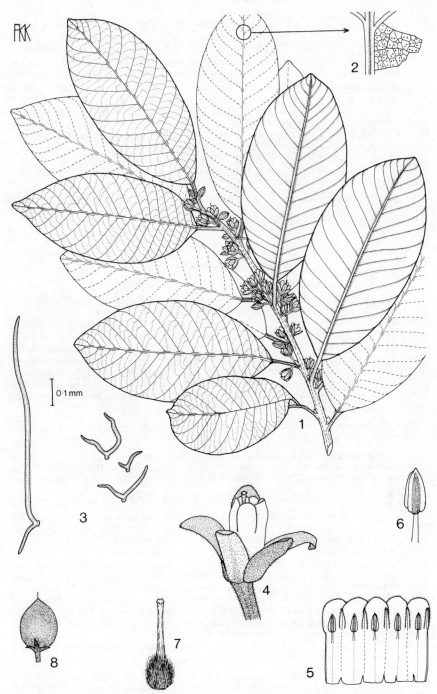

Tab. 44. ANINGERIA ALTISSIMA. 1, habit (×⅔); 2, magnified detail of leaf showing punctate appearance; 3 hairs of ovary (left) and pedicel (right) showing the lateral attachment characteristic of the family Sapotaceae; 4, flower (×4); 5, corolla opened out showing stamens and staminodes (×4); 6, stamen, ventral view (×12); 7, gynoecium (×4), 1–7 from *Chandler* 1994; 8, young fruit (×1) *Harris* 838.

Sideroxylon altissimum (A. Chev.) Hutch. & Dalz. in F.W.T.A. **2**: 12 (1931). Syntypes as above.
Pouteria altissima (A. Chev.) Baehni in Candollea **9**: 292 (1942). Syntypes as above.

Tall tree, up to 50 m., with deeply fluted bole and spreading crown; bark of trunk pale creamy grey, irregularly and shallowly fissured. Young stems yellowish puberulous, soon glabrescent, older twigs dark grey with wrinkled surface. Leaves membranous, pellucid-punctate, not confined to branch ends. Lamina 4–16 × 2·5–6 cm., ovate or obovate-oblong, apex rounded or more often cuspidate with blunt to emarginate apiculum, base rounded or cuspidate; petioles up to 1 cm. long. Lateral nerves 11–23 on each side. Leaves completely glabrous or with midrib and lateral nerves of lower surface lanate-puberulous. Pedicels up to 1 cm. long. Calyx c. 4 mm. long, sepals elliptic or oblong. Corolla 5–6 mm. long, greenish-yellow. Anthers c. 1 mm. long. Style up to 4 mm. long. Fruit up to 2 cm. in diameter and red when ripe, obovoid to subglobose. Seed up to 1·5 cm. long.

Zambia. N: Mbala (Abercorn) to Kalambo Falls road, st. 30.vii.1949, *Greenway & Hoyle* 1100 (FHO).
Tropical Africa from Gabon to Ethiopia. Dominant tree in evergreen forest.

4. MALACANTHA Pierre

Malacantha Pierre, Notes Bot. Sapot.: 60 (1891).

Deciduous tree. Leaves exstipulate, membranous, pellucid-punctate, villous on lower surface; lateral nerves extending to the leaf edge to form the thickened margin, spanned at right angles by conspicuous tertiary veins. Flowers 5-merous, sessile, in dense fascicles in axils of current or fallen leaves, the buds accompanied by hairy sepal-like bracts. Calyx densely pilose; sepals free to base, strongly imbricate. Corolla with tube longer than lobes. Stamens inserted in corolla-tube, anthers introrse, included. Staminodes absent. Ovary subglobose, densely pilose; style cylindrical, slightly expanded towards apex. Fruit a sessile subglobose berry with persistent style. Seed solitary, compressed-ellipsoid; testa shiny, brown, with long lateral scar; endosperm absent, cotyledons fleshy.

Monotypic; related to *Aningeria*, with which it shares the pellucid-punctate leaf-character, but distinguished by the lack of staminodes, introrse anthers and narrow seed-scar.

Malacantha alnifolia (Baker) Pierre, Notes Bot. Sapot.: 61 (1891). — Engl., Mon. Afr. Pflanzen. **8**: 49 (1904). — J. H. Hemsley in Kew Bull. **15**: 284 (1961); in F.T.E.A., Sapotaceae: 24, fig. 3 (1968). — Heine in F.W.T.A. ed. 2, **2**: 24 (1963). TAB. **45**. Type from Nigeria.
Chrysophyllum? alnifolium Baker in F.T.A. **3**: 499 (1877). Type as above.

Sparsely branched tree up to 25 m. or more with fluted bole slightly buttressed at the base; bark brown, flaking; slash pink; timber white and soft. Apical buds and young stems densely brownish-pubescent, soon glabrescent; bark of twigs dark grey, finely wrinkled. Leaves 14–38 × 7–18 cm., obovate or elliptic, the apex rounded, minutely cuspidate and with the midrib sometimes prolonged into a deciduous hair-point, the base tapering; petiole 12–15 mm. long; lateral nerves many-paired. Upper leaf surface with channelled midrib and impressed lateral nerves, subglabrous except along the midrib. Lower surface with prominent nerves, villous with long-stalked Y-shaped brownish hairs especially on midrib and nerves. Flowers slightly aromatic. Calyx c. 4·5 mm. long, lobes broadly elliptic, strongly concave, brownish appressed-pubescent on exterior surface. Corolla white or cream, slightly exceeding calyx at anthesis. Anthers c. 1·4 mm. long. Style equalling corolla at anthesis. Fruit up to 2·5 cm. in diameter, red and fleshy when ripe. Seed up to 1·7 cm. long.

Mozambique. Z: Lugela Distr., Namagoa, Lugela to Mocuba, fl., *Faulkner* 91 (K).
Widespread in tropical Africa from Senegal to the Sudan; known from Mozambique only from the specimen cited. In E. Africa an understorey tree in lowland rainforest and riverine forest, and in deciduous forest.

M. alnifolia has been divided by J. H. Hemsley (op. cit.: 287, 1961) into two varieties which differ in indumentum characters. Var. *sacleuxii* (Lecomte) J. H. Hemsley is restricted to Zanzibar.

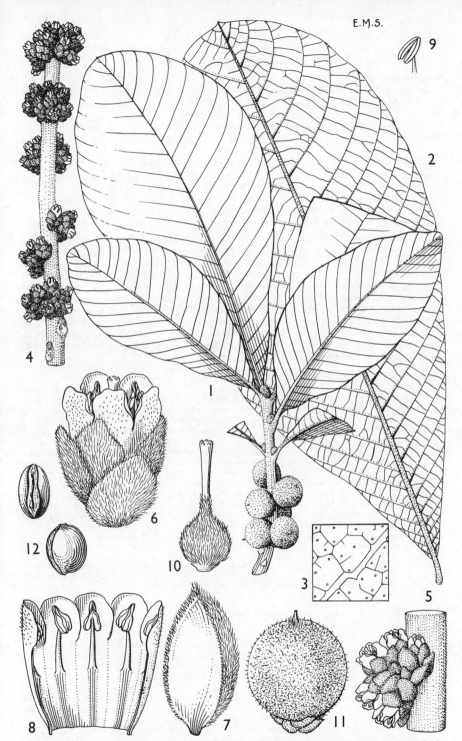

E.M.S.

Tab. 45. MALACANTHA ALNIFOLIA. 1, fruiting branchlet (× ½), from *Semsei* 1954; 2, leaf (× 1½), from *Semsei* 866; 3, lower surface of leaf (× 40), from *Drummond & Hemsley* 4035; 4, flowering branch of var. SACLEUXII (× 1); 5, flower-cluster of same (× 2), 4, 5, from *Vaughan* 1967; 6, flower (× 6); 7, sepal (× 6); 8, section of corolla (× 6); 9, anther (× 6); 10, ovary (× 6), 6–10, from *Mshatshi* in Herb. Amani 3062; 11, fruit (× 1½); 12, seeds (× 1), 11, 12, from *Semsei* 1954. From F.T.E.A.

5. BEQUAERTIODENDRON De Wild.

Bequaertiodendron De Wild. in Rev. Zool. Afr. **7**, Suppl. Bot.: 22 (1919); Pl. Bequaert. **4**: 143 (1926). — Heine & J. H. Hemsley in Kew Bull. **14**: 306 (1960).
Englerophytum Krause in Engl., Bot. Jahrb. **50**, Suppl.: 344 (1914) (see discussion after genus description).
Chrysophyllum Sect. *Zeyherella* Engl., Mon. Afr. Pflanzen. **8**: 46 (1904).
Pachystela Sect. *Zeyherella* (Engl.) Lecomte in Bull. Mus. Hist. Nat. Paris **25**: 193 (1919).
Zeyherella (Engl.) Aubrév. & Pellegrin in Bull. Soc. Bot. Fr. **105**: 37 (1958).
Boivinella Aubrév. & Pellegrin, loc. cit. non A. Camus (1925), nom. illegit.
Neoboivinella Aubrév. & Pellegrin, op. cit. **106**: 23 (1959).
Pseudoboivinella Aubrév. & Pellegrin in Notul. Syst. **16**: 260 (1960).
Pouteria sensu A. Meeuse in Bothalia **7**: 332 (1960) pro parte, non Aubl.
Amorphospermum sensu Baehni in Boissiera **11**: 102 (1965) pro parte, non F. Muell.

Shrubs and trees. Stipules present or absent. Leaves coriaceous, characteristically sericeous on lower surface with a dense appressed indumentum; lateral nerves very numerous, closely spaced, diverging at a wide angle from the midrib. Flowers ⚥ or rarely unisexual, 5-merous, subsessile to pedicellate, solitary or in fascicles, plants sometimes cauliflorous. Sepals almost free to base. Corolla longer or shorter than calyx; lobes longer or shorter than tube. Stamens inserted at mouth of corolla-tube; filaments short to absent, sometimes shortly monadelphous at base; anthers extrorse, dehiscing laterally or introrse. Staminodes (in F.Z. area) absent or small and irregular in shape and number. Ovary ovoid to globose, densely pilose, with short to long style. Fruit a 1-seeded berry. Seed compressed-ellipsoid, with broad or narrow lateral scar; endosperm absent, cotyledons fleshy.

A tropical African genus of perhaps half a dozen species.

In their discussion on the genus *Bequaertiodendron*, Heine & Hemsley stated that the use of this name must remain in some doubt since the earlier name *Englerophytum* might be a synonym. *E. stelachantha* Krause, the type of *Englerophytum*, seemed to be closely related to *Bequaertiodendron magalismontanum*, but the type specimen had apparently been destroyed and its identity was not completely certain.

Since then, Aubréville (in Adansonia **1**: 434, 1971) has reported the discovery of an isotype of *E. stelachantha*, which proved that the species does indeed fall within Heine & Hemsley's concept of *Bequaertiodendron*. (Aubréville himself divides *B. magalismontanum* between *Zeyherella* and *Englerophytum* and places *B. natalense* in *Neoboivinella* and the tropical African *B. oblanceolatum* in *Pseudoboivinella*!) Following Heine & Hemsley, therefore, I should transfer species of *Bequaertiodendron* to *Englerophytum*. However, I am unwilling to take this step because in my opinion the delimitation of genera nos. 2–7 in this account is unsatisfactory and I hope that eventually most or all of them will be amalgamated; this being so, it seems futile to create yet more combinations, especially when the widespread *B. magalismontanum* is now generally recognised by this name in southern Africa.

Branching-pattern not obviously subterminal; flowers several to many in fasciculate clusters
                                                1. *magalismontanum*
Branching-pattern distinctly subterminal; flowers solitary or 2–3 per node - 2. *natalense*

1. **Bequaertiodendron magalismontanum** (Sonder) Heine & J. H. Hemsley in Kew Bull. **14**: 307 (1960). — A. Meeuse in Fl. Southern Afr. **26**: 37, fig. 6, 1 (1963). — Heine in F.W.T.A. ed. 2, **2**: 25 (1963). — J. H. Hemsley in Kew Bull. **20**: 469 (1966); in F.T.E.A., Sapotaceae: 21 (1966). — Palmer & Pitman, Trees of Southern Afr. **3**: 1741, with fig. (1972). — R. B. Drumm. in Kirkia **10**: 266 (1975). Type from S. Africa (Transvaal).
    Chrysophyllum magalismontanum Sonder in Linnaea **23**: 72 (1850) ("magalis-montana"). — Baker in F.T.A. **3**: 498 (1877). — Engl., Mon. Afr. Pflanzen. **8**: 47, t. 16 fig. C (1904); in Fl. Pl. S. Afr. **3**: 98 (1923). — Harv. & Wright in Harv. & Sond., F.C. **4**: 437 (1906). — Gerstner in Journ. S. Afr. Bot. **12**: 49, fig. 4 (1946). — O. B. Miller, B.C.L.: 46 (1948). — Brenan in Mem. N.Y. Bot. Gard. **8**: 498 (1954). — F. White, F.F.N.R.: 321 (1962). — Fanshawe, Check-list Woody Pl. Zambia: 10 (1973). Type as above.
    Chrysophyllum argyrophyllum Hiern, Cat. Afr. Pl. Welw. **3**: 641 (1898). — Engl., op. cit.: 46, t. 16 fig. A (1904). — S. Moore in Journ. Linn. Soc., Bot. **40**: 131 (1911). — Eyles in Trans. Roy. Soc. S. Afr. **5**: 437 (1916). — Steedman, Trees etc. S. Rhod.: 62 (1933). — Burtt Davy & Hoyle, N.C.L.: 70 (1936). — Pardy in Rhod. Agric. Journ. **53**: 51 (1956). Syntypes from Angola.
    Sideroxylon randii S. Moore in Journ. Bot., Lond. **41**: 402 (1903). — Harv. & Wright, tom. cit.: 439 (1906). Type from S. Africa (Transvaal).

Chrysophyllum? carvalhoi Engl., op. cit.: 47 (1904). Type: Mozambique, between Moçambique and Gorungosa, 1884–85, *Carvalho* (B, holotype†; COI, isotype).

Chrysophyllum wilmsii Engl., op. cit.: 46, t. 16 fig. B (1904). — Harv. & Wright, tom. cit.: 437 (1906). Type from S. Africa (Transvaal).

Chrysophyllum lujae De Wild., Pl. Bequart. **4**: 133 (1926). Type: Mozambique, Morrumbala, *Luja* 330 (BR, holotype).

Zeyherella magalismontana (Sonder) Aubrév. & Pellegrin in Bull. Soc. Bot. Fr. **105**: 37 (1958). — Baehni in Boissiera **11**: 69 (1965). Type as for *Bequaertiodendron magalismontanum*.

Pouteria magalismontana (Sonder) A. Meeuse in Bothalia **7**: 335, fig. 7 (1960). Type as above.

For a fuller synonymy, see Heine & Hemsley, loc. cit.

A bush or medium-sized evergreen tree 1–15 m. tall, often with 2–4 main stems. Trunk (in larger specimens) slightly fluted and buttressed; bark smooth and scaling in smaller specimens, becoming roughly fissured. Young branchlets densely ferrugineous-tomentose. Stipules 2–7 mm. long, subulate, soon deciduous. Leaf-lamina 5·5–22 cm. long, 2·5–7 times as long as broad, elliptic to narrowly obovate, apex emarginate or shortly acuminate or rounded and apiculate; midrib often produced into a mucro c. 1 mm. long. Petiole 1–2 cm. long. Upper leaf surface dark green, smooth and glabrous, with impressed midrib and inconspicuous lateral nerves. Lower leaf surface densely velutinous, usually rusty-brown when young and becoming silvery with age, but occasional specimens with mature leaves vividly reddish-brown below. Flowers borne in dense fascicles on new and old wood throughout the plant: from axils of current leaves to near the ground on the main trunk. Pedicels (0) 1–12 mm. long. Calyx 2·2–5 mm. long, cup-shaped, dorsally pilose, segments broadly ovate. Corolla 3·6–5·4 mm. long, pinkish, red or purplish-brown, campanulate, longer or shorter than calyx, divided to ⅔ into ovate-acute lobes auriculate at the base. Stamens with filaments of very variable length; anthers 1·6–2·2 mm. long, included in or somewhat exserted from corolla. Staminodes usually absent, when present represented by ovate scales c. 0·6 mm. long. Gynoecium 4–4·4 mm. long. Fruit up to 2·5 × 1·8 cm., ellipsoid, red with edible purple pulp. Seed solitary (2), c. 1·4 cm. long, whitish or pale brown.

Botswana. SE: Mannyelanong Hill, 32 km. SW. of Gaberone, 1475 m., fl. 30.ix.1974, *Mott* 385 (SRGH). **Zambia.** B: 5 km. S. of Kalabo, fr. 17.xi.1959, *Drummond & Cookson* 6565 (PRE; SRGH). N: Mporokoso Distr., Kundabwika Falls, Kalungwishi R., 940 m., fl. 19.ix.1957, *Whellan* 1407 (COI; SRGH). W: Mwinilunga Distr., W. of Kalene Mission, fr. 23.ix.1952, *Angus* 521 (BM; COI; FHO; K; PRE). C: Lusaka Distr., between Luangwa Bridge and Rufunsa, fl. 6.ix.1947, *Brenan & Greenway* 7818 (BM; COI; FHO; K). E: near Nyimba R., fl. 24.viii.1929, *Burtt Davy* 20909 (BM; FHO; K). S: Gwembe, Lowe R., fr. 23.xi.1955, *Bainbridge* 206/55 (FHO). **Zimbabwe.** N: Mazoe Distr., 1310 m., fl. x.1906, *Eyles* 378 (BM; SRGH). W: Matopos Distr., fl. 1932, *Mundy* 5579 (BM; SRGH). C: Mermaid's Pool E. of Salisbury, fl. 29.viii.1931, *Gilliland* 8 (BM; SRGH). E: Umtali Distr., near Umwindsi R., 910 m., st. 5.ii.1957, *Chase* 6323 (COI; K). S: Victoria Distr., Kapota Hill, 1130 m., fl. ix.1956, *Miller* 3647 (SRGH). **Malawi.** C: Ntchisi Mt., 1450 m., fr. 20.ii.1959, *Robson & Steele* 1688 (BM; K; LISC; PRE). S: Mulanje Mt., fl. 1.x.1957, *Chapman* 447 (BM; FHO; K; MAL; PRE). **Mozambique.** N: Eráti, c. 12 km. from Namapa to Alua, Mt. Geovi, c. 500 m., fr. 8.i.1964, *Torre & Paiva* 9891 (COI; LISC; LMU). Z: Mt. Gúruè, c. 3 km. from waterfall on R. Licungo, c. 1200 m., fl. 24.ii.1966, *Torre & Correia* 14814 (COI; EA; LISC; MO; WAG). T: Maravia, Fingoè, road to Vila Vasco da Gama, st. 27.vi.1949, *Andrada* 1675 (LISC). MS: Manica, Dombe, fr. 24.xi.1965, *Pereira & Marques* 890 (BM; COI; LISC; LMA; LMU; PRE; SRGH).

Widespread in tropical and southern Africa. In riverine fringing forest, on rocky hillsides.

A. Meeuse (in Fl. Southern Afr. **26**: 39, 1963) discusses the occurrence of staminodes and the occasional appearance of ♀ flowers in *B. magalismontanum*. The forms which he described have all been found among material from the F.Z. area.

2. **Bequaertiodendron natalense** (Sonder) Heine & J. H. Hemsley in Kew Bull. **14**: 308 (1960). — A. Meeuse in Fl. Southern Afr. **26**: 39 (1963). — J. H. Hemsley in F.T.E.A., Sapotaceae: 19, fig. 2 (1968). — Palmer & Pitman, Trees of Southern Afr. **3**: 1742, with fig. (1972). — R. B. Drumm. in Kirkia **10**: 266 (1975). TAB. **46**. Type from S. Africa (Natal).

Chrysophyllum natalense Sonder in Linnaea **23**: 72 (1850). — Engl., Mon. Afr. Pflanzen. **8**: 43, t. 34 fig. C (1904). — Harv. & Wright in Harv. & Sond., F.C. **4**: 437 (1906). — S. Moore in Journ. Linn. Soc., Bot. **40**: 131 (1911). — Eyles in Trans. Roy. Soc. S. Afr. **5**: 437 (1916). — Burtt Davy & Hoyle, N.C.L.: 70 (1936). — Gerstner in Journ. S. Afr. Bot. **12**: 48, fig. 2 (1946). Type as above.

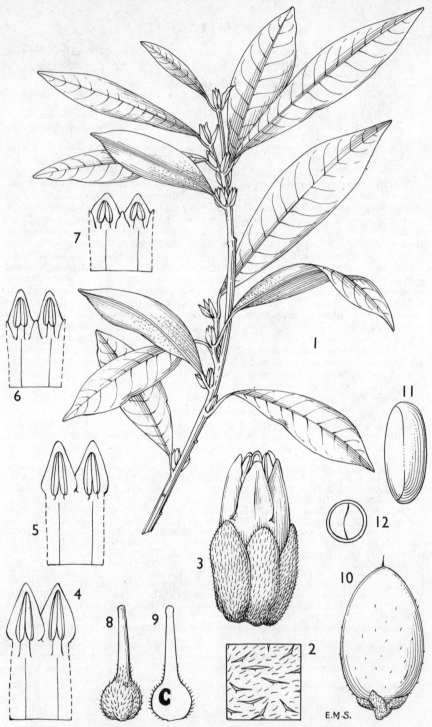

Tab. 46. BEQUAERTIODENDRON NATALENSE. 1, flowering branch ($\times \frac{2}{3}$); 2, lower surface of leaf ($\times 20$), 1, 2, from *Eggeling* 6821; 3, flower ($\times 6$), from *Faulkner* 1080; 4–7 sections of corolla with stamens and staminodes variously developed ($\times 6$), 4 from *Faulkner* 1080, 5 from *Drummond* & *Hemsley* 3171, 6 from *Purseglove* 834, 7 from *Eggeling* 3164; 8, ovary ($\times 6$); 9, section of ovary ($\times 6$); 10, fruit ($\times 1\frac{1}{2}$); 11, seed ($\times 1\frac{1}{2}$); 12, transverse section of seed ($\times 1\frac{1}{2}$), 8–12 from *Faulkner* 1080. From F.T.E.A.

Chrysophyllum kilimandscharicum G. M. Schultze in Notizbl. Bot. Gart. Berl. **12**: 196 (1934). Type from Tanzania.
Boivinella natalensis (Sonder) Pierre ex Aubrév. & Pellegrin in Bull. Soc. Bot. Fr. **105**: 37 (1958). Type as for *Bequaertiodendron natalense*.
Neoboivinella natalensis (Sonder) Aubrév. & Pellegrin, op. cit. **106**: 23 (1959). Type as above.
Pouteria natalensis (Sonder) A. Meeuse in Bothalia **7**: 339 (1960). Type as above.
Amorphospermum natalense (Sonder) Baehni in Boissiera **11**: 103 (1965). Type as above.

Evergreen shrub or tree up to 12 m. tall, with characteristic repeated sub-apical branching. Trunk (in large specimens) somewhat fluted; bark brown, flaking. Young branchlets densely dark brown appressed-pubescent; older twigs with smooth, grey and brown striated bark. Stipules absent or very soon deciduous. Leaves tending to be clustered at branch ends. Lamina 5–16 × 2–5 cm., oblanceolate to narrowly elliptic, the apex cuspidate-acuminate, the base narrowly acute. Petiole 5–10 mm. long. Upper surface smooth, glabrous, greyish-green; midrib narrow, impressed, lateral nerves inconspicuous. Lower surface densely silvery-sericeous, mottled due to presence of scattered larger brownish hairs. Flowers solitary or in groups of 2 or 3, sessile, borne in leaf axils. Calyx 4·5–6 mm. long, narrowly ovoid, thick and leathery, deeply divided into ovate lobes but these remaining closely appressed and almost completely enclosing corolla; outer surface with brownish indumentum. Corolla 3·4–4 mm. long, whitish or yellowish; lobes $\frac{1}{3}$–$\frac{1}{2}$ as long as tube, ovate, auricled at the base. Anthers 1·2–1·4 mm. long, not exserted. Staminodes 0–5, when present represented by petaloid scales c. 0·5 mm. long. Gynoecium 3–4 mm. long. Fruit 2–2·5 × 1–1·5 cm., narrowly ovoid to cylindrical, puberulous, deep red and edible when ripe. Seed c. 20 × 8 mm.

Zimbabwe. E: Melsetter Distr., Mutsangazi R., W. of Mt. Peni, 900 m., fr. 13.xii.1972, *Müller & Goldsmith* 2064 (K; SRGH). **Malawi.** S: Mulanje Distr., Machemba, 1310 m., st. 2.iv.1963, *Chapman* 1833 (MAL; SRGH). **Mozambique.** N: Mt. Murripa, 40 km. from Entre-Rios to Ribáuè, c. 1100 m., fr. 15.xii.1967, *Torre & Correia* 16512 (BR; COI; EA; LISC; MO; PRE; SRGH). Z: S. side of Mt. Tumbini, fl. 30.viii.1949, *Andrada* 1804 (COI). T: Mt. Zóbuè, fl. & fr. 3.x.1942, *Mendonça* 614 (LISC). MS: Garuso, Bandula, st. 28.i.1949, *Chase* 1704 (BM; K; SRGH). M: Namaacha, R. Impamputo, fl. 26.vii.1967, *Marques* 2096 (BM; COI; LISC; LMU; PRE; SRGH).
Occurring in Uganda, Kenya, Tanzania and S. Africa. In shade in hygrophilous mixed evergreen forest.

6. PACHYSTELA Engl.

Pachystela Engl., Mon. Afr. Pflanzen. **8**: 35 (1904).
Pouteria sensu Baehni in Candollea **7**: 472 (1938) et sensu A. Meeuse in Bothalia **7**: 332 (1960), pro parte, non Aubl. (1775).

Shrubs and trees. Stipules present, rigid, persistent. Leaves coriaceous; lateral nerves fairly distant, curving towards leaf apex; tertiary venation inconspicuous. Flowers 5-merous, fascicled in axils of current and fallen leaves, sessile or pedicellate. Calyx segments shortly united at the base. Corolla lobes longer than tube. Stamens inserted at level of base of corolla lobes, with long flexuous filaments; anthers extrorse, included in to considerably exserted from corolla. Staminodes absent or present and then often irregular in shape and number. Ovary ± ovoid, pilose; style long, cylindrical. Fruit a 1-seeded berry. Seed compressed-ellipsoid, the scar covering more than half its surface; endosperm absent, cotyledons fleshy.

A genus of c. 6 species confined to tropical Africa.

Pachystela brevipes (Baker) Engl., Mon. Afr. Pflanzen. **8**: 37 (1904). — Burtt Davy & Hoyle, N.C.L.: 71 (1936). — Topham, N.C.L. ed. **2**: 94 (1958). — F. White, F.F.N.R.: 322 (1962). — Heine in F.W.T.A. ed. 2, **2**: 28, fig. 206 (1963). — Baehni in Boissiera **11**: 99, fig. 156 (1965). — J. H. Hemsley in F.T.E.A., Sapotaceae: 36, fig. 6 (1967). — Fanshawe, Check-list Woody Pl. Zambia: 32 (1973). — R. B. Drumm. in Kirkia **10**: 266 (1975). TAB. **47**. Syntypes from Tanzania and Malawi: N. end of L. Malawi, *Kirk* (K, isotype).
Sideroxylon brevipes Baker in F.T.A. **3**: 502 (1877). Syntypes as above.
Chrysophyllum cinereum Engl., Bot. Jahrb. **12**: 522 (1890). Type from Angola.
Chrysophyllum stuhlmannii Engl., Pflanzenw. Ost-Afr. **C**: 306 (1895). Syntypes: Mozambique, Zambezia, Quelimane, *Stuhlmann* (B†; HBG); Malawi, *Buchanan* 793 (BM; K, isosyntypes).

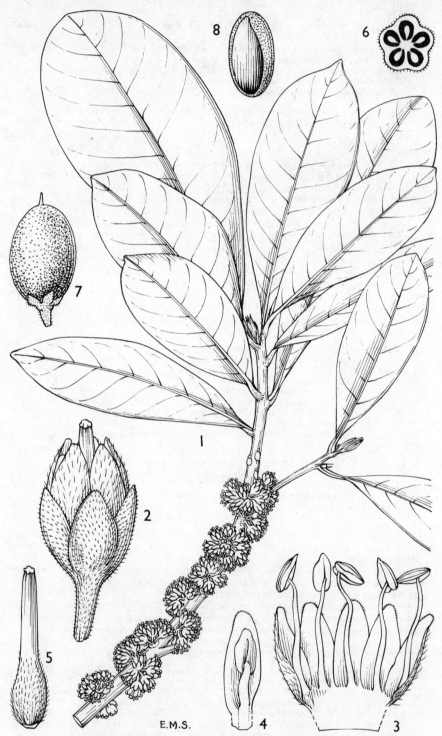

Tab. 47. PACHYSTELA BREVIPES. 1, flowering branch (× ⅔), from *Lyne* 83; 2, young flower (× 6); 3, corolla dissected to show stamens (× 6); 4, corolla-lobe and stamen of young flower (× 6); 5, ovary (× 6); 6, diagrammatic transverse section of ovary; 2–6, from *Eggeling* 440 in F.D. 744; 7, fruit (× ½); 8, seed (× 1½), 7, 8, from *Williams* 102. From F.T.E.A.

Pachystela cinerea (Engl.) Engl., op. cit.: 36 (1904). — S. Moore in Journ. Linn. Soc., Bot. **40**: 133 (1911). Type as for *Chrysophyllum cinereum*.
Pouteria brevipes (Baker) Baehni in Candollea **9**: 290 (1942). — A. Meeuse in Bothalia **7**: 333, fig. 5 (1960). Type as for *Pachystela brevipes*.
For a fuller synonymy, see Heine, loc. cit.

A bush or much-branched evergreen tree 4–20 m. tall; bark greyish-brown to blackish, rather rough; bole (in large specimens) deeply fluted and slightly buttressed at the base. Young branchlets and apical buds greyish-pubescent; older twigs glabrescent with smooth grey bark. Stipules 3–5 mm. long, subulate, caducous. Leaves tending to be crowded at branch ends. Lamina 9–25 × 6·5–10 cm., narrowly obovate, the apex obtuse and apiculate or shortly cuspidate-acuminate, the base tapering, acute. Petiole 5–12 mm. long. Upper leaf surface glossy, glabrous, drying pale to dark brown or muddy green; midrib slightly raised, lateral nerves impressed, 6–9 on each side, curving and becoming indistinct near the margin. Lower leaf surface minutely silvery-sericeous, drying pinkish-brown, becoming glabrescent and then brown; midrib and lateral nerves very prominent. Flowers in dense few- to many-flowered fascicles on old wood and in leaf axils, very sweetly scented; pedicels up to 3 mm. long. Calyx 3–4 mm. long, cup-shaped, the lobes narrowly ovate, not differentiated into inner and outer sepals (i.e. not strongly imbricate), dorsally appressed-pubescent. Corolla c. 6 mm. long, ± tubular, divided to c. ⅔ into narrowly ovate lobes, pale greenish or cream-coloured. Filaments 2–5 mm. long; anthers 1·7–2·2 mm. long, often exserted from corolla. Staminodes absent or represented by small fleshy ovate or narrowly linear scales. Gynoecium 4·5–5·2 mm. long. Fruit up to 2·5 cm. long, ellipsoid, crowned with the persistent style, yellow or orange when ripe, with sweet edible mucilaginous flesh; outer surface appressed-pubescent, the indumentum rubbing off. Seed up to 2 cm. long, with shiny brown testa and large pale lateral scar covering up to ⅔ of the surface area.

Zambia. N: Lunzua R., fr. 21.vii.1930, *Hutchinson & Gillett* 3978 (BM; COI; LISC; SRGH). W: Chingola, fr. 25.viii.1954, *Fanshawe* 1488 (FHO; K). **Zimbabwe.** E: Melsetter Distr., path from Hayfield to junction of Lusitu and Haroni Rs., 400 m., st. 25.xi.1955, *Drummond* 50001 (K; LISC; PRE; SRGH). **Malawi.** N: Nkhata Bay, Chintece Beach, Lakeshore road, 40 km. S. of Nkhata Bay, 475 m., fl. 28.v.1972, *Pawek* 5401 (K; MAL; SRGH). S: Zomba Distr., Likangala Bridge, fl. 24.i.1964, *Salubeni* 212 (MAL; SRGH). **Mozambique.** N: Cova Chaves, fl. 13.viii.1948, *Andrada* 1257 (COI; LISC; LMA). Z: Namacurra, road to Quelimane, fl. & fr. immat. 28.v.1949, *Andrada* 1532 (COI). MS: Búzi R. valley, c. 610 m., fr. ix.1963, *Goldsmith* 41/63 (BM; K; LISC; SRGH).
Widespread in tropical Africa. In riverine forest and in dry evergreen forest fringing lakes.

Closely related to the similarly widespread *P. msolo* (Engl.) Engl. According to Fanshawe, loc. cit., *P. brevipes* also occurs in the Central province of Zambia.

7. VINCENTELLA Pierre

Vincentella Pierre, Notes Bot. Sapot.: 37 (1891).
Sideroxylon Sect. *Bakerisideroxylon* Engl., Bot. Jahrb. **12**: 518 (1890); in Engl. & Prantl, Pflanzenfam. **IV**, 1: 144 (1890).
Bakerisideroxylon (Engl.) Engl., Mon. Afr. Pflanzen. **8**: 33 (1904).
Pouteria Aubl. Sect. *Bakerisideroxylon* (Engl.) Baehni in Candollea **9**: 382 (1942).

Trees and shrubs. Stipules present, persistent or caducous. Leaves moderately coriaceous, lateral nerves fairly conspicuous, distant, curving towards leaf apex. Flowers 5-merous, fascicled or solitary, borne in axils of current and fallen leaves, long-pedicellate or sessile. Calyx of small sepals ± free to the base. Corolla much exceeding calyx, lobes longer or shorter than tube. Stamens inserted at level of base of corolla lobes, with long slender filaments or subsessile; anthers extrorse or introrse. Staminodes present or absent. Ovary ovoid, densely long-pilose, with slender style. Fruit a 1-seeded berry. Seed compressed-ellipsoid; scar long, occupying lateral face; endosperm absent, cotyledons fleshy.

A genus of c. 7 species confined to tropical Africa.

Leaf apex obtuse or rounded; flowers long-pedicellate; corolla lobes much longer than tube, reflexed at anthesis; anthers borne on long filaments, extrorse - - 1. *passargei*
Leaf apex cuspidate-acuminate; flowers sessile; corolla lobes much shorter than tube, not or scarcely reflexed at anthesis; anthers subsessile, introrse - - - - 2. *muelleri*

1. **Vincentella passargei** (Engl.) Aubrév., Fl. For. Soud.-Guin.: 427, t. 93, 1 (1950). — F.
White, F.F.N.R.: 322 (1962). — Heine in F.W.T.A. ed. 2, **2**: 23 (1963). — J. H. Hemsley
in F.T.E.A., Sapotaceae: 40, fig. 7 (1968). — Fanshawe, Check-list Woody Pl. Zambia: 46
(1973). Type from Cameroon.

    Bakerisideroxylon passargei Engl., Mon. Afr. Pflanzen. **8**: 35, t. 11A (1904). Type as
above.

    Bakerisideroxylon sapinii De Wild. in Rev. Zool. Afr. **7**, Suppl. Bot.: 16 (1919). Type
from Zaire (Katanga).

    Pouteria tridentata Baehni in Candollea **9**: 386 (1942); in Boissiera **11**: 56 (1965). Type
from Tanzania.

    Vincentella sapinii (De Wild.) Brenan in Mem. N.Y. Bot. Gard. **8**: 498 (1954). — A.
Meeuse in Bothalia **7**: 342 (1960). Type as for *Bakerisideroxylon sapinii*.

Shrub or much-branched evergreen tree 2–8 m. tall, with spreading crown and
drooping branches. Trunk slightly fluted at base, bark smooth, light brown or grey.
Young branchlets densely ferrugineous-pubescent. Stipules 5–9 mm. long,
subulate, persistent. Leaves crowded at branch ends. Lamina 5–12 × 2·5–5 cm.,
narrowly obovate with obtuse or rounded apex and acute base; margin slightly
revolute and undulate; lateral nerves 7–11 on each side. Petiole 6–10 mm. long.
Upper leaf surface smooth, glossy, glabrous, drying brownish- or greyish-green,
with impressed nerves; lower surface mat, glabrous, greyish-green to pinkish-brown
with prominent hirsute to glabrescent pale brown to orange nerves. (Very young
leaves may be ephemerally lanate on both sides.) Flowers fascicled in axils of current
and recently fallen leaves, on slender pedicels 3–6 mm. long. Calyx c. 1·6 mm. long,
pubescent, divided almost to the base into narrowly oblong segments. Corolla c. 3·5
mm. long, greenish-white; petals almost free, narrowly obovate, becoming reflexed
at anthesis. Filaments slender, c. 3 mm. long; anthers 0·6 mm. long, extrorse.
Staminodes a little shorter than to equalling petals, lanceolate in outline, with ±
lacerate margin. Gynoecium c. 2·5 mm. long; ovary hemispherical, densely and
conspicuously hirsute, crowned with the slender style. Fruit up to 1·5 cm. long,
ellipsoid, yellow or orange, puberulous to ± glabrous, with withered flower parts
persisting at base; flesh edible. Seed up to 12 × 8 mm.; testa shiny pale greyish-
brown with paler ± elliptic lateral scar.

Zambia. N: between Mpika and R. Chambesi, fl. 17.vii.1930, *Hutchinson & Gillett* 3790
(BM; K; SRGH). W: 110 km. from Mwinilunga to Solwezi, near R. Kabompo, fl. 17.iv.1952,
White 3276 (BM; FHO; K; PRE; SRGH). **Malawi.** N: Nkhata Bay, border of Chombe Tea
Estate near Mweza Village, fl. 13.iv.1960, *Adlard* 338 (MAL; SRGH). C: Nkhota Kota Distr.,
Chia area, fl. 3.ix.1946, *Brass* 17510 (BM; K; PRE; SRGH). **Mozambique.** N: Ribáuè,
M'Puipe R., 650 m., fl., *Gomes e Sousa* 2305 (K; PRE). Z: Lugela Distr., Munguluni Mission,
fl. & fr. immat., *Faulkner* 69 (K). MS: Beira, Nyamaruza Camp, Zuni to Chiniziua area, fl.
v.1973, *Tinley* 2825 (SRGH).

Tropical Africa from Guinea south-eastwards to southern Tanzania, Zaire and Angola. In
fringing forest.

According to Fanshawe, loc. cit., *V. passargei* also occurs in the Barotseland and Central
provinces of Zambia.

I do not find that the anthers and filaments in *V. passargei* are hispidulous, as shown in the
figure in F.T.E.A., loc. cit.

2. **Vincentella muelleri** Kupicha in Candollea **33**: 37, fig. 4 (1978). TAB. **48**. Type from
Malawi, Mulanje Distr., Ruo Gorge, c. 900 m., fl. 1.ix.1970, *Müller* 1463 (K, holotype;
SRGH, isotype).

Tree or shrub 4–7 m. tall; bark smooth, slash pinkish. Branching-pattern
distinctly subterminal; branch apices with densely crowded nodes rough with
persistent subulate stipules 2–7 mm. long; tips of branches ferrugineous-pubescent,
older parts glabrous. Leaf-lamina 6–16·5 × 1·8–4 cm., oblong-elliptic to narrowly
obovate-oblong, the apex cuspidate-acuminate with rounded tip, the base
acuminate; petiole 6–16 mm. long. Upper leaf surface glabrous, with impressed
midrib and lateral nerves, these 7–11 on each side. Lower surface with raised nerves,
glabrous except for the presence of appressed hairs on the midrib and sometimes also
on the lateral nerves. Flowers (4)5-merous, solitary, sessile, borne in leaf axils and on
older wood (plants said to be cauliflorous). Calyx 3–4 mm. long, cup-shaped; lobes
ovate to suborbicular, dorsally appressed-pubescent. Corolla 7–10 mm. long, white,
tubular, lobed to c. 3 mm. deep; lobes with ovate median area of thicker texture and
lateral wings of thinner tissue by which adjacent lobes are sometimes united. Anthers

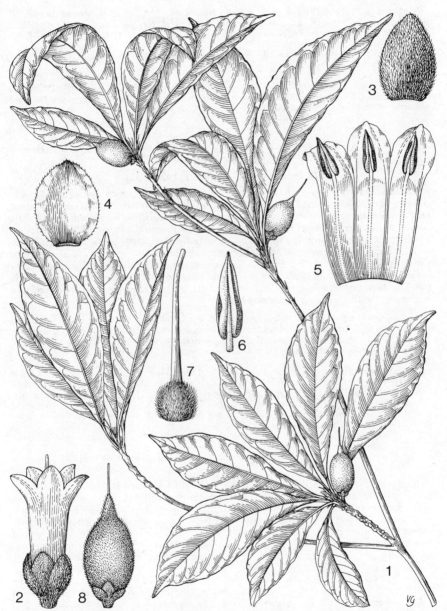

Tab. 48. VINCENTELLA MUELLERI. 1, habit (× ⅔) *Müller* 1463 & *Torre* & *Paiva* 10301; 2, flower
(× 3); 3, sepal, dorsal view (× 6); 4, sepal, ventral view (× 6); 5, part of flower opened out
(× 4); 6, stamen, from inside of flower (× 8); 7, gynoecium (× 4), all from *Müller* 1463; 8,
young fruit (× 1½) *Torre & Correia* 16403. From Candollea **33**: 38 (1978).

1·75–2·1 mm. long, introrse, sessile or borne on filaments up to 1 mm. long. Staminodes absent. Ovary c. 2 mm. long; style 5–6 mm. long. Mature fruit not seen but stated to be pink with white flesh and containing 1(?) shiny brown seed.

Malawi. S: Mulanje Plateau, st. 24.x.1929, *Burtt Davy* 22131 (FHO). **Mozambique.** N: Ribáuè, serra de Mepaluè, c. 1500 m., fl. bud 28.i.1964, *Torre & Paiva* 10301 (K; LISC; LMU). Z: Gúruè, confluence of Malema and Cocossi Rs., c. 1650 m., fl. & fr. immat. 6.xi.1967, *Torre & Correia* 15921 (LISC).

Unknown outside the F.Z. area. In *Newtonia* medium-altitude montane forest and dense humid riverine forest.

In its vegetative characters *V. muelleri* is very similar to the rare Kenyan species *Pachystela subverticillata* E. A. Bruce, but their flowers are quite different. The systematic position and phenetic relationships of *V. muelleri* are discussed in Candollea, loc. cit.

8. SIDEROXYLON L.

Sideroxylon L., Sp. Pl. **1**: 192 (1753).

Shrubs and small trees. Leaves exstipulate, coriaceous, with characteristic fine tertiary vein reticulation. Flowers subsessile to petiolate, fascicled in axils of current and/or recently fallen leaves, usually pentamerous. Calyx lobes almost free. Petals united at base into a short to long tube. Stamens inserted at level of base of lobes; anthers extrorse, included to somewhat exserted. Staminodes large, petaloid, entire to lacerate. Gynoecium with subglobose, densely pilose ovary tapering at apex into the cylindrical style. Fruit a 1-seeded subglobose berry with very sticky mesocarp; style persistent. Seed subglobose to ovoid, often faintly pentangular; testa shiny, pale to dark brown; scar small, basal, subcircular; endosperm copious; embryo horizontal, with thin leafy cotyledons.

Authorities disagree about the delimitation of *Sideroxylon*, some including in it species from the Far East and America, others taking a much narrower view. Aubréville (in Adansonia **3**: 29–38, 1963) restricts *Sideroxylon* to the small group of species closely related to the type, *S. inerme*, found in E. Africa, Madagascar, the Mascarenes and Macronesia.

Sideroxylon inerme L., Sp. Pl. **1**: 192 (1753). — Engl., Mon. Afr. Pflanzen. **8**: 27, t. 7 fig. B (1904). — Harv. & Wright in Harv. & Sond., F.C. **4**: 438 (1906). — Gerstner in Journ. S. Afr. Bot. **12**: 47, fig. 1 (1946). — A. Meeuse in Bothalia **7**: 323 (1960); in Fl. Southern Afr. **26**: 34 (1963). — J. H. Hemsley in Kew Bull. **20**: 472 (1966); in F.T.E.A., Sapotaceae: 33 (1968). — Palmer & Pitman, Trees of Southern Afr. **3**: 1735 with fig. (1972). Type probably from S. Africa (Cape Prov.).

    Sideroxylon inerme var. *schlechteri* Engl., loc. cit. Type: Mozambique, Maputo (Lourenço Marques), *Schlechter* 11710 (K, isotype).

    Calvaria inermis (L.) Dubard in Ann. Mus. Col. Marseille **20**: 86 (1912). Type as for *Sideroxylon inerme*.

Evergreen shrub or small tree 2–15 m. tall with brown or blackish bark. Apical buds ferrugineous pubescent; distal leafy part of branchlets pubescent or glabrous; older leafless twigs with grey striated bark. Leaf lamina 3·5–9·5 × 2–4·5 cm., elliptic to obovate, sometimes elliptic-ovate or suborbicular, the apex rounded, slightly emarginate or acute, the base cuspidate to acuminate; petiole 0·5–2 cm. long. Upper leaf surface glabrous, midrib level with surface. Lower surface often ferrugineous-pubescent, glabrescent; midrib fairly prominent, lateral nerves indistinct, tertiary veins forming a fine even reticulation. Flowers 4- or 5-merous, scented, in few- to many-flowered fascicles in leaf axils or on older wood; pedicels 0–4 mm. long. Calyx 1·6–2 mm. long, cup-shaped; lobes suborbicular, imbricate, ± glabrous. Corolla 2·4–3·6 mm. long, campanulate, whitish, cream-coloured or greenish; lobes 2–3 times as long as tube, ovate, auricled at base. Anthers 1–1·4 mm. long. Staminodes c. 1·8 mm. long. Fruit up to 1 cm. in diameter, purple to blackish. Seed up to 6 mm. in diameter, glossy, cream-coloured or dark brown.

Flowers usually borne in axils of current leaves, rarely on older wood; pedicels 2–4 mm. long; leaves obovate to elliptic; leafy branchlets usually glabrous; young leaves glabrous

                                      subsp. *inerme*
Flowers usually borne proximally to current year's leaves but sometimes also in leaf axils; pedicels 0–2 mm. long (in F.Z. area; note that pedicels elongate in fruit); leaves usually

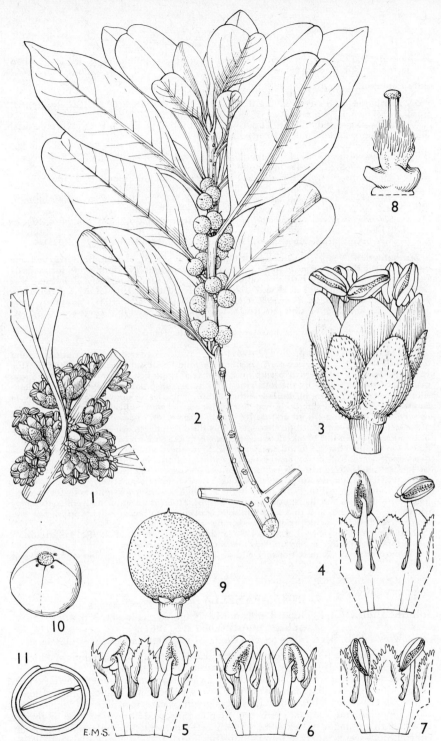

Tab. 49. SIDEROXYLON INERME subsp. DIOSPYROIDES. 1, part of flowering branchlet (× 2), *Faulkner* 727; 2, fruiting branch (× ⅔), *Drummond & Hemsley* 3246; 3, flower (× 10), *Faulkner* 727; 4–7, sections of corollas with stamens and staminodes variously developed (× 10), 4 from *Faulkner* 727, 5 from *Gillet* 4516, 6 from *Kassner* 422, 7 from *Boivin*; 8, ovary and receptacle (× 10); 10, seed in basal-lateral view (× 3); 11, transverse section of seed (× 3). 9–11. from *Drummond & Hemsley* 3534. From F.T.E.A.

obovate; leafy branchlets often ferrugineous- or greyish-pubescent; young leaves often ferrugineous-pubescent below - - - - - - - subsp. *diospyroides*

Subsp. **inerme**.

Mozambique: GI: Praia de Zavora, fl. 27.ii.1955, *E. M. & W.* 689 (BM; LISC; SRGH). M: Costa do Sol, fl. 6.ii.1962, *Balsinhas* 535 (BM; COI; K; LISC; LMA; PRE).
Also in Swaziland and S. Africa (Transvaal, Natal and Cape Prov.). In thickets on coastal dunes and in mixed woodland, often on termitaria.

Subsp. **diospyroides** (Baker) J. H. Hemsley in Kew Bull. **20**: 476 (1966); in F.T.E.A., Sapot.: 35, fig. 5 (1968). TAB. **49**. Type from Tanzania.
    Myrsine querimbensis Klotzsch in Peters, Reise Mossamb., Bot. **1**: 185 (1861). Type: Mozambique, Cabo Delgado, Querimba I., *Peters* (B, holotype†).
    Sideroxylon diospyroides Baker in F.T.A. **3**: 502 (1877). — Engl., Mon. Afr. Pflanzen. **8**: 27, t. 7 fig. A (1904). Type as for *Sideroxylon inerme* subsp. *diospyroides*.
    Sideroxylon inerme var. *schlechteri* sensu S. Moore in Journ. Linn. Soc., Bot. **40**: 133 (1911).
    Sideroxylon inerme subsp. *inerme* sensu R. B. Drumm. in Kirkia **10**: 265 (1975).

Zimbabwe. S: Nuanetsi Distr., near tributary of Guluene R., c. 5 km. from Mozambique border, fl. 18.x.1975, *Drummond* 10404 (SRGH). **Mozambique.** N: 5 km. N. of Mocimboa da Praia, bank of R. Mepanga, fl. 10.xi.1960, *Gomes e Sousa* 4625 (COI; K; PRE). Z: Pebane, near the lighthouse, c. 30 m., fr. 8.iii.1966, *Torre & Correia* 15066 (B; LISC; LUAI; P). MS: 9 km. N. of Maringua, Sabi R., st. 23.vi.1950, *Chase* 2535 (BM; PRE; SRGH). GI: Gaza Prov., near Kapateni, 65 km. NE. of Chicualacuala (Malvérnia), st. 25.iv.1962, *Drummond* 7716 (K; SRGH).
Somalia, Kenya and Tanzania. Habitat as for type subspecies.

J. H. Hemsley, in Kew Bull. **20**: 472–478 (1966), has discussed in detail the relationship and delimitation of the subspecies of *S. inerme*. He found that subsp. *diospyroides* from the F.T.E.A. area and northern Mozambique could readily be distinguished from S. African material of subsp. *inerme* by the following characters. Subsp. *diospyroides* has subsessile to shortly pedicellate flowers (pedicels less than 3 mm. long), leaves obovate to elliptic-obovate and seeds 5–7 mm. in diameter. Subsp. *inerme* has flowers with pedicels 3 mm. long or more, leaves elliptic or oblong-elliptic and seeds 6–9 mm. in diameter. My own survey has shown that of these three pairs of contrasting characters, only the difference in leaf-shape proves a reliable distinction and even this breaks down in specimens from southern Mozambique, as Hemsley noted. On the other hand, I found that in the F.Z. area other features can be used to separate the two subspecies: subsp. *diospyroides* usually has flowers borne in axils of fallen leaves and the young stems and foliage are often pubescent, whereas subsp. *inerme* usually has flowers borne in the current leaf axils and young shoots glabrous. These characters hold true for subsp. *inerme* in S. Africa but not for subsp. *diospyroides* in the F.T.E.A. area. It thus appears that within the long narrow north-south distribution range of *S. inerme* there is variation in several, only partly correlated, characters, and that although it is convenient to recognise subspp. *inerme* and *diospyroides* the latter includes rather a wide range of variation.
Burtt Davy and Hoyle (N.C.L.: 71, 1936) and Topham, N.C.L. ed. 2: 95 (1958) record *S. inerme* for Malawi, but it is very unlikely that this species occurs there.
The description of Gomes e Sousa (Dendrol. Moçamb. **2**: 627, 1967) applies to the species as a whole.

9. INHAMBANELLA (Engl.) Dubard

Inhambanella (Engl.) Dubard in Ann. Mus. Col. Marseille, sér. 3, **3**: 42 (1915). — Aubrév. in Adansonia **1**: 6 (1961).
    Mimusops L. Sect. *Inhambanella* Engl., Mon. Afr. Pflanzen. **8**: 80 (1904).

Small to medium-sized evergreen tree. Leaves with inconspicuous, caducous stipules, moderately coriaceous in texture; lamina glabrous; lateral nerves conspicuous, rather widely spaced and curving towards leaf-apex, tertiary vein reticulation regular, fine and distinct. Flowers pedicellate, fascicled in leaf axils. Calyx 5-merous, the lobes almost free, strongly imbricate. Corolla with 5 lobes, each with 1 median and 2 lateral segments. Stamens inserted at level of base of corolla lobes; anthers included, extrorse. Staminodes present, entire, petaloid. Ovary conical, pubescent, tapering into the short cylindrical style. Fruit a large 1-seeded berry. Seed compressed oblong-ellipsoid, with oblong lateral scar; endosperm absent, cotyledons fleshy.

A monotypic genus confined to E. tropical Africa. *Inhambanella* is one of the few examples in Sapotaceae where the distinctive characters separating the two subfamilies Mimusopoideae and Sideroxyloideae break down: it has corolla lobes with lateral segments, as in the former, but a 5-merous, uniseriate calyx, as in the latter. According to Aubréville (loc. cit.), the other African genera with this character-combination are *Lecomtedoxa* (Engl.) Dubard, *Gluema* Aubrév. & Pellegrin and *Kantou* Aubrév. & Pellegrin; he argues that the first two are very different from *Inhambanella* in having dry, dehiscent fruits, and *Kantou* is probably its closest relative. A. Meeuse, on the other hand, decided that *Inhambanella* is synonymous with *Lecomtedoxa* (see species synonymy).

Inhambanella henriquesii (Engl. & Warb.) Dubard in Ann. Mus. Col. Marseille, sér. 3, **3**: 43 (1915). — Baehni in Boissiera **11**: 88 (1965). — J. H. Hemsley in F.T.E.A., Sapotaceae: 45, fig. 9 (1968). — Palmer & Pitman, Trees of Southern Afr. **3**: 1763 with fig. (1972). — R. B. Drumm. in Kirkia **10**: 266 (1975). TAB. **50**. Type: Mozambique, Inhambane, *Ferreira* s.n. (COI, holotype).
    Mimusops henriquesii Engl. & Warb. in Engl., Mon. Afr. Pflanzen. **8**: 80, t. 25A (1904) ('henriquezii', but corrected, op. cit.: 88). Type as above.
    Lecomtedoxa henriquesii (Engl. & Warb.) A. Meeuse in Bothalia **7**: 344, fig. 8 (1960); in Fl. Southern Afr. **26**: 40 (1963). — Gomes e Sousa, Dendrol. Moçamb. **2**: 619 (1967). Type as above.

Evergreen tree 6–18 m. tall, notable for the brilliant coppery-red colour of its young leaves. Bark dark brown, rough and flaking; slash very fibrous, orange, exuding copious white latex. Young branchlets glabrescent, with rough longitudinally striate grey bark; apical buds surrounded by subulate reddish-brown tomentose scales. Leaves clustered at branch apices. Lamina 4·5–14(23) × 2·75–5(8) cm., obovate or obovate-oblong; apex obtuse to emarginate, usually much reflexed (so that the leaf cannot be pressed flat), base acute to acuminate; margin thickened, pellucid, undulate. Petiole 2–4·5 cm. long. Upper leaf surface shiny, drying grey or brownish, with slightly raised midrib and 9–11 curved lateral nerves on each side, with a fine reticulation between. Lower surface mat, drying yellowish-green, with midrib and nerves more conspicuous than above. Pedicels 8–10 mm. long. Calyx c. 4·8 mm. long, cylindrical, dorsally pubescent, lobes spreading slightly during anthesis. Corolla c. 5·5 mm. long, yellow, strongly honey-scented, divided for ⅓–⅔ its length into 5 lobes; each lobe comprising an ovate median segment and two small lanceolate lateral segments. Anthers c. 2·3 mm. long. Staminodes c. 3·4 mm. long, lanceolate. Gynoecium c. 3 mm. long. Fruit 2–4 cm. in diameter, subglobose or ellipsoid, orange or crimson when ripe, with edible pulp. Seed up to 3 cm. long.

Zimbabwe. E: Chipinga Distr., Nyangamba R. area, 610 m., fl. viii.1962, *Goldsmith* 178/62 (BM; COI; FHO; K; LISC; SRGH). **Malawi. S:** Nsanje Distr., Malawi Hills, 640 m., st. 3.i.1964, *Chapman* 2179 (FHO; SRGH). **Mozambique. N:** 20 km. from Mocimboa da Praia to Palma, near road, fl. 14.xi.1960, *Gomes e Sousa* 4621 (COI; K; PRE). **Z:** near Maganja da Costa, fl. 31.vii.1943, *Torre* 5730 (K; LISC; LMU; PRE; SRGH). **MS:** 25 km. from Lacerdonia, N. of new railway, 200 m., st. 6.xii.1971, *Müller & Pope* 1913 (K; LISC; SRGH). **GI:** Chidenguele, Marie, st. 15.viii.1947, *Pedro & Pedrogão* 1762 (LMU). **M:** Bela Vista, Licuati, fr. 6.xii.1961, *Lemos & Balsinhas* 264 (BM; COI; K; LISC; LMU; PRE).
    Also known from coastal regions of Kenya, Tanzania and S. Africa (Natal). In mixed evergreen forest.

10. MIMUSOPS L.

Mimusops L., Sp. Pl. **1**: 349 (1753).

Trees and shrubs. Leaves exstipulate (or stipules soon caducous), thinly to strongly coriaceous, not clustered at ends of branches. Flowers borne in axils of current leaves, pedicellate. Sepals 8, in two slightly dissimilar whorls of 4, ± free; inner sepals smaller and paler than outer, and with ciliate margins. Corolla of 8 members joined in a short tube; each member comprising a median and two lateral segments, the latter usually entire, sometimes further divided. Stamens 8, opposing median segments, adnate to corolla tube; anthers extrorse, apiculate. Staminodes 8, alternating with stamens, triangular, dorsally and marginally pilose, bending into the centre of the flower to form a conical sheath round the gynoecium. Ovary shortly cylindrical or subglobose, ribbed, 8-locular, densely brownish appressed-pubescent; style cylindrical, ± truncate. Fruits baccate, fleshy to ± coriaceous, sometimes edible, 1-several-seeded, with persistent calyx at base. Seeds ± ellipsoid, laterally compressed, with hard, highly polished testa; scar small, subcircular, basal or basi-lateral; endosperm abundant, cotyledons thin.

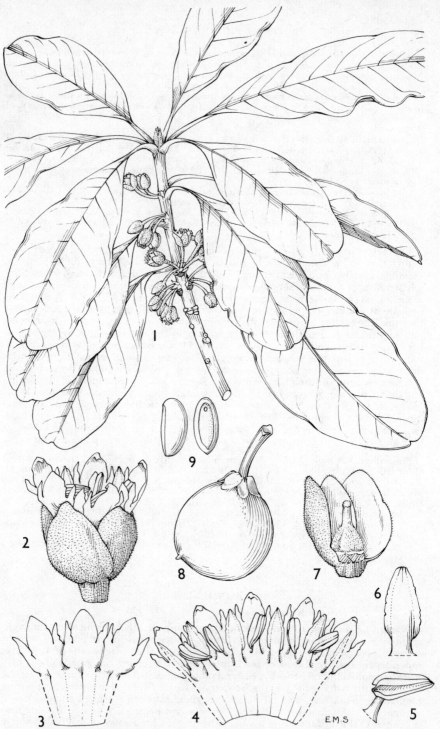

Tab. 50. INHAMBANELLA HENRIQUESII. 1, flowering branch (×⅔); 2, flower (×4); 3, part of corolla viewed from outside (×4); 4, corolla opened out showing stamens and staminodes, viewed from inside (×4); 5, stamen (×8); 6, staminode (×8); 7, flower with two sepals and corolla removed to show gynoecium (×4), 1–7 from *Rawlins* H25/58; 8, fruit (×1), from *Gomes e Sousa* 4403; 9, seeds (×⅔) after *J. Sausotte-Guérel* in Adansonia, mém. 1, t. 18/2. From F.T.E.A.

A genus of 20–30 species, centred in tropical Africa but extending into tropical Asia.

M. _elengi_ L., a S.E. Asian species with oblong-apiculate, undulate-margined leaves, is occasionally grown in Africa as an ornamental tree, and specimens have been seen from Mozambique (Maputo).

Burtt Davy & Hoyle (N.C.L.: 70, 1936) cite the Angolan species M. _mayumbensis_ Greves from Malawi, but this is not confirmed.

1. Lateral segments of corolla each divided almost to the base into 2 narrowly triangular laciniae; shrub with strongly coriaceous cordate to narrowly obovate leaves, inhabiting coastal scrub   -   -   -   -   -   -   -   -   -   -   -   -   - 1. _caffra_
- Lateral segments of corolla simple (or sometimes divided in M. _kummel_, but this species with habitat different from above); trees or shrubs with thinly to strongly coriaceous leaves which may be obovate but not cordate   -   -   -   -   -   -   -   -   -   2
2. Flowering pedicels distinctly longer than petioles   -   -   -   -   -   -   -   3
- Flowering pedicels shorter than to slightly exceeding petioles   -   -   -   -   4
3. Flowers 1–2 in leaf axils; anthers c. 5·5 mm. long; leaves up to 5 × 2·3 cm., thinly coriaceous
2. _obovata_
- Flowers 1–4 in leaf axils; anthers 2·5–4 mm. long; leaves up to 14 × 5·5 cm., coriaceous
3. _kummel_
4. Style 1·5–4 mm. long; leaves narrowly oblong to oblanceolate, with acute to acuminate tip; flowering pedicels much shorter than petioles   -   -   -   -   - 5. _aedificatoria_
- Style 5–9 mm. long; plants not as above   -   -   -   -   -   -   -   -   5
5. Leaves broadest above the middle, the apex usually rounded; lower surface glabrous or sparsely greyish appressed pubescent   -   -   -   -   -   -   - 6. _obtusifolia_
- Leaves usually broadest at or below the middle, rarely above, the apex usually acute; lower surface usually densely appressed ferrugineous pubescent at first, becoming greyish and glabrescent   -   -   -   -   -   -   -   -   -   -   -   - 4. _zeyheri_

1. **Mimusops caffra** E. Mey. ex A.DC. in DC., Prodr. **8**: 203 (March 1844). — Engl., Mon. Afr. Pflanzen. **8**: 72, t. 27 fig. B (1904). — Sim, For. Fl. Port. E. Afr.: 80, t. 75 (1909). — Gerstner in Journ. S. Afr. Bot. **12**: 52, fig. 7 (1946). — A. Meeuse in Bothalia **7**: 356, fig. 12 (1960); in Fl. Southern Afr. **26**: 45 (1963). — Gomes e Sousa, Dendrol. Moçamb. **2**: 623 (1967). — Palmer & Pitman, Trees of Southern Afr. **3**: 1747, with fig. (1972). TAB. **51**. Type from S. Africa (Cape Prov. or Natal).

Mimusops revoluta Hochst. apud Krauss in Flora **27**: 825 (Dec. 1844). Type from S. Africa (Natal).

Much-branched evergreen shrub or small tree 1–5(12) m. tall. Young stems greyish or ferrugineous appressed-pubescent, older branches glabrous with striated grey bark. Leaves very leathery. Lamina 3–9 × 1·8–4·3 cm., cordate or broadly to narrowly obovate with apex slightly emarginate or rarely rounded-acute; base acute; margin thickened, slightly revolute; petiole 7–16 mm. long. Midrib slightly raised on both surfaces, other nerves faint. Upper leaf surface appressed-pubescent when young, soon glossy, glabrous, lower surface densely and minutely brownish or ferrugineous appressed-pubescent, rarely subglabrous. Flowers white, borne 1–8 in leaf axils; pedicels 1·5–3 cm. long. Calyx c. 8 mm. long; lobes of outer whorl narrowly triangular, brownish-pubescent, those of inner whorl lanceolate, paler and more delicate than outer, with distinct median dorsal groove. Corolla equalling calyx, connate at base into a tube c. 1·5 mm. long; median segments lanceolate with involute margins, lateral segments slightly shorter, each divided almost to the base into two narrowly triangular laciniae. Stamens c. 5 mm. long; filaments subulate, anthers c. 3·5 mm. long. Staminodes c. 3·5 mm. long. Ovary c. 2·5 mm. long; style c. 6·5 mm. long. Fruit an edible plum-shaped berry up to 2·5 cm. long, orange-red when ripe; fruiting calyx not reflexed; style persistent. Seed 10–13 mm. long.

Mozambique. GI: Inhambane Prov., Inharrime, fr. 22.vi.1960, _Lemos & Balsinhas_ 154 (BM; COI; K; LISC; PRE; SRGH). M: Polana, fl. 20.i.1960, _Lemos & Balsinhas_ 16 (COI; K; LISC; PRE; SRGH).
Also found in S. Africa, from Natal to the Cape Prov. In coastal thickets.

M. _caffra_ can be confused with M. _obtusifolia_ (no. 6), since both occur in coastal scrub and have rounded leaves which are broadest at the apex. Useful characters for distinguishing them are their different relative lengths of petiole and pedicel and different colour of indumentum on the lower leaf surface.

The following specimen has been named M. _caffra_: **Mozambique.** M: camping site of Polana beach, fl. 17.viii.1974, _Balsinhas_ 2642 (K; SRGH). Although probably correctly assigned to this species it is anomalous in several respects, having leaves unusually large and less

Tab. 51. MIMUSOPS CAFFRA. 1, habit (× ⅔); 2, flower (× 2); 3, lobe of outer calyx whorl, dorsal view (× 4); 4, lobe of inner calyx whorl, dorsal view (× 4); 5, corolla member comprising an inner median segment and two bipartite outer lateral segments; flower contains 8 such units, joined into a ring at "x" (× 4); 6, stamen and two staminodes; the region "x" is united with the ventral surface of the corolla tube (× 4); 7, anther, dorsal view (× 6); 8, gynoecium (× 4), 1–8 from *Lemos & Balsinhas* 16; 9, fruit (× ⅔) *Gomes e Sousa* 1908; 10, seed, ventral (right) and lateral (left) views (× ⅔) *Mogg* 27471.

tapered at the base than is normal, the leaf apex apiculate, and the lateral corolla segments simple.

Wild and Barbosa, in the F.Z. supplement, Vegetation Map of the F.Z. Area: 17 (1968), state that *M. caffra* is one of the commonest shrub species in the littoral scrub of central Mozambique (Zambezia), but this seems doubtful in view of the absence of any herbarium specimens from further north than the Sul do Save province.

2. **Mimusops obovata** Sonder in Linnaea **23**: 17 (1850). — Engl., Mon. Afr. Pflanzen. **8**: 72, t. 27 fig. D (1904). — Wright in Harv. & Sond., F.C. **4**: 442 (1906). — Gerstner in Journ. S. Afr. Bot. **12**: 54, fig. 10 (1946). — A. Meeuse in Bothalia **7**: 358 (1960); in Fl. Southern Afr. **26**: 46 (1963). — Palmer & Pitman, Trees of Southern Afr. **3**: 1749, with fig. (1972). Type from S. Africa (Cape Prov.).

    Mimusops oleifolia N.E. Br. in Kew Bull. **1895**: 109 (1895). — Engl., op. cit.: 73, t. 34 fig. B (1904). — Wright, loc. cit. Type from S. Africa (Natal).

    Mimusops woodii Engl., op. cit.: 65, t. 26 fig. A (1904). — Wright, tom. cit.: 440 (1906). Type from S. Africa (Natal).

    Mimusops rudatisii Engl. & Krause in Engl., Bot. Jahrb. **49**: 395 (1913). Type from S. Africa (Natal).

Shrub or tree up to 20 m. tall with slender, much-branched, very leafy twigs; young branches appressed greyish- or ferrugineous-pubescent at apex, soon glabrous; bark of twigs pale brown, finely longitudinally fissured. Leaves thinly coriaceous, glabrous; lamina 1·8–5 × 1–2·3 cm., elliptic or elliptic-obovate, the apex acute or slightly apiculate or rounded, the base acute; margin slightly undulate; midrib and vein reticulation slightly raised on both surfaces; lamina glossy above, mat below; petiole 4–7 mm. long. Flowers white to pale yellowish, scented, borne 1–2 in leaf axils; pedicels 1–3 cm. long. Calyx c. 1 cm. long; segments free, triangular, the inner slightly narrower and paler than the outer. Corolla ± equalling calyx; tube c. 1 mm. long; median and lateral segments subequal in shape and size, ± lanceolate. Anthers c. 5·5 mm. long. Staminodes about as long as stamens and corolla. Ovary c. 3 mm. long, style c. 8 mm. Fruit 2–3·5 × 1–2 cm., ovoid, beaked, often 1-seeded, yellow to orange-red when ripe. Seed up to 2·5 cm. long.

    Mozambique. GI: Guijá, Massingir, near Rio dos Elefantes, st. 2.xii.1944, *Mendonça* 3234 (COI; LISC; LMU; PRE; SRGH). M: Namaacha, Mt. M'Ponduine, fl. 28.x.1971, *Marques* 2331 (BM; COI; LISC; LMA; LMU; PRE; SRGH).

    Also in eastern S. Africa, from the Cape northwards. In gallery forest and remnant moist evergreen forest.

3. **Mimusops kummel** A.DC. in DC., Prodr. **8**: 203 (1844). — Baker in F.T.A. **3**: 508 (1877). — Engl., Mon. Afr. Pflanzen. **8**: 75, t. 30 fig. A (1904). — Heine in F.W.T.A. ed. 2, **2**: 20 (1963). — J. H. Hemsley in F.T.E.A., Sapotaceae: 54 (1968). Type from Ethiopia.

    Mimusops langenburgiana Engl., op. cit.: 70, t. 28 fig. D (1904). — Burtt Davy & Hoyle, N.C.L.: 70 (1936). Type from Tanzania.

Shrub or small to large tree; bark dark grey, fissured longitudinally or into square flakes, slash pink. Young stems yellowish, whitish or ferrugineous pubescent, soon glabrescent, older twigs with longitudinally wrinkled grey bark. Leaves 4·5–14 × 2–5·5 cm., ovate-oblong to obovate, the apex rounded, acute or acuminate, the base acute; petiole 0·5–1(1·5) cm. long. Upper surface mat or glossy, with prominent midrib but usually obscure vein reticulation, glabrous. Lower surface mat, with prominent midrib and obscure reticulation, usually glabrous, rarely ferrugineous appressed-pubescent. Flowers 1–4 in leaf axils; pedicels 2–8 cm. long. Calyx 9–12 mm. long. Corolla equalling or slightly shorter than calyx, white; lateral segments simple, lanceolate or sometimes divided into 2–3 laciniae, median segments elliptic; tube 1–2 mm. long. Anthers 2·5–4 mm. long. Gynoecium 8–12 mm. long. Fruit up to 2·5 cm. long, an edible plum-shaped 1–2-seeded berry yellowish-orange to red when ripe. Seed up to 1·8 cm. long.

    Malawi. N: Chitipa Distr., Misuku Hills, Kalenga R., 1095 m., fr. 8.vii.1973, *Pawek* 7147 (K; MAL).

    Widely distributed in tropical Africa. In riverine vegetation.

    The specimen cited is the only example of this species known from the F.Z. area, and it is somewhat unusual in having 2-seeded fruits which were probably 3 cm. long when fresh. *M. kummel* occurs in neighbouring parts of Tanzania and so may be expected from northern Malawi.

    M. kummel is closely similar to both *M. zeyheri* (no. 4) and *M. obovata*, its distribution overlapping that of the former. Sterile material of *M. kummel* can be distinguished from that of *M. zeyheri* by several differential, though not completely diagnostic, leaf characters: in *M. kummel* the petioles tend to be short, the lower leaf surface is normally glabrous and the venation inconspicuous; in *M. zeyheri* the petioles are longer, the lower leaf surface is often ferrugineous-pubescent and the tertiary venation raised.

4. **Mimusops zeyheri** Sonder in Linnaea **23**: 74 (1850). — Engl., Mon. Afr. Pflanzen. **8**: 73, t. 27 fig. C (1904). — Steedman, Trees etc. S. Rhod.: 62, t. 58 (1933). — O. B. Miller, B.C.L.: 46 (1948). — A. Meeuse in Fl. Pl. Afr. **30**: t. 1164 (1954); in Bothalia **7**: 361 (1960); in Fl. Southern Afr. **26**: 47 (1963) pro parte excl. syn. *M. kirkii*. — Pardy in Rhod. Agric. Journ. **53**: 517, with photogr. (1956). — F. White, F.F.N.R.: 321 (1962). — Gomes e Sousa, Dendrol. Moçamb. **2**: 624 (1967) pro parte. — J. H. Hemsley in F.T.E.A., Sapotaceae: 56 (1968). — Palmer & Pitman, Trees of Southern Afr. **3**: 1751, with fig. (1972) excl. syn. — Fanshawe, Check-list Woody Pl. Zambia: 29 (1973). — R. B. Drumm. in Kirkia **10**: 266 (1975). Type from S. Africa (Transvaal).

   Mimusops zeyheri var. *laurifolia* Engl., loc. cit. — Eyles in Trans. Roy. Soc. S. Afr. **5**: 438 (1916). — Burtt Davy & Hoyle, N.C.L.: 71 (1936). — Topham, N.C.L. ed. 2: 94 (1958). Type: Malawi, 1895, *Buchanan* 304 (B†, holotype; BM, K, isotypes).

   ?*Mimusops blantyreana* Engl., op. cit.: 83 (1904). — Burtt Davy & Hoyle, op. cit.: 70 (1936). Type: Malawi, Blantyre, *Buchanan* 7024 (not seen).

   Mimusops monroi S. Moore in Journ. Bot., Lond. **49**: 154 (1911). — Eyles, tom. cit.: 437 (1916). Type: Zimbabwe, Victoria, 1909, *Monro* 761 (BM, holotype; SRGH, isotype).

   Mimusops decorifolia S. Moore, loc. cit. — Eyles, loc. cit. Type: Zimbabwe, Victoria, 1909, *Monro* 811 (BM, holotype).

   Mimusops sp. — Brenan in Mem. N.Y. Bot. Gard. **8**: 499 (1954).

Evergreen shrub branching near the ground to tall tree with long clean trunk, 2–23 m. tall; slash white, pink or crimson; bark grey to black, smooth or shallowly grooved or roughly reticulately fissured; buttresses absent; crown (in larger specimens) wide-spreading, rounded. Young stems and leaves often densely ferrugineous appressed-pubescent, glabrescent, older twigs with longitudinally wrinkled grey bark. Leaf-lamina 3·5–11·5 × 1·4–5·5 cm., elliptic, oblong-elliptic, obovate-elliptic or lanceolate; apex tapering or acute or rounded and slightly emarginate, but most often bluntly apiculate; base acute to acuminate; petiole 0·6–3·5 cm. long. Upper leaf surface glossy, usually with prominent midrib and vein reticulation but sometimes the latter obscured beneath the thick leathery cuticle, initially or soon becoming completely glabrous. Lower surface mat, with prominent midrib and fairly conspicuous reticulation, more tardily glabrescent than above, indumentum sometimes becoming grey, sometimes persisting along midrib. Flowers sweetly scented, borne in axillary fascicles of 1–7; pedicels 10–17 mm., but elongating in fruit. Calyx 6·5–9·5 mm. long. Corolla equalling or slightly longer or shorter than calyx, white to pale brownish-yellow; lateral segments narrowly triangular or lanceolate, median segment equalling, longer or shorter than laterals, elliptic with narrow basal attachment; tube 1–2 mm. long. Anthers 2·5–3·5 mm. long. Staminodes half as long as to almost equalling corolla, shorter or longer than stamens. Gynoecium 6·5–9 mm. long. Fruit up to 4·5 cm. long, edible, plum-shaped, 1-seeded, with brittle yellow or orange epicarp and floury astringent orange pulp; fruiting calyx cupped round fruit or spreading. Seed c. 1·8 mm. long.

Botswana. N: Chobe Distr., Serondela, near Chobe R., 915 m., x.1951, *Miller* B/1191 (FHO). SE: Ootse Mt., 40 km. SE. of Gaberone, 1370 m., fr. 23.vi.1974, *Woollard* 3 (SRGH). **Zambia.** B: Zambesi Distr., Kalambosa, fr. 14.iv.1949, *West* 2900 (SRGH). N: Mansa (Fort Rosebery), fl. buds 16.viii.1952, *White* 3068 (BM; COI; FHO; K). W: Ndola Distr., Ndola Golf Course, fl. 23.x.1952, *Angus* 653 (BM; FHO; K; PRE). C: 142 km. N. of Kabwe (Broken Hill), fr. vii.1909, *Rogers* 8325 (SRGH). E: Nsadzu to Chipata (Fort Jameson) road, 900 m., fl. 25.xi.1958, *Robson* 704 (BM; K; LISC; PRE; SRGH). S: Mazabuka Distr., Kafue Gorge, c. 1000 m., fl. 4.xi.1972, *Strid* 2437 (C). **Zimbabwe.** N: Shamva Distr., Mazoe R., Golden Scar, immat. fr. 29.xii.1963, *Masterson* 296 (SRGH). W: Matopos, Farm Besna Kobila, 1430 m., fl. xii.1953, *Miller* 1985 (PRE; SRGH). C: Gwelo Distr., 16 km. E. of Gwelo, 1460 m., fr. 21.xi.1965, *Biegel* 576 (SRGH). E: Umtali Commonage, Meikle's Jungle, fr. 27.viii.1948, *Chase* 869 (BM; COI; LISC; PRE; SRGH). S: Victoria Distr., Zimbabwe, Acropolis Hill, fr. 19.viii.1965, *West* 6733 (SRGH). **Malawi.** N: Rumphi Distr., near Rumphi Bridge on Livingstonia road, fl. 20.x.1962, *Adlard* 507 (FHO; MAL; SRGH). C: Dedza Distr., Ciwau, fr. 16.x.1960, *Chapman* 992 (FHO; K; MAL; SRGH). S: Ntcheu Distr., confluence of Dombole and Livulezi Rs., fr. 1.ix.1971, *Salubeni* 1708 (MAL; SRGH). **Mozambique.** N: 10 km. from Mutuali to Malema, fl. 21.x.1953, *Gomes e Sousa* 4150 (COI; K; PRE). T: between Marueira and Songo, fr. 7.ii.1972, *Macedo* 4806 (K; LISC; LMA; LMU; SRGH). MS: Plateau E. of

Haroni R., above its confluence with Lusitu, 450 m., fl. 14.i.1969, *Biegel* 2837 (K; LISC; PRE; SRGH). M: Namaacha, by R. Impamputo, fr. 20.vii.1967, *Marques* 2065 (BM; COI; K; LISC; LMU; PRE; SRGH).

Also known from Tanzania, Angola and S. Africa (Transvaal and Natal). In riverine fringe vegetation, also on termitaria and on rocky hill slopes.

5. **Mimusops aedificatoria** Mildbr. in Notizbl. Bot. Gart. Berl. **14**: 109 (1938). — J. H. Hemsley in F.T.E.A., Sapot.: 58 (1968). Type from Tanzania.

Very closely related to *M. zeyheri*. It differs in having the following combination of characters: leaves narrowly oblong to oblanceolate, lamina up to 12 × 3·5 cm. and petiole up to 3 cm. long; pedicels distinctly shorter than petioles, up to 1(1·5) cm. long; style short, 1·5–4 mm. long.

Malawi. S: Lake Malawi (L. Nyasa), Karonga, fl. xi.1893, *Scott Elliot* 6404 (BM; K). Otherwise known from eastern Kenya and Tanzania. In humid forest.

6. **Mimusops obtusifolia** Lam., Encycl. Méth., Bot. **4**: 186 (1797), excl. syn. Type a specimen at P., without indication of locality or collector.

    Mimusops fruticosa A.DC. in DC., Prodr. **8**: 202 (1844). — Baker in F.T.A. **3**: 508 (1877). — Engl., Mon. Afr. Pflanzen. **8**: 66, t. 23 fig. B (1904). — Dubard in Ann. Mus. Col. Marseille, sér. 3, **3**: 50, fig. 18 (1915). — J. H. Hemsley in F.T.E.A., Sapotaceae: 53, fig. 11 (1968). — R. B. Drumm. in Kirkia **10**: 266 (1975). — Hall-Martin in Kirkia **12**: 175 (1980). Type from E. Africa.

    Mimusops kirkii Baker, tom. cit.: 507 (1877). Syntypes from Mozambique: Lower Shire Valley, Shamo, *Kirk* (K, four gatherings).

    Mimusops busseana Engl., op. cit.: 79 (1904). Type from Tanzania/Mozambique border.

    Mimusops kilimanensis Engl., op. cit.: 67 (1904). Type: Mozambique, Zambezia, near Quelimane (Kilimane), Puguruni, *Stuhlmann* 1007 (B, holotype†; HBG, isotype).

    Mimusops zeyheri sensu A. Meeuse in Bothalia **7**: 361 (1960); in Fl. Southern Afr. **26**: 47 (1963). — Sensu Gomes e Sousa, Dendrol. Moçamb. **2**: 624 (1967), pro parte quoad syn. *M. kirkii*, non Sonder.

Many-stemmed shrub, or tree, 1·5–20 m. tall; bark rough; crown (in large specimens) hemispherical. Young stems appressed brownish pubescent, soon glabrescent; bark of older twigs grey, finely longitudinally fissured and wrinkled. Leaves 3·5–9·5 × 2·2–7 cm., elliptic, broadly elliptic, obovate or suborbicular; apex rounded, often slightly emarginate or apiculate, base acute; petiole 1·3–4 cm. long. Upper leaf surface glossy, usually with prominent midrib and vein reticulation but sometimes the latter obscured by the thick leathery cuticle, glabrous. Lower surface mat, with prominent midrib and level or raised reticulation, glabrous or sparsely greyish appressed-pubescent. Flowers sweetly scented, in axillary fascicles of 1–3; pedicels 8–20 mm. long, lengthening considerably from bud to fruit. Calyx up to 9 mm. long. Corolla up to 8·5 mm. long, white, cream-coloured or pale pink; lateral segments narrowly triangular or lanceolate, median segment elliptic with narrow attachment, longer or shorter than laterals; tube c. 1 mm. long. Anthers 2·5–3 mm. long. Staminodes longer or shorter than stamens. Gynoecium c. 10 mm. long. Fruit up to 2·5 cm. long, edible, plum-shaped, 1–5(6)-seeded, with brittle orange or red epicarp; fruiting calyx spreading to reflexed. Seed up to 2 cm. long.

Zimbabwe. E: Chipinga Distr., Dinde R., fl. 28.ix.1958, *Phelps* 265 (K; LISC; SRGH). S: Ndanga Distr., Chitsa's Kraal, fl. 5.vi.1950, *Chase* 2338 (BM; COI; LISC; PRE; SRGH). **Malawi.** S: Chikwawa Distr., Lengwe Game Reserve, 200 m., st. 18.viii.1970, *Hall-Martin* 928 (K; SRGH). **Mozambique.** N: Nampula area, Mureveia, Chefe Niaro, fl. 22.xi.1948, *Andrada* 1470 (COI; LISC; LMA). Z: Namagoa, fl. x–xii, 1944, *Faulkner* 56 (BM; COI; K; PRE; SRGH). T: margin of R. Zambesi, 7 km. W. of Tete, fl. 20.xi.1965, *Neves Rosa* 126 (LISC; LMA). MS: Sofala Prov., Chiniziua, fr. 17.iv.1957, *Gomes e Sousa* 4365 (COI; FHO; K; LISC; LMU; PRE). GI: Gaza, Caniçado, 18 km. from Mabalane (Vila Pinto Teixeira) to Combomune, fr. 21.viii.1969, *Correia & Marques* 1105 (COI; LMU; PRE; SRGH). M: Magude, between Mapulanguene and Rio dos Elefantes, fl. 29.i.1948, *Torre* 7236 (BR; LISC; MO; WAG).

Coastal regions of Kenya and Tanzania, S. Africa (N. Transvaal), Comoro Is. and Madagascar. In riverine forest, in coastal scrub, on termitaria.

After his detailed description of *M. obtusifolia*, Lamarck cited a specimen from Mauritius, collected by Joseph Martin. The specimen identified in Paris as the type of this species, however, has no locality or collector. *M. obtusifolia* is not known from Mauritius (Dr. H. Heine, pers. comm.), and it is possible that the description was based on a cultivated plant which came originally from E. Africa. This seems the more likely since A. De Candolle based his description of *M. fruticosa* (syn. *M. obtusifolia*) on a *Bojer* specimen cultivated in Mauritius but originating from E. Africa. I am most grateful to Dr. Heine for his help in typifying this species.

M. obtusifolia varies throughout its range in respect to the relative lengths of petiole and pedicel. In material from the F.Z. area the petioles are longer than pedicels, whereas further north the pedicels usually exceed the petioles (cf. J. H. Hemsley in F.T.E.A., Sapotaceae: 53, fig. 11, 1968). The leaves of *M. obtusifolia* are usually relatively broader and more rounded at the apex than shown in this figure.

11. VITELLARIOPSIS (Baill.) Dubard

Vitellariopsis (Baill.) Dubard in Ann. Mus. Col. Marseille, sér. 3, **3**: 44 (1915). — Aubrév. in Adansonia **3**: 41 (1963). — Kupicha in Candollea **33**: 29 (1978).
  Mimusops Sect. *Vitellariopsis* Baill. in Bull. Soc. Linn. Paris **2**: 942 (1891).
  Austromimusops A. Meeuse in Bothalia **7**: 347 (1960).

Shrubs and trees, with leaves crowded at tips of branches. Leaves stipulate, stipules persistent or caducous; vein reticulation characteristically fine and tessellate. Flowers borne in axils of current leaves, pedicelled. Sepals 8, in two dissimilar whorls of 4, shortly united at base. Corolla of 8 members fused at the base into a short tube, each member comprising a median and two lateral segments. Stamens 8, opposed to median segments; anthers extrorse. Staminodes alternating with stamens, triangular, dorsally and marginally pilose, bent inwards to form a cone sheathing the style. Ovary densely hairy, 8-locular; style slender. Fruit baccate, dryish, 1 (few)-seeded. Seed ellipsoid, with large hilum occupying up to half the seed surface; testa crustaceous, dull; endosperm absent, cotyledons fleshy.

A genus of 5 species from E. tropical and southern Africa. Closely related to *Mimusops,* but differing in having leaves clustered at branch apices, in its characteristic leaf venation, and in seed morphology.

Leaves elliptic to obovate-elliptic, rounded at apex; lower surface ferruginous crispate-
    pubescent (indumentum wearing off in older leaves)   -   -   -   - 1. *ferruginea*
Leaves narrowly obovate, acute to acuminate at apex; lower surface glabrous   2. *marginata*

1. **Vitellariopsis ferruginea** Kupicha in Candollea **33**: 30, fig. 1, 2 (1978). TAB. **52**. Type: Zimbabwe, Umtali Distr., Dora Farm, 16 km. S. of Umtali on Melsetter road, fl. 21.xi.1948, *Chase* 964 (BM, holotype; K, isotype).
    Austromimusops sylvestris (S. Moore) A. Meeuse in Bothalia **7**: 354, fig. 11 (1960), excl. typ.
    Vitellariopsis sylvestris (S. Moore) Aubrév. sensu R. B. Drumm. in Kirkia **10**: 266 (1975) quoad specim. *Chase* 964 excl. typ.

Tree or large bush 3–7 m. high with smooth bole and pink slash. Branching-pattern divaricate or verticillate, the proximal part of each twig smooth, with distant leaf scars, the distal part rough with very crowded scars; leaves in "rosettes" at branch ends; leafy part of twigs and petioles with dense reddish indumentum. Stipules soon caducous. Leaf lamina 3·5–11 × 1·3–6 cm., elliptic to obovate-elliptic, the apex rounded or slightly emarginate, the base rounded to cuspidate; petiole 2–7 mm. long. Upper leaf surface glabrous, drying dark greyish-brown, with slightly raised midrib and fine vein reticulation; lower surface ferruginous crispate-pubescent, glabrescent, with prominent midrib and conspicuous reticulation. Flowers white to pale pink, scented, attractive, borne 1–3 in leaf axils; pedicels 1–2 cm. long; buds and pedicels densely ferruginous-pubescent. Calyx c. 7·5 mm. long, outer sepals of thicker texture than inner. Corolla c. 7·5 mm. long, connate at base for c. 1·5 mm.; median segments elliptic, slightly shorter than lateral segments, these ovate, tapering at apex. Stamens c. 4 mm. long; filaments subulate; anthers c. 3·5 mm. long. Staminodes equalling stamens in length. Gynoecium c. 8 mm. long. Fruit up to 4·5 cm. long, ovoid-ellipsoid, apiculate, ferruginous-tomentose when young, glabrescent, 1–2-seeded. Seed up to 3 cm. long.

Zimbabwe. E: Umtali Distr., Bushman's Haunt, Zimunya's T.T.L., 1010 m., fr. 13.i.1952, *Chase* 4328 (BM; COI; K; LISC; PRE; SRGH). S: Buhera Distr., Chironga Ruins near Matandera, fl. buds 19.xi.1963, *Masterson* 274 (SRGH).
    Known only from Zimbabwe. On granite hills, among rocks.

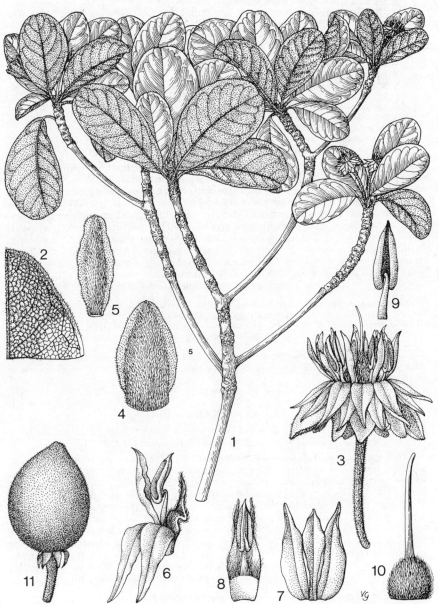

Tab. 52. VITELLARIOPSIS FERRUGINEA. 1, habit (× ⅔) *Chase* 965 & 4328; 2, part of lower leaf surface (× 3); 3, flower (× 3); 4, outer sepal, dorsal view (× 4); 5, inner sepal, dorsal view (× 4); 6, part of corolla with stamen and staminode attached (× 4); 7, same part of corolla showing the 3-segment petal unit (× 4); 8, a stamen and its two adjacent staminodes, from outside of flower (× 4); 9, stamen, from inside of flower (× 5); 10, gynoecium (× 4), all from *Chase* 965; 11, fruit (× 1) *Chase* 4328. From Candollea **33**: 31 (1978).

2. **Vitellariopsis marginata** (N.E. Br.) Aubrév. in Adansonia **3**: 42 (1963). — Palmer &
Pitman, Trees of Southern Africa **3**: 1759, with fig. (1972). — Kupicha in Candollea **33**:
37, fig. 2, 3 (1978). Type from S. Africa (Cape Prov.).
> *Mimusops marginata* N.E. Br. in Kew Bull. **1895**: 108 (1895). — Engl., Mon., Afr.
> Pflanzen. **8**: 71 (1904). — Gerstner in Journ. S. Afr. Bot. **12**: 54, figs. 8 & 9 (1946).Type as
> above.
> *Mimusops natalensis* Schinz in Bull. Herb. Boiss. **4**: 441 (1896). Type from S. Africa
> (Cape Prov.).
> *Mimusops schinzii* Engl., op. cit.: 70, t. 29 fig. A (1904). — Gerstner, loc. cit., *nom.
> illegit.* Type as for *Mimusops natalensis.*
> *Mimusops sylvestris* S. Moore in Journ. Linn. Soc., Bot. **40**: 132 (1911). Type:
> Mozambique, Madanda Forests, c. 120 m., fl. 6.xii.1906, *Swynnerton* 570 (BM, holotype;
> K, isotype; SRGH).
> *Inhambanella natalensis* (Schinz) Dubard in Ann. Mus. Col. Marseille **23**: 43 (1915).
> Type as for *Mimusops natalensis.*
> *Austromimusops marginata* (N.E. Br.) A. Meeuse in Bothalia **7**: 348 (1960); in Fl.
> Southern Afr. **26**: 43, fig. 7, 2 (1963). Type as for *Vitellariopsis marginata.*

Shrub or tree 3–7 m. tall. Young stems glabrous or ferrugineous-pubescent and
soon glabrescent; bark grey, striated. Leaves glabrous; lamina 4·5–10 × 1·5–4 cm.,
narrowly obovate, apex acute to cuspidate-apiculate, base tapering gradually; midrib
and the very fine vein reticulation raised on both surfaces; margin revolute; petiole
6–10 mm. long. Flowers white, in 1–4-flowered axillary fascicles, on ferrugineous-
pubescent pedicels up to 3 cm. long, each subtended by a minute, broadly triangular,
caducous bract. Calyx c. 9 mm. long. Corolla c. 8·5 mm. long; tube c. 1 mm. long;
median and lateral lobes similar, narrowly oblong-elliptic. Anthers c. 2·5 mm. long.
Staminodes c. 4·5 mm. long. Gynoecium c. 13 mm. long. Fruit up to 5 cm. long and
3·5(5) cm. in diameter, ovoid, ellipsoid or globose, with pointed apex, glabrous and
purplish-red when ripe; fruiting calyx spreading to reflexed. Seed 20–25 mm. long.

Mozambique. MS: Madanda Forests, c. 120 m., fl. 6.xii.1906, *Swynnerton* 570 (BM; K).
GI: São Sebastião peninsula, fl. 10.xi.1958, *Mogg* 29138 (BM; K; LISC; LMU; SRGH). M:
Maputo, near Goba, fl. 18.xii.1948, *Barbosa* 744 (K; LISC; LMU; SRGH).
Also in S. Africa (Cape Prov., Natal and Transvaal). Woodland on sandy soil, ravine forest.

The identity of *Swynnerton* 570, the type of *M. sylvestris* from Mozambique (MS), is a little
doubtful, and a fresh collection from the same region is most desirable; see discussion in
Candollea **33**: 37 (1978).

12. MANILKARA Adans.

Manilkara Adans., Fam. Pl. **2**: 166 (1763). — H. J. Lam in Bull. Jard. Bot. Buitenz.
sér. 3, **7**: 234 (1925). — Aubrév. in Adansonia **11**: 251 (1971).

Trees and shrubs, with leaves often borne in terminal clusters. Stipules absent or
soon caducous. Leaves often with appressed silvery indumentum on lower surface.
Flowers often numerous, borne in axils of current and recently-fallen leaves, long-
pedicellate. Sepals 6, in two dissimilar whorls of 3, the inner ones thinner in texture
and paler than the outer. Corolla of 6 members joined at the base into a short tube,
each member comprising a median and two lateral segments. Stamens 6*,
epipetalous; anthers extrorse. Staminodes 6*, alternating with stamens, glabrous,
petaloid to ligulate, dentate or laciniate, erect, not forming a sheath round the
gynoecium. Ovary with 6–16 locules, densely pilose, with long slender glabrous
style. Fruit a fleshy or leathery berry, 1–several-seeded. Seeds ellipsoid to obovoid,
with narrow linear or large and broad lateral scar; endosperm abundant.

A pantropical genus of c. 50 species, with some 15 in Africa.

Specimens of two exotic species of *Manilkara* have been seen from Mozambique, the Asian
M. kauki (L.) Dubard and the Central American *M. zapota* (L.) van Royen. The latter is
cultivated for its edible fruit called Sapodilla or Nispero, and is also a source of chicle, used in
the manufacture of chewing gum.

1. Lower leaf surface never with an appressed silvery indumentum, either completely glabrous
   or crispate-pubescent; petioles up to 1 cm. long - - - - - - - 2
 - Lower leaf surface ± obscured by a dense matted appressed indumentum of minute hairs

---

* Aberrant forms of androecium occur in *M. discolor.*

scarcely visible with a hand-lens, often giving a silvery sheen, not crispate-pubescent; or if
leaves apparently glabrous then petiole 1 cm. long or more - - - - - 3
2. Leaves clustered on short warty branches (these actually the overtopped stem apices)
 1. *mochisia*
 - Leaves not clustered as above, but borne at moderately spaced nodes on main branches
 2. *concolor*
3. Vein reticulation on upper leaf surfaces very fine, giving the lamina a minutely granular
 appearance - - - - - - - - - - - - - - 4
 - Vein reticulation on upper leaf surface relatively coarse, giving the lamina an elongate-
 reticulate, striate appearance - - - - - - - - 5. *sansibarensis*
4. Leaves 7–14 cm. long; flowers with 6 stamens and 6 staminodes, the petals always
 comprising 3 segments - - - - - - - - - - 3. *obovata*
 - Leaves 3·5–7(9) cm. long; flowers usually with 12 stamens and no staminodes and petals of 3
 segments, but sometimes androecium comprising 12 staminodes or 6 stamens and 6
 staminodes, and then petals only shortly 3-lobed - - - - - - 4. *discolor*

1. **Manilkara mochisia** (Baker) Dubard in Ann. Mus. Col. Marseille, sér. 3, **3**: 26 (1915). —
 A. Meeuse in Bothalia **7**: 369, fig. 16 (1960); in Fl. Southern Afr. **26**: 50, fig. 8 (1963). — F.
 White, F.F.N.R.: 321 (1962). — J. H. Hemsley in Kew Bull. **20**: 483 (1966); in F.T.E.A.,
 Sapot.: 64 (1968). — Palmer & Pitman, Trees of Southern Afr. **3**: 1755, with fig. (1972). —
 Fanshawe, Check-list Woody Pl. Zambia: 28 (1973). — R. B. Drumm. in Kirkia **10**: 266
 (1975). — Hall-Martin in Kirkia **12**: 175 (1980). Type: Mozambique, Tete, xi.1859, *Kirk*
 (K, lectotype).
 Mimusops mochisia Baker in F.T.A. **3**: 506 (1877). — Engl., Mon. Afr. Pflanzen. **8**: 63, t.
 22 fig. B (1904). — Burtt Davy & Hoyle, N.C.L.: 71 (1936). — Gerstner in Journ. S. Afr.
 Bot. **12**: 52 (1946). Type as above.
 Mimusops densiflora Engl., Pflanzenw. Ost-Afr. **C**: 307 (1895) non Baker, *nom. illegit.*
 Type from Tanzania.
 Mimusops menyhartii Engl., op. cit.: 63, t. 22 fig. D (1904). Type: Mozambique,
 Boroma, *Menyhart* 771 (C, isotype; Z, holotype).
 Mimusops macaulayae Hutch. & Corb. in Kew Bull. **1920**: 329, fig. A (1920). —
 Roessler in Prodr. Fl. SW. Afr. **106**: 1 (1967). Type: Zambia, Mumbwa, *Macaulay* 1002
 (K, holotype).
 Mimusops spiculosa Hutch. & Corb., tom. cit.: 331, fig. B (1920). Type: Zimbabwe or
 Zambia, Victoria Falls, xi.1905, *Allen* 185 (K, holotype).
 Mimusops umbraculigera Hutch. & Corb., loc. cit.: fig. C (1920). — Steedman, Trees
 etc. S. Rhod.: 62 (1933). Type: Zimbabwe, without definite locality or collector, *Herb.
 Dept. of Agric. S. Rhod.* 2639 (K, holotype).
 Manilkara densiflora Dale, Addit. & Correct. "Trees & Shrubs Kenya Col.": 25 (1939).
 Type as for *Mimusops densiflora.*
 Manilkara macaulayae (Hutch. & Corb.) H. J. Lam in Blumea **4**: 356 (1941). — A.
 Meeuse, tom. cit.: 373, fig. 18 (1960); op. cit.: 51 (1963). Type as for *Mimusops
 macaulayae.*
 Manilkara menyhartii (Engl.) H. J. Lam, loc. cit. Type as for *Mimusops menyhartii.*
 Manilkara spiculosa (Hutch. & Corb.) H. J. Lam, loc. cit. Type as for *Mimusops
 spiculosa.*
 Manilkara umbraculigera (Hutch. & Corb.) H. J. Lam, loc. cit. Type as for *Mimusops
 umbraculigera.*

Evergreen tree or large bush 3–20 m. tall, with rough dark bark and pendulous
branches, said to spread by suckering. Branching subterminal, strongly divaricate,
often subverticillate; leaves densely clustered at stem apices which are then
overtopped by younger branches. Leaves 1·8–9 × 1–3·4 cm., oblong, oblong-
obovate or cordate, the apex rounded and emarginate, the base acute; petiole 1–10
mm. long. Upper leaf surface with midrib slightly raised and lateral veins level or
impressed; lower surface with prominent midrib and veins level to strongly
impressed; leaves often crispate-pubescent below and sometimes also sparsely
pubescent above, patchily glabrescent. Flowers abundant, strongly scented, in
fascicles of 3–4 per node; pedicels 5–7 mm. long. Calyx c. 3·5 mm. long, the sepals
elliptic-ovate. Corolla 3·6–4 mm. long, greenish- or brownish-yellow; basal tube c.
0·4 mm. long; median and lateral segments ± equal in shape and size, narrowly
elliptic. Anthers 1·3–1·9 mm. long. Staminodes very variable even within a single
flower, ranging in shape from a small triangular or ovate-dentate scale to a long
narrow process exceeding the stamen. Gynoecium 2·8–3·6 mm. long. Fruit up to 18
mm. long when ripe, ellipsoid, yellow with red pulp, edible, 1–3-seeded. Seed 8–13
mm. long.

Botswana. N: Linyanti R. at Hunter's Camp, c. 17 km. SW. of Hyena Camp, fl. 27.x.1972,
Pope, Biegel & Russell 878 (K; LISC; SRGH). **Zambia.** C: Luangwa Valley, near Kapampa

R., c. 610 m., st. 11.i.1966, *Astle* 4340 (SRGH). E: Chipata (Fort Jameson), 4.xii.1961, *Grout* 269 (FHO). S: Mazabuka Drift, near Kariba Hills, st. 4.xii.1957, *Goodier* 440 (COI; K; PRE; SRGH). **Zimbabwe.** N: Gokwe Distr., Copper Queen African Purchase Area, fl. 8.xi.1963, *Bingham* 899 (K; SRGH). W: Wankie Distr., Devil's Cataract, Victoria Falls, fl. xi.1959, *Armitage* 166/59 (SRGH). C: Gatooma Distr., Mornington, st. 1.ii.1951, *Golding* 31267 (K; LISC; SRGH). E: Inyanga North, Lawley's Concession, st. 19.ii.1954, *West* 3373 (K; SRGH). S: Ndanga Distr., Chitsa's Village, Sabi R., st. 6.vi.1950, *Chase* 2367 (BM; SRGH). **Malawi.** C: Kasungu Game Reserve, Lifupa, fl. 16.xii.1962, *Chapman* 1766 (FHO; K; LISC; MAL; SRGH). S: Chikwawa Distr., Lengwe National Park, fl. xi.1970, *Hall-Martin* 1133 (SRGH). **Mozambique.** Z: Mopeia, st. iv.1972, *Earle* P111C (SRGH). T: Mutara, st. 29.xi.1971, *Haffern* 100 (SRGH). MS: Gorongosa, National Game Park, c. 40 m., fl. 10.xi.1963, *Torre &* *Paiva* 9166 (LISC). GI: Guijá Distr., Missão de S. Vicente de Paula, st. 18.vi.1947, *Pedrogão* 348 (COI; K; PRE; SRGH). M: between Boane and Moamba, fr. 23.xii.1954, *Myre &* *Carvalho* 2008 (LMA; PRE; SRGH).

Also found in S. Africa (Natal, Transvaal), SW. Africa, Angola, Tanzania, Kenya and Somali Republic. In savannah woodland, often on termitaria.

Fanshawe, loc. cit. also records *M. mochisia* from the Barotseland and Northern Provinces of Zambia.

There is some degree of correlation between geography and presence or absence of leaf pubescence: specimens with crispate-pubescent leaves are found in Botswana, Zambia, Zimbabwe (except E) and central Malawi, while plants with glabrous leaves occur throughout Mozambique, Zimbabwe and Malawi. My findings in the F.Z. area thus agree with those of J. H. Hemsley (1968: 65) for F.T.E.A. and of A. Meeuse (1963: 48–52) for Fl. Southern Afr.: the pubescent form tends to occur in the inland part of the distribution range, the glabrous form in the coastal part. (Meeuse treats pubescent specimens as a separate species, *M. macaulayae*.)

2. **Manilkara concolor** (Harv. ex Wright) Gerstner in Journ. S. Afr. Bot. **14**: 171 (1948). — A. Meeuse in Bothalia **7**: 367, fig. 15 (1960); in Fl. Southern Afr. **26**: 48 (1963). — Gomes e Sousa, Dendrol. Moçamb. **2**: 621 (1967). — Palmer & Pitman, Trees of Southern Afr. **3**: 1753, with fig. (1972). Type from S. Africa (Natal).

　　Mimusops concolor Harv. ex Wright in Dyer, Fl. Cap. **4**: 443 (1906). — Gerstner, op. cit. **12**: 52, fig. 6 (1946). Type as above.

Tree 3–12 m. tall. Leaves not clustered at or confined to branch ends, glabrous; lamina 2·5–8·5 × 1–3·5 cm., elliptic, oblong-elliptic, oblong-obovate or narrowly oblong, apex very slightly to moderately emarginate, base acute; petiole 2·5–8 mm. long. Midrib slightly raised on upper leaf surface, prominent below, nerve reticulation level to quite deeply impressed on both surfaces. Flowers white or greenish-yellow, nectariferous, borne in fascicles of up to 8 flowers in leaf axils; pedicels 5–13 mm. long. Calyx 2·8–3·2 mm. long, sepals triangular to ovate. Corolla c. 4 mm. long, connate at base for c. 0·6 mm.; median segments narrowly elliptic, slender at base, lateral segments broader. Anthers 1·5 mm. long. Staminodes triangular to dentate-acuminate, variable in size. Gynoecium c. 3·5 mm. long. Fruit up to 1·5 cm. in diameter, subglobose, 1–2-seeded, orange and edible when ripe. Seeds up to 11 mm. long.

Zimbabwe. S: Nuanetsi Distr., Gona-re-Zhou National Park, Naivasha Camp, st. 17.x.1975, *Drummond* 10399 (SRGH). **Mozambique.** GI: Inhambane Prov., Inharrime, Ponta Zavora, fr. 16.x.1957, *Barbosa & Lemos* 8078 (COI; K; LISC; LMA). M: Inhaca I., st. 15.vii.1957, *Mogg* 27207 (K; PRE; SRGH).

Also in S. Africa (Natal). In coastal scrub, in tree savannah, often on termitaria.

3. **Manilkara obovata** (Sabine & G. Don) J. H. Hemsley in Kew Bull. **17**: 171 (1963); op. cit. **20**: 468 (1966); in F.T.E.A., Sapotaceae: 68 (1968). — Heine in F.W.T.A. ed. 2, **2**: 20 (1963). — Fanshawe, Check-list Woody Pl. Zambia: 28 (1973). Type from Sierra Leone.

　　Chrysophyllum obovatum Sabine & G. Don in Trans. Hort. Soc. Lond. **5**: 458 (1824). Type as above.

　　Mimusops multinervis Baker in F.T.A. **3**: 506 (1877). — Engl., Mon. Afr. Pflanzen. **8**: 57, t. 20 fig. A (1904). Type from Nigeria.

　　Mimusops cuneifolia Baker, loc. cit. — Engl., op. cit.: 64 (1904). Type from Angola.

　　Mimusops lacera Baker, tom. cit.: 507 (1877). — Engl., op. cit.: 59, t. 20 fig. B (1904). Syntypes from Nigeria.

　　Mimusops schweinfurthii Engl., Bot. Jahrb. **12**: 523 (1890); op. cit.: 87, t. 20 fig. D (1904). Syntypes from the Sudan.

　　Mimusops welwitschii Engl., tom. cit.: 524 (1890); op. cit.: 58 (1904). Type from Angola.

　　Mimusops densiflora Baker in Kew Bull. **1895**: 148 (1895). Type from Nigeria.

　　Mimusops chevalieri Pierre in Bull. Mus. Hist. Nat. Paris **7**: 139 (1901). Type from Mali.

Mimusops kerstingii Engl., op. cit.: 78, t. 26 fig. D (1904). Type from Togo.

Mimusops propinqua S. Moore in Journ. Linn. Soc., Bot. **37**: 177 (1905). Type from Uganda.

Chrysophyllum holstii Engl., Bot. Jahrb. **49**: 390 (1913). Type from Tanzania.

Manilkara multinervis (Baker) Dubard in Ann. Mus. Col. Marseille, sér. 3, **3**: 24 (1915). — Heine, loc. cit. — J. H. Hemsley, op. cit. **20**: 490–500, fig. 4 (1966); op. cit.: 69 (1968). Type as for *Mimusops multinervis*.

Manilkara cuneifolia (Baker) Dubard, tom. cit.: 23, fig. 11 (1915). Type as for *Mimusops cuneifolia*.

Manilkara lacera (Baker) Dubard, tom. cit.: 24 (1915). Type as for *Mimusops lacera*.

Manilkara schweinfurthii (Engl.) Dubard, tom. cit.: 25 (1915). Syntypes as for *Mimusops schweinfurthii*.

Manilkara propinqua (S. Moore) H. J. Lam in Blumea **4**: 356 (1941). Type as for *Mimusops propinqua*.

Manilkara sp. 1 and *Manilkara sp. 2* — F. White, F.F.N.R.: 321 (1962).

Manilkara multinervis subsp. *schweinfurthii* (Engl.) J. H. Hemsley, tom. cit.: 497 (1966); op. cit.: 70 (1968). Syntypes as for *Mimusops schweinfurthii*.

Tree up to 14 m. or more. Leaves coriaceous, exstipulate or with subulate stipules soon falling, not clustered at ends of branches. Leaf lamina 7–14 × 2·5–7 cm., obovate to oblong, the apex obtuse to cuspidate-acuminate, the base cuspidate; petiole 1–4 cm. long. Upper leaf surface glabrous; midrib slightly impressed, vein reticulation very fine. Lower surface with dense appressed indumentum of minute hairs giving a silvery sheen; midrib prominent, lateral nerves inconspicuous. Flowers clustered in leaf axils, with pedicels c. 1 cm. long. Calyx c. 5 mm. long, sepals ± elliptic. Corolla c. 6 mm. long, with tube c. 1·5 mm. long, white to creamy yellow; median segments c. 4 mm. long, narrowly elliptic, lateral segments a little shorter, narrowly triangular. Anthers c. 1·5 mm. long. Staminodes c. 3 mm. long, oblong in outline, the apex irregularly laciniate. Gynoecium up to 8 mm. long. Fruit up to 2·5 cm. long, obovoid to subglobose. Seed c. 1 cm. long.

Zambia. W: Solwezi R., 3 km. SE. of Solwezi Boma, st. 13.ix.1952, *White* 3245 (BM; FHO; K).

Widely distributed in tropical Africa. In riverine forest.

Fanshawe, loc. cit., cites *M. obovata* from all provinces of Zambia except E, but this must remain doubtful in the absence of corroborating herbarium material.

The characters used by Heine (in F.W.T.A. ed. 2, **2**: 19, 1963) and J. H. Hemsley (in F.T.E.A., Sapotaceae: 63, 1968) in their keys to separate *M. obovata* and *M. multinervis* are unconvincing, and the two species are said to be sympatric over much of their ranges and to occupy similar habitats. I find it impossible to distinguish the two taxa in the herbarium, and so have treated *M. multinervis* as a synonym of *M. obovata*. F. White (in Keay & Onochie, Nigerian Trees **2**: 347, 1964) has previously arrived at the same conclusion.

4. **Manilkara discolor** (Sonder) J. H. Hemsley in Kew Bull. **20**: 510 (1966); in F.T.E.A., Sapot.: 70, fig. 13 (1968). — Palmer & Pitman, Trees of Southern Afr. **3**: 1755, with fig. (1972). — Drummond in Kirkia **10**: 266 (1975). TAB. **53**. Type from S. Africa (Natal).

Labourdonnaisia discolor Sonder in Linnaea **23**: 73 (1850). — Gerstner in Journ. S. Afr. Bot. **12**: 49, fig. 5 (1946). Type as above.

Muriea discolor (Sonder) Hartog in Journ. Bot. **7**: 145 (1879). — Dubard in Ann. Mus. Col. Marseille, sér. 3, **3**: 29 (1915). — A. Meeuse in Bothalia **7**: 376 (1960); in Fl. Southern Afr. **26**: 52 (1963). Type as above.

Mimusops discolor (Sonder) Hartog, tom. cit.: 358 (1879). — Engl., Mon. Afr. Pflanzen. **8**: 55 (1904). Type as above.

Mahea natalensis Pierre, Notes Bot. Sapot.: 10 (1890). Type from S. Africa (Natal).

Mimusops buchananii Engl., Pflanzenw. Ost.-Afr. **C**: 307 (1895); op. cit.: 56, t. 19 fig. B (1904). Type: Malawi, Shire Highlands, 1891, *Buchanan* 684 (B, holotype †; BM, K, isotypes).

Mimusops altissima Engl., op. cit.: 55 (1904). Type from Tanzania.

Manilkara natalensis (Pierre) Dubard, tom. cit.: 28 (1915). Type as for *Mahea natalensis*.

Manilkara altissima (Engl.) H. J. Lam in Blumea **4**: 354 (1941). — Wild & Barbosa in F.Z. Supplement Veg. Map of F.Z. area: 15 (1968). Type as for *Mimusops altissima*.

For a fuller synonymy see J. H. Hemsley in Kew Bull. **20**: 510 (1966).

Tree 3–40 m. high; bole of larger specimens shallowly fluted near the base; slash pink; bark brown or blackish grey, longitudinally fissured. Young stems glabrous to densely greyish-pubescent but soon glabrescent, twigs with granular grey bark.

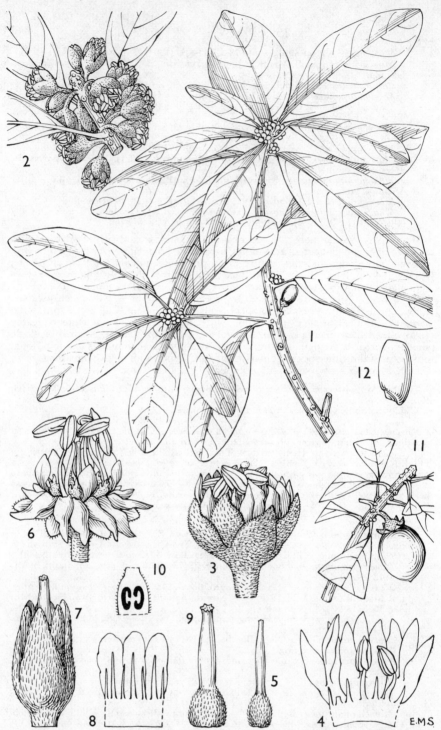

Tab. 53. MANILKARA DISCOLOR. 1, branch (× ⅔), from *Williams* 420; 2, inflorescences (× 1); 3, flower (× 4); 4, corolla section of the same (× 4); 5, ovary of same (× 4), 2–5 from *Gardner* in F.D. 1024; 6, more mature flower (× 4), from *Williams* 663; 7, reduced female flower (× 8); 8, corolla-section of same (× 8); 9, ovary of same (× 8); 10, section of ovary (× 8). 7–10 from *Timothy* 548 in C.M. 15954; 11, young fruit, attached (× 1); 12, seed (× 2), 11, 12 from *Nicholson* 49. From F.T.E.A.

Leaves tending to be clustered at branch ends. Lamina 3·5–7(13) × 1·5–5 cm., elliptic-obovate to obovate, apex emarginate, acute or apiculate, base acute; petiole 6–18 mm. long. Upper leaf surface glabrous, dark green, midrib impressed, very fine reticulation visible but veins not raised or impressed. Lower surface silvery-sericeous with a dense appressed indumentum, midrib prominent, main lateral nerves faintly visible, numerous, reticulation obscured by indumentum. Flowers yellow, in axillary fascicles on pedicels 3–6 mm. long. Form of flowers variable (see J. H. Hemsley, tom. cit.: 500–510, 1966); normal form as follows. Calyx c. 4 mm. long, sepals ovate. Corolla equalling calyx, tube c. 1 mm. long; each petal 3-lobed, the median segment obovate, tapering to a slender basal attachment, the lateral segments elliptic. Stamens 12 in 2 whorls; anthers c. 1·5 mm. long. Staminodes absent. Gynoecium c. 3 mm. long. Other flower forms: corolla simpler, petals entire (lateral segments absent) or trifid, or lateral segments present but united to near the apex; stamens all, or only the inner whorl, replaced by staminodes, these filament-like or represented by dentate flaps of tissue. (Flowers with a reduced androecium, i.e. less than 12 stamens, have a reduced corolla.) Fruit up to 1·3 × 0·8 cm., ovoid or ellipsoid, initially puberulous, glabrescent and yellow when ripe, edible, 1 (2)-seeded. Seeds up to 1 cm. long.

Zimbabwe. E: Umtali Distr., Tshakwe Mt., Burma Farm, 1400 m., fl. 21.viii.1955, *Chase* 5722 (BM; K; LISC; PRE; SRGH). **Malawi.** N: Viphya, Champoyo, 1830 m., st. 15.vii.1964, *Chapman* 2259 (FHO). S: Shire Highlands, fl. 1891, *Buchanan* 684 (BM; K). **Mozambique.** N: Pemba (Porto Amélia), fl. buds 9.ix.1948, *Andrada* 1326 (COI; LISC; LMA). Z: Bajone, 3·2 km. from Murroa to Namuera, fl. buds 2.x.1949, *Barbosa & Carvalho* 4275 (K). MS: Serra Macuta, 1040 m., fl. buds 3.vi.1971, *Müller & Gordon* 1786 (LISC; SRGH). GI: Macia, Incaia, st. 16.vii.1947, *Pedro & Pedrogão* 1499 (COI; K; SRGH). M: Maputo, Inhaca I., st. 2.vi.1970, *Correia & Marques* 1592 (BM; COI; LMU).

Also known from S. Africa (Natal), Tanzania and Kenya. In lowland mixed evergreen forest and mountain rainforest, usually in situations with good drainage.

5. **Manilkara sansibarensis** (Engl.) Dubard in Ann. Mus. Col. Marseille, sér. 3, **3**: 26 (1915). — A. Meeuse in Bothalia **7**: 367 (1960). — J. H. Hemsley in F.T.E.A., Sapotaceae: 66 (1968). Type from Tanzania (Zanzibar).

Mimusops sansibarensis Engl., Pflanzenw. Ost.-Afr. **C**: 307 (1895); Mon. Afr. Pflanzen. **8**: 58, t. 21 fig. B (1904). Type as above.

Small tree or shrub 3–8 m. tall. Young stems glabrous; twigs flexuous, rather sparsely branched, with roughly striated bark. Leaves not confined to branch apices, exstipulate. Lamina 5–11·5 × 2·5–7 cm., very variable in shape: oblong, elliptic, oblong-obovate or broadly obovate, apex bluntly acute or rounded and emarginate, base acute; petiole 1–3 cm. long; margin revolute. Upper surface glabrous, midrib and vein reticulation impressed, the latter relatively coarse; lower surface very minutely sericeous, midrib prominent, vein reticulation impressed. Flowers white, in dense axillary fascicles; pedicels up to 12 mm. long. Calyx c. 4 mm. long; sepals triangular, strongly reflexed at anthesis. Corolla c. 5 mm. long, with basal tube c. 1·5 mm. long; median segment elliptic, narrowed to a slender base, lateral segments ovate with narrowly triangular apex. Anthers c. 1·5 mm. long. Staminodes equalling stamens, petaloid, narrowly oblong, ± dentate at apex. Gynoecium c. 13 mm. long. Fruit up to 1·3 cm. long, elliptic to subglobose, 1–4-seeded. Seeds 7–11 mm. long.

Mozambique. N: Pundanhar, fl. 2.x.1960, *Gomes e Sousa* 4572 (COI; K; PRE).

Extending north to coastal regions of Tanzania and Kenya. Deciduous woodland on sandy soil, often in areas of poor drainage.

107. EBENACEAE

By F. White

The morphology, anatomy and cytology of the Ebenaceae have been studied more fully than those of many tropical families and have yielded much information of taxonomic interest. Hence, the following account is somewhat more detailed than is usual in Flora Zambesiaca. I am indebted to F. S. P. Ng, A. N. Caveney and J. Cassells for providing unpublished information, mostly on anatomy, and to R. B. Drummond, T. Müller, G. Pope and H. Wild for their comments on an early draft.

Trees, shrubs or suffrutices without milky latex. Heartwood sometimes black. Leaves simple, exstipulate, usually alternate and entire. Inflorescence sometimes cauline, determinate, usually cymose or fasciculate but sometimes a simple or branched false raceme, or reduced to a solitary flower which usually terminates a bracteate peduncle. Flowers actinomorphic, hypogynous, unisexual, but usually with rudiments of the other sex, 3–8-merous. Calyx gamosepalous, entire and truncate to deeply lobed, always persistent in fruit and usually accrescent. Corolla gamopetalous, shortly to deeply lobed; tube sometimes fleshy and constricted at the throat; lobes contorted sintrorsely in bud. Disk present or absent. Stamens from (2) 3 to more than 100, epipetalous or borne on receptacle, exserted or included; filaments often very short; anthers basifixed, usually apiculate and setulose, often of unequal size and 2 or more arising from a single filament. Pistillode very variable in development, rarely completely absent. Ovary syncarpous, 2–8-carpellary, each carpel bi-ovulate and usually completely or incompletely divided by a false septum into 2 uni-ovulate locules; styles equal in number to the carpels, distinct or basally connate, very rarely completely united; stigmas usually large and conspicuous; ovules apical, pendulous, anatropous, with 2 integuments. Staminodes very variable in development, rarely absent. Fruit usually a berry, rarely showing tardy dehiscence. Seeds large, with a distinct circum-peripheral vascular loop. Hilum small, apical, inconspicuous. Testa coriaceous (parenchymatous). Endosperm abundant, hard, horny, smooth or ruminate; embryo usually about half as long as the seed; radicle large, superior; cotyledons foliaceous.

A medium-sized family of 2 genera and c. 500 species, distributed throughout the tropics and subtropics of both hemispheres, with most species in the Indo-Pacific region but with the centre of variation in Africa (including Madagascar). In Africa the family is represented in most terrestrial vegetation types below 2000 m. except in the driest areas. The majority of species are neither very rare nor particularly abundant. D. kaki L., the persimmon, is sometimes cultivated.

In recent years the number of genera recognized has been reduced from 6 to 2. Those no longer upheld were based on erroneous observation (sex distribution, aestivation) or on characters which have proved to be very variable (number of floral parts).

Some species produce Ebony of commerce, and the wood of others is locally used for making handles of spears, axes and other tools, or for building. The fruit of many species is edible, though very astringent when unripe. Palatability varies greatly, and a few species are toxic. The latter, including Diospyros mweroensis in our area, are sometimes used as fish poison.

Most tropical African species of Ebenaceae are relatively uniform, and very few can profitably be subdivided. In southern Africa, however, several species show complicated patterns of ecogeographical replacement. Most of this variation is concentrated in South Africa, but of the species involved, 4, namely Diospyros lycioides, Euclea crispa, E. natalensis, and E. racemosa, are widely distributed in the Flora Zambesiaca area and their relevant subspecies are described in the present account. In some cases, at least locally, e.g. E. crispa subsp. crispa and subsp. linearis, E. natalensis subsp. natalensis and subsp. rotundifolia, and E. racemosa subsp. sinuata and subsp. zuluensis, it appears at first sight that distinct species are involved. Consideration of the overall pattern, however, shows that specific rank is not justified. Since those with only local knowledge, especially field workers, might be sceptical when told that such strikingly dissimilar variants are conspecific, they are described and illustrated in some detail. It must be remembered, however, that the fundamental unit of taxonomy is the species. Hence, this extra information is strictly subordinated to the general treatments of the species concerned.

Unicellular trichomes are found in all species and are simple or two-armed. Club-shaped, multicellular glandular hairs occur sporadically, possibly more widely than at present known, but are conspicuous in only a few species including *Euclea natalensis* and some species in *Diospyros* section *Royena*. Peltate scales occur in some species of *Euclea* and a few non-African species of *Diospyros*. These scales have sometimes been referred to in the literature as a "rusty exudate". Nearly all the African species are distinct in cuticular characters including microscopic details of the indumentum, but these features have not been used in the present account. Most of the African sections of *Diospyros* can be characterized using the cuticle (Cassells & White, in prep.).

In most species, possibly all, hydathode glands occur on the lower surface of the leaf. Especially in young leaves, they secrete droplets of a sugary fluid. Apparently, they were first described by Busch (*Anatomisch-systematische Untersuchung der Gattung* Diospyros, Thesis, Erlangen, 1913), who did not discuss their potential taxonomic importance in any detail. He suggested that their function might be associated with ants. It seems that their precise distribution on the leaf is often species-specific. Letouzey & White (Fl. Gabon, **18**, Ebenaceae 1970; Fl. Cameroun, **11** Ebenaceae, 1970) describe and illustrate them for several Guineo-Congolian species. Unfortunately these glands cannot always be easily detected, especially on mature leaves of herbarium specimens, and living material of relatively few Zambezian species has been examined to date. For these reasons they are not mentioned further in the present account. Larger glandular areas, the details of which also seem to be species-specific, also occur towards the base of the lamina in some species.

Since the inflorescence is fundamentally cymose, no useful distinction can be drawn between bracts and bracteoles, and all "leaf-like" structures associated with the inflorescence are referred to as bracts in the present account.

In *Diospyros* the female flowers are often larger than the male and occur in fewer-flowered inflorescences, but in *Euclea* the female flowers are distinctly smaller.

It was formerly thought that the genus *Royena* differs from *Diospyros* in having hermaphrodite flowers, but this has been shown to be false (White & Barnes, in Oxford Univ. For. Soc. Journ., ser. 4, **6**: 31–34, 1958; de Winter in Bothalia, **7**: 17–18, 1958). No member of the family is known to be normally hermaphrodite. Male plants, however, of some species, both of *Diospyros* and *Euclea*, can occasionally produce functional female flowers and viable seed (Wright, Ann. Roy. Bot. Gard. Peradeniya, **2**(1): 1–106, **2**(2): 1–78, 1904; Ng, unpublished thesis, 1971; White, pers. obs.). The precise mechanism remains to be studied.

The stamens are variously arranged and even within a single individual there is often variation. In the section *Royena* of *Diospyros* there are 8 or 10 stamens which are inserted at the base of the corolla-tube, one opposite to and one alternating with each corolla lobe. In other species there are four times as many stamens as corolla-lobes and they are united in radial pairs by their filaments. However, departures from these conditions involving increase or decrease in number and irregularities of adnation are frequent.

In the fruit a hypodermis of varying thickness is differentiated. It is nearly always made up of stone cells, very rarely of fibres or a mixture of fibres and stone cells. The remainder of the pericarp forms a pulp of varying degrees of fibrousness. In several species a somewhat gelatinous pulpy "endocarp" is differentiated from a fibrous mesocarp. This endocarp, which forms a distinct layer round each seed, adheres closely to it and resembles a sarcotesta.

In some species of *Diospyros* section *Royena* the septa form lines of weakness and the fruit is sometimes described as capsular or tardily dehiscent. In no species, however, is the fruit completely and spontaneously dehiscent. Dehiscence in herbarium specimens in some species, e.g. *D. lycioides*, seems to be the result of mechanical pressure applied during drying. In the Flora Zambesiaca area incomplete dehiscence occurs in *D. dichrophylla* and *D. villosa*, and appears to be associated with dispersal (White, pers. obs.). In these species the fruits normally "dehisce" only if firm pressure is applied at the apex. This causes the seeds with their firmly adhering pulpy endocarp to be exposed and easily removed. Both species are said to be dispersed by primates, including monkeys, baboons and man. In *D. dichrophylla* the fruits often fall to the ground before they are dispersed and are then sometimes eaten by other mammals. Seeds of *D. dichrophylla* have been found, not only in the droppings of baboons, but also of bush pigs, and in the rumen of the Blue Duiker (*Cephalophus monticola*), where they are not associated with the fibrous, outer pericarp, which presumably is not eaten (K. von Gadow, pers. comm.). This means, either that the Blue Duiker, which is a fruit specialist, can liberate the seeds from the undehisced fruit, or it feeds on seeds which have been dropped by foraging monkeys.

In *D. dichrophylla*, birds also remove one or more seeds with their gelatinous envelope from fruits still attached to the parent plant. Partial "dehiscence" of the remainder of the fruit sometimes follows, presumably because of drying out. Small fallen fruits often show dehiscence, but invariably they contain parasitized seeds.

The seeds of Ebenaceae and Sapotaceae differ much more than was previously supposed (Ng, Thesis, 1971). Seed structure also provides the best means of separating *Diospyros* and *Euclea* (TAB. **54**).

The testa consists of the outer integument only. After fertilization the inner integument disintegrates except for its inner epidermis which persists as an endothelium and behaves as if it were the surface layer of the embryo-sac. The testa is structurally simple but the outer epidermis is variable from species to species in shape, size and degree of cell-wall thickening, though few species have yet been studied in detail. In some species the endosperm is ruminate

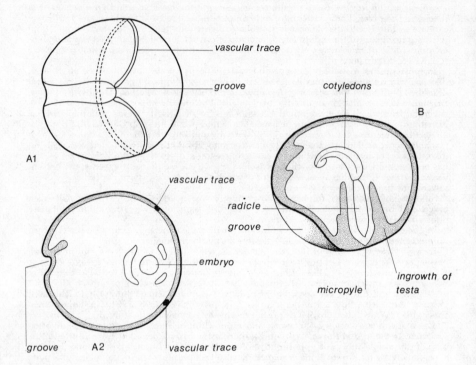

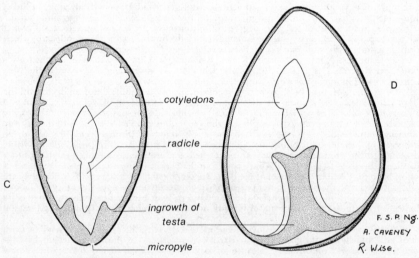

Tab. 54. Seeds of Ebenaceae. A. — EUCLEA CRISPA subsp. LINEARIS. A1, apical view; A2, transverse section through radicle. B. — EUCLEA DIVINORUM, longitudinal section. C. — DIOSPYROS MESPILIFORMIS, longitudinal section. D. — DIOSPYROS WHYTEANA, longitudinal section. A–C × 5, D × 10.

because of ingrowths from the testa due to localized meristematic activity. Whether the endosperm is ruminate or not is sometimes variable within species. The deep invagination surrounding the radicle of *Euclea* (see below) and the similar but shallower ingrowth in *Diospyros* section *Royena*, are, however, taxonomically significant.

In the great majority of species of *Diospyros* a single vascular strand runs round the seed in the raphe anti-raphe plane, which is also the median radial plane of the seed. Its passage is marked by a groove, ridge or line of lighter pigmentation. The embryo is completely surrounded by endosperm and has a strongly developed radicle. The flat, leafy cotyledons are normally orientated in the plane of the vascular loop.

The seed of *Euclea* differs in several specialized departures from the basic plan. In *Euclea* the seed is displaced in growth so that the plane of the vascular loop fails to coincide with the median radial plane. This condition also occurs in a few species of *Diospyros*, e.g. *D. ferrea*, where the vascular loop is at right-angles to the median radial plane, but in those species the other peculiar features of the *Euclea* seed are absent.

In *Euclea* only one seed normally develops. During development the abortive ovules and the central axis are pushed to one side of the fruit and the axis leaves an impression on the seed in the form of a lateral groove down one side. Viewed from the apex the *Euclea* seed shows 3 radiating lines consisting of the lateral groove (impression of the displaced axis) and the two ends of the vascular loop meeting at the apex. The embryo is orientated in the plane of the vascular loop but the cotyledons are strongly curved towards the axis. This is a peculiarity not found in *Diospyros* where the embryo is straight or curved in its own plane.

In *Euclea* the testa is invaginated to form a sheath surrounding the radicle for the whole of its length.

In *Diospyros* the testa occasionally, as in section *Royena*, forms a sheath round the distal half of the radicle but the other peculiarities of the *Euclea* seed are absent.

On germination the cotyledons are emergent (except in a few Asiatic species). They are either non-photosynthetic and are soon shed or are photosynthetic and more persistent. In *Diospyros* section *Royena* the hypocotyl is short and the cotyledons do not always emerge above the surface of the soil. Three cotyledons are sometimes present in *D. lycioides* subsp. *sericea*.

Chromosome numbers form a single euploid series with 2n = 30, 60, 90 (White & Vosa in Bol. Soc. Brot., Sér. 2, **53**: 269–291, 1980). Polyploids are rare. Apart from *Diospyros lycioides* subsp. *sericea*, which is both diploid and tetraploid, polyploids are unknown in the Flora Zambesiaca area.

Inflorescence cymose or fasciculate or flowers solitary, fruit usually 2- or more seeded; seed very rarely (sometimes in *D. abyssinica* and *D. ferrea*) subglobose; with 2 lines radiating from the apex; embryo straight or curved in its own plane; radicle not surrounded by invagination of the testa or only in lower half - - - - - **1. Diospyros**
Inflorescence a simple or branched pseudo-raceme; fruit nearly always 1-seeded; seed subglobose with 3 radiating lines from the apex, namely the two ends of the vascular loop and the impression of the displaced axis of the fruit; embryo with cotyledons flexed at right-angles to the radicle; radicle surrounded by invagination of the testa for whole of length - - - - - - - - - - - - - **2. Euclea**

1. DIOSPYROS L.

Diospyros L., Sp. Pl.: 1057 (1753); Gen. Pl., ed. 5: 478 (1754).

Trees, shrubs or suffrutices. Leaves nearly always alternate, margin entire. Inflorescence usually cymose or fasciculate or reduced to a solitary flower. Flowers dioecious, the female usually larger than the male. Calyx very variable, 3–8-lobed or cup-shaped and entire, usually accrescent. Corolla 3–8-lobed, very variable. Stamens 2–100 or more, included or exserted, solitary or united in pairs or larger groups, adnate to corolla-tube or not; anthers (in our area) dehiscing by longitudinal slits. Pistillode variable, absent to very well-developed. Disk well-developed or not, sometimes fimbriate. Staminodes variable, absent to well-developed. Ovary globose, ovoid or conoidal, glabrous or hairy, 3–8 carpellary; carpels (except in section *Ferrea*) completely or incompletely septate. Styles usually at least partly free, rarely completely united; stigmata usually fleshy and expanded. Fruit usually a 1- to many-seeded berry, rarely tardily dehiscent. Seed usually ellipsoid or shaped like a segment of an orange, very rarely subglobose, with 2 lines radiating from the apex; embryo straight or curved in its own plane, radicle not completely surrounded by an invagination of the testa.

Nearly 500 species. Widely distributed in the tropics and subtropics with a few species in the warm temperate zones both north and south of the equator.

The 90 species occurring on the African mainland belong to 18 sections, several of which have been previously regarded as genera. Most of the sections appear to be confined to Africa.

They are well characterized both by morphological and anatomical features. By contrast, many of the sections currently recognized in the Indo-Pacific Region (Bakhuizen van den Brink in Bull. Jard. Bot. Buitenzorg, sér. 3, 15, 1-515, pl. i–xx, 1936–1955) seem to be artificial.

Key to the sections

Anthers distinctly exserted:
  Locules 3, biovulate - - - - - - - - - 1. FORSTERIA
  Locules 4 or more, uniovulate:
    Fruiting calyx cyathiform, truncate - - - - - - 2. KATULA
    Fruiting calyx distinctly lobed:
      Male calyx closed in bud, truncate or shallowly lobed at anthesis; female calyx with distinct sinuses alternating with the lobes - - - - 3. FORBESIA
      Male calyx with open or slightly imbricate aestivation, deeply lobed at anthesis; female calyx without sinuses - - - - - - - - 4. BREVITUBA
Anthers included:
  Flowers either solitary on long pedicels, or in long-pedunculate, few-flowered cymes; corolla-lobes at first ascending and forming a continuation of the corolla-tube, then spreading or reflexed; pistillodes and antherodes usually well-developed so that the flowers appear bisexual - - - - - - - - 5. ROYENA
  Flowers etc. not as above:
    Male corolla more than 1·3 cm. long, lobed at least to ¼:
      Male flowers subsessile, 3–5 together at ends of distally expanded and flattened supra-axillary peduncles; calyx botuliform, completely concealing corolla before anthesis; fruiting calyx cyathiform, truncate - - - - - 6. ERIKESI
      Male flowers etc. not as above - - - - - - - 7. TABONACA
    Male corolla up to 1·2 cm. long, lobed for ¼ or less (except in sect. *Marsupium*):
      Female flowers with up to 2 bracts; male corolla lobed to ¼ or less:
        Male calyx lobes acute, usually narrowly deltate; male corolla narrowly conoidal in bud, lobes acute, usually narrowly deltate - - - - 8. NOLTIA
        Not as above:
          Corolla glabrous or minutely puberulous towards apex - 9. CALVITIELLA
          Corolla tomentose - - - - - - - 10. BREVISTYLA
      Female flowers with 6 or more deltate bracts immediately below calyx; male corolla lobed to c. ⅓, tomentose, conoidal in bud - - - - 11. MARSUPIUM

1. Sect. FORSTERIA Bakh. in Bull. Jard. Bot. Buitenz., Sér. 3, **15**: 8 (1937). — F. White in Bull. Jard. Bot. Nat. Belg. **50**: 450 (1980).
    Maba J. R. & G. Forst., Char. Gen. Pl.: 121, t. 61 (1776).

Male calyx closed in bud, at anthesis cyathiform, truncate or irregularly lobed. Male corolla small, 3-lobed to below the middle, densely strigulose. Stamens 8–16, inserted on receptacle or base of corolla-tube, slightly exserted. Ovary ovoid-conoidal, glabrous or strigulose-tomentose. Locules 3, biovulate. Fruit (in Africa) up to 1·5 cm. diameter, globose or ellipsoid. Seeds 1–2; endosperm smooth. Fruiting calyx scarcely accrescent, patelliform or shallowly cyathiform.

Bud scales absent; calyx of flower and fruit irregularly lobed; ovary densely strigulose
                                       1. *ferrea*
Bud scales present; calyx truncate or denticulate; ovary glabrous - - - 2. *natalensis*

1. **Diospyros ferrea** (Willd.) Bakh. in Gard. Bull. Str. S. 7: 162 (1933); in Bull. Jard. Bot. Buitenz., Sér. 3, 15: 50 (1936); tom. cit.: 431 (1941). — F. White in F.W.T.A., ed. 2, **2**: 11 (1963); in Bull. Jard. Bot. Nat. Belg. **48**: 273, tt. 5–6 (1978); in Distr. Pl. Afr. **14**: 455 (1978). — Letouzey & F. White in Fl. Cam. **11**, & Fl. Gab. **18**: 69, t. 16 fig. 13–22 (1970). — Drummond in Kirkia, **10**: 266 (1975). — K. Coates Palgrave, Trees of Southern Afr.: 745 (1977). TAB. **55**, fig. A. Type from SW. India.
    Maba buxifolia (Rottb.) A. L. Juss. in Ann Mus. Hist. Nat. **5**: 418 (1804). — Hiern in Trans. Camb. Phil. Soc. **12**: 116 (1873). Type from SW. India.
    Maba buxifolia var. *ebenus* Thwaites, Enum. Pl. Zeylan.: 183 (1960). — Hiern., loc. cit. Type from Sri Lanka.
    Ebenus buxifolia (Rottb.) O. Kuntze, Rev. Gen. Pl. **2**: 408 (1891). Type as for *M. buxifolia*.

Evergreen shrub or small tree up to 18 m. high. Bud-scales absent or minute and fugaceous. *Leaves* subcoriaceous, drying pale brown; lamina 3·5 × 1·8–8 × 3·5 cm., broadly elliptic to lanceolate-elliptic, apex usually shortly acuminate or sub-acuminate, sometimes obtuse or rounded, lower surface glabrous except for a few setulose hairs, chiefly on the midrib and margin; secondary nerves in about 5–9 pairs,

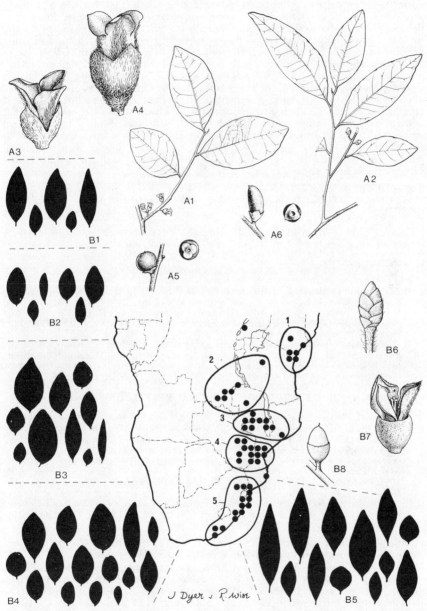

Tab. 55. DIOSPYROS sect. FORSTERIA. A, — D. FERREA. A1, flowering branchlet (× ½) *Goldsmith* 109/67; A2, flowering branchlet (× ½), *Le Testu* 863; A3, male flower (× 4), *Perrotet* 453, A4, female flower (× 4), *Welwitsch* 2527; A5, fruits (× ⅔) *Gossweiler* 2789; A6, fruits (× ⅔) *Vigne* 2518. B. — D. NATALENSIS, B1–5, variation in leaf-shape and size from areas 1–5 respectively (× ½); B6, terminal bud (× 5); B7, female flower (× 4) *Wood* 1414; B8, fruit (× 1) *Fanshawe* 2035.

together with the tertiary nerves and veins forming an indistinct reticulum. *Male flowers* subsessile in 1–3(4)-flowered cymules in axils of leaves or towards base of current year's growth in axils of caducous bracts; peduncle 0·1–0·3 cm. long, densely setulose. Calyx 0·3 cm. long, densely strigulose outside, glabrous inside; lobes 3, ± deltate, very variable in shape and size. Corolla 0·5 cm. long, lobed to just below the middle, lobes 3, elliptic, somewhat fleshy, ascending, with densely strigulose mid-petaline lines, otherwise glabrous. Stamens 8–14, in pairs; filaments 0·05–0·1 cm. long, inserted on the receptacle, glabrous; anthers 0·1–0·15 cm. long, lanceolate, apiculate, glabrous or with a few apical hairs. Pistillode represented by a tuft of hairs. *Female flowers* solitary in axils of leaves and caducous foliaceous bracts towards base of current year's shoot, subsessile, somewhat larger than the male and with calyx and corolla less deeply divided. Staminodes absent. Ovary 0·3 × 0·25 cm., conoidal, densely strigulose, terminating in 3 fleshy, bilobed stigmas. *Fruit* c. 1·3 × 0·9 cm., broadly ellipsoid, glabrous except for a few strigulose hairs towards the base of the persistent style. Seeds 1–2, 0·9 × 0·4 × 0·3 cm., sub-ellipsoid. Fruiting calyx patelliform, 0·3 cm. long, irregularly 3-lobed, lobes broadly deltate to hemi-orbicular. *Chromosome number*: 2n = 30.

Zimbabwe. E: Chirinda Forest, ♂ fl. xi.1967, *Goldsmith* 109/67 (FHO; SRGH).
Mozambique: N: Pemba (Porto Amelia), Nangororo–Quissanga 15 km., *Gomes e Sousa* 4526 (BR; COI; K). MS: Serra da Gorongosa, 1050 m., ♂ fl. 24.x.1965, *Torre & Pereira* 12583 (COI; K; LISC; LMU).
Widespread in the Old World tropics. Interpreted widely it has a sporadic distribution from Senegal to Hawaii. *D. ferrea* is an ecological and chorological transgressor, but in our area its ecology seems to be rather uniform and its most characteristic habitat is transitional rain forest between 1000 and 1200 m.

2. **Diospyros natalensis** (Harv.) Brenan in Mem. N.Y. Bot. Gard. **8**: 501 (1954). — F. White, F.F.N.R.: 324, t. 58 fig. a (1962). — de Winter in F.S.A. **26**: 58, t. 9 fig. 1 (1963). — Palmer & Pitman, Trees of Southern Afr. **3**: 1789 cum tab. & photogr. (1972). — Drummond in Kirkia, **10**: 267 (1975). — K. Coates Palgrave, Trees of Southern Afr.: 748 (1977). TAB. **55**, fig. B. Type from South Africa.
    Maba natalensis Harv., Thes. Cap. **2**: 7, t. 110 (1863). — Hiern in Trans. Camb. Phil. Soc. **12**: 131 (1873). — Sim, For. Fl. Port. E. Afr.: 82 (1909). — S. Moore in Journ. Linn. Soc. Bot. **40**: 134 (1911). Type as above.
    Maba dawei Hutch. in Kew Bull. **1921**: 330 (1912). Type: Mozambique, Chimoio, Garuso, ♂ fl. i.1912, *Dawe* 524 (K, holotype).
    Diospyros dawei (Hutch) Brenan in Kew Bull. **1948**: 111 (1948), syn. nov. Type as above.
    Diospyros nummularia Brenan, loc. cit. — K. & O. Coates Palgrave, Trees of Central Afr.: 165 cum tab. et photgr. (1957). — F. White, F.F.N.R.: 326, t. 58 fig. b (1962). — de Winter in tom. cit.: 60 (1963). Types: Zimbabwe, Salisbury, 1370 m., fr. iv.1922, *Eyles* 3414 (K, holotype). 11 paratypes, all from Zimbabwe, are also cited in the protologue.
    Diospyros nyasae Brenan in Mem. N.Y. Bot. Gard. **8**: 500 (1954), syn. nov. Types; Malawi, without precise locality, ♂, ♀ fl. fr. 1891, *Buchanan* 957 (K, holotype); id., ♂ fl., 1891, *Buchanan* 977 (K, paratype); Mt. Mulanje, Likabula Gorge, 840 m., fr. 20.vi.1946, *Brass* 16385 (K, paratype; MO).
    Diospyros natalensis subsp. *nummularia* (Brenan) F. White ex auct.; Palmer & Pitman, loc. cit. — Wild, Biegel & Mavi, Rhod. Bot. Dict., ed. 2: 151 (1972). — K. Coates Palgrave, tom. cit., t. 246 *nom. illegit.* (*sine relat. nom.*).

Much-branched evergreen shrub or small tree up to 25 m. high. Young shoots and inflorescences protected by (4) 6–8 bud-scales. *Leaves* subcoriaceous, drying pale brown above; lamina 1·3 × 0·6–3·6 × 2 cm., very variable, suborbicular, elliptic, lanceolate-elliptic, ovate-elliptic or ovate, apex obtuse or subacute; lower surface glabrous, but margin with a few ciliolate hairs; secondary nerves in 3–5 pairs, darker than the lamina, tertiary nerves and veins indistinct. *Male flowers* solitary or in 2–4-flowered cymes in axils of leaves or of fallen leaves on second-year branchlets; pedicels c. 0·1 cm. long, almost glabrous. Calyx truncate or denticulate, 0·2–0·3 cm. long, usually glabrous except for the sparsely ciliate margin. Corolla 0·5 cm. long, widely open at the throat, shortly strigulose–tomentellous outside except near the base, glabrous inside; lobes 3, broadly deltate, 0·25 cm. long. Stamens c. 16, exserted, 0·2–0·3 cm. long; filaments very short, glabrous, inserted at base of corolla-tube; anthers lanceolate, apiculate, glabrous. Pistillode absent. *Female flowers* solitary, similar to male. Staminodes 9, inserted half-way up the corolla-tube, 3 alternating with the lobes and 2 opposite each lobe, 0·15 cm. long, glabrous. Ovary 0·3 × 0·25 cm., ovoid-conoidal, glabrous; locules 3, 2-ovulate; styles 3, 0·15 cm. long, united in lower half, bifid at apex. *Fruit* 0·8 × 0·45 cm., obovoid-ellipsoid or ellipsoid, apiculate, surrounded at the base by the slightly accrescent calyx and looking like an acorn in its cupule. Seed 1, 0·7 × 0·3 × 0·3 cm., sub-ellipsoid. *Chromosome number*: 2n = 30.

Zambia. N: Mporokoso Distr., L. Mweru, Chienge, fr.18.viii. 1958, *Fanshawe* 4728 (FHO; K). C: Luangwa Valley, Kapampa R., foot of escarpment, 765 m., st. 19.i.1966, *Astle* 4460 (SRGH). E. Petauke Distr., Chipata to Lusaka 130 km., st. 24.v.1952, *White* 2880 (BR; FHO; K); Chadiza, Chapiri Hill, 900 m., fl. 29.xi.1958, *Robson* 786 (FHO; K). **Zimbabwe.** N: Sipolilo, Nyamnyetsi Estate, st. 9.v.1978, *Nyariri* 102 (SRGH). C: Salisbury Distr., Goromonzi, ♀ fl. 1.i.1927, *Eyles* 4601 (K; SRGH). E: Umtali Distr., Mandini Mt., 25 km. S of Umtali, ♂ fl. 19.vi.1960, *Chase* 7352 (BM; FHO; K; SRGH). S: near Tokwe Dam, *White* 10109, st. 4.ii.1973, (FHO). **Malawi.** C: Chongoni Forest, Cibenthu Hill, fr. 25.ix.1960, *Chapman* 926 (FHO; K; PRE; SRGH). S: Mt. Mulanje, Lukulezi, Ngono R., fr. 15.x.1957, *Chapman* 471 (BR; FHO; K). **Mozambique.** Z: serra de Morrumbala, 700 m., ♂ fl. 9.xii.1971, *Müller & Pope* 1982 (FHO; LISC; SRGH). T; Chicoa, serra de Songo, 900 m., fr. 31.xii.1965, *Torre & Correia* 13945 (C; COI; FHO; K; LISC; LMU). MS; Cheringoma Distr., Chiniziua, fr. 12.iv.1957, *Gomes e Sousa* 4352 (COI; EA; FHO; K; PRE; SRGH). SS: Gaza, Bilene, Praia de S. Martinho, ♂ fl. 7.xi.1969, *Correia & Marques* 1451 (BM; LISC; SRGH). LM: Namaacha Falls, fr. 22.ii.1955, *E. M. & W.* 542 (BM; LISC; SRGH).

Extending northwards to Zaire (Upper Shaba), Uganda and Kenya, and southwards to the eastern Cape Province, South Africa. Also in Madagascar. In evergreen and semi-evergreen forest, bushland and thicket, and on rocky outcrops. From near sea-level to 1600 m.

D. ferrea and *D. natalensis* are possibly the two most variable species of *Diospyros* on the African mainland. Although their overall ranges overlap in East Africa, they never seem to occur in close proximity and there is no difficulty in distinguishing them. By contrast, in Madagascar, where *D. natalensis* is excessively variable, some of the specific differences appear to break down.

Experiments have shown (White, unpublished) that leaf-shape and -size remain constant in cultivation. Nevertheless the overall pattern of variation is too complicated (tab. 55 fig. B) to justify the recognition of segregate species nor of infra-specific taxa. Some variants appear to have arisen polytopically in response to specialized local conditions, e.g. the narrow-leaved rheophytes (plants of stream banks which are frequently submerged by rapidly flowing floodwater). The latter, which occur in Upper Shaba, on Mt. Mulanje (*Brass* 16385) and at the base of the Chimanimani Mts. in Mozambique (*Müller & Pope* 2115), are quite different from other populations of *D. natalensis* growing nearby, but are connected to them by intermediates growing in the same general area.

2. Sect KATULA F. White in Bull. Jard. Bot. Nat. Belg. **50**: 452 (1980).

Male calyx cyathiform, very shallowly lobed. Male corolla 0·8 cm. long, rotate, 4-lobed to below the middle. Stamens c. 50. Female flowers with foliaceous bracts. Ovary ovoid-conoidal, glabrous. Locules 8, uniovulate. Styles 4, longer than ovary, united in lower half; stigmas fleshy, irregularly lobed. Fruit ovoid or broadly ellipsoid, c. 2·5 × 2·2 cm. Seeds 8; endosperm smooth. Fruiting calyx markedly accrescent, cyathiform.

3. **Diospyros mweroensis** F. White in Bull. Jard. Bot. État. Brux. **27**: 530, t. 56 (1957); F.F.N.R.: 326, t. 60 (1962). — Malaisse, Bull. Trim. CEPSI, Problèmes Sociaux Congolais, **90–91**: 321, cum photogr. 1–3 & cart. (1970). TAB. **56**. Types: holotype from Zaire. 13 paratypes are also cited in the protologue including, *inter alia*, the following from the FZ area: Zambia, Mpulungu to Mbala 30 km., ♂ fl. 17.xi.1952, *White* 3683 (FHO).

Small, bushy, semi-deciduous tree up to 10 m. high. Bark smooth, shed in large irregular scales. *Leaves* chartaceous, drying dark brown above, pale yellowish-brown beneath; lamina 3·2 × 1·2–9·5 × 3·5 cm., oblanceolate, oblanceolate-oblong or obovate-oblong, apex usually shortly and abruptly acuminate, rarely acute or obtuse, base cuneate or rounded; lower surface densely strigulose-setulose on nerves and veins, less densely so on lamina; lateral nerves in 5–9 pairs, prominent beneath, arcuate and anastomosing near the margin; tertiary nerves and veins forming a lax reticulum, slightly prominent beneath. *Male flowers* subfasciculate or in shortly pedunculate cymes, 6–9 together in axils of fallen leaves or on older branchlets; pedicels 0·2–0·3 cm. long, densely setulose-pubescent. Calyx at first closed in bud, at anthesis cyathiform, 0·5 × 0·6 cm., strigulose-tomentellous with dark brown hairs, margin subtruncate with 4 broadly deltate lobes scarcely 0·1 cm. long. Corolla

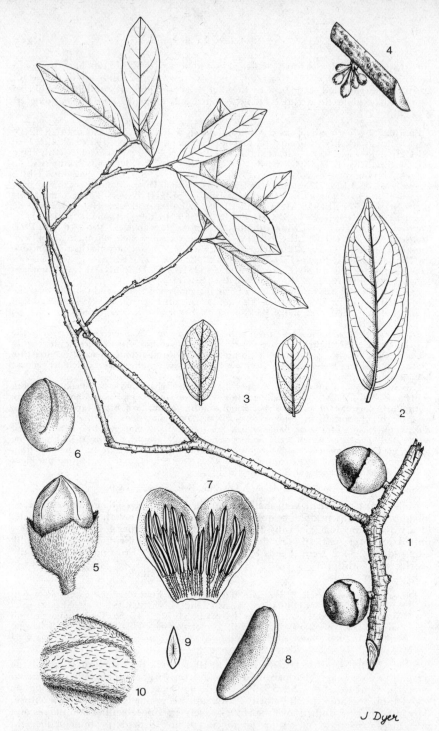

Tab. 56. DIOSPYROS MWEROENSIS (sect. Katula). 1, fruiting branchlet (× ½) *Schmitz* 683; 2 and 3, leaves to show range of variation (× ½) *Schmitz* 1865; 4, male flower buds (× 1) *White* 3683; 5, male flower bud (× 4) *Schmitz* 1166; 6, male corolla, unopened (× 4) *Schmitz* 1166; 7, part of corolla from inside and stamens (× 4) *Schmitz* 1210; 8, seed (× 2) *Duvigneaud* 1259d; 9, seed, transverse section (× 2) *Duvigneaud* 1259d; 10, indumentum of lower leaf surface (× 10) *Schmitz* 1865. By permission of the Editor of Bull. Jard. Bot. Nat. Belg.

coriaceous, subrotate, 0·7–0·8 cm. long, fulvous-strigulose-tomentellous; tube 0·15–0·2 cm. long; lobes 4, lingulate. Stamens c. 50, far-exserted, up to 0·6 cm. long, filaments up to 0·25 cm. long, setulose, inserted on corolla-tube; anthers narrowly lanceolate, apiculate, connective sparsely strigulose on abaxial surface. *Female flowers* in 1–5-flowered fascicles on older leafless branchlets; pedicels 0·5–0·6 cm. long, tomentellous; bracts foliaceous, 0·9 × 0·5 cm., caducous. Calyx 0·8 cm. long, cyathiform, brown-velutinous-tomentellous outside, paler tomentellous inside, very shallowly and irregularly lobed at apex. Corolla 1·3 cm. long, rotate; tube 0·4 × 0·4 cm., lobes 4, ovate-lingulate, 0·9 × 0·6 cm., outer surface tomentellous where exposed in bud, glabrous where overlapped in bud. Staminodes c. 16, 0·4 cm. long, inserted on corolla-tube; antherodes densely setulose. Ovary 0·3 × 0·3 cm.; common style 0·3 cm. long, densely strigulose, ending in 4 fleshy, glabrous, 0·2 cm. long, irregularly lobed stigmata. *Fruit* apiculate, yellow at first, black when ripe, c. 2·5 × 2·2 cm., glabrous. Seeds 8, black, 1·6 × 0·6 × 0·4 cm. Fruiting calyx 1·4 × 2·4 cm., glabrescent in patches outside, persistently and minutely strigulose inside.

Zambia. N: Mbala Distr., Kalambo Falls, fr. 7.vii.1941, *Greenway* 6206 (EA; K); Luwingu Distr., Nsombo to Luwingu 7 km., fr. 6.iv.1961, *Angus* 2718 (FHO; SRGH).
Also in the Upper Shaba (Katanga) Province of Zaire. In miombo woodland and on termite mounds. 800–1500 m.

The fruits of *D. mweroensis* are widely used as a fish poison. According to Malaisse (loc. cit.), who describes the method used, the active principle is a binaphthoquinone closely related to diospyrine and similar compounds which are widely distributed in the family. The poison causes the fish to lose their sense of direction, enabling them to be more easily caught. Because it is also toxic to molluscs, the latter are completely absent from rivers which are regularly fished in this way.

3. Sect. FORBESIA F. White in Bull. Jard. Bot. Nat. Belg.**50**: 453 (1980).
Sect. *Ebenus* Endl., Gen. Pl.: 742 (1839) pro parte. — Hiern in Trans. Camb. Phil. Soc. **12**: 146, 148 (1873) pro parte.

Male calyx closed in bud, at anthesis forming a truncate or denticulate cup. Corolla 0·8–1·3 cm. long, glabrous or minutely puberulous outside, widely open at the throat, lobes 4–6. Stamens c. 30, glabrous, attached to corolla-tube, exserted. Female calyx cyathiform with a shallowly lobed, venose margin alternating with 4–5 decurrent sinuses. Ovary globose, glabrous or minutely puberulous; locules 6–10, uniovulate; styles ± as long as ovary, united only at the base, ending in fleshy, irregularly lobed stigmata. Seeds 10 or fewer, endosperm smooth. Fruiting calyx patelliform, shallowly to deeply lobed.

Leaves elliptic or obovate; male flowers 1–3; calyx irregularly lobed; corolla c. 0·8 cm. long, tube no longer than the calyx, lobes hemi-orbicular; female flowers subsessile; calyx 0·7 cm. long, lobes slightly decurrent; corolla c. 0·8 cm. long; fruit c. 1·5 × 1·5 cm; fruiting calyx indistinctly lobed   -   -   -   -   -   -   -   -   4. *consolatae*
Leaves suborbicular or orbicular; male flowers solitary; calyx truncate or denticulate; corolla 0·9–1·3 cm. long; tube longer than the calyx, lobes broadly lingulate; female flowers with pedicels c. 0·3 cm. long; calyx 0·9–1·2 cm. long, lobes markedly decurrent; corolla 1·2–1·5 cm. long; fruit c. 2·5 × 2·5 cm.; lobes of fruiting calyx c. 0·8 cm. long   -   5. *rotundifolia*

4. **Diospyros consolatae** Chiov., Racc. Bot. Miss. Consol. Kenya: 77 (25.iii.1935). Type from Kenya.
Diospyros vaughaniae Dunkley in Kew Bull. **1935**: 262 (4.ix.1935). Types from Zanzibar.

Much and stiffly branched, erect evergreen shrub or small bushy tree up to 7 (9) m. tall. *Leaves* subcoriaceous, drying pale brown or yellowish-brown, often wilted on the tree or wilting soon after picking, and consequently in herbarium specimens usually with revolute or undulate margins; lamina 4·5 × 2·5–10 × 4·5 cm., elliptic or obovate, apex usually subacute to subacuminate, sometimes broadly rounded or emarginate; lower surface glabrous, lateral nerves in 5–9 pairs, indistinct. *Male flowers* 1–3 together in leaf-axils and in axils of fallen leaves; pedicels 0·1–0·3 cm. long. Calyx c. 0·4 cm. long, minutely puberulous outside. Corolla c. 0·8 cm. long, campanulate; tube no longer than the calyx, 0·4–0·4 cm.; lobes 4, 0·4 × 0·5 cm., hemi-orbicular. Stamens c. 30, 0·3 cm. long, glabrous; filaments 0·05 cm. long. *Female flowers* solitary or in pairs in leaf axils and in axils of fallen leaves, subsessile. Calyx c. 0·7 cm. long, minutely puberulous outside, tube cup-shaped, 0·4 cm. long; lobes 4(6), reduplicate and alternating with slightly decurrent sinuses. Corolla c. 0·8 cm. long, infundibuliform; minutely puberulous outside; tube 0·4 cm. long, 0·2 cm. wide below, expanded above; lobes 4(–6), 0·4 × 0·4 cm., suborbicular. Staminodes c. 15, 0·3 cm. long, glabrous, attached to the corolla-tube, exserted and blocking the throat. Ovary 0·2 × 0·2 cm., minutely puberulous; styles 3–4, 0·3 cm. long; locules 6–8. *Fruit* c. 1·5 × 1·5 cm., globose, glabrous. Seeds 6 or 8 or fewer by abortion, 1·2 × 0·6 × 0·4 cm., very dark brown. Fruiting calyx up to 1·2 cm. long, indistinctly lobed.

Mozambique. N: Cabo Delgado, Palma, near the lighthouse, fr. 16.ix.1948, *Andrada* 1355 (C; COI; FHO; K; LISC; LMA; LMU). Z: Pebane, fr. 24.x.1942, *Torre* 4662 (BR; FHO; LISC; MO; WAG).

From Kenya southwards to Mozambique. Mostly in coastal thicket, more rarely in woodland further inland. 0–100 m.

The taxonomic relationships of *D. consolate* and *D. rotundifolia* are discussed under the latter.

5. **Diospyros rotundifolia** Hiern in Trans. Camb. Phil. Soc. **12**: 181 (1873). — de Winter in F.S.A. **26**: 62 (1963). — Palmer & Pitman, Trees of Southern Africa, **3**: 1795 cum tab. & photogr. (1972). — K. Coates Palgrave, Trees of Southern Africa: 750 (1977). TAB. **57**. Type: Mozambique, Delagoa Bay, fr., *Forbes* 34 (CGE, holotype; P).

Evergreen shrub or rigidly branched tree with wide-spreading branches, 2–8 (11) m. tall. *Leaves* coriaceous, drying pale brown or greenish-brown; lamina 3 × 2·6–4·5 × 3·8 cm., suborbicular or orbicular, apex broadly rounded to subtruncate, margin recurved; lower surface glabrous; lateral nerves in c. 5 pairs, indistinct. *Male flowers* solitary in leaf-axils and in axils of fallen leaves; pedicels 0·3–0·6 cm. long. Calyx 0·3–0·5 cm. long, minutely puberulous outside. Corolla 0·9–1·3 cm. long, campanulate; tube much longer than the calyx, 0·6 × 0·4–0·7 × 0·5 cm., widely open at the throat; lobes 4–6, up to 0·6 × 0·45 cm., broadly lingulate, spreading. Stamens c. 30, 0·3–0·4 cm. long, glabrous; filaments c. 0·1 cm. long. *Female flowers* solitary in leaf-axils and in axils of fallen leaves; pedicels c. 0·3 cm. long. Calyx 0·9–1·2 cm. long, glabrous; tube cup-shaped, very thick and fleshy, contrasting with the wide, strongly venose, membranous marginal frill which forms (4) 5 plicate lobes alternating with (4) 5 decurrent sinuses. Corolla 1·2–1·5 cm. long, hypocrateriform, widely open at the throat, glabrous; tube 0·6–0·8 × 0·4–0·5 cm., much shorter than the calyx; lobes (4) 5–6, 0·5–0·7 × 0·5–0·7 cm., spreading, suborbicular. Staminodes c. 15, c. 0·4 cm. long, attached to base of corolla-tube, glabrous. Ovary 0·3 × 0·3 cm., glabrous; styles (4) 5, 0·275 cm. long; locules (8) 10. *Fruit* c. 2·5 × 2·5 cm., glabrous, globose. Seeds 8 or 10 or fewer by abortion, 1·5 × 0·8 × 0·3 cm., very dark brown. Fruiting calyx scarcely accrescent but becoming flattened and patelliform as the fruit matures, lobes c. 0·8 cm. long.

Mozambique. GI: Vilanculos, Santa Carolina Isl., fr. iv.1962, *Guy* in GHS 133188 (K; LISC; SRGH); Gaza, beach of Xai-Xai (Vila João Belo), fr. 23.ii.1941, *Torre* 2609 (LISC). M: Ponta do Ouro, ♂ fl. 7.ix.1948, *Gomes e Sousa* 3873 (COI, K).

Also in northern Natal. In coastal thicket on sand dunes, where it is often co-dominant with *Mimusops caffra* and *Sideroxylon inerme*. 0–200 m.

Over the greater part of their ranges *D. consolatae* and *D. rotundifolia* seem to differ consistently in the characters mentioned in the key. A few specimens however collected from near the southern limit of *D. consolatae* and the northern limit of *D. rotundifolia* are intermediate.

Intermediates between **D. consolatae** and **D. rotundifolia**.

Mozambique. Z: Pebane, near the lighthouse, 20 m., fr. 8.iii.1966, *Torre & Correia* 15070 (LISC). SS: Vilanculos, fr. 1.ix.1944, *Mendonça* 1931 (FHO; LISC; MO); between Vilanculos and Mapinhane, fr. 30.viii.1944, *Mendonça* 1898 (C; COI; FHO; K; LISC; LMU; WAG); Massinga, rio das Pedras, ♀ fl. 17.xi.1941, *Torre* 3848 (COI; FHO; K; LISC; LMU), ♂ fl. 17.xi.1941, *Torre* 3843 (LISC).

These specimens are variously intermediate. Thus: *Torre & Correia* 15070 has the fruiting calyx of *consolatae* and the habit and habitat of *rotundifolia*; the leaves are intermediate. *Mendonça* 1898 and 1931 have the leaves of *consolatae* and the fruiting calyx of *rotundifolia*. *Torre* 3843 and 3848 have the leaves of *consolatae*, but the female flowers of *Torre* 3848 approach *rotundifolia*.

4. Sect. BREVITUBA F. White in Bull. Jard. Bot. Nat. Belg. **50**: 453 (1980).

Male calyx cyathiform, lobed to the middle or beyond. Male corolla small, less than 1 cm. long, 3–6-lobed, usually subrotate and divided to beyond the middle,

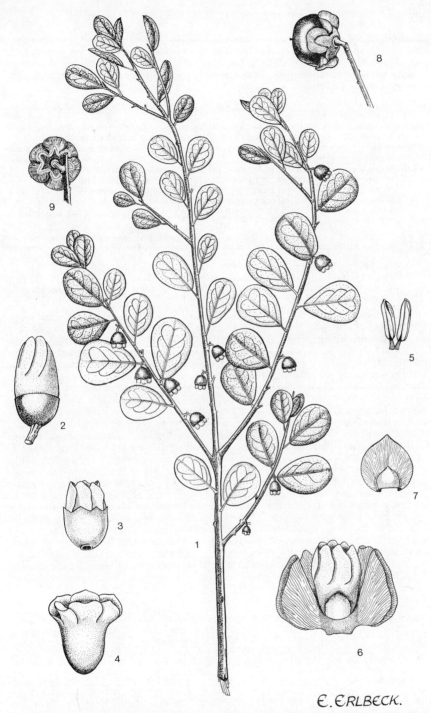

E. ERLBECK.

Tab. 57. DIOSPYROS ROTUNDIFOLIA (sect. Forbesia). 1, flowering branchlet (×½) *Mendonça* 3389; 2, male flower-bud (×2) *Mendonça* 3288; 3, male flower (×2) *Mendonça* 3288; 4, male corolla (×2) *Mendonça* 3288; 5, stamens (×3) *Mendonça* 3288; 6, female flower with one calyx-lobe removed (×2) *Mendonça* 3288; 7, female calyx-lobe, adaxial surface (×2) *Mendonça* 3288a; 8, fruit (×½) *Torre* 2609; 9, fruiting calyx (×½) *Barbosa* 7488.

usually glabrous or almost so. Stamens (5) 10–120, inserted on corolla-tube, exserted. Ovary glabrous. Locules 4, 6, 8 or 10, uniovulate. Styles 3–5, or completely united, usually much shorter than the ovary. Fruit small, up to 2 cm. diameter, usually globose, 1–10-seeded. Endosperm smooth or ruminate. Fruiting calyx usually scarcely accrescent.

Leaves broadest in upper half, more than 3 cm. in diameter, drying yellowish or pale brown;
    calyx-lobes in fruit 0·8–1 × 0·5–0·6 cm.   -   -   -   -   -   -   -   6. *squarrosa*
Leaves broadest at or below the middle, or if broadest in upper half then usually less than 2 cm.
    in diameter, and drying dark brown or blackish; calyx-lobes in fruit up to 0·6 cm. long:
    Leaves mostly crowded at ends of very short spur-shoots, oblanceolate to spathulate; male
      flowers in lax cymes   -   -   -   -   -   -   -   -   -   7. *quiloensis*
    Leaves confined to long shoots, widest near the middle; male flowers in contracted cymes:
      Leaf-apex usually emarginate; fruit fusiform: lobes of fruiting calyx acute:
        Petiole c. 0·6 cm. long; leaf margin plane; fruits distinctly pedicellate -   -   10. *sp. 1*
        Petiole c. 0·2 cm. long; leaf margin undulate; fruits subsessile   -   -   8. *inhacaensis*
      Leaf-apex not emarginate; fruit ellipsoid or subglobose; lobes of fruiting calyx rounded
                                     9. *abyssinica*

6. **Diospyros squarrosa** Klotzsch in Peters Reise Mossamb. Bot.: 184 (1861). — Hiern in Trans. Camb. Phil. Soc. **12**: 190 (1873). — Sim, For. Fl. Port. E. Afr.: 83 (1909). — F. White, F.F.N.R.: 332, t. 59 fig. f. (1962). — Drummond in Kirkia, **10**: 267 (1975). — K. Coates Palgrave, Trees of Southern Afr.: 752 (1977). TAB. **58** fig. A. Types: Mozambique, Sena, ♀ fl., *Peters* s.n. (B†, holotype); Sena, fr. 7.i.1860, *Kirk* s.n. (K, neotype, here designated).

Small deciduous tree 3–9 m. high, sometimes flowering as a shrub c. 2 m. high. Bark grey or brown, with close and shallow longitudinal fissures. *Leaves* subcoriaceous, drying pale brown with pale yellow nerves above and beneath; lamina 5·5 × 3–11 × 6·5 cm., obovate; apex shortly cuspidate; lower surface densely setulose on midrib, sparsely so elsewhere; secondary nerves in 6–9 pairs, prominent beneath, tertiary nerves subscalariform, venation closely reticulate, slightly prominent above, more so beneath. *Male flowers* in (1) 3(7)-flowered lax cymules borne towards the base of the current year's shoot in the axils of caducous reduced leaves and the first-formed foliage leaves; peduncle c. 0·6 cm. long; pedicels 0·1–0·3 cm. long, fulvous-tomentellous. Calyx 0·25 cm. long, strigulose-tomentellous outside, glabrous inside, 4-lobed to about the middle; lobes suborbicular. Corolla 0·7 cm. long, subrotate, strigulose on mid-petaline lines, otherwise glabrous; tube 0·15 cm. long; lobes 4, ovate-lingulate. Stamens 24–28, 0·4–0·5 cm. long; filaments 0·1–0·15 cm. long, glabrous, inserted singly or in pairs on corolla-tube; anthers linear-lanceolate, apiculate, glabrous or strigulose. Pistillode 0·1 × 0·1 cm., sparsely setulose. *Female flowers* solitary, borne as in male; pedicel 0·4 cm. long. Calyx as in male but 0·6 cm. long and more deeply-lobed; lobes markedly imbricate. Corolla as in male but 0·7–0·85 cm. long. Staminodes 4, inserted on corolla-tube and alternating with the lobes, 0·2 cm. long, completely glabrous except for a single apical seta. Ovary 0·25 × 0·25 cm., subglobose; locules 8 or 10, uniovulate; styles 4(5), as long as ovary, ascending, expanded distally, fleshy and stigmatic on inner surface. *Fruit* c. 2 × 2 cm., subglobose, glabrous. Seeds 8 or 10 or fewer by abortion, dark brown, 1·2 × 0·6 × 0·5 cm., surface wrinkled, endosperm ruminate. *Fruiting calyx* slightly accrescent, c. 1·2 cm. long, lobes 0·8 × 0·6 cm., reflexed, with conspicuous, ± flabellate, longitudinal veins.

Zambia. C: Katondwe, fr. 24.ii.1965, *Fanshawe* 9209 (K). E: Chipata Distr., Mwangazi R., ♂ fl. 26.xi.1958, *Robson* 714 (BM; FHO; SRGH). S: Choma Distr., Mochipapa Agricultural Station, ♀ fl. 11.i.1960, *White* 6227 (FHO; SRGH). **Zimbabwe.** N: Kariba Gorge, fr. vi.1960, *Goldsmith* 70/60 (BM; FHO; SRGH). W: Wankie Distr., Victoria Falls Village, 975 m., fr. 17.vii.1975, *Gonde* 43 (SRGH). E: Chipinga Distr., E bank of Sabi R, fr. 8.vi.1950, *Wild* 3445 (BM; FHO; SRGH). S: Nuanetsi Distr., Chipinda Pools, fr. 6.iv.1970, *Sherry* 85/70 (FHO; SRGH). **Malawi:** N: Rumpi Distr., Njakwa, fr. 14.v.1952, *White* 2849 (BR; FHO; K; SRGH). C: Dedza Distr., Chipoka, fr. 7.vii.1954, *Banda* 42 (BM; BR; FHO). S: Nsanje Distr., Chiromo, ♂ fl. i.1894, *Scott Elliot* 2789 (BM; CGE; K). **Mozambique.** N: Memba, 30 m., ♀ fl. 6.xii.1963, *Torre & Paiva* 9441 (LISC). Z: Gúruè to Lioma 24 km., fr. 25.ii.1966, *Torre & Correia* 14865 (C; COI; FHO; K; LISC; LMU). T: between Estima and Cahó, fr. i.1972, *Macêdo* 4722 (LISC; SRGH). MS: Manica to Chimoio (Vila Pery) 2 km., 900 m., ♂ fl. 25.xi.1965, *Torre & Correia* 13249 (LISC).

From Kenya to Mozambique, penetrating inland along river valleys, and with a distant outlying population in Zaire (Shaba). Virtually absent from the Central African Plateau. In deciduous woodland and thicket. At higher altitudes only on termite mounds. From near sea-level to 1200 m.

7. **Diospyros quiloensis** (Hiern) F. White in Bull. Jard. Bot. État. Brux. **26**: 244 (1956); F.F.N.R.: 332, t. 59 fig. h (1962). — K. & O. Coates Palgrave, Trees of Central Afr.: 168 cum tab. & photogr. (1957). — Wild, Biegel & Mavi, Rhod. Bot. Dict., ed. 2: 151 (1972). — Drummond in Kirkia, **10**: 267 (1975). — K. Coates Palgrave, Trees of Southern Afr.: 749 (1977). TAB. **58**, fig. B. Syntypes from Tanzania.
   Maba quiloensis Hiern in Trans. Camb. Phil. Soc. **12**: 132 (1873). Types as above.

Small deciduous tree up to 8 m. tall. Bark on bole and branches dark grey, very thick, deeply fissured longitudinally and cracked transversely and looking like Crocodile skin, the fissures 1–2 cm. deep and 1–3 cm. apart; sometimes a multiple-stemmed shrub. *Leaves* subcoriaceous, dark green, glossy, usually drying blackish, scattered on long shoots, but mostly crowded at ends of very short spur-shoots; lamina up to 6 × 2 (7 × 3·5) cm., oblanceolate to spathulate, apex rounded to subacute, base cuneate, margin plane; lower surface sparsely puberulous especially on the midrib; lateral nerves in c. 7–9 pairs, indistinct; tertiary nerves and veins indistinct on both surfaces; petiole c. 0·2 cm. long. *Male flowers* in 2–12-flowered, rather lax cymes, borne below the leaves towards base of spur shoots, and in axils of fallen leaves on long shoots; peduncle c. 0·1 cm. long, pedicels very slender, up to 0·4 cm. long. Calyx 0·15–0·2 cm. long, glabrous except for minute marginal setulae; lobes 3–4, deltate, 0·1–0·15 cm. long. Corolla 0·6–0·8 cm. long, rotate, glabrous; tube c. 0·15 cm. long; lobes 3–4, ovate-deltate. Stamens 0·4–0·5 cm. long, usually distinctly exserted, 9 or 12, inserted on corolla-tube, one opposite to and 2 alternating with each lobe; anthers 0·15–0·2 cm. long, lanceolate, not apiculate, densely setulose. Pistillode minute, glabrous. *Female flowers* subsessile, solitary or in fascicles of 2–5, at base of spur shoots and in leaf-axils and axils of fallen leaves on long shoots. Calyx and corolla similar to male but up to 0·4 and 0·9 cm. long respectively. Staminodes 3, c. 0·3 cm. long, attached to corolla-tube and alternating with the lobes, minutely setulose at apex. Ovary 0·3 × 0·25 cm, obovoid, glabrous; style 0·15 cm. long, stout, ending in a large, irregularly lobed fleshy stigma; locules 6, uniovulate. *Fruit* up to 2 × 0·9 cm., glabrous, fusiform, style persistent. Seed 1, ellipsoid; endosperm smooth. Fruiting calyx c. 0·5 cm. long, the tube shorter than the lobes, much wider than the pedicel, lobes rounded or obtuse.

Zambia. C: Feira Distr., Katondwe, fr. 11.ii.1963, *Grout* 288 (FHO). S: Zambezi Valley, Sinazongwe to Choma, 10 km., fr. 24.vi.1961, *Bainbridge* 448 (FHO). **Zimbabwe.** N: Urungwe Distr., Chirundu, fr. 20.iii.1952, *Whellan* 640 (FHO; SRGH). W: Victoria Falls, banks of Zambezi R., fr. 24.vii.1950, *Robertson & Elffers* 41 (FHO; SRGH). C: Gatooma Distr., Umniati R., 48 km. W. of Gatooma, 951 m., fr. 6.iv.1969, *Burrows* 326 (SRGH). E: Melsetter Distr., Sabi Valley, between Hot Springs and Birchenough Bridge, ♂ fl. 23.x.1948, *Chase* 930 (BM; BR; K; SRGH). S: Buhera Distr., Birchenough Bridge area, 480 m., fr. 21.iv.1969, *Biegel* 2943 (SRGH). **Malawi.** S: Mangochi Distr., Lundwe, st. 10.vii.1954, *Jackson* 1361 (FHO). **Mozambique.** N: 5 km. S. of Pemba (Porto Amélia), fr. 3.iii.1961, *Gomes et Sousa* 4642 (K). T: Tete to Changara 6 km., ♂ fl. 26.xii.1965, *Torre & Correia* 13814 (C; COI; FHO; K; LISC; LMU); fr. iii.1966, *Torre & Correia* 15228 (LISC). MS: Chemba, fr. vii.1947, *Simão* 1365 (EA).

Only known from the Zambezi, Shire, Luangwa and Sabi River valleys and from a few localities on the coastal plain of northern Mozambique and Tanzania. In deciduous woodland, especially *Colophospermum mopane* woodland, and deciduous thicket. From near sea-level to 900 m.

The northern populations in Tanzania and northern Mozambique differ from the southern populations, which are centred on the Zambezi Valley and its tributaries, in having broader leaves which are less prominently crowded on spur shoots, and in a few minor floral features. The material, however, is extremely exiguous. Of the 3 available gatherings 2 are types, allegedly from Tanzania, but since they are only vaguely localized and the species has not been re-collected in Tanzania, further investigation is called for.

A puzzling specimen collected from within the geographical range of *D. quiloensis* but with some features of *D. inhacaensis* though differing from both, is treated below as *Diospyros* sp. 1.

8. **Diospyros inhacaensis** F. White in Bol. Soc. Brot., Sér. 2, **36**: 97, t. 1 (1962). — de Winter in F.S.A. 26: 60 (1963). — Palmer & Pitman, Trees of Southern Afr. 3: 1791 cum tab. & photogr. (1972). — K. Coates Palgrave, Trees of Southern Afr.: 746 (1977). TAB. **58**, fig. C. Type: Mozambique, Inhaca Isl., fr. 14.vii.1957, *Mogg* 27221 (J; K, holotype; PRE; SRGH). 6 paratypes are also cited in the protologue, including 2 from Mozambique.

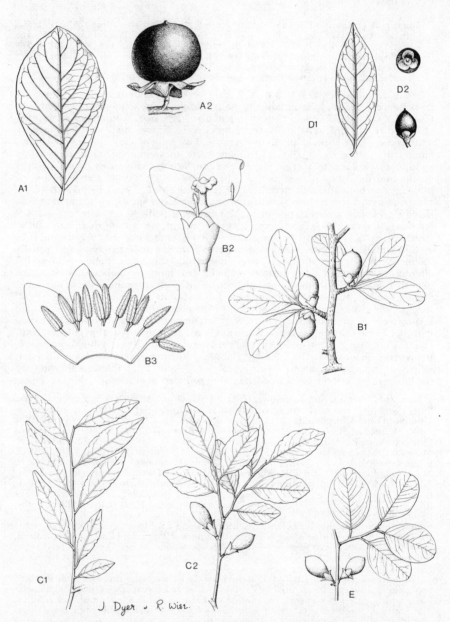

Tab. 58. DIOSPYROS sect. BREVITUBA. A. — D. SQUARROSA. A1, leaf (× ½); A2, fruit (× 1). B. — D. QUILOENSIS. B1, fruiting branchlet (× ½) *Bainbridge* 448; B2, female flower (× 4) *Chase* 4700; B3, male corolla and stamens (× 4), *Chase* 930. C. — D. INHACAENSIS. C1, leaves (× ½) *Mogg* 28029; C2, fruiting branchlet (× ½) *Mogg* 28135. D. — D. ABYSSINICA subsp. ABYSSINICA. D1, leaf (× ½), D2, fruits (× 1). E. — D. SP. 1, fruiting branchlet (× ½) *Torre, Carvalho & Ladeira* 18965. A & D from F.F.N.R.

Evergreen shrub or tree up to 20 m tall. Bark dark grey, rough but less so than in *D. quiloensis*. *Leaves* subcoriaceous, drying pale brown or blackish; lamina up to 7 × 3·6 cm., narrowly or broadly rhombic-elliptic, apex usually subacuminate, rarely acute, the tip itself usually slightly emarginate, base cuneate, margin, both in living and dried leaves, conspicuously undulate; lower surface glabrous; lateral nerves in 5–7 pairs, slightly prominent beneath; tertiary nerves and veins scarcely visible; petiole about 0·2 cm. long. *Male flowers* axillary and ramuligerous, in contracted 7–12-flowered cymes; peduncle 0·1 cm. long; pedicels 0·1–0·2 cm. long, glabrous. Calyx 0·2 cm. long, with 3 narrowly deltate, acute lobes, 0·15 × 0·15 cm., glabrous outside except for the sparsely setulose margin, glabrous inside. Corolla subrotate, 0·6 cm. long, glabrous; tube 0·15 cm. long; lobes 3, 0·45 × 0·3 cm., apex obtuse. Stamens 5–11, exserted, up to 0·5 cm. long; anthers 0·1–0·15 cm. long, lanceolate, obtuse, densely setulose. Pistillode 0·1 cm. long, glabrous, *Female flowers* unknown. *Fruits* subsessile, solitary or 2–3 together in the axils of fallen leaves on second and third year branchlets, up to 2 × 1·2 cm., glabrous, fusiform, apiculate, style persistent. Seed 1, obovoid, 1·1 × 0·6 cm., black, endosperm smooth. Fruiting calyx usually c. 0·7 cm. long, glabrous, tube longer than the lobes, gradually narrowed to the pedicel; lobes ovate-deltate. *Chromosome number*: 2n = 30.

Mozambique. GI: Vilanculos, Mapinhane, st. 31.viii.1942, *Mendonça* 51 (FHO; LISC; LMU). M: Maputo, Ponta do Ouro, *Torre & Correia* 17791, ♂ fl. 4.iii.1968 (C; FHO; K; LISC); Inhaca Isl., fr. 8.vii.1958, *Mogg* 28029 (BM; J; SRGH).
Also in northern Natal. In coastal dune forest, occasional to locally dominant. 10–200 m.

9. **Diospyros abyssinica** (Hiern) F. White in Bull. Jard. Bot. État Brux. **26**: 241 (1956); tom. cit.: 294, t. 76 fig. f-l, t. 77 fig. c-d; F.F.N.R.: 332, t. 59 fig. g (1962); in F.W.T.A., ed. 2, **2**: 10 (1963). — Letouzey & F. White in Fl. Cam. **11** & Fl. Gab. **18**: 29, t. 19 fig. 12–18 (1970). — Wild, Biegel & Mavi, Rhod. Bot. Dict., ed. 2: 150 (1972). — Drummond in Kirkia, **10**: 266 (1975). — K. Coates Palgrave, Trees of Southern Afr.: 743 (1977). Types from Ethiopia.
Maba abyssinica Hiern in Trans. Camb. Phil. Soc. **12**: 132 (1873). Types as above .
Ebenus abyssinica (Hiern) O. Kuntze, Rev. Gen. Pl. **2**: 408 (1891). Types as above.

Small, medium-sized or large tree up to 36 m. high, but sometimes flowering as a shrub 2 m. high. Bole long, straight, slender. Bark dark grey or blackish, rough, reticulate and exfoliating on old trees. *Leaves* subcoriaceous, drying grey-green or blackish; lamina 3 × 1·2–12 × 4 cm., elliptic, oblanceolate-elliptic or lanceolate-elliptic, apex obtuse to shortly and bluntly subacuminate, rarely distinctly acuminate; lower surface glabrescent; lateral nerves in 5–12 pairs; venation prominent and closely reticulate on both surfaces. *Male flowers* axillary and ramuligerous, in contracted 10–18-flowered cymes; peduncle 0·1 cm. long; pedicels 0·1 cm. long, fulvous-setulose. Calyx 0·2 cm. long, shallowly cyathiform, with 3–4 short, broadly deltate lobes, glabrous outside except for a few minute marginal hairs, glabrous inside. Corolla 0·5–0·6 cm. long, sub-rotate, glabrous; tube 0·15 cm. long; lobes 3–4, 0·45 × 0·3 cm., broadly elliptic, apex obtuse. Stamens 10–15, 0·2–0·4 cm. long; anthers lanceolate-apiculate, sparsely setulose towards the apex. Pistillode 0·1 cm. long or absent, glabrous. *Female flowers* axillary or ramuligerous, in (1–2) 3–5(8)-flowered fascicles; pedicels 0·2 cm. long, fulvous-setulose. Calyx 0·6 cm. long, cyathiform, glabrous outside, finely strigulose towards the base inside, divided almost to the base; lobes 3–4, up to 0·6 × 0·6 cm., suborbicular, sometimes apiculate, strongly imbricate. Corolla slightly shorter than the calyx, otherwise as in male. Staminodes 3–4, glabrous, 0·2 cm. long, filiform, exserted, attached to the throat of the corolla and alternating with the lobes. Ovary 0·4 × 0·2 cm., conoidal, glabrous, gradually merging into the short undivided style; locules 6, uniovulate; stigmatic lobes, 3, about 0·1 cm. long, ascending. *Fruit* up to 1·4 × 0·9 cm., glabrous, ellipsoid or subglobose, style persistent. Seed(s) 1 (very rarely 2), 0·9 × 0·6 cm., globose to sub-ellipsoid, black; endosperm smooth. Fruiting calyx scarcely accrescent, c. 0·7 cm. long, becoming patelliform.

D. abyssinica, which is widespread in tropical Africa, is moderately variable in leaf-shape and size. This variation is too diffuse to justify the recognition of infra-specific taxa except for subsp. *chapmaniorum*, a consistently small-leaved variant in Malawi and adjacent parts of Zambia and Mozambique.
In our area *D. abyssinica* does not appear to be of economic importance, but in Tanzania the boles of straight-stemmed trees are used for the masts of dhows, and the tough wood is

considered suitable for the handles of pick-axes and other tools; up to 20000 were formerly produced each year.

Subsp. **abyssinica**. — TAB. **58**, fig. D.

    Maba mualala Welw. ex Hiern in Trans. Camb. Phil. Soc. **12**: 111 (1873). — S. Moore in Journ. Linn. Soc., Bot. **40**: 133 (1911). — Steedman, Trees etc. S. Rhod.: 64 (1933). Types from Angola.

    Ebenus mualata (sphalm) (Hiern) Kuntze, Rev. Gen. Pl. **2**: 408 (1891). Types as above.

    Diospyros welwitschii Hiern, Cat. Afr. Pl. Welw. **3**: 653 (1898). Types from Angola.

Evergreen tree up to 36 m. high but sometimes flowering as a shrub 2 m. high. Leaves 7 × 2·5–12 × 4 cm., elliptic or lanceolate-elliptic, apex shortly and bluntly subacuminate to acuminate.

Zambia. W: Chingola, ♀ fl. 5.ii.1957, *Fanshawe* 2997 (BR; FHO; K). C: Broken Hill Forest Reserve, st. 21.ix.1947, *Brenan & Trapnell* 7899 (EA; FHO; K). **Zimbabwe.** E: Chirinda Forest, ♂ fl. x.1905, *Swynnerton* 2 (BM; K; SRGH). **Mozambique.** MS: 30 km. NE. of Inhamitanga, st. 5. xii.1971, *Müller & Pope* 1884 (FHO; LISC; SRGH).

Widespread in tropical Africa from Guinea Republic to Eritrea and southwards to the Flora Zambesiaca area and Angola, but absent from the Guineo-Congolian rain forests except locally near their northern and southern fringes. A striking hiatus in its range in Malawi is occupied by subsp. *chapmaniorum*. In evergreen forest and thicket and on termite mounds. 200–1350 m.

Subsp. **chapmaniorum** F. White in Bull. Jard. Bot. Nat. Belg. **50**: 393 (1980).

    Types: Malawi, Ntchisi Mt., ♂ fl. 23.iii.1963, *Chapman* 1824 (BR; FHO; K, holotype; LISC; SRGH). 11 paratypes are also cited in the protologue.

Evergreen tree 18–30 m. tall. Leaves, 3 × 1·2–5 × 2 cm., usually oblanceolate-elliptic, sometimes elliptic, apex obtuse.

Zambia. E: Nyika Plateau, headwaters of Chire R., st. 7.v.1952, *White* 2758 (BM; FHO). **Malawi.** N: Misuku Hills, Mugesse Forest, st. 20.iv.1963, *Chapman* 1894 (FHO; SRGH). C: Ntchisi Mt., st. 3.v.1961, *Chapman* 1268 (FHO; K; SRGH). S: Kirk Range, Cirobwe Mt., st. 22.v.1961, *Chapman* 1323 (FHO; SRGH). **Mozambique.** MS: Chimoio, Garuso forest, st. iv. 1935, *Gilliland* 1814 (BM; FHO).

Only known from our area. In montane rain forest. 1400–2220 m.

The leaves of young plants are more variable than those of mature individuals and may be narrower or broader than those described above.

10. **Diospyros** sp. 1. — TAB. **58**, fig. E.

Small tree 6 m. tall. *Leaves* drying blackish, scattered on long shoots; lamina up to 3·8 × 3 cm., suborbicular to very broadly elliptic, broadest at or near the middle, apex broadly rounded and shallowly emarginate, base broadly cuneate to subtruncate, margin plane; lower surface glabrous; lateral nerves in about 8 pairs, together with the veins forming a subprominent reticulum on upper surface, less distinct beneath; petiole c. 0·6 cm. long. *Fruit* c. 1·5 × 1 cm., broadly ellipsoid, apiculate, solitary or in fascicles of 2–3 borne below the foliage leaves towards base of current year's growth, distinctly pedicellate; pedicels c. 0·2 cm. long. Fruiting calyx 0·6 cm. long; lobes ovate-deltate.

Mozambique. T: Cahora Bassa, rio Mucangádzi, encosta do monte, próx. do Posto Policial no. 3, ao km. 5 da barragem, 867 m., fr. i.1973. *Torre, Carvalho & Ladeira* 18965 (C; FHO; K; LISC; LMU).

Not known elsewhere. In rocky places. 870 m.

The specimen cited above is closely related to *D. quiloensis* and *D. inhacaensis*. It was collected from well inside the geographical range of the former but 700 km. from the nearest known occurrence of the latter. Nevertheless in the insertion of its leaves and the shape of its lamina and fruiting calyx it is much closer to *D. inhacaensis*. It is very different from both, however, in its long petiolate leaves and pedicellate fruits. ·

5. Sect. ROYENA (L.) F. White in Bol. Soc. Brot., Sér. 2, **53**: 281 (1980); in Bull. Jard. Bot. Nat. Belg. **50**: 455 (1980).

    Royena L., Sp. Pl.: 297 (1753); Gen. Pl., ed. 5: 188 (1754).

Flowers solitary on long pedicels or in long-pedunculate, few-flowered cymes. Calyx usually deeply divided; lobes 4–5(6), deltate, ovate or lanceolate, aestivation valvate or open Corolla glabrous to densely pubescent, urceolate or campanulate, up to 1 cm. long, deeply divided; lobes proximally ascending and continuing the tube, then spreading or reflexed. Stamens 8 or 10, inserted in a single row at base of corolla-tube, alternating with and opposite to the corolla-lobes, included. Pistillode similar to functional gynoecium but styles not bifid and stigmatic surfaces not developed. Ovary pubescent; locules 4–10, uniovulate. Styles 2–5, more or less equalling ovary, united in lower half, bilobed at apex, stigmatic surface only slightly expanded. Antherodes well-developed but usually concealed by long, strigose-sericeous hairs. Fruit a berry, sometimes tardily dehiscent, 1–4 cm. in diameter. Seeds 10 or fewer; endosperm smooth or ruminate. Fruiting calyx slightly to markedly accrescent.

Suffrutex; stems and leaves with pilose hairs up to 0·3 cm. long -   -   -   - 11. *anitae*
Shrubs or trees, or if suffrutose then lacking long weak hairs:
   Flowers in 2–5(7)-flowered cymes; calyx lobes valvate with white-tomentellous, reduplicate
      margins contrasting with the less densely hairy outer surface; corolla up to 0·6 cm. long:
     Bracts foliaceous; calyx completely concealing the fruit:
       Lamina sub-obdeltate, apex both cuspidate and truncate or emarginate; endosperm
        smooth -   -   -   -   -   -   :   -   -   - 12. *truncatifolia*
       Lamina obovate or obovate-elliptic, apex obtuse to shortly acuminate; endosperm
        shallowly ruminate -   -   -   -   -   -   - 13. *usambarensis*
     Bracts not foliaceous; calyx lobes reflexed in fruit   -   -   -   - 14. *zombensis*
   Flowers solitary (in *D. villosa* rarely in 2–3-flowered cymes, but then corolla more than 0·7
    cm. long):
    Leaf-margin with long hyaline hairs; calyx in flower and fruit urceolate, shortly toothed
                                   15. *whyteana*
    Leaf-margin without long hyaline hairs; calyx deeply cleft:
     Leaves cordate or rounded at the base, venation deeply impressed above, very
      prominent beneath; fruit densely hispid-tomentose, tardily dehiscent - 16. *villosa*
     Leaves cuneate, venation otherwise; fruit glabrescent to tomentellous:
      Leaf-margin revolute; fruit depressed-globose, tomentellous, tardily dehiscent
                                   17. *dichrophylla*
      Leaf-margin plane; fruit ellipsoid or globose, puberulous or glabrescent, indehiscent
                                   18. *lycioides*

11. **Diospyros anitae** F. White in Bol. Soc. Brot., Sér. 2, **54**: 1 (1980). Type: Mozambique, Imala to Mecuburi 30 km., ♀ fl., fr. 16.i.1964, *Torre & Paiva* 10008 (LISC, holotype).

Rhizomatous suffrutex up to 40 cm. tall. Branchlets and leaves with pilose hairs up to 0·3 cm. long. *Leaves* chartaceous; lamina up to 10·5 × 6·5 cm., obovate, apex cuspidate, base rounded or subcordate; secondary nerves in 5–6 pairs, prominent beneath; tertiary nerves and veins closely reticulate, not prominent, but on lower surface darker than the lamina. *Male flowers* unknown. *Female flowers* solitary or in 2-flowered cymules, pentamerous. Calyx 0·2 cm. long; lobes deltate, 0·15 cm. long, tomentellous inside, glabrous outside except for a few minute, glandular hairs; margin densely ciliolate. Corolla 0·35 cm. long, glabrous except for a few minute glandular hairs; lobes suborbicular, 0·2 cm. long. *Fruit* (immature) globose, 2 cm. diameter, tomentellous. Fruiting calyx accrescent, c. 1 cm. long; lobes lingulate, obtuse, reflexed.

Mozambique. N: Imala to Mecuburi 30 km., ♀ fl., fr. 16.i.1964, *Torre & Paiva* 10008 (LISC).
Only known from the type locality. In *Brachystegia* woodland. 450 m.

12. **Diospyros truncatifolia** A. N. Caveney in Bull. Jard. Bot. Nat. Belg. **50**: 395 (1980). TAB. **59**, fig. A. Type: Mozambique, Malema Distr., Mutuali, near Catholic Mission, fr. 12.iv.1953, *Gomes e Sousa* 4099 (COI; FHO; K, holotype; LMJ).
   Royena amnicola sensu Gomes e Sousa, Dendrol. Moçamb. Estudo Geral, **2**: 634, t. 197 (1966).

Shrub or small tree 3–8 m. high. Bark rough. Branchlets spreading-pubescent with dense, short, setose hairs and longer, fulvous, pilose hairs. *Leaves* chartaceous, drying dull brown or grey-green; lamina sub-obdeltate, gradually tapered from near the apex to the slightly cordate base, apex both cuspidate, and truncate or emarginate; margin ciliate: lower surface velutinous-tomentose; secondary nerves in 5–6 pairs, prominent below. *Male flowers* usually in 3-flowered axillary cymes, tetramerous; peduncle up to 1·5 cm. long; pedicels up to 0·8 cm. long; bracts c. 0·9 × 0·3 cm., foliaceous, opposite, lanceolate, abruptly acuminate; indumentum of inflorescence consisting of long spreading, short setose, and reddish glandular hairs.

Calyx 0·4 cm. long; lobes tomentellous on both surfaces, deltate or ovate-acuminate, the margins reduplicate. Corolla 0·4-0·5 cm. long, lobed almost to the base; lobes subacute, glabrous. Disk fimbriate. Stamens 8, 0·25-0·3 cm. long; filaments very short, glabrous; anthers 0·2-0·25 cm. long, lanceolate, densely sericeous-pubescent. Pistillode 0·1 cm. long, conoidal, minutely tomentellous; locules 8; stylodes 4, glabrous. *Female flowers* unknown. *Fruit* 2-2·8 cm. in diameter, globose, fulvous-velutinous. Seeds up to 8, 1·1-1·7 cm. long, reddish-brown; testa smooth or minutely pitted; endosperm smooth. Fruiting calyx accrescent, completely concealing the fruit; lobes 2 × 1·5-4·1 × 2·2 cm., ovate, acute or acuminate, velutinous-tomentose.

Malawi. S: Mangoche Distr., Kawinga, Ntaja, fr. 29.iv.1955, *Jackson* 1652 (FHO).
Mozambique. N: Cabo Delgado, between Montepuez and Balama, fr. 29.viii.1948, *Barbosa* 1918 (LISC; LMA; LMU).
Also in southern Tanzania. In rocky places and on termite mounds, otherwise ecology unknown, 500–900 m.

The fruits are edible.

13. **Diospyros usambarensis** F. White in Bull. Jard. Bot. Etat Brux. **33**: 366 (1963). — Drummond in Kirkia, **10**: 267 (1975). — K. Coates Palgrave, Trees of Southern Africa: 753 (1977). TAB. **59**, fig. B. Types: Mozambique, Sena, ♀ fl., *Peters* s.n. (B†, holotype of *D. macrocalyx*, see below).
   Diospyros macrocalyx Klotzsch in Peters Reise Mossamb. Bot.: 182 (1862) non *D. macrocalyx* A.DC. (1844). Type: *Peters* s.n. (B†, holotype) see above.
   Royena macrocalyx Gürke in Engl., Pflanzenw. Ost. Afr. **C**: 305 (1895). — Brenan, T.T.C.L.: 189 (1949); in Mem. N.Y. Bot. Gard. **8**: 499 (1954). — Gomes e Sousa, Dendrol. Moçamb. Estudo Geral, **2**: 635, t. 198 (1966). Type as above.
   Diospyros loureiriana auct. non G. Don, Hiern in Trans. Camb. Phil. Soc. **12**: 194 (1873) pro parte. — Sim, For. Fl. Port. E. Afr.: 83 (1909). — S. Moore in Journ. Linn. Soc., Bot. **40**: 134 (1911).
   Royena usambarensis Gürke ex Engl. in Abh. Preuss. Akad. Wiss.: 34 (1894); Pflanzenw. Ost. Afr., **A**: 73 (1895) *nom. nud.*

Semi-deciduous shrub or tree (1) 2–6(10) m. tall. Bark dark grey or black, rough, deeply fissured. Branchlets reddish with spreading pilose, setulose and glandular hairs. *Leaves* chartaceous, drying greenish-black or reddish brown; lamina 2·9 × 1·6–11 × 6·8 cm., obovate or obovate-elliptic; apex obtuse, acute or shortly acuminate, base rounded to subcordate, margin ciliate or not; lower surface almost glabrous to densely pubescent on the lamina and puberulous to tomentellous or pilose-tomentose on the nerves; lateral nerves in 5–6 pairs, tertiary nerves and veins forming a reticulum darker than the lamina on lower surface. *Male flowers* tetramerous, usually in 3–7-flowered cymes in axils of reduced leaves and proximal true leaves towards base of current year's growth, rarely solitary; peduncle up to 2·5 cm. long; pedicels up to 1·5 cm. long, puberulous; bracts up to 0·5 × 0·3 cm., opposite, ovate-lanceolate. Calyx 0·3-0·4 cm. long, valvate-reduplicate, deeply lobed; lobes deltate or ovate-acuminate, pallid-tomentellous on both sides. Corolla 0·4-0·6 cm. long, lobed almost to the base; lobes rounded to acute, minutely setulose, reflexed at apex. Disk undulate, glabrous. Stamens 8, 0·3 cm. long; filaments very short, hairy distally; anthers 0·25 cm. long, lanceolate, densely sericeous-pubescent. Pistillode conoidal, minutely tomentellous to almost glabrous, locules 8, uniovulate; stylodes 4, glabrous. *Female flowers* similar to male. Staminodes 8, 0·15-0·2 cm. long; antherode densely hairy. Ovary 0·15-0·2 cm. long, conoidal, tomentellous to almost glabrous; locules 8; styles united only at base, glabrous. *Fruit* yellowish, 1·5-2·5 cm. in diameter, globose, setulose-puberulous. Seeds 8 or fewer, 1·1 cm. long, dull brown; testa smooth or slightly reticulate; endosperm shallowly ruminate. Fruiting calyx strongly accrescent, concealing the fruit, lobes up to 4 × 2·5 cm., broadly ovate, membranous with prominent venation, puberulous.

This species was originally described by Klotzsch as *D. macrocalyx*, a name already used by A. De Candolle. When Gürke transferred it to *Royena* he was able to use the same epithet, and the name *R. macrocalyx* has been widely used during this century. Gürke cited his previously published nomen nudum *R. usambarensis* in synonymy. When White returned *R. macrocalyx* to *Diospyros* he adopted the name *usambarensis* in order to retain bibliographic continuity, not then realizing that a strong case could be made for recognizing two subspecies. This means, however, that the southern subspecies must be called *usambarensis*, notwithstanding the fact that the epithet when it was originally published illegitimately referred to the northern

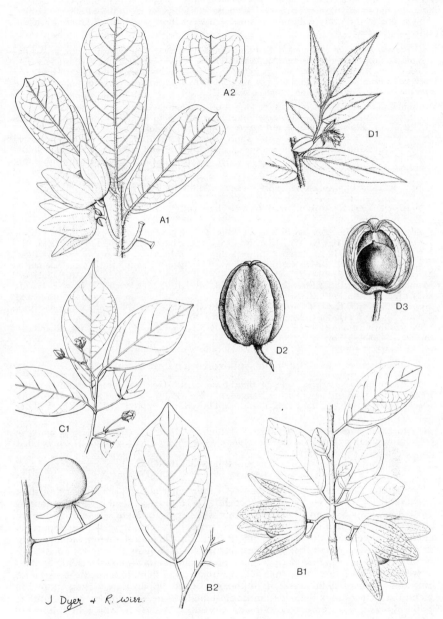

J Dyer + R. wise.

Tab. 59. DIOSPYROS sect. ROYENA (pro parte). A. — D. TRUNCATIFOLIA. A1, fruiting branchlet (× ½) *Richards* 18026; A2, leaf apex (× ½) *Gomes* e *Sousa* 4099. B. — D. USAMBARENSIS subsp. USAMBARENSIS. B1, fruiting branchlet showing proximal leaves (× ½) *Torre* & *Paiva* 9760; B2 distal leaf (× ½) *Pedro* 2184. C. — D. ZOMBENSIS. C1, flowering branchlet (× ½) *Lawton* 1555; C2, fruit (× ½) *Torre* 5777. D. — D. WHYTEANA. D1, flowering branchlet (× ½); D2, fruiting calyx (× 1); D3, fruit with half of fruiting calyx removed. D from F.F.N.R.

subspecies. A new epithet, *rufescens*, has been provided for the northern subspecies, which, although not strictly occurring in the Usambara Mts., is plentiful in the adjacent lowlands. This is one of the many unfortunate consequences of the application of Article 26 of the International Rules.

D. usambarensis is represented in the Flora Zambesiaca area by the typical subspecies, which is replaced further north in the coastal parts of Tanzania and Kenya by subspecies, *rufescens* A. N. Caveney. Although typical subsp. *rufescens* appears to be absent from our area, certain specimens from northern Mozambique, which are cited below, are intermediate between the two subspecies. For this reason a key is provided.

Branchlets and inflorescence-axes with long spreading hairs, easily seen by naked eye; leaves
　　drying reddish-brown, obovate-elliptic, apex obtuse to broadly acute, lower surface
　　densely pubescent on lamina and nerves with long spreading hairs - subsp. *rufescens*
Branchlets and inflorescence-axes with short, mostly spreading hairs; leaves nearly always
　　drying greenish-black, obovate, apex usually shortly and abruptly acuminate, lower
　　surface virtually glabrous on lamina, sparsely puberulous to tomentellous with short
　　spreading hairs on main nerves - - - - - - - - subsp. *usambarensis*

Subsp. **usambarensis**— F. White & Caveney in Bull. Jard. Bot. Nat. Belg. **50**: 396 (1980).

Semi-deciduous shrub or small tree 1–10 m. tall.

Zimbabwe. E: Inyanga North Reserve, World's View, near confluence of Ruenya R., fl. x.1958, *Davies* 2520 (K; SRGH). S: Sabi R., Chitsa's Kraal, fr. 9.vi.1950, *Chase* 2400 (BM; BR; EA; FHO; K; SRGH). **Malawi.** S: Shire Valley below Blantyre, Chikwawa, fr. 19.v.1952, *White* 2867 (FHO; K). **Mozambique.** N: Cabo Delgado, Montepuez Distr., Balama, fl. 31.viii.1948, *Andrada* 1313 (C; COI; FHO; K; LISC; LMA; LMU). Z: Alto Molócuè, Mamala, ♀ fl. 20.xii.1967, *Torre & Correia* 16658 (BR; LISC; MO; WAG). T: between Casula and Chiúta, fr. 8.vii.1949, *Barbosa & Carvalho* 3503 (EA). MS: Báruè, Mungári to Tambara 54 km., fl. 16.xii.1965, *Torre & Correia* 13695 (EA; FHO; LISC; M; P). GI: Inhambane, Govuro, Mambone, 28.v.1941, *Torre* 2765 (C; FHO; LISC; WAG).

Only known from the Flora Zambesiaca area. Ecology imperfectly known; in woodland and thicket. 20–500 (915) m.

The fruits are edible and the roots are used locally to dye the lips and teeth red.

Intermediates between subsp. **usambarensis** and subsp. **rufescens**

Mozambique. N: Palma, near the lighthouse, fl., fr. 22.x.1942, *Mendonça* 1037 (COI; FHO; LISC; LMU); Palma Distr., Nangade, fl. 26.ix.1948, *Pedro & Pedrogão* 5389 (EA); Mecúfi Distr., Lúrio, ♀ fl. 29.x.1942, *Mendonça* 1115 (BR; C; FHO; K; LISC; PRE; WAG).

14. **Diospyros zombensis** (B. L. Burtt) F. White, [F.F.N.R.: 331 (1962) sine relat. nom.]; in Bull. Jard. Bot. Etat Brux. **33**: 366 (1963). TAB. **59**, fig. C. Type: Malawi, Zomba, fl. & fr. xi.1915, *Purves* 260 (K, holotype).
　　Royena zombensis B. L. Burtt in Kew Bull. **1935**: 289 (1935). Type as above.
　　Royena amnicola B. L. Burtt, tom. cit.: 290. Type from Tanzania.

Evergreen shrub or tree 2–15 m. tall. *Leaves* chartaceous, usually drying dull, dark reddish-brown above, paler beneath, more rarely blackish-green; lamina 5·6 × 3·3–13·7 × 8·3 cm., obovate, apex shortly acuminate, base rounded to subcordate; margin pilose; lower surface sparsely to densely setulose, especially along the nerves, and also usually with scattered small glandular hairs; lateral nerves in c. 7 pairs, prominent beneath, tertiary nerves and veins closely reticulate, not prominent but on lower surface darker than the lamina. *Male flowers* usually in 2–5-flowered axillary cymes, pentamerous; peduncle up to 1 cm. long; pedicels up to 1·8 cm. long; bracts 0·3–0·5 cm. long, opposite, linear, puberulous with setulose and glandular hairs. Calyx 0·25–0·4 cm. long; lobes deltate, strigulose-and glandular-puberulous outside, tomentellous inside. Corolla up to 0·6 cm. long, lobed almost to the base; lobes obtuse, spreading or reflexed; glabrous or with a few mid-petaline downwardly directed hairs and scattered small glandular hairs outside. Disk thin, fimbriate. Stamens 10, 0·3 cm. long; filaments 0·05 cm. long, glabrous; anthers 0·25 cm. long, lanceolate, densely strigose. Pistillode conoidal, tomentellous; locules 8 or 10; stylodes 4–5, mostly glabrous. *Female flowers* similar to male. Staminodes 10, 0·15 cm. long; antherodes densely hairy. Ovary 0·2 × 0·2 cm., conoidal, tomentellous; locules 8 or 10; styles 4–5, glabrous, united only at the base. *Fruit* subglobose, 2–3 cm. in diameter, setulose. Seeds 10 or fewer, 1–1·4 cm. long, reddish-brown, shining; testa deeply and reticulately grooved; endosperm deeply ruminate. Fruiting calyx accrescent; lobes 1·2–1·8 cm. long, lanceolate, acute, reflexed.

Zambia. N: Mupamadzi R, 7 km. from Luangwa escarpment, ♀ fl., imm. fr. 7.ix.1968, *Lawton* 1555 (FHO). E: Great East Road, 12 km. W. of Kachalola, fr. 17.iii.1959, *Robson* 1747 (BM; FHO; SRGH). **Malawi.** N: Nkhata Bay, fr. 23.vi.1960, *Adlard* 378 (FHO; K; SRGH). C: Mchinji Distr., Calansano Estate, fl. 15.xii.1955, *Adlard* 135 (FHO). S: Mt. Mulanje, lower slopes between Likabula and Palombe, fr. iv.1958, *Chapman* 559 (FHO). **Mozambique.** T: Moatize to Vila Coutinho 25 km., fr. 13.i.1966, *Correia* 442 (FHO; K; LISC; LMU).

Also in Kenya and Tanzania. In various types of forest, woodland, thicket and wooded grassland. 350–900 m.

15. **Diospyros whyteana** (Hiern) F. White in Bothalia, **7**: 458 (1961); F.F.N.R.: 326, t. 58 fig. d-e (1962); in Mitt. Bot. Staatssamml. München, **10**: 94, t. 3 (1971). — de Winter in F.S.A. **26**: 69 (1963). — Palmer & Pitman, Trees of Southern Afr. **3**: 180 cum tab. & photogr. (1972). — Wild, Biegel & Mavi, Rhod. Bot. Dict., ed. 2: 151 (1972). — Drummond in Kirkia, **10**: 267 (1975). — K. Coates Palgrave, Trees of Southern Afr.: 753, t. 247 (1977). TAB. **59**, fig. D. Type: Malawi, Mt. Mulanje, ♂ fl. x-xi.1891, *Whyte* s.n. (BM, holotype).

Royena whyteana Hiern in Trans. Linn. Soc. Bot., Sér. 2, **4**: 25 (1894). Type as above.
Royena lucida L., Sp. Pl.: 397 (1753) non *Diospyros lucida* Loud. (1841). — Steedman, Trees, etc. S. Rhod.: 64 (1933). Type from South Africa.
Royena lucida var. *whyteana* (Hiern) de Winter & Brenan in Mem. N.Y. Bot. Gard. **8**: 499 (1954). Type as for *D. whyteana*.

Evergreen shrub or tree up to 13 m. high. *Leaves* subcoriaceous, drying pale reddish-brown; lamina $3 \times 1 \cdot 4$–8×3 cm., lanceolate, lanceolate-elliptic or elliptic; apex broadly acute to subacuminate; base rounded to cordate; margin fringed with long hyaline hairs; lower surface sparsely hairy with long, weak, appressed hairs; lateral nerves in c. 6 pairs, venation on lower surface indistinct, but darker than the lamina. *Male flowers* solitary, axillary, pentamerous (rarely hexamerous); pedicel $0 \cdot 5$–2 cm. long; bracts 2, foliaceous, c. $0 \cdot 6 \times 0 \cdot 25$ cm., separated by an internode, ovate-lanceolate; indumentum of unicellular hairs and shorter, multicellular, glandular hairs. Calyx $0 \cdot 35$–$0 \cdot 75$ cm. long, urceolate, shortly toothed, indumentum appressed to spreading. Corolla $0 \cdot 5$–$1 \cdot 2$ cm. long, campanulate, minutely puberulous, lobed to about the middle. Disk fimbriate. Stamens 10, all fertile or some replaced by staminodes, $0 \cdot 4$–$0 \cdot 5$ cm. long; filaments $0 \cdot 05$–$0 \cdot 1$ cm. long, glabrous, anthers $0 \cdot 3$–$0 \cdot 45$ cm. long, lanceolate, setulose on the distal half of the connective, otherwise glabrous. Pistillode conoidal, tomentellous; locules 4 or 6; styles 2–3, hairy in the united lower half. *Female flowers* similar to male. Staminodes 6–10, $0 \cdot 12$ cm. long, glabrous or sparsely setulose at apex. Ovary $0 \cdot 2 \times 0 \cdot 15$ cm., conoidal; locules 4 or 6; styles 2 or 3. *Fruit* subglobose, up to 2 cm. diameter, glabrescent. Seeds 4 or fewer, $0 \cdot 7$ cm. long; endosperm smooth. Fruiting calyx accrescent, papery, inflated, bladder-like, completely concealing the fruit. *Chromosome number*: 2n = 30.

Zambia. N: Mbala Distr., D'hlumiti Kloof, imm. fr. 19.ii.1955, *Richards* 4580 (K). E: Lundazi Distr., Nyika Plateau, 9 km. S.W. of Rest House, imm. fr. 29.x.1958, *Robson* 355 (BM; FHO; K; SRGH). **Zimbabwe.** N: Sipolilo Distr., Nyamunyeche Estate, Great Dyke, st. 3.xi.1978, *Nyariri* 459 (SRGH). C: Selukwe Distr., Ferny Creek, fr. 8.xii.1953, *Wild* 4285 (BM; BR; FHO; P; SRGH). E: Umtali Distr., S.W. Vumba Mts., ♂ fl. 18.ix.1955, *Chase* 5799 (BM; BR; FHO; K; SRGH). S: Belingwe Distr., Mwembe stream, st. vii.1965, *West* 6697 (SRGH). **Malawi.** N: Karonga Distr., Misuku Hills, Matipa Forest, Khunguruwe Hill, ♂ fl. x.1954, *Chapman* 247 (FHO; BR; K). C: Ntchisi Distr., Nchisi Mt., fr. 31.vii.1946, *Brass* 17055 (BM; BR; K; SRGH). S: Mulanje Mt., Lukulezi Valley, ♂ fl. 2.x.1957, *Chapman* 452 (BM; FHO). **Mozambique.** N: Lichinga Distr., Massangulo Hills, 1400 m., fr. 25.ii.1964, *Torre & Paiva* 10820 (COI; FHO; K; LISC; LMU). Z: Gúruè hills, base of Pico Namúli, 1700 m., fr. 9.iv.1943, *Torre* 5126 (BR; C; EA; FHO; LISC; MO; SRGH; WAG). T: Angónia Distr., Dómuè Mt. 1800 m., fr. 16.x.1943, *Torre* 6047 (C; COI; FHO; LISC; M; MO; WAG). MS: Manica Prov., Tsetsera, Zuira Hills, 1980 m., ♂ fl. 11.xi.1965, *Torre & Pereira* 12888 (LISC).

From Tanzania to Cape Province, South Africa. Mostly in and at the edges of evergreen montane forest and scrub forest, sometimes (at lower altitudes) in bushland in rocky places, descending almost to sea-level in South Africa. 1220–2400 m.

16. **Diospyros villosa** (L.) de Winter in Bothalia, **7**: 458 (1961); in F.S.A. **26**: 78 (1963). — Palmer & Pitman, Trees of Southern Afr. **3**: 1811 cum tab. (1972). — K. Coates Palgrave, Trees of Southern Afr.: 753 (1977). TAB. **60**, fig. A. Type from South Africa.

Royena villosa L., Syst. Nat. **12** (2): 302 (1767). Type as above.

Erect or scandent shrub 1–4 m. high (elsewhere sometimes much taller) or a rhizomatous suffrutex c. $0 \cdot 5$ m. high. Leaves chartaceous, drying dull brown above, much paler beneath; lamina 3 × 1·5–6·5 × 3·5 cm., obovate-oblong, apex usually broadly rounded and slightly emarginate, sometimes obtuse, base usually cordate to rounded; margin slightly revolute, lateral nerves in c. 5 pairs, together with the reticulate tertiary nerves and veins deeply impressed above and prominent beneath; lower surface sparsely to densely pubescent with long, slender, appressed or spreading hairs, rarely tomentose. *Male flowers* solitary or in 2–3 flowered cymes, axillary, pentamerous; peduncle up to 1·5 cm. long; pedicel (including peduncle of solitary flowers) 0·5–2 cm. long; bracts 2, 0·4–0·8 cm. long, ovate to linear-oblanceolate, subopposite. Calyx up to 0·7 cm. long, cleft almost to the base; lobes ovate-lanceolate, fulvous-tomentose, the margins often reduplicate. Corolla up to 1·2 cm. long, deeply divided; lobes oblong, reflexed, broadly acute at apex, tomentose outside except at the base and where overlapped in bud. Disk indistinct. Stamens 10, 0·35–0·4 cm. long; filaments up to 0·05 cm. long, glabrous; anthers up to 0·35 cm. long, lanceolate, densely strigose-setose. Pistillode conoidal, fulvous-tomentose; locules 10; common style short, stout, densely puberulous, the 5 branches long and glabrous distally. *Female flowers* similar to the male but slightly smaller. Staminodes 10, 0·2 cm. long; filaments glabrous; antherodes densely strigose. Ovary 0·3 cm. long, ovoid-conoidal, 5-angled, densely strigose-tomentose, especially on the angles; locules 10, style short, densely pubescent, the 5 branches glabrous distally. *Fruit* up to 3 × 3·8 cm., depressed-globose, tardily dehiscent, densely hispid-tomentose with long, deciduous bristles. Seeds 3–8, up to 1·5 cm. long, dull brown, endosperm smooth. Fruiting calyx strongly accrescent, closely surrounding but not completely concealing the fruit except when young; lobes up to 3 × 2 cm., ovate, densely puberulous, with prominent longitudinal nerves. *Chromosome number*: 2n = 30.

Mozambique. GI: Gaza, near Chipenhe, Régulo Chiconela, floresta de Chirindzeni, fr. 13.x.1957, *Barbosa & Lemos* 8031 (FHO; K; LISC). M: Maputo Prov., between Marracuene and Manhiça, ♂ fl. 29.iii.1958, *Barbosa & Lemos* 8277 (K; LISC; SRGH).

Also in South Africa extending from northern Transvaal to southern Cape Province. Ecology in Mozambique poorly known. In woodland and grassland and at forest edges. From sea-level to 300 m.

The suffruticose specimens resemble the closely related South African spcies *D. galpinii* Hiern in habit but not in other respects. Similar variants occur on the coastal plain of northern Natal (de Winter in F.S.A. **26**: 78, 1963).

17. **Diospyros dichrophylla** (Gand.) de Winter in F.S.A. **26**: 75 (1963). — Palmer & Pitman, Trees of Southern Afr. 3: 1809 cum tab. & photogr. (1972). — K. Coates Palgrave, Trees of Southern Afr.: 745 (1977). TAB. **60**, fig. B. Type from South Africa.
    Royena dichrophylla Gand. in Bull. Soc. Bot. Fr. **65**: 56 (1918). Type as above.

Erect, densely leafy evergreen shrub up to 6 m. tall, elsewhere sometimes arborescent and up to 16 m.; branches erect or ascending; branchlets shortly and densely yellowish-pubescent. *Leaves* coriaceous, drying dull pale brown or yellowish-brown beneath; lamina 1·7 × 0·6–8·8 × 2·5 cm., narrowly obovate to oblanceolate, apex rounded to broadly acute, base cuneate, margin usually revolute; lower surface appressed-pubescent; lateral nerves in c. 5 pairs, indistinct. *Male flowers* solitary, axillary, pentamerous; pedicel 0·8–2 (2·8) cm. long; bracts 0·25–0·4 cm. long, linear or oblanceolate, distant, deciduous. Calyx 0·2–0·4 cm. long strigulose-tomentellous; lobes 0·15–0·3 cm. long, deltate. Corolla 0·7–1·2 cm. long, densely strigulose outside; lobes 0·5–1 cm. long. Disk fimbriate. Stamens 10, 0·4–0·45 cm. long; filaments 0·05–0·07 cm. long, hairy towards apex; anthers 0·3–0·35 cm. long, densely strigose. Pistillode subglobose, sericeous-tomentellous, slightly ridged; locules 10; stylode puberulous, divided into 5 branches. *Female flowers* similar to male. Staminodes 10, 0·2 cm. long; antherodes densely strigose. Ovary 0·25 × 0·2 cm., subglobose, slightly ridged; densely sericeous-tomentellous; locules 10; style puberulous, the 5 branches glabrous towards the tips. *Fruit* up to 2 × 2·5 cm., depressed-globose, densely yellowish-velvety-pubescent, tardily dehiscing from above into 5 valves. Seeds 3–8, 1·3 cm. long, shining-brown; endosperm smooth. Fruiting calyx strongly accrescent, lobes up to 2 cm. long, with many longitudinal nerves, usually strongly reflexed but occasionally only partly reflexed. *Chromosome number*: 2n = 30.

Mozambique. M: Namaacha, fr. 20.i.1958, *Barbosa & Lemos* 8243 (FHO; LISC).

From Mozambique to Cape Province, South Africa. In semi-deciduous bushland and scrub forest. 50–200 m.

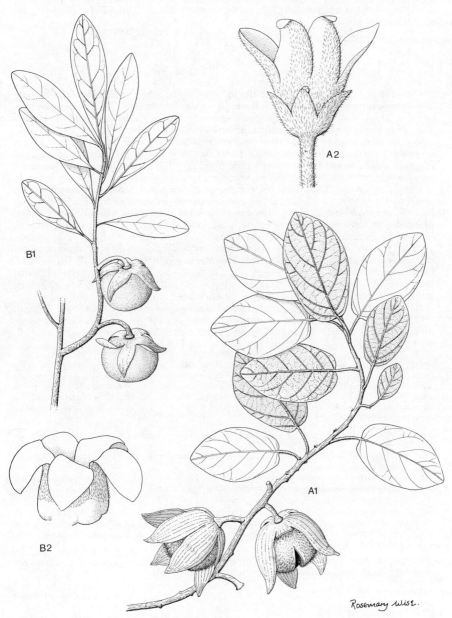

Rosemary Wise.

Tab. 60. DIOSPYROS sect. ROYENA (pro parte). A. — D. VILLOSA. A1, fruiting branchlet (× ½)
White 10323; A2, female flower (× 4) *Pegler* 468. B. — D. DICHROPHYLLA. B1, fruiting
branchlet (× ½) *Toms* 58; B2, male corolla (× 4) *Wells* 2049.

18. **Diospyros lycioides** Desf. in Ann. Mus. Par. **6**: 448, t. 62 fig. 1 (1805). — de Winter in
F.S.A. **26**: 7 (1963). — Wild in Kirkia, **7**: 34 (1968). — Palmer & Pitman, Trees of
Southern Afr. **3**: 1805, cum tab. and photogr. (1972). — Drummond in Kirkia, **10**: 267
(1975). — K. Coates Palgrave, Trees of Southern Afr.: 747, t. 244 (1977). — F. White &
Vosa in Bol. Soc. Brot., Sér. 2, **53**: 276, 285, tab. 4 (1980). TAB. **61**. Type from a plant
cultivated in Paris.
  Royena pallens sensu Hiern in Trans. Camb. Phil. Soc. **12**: 85 (1873) pro parte. — S.
Moore in Journ. Linn. Soc., Bot. **40**: 134 (1911). — R.E. Fr., Wiss. Ergebn. Schwed.
Rhod. -Kongo -Exped. **1**: 256 (1916). — Steedman, Trees, etc. S. Rhod.: 64 (1933). — O.
B. Mill, in Journ. S. Afr. Bot. **18**: 68 (1952).
  Diospyros pallens sensu F. White, F.F.N.R.: 331, t. 59 fig. e (1962).

Deciduous shrub or small tree up to 6 m. high, sometimes suffruticose. Bole
sometimes spinescent towards the base. Bark grey, ± smooth. Leaves chartaceous,
drying dull dark brown or grey-green above, paler beneath; lamina
1·5 × 0·7–11 × 2·5 cm., obovate to oblanceolate, apex rounded to acute, base cuneate;
lower surface sparsely to densely sericeous-pubescent, especially on the nerves;
secondary nerves in 5–6 pairs. *Male flowers* solitary, axillary or in axils of reduced
leaves towards base of current year's shoot; pedicels 0·7–1·7 cm. long. Calyx up to 0·8
cm. long, densely sericeous-pubescent, deeply cleft almost to the base; lobes 5,
narrowly deltate or lanceolate-acuminate. Corolla up to 1 cm. long, campanulate,
widely open at the throat, densely strigulose outside, lobed to just below the middle;
lobes 5, ovate-oblong, obtuse. Stamens 10, 0·3–0·45 cm. long; filaments very short,
glabrous; anthers narrowly lanceolate, apiculate, densely setose with long hairs at
base of connective and shorter hairs along its length on both surfaces. Pistillode
similar to functional gynoecium but rudimentary ovary conoidal and styles not bifid
and stigmatic at the apex. *Female flowers* similar to male. Staminodes 10, 0·1–0·2 cm.
long, densely setose. Ovary subglobose, ridged, 0·25 × 0·25 cm., tomentellous;
locules 6, 8 or 10; styles (3–4)5, common style puberulous, branches glabrous,
ending in a shallowly bi-lobed stigma. *Fruit* red, becoming black, ovoid or globose,
apiculate, up to 2 × 1·5 cm., puberulous or glabrescent. Seeds 1–6 or more, brown,
up to 1·3 cm. long; endosperm smooth. Fruiting calyx accrescent, up to 1·5 cm. long,
lobes narrowly deltate, ultimately strongly reflexed. *Chromosome numbers*: 2n = 30,
60.

D. lycioides is one of the most widely distributed woody species in southern Africa, extending
from Upper Shaba and Mbala District, Zambia southwards to the Cape Province, South
Africa. Three of its four subspecies, which show ecogeographical replacement, occur in the
Flora Zambesiaca area. Intermediates between subsp. *lycioides* and subsp. *sericea* are found in a
narrow zone of contact. In Zimbabwe, *D. lycioides* tolerates relatively low concentrations of
copper in the soil.

Mature leaves with nerves not raised, both surfaces usually glabrous or sparsely appressed-
  hairy - - - - - - - - - - - - - subsp. *lycioides*
Mature leaves with secondary nerves and main veins at least slightly raised on lower surface,
  usually densely hairy:
  Secondary nerves and main veins prominently raised - - - - subsp. *guerkei*
  Secondary nerves only prominently raised - - - - - subsp. *sericea*

Subsp. **guerkei** (Kuntze) de Winter in Bothalia, **7**: 458 (1961); in F.S.A. **26**: 75 (1963). — F.
White & Vosa in Bol. Soc. Brot., Sér. 2, **53**: 283, 291, tab. 4 fig C (1980). Type from Natal.
  Royena guerkei Kuntze, Rev. Gen. **3** (2): 196 (1898). — de Winter in Fl. Pl. Afr. tab.
1262 (1958). Type as above.

Shrub 1 m. high (elsewhere up to 5 m.). Branchlets ascending, not spinescent. Leaves not
crowded. *Chromosome number*: 2n = 30.

Botswana. SE: Mannyelanong Hill, 32 km. S. of Gaborone, 1465 m., fr. 9.ii.1975, *Mott* 654
(SRGH).
  Also in South Africa.

Subsp. **lycioides** — de Winter in F.S.A. **26**: 73 (1963). — de Winter & F. White, Prodr. Fl.
S.W. Afr. **107**: 3 (1967). — Drummond in Kirkia, **10**: 267 (1975). — F. White & Vosa in
Bol. Soc. Bot., Sér. 2, **53**: 282, 291, tab. 4 fig. a (1980). TAB. **61**, fig. A.

Shrub or small tree up to 6 m. high. Branchlets spreading at right-angles or
slightly ascending at the ends, occasionally spinescent, mostly naked towards the
base with the leaves crowded towards the tips. *Chromosome numbers*: 2n = 30, 60.

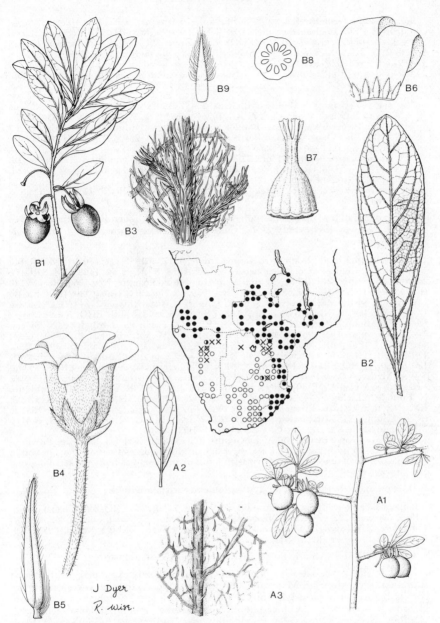

Tab. 61. Pictorialized distribution map of DIOSPYROS LYCIOIDES (sect. Royena). A. — Subsp. LYCIOIDES. A1, fruiting branchlet (× ½) *White* 10089; A2, leaf (× 1½) *Esterhuysen* 2430; A3, lower leaf surface (× 12) *Thode* A1439. B. — Subsp. SERICEA. B1, fruiting branchlet (× ½) *Mendonça* 309; B2, leaf (× 1½) *Edwards* 1563; B3, lower leaf surface (× 12) *Adlard* 146; B4, male flower (× 4) *Greenway* 8815; B5, stamen (× 10) *Greenway* 8815; B6; part of female corolla and staminodes (× 4), *Lawton* 647; B7, ovary (× 8) *Lawton* 647; B8, transverse section of ovary (× 8) *Lawton* 647; B9, staminode (× 10) *Lawton* 647.

Hollow circles indicate subsp. *lycioides,* solid circles subsp. *sericea,* and half solid circles degree squares in which both subspecies occur. Crosses represent intermediates between the two subspecies.

Botswana. N: Ngamiland, 6 km. S. of Shakawe, 1000 m., fr. 24.iv.1975, *Müller & Biegel* 2261 (SRGH). SW: Bohelabatho pan, fl. 1.xi.1978, *Skarpe* 293 (SRGH). SE: Mochudi, Marico R., ♂ fl. x.1946, *O. B. Miller* B/467 (FHO). **Zimbabwe.** W: Kezi to Sun Yat Sen 38 km., Kaholi R. fr. 31.i.1973, *White* 10079 (FHO; SRGH). S: Shashi R. near Lindi Rest Camp, fr. 1.ii.1973, *White* 10089 (FHO; SRGH).

Also widespread in the drier parts of Namibia and South Africa. In riparian forest and thicket, 600–1000 m.

Subsp. **sericea** (Bernh.) de Winter in Bothalia, **7**: 457 (1961); in F.S.A. **26**: 74 (1963). — de Winter & F. White, Prodr. Fl. S.W. Afr. **107**: 4 (1967). — F. White, F.F.N.R.: 435 (1962). — Wild, Biegel & Mavi, Rhod. Bot. Dict.: 150 (1972). — Drummond in Kirkia, **10**: 267 (1975). — F. White & Vosa in Bol. Soc. Brot, Sér. 2, **53**: 284, 292, tab. 4 fig. b (1980). TAB. **61**, fig. B. Type from Natal.

Royena sericea Bernh. in Flora, **27**: 824 (1844). — O. B. Mill. in Journ. S. Afr. Bot. **18**: 68 (1952). — Wild, Guide Fl. Vict. Falls: 152 (1953). — Brenan in Mem. N.Y. Bot. Gard. **8**: 499 (1954). Type as above.

Royena lycioides subsp. *sericea* (Bernh.) de Winter in Bothalia, **7**: 17 (1958). Type as above.

Shrub or small tree up to 6 m. high. Branchlets usually ascending, with the leaves inserted more or less evenly along their length, rarely spinescent. *Chromosome numbers*: 2n = 30, 60.

Caprivi Strip. Lisikili, 24 km. E. of Katima Mulilo, fl. 17.vii.1952, *Codd* 7039 (BM; BR; K). **Botswana.** N: Chobe-Zambezi confluence, fr. 11.iv.1955, *E. M. & W.* 1469 (BM; SRGH). **Zambia.** B: South Kashishi R., 34 km. W. of Balovale Pontoon, fr. 26.v.1960, *Angus* 2280 (FHO; SRGH). N: 48 km. N. of Kasama, fr. vii.1930, *Hutchinson & Gillett* 3809 (BM, K). W: Mwinilunga Distr., Matonchi Farm, ♂ fl., fr. 16.x.1937, *Milne-Redhead* 2814 (K). C: Lusaka Distr., Mt. Makulu Research Station, ♀ fl. 7.xi.1956, *Angus* 1438 (BR; FHO; K). S: Choma Distr., Siamambo F.R., fr. 12.xii.1952, *Angus* 925 (BR; FHO; K). **Zimbabwe.** N: Sinoia, Dichwe Forest, fr. 16.ii.1965, *West* 6344 (FHO; SRGH). W: Victoria Falls, fl. ix.1905, *Gibbs* 112 (BM). C: Salisbury, ♂ fl. 25.x.1936, *Eyles* 8813 (BR; K). E: Chirinda, fr. iii.1950, *Hack* 179/50 (SRGH). S: Fort Victoria to Beit Bridge, 2 km. S. of Lundi Bridge, ♂ fl. 4.xii.1961, *Leach* 11287 (FHO; SRGH). **Malawi.** N: Nkhata Bay, Vipya Plateau, Elephant Rock, 56 km. S.W. of Mzuzu, fr. 15.v.1971, *Pawek* 4800 (K). C: Dedza, ♂ fl. 30. ix.1954, *Greenway* 8815 (EA; FHO; K). S: Mulanje, ♂ fl. 30.ix.1946, *Brass* 17882 (BM; BR; EA; K; SRGH). **Mozambique.** N: Litunde to Lichinga (Vila Cabral), ♀ fl. 10.x.1942, *Mendonça* 713 (COI; FHO; K; LISC; LMU). Z: montes de Ile, Errego, imm. fr. iii.1966, *Torre & Correia* 14979 (LISC). T: Calóbue to Vila Coutinho, 1550 m., fr. 5.iii.1964, *Torre & Paiva* 11049 (C; FHO; LISC; MO; WAG). MS: Espungabera, ♀ fl., fr. 31.x.1944, *Mendonça* 2701 (LISC).

From Zaire (Upper Shaba) and Angola southwards to Namibia and eastern Cape Province, South Africa. In bushland and thicket, especially on riverbanks and termite mounds and in rocky places. Sometimes at edges of riparian forest or forming secondary thickets following over-grazing. 600–1525 m.

Intermediates between subsp. **lycioides** and subsp. **sericea**.

Botswana. N: Lake Ngami, ♂ fl. ix.1896, *F. D. & E. J. Lugard* 19 (K). SW: Ghanzi Pan, fl. x.1969, *Brown* 6733 (SRGH). **Zimbabwe.** W: Matopos Research Station, fl. ix.1927, *Darbyshire* 2376 (SRGH). S: Gwanda Distr., Mjingwe Weir, Liebig's Ranch, 685 m., fl. x.1963, *Norris-Rogers* 142 (SRGH).

6. Sect. ERIKESI F. White in Bull. Jard. Bot. Nat. Belg. **50**: 456 (1980).

Male flowers subsessile, 3–5 together at ends of distally expanded and flattened supra-axillary peduncles. Male calyx botuliform, papery, completely concealing the corolla before anthesis. Male corolla 1·4–2·2 cm. long, 3–5-lobed to $\frac{1}{3}$ or more, glabrous. Stamens c. 20, inserted on corolla-tube, included. Ovary conoidal, glabrous. Locules 8, uniovulate. Style undivided. Stigmata 4, fleshy, sessile. Fruit 2–3·5 cm. in diameter. Seeds 8 or fewer. Endosperm shallowly ruminate. Fruiting calyx markedly accrescent, cyathiform, truncate.

19. **Diospyros senensis** Klotzsch in Peters Reise Mossamb. Bot.: 183 (1861). — Hiern in Trans. Camb. Phil. Soc. **12**: 181 (1873) pro parte. — Sim, For. Fl. Port. E. Afr.: 83 (1909. — S. Moore in Journ. Linn. Soc., Bot. **40**: 134 (1911). — F. White in Bull. Jard. Bot. Etat Brux. **27**: 522, t. 54 fig. g-1 (1957); F.F.N.R.: 326, t. 58 fig. c (1962). — Drummond in Kirkia, **10**: 267 (1975). — K. Coates Palgrave, Trees of Southern Afr.: 751 (1977). TAB. **62**, fig. A. Types: Mozambique, Sena, ♂ fl., *Peters* s.n. (B†, holotype); Chioco to Tete 6 km., ♀ fl. ii.1968, *Torre & Correia* 17691 (LISC, neotype, here designated).

Diospyros shirensis Hiern in Journ. Bot. Lond. **33**: 179 (1895). — S. Moore loc. cit.

Syntypes: Malawi, Mangoche Flats, ♂ fl. xii.1894, *Scott Elliot* 8433 (BM; K); Ruo R., ♂ fl. xii.1894, *Scott Elliot* 8674 & 8681 (BM; K).

Shrub, or small tree 2–9 m. high with one or many stems; bole sometimes fluted at base; older branchlets and branches with large, sharp-pointed spines. Bark grey, smooth and peeling to reveal cream underlayer, or rough and blackish (? fire damage). Branchlets densely grey- or fulvous-pubescent or tomentose with spreading flexuose hairs. *Leaves* chartaceous, drying pale grey-green above and grey beneath with conspicuous yellowish nerves and veins; lamina 4 × 2–12 × 5 (16·3 × 9) cm., obovate or narrowly obovate, apex mostly rounded or obtuse, sometimes subacute or very shortly and bluntly subacuminate, base cunate, rarely rounded to subcordate; lower surface pubescent with spreading flexuose hairs; lateral nerves in 5–9 widely spaced pairs; tertiary nerves and veins forming a lax reticulum. *Male flowers* at ends of 0·2–0·7 cm. long, very hairy peduncles; bracts usually inconspicuous, 0·25 × 0·1 cm., very rarely foliaceous. Calyx 0·4–0·7 cm. long, botuliform, splitting at apex into 2–4 short teeth, densely pubescent. Corolla c. 1·3 cm. long, rotate, almost glabrous; tube 0·6–0·7 cm. long; lobes 4, ovate, acute and apiculate. Stamens c. 20; anthers apiculate, glabrous. Pistillode absent. *Female flowers* solitary or in 2–3-flowered cymules; peduncle densely hairy, 0·3–0·6 cm. long, upper part slightly swollen and greatly increasing in thickness as fruit develops. Calyx 0·45 cm. long, unlobed, urceolate in flower, becoming cyathiform in fruit, coriaceous and rugose in lower half, membranaceous distally, densely velutinous-tomentose outside especially towards the base, densely hairy within. Corolla similar to male. Staminodes absent. Ovary 0·3 × 0·2 cm., glabrous, style 0·2 cm. long, stout, simple, hairy in upper half, obscurely lobed at apex. *Fruit* c. 2·2 × 2 cm., ellipsoid to subglobose, glabrous except at base of persistent style. Seeds 8 or fewer.

Zambia. C: tributary or Mwomboshi R, where it crosses Lusaka-Kabwe road, ♀ fl. 6.xii.1957, *Fanshawe* 4118 (FHO). E: Petauke, Old Boma Road, 800 m., ♂ fl. 5.xii.1958, *Robson* 851 (SRGH). S: Gwembe Valley, 10 km. E. of Sinazeze, fr. 17.vi.1961, *Angus* 2922 (FHO; SRGH). **Zimbabwe.** N: Kessesse R, near Kariba, fr. i.1956, *Goodier* 26 (K; SRGH). W: Wankie, fr. vii.1954, *Orpen* 45/54 (SRGH). C: Gatooma Distr., Umniati R., 48 km. W. of Gatooma, 915 m., st. 6.iv.1969, *Burrows* 309 (SRGH). E: Inyanga Distr., Lawley's Concession, fr. 19.ii.1954, *West* 3370 (SRGH). **Malawi.** C: Dowa Distr., Lake Nyasa Hotel, fr. 3.viii.1951. *Chase* 3866 (BR; FHO; SRGH). S: Lengwe Game Reserve, fr. 18.viii.1970, *Hall-Martin* 1510 (FHO). **Mozambique.** N: Chalaua to Moma 16 km., ♂ fl. 20.i.1968, *Torre & Correia* 17282 (C; COI; FHO; K; LISC; LMU; WAG). Z: Mocuba to Maganja da Costa 22 km., 200 m., fr. 8.xi.1966, *Torre & Correia* 14474 (BR; EA; FHO; LISC; M; P; SRGH). T: Chioco to Tete 6 km., 250 m., ♀ fl. 16.ii.1968, *Torre & Correia* 17691 (C; FHO; LISC; MO; PRE; WAG). MS; Manica Distr., Dombe to Chuaca 16 km., ♂ fl. x.1953, *Pedro* 4422 (LMA).

Also in Zaire. In deciduous woodland and thicket. 30–1000 m.

The tough wood is used for hoe-handles and building.

7. Sect. TABONACA F. White in Bull. Jard. Bot. Nat. Belg. **50**: 457 (1980).

Leaves usually hairy. Male calyx cyathiform, shallowly lobed to deeply cleft, usually much shorter than the corolla-tube. Male corolla 1·3–3·5 cm. long, hypocrateriform, usually hairy, lobed to ⅓ or more. Stamens 12–30, inserted on corolla-tube or receptacle, included. Female, calyx not completely concealing corolla-tube. Ovary hairy. Locules 6–10, uniovulate. Styles 3–5, free, usually at least one third as long as ovary. Fruit 2·5 × 3–5 × 5 cm., subglobose or subcylindric, hairy. Seeds 10 or fewer. Fruiting calyx accrescent or not, rarely completely concealing the developing fruit.

Calyx lobed almost to the base, lobes strongly transversely undulate in fruit
                                                                20. *pseudomespilus*
Calyx lobed only in upper half, lobes not undulate:
  Leaf-lamina with tertiary nerves and veins forming a close reticulum prominent on both
    surfaces; fruiting calyx concealing the lower half to two thirds of the fruit
                                                                21. *gabunensis*
  Leaf-lamina with invisible or scarcely visible tertiary nerves and veins; fruiting calyx only
    concealing base of fruit:
    Small cauliflorous tree; male calyx 0·6 cm. long, blackish-tomentose; female flowers
      with inconspicuous bracts   -   -   -   -   -   -   -   22. *batocana*
    Rhizomatous geoxylic suffrutex with mostly axillary flowers; male calyx 1·3 cm. long,
      fulvous-tomentose; female flowers with conspicuous bracts   -   23. *chamaethamnus*

20. **Diospyros pseudomespilus** Mildbr. in Notizbl. Bot. Gart. Berl. **9**: 1050 (1926). — Letouzey & F. White in Fl. Cam. **11** & Fl. Gab. **18**: 140, t. 22 (1970). — F. White in Bull. Jard. Bot. Nat. Belg. **48**: 311, t. 7 (1978); in Distr. Pl. Afr. **14**: 479–482 (1978). Type from Cameroon.

   Diospyros undabunda auct. non Hiern ex Greves in Journ. Bot. Lond. **67** (Suppl.): 80 (1921) sens. strict. — F. White, F.F.N.R.: 328, t. 58 fig. 5 (1962); in F.W.T.A., ed. 2, **2**: 12 (1963); in Nig. Trees, **2**: 332 (1964). Types from Angola.

Tree, but sometimes flowering as a shrub 4 m. high. Bole slender. Bark dark grey, finely and closely furrowed. *Leaves* chartaceous to subcoriaceous, drying dull red-brown above, paler beneath; lamina 7 × 2·5–26 × 9 cm., oblong-elliptic or oblanceolate-elliptic, apex shortly and bluntly acuminate, base rounded to subcordate; lower surface with a variable indumentum; secondary nerves in 7–10 pairs, prominent beneath, arcuate, anastomosing close to margin, tertiary nerves and veins forming a lax reticulum, impressed or not impressed above, prominent beneath. *Male flowers* axillary and ramuligerous, in 3–5-flowered congested cymes; peduncle ± 0·1 cm. long; pedicels up to 0·2 cm. long. Calyx 0·7–0·8 cm. long, fulvous-tomentose on both surfaces, lobed almost to the base; lobes (4)5, narrowly deltate. Corolla 1·5 cm. long, sparsely strigulose outside, more densely so on the mid-petaline lines; tube 0·7 cm. long, only slightly thickened at the throat; lobes 5, broadly elliptic. Stamens (16) 20–25, up to 0·5 cm. long; filaments less than 0·1 cm. long, setulose, inserted on the receptacle; anthers lanceolate, scarcely apiculate, setulose. Pistillode 0·2 cm. long, setulose. *Female flowers* solitary or 2–3 together, borne as in male; pedicels stout, c. 0·1 cm. long. Calyx as in male but more coriaceous and lobes with markedly undulate margins. Corolla slightly longer than the male. Staminodes (10) 12–16, glabrous except for the setulose apex and base of the antherode, 0·5 cm. long, inserted on corolla-tube. Ovary 0·3 × 0·2 cm., ovoid, densely setose; locules (6) 8 or 10, styles (3) 4 or 5, united in lower half, each ending in a 2-lipped, fleshy stigma. *Fruit* up to 4 cm. in diameter, bright red, looking like a tomato, depressed-globose, glabrescent except near the style and towards the base. Seeds (4) 8 or 10. Fruiting calyx 2–4·2 cm. long; lobed nearly to the base, markedly accrescent but only concealing the fruit when very young; lobes narrowly deltate, tomentose inside, variously hairy outside, margin strongly transversely undulate.

Widespread in the rain forests of the Guineo-Congolian Region from southern Nigeria to eastern Zaire. The three subspecies recognized by White (loc. cit.) differ chiefly in the size and indumentum of their leaves and fruiting calyces and in leaf venation. Subsp. *pseudomespilus* and subsp. *undabunda* are confined to the Guineo-Congolian Region whereas subsp. *brevicalyx*, which is an ecological and chorological transgressor, extends south into the northern half of the Zambezian Region.

Subsp. **brevicalyx** F. White in Bull. Jard. Bot. Nat. Belg. **48**: 314, t. 7 fig. g-1 (1978); in Distrib. Pl. Afr. **14**: 481 (1978). TAB. **62**, fig. B. Type: Zambia, Mwinilunga Distr., 6 km. N. of Chikundulu Stream, fr. 4.x.1952, *White* 3448 (BM; BR; FHO; K, holotype).
   Diospyros undabunda sensu F. White, F.F.N.R.: 328, t. 58 fig. 5 (1962).

Small tree up to 10 m. tall (elsewhere said to reach 35 m.). Leaves coriaceous, 9·8 × 3·4–14 × 5 cm.; midrib densely puberulous beneath, hairs up to 0·1 cm. long, scarcely visible to naked eye; intercostal nerves and veins not or scarcely impressed above, subprominent beneath. Fruiting calyx 2–2·8 cm. long, persistently tomentose outside at least along the median axis.

Zambia: B: Balovale, fr. vii.1933, *Trapnell* 1295 (K). W: Mwinilunga to Kabombo 48 km., ♂ fl. 3.x.1952, *White* 3439 (BM; BR; FHO; K)
Extending southwards from the Central African Republic to Angola and Zambia. In the Flora Zambesiaca area almost confined to Kalahari Sand where it is one of the most characteristic species of dry evergreen forest (mavunda) dominated by *Cryptosepalum pseudotaxus*. 1000–1450 m.

21. **Diospyros gabunensis** Gürke in Engl., Bot. Jahrb., **26**: 72 (1898). — F. White in F.W.T.A., ed. 2, **2**: 12 (1963); in Bull. Jard. Bot. Nat. Belg. **48**: 317 (1978); in Dist. Pl. Afr. **14**:457 (1978). — Letouzey & F. White in Fl. Cam. **11** & Fl. Gab. **18**: 77, t. 9 (1970). TAB. **62**, fig. C. Type from Gabon.
   Diospyros sp. 1, F. White, F.F.N.R.: 435 (1962).

Tree (6) 10–20 m. high. Bole slender. Bark black, hard and brittle, usually smooth. Young branchlets tomentellous with black or dark brown, spreading hairs; older branchlets smooth and black. *Leaves* subcoriaceous, usually drying greenish-brown,

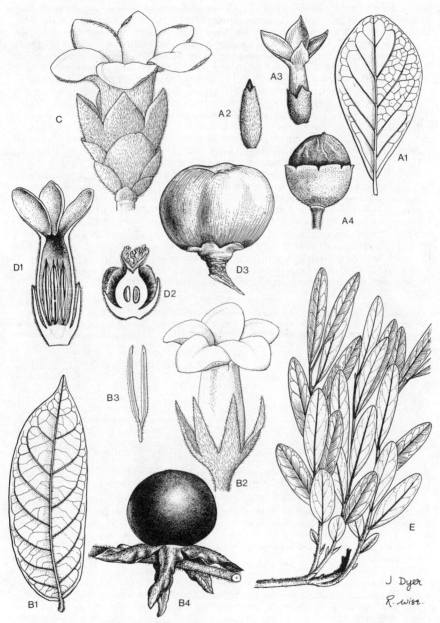

Tab. 62. DIOSPYROS sect. ERIKESI and sect. TABONACA. A. — D. SENENSIS. A1, leaf (× ½); A2, male flower bud (× 2); A3, male flower (× 2); A4, fruit (× 1). B. — D. PSEUDOMESPILUS Subsp. BREVICALYX. B1, leaf (× ½); B2 male flower (× 4) *Angus* 602; B3 stamens (× 10) *Angus* 602; B4, fruit (× ¾). C. — D. GABUNENSIS, female flower (× 2½) *Le Testu* 7759. D. — D. BATOCANA. D1, male flower, longitudinal section (× 3); D2, ovary, longitudinal section (× 3); D3, fruit (× ¾). E. — D. CHAMAETHAMNUS, habit (× ½). A, B1, B4, D & E from F.F.N.R.

shining on both surfaces; lamina c. 18 × 6·5 cm., very variable in shape and size, lanceolate, elliptic, oblanceolate-elliptic or oblanceolate, apex acuminate, base rounded or cuneate, slightly decurrent, margin often revolute, especially at the base; acumen 1·8–2·5 × c. 0·5–0·8 cm. at the base, deltate, apex rounded; lower surface strigulose-puberulous on the nerves, sparsely so on the lamina; lateral nerves in 8–10 pairs, prominent beneath, anastomosing in loops near the margin; intermediate nerves, tertiary nerves and veins forming a close reticulum prominent on both surfaces. *Male flowers* subsessile in 1–12-flowered fascicles in the leaf-axils and on the older branchlets but not on trunk. Calyx 0·7–0·8 cm. long, cyathiform, tomentose outside with chocolate-brown hairs, tube 0·5–0·6 cm. long, lobes 5(6), deltate. Corolla 1·8–2·2 cm. long, fulvous-tomentellous outside; tube up to 1·2 × 0·5 cm., cylindric, not constricted at the throat; lobes (4) 5 (6), ovate-oblong, obtuse. Stamens 20–30, c. 1 cm. long, filaments 0·2–0·3 cm. long, completely fused to base of corolla-tube; anthers linear, slightly apiculate, densely setulose for whole length. Pistillode 0·1 × 0·2 cm., depressed-globose, glabrous or bearing a tuft of hairs. *Female flowers* subsessile, solitary or 2–5 together in leaf-axils or on older branchlets, surrounded at the base by several bracts up to 1 cm. long, similar to the male, but larger and calyx sometimes more deeply divided. Staminodes 5–10, as long as the corolla-tube, similar to the stamens but more slender. Ovary 0·5–0·6 × 0·3–0·4 cm., ovoid-conic, brown-tomentose; styles 4–5, 0·2 cm. long, united at the base, each ending in a bilobed, fleshy stigma; locules 8 or 10, uniovulate, but sometimes imperfectly separated near apex of ovary. *Fruit* c. 2·5(3) cm. in diameter, subglobose or broadly ellipsoid, with persistent styles, brown- or black-tomentose in patches; seeds 8 or 10. Fruiting calyx accrescent, persistently fulvous-brown- or black-tomentose, cup-shaped and concealing the lower half to two thirds of the fruit; lobes 5, 0·4–2·2 cm. long.

Zambia. N: Mbala Distr., Lunzua R., below falls, fr. 2.iv.1960, *Fanshawe* 5626 (FHO).
Widespread in the rain forests of the Guineo-Congolian Region from Sierra Leone to Rwanda, and extending south along river valleys to northern Zambia, where it is only known from the locality cited. 1220–1525 m.

22. **Diospyros batocana** Hiern in Trans. Camb. Phil. Soc. **12**: 174 (1873). — O. B. Mill. in Journ. S. Afr. Bot. **18**: 67 (1952). — Wild, Guide Fl. Vict. Falls: 152 (1953). — Pardy in Rhod. Agric. Journ. **53**: cum photogr. (1956). — F. White, F.F.N.R.: 330, t. 57 fig. a-b (1962). — Drummond in Kirkia, **10**: 266 (1975). — K. Coates Palgrave: 744 (1977). TAB. **62** fig. D. Type: Zambia, "Batoka country", ♂ fl. vii.1860, *Kirk* s.n. (K, holotype).
   Diospyros xanthocarpa Gürke in Warb., Kunene-Samb. Exped. Baum: 328 (1903). — R.E. Fr., Wiss. Ergebn. Schwed. Rhod.-Kongo-Exped. **1**: 257 (1916). Type from Angola.
   Diospyros odorata Hiern ex Greves in Journ. Bot. Lond. **67** Suppl. 2: 78 (1929). Types from Angola.
   Diospyros odorata var. *rhodesiana* Rendle in Journ. Bot. Lond. **70**: 93 (1932). Type: Zimbabwe, Victoria Falls, fl. 7.viii.1929, *Rendle* 318 (BM, holotype).

Small, evergreen sclerophyllous tree 3–8(12) m. high, very rarely flowering as a shrub less than 1 m. high. Crown dense, rounded. Bark blackish, rough, deeply fissured. *Leaves* rigid, coriaceous, glossy above, pale glaucous-green beneath; lamina 6 × 2–12 × 4 cm., usually narrowly to broadly elliptic and rounded or subacute at both ends, but sometimes lanceolate or lanceolate-elliptic and acute at both ends; lower surface glabrous except for a few strigulose hairs; lateral nerves in about 7 pairs, indistinct above, subprominent beneath; venation indistinct. *Male flowers* rarely axillary, usually in subsessile clusters on the older branchlets and on branches up to 10 cm. in diameter, the persistent axes forming prominent coralloid bosses. Calyx 0·6 cm. long, cyathiform, with 4–6 very short, broadly deltate or rounded lobes, strigulose-tomentellous outside with blackish, upward-directed hairs. Corolla c. 1·4 cm. long, hypocrateriform, white and waxy inside, strigulose-tomentellous outside with blackish, downward-directed hairs; tube c. 1 cm. long, broadest at the middle, thickened and constricted inside at the throat; lobes 4–6, 0·4 × 0·3 cm., suborbicular. Stamens 15–17, included, inserted on the receptacle, 0·6 cm. long; filaments c. 0·1 cm. long; anthers linear-apiculate, glabrous. Pistillode represented by a tuft of glandular hairs. *Female flowers* similar to the male in structure and position, but calyx more deeply lobed and corolla with broader tube and shorter lobes. Bracts inconspicuous, up to 0·3 cm. long. Staminodes 6–8, densely setulose, attached to base of corolla-tube. Ovary 0·5 × 0·5 cm., subglobose, merging into the short, stout style, strigulose-tomentellous; locules 6, styles 3, short; stigmatic lobes, large, fleshy, deeply and irregularly lobed. *Fruit* up to 4 cm. in diameter, obovoid or oblate, orange, tomentose when young with chocolate-brown or rufous appressed hairs, glabrescent in patches. Seeds 5–6, c. $2 \times 1 \times 0.8$ cm., endosperm smooth. Fruiting calyx 0·5–0·6 cm. long, scarcely accrescent, patelliform.

Botswana. N: Lesuma Valley, Chobe, fr. ix.-x. *O. B. Miller* B/480 (FHO). **Zambia.** B: Kalabo Distr., Kalabo to Sihole 8 km., fr. 16.ii.1952, *White* 2088 (FHO; K). N: L. Bangweulu, Samfya, fr. 21.viii.1952, *Angus* 270 (BM; BR; BRLU; FHO; K). W: Solwezi, ♂ fl. 15.vi.1960, *Milne-Redhead* 515 (BR; K). C: Mpika Distr., 3 km. W. of Kateti-Luangwa confluence, ♂ fl. 4.v.1965, Mitchell 2820 (SRGH). S: Namwala, fr. 11.vi.1949, *Hornby* 3011 (FHO; SRGH). **Zimbabwe.** W: Victoria Falls Hotel, ♂ fl. 1.ix.1955, *Chase* 5755 (BM; FHO; SRGH).

Also in Angola and Zaire. Widespread in the northern, wetter half of the Zambezian Region, but extending into the drier southern half only on Kalahari Sand. In various types of woodland. 900–1525 m.

The relationships of *D. batocana* and *D. chamaethamnus* are discussed under the latter species.

The specific epithet implies that Kirk collected the type-specimen in the Batoka Highlands in the Southern Province of Zambia, but *D. batocana* is not known to grow there. It is much more likely to have been collected at or near Sesheke (White in Comptes Rendus IVe Réunion AETFAT: 179, 1962).

23. **Diospyros chamaethamnus** Dinter ex Mildbr. in Notizbl. Bot. Gart. Berl. **15**: 757 (1942). — F. White, F.F.N.R.: 330, t. 58, fig. j (1962). — de Winter in F.S.A. **26**: 60 (1963). — de Winter & F. White, Prod. Fl. SW. Afr. **107**: 2, 3 (1967). TAB. **62**, fig. E. Type from Namibia.

Rhizomatous suffrutex forming extensive colonies up to several metres across. Stems usually unbranched and less than 30 cm. high, but in the absence of fire up to 45 cm. high and sometimes sparsely branched. Leaves similar to those of *D. batocana* but often much smaller. *Flowers* subsessile, usually in clusters of 1–7 in the axils of leaves or reduced scale leaves, or, on second-year shoots, in the axils of fallen leaves, similar in structure to those of *D. batocana* but much larger and of different proportions, and the female flowers with conspicuous bracts. *Male flowers*: Calyx 1·3 cm. long, fulvous-tomentose, lobes c. 0·6–0·7 cm. long. Corolla c. 1·6 cm. long, lobes 0.8×0.6 cm. *Female flowers* with several conspicuous bracts at the base up to 0·6 cm. long. Calyx 1–1·5 cm. long, fulvous-tomentose; lobes 0·5–0·8 cm. long. Corolla 1·8 cm. long, lobes 0·7 cm. long. *Fruit* subglobose, c. 3.5×3.5 cm., fulvous-tomentose, at least when young. Seeds 6, mature seeds unknown.

Botswana: N: 29 km. S. of Khardoum Valley, fl. buds 14.iii.1965, *Wild & Drummond* 7029 (FHO; LISC; SRGH). **Zambia.** B: Sesheke Distr., Masese, ♀ fl. 20.vi.1960, *Fanshawe* 5760 (BR; FHO; K).

Also in Namibia and Angola. Apparently confined to Kalahari Sand. In sparse grassland with suffrutices at the edges of dambos and in various types in woodland. 900–1000 m.

The fruit, although insipid, is an important item in the diet of certain tribes of bushmen (R. B. Lee, "Subsistence ecology of Kung bushmen", 1965; B. Maguire, unpublished thesis, University of Witwatersrand, 1978).

D. chamaethamnus is very closely related to *D. batocana*. Their geographical ranges overlap slightly, but that of *D. chamaethamnus* extends much further south.

Throughout its range *D. chamaethamnus* consistently differs from *D. batocana* in the size and proportions of its flowers and the colour of the indumentum on the flowers and fruit. It seems, however, that *D. batocana* very locally produces a dwarf shrubby variant which can be confused with *D. chamaethamnus*. Thus, *Mutimushi* 3541 (K) from Dambwa Forest Reserve, Livingstone Distr., Zambia, was collected from a 0·6 m. high plant forming small colonies at the margin of a dambo on Kalahari Sand. Although in habit it is closer to *D. chamaethamnus*, the blackish indumentum, small inconspicuous bracteoles and small shallowly lobed fruiting calyx clearly indicate a close relationship with *D. batocana*.

8. Sect. NOLTIA (Schum.) Hiern in Trans. Camb. Phil. Soc. **12**: 149 (1873) quoad *D. tricolor* et *D. barteri* tantum. — F. White in Bull. Jard. Bot. Nat. Belg. **50**: 457 (1980).
Noltia Schum., Beskr. Guin. Pl.: 189 (1827); K. Danske Vid. Selsk. Nat. Math. Afhandl. **3**: 209 (1828).

Male calyx cyathiform, lobes deltate. Male corolla narrowly conoidal in bud, glabrous to tomentose, lobed to ¼ or less, lobes narrowly deltate. Stamens 3–15, inserted on receptacle, included. Ovary usually conoidal or fusiform, tomentose at least towards the apex; locules usually 4, uniovulate. Style scarcely differentiated from ovary. Stigmata 2(3) sessile, usually very short. Fruit 2·5 × 1·5–7 × 3 cm., various. Seeds 6 or fewer; endosperm smooth or ruminate. Fruiting calyx scarcely accrescent.

24. **Diospyros mafiensis** F. White in Bull. Jard. Bot. Nat. Belg. **50**: 394 (1980). TAB. **63**.
   Type: from Mafia Isl.

Evergreen shrub or small tree up to 6 m. tall. Branchlets very slender, blackish, densely and minutely setulose. *Leaves* subcoriaceous, drying dull dark reddish-brown above, pale pinkish-grey beneath; lamina usually less than 5·5 × 2·2 cm., occasionally up to 7·5 × 3·2 cm., elliptic or lanceolate-elliptic, apex shortly and bluntly subacuminate, base cuneate, margin slightly revolute; lower surface sparsely strigulose-puberulous; lateral nerves in 4–5 pairs, inconspicuous, tertiary nerves and veins invisible. *Male flowers* solitary, subsessile, axillary and in axils of fallen leaves on second-year shoots. Calyx 0·15 cm. long, lobed to two thirds, lobes 3, narrowly deltate, densely strigulose outside, glabrous inside. Corolla 0·6 cm. long, glabrous, lobes 3, 0·2 cm. long, mature corolla unknown. Stamens 3, 0·25 cm. long; filaments 0·05 cm. long, setose; anthers 0·2 cm. long, glabrous, apiculate. Rudimentary ovary absent. *Female* flowers borne as in male. Calyx glabrous outside except near margins, densely strigulose inside. Corolla narrowly conoidal in bud, mature corolla unknown. Ovary 0·2 cm. long, obovoid-cylindric, densely strigulose; locules 6. *Fruit* up to 2 × 2 cm., obovoid-cylindric, broadly cylindric or subglobose, sulcate between the seeds when young, verruculose, apiculate, glabrous except near the base of the persistent styles. Seeds (?6) 4 or fewer by abortion, 1·5 × 0·6 × 0·6 cm., dull blackish-brown; endosperm smooth. Fruiting calyx not accrescent.

   Mozambique. N: Niassa, Mocímboa da Praia, ♀ fl., immat. fr. 12.ix.1943, *Andrada* 1345 (COI; LISC; LMA); Palma to Nangade 25 km., fr. 5.xi.1960, *Gomes e Sousa* 4602 (FHO; K). Also in Tanzania and on Mafia Isl. In coastal forest. 0–250 m.

9. Sect. CALVITIELLA F. White in Bull. Jard. Bot. Nat. Belg. **48**: 246 (1978); op. cit. **50**: 459 (1980).

Male calyx shallowly cyathiform, truncate or distinctly lobed. Male corolla narrowly ellipsoid in bud, glabrous or minutely puberulous towards apex, lobed to one quarter or less, lobes usually rounded or obtuse. Ovary subglobose or ovoid-conoidal, glabrous or strigulose; locules 4 or 6, uniovulate. Style either not differentiated from ovary and with 2–3, short, sessile, fleshy and irregularly lobed stigmata, or styles 3, free, fleshy and stigmatic or inner face. Fruit up to 3·5 cm. diameter, subglobose or ovoid-conoidal. Seeds 6 or fewer; endosperm smooth or ruminate. Fruiting calyx not accrescent.

Leaves obliquely trullate, apex deeply emarginate   -   -   25. *hoyleana* subsp. *hoyleana*
Leaves lanceolate-elliptic, apex attenuate-acuminate -   -   -   -   -   -   26. *sp. 2*

25. **Diospyros hoyleana** F. White in Bull. Jard. Bot. Brux. **26**: 245 (1956); F.F.N.R.: 332, t. 59 fig. j–k (1962); in F.W.T.A., ed. 2, **2**: 15, t. 202a fig. 7 (1963); in Bull. Jard. Bot. Nat. Belg. **48**: 343, t. 9 fig. a–b (1978); in Distr. Pl. Afr. **14**: 463–465 (1978); in Afr. J. Ecol. **19**: 44, tab. 7 fig. a–b (1981). — Letouzey & F. White in Fl. Cam. **11** & Fl. Gab. **18**: 88, t. 11 fig. 1–14 (1970). TAB. **64**, fig. A. Type from Cameroon.
   Maba kamerunensis Gürke in Engl. Bot. Jahrb. **46**: 150 (1912) non *D. kamerunensis* Gürke (1898). Type as above.

Small tree usually 8–15 m. high, but sometimes flowering as a shrub c. 3 m. high. *Leaves* subcoriaceous, drying blackish-brown above, paler brown beneath; lamina up to 7 × 2·2 cm, very variable in shape; lower surface glabrous except for a few setulose hairs towards the base of the midrib; midrib prominent above, often impressed beneath; nerves almost or quite invisible. *Male flowers* subsessile, axillary or ramuligerous, solitary or in fascicles of 2–7; pedicels 0·05 cm. long, densely setulose. Calyx 0·125–0·15 cm. long, sparsely setulose outside, margin ciliolate, glabrous inside; lobes 3, up to 0·06 cm. long, hemi-orbicular. Corolla c. 0·9–1 × 0·275 cm.; tube 0·8 cm. long; lobes 0·15 cm. long, ascending, scarcely longer than broad. Stamens c. 9, included, c. 0·45 cm. long; filaments 0·05 cm. long, inserted on the receptacle, densely setulose; anthers 0·4 cm. long, linear-lanceolate, markedly apiculate; connective setulose towards base on both surfaces, otherwise glabrous. Pistillode represented by a tuft of hairs. *Female flowers* as in male but

Tab. 63. DIOSPYROS MAFIENSIS (sect. Noltia). 1, branchlet (× ½) *Andrada* 1345; 2, leaves (× ½) *Semsei* 1344; 3, female flower-bud (× 5) *Andrada* 1345; 4, immature fruits (× 1) *Semsei* 1344; 5, mature fruit (× 1) *Holtz* 389. 6, seed (× 1) *Semsei* 1344.

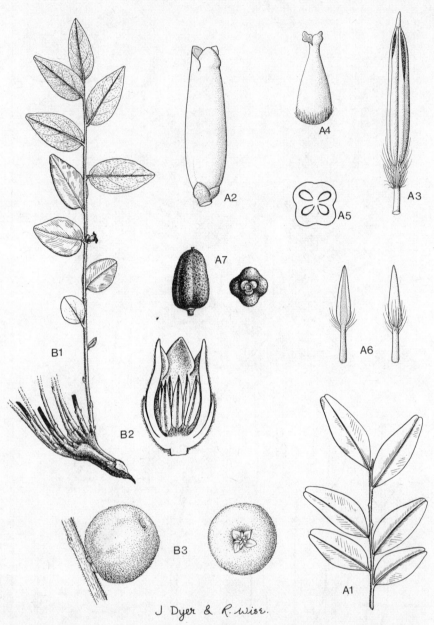

J Dyer & R. Wise.

Tab. 64. DIOSPYROS sect. CALVITIELLA and sect. MARSUPIUM. A. — D. HOYLEANA. A1, habit
($\times \frac{1}{2}$); A2, male flower ($\times 4$) *Brenan* 9471; A3, stamen ($\times 10$); A4, pistil ($\times 10$) *Devred* 2696;
A5, t.s. ovary ($\times 15$) *Devred* 2696; A6, staminodes ($\times 10$), *Devred* 2696, A7, fruits ($\times 1$). B.
— D. VIRGATA. B1, habit ($\times \frac{1}{2}$); B2, male flower, longitudinal section ($\times 3$); B3, fruits ($\times \frac{2}{3}$),
Gossweiler 4095. A1, A7, B1 & B2 from F.F.N.R.

corolla only 0·6 cm. long. Staminodes 4–8, lanceolate, 0·25 cm. long, with a tuft of strigulose hairs at base of antherode. Ovary narrowly ovoid-conoidal, together with the style 0·3 × 0·15 cm., glabrous, encircled at the base by a fringe of hairs; locules 4, uniovulate; stigmata 2, fleshy, sessile, irregularly lobed, 0·1 cm. long. *Fruit* up to 2 × 1·8 cm., red then black, globose, conoidal or ellipsoid-conoidal, verruculose, glabrous. Seeds 1–4, reddish-brown, 1 × 0·8 × 0·6 cm.; endosperm usually deeply ruminate, occasionally shallowly ruminate or smooth. Fruiting calyx scarcely accrescent, 0·2 cm. long.

Widely distributed in the understorey of rain forest in the Guineo-Congolian Region from SE. Nigeria to eastern Zaire. Of its 2 subspecies, subspecies *augustifolia* F. White occurs in north-eastern and eastern Zaire, whereas the typical subspecies, which has a more westerly distribution, is an ecological and chorological transgressor which extends southwards into the northern part of the Zambezian Region and reaches its southern limits in the Flora Zambesiaca area.

The taxonomy of *D. hoyleana* and its relatives is the most troublesome among African members of the genus, and certain problems remain unresolved. In particular, some incompletely collected, small, geographically isolated populations cannot yet be satisfactorily accounted for. One of them (*Diospyros* sp. 2, below) is very similar in leaf-shape to *D. hoyleana* subsp. *angustifolia*, but ecogeographical considerations, combined with the complex pattern of relationships in the group it belongs to, suggest that it might have originated independently of subsp. *augustifolia*.

Subsp. **hoyleana** — F. White in Bull. Jard. Bot. Nat. Belg. **48**, 344, t. 9 fig. a (1978).

Evergreen shrub or small tree up to 7 m. tall. Leaves obliquely trullate, apex deeply emarginate, base asymmetric.

Zambia. W: Mwinilunga Distr., Zambezi R., near Kalene Hill, st. 21.ix.1952, *Angus* 506 (BM; BR; FHO; K).

From SE. Nigeria to northern Angola and Zambia. In the Flora Zambesiaca area only known from riparian forest but in the Upper Shaba Province of Zaire it is a characteristic member of dry evergreen forest (muhulu). 1400 m.

26. Diospyros sp. 2.

Tree 5–12 m. tall. Bark smooth or slightly rough, minutely fissured. *Leaves* subcoriaceous, drying blackish-brown above, paler beneath; lamina up to 9 × 3 cm., lanceolate-elliptic, apex attenuate-acuminate, base slightly asymmetric, lower surface glabrous or very sparsely and minutely strigulose; midrib slightly prominent above, more distinctly so beneath; lateral nerves indistinct, other nerves and veins invisible. *Flowers* unknown, probably borne in fascicles; persistent pedicels forming coralloid bosses on 2–4-year old branchlets. *Fruit* unknown.

Zimbabwe. E: Melsetter Distr., Haroni-Macurupini Forest, 1 km. E. of Haroni R., 460 m., st. 17.viii.1975, *Müller* 2388 (FHO; SRGH). **Mozambique.** MS: Chimanimani Mts., E. slopes of Mevumosi R. on way to Gossamer Falls, 600 m., st. 24.iv.1974, *Müller & Pope* 2120 (FHO; SRGH); Serra Macuta, southern slopes, 700 m., st. 1.vi.1971, *Müller & Gordon* 1767 (FHO; LISC; SRGH).

In lowland rain forest dominated by *Newtonia buchananii*. 460–700 m.

The relationships of this taxon to the *D. hoyleana* complex are discussed by White in Afr. J. Ecol. **19**: 44 (1981).

10. Sect. BREVISTYLA F. White in Bull. Jard. Bot. Nat. Belg. **50**: 459 (1980).

Male calyx tomentose, cyathiform, deeply lobed. Male corolla ellipsoid or narrowly conoidal in bud, tomentose, lobed to ⅓ or less. Ovary ovoid-conoidal, tomentose, locules 4–8, uniovulate. Style not or scarcely differentiated from the ovary and ending in a lobed, fleshy stigma, or styles 2–4, very short and ending in capitate stigmata. Fruit 2·5 × 2·5–4·5 × 4 cm., globose or ellipsoid. Seeds 8 or fewer; endosperm smooth or ruminate. Fruiting calyx not or scarcely accrescent.

Lower leaf-surface appressed sericeous-tomentose when young, lower epidermis glaucous and
    papillose; female calyx up to 0·4 cm. long -   -   -   -   -   -   27. *verrucosa*
Lower leaf-surface lanate-tomentose or minutely puberulous when young, lower epidermis not
    glaucous, not papillose; female calyx at least 0·8 cm. long:
    Leaves less than twice as long as broad, apex and base usually rounded, lower surface with
      very prominent venation, secondary nerves in 6–8 pairs; lobes of female calyx flat
                                     28. *kirkii*

Leaves more than 2·5 times as long as broad, apex usually acute to subacuminate, lower
surface with scarcely prominent venation, secondary nerves in 15–20 pairs; lobes of female
calyx strongly undulate -   -   -   -   -   -   -   -   - 29. *mespiliformis*

27. **Diospyros verrucosa** Hiern in Trans. Camb. Phil. Soc. **12**: 167 (1873). — Sim, For. Fl.
Port. E. Afr.: 83 (1909). TAB. **65**, fig. A. Syntypes: "Prov. Zanguebar", *Kirk*, s.n. (K);
Rovuma R., 20 miles above the mouth , viii.1862, *Kirk* s.n. (K).

Evergreen shrub or small tree usually less than 5 m. high, rarely up to 15 m. Bark
blackish with close longitudinal fissures. Young branchlets tomentellous. *Leaves*
subcoriaceous, drying blackish-brown above, paler beneath; lamina up to 12·5 × 5·5
cm., ovate or ovate-oblong to lanceolate, apex subacute to shortly acuminate, base
usually rounded; lower surface appressed sericeous-tomentose, somewhat glabrescent except on nerves; lower epidermis glaucous, papillose; secondary nerves in 5–7
pairs, prominent beneath, tertiary nerves and veins usually indistinct. *Male flowers*
in 3–7-flowered cymules, borne towards the base of the current year's growth in the
axils of caducous reduced leaves or the first formed foliage leaves; peduncle 0·1–0·2
cm. long; pedicels 0·1–0·2 cm. long. Calyx c. 0·25 cm. long, lobes 4, 0·15 cm. long,
deltate. Corolla 0·6 cm. long, ellipsoid or ovoid-ellipsoid in bud, lobes 4, c. 0·1 cm.
long. Stamens c. 12, included, inserted on receptacle; anthers linear-lanceolate,
apiculate, strigose. Pistillode minute, tomentose. *Female flowers* usually solitary,
rarely in 2–3-flowered cymules, in axils of leaves or of bracts below the leaves;
peduncle (including pedicel) 0·3–1·2 cm. long. Calyx 0·3–0·4 cm. long, lobes 4, c. 0·2
cm. long, broadly deltate. Corolla urceolate, 0·7 × 0·3 cm., lobes 4, less than 0·1 cm.
long. Ovary 0·4 × 0·25 cm.; locules 4, uniovulate; style scarcely differentiated from
ovary, less than 0·1 cm. long, bifid at apex, branches ending in 2 fleshy, irregularly
lobed stigmata. *Fruit* up to 4·5 × 4 cm., orange, subglobose or broadly ellipsoid,
verrucose, especially when young, glabrescent. Seeds 4 or fewer, blackish,
endosperm ruminate. Fruiting calyx sometimes with reflexed lobes.

Mozambique. N: Cabo Delgado, Pemba (Porto Amélia), near Ancuabe, fr. 25.viii.1948,
Barbosa 1888 (COI; FHO; LISC; LMU). Z: Cundine, ♂ fl. 4.ii.1905, *Le Testu* 635 (BM; FHO;
P). MS: between Dondo and Beira, fr. 31.xii.1943, *Torre* 6326 (K; LISC).
Also in Tanzania. In coastal forest and forest regrowth. 30–200 m.

28. **Diospyros kirkii** Hiern in Trans. Camb. Phil. Soc. **12**: 199 (1873). — Brenan in Mem.
N.Y. Bot. Gard. **8**: 499 (1954). — F. White, F.F.N.R.: 331 (1962); in Systematics
Association Publ. **4**, 88–96, t. 9 fig. a (1962). — Gomes e Sousa, Dendrol. Moçamb.
Estudo Geral, **2**: 630, t. 195, cum phot. (1966). — Wild in Kirkia, **7**: 11, 13, 16, 17, 21, 23,
34, 54 (1968); op. cit., **9**: 245 (1974). — Wild, Biegel & Mavi, Rhod. Bot. Dict., ed. 2: 150
(1972). — Drummond in Kirkia, **10**: 266 (1975). — K. Coates Palgrave, Trees of
Southern Africa: 747 (1977) TAB. **65**, fig. B. Type: Mozambique, above Tete, Cahora
Bassa ("Kebrabassa"), ♂ fl. xi.1860 *Kirk* s.n. (K, holotype).
   Diospyros platyphylla Welw. ex Hiern, tom. cit. (1873): 266. Type from Angola.
   Diospyros latifolia Gürke in Engl. Bot. Jahrb. **26**: 63 (1898). Lectotype from Tanzania.
   Diospyros baumii Gürke in Warb., Kunene-Samb.-Exped. Baum: 328 (1903). — R.E.
Fr., Wiss. Ergebn. Schwed. Rhod.-Kongo-Exped. **1**: 257 (1916). Type from Angola.
   Diospyros flexilis Hiern ex Greves in Journ. Bot. Lond. **67**, Suppl. 2: 77 (1929). Types
from Angola.

Small, deciduous or semi-deciduous tree 4–8 m. high. Bole often twisted. Crown
rounded. Bark dark grey or blackish, rough with deep longitudinal fissures. Young
branchlets, leaves and inflorescence-axes densely covered with pinkish lanate
indumentum. *Leaves* subcoriaceous, drying pale reddish-brown; lamina
8 × 5–18 × 12 cm., broadly elliptic, apex and base usually rounded; lower surface
persistently pubescent with spreading flexuose hairs, rarely glabrescent; secondary
nerves in 6–8 pairs, prominent beneath, tertiary nerves and veins forming a
prominent reticulum beneath. *Male flowers* in 3-or more flowered cymules borne
towards the base of the current year's growth in the axils of caducous reduced leaves
or the first-formed foliage leaves; peduncles 0·5–1·5 cm. long; pedicels ç. 0·1 cm.
long. Calyx 0·4–0·7 cm. long, lobes 4–5, 0·3–0·5 cm. long, deltate. Corolla 1 cm. long,
urceolate, velutinous-tomentose outside, glabrous inside and thickened at the throat;
tube 0·8 cm. long; lobes 4–5, 0·2 cm. long, ovate-apiculate. Stamens 14–16, included,
0·5–0·6 cm. long; filaments very short, inserted on the receptacle; anthers linear-
lanceolate, apiculate, glabrous. Pistillode minute, tomentose. *Female flowers* similar
to male but solitary or in pairs and much larger, especially the calyx which is 0·8–1

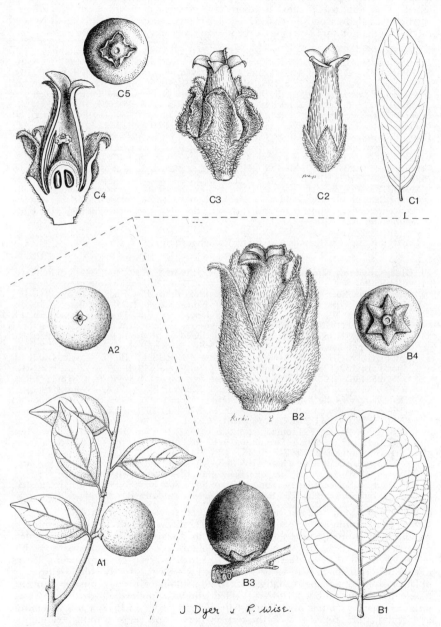

Tab. 65. DIOSPYROS sect. BREVITUBA. A. — D. VERRUCOSA. A1, fruiting branchlet (× ½), *Torre &
Correia* 16169; A2, fruit (× ½), *Torre & Correia* 16169. B. — D. KIRKII. B1, leaf (× ½), *Keay*
FHI 21353; B2, female flower (× 4), *Duff* 51; B3, B4, fruits (× ¾), *Chapman* 205. C. — D.
MESPILIFORMIS. C1, leaf (× ½), *Brenan* 7730; C2, male flower (× 4), *Mendes* 956; C3,
female flower (× 4), *Townsend* 268; C4, female flower, longitudinal section (× 5); C5, fruit
(× 1), C4 and 5 from F.F.N.R.

cm. long and is more deeply-lobed. Staminodes c. 8, filiform. Ovary 0·6 × 0·4 cm.; locules 4; style not differentiated from the ovary; stigma large, fleshy, sessile, 3–4-lobed. *Fruit* up to 3·5 cm. in diameter, golden-yellow or orange, globose, verruculose, glabrescent. Seeds 3–4, reddish-brown; endosperm ruminate. Fruiting calyx with flat not undulate lobes, except sometimes in very old specimens.

Zambia. B: near Kamanga, fr., *Martin* 857 (FHO). N: Mbala Distr., Kalambo Falls, ♂ fl. 21.x.1947, *Brenan & Greenway* 8178 (BR; EA; FHO; K). W: Ndola, ♀ fl., fr. xii.1932, *Duff* 50 & 51 (FHO). C: Lusaka Distr., Mt. Makulu, ♂ fl. 31.x.1956, *Angus* 1430 (BM; BR; EA; FHO; K). E: 48 km. E. of Petauke, fr. 21.iv.1952, *White* 2434 (BR; FHO; K). S: Gwembe Valley, ♂ fl. 18.xi.1955, *Bainbridge* 199/55 (FHO; K). **Zimbabwe.** N: Sebungwe Distr., Kariyangwe Camp, 915 m., ♂ fl. x.1955, *Davies* 1570 (BR; FHO; SRGH). W: Wankie, 31.vii.1930, *Pole-Evans* 3074 (K, P; SRGH). C: near Hartley, fr. i.1936, *Edwards* 1/36 (FHO). **Malawi.** N: Rumphi Distr., Vintukutu F.R., fr. vi.1954, *Chapman* 205 (FHO). C.: Mua-Livulezi F.R., ♂ fl. 17.xi.1954; *Adlard* 198 (FHO). S: Chikwawa, 300 m., fr. 5.x.1946, *Brass* 17994 (K; MO; SRGH). **Mozambique.** N: near Montepuez, ♂ fl. 17.x.1942, *Mendonça* 932 (C; COI; FHO; K; LISC; LMU). Z: Gúruè near Lioma, Monte Comé, 700 m., fl. buds 10.xi.1967, *Torre & Correia* 16047 (FHO; LISC; MO; SRGH; WAG). T: Ruenya R., near Massanga, 550 m., fr. 25.ix.1948, *Wild* 2626 (BR; K; SRGH).

Also in Tanzania, Zaire (Upper Shaba) and Angola. In various types of woodland, especially open woodland on stony soils and at the edges of bamboo. In Zimbabwe it tolerates relatively low concentrations of copper in the soil. It also occasionally occurs on arsenical sóils but is not common and is dwarfed. 350–1370 m.

The fruit is eaten and the wood is used for making furniture.

29. **Diospyros mespiliformis** Hochst. ex A.DC., Prodr. **8**: 672 (1844). — Hiern in Trans. Camb. Phil. Soc. 12: 165 (1873). — Sim, For. Fl. Port. E. Afr.: 82 (1909). — S. Moore in Journ. Linn. Soc., Bot. **40**: 134 (1911). — Eyles in Trans. Roy. Soc. S. Afr. **5**: 439 (1916). — R.E. Fr., Wiss. Ergebn. Schwed. Rhod.-Kongo-Exped. **1**: 257 (1916). — Steedman, Trees etc. S. Rhod.: 63 (1933). — Mendonça, Est. Ens. Docum., **1**, Contrib. Fl. Moçamb.: 53 (1950). — Gomes e Sousa, Dendrol. Moçamb. **1**, Commerc. Timbers: 154 cum tab. (1951). excl. syn. *D. bicolor* Winkl.; Dendrol. Moçamb. Estudo Geral, **2**: 629, t. 194, cum photogr. (1966). — Codd, Trees & Shrubs Kruger Nat. Park: 143, t. 133, cum photogr. (1951). — O. B. Mill. in Journ. S. Afr. Bot. **18**: 67 (1952). — Williamson, Useful Pl. Vict. Falls: 152 (1953). — Williamson, Useful Pl. Nyasal.: 49 (1955). — White, F.F.N.R.: 328, t. 58 fig. g–h (1962); in Systematics Association Publ. **4**: 88–96, tt. 8, 10 & 11 (1962); in F.W.T.A., ed. 2, **2**: 12, t. 203 (1963). — de Winter in F.S.A. **26**: 62, t. 9 fig. 2 (1963). — B. de Winter, M. de Winter & Killick, Sixty-six Transv. Trees: 140, cum cart. & photogr. (1966). — de Winter & F. White, Prodr. Fl. SW. Afr. **107**: 4 (1967). — Letouzey & F. White in Fl. Cam. **11** & Fl. Gab. **18**: 110, t. 16 fig. 1–12 (1970). — Palmer & Pitman, Trees of Southern Afr. **3**: 1793 cum tab. & photogr. (1972). — Wild, Biegel & Mavi, Rhod. Bot. Dict., ed. 2: 150 (1972). — Drummond in Kirkia, **10**: 267 (1975). — K. Coates Palgrave, Trees of Southern Africa: 748: t. 245 (1977). TAB. **65**, fig. C. Types from Ethiopia.

Diospyros bicolor Klotzsch in Peters Reise Mossamb. Bot.: 184 (1864) non *D. bicolor* H. Winkl. (1908). Type: Mozambique, Sena, imm. fr., *Peters* s.n. (B†, holotype).

Diospyros sabiensis Hiern in Journ. Linn. Soc., Bot. **40**: 135 (1911). — Eyles, loc. cit. Type: Zimbabwe, Sabi R., ♂ fl.9.xi. 1906, *Swynnerton* 1209 (BM, holotype).

Evergreen tree up to 25 m. high or more, rarely flowering as a shrub c. 3 m. high. Crown dense, rounded. Bole sometimes fluted at the base. Bark dark brown or black, rough, longitudinally fissured, exfoliating in square scales. Young branchlets tomentellous with pinkish appressed hairs, glabrescent. *Leaves* subcoriaceous, old leaves drying dull grey-green above and yellowish-green beneath, young leaves drying pale reddish-brown; lamina 6 × 2·2–14 × 4·5 cm., mostly oblong-elliptic or oblanceolate-elliptic; apex usually acute to subacuminate, rarely obtuse or rounded; base cuneate or rounded; lower surface minutely puberulous with appressed strigulose hairs or spreading flexuose hairs; lateral nerves in 15–20 pairs, ascending at 45°, indistinct, together with the tertiary nerves and veins forming a reticulum which is slightly prominent on both surfaces. *Male flowers* sub-sessile in 3s at the end of 0·4–0·6 cm. long peduncles arising from the axils of caducous reduced leaves at the base of the current year's growth or of the first-formed normal leaves. Calyx 0·3 cm. long, lobes 4–5, deltate, 0·15 cm. long. Corolla 0·6 cm. long, narrowly urceolate, sericeous-tomentose outside; tube 0·5 cm. long, glabrous inside and thickened at the throat; lobes 4–5, deltate, 0·1 cm. long. Stamens c. 14, included, 0·4 cm. long; filaments 0·1 cm. long, glabrous, inserted on the receptacle; anthers narrowly lanceolate, apiculate, glabrous, except for a few short hairs on the connective.

Pistillode minute, tomentose. *Female flowers* subsessile, solitary, or rarely 2–3 together, in the axils of reduced leaves at the base of the current year's shoot. Calyx and corolla similar to the male but calyx 0·8 cm. long and lobes cordate-deltate with undulate-plicate reflexed margins. Corolla 1–1·2 cm. long. Staminodes 6–12, filiform, 0·4 cm. long, glabrous, inserted at base of corolla. Ovary 0·3 cm. in diameter; style scarcely differentiated from the ovary; stigma sessile; locules 4 or 6. *Fruit* up to 2·5 cm. in diameter, yellow, globose, verruculose, glabrescent, but a few hairs persisting near the base of the persistent style. Seeds 3–6, reddish-brown; endosperm deeply ruminate. Fruiting calyx patelliform or shallowly cyathiform, lobes with recurved, strongly undulate margins. *Chromosome number*: 2n = 30.

Caprivi Strip: Katima Mulilo to Ngoma 9 km., fr. 5.i.1959, *Killick & Leistner* 3304 (M; SRGH). **Botswana**. N: Okavango Swamp, Moremi Wildlife Reserve, Gobega lagoon, 900 m., fr. 5.iii.1972, *Biegel & Russell* 3861 (FHO; SRGH). **Zambia**. N: 22 km. S.E. of Kafulwe, ♀ fl. 3.xi.1952, *White* 3594 (BM; BR; FHO; K). W: Solwezi Distr., Meheba R, fr. 25.vii.1930, *Milne-Redhead* 771 (K). C: Kafue Gorge, below Kafue Dam, 1000 m., ♂ fl. 4.xi.1972, *Strid* 2434 (FHO). E: Petauke to Sasare, 800 m., ♂ fl. 4.xii.1958, *Robson* 830 (FHO; SRGH). S: Katambora, fr. 22.viii.1947, *Brenan* 7730 (BR; EA; K). **Zimbabwe**. N: Sebungwe Distr., Kariyangwe Camp, fr. 26.vi.1951, *Lovemore* 76 (FHO; SRGH). W: Victoria Falls, fr. 31.vii.1941, *Greenway* 6249 (K). C: Hartley, ♂, ♀ fl. 11.xi.1932, *Eyles* 7266 (K; SRGH). E: Umtali, ♀ fl. xi.1944, *Chase* 129 (BM; SRGH). S: Chiredzi R, ♂ fl. 11.x.1951, *McGregor* 65/51 (FHO; SRGH). **Malawi**. N: Likoma Isl., viii.1887, *Bellingham* s.n. (BM). C: Balaka to Ncheu 18 km., fr. 19.iii.1964, *Chapman* 2239 (FHO). S: Lengwe National Park, imm. fr. 14.xii.1970, *Hall-Martin* 1115 (FHO; SRGH). **Mozambique**. N: between Pemba (Porto Amélia) and Ancuabe, fr. 24.viii.1948, *Barbosa* 1869 (K; LISC; LMA; LMU). T: between Tete and Changara, Mazoi R., 300 m., fr. 6.i.1966, *Torre & Correia* 14021 (C; COI; FHO; LISC; MO; WAG). MS: Vila Machado, base of serra de Chiluvo, fr. 14.iv.1948, *Mendonça* 3927 (C; FHO; LISC; MO; WAG). GI: between Funhalouro and Mavume, fr. 17.v.1941, *Torre* 2669 (BR; FHO; LISC; M; SRGH). M: Maputo R. Salamanga, ♀ fl. 19.xi.1948, *Gomes e Sousa* 3877 (COI; FHO; MO; SRGH).

Widespread in tropical Africa from Senegal to Eritrea and the Yemen and southwards to Namibia, the Transvaal and southern Mozambique, but absent from the rain forests of the Guineo-Congolian Region except locally along their northern fringes. In our area most frequently in riparian forest, more rarely on termite mounds or rocky outcrops or in dry semi-evergreen forest. 60–1370 m.

The fruit is eaten and the wood is used for making canoes, drums, pestles, mortars and rifle butts.

Intermediates between **D. kirkii** and **D. mespiliformis**. — F. White in Systematics Association Publ. **4**: 89–92, t. 9 fig. b–j.

Zambia. N: Mbala Distr., 18 km. S. of L. Tanganyika, fr. 22. vii.1930, *Hutchinson & Gillett* 3994 (BM; K). S: Namwala Distr., Kafue National Park, Burning Experimental Plots, ♂ fl. 9.xi.1962, *Mitchell* 15/21 (FHO). **Zimbabwe**. N: Sebungwe Distr., Kariyangwe R. crossing on way to Subu R., 915 m., ♂ fl.x.1955, *Davies* 1569 (FHO; K; SRGH). W: Nyamandhlovu Distr., Tjolotjo Tribal Trust Land, near Jalume, ♂ fl. 5.xi.1964, *Crozier* 2/64 (FHO; SRGH). E: Inyanga Distr., Van Niekerk ruins, Inyangambe R., 1220 m., fr. 2.viii.1950, *Wild* 3515 (BR; FHO; K; SRGH). **Malawi**. S: Chikwawa, Njombo, fr. 12.v.1959, *Jackson* 2321 (FHO; SRGH). **Mozambique**. N: Malema Distr., Malema Plain, ♂ fl. 26.ix.1942, *Hornby* 2281 (EA, K). Z: Mocuba Distr., Namagoa Estate, 60–120 m, ♀ fl. ix-x.1944, *Faulkner* 219 (K; SRGH); ♂ fl. 1.xi.1948, *Faulkner* 319 (BR; EA; K; SRGH); ♀ fl. i. xi.1948, *Faulkner* 320a (K; SRGH); fr. 29.xii.1948, *Faulkner* 320b (K). T: Tete to Changara 20 km., fr. v.1971, *Torre & Correia* 18343 (LISC). MS: Chemba, Chiou, fr. 18.iv.1960, *Lemos & Macuácua* 118 (BM; K; SRGH).

Also in Tanzania and Angola. In the ecotone between riparian forest and escarpment woodland, and in various types of woodland, 60–1000 (? 1300) m.

The geographical range of *D. kirkii* lies entirely within that of the much more widely distributed *D. mespiliformis*. At various places within their area of overlap individuals occur which collectively bridge the morphological gap between them. It is possible that some of the intermediates cited above, especially those that are not very different from *D. kirkii* or *D. mespiliformis*, are merely extreme variants of those species. For others, however, there is strong evidence in support of a hybrid origin.

In habit and ecology *D. kirkii* and *D. mespiliformis* are normally very different. It seems that they come into contact only very locally, but then may produce hybrid progeny. Thus, in Sebungwe District, Zimbabwe, *D. kirkii* is common in escarpment *Brachystegia-Julbernardia globiflora* woodland, and *D. mespiliformis* occurs in the riparian fringe of the Kariangwe stream. Putative hybrids occur in the ecotone between them (D. Lovemore, pers. comm.). Similarly, of the 5 specimens collected by Faulkner on the Namagoa Estate, Mocuba District, Mozambique, no. 219 is close to *D. kirkii*, but not identical, 220 is close to *D. mespiliformis*, and 319, 320a and 320b are variously intermediate.

No specimens resembling the intermediates cited above have been collected outside the area where both putative parents are present.

11. Sect. MARSUPIUM F. White in Bull. Jard. Bot. Nat. Belg. **50**: 460 (1980).

Male calyx cyathiform, deeply divided, lobes narrowly deltate. Male corolla conoidal in bud, tomentose, lobed to c. ⅓, lobes acute. Ovary conoidal, densely strigulose; locules 6, uniovulate. Styles 3, 0·15 cm. long, erect, glabrous distally, stigmatic on inner surface. Fruit subglobose, 2·5 × 2·5 cm., tomentose. Seeds 6 or fewer. Endosperm smooth. Fruiting calyx not accrescent.

30. **Diospyros virgata** (Gürke) Brenan in Kew Bull. **1953**: 437 (1953). — F. White, F.F.N.R.: 330, t. 59 fig. c–d (1962). — de Winter in F.S.A. **26**: 61 (1963). — de Winter & F. White, Prodr. Fl. SW. Afr. **107**: 4 (1967). TAB. **64**, fig. B. Type from Angola.
  Maba virgata Gürke in Warb., Kunene-Sambesi Exped. Baum: 327 (1903). — R.E. Fr., Wiss. Ergebn. Schwed. Rhod.-Kongo-Exped. **1**: 256 (1916). Type as above.
  Maba fragrans Hiern ex Greves in Journ. Bot. Lond. **67**, Suppl. 2: 75 (1929). Types from Angola.

Usually a rhizomatous suffrutex up to 0·45 m. high and forming mats a few metres in diameter; when fire-protected a shrub up to 2·5 m. high. *Leaves* drying pale yellow-brown; lamina 4 × 1·5–7 × 3 cm., lanceolate, apex acute or subacute, the tip usually mucronate; lower surface densely fulvous-sericeous; secondary nerves in (3) 4–5 pairs, inconspicuous; tertiary nerves and veins invisible. *Male flowers* pale yellow, sessile in 1–7-flowered clusters in the axils of leaves or of fallen leaves. Calyx 0·4 cm. long, strigulose-tomentellous outside, lobed almost to the base; lobes 3(4) narrowly deltate. Corolla 0·7–0·8 cm. long, cream, urceolate, strigulose-tomentellous almost to the base outside; lobes 3(4), ovate-deltate, 0·25 cm. long. Stamens 12–14, visible at the throat but included, 0·3–0·4 cm. long; filaments very short, inserted on the receptacle; anthers lanceolate, apiculate, glabrous. Pistillode represented by a tuft of hairs. *Female flowers* solitary, shortly pedicellate, with 6 or more deltate-bracts, otherwise similar to the male but corolla-tube slightly wider and more contracted at the throat. Staminodes 3(4), filiform, inserted on the receptacle. Ovary 0·5 × 0·4 cm., locules 6. Styles 3, 0·15 cm. long, erect, glabrous distally, stigmatic on inner surface. Fruit 2·5 × 2·5 cm., subglobose, fulvous-tomentose, outer pericarp somewhat woody. Seeds 6 or fewer. Fruiting calyx not accrescent, 0·4 cm. long.

Caprivi Strip. Katima Mulilo, fl. vii.1974, *Vahrmeijer & du Preez* 456 (SRGH). **Zambia.** B: Kalabo, ♀ fl., imm. fr. 18.vii.1957, *Angus* 1641 (BR; FHO). W: Mwinilunga Distr., near source of Matonchi R., fr. 7.x.1937, *Milne-Redhead* 2625 (BM; BR; K). C: Serenje Corner, ♂ fl., 16.vii.1930, *Hutchinson & Gillett* 3710 (BM; BR; K; SRGH). S: 24 km. W. of Namwala, ♂ fl. 25.vii.1952, *White* 2988 (FHO; K).
Also in Zaire, Angola and Namibia. Almost confined to Kalahari Sand but with an eastern extension along the Zambezi-Zaire watershed to Serenje District in Zambia. In dry evergreen forest and various types of woodland, persisting in fire-induced communities. 900–1700 m.

2. **EUCLEA** Murray

Euclea Murray, Syst. Veg. ed. 13: 747 (1774).

Trees, shrubs or suffrutices, usually evergreen. Leaves alternate, subopposite or subverticillate, coriaceous, margin sometimes crenulate, otherwise entire. Inflorescence a simple or branched false raceme, or rarely (not in our area) flowers solitary. Flowers dioecious, the male usually larger than the female. Calyx 4–5-lobed shallowly cyathiform or patelliform, not accrescent. Corolla campanulate, 4–5-lobed, usually to beyond the middle. Stamens 10–30, basically 2 opposite to and 2 alternating with each corolla lobe, but there are many modifications of this arrangement; anthers slightly exserted or at least clearly visible at the throat, lanceolate or narrowly oblong, usually strigulose, often dehiscing at first by large ellipsoidal apical pores which later become longitudinal slits; filaments usually shorter than the anthers. Pistillode usually very reduced, with or without stylodes. Disk fleshy, fimbriate, sometimes undulate. Staminodes absent or very reduced. Ovary globose, hairy or covered with peltate scales, locules usually 4–6 and uniovulate, occasionally 2–3 and incompletely septate and biovulate. Styles 2–3, free or united in lower half, often as long as ovary, usually glabrous, ending in a slightly expanded bilobed stigmatic surface. Fruit a small, globose, 1 (3)-seeded berry up to 1 cm. diameter; mesocarp exiguous. Seed subglobose, with 3 radiating lines from the apex; embryo with cotyledons flexed at right-angles to the radicle; radicle entirely surrounded by an invagination of the testa.

About 12 species, confined to Africa, Arabia, Socotra and the Comoro Islands. All species occur in South Africa including Namibia, and 6 extend into the tropics, of which 5 are found in the Flora Zambesiaca area.

The description given above applies only to subgenus *Euclea*. Subgenus *Rymia*, which is confined to South Africa, Namibia and south-west Angola, differs in its very shallowly lobed corolla with 5–8 lobes. The flowers of subgenus *Euclea* are predominantly tetramerous. Some individuals of *E. undulata* are consistently pentamerous, however. Otherwise pentamerous flowers are of sporadic occurrence.

The traditional differences between *Euclea* and *Diospyros* at first sight appear to be small, especially in view of the considerable variation within the latter. As explained after the family description, however, the structure of the fruit and seed is fundamentally different. *Euclea*, in appearance, is also very distinct, though the differences are difficult to express concisely in words. No serious student of the family has ever confused them, except for a rare slip of the pen and the biased judgement of a single anatomist (Parmentier, Ann. Univ. Lyon, **6** (2): 81, 1892).

In parts of Kenya, *Euclea divinorum* and *E. racemosa* subsp. *schimperi* are important weeds, especially in tsetse-barrier clearings (Ivens, *East African weeds and their control*, p. 56, 1967). When natural forest or thicket is cleared they often become dominant because of their apparently inexhaustible capacity for coppice and root sucker production. They also appear to invade existing pasture.

Fruits are normally produced in abundance, but, although they are eaten by birds and mammals, including squirrels and man, they do not seem to be much sought-after, and frequently remain on the plant or on the ground beneath it for a year or more after ripening.

Leaves at least 14 times as long as broad; branchlets and inflorescence-axes usually glabrous
    except for peltate scales - - - - - 2. *crispa* (subsp. *linearis* sens. strict.)
Leaves less than 14 times as long as broad:
  Branchlets and inflorescence-axes (at least when young) puberulous to tomentellous; leaves
    with prominent reticulate venation, especially on upper surface:
    Inflorescence usually branched; calyx cyathiform, lobed to the middle or beyond, lobes
      narrowly deltate; corolla-lobes with a conspicuous mid-petaline keel; petiole (0·4) 0·5–1
      cm. long; peltate scales absent - - - - - - 1. *natalensis*
    Inflorescence unbranched; calyx patelliform, denticulate or lobed only in upper half, lobes
      broadly deltate; corolla-lobes unkeeled; petiole up to 0·2 cm. long; peltate scales present
      on branchlets, leaves and calyx at least when young - - - - 2. *crispa*
  Branchlets and inflorescence-axes glabrous, except for peltate scales in *E. divinorum* and *E.*
    undulata; leaves with inconspicuous to subprominent venation:
    Branchlets, leaves and inflorescence-axes completely glabrous; corolla (in our area) mostly
      urceolate, lobed only in upper half - - - - - 3. *racemosa*
    Branchlets, leaves and inflorescence-axes with peltate scales; corolla campanulate, lobed to
      beyond the middle:
      Leaves broadest at or below the middle, petiole at least 0·4 cm. long, inflorescences
        paired in at least some leaf-axils; each corolla-lobe always with more than 25 hairs;
        ovary strigulose - - - - - - - - 4. *divinorum*
      Leaves broadest in upper half, petiole up to 0·3 cm. long; inflorescences solitary; each
        corolla-lobe nearly always with fewer than 20 hairs; ovary covered with whitish scales
              5. *undulata*

1. **Euclea natalensis** A.DC. in DC., Prodr. **8**: 218 (1844). — Hiern in Trans. Camb. Phil. Soc. **12**: 101 (1873). — S. Moore in Journ. Linn. Soc., Bot. **40**: 134 (1911). — O. B. Mill. in Journ. S. Afr. Bot. **18**: 68 (1952). — de Winter in F.S.A. **26**: 88 (1963). — Gomes e Sousa, Dendrol. Moçamb. Estudo Geral, **2**: 632, t. 196 cum photogr. (1966). — White in Mitt., Bot. Staatssamml. München, **10**: 100–102, 104–105, t. 4 (1970); in Bull. Jard. Bot. Nat. Belg. **50**: 397 (1980). — Wild in Kirkia, **7**, Suppl.: 42 (1970). — Palmer & Pitman, Trees of Southern Afr. **3**: 1773 cum tab. & photogr. (1972). — Wild, Biegel & Mavi, Rhod. Bot. Dict., ed. 2: 163 (1972). — Drummond in Kirkia, **10**: 266 (1975). — K. Coates Palgrave, Trees of Southern Afr.: 738, tab. 243 (1977). TAB. **66**, fig. A–E; **68**, fig. F. Type from Natal.
    Euclea multiflora Hiern sens. lat. tom. cit.: 100, t. 3 (1873). — Codd, Trees & Shrubs Kruger Nat. Park: 144 (1951). — O. B. Mill. in Journ. S. Afr. Bot. **18**: 68 (1952). — F. White, F.F.N.R.: 333 (1962). Types from Angola and South Africa.
    Euclea fructuosa, sensu Steedman, Trees, etc. S. Rhod.: 63 (1933). — Williamson, Useful Pl Nyasal.: 57 (1955) non Hiern sens. strict.

Shrub or small tree (0·5) 2–12 (18) m. high. Foliage very dark green. Young shoots densely puberulous. *Leaves* alternate, very rarely subopposite; petiole (0·4) 0·5–1 cm. long, lamina very variable in shape and size, up to 12 × 4·5 cm.; lower surface sparsely and minutely setulose or strigulose to densely tomentose, venation prominent and reticulate on both surfaces, especially the upper. *Inflorescence* 1·5–4·5 cm. long, usually branched, with densely puberulous to tomentellous axes. *Male flowers* up to 0·5 cm. long. Calyx cup-shaped, densely hairy, lobed to the middle or beyond, lobes narrowly deltate. Corolla glabrous except for the densely strigulose mid-petaline lines and a few marginal setae, lobed to about the middle; lobes with a conspicuous mid-petaline keel. Stamens usually 16. Pistillode minute, setulose with 2 stylodes. *Female flowers* without staminodes. Ovary densely strigose. *Fruit* c. 0·9 cm. diameter, sparsely strigose, glabrescent. *Chromosome number*: 2n = 30.

From Kenya and Zaire southwards to the Cape Province, South Africa. Sea level to 1525 m.
E. natalensis shows a complicated pattern of ecogeographical variation. In an earlier work (op. cit., 1970) I suggested that the recognition of 6 more or less allopatric subspecies might be justified, but that their precise delimitation would be arbitrary because of the occurrence of intermediates. Subsequent extensive field work, coupled with experimental studies, has shown that intermediates are sufficiently localized and infrequent not to pose practical problems and that the differentiating features of leaf-shape and indumentum are constant in cultivation. Accordingly, the species has been divided into 8 subspecies (White, op. cit., 1980), five of which occur in the Flora Zambesiaca area.
In East Africa a black dye, which is obtained from the roots, is used for dying mats. The roots are also chewed by women to impart a red colour to their mouths, and a decoction of the roots is used as an anthelmintic. The twigs are used for tooth brushes.

Leaves suborbicular to broadly oblong-elliptic   -   -   -   -   -   subsp. *rotundifolia*
Leaves otherwise:
  Leaves obovate to broadly obovate, apex broadly rounded or obtuse -   - subsp. *obovata*
  Leaves otherwise:
    Leaves densely pubescent beneath, broadest at the middle, apex acute:
      Leaves more than 2·5 cm. wide   -   -   -   -   -   - subsp. *acutifolia*
      Leaves less than 1·6 cm. wide -   -   -   -   -   - subsp. *angustifolia*
    Leaves sparsely and minutely setulose beneath, broadest in upper half or apex obtuse
                                      subsp. *natalensis*

Subsp. **rotundifolia** F. White in Bull. Jard. Bot. Nat. Belg. **50**: 398 (1980). TAB. **66**, fig. B. Type from Natal.

Shrub or bushy tree, often with a leaning and crooked bole and wide-spreading branches. Leaf-lamina 5·5 × 3·5–8 × 6 cm., thickly coriaceous, suborbicular to broadly oblong-elliptic, apex very broadly rounded, margin strongly revolute, lower surface sparsely strigulose. *Chromosome number*: 2n = 30.

Mozambique. M: Inhaca Isl., Ponta Torres, fr. 12.vii.1957, *Barbosa* 7683 (FHO). Also in Natal. In coastal thicket, often subjected to salt spray. 0–100 m.

Intermediates between subsp. *rotundifolia* and subsp. *natalensis* occur at the northern (Inhaca Island) and southern (near Durban) limits of the range of the former. Elsewhere (White, field obs.), the two subspecies seem to remain distinct although they often occur in close proximity. Juvenile plants of subsp. *natalensis* sometimes occur in gaps in wind-trimmed coastal thicket but do not reach maturity there. Similarly, young plants of subsp. *rotundifolia* are occasionally found in the pioneer stages of coastal forest but are never seen as adults.

Subsp. **obovata** F. White in Bull. Jard. Bot. Nat. Belg. **50**: 398 (1980). TAB. **66**, fig. C. Type from Kenya.
    Euclea fructuosa Hiern in Trans. Camb. Phil. Soc. **12**: 101 (1873). — Sim, For. Fl. Port. E. Afr.: 81 (1909). Types: Mozambique, Luame R. mouth, fr. 8.ii.1861, *Kirk* s.n. (K, syntype); between Tete and the sea coast, fr. 16.iii.1860, *Kirk* s.n. (K, syntype).

Shrub or small bushy tree, usually 2–7 m. tall. Leaf-lamina 6 × 2·5–11 × 4·5 cm., obovate to broadly oblanceolate, apex broadly rounded or obtuse, lower surface sparsely and minutely strigulose-puberulous.

Malawi. N: Lukoma, fl. viii.1887, *Bellingham* s.n. (BM). C: Nchisi Mt., st. 7.ix.1929, *Burtt Davy* 21364 (FHO). S: foot of Mt. Mulanje, Nakaye stream, fl. 15.vi.1958, *Chapman* 590 (BR; FHO; K; SRGH). **Mozambique.** N: Cabo Delgado lighthouse, fl. 16.ix.1948, *Barbosa* 2168 (K; LISC; LMA; LMU). Z: Pebane, fr. 8.iii.1961, *Torre & Correia* 15103 (FHO; LISC; MO). MS: Chemba, near Maringuè, fl. 19.vii.1941, *Torre* 3106 (C; COI; FHO; LISC; WAG). GI: Ilha de Bazaruto, fr. viii.1936, *Gomes e Sousa* 1840 (K).
Extending from Kenya to southern Mozambique, and with an outlying population in southern Natal and Eastern Cape Province, South Africa. A characteristic plant of bushland, thicket and scrub forest. In inland localities often in riparian forest, more rarely in *Brachystegia* woodland and on termite mounds. 0–900 m.

In southern Mozambique at about 24°S subsp. *obovata* is suddenly replaced by subsp.

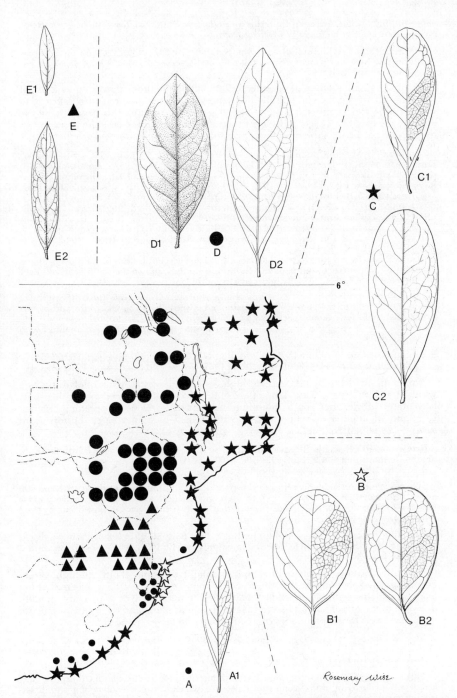

Tab. 66. Leaf-shape in EUCLEA NATALENSIS (all × ½). A. — Subsp. NATALENSIS. A1, *Barbosa* & *Lemos* 7786. B. — Subsp. ROTUNDIFOLIA. B1, *White* 10358 bis; B2, *White* 10452. C. — Subsp. OBOVATA. C1, *Clements* 548; C2, *Torre* 5746. D. — Subsp. ACUTIFOLIA. D1, *Burtt Davy* 20699; D2, *White* 10108. E. — Subsp. ANGUSTIFOLIA. E1, *Yalala* 178; E2, *Yalala* 159.

natalensis but a single specimen from this area (*Barbosa* 7838, Xai-Xai (Vila de Josão Belo), typologically at least, conforms to subsp. *obovata*.

Subsp. **acutifolia** F. White in Bull. Jard. Bot. Nat. Belg. **50**: 397 (1980). TAB. **66**, fig. D. Type: Zambia, Mazabuka Distr., Siamambo Forest Reserve, fl. 30.vii.1952, *Angus* 98 (K, holotype; BR; FHO).

Shrub or small bushy tree, usually 2–7 m. tall, sometimes flowering as a dwarf shrub or suffrutex c. 0·6 cm. tall. Bark dark grey, rough and fissured with flakes up to 1 cm. thick. Leaf-lamina 6·6 × 2·4–11·5 × 4 cm., subrhombic, broadest near the middle, apex acute, densely pubescent beneath with spreading crispate hairs and an understorey of small, shortly stalked, glandular hairs.

Zambia. N: Mbala-Mpulungu 7 km., fl. 2.iv.1961, *Angus* 2625 (FHO; SRGH). W: Kasempa, fl. 19.viii.1961, *Fanshawe* 6687 (FHO). C: Fiwila, fl. 29.ix.1957, *Fanshawe* 3763 (BR; K; SRGH). E: Chipata, fr. 1.vi.1958, *Fanshawe* 4496 (K). S: Choma Distr., Siamambo Forest Reserve, fr. 30.vii.1952, *Angus* 98 (BR; FHO; K). **Zimbabwe.** N: Msonedi, 26.ix.1937, *McGregor* 122/37 (BM; FHO; SRGH). W: Matopos Hills, Maleme Rest Camp, fr. 30.i.1973, *White* 10061 (FHO; SRGH). C: Selukwe, Ferny Creek, fr. 8.xii.1966, *Biegel* 1528 (BR; SRGH). E: between Chirinda andChipinga, fl. vii.1962, *Goldsmith* 161/62 (K; SRGH). S: Tokwe Dam, fr. 4.ii.1973, *White* 10108 (FHO; SRGH). **Malawi.** C: 64 km. N. of Kasungu, fl. 27.ix.1950, *Jackson* 175 (BR; K). **Mozambique.** MS: Serra de Garuso, fr. 4.iii.1948, *Barbosa* 1103 (FHO; K; LISC; LMU).

Also in Zaire (Upper Shaba) and Tanzania. In various types of scrub forest, bushland and thicket, especially on rocky outcrops and on termite mounds, and on the banks of rivers. Very local in miombo woodland.'Mostly between 915 and 1525 m.

The distribution of subsp. *acutifolia* is parapatric with that of subsp. *obovata* to the east and of subsp. *angustifolia* to the south. The transitions between them, however, are abrupt and intermediates are few. Some specimens from northern Zambia (*Bredo* 5191, *Lawton* 1519, *Mutimushi* 900) have a similar leaf-shape to subsp. *obovata*, but agree in their indumentum with subsp. *acutifolia*.

Subsp. **angustifolia** F. White in Bull. Jard. Bot. Nat. Belg. **50**: 397 (1980). TAB. **66**, fig. E. Type: Botswana, 24 km. E. of Kanye, fl. vi.1962, *Yalala* 159 (K, holotype; FHO; SRGH).

Shrub or small bushy tree 2–5 m. high, rarely (in South Africa) flowering as a shrublet 0·6 m. high. Leaf-lamina 3·2 × 0·9–7 × 1·6 cm., very rarely broader, lanceolate-elliptic, apex acute, densely pubescent beneath, with spreading crispate hairs and an under storey of small, shortly stalked glandular hairs. Differs from subsp. *acutifolia* in its smaller, proportionally narrower leaves.

Botswana. SE: Bakhatla Reserve, Mochudi, fl. ii.1914, *Rogers* 6630 (K). **Zimbabwe.** S: Shashi-Limpopo junction, fl. 2.ii.1973, *White* 10096 (FHO).
Also in the Transvaal. In bushland on rocky outcrops. 300–900 m.

Subsp. *angustifolia* suddenly replaces subsp. *acutifolia* in Zimbabwe at about 20°S and 900 m. alt. Although there is a slight overlap or interdigitation of their ranges subsp. *angustifolia* is very local in the contact zone, and it seems that the two subspecies do not occur in close proximity.

Subsp. **natalensis** — F. White in Bull. Jard. Bot. Nat. Belg. **50**: 397 (1980). TAB. **66**, fig. A. *Euclea multiflora* Hiern in Trans. Camb. Phil. Soc. **12**: 100, t. 3 fig. 1 (1873) sens. strict. Type from Natal.

Multiple stemmed shrub 2–4 m. high or small tree up to 12 (18) m. high, rarely flowering when only 60 cm.–1 m. high. Leaf-lamina 4·5 × 1·3–10·5 × 3·5 cm. Differs from subsp. *angustifolia* in its less hairy, less tapered leaves which are upually broadest in the upper half and seem to lack glandular hairs, and from subsp. *obovata* in its narrower, more pointed leaves. *Chromosome number*: 2n = 30.

Mozambique. GI: Panda Distr., Mauéele, st. ix.1936, *Gomes e Sousa* 1820 (K). M: Maputo Distr., Santaca, fl. 15.viii.1947, *Gomes e Sousa* 3600 (EA; K; LISC; SRGH).
Also in Swaziland and South Africa. In coastal forest and scrub forest. 10–300 m.

This is the most variable subspecies. It is both intermediate and bridges the morphological gap between subsp. *angustifolia* and subsp. *obovata*, which, if subsp. *rotundifolia* is excluded from consideration, are more different than any other pair of subspecies. Throughout almost the whole of its geographical range subsp. *natalensis* is the only subspecies to occur. Locally, however, it comes into contact with subspp. *acutifolia, obovata* and *rotundifolia*, and a few specimens are difficult to place or are genuinely intermediate. Nevertheless, the great majority of specimens of subsp. *natalensis* lie outside the range of variation of the three subspecies named above.

2. **Euclea crispa** (Thunb.) Gürke in Engl. & Prantl, Pflanzenfam. **4** (1): 158 (1891). — de Winter in F.S.A. **26**: 90 (1963). — Wild in Kirkia, **7**, Suppl.: 42 (1970). — Palmer & Pitman, Trees of Southern Afr. **3**: 1775 cum. tab. & photogr. (1972). — Wild, Biegel & Mavi, Rhod. Bot. Dict., ed. 2: 162 (1972). — Drummond in Kirkia, **10**: 266 (1975). — White in Gardens' Bull. Singapore, **29**: 68, t. 1 (1976); in Bull. Jard. Bot. Nat. Belg. **50**: 396 (1980). — K. Coates Palgrave, Trees of Southern Afr.: 736 (1977). TAB. **67**, fig. A–C. Type from South Africa.

Celastrus crispus Thunb. in Hoffm., Phyt. Blatt. **1**: 23 (1803). Type as above.

Shrub or small bushy tree up to 8 m. high, or a rhizomatous suffrutex or virgate shrub 0·3–3 m. high. Young shoots densely puberulous to glabrous, except for peltate, usually rusty, scales. *Leaves* alternate or subopposite; petiole up to 0·2 cm. long; lamina very variable, mostly obovate to oblanceolate or linear, margin often crenulate, usually revolute, or markedly undulate (subsp. *ovata*); lower surface always with peltate, usually rusty, scales, at least when young, nearly always (except in subsp. *linearis* sens. strict) also sparsely to densely puberulous with spreading crispate hairs, or tomentose (subsp. *ovata* when young), often glabrescent; venation prominent and reticulate, especially on upper surface, very rarely impressed above. *Inflorescence*, solitary, axillary, unbranched, up to 3 cm. long, always with peltate scales, and (except in subsp. *linearis* sens. strict) sparsely to densely puberulous. *Male flowers* c. 0·25 cm. long. Calyx patelliform, with very short, broadly deltate teeth. Corolla lobed to beyond the middle, glabrous except for the sparsely strigulose mid-petaline lines. Stamens 10–18, mostly in pairs; anthers strigulose. Pistillode minute, setulose, with or without well-developed stylodes. *Female flowers* with c. 8 filiform, setulose staminodes or staminodes absent. Ovary densely strigose. *Fruit* c. 0·6 cm. diameter, sparsely strigose, glabrescent. *Chromosome number*: 2n = 30.

From Angola and Zambia south to the Cape Province, South Africa. From 1220–2830 m. in our area, but descending almost to sea-level in South Africa.

E. crispa has an extremely scattered distribution north of the Zambezi, but is much more abundant further south. It occurs in a wide range of vegetation types and in no fewer than 5 of the phytogeographic regions recognized by White (Boissiera, **24**, p. 659–666, 1976), and hence is an ecological and chorological transgressor.

In leaf-shape it is one of the most variable of all African trees. Within this one species there is almost as much variation in leaf-shape as in the whole of the genus *Salix*. Compared with *E. natalensis* and *E. racemosa*, relatively little of this variation appears to be sufficiently well-correlated with ecology and geography to justify taxonomic recognition, and only two extreme variants are given subspecific rank.

Subsp. *ovata*, which occupies parts of the arid interior of South Africa, and extends only a short way into the Flora Zambesiaca area in south-east Botswana, behaves as a normal subspecies. Subspecies *linearis*, however, which has a disjunct distribution, occurs in three widely separated areas which are somewhat different in their ecology. The possibility of a polytopic origin from *E. crispa* subsp. *crispa* or a common ancestor cannot entirely be discounted, but requires further investigation.

Leaf margin strongly undulate - - - - - - - - subsp. *ovata*
Leaf margin not or slightly undulate:
  Leaf-lamina less than 7 times as long as broad - - - - - subsp. *crispa*
  Leaf-lamina at least 7 times as long as broad:
    Leaf-lamina 7–13·9 times as long as broad - - - - subsp. *linearis* sens. lat.
    Leaf-lamina at least 14 times as long as broad - - - subsp. *linearis* sens. strict.

Subsp. **ovata** (Burch.) F. White in Bull. Jard. Bot. Nat. Belg. **50**: 397 (1980). TAB. **67**, fig. B. Type from South Africa.

Euclea ovata Burch., Trav. **1**: 387 (1822). — A.DC. in DC., Prodr. **8**: 218 (1844). — Hiern in Trans. Camb. Phil. Soc. **12**: 98 (1873) excl. syn. *Celastrus crispus* Thunb. Type as above.

Euclea crispa var. *ovata* (Burch.) de Winter in Bothalia **7**: 403 (1960); in F.S.A. **26**: 92 (1963). Type as above.

Bushy tree up to 5 m. tall. Leaf-lamina up to 3·6 × 2 cm., rhombic-elliptic or trullate-elliptic, tapering from about the middle to the narrowly acute and usually mucronate apex, margin strongly undulate, lower surface densely pubescent with long spreading hairs, at least when young.

Botswana. SE: Pharing, Mokgodumo valley, ♀ fl. xii.1947, *Miller* B/550 (K).
Also in the Orange Free State and northern Cape Province. In bushland. 1200 m.

Subsp. **crispa** — F. White in Bull. Jard. Bot. Nat. Belg. **50**: 396 (1980). TAB. **67**, fig. A.

Euclea crispa var. *crispa* — de Winter in Bothalia, **7**: 403 (1960); in F.S.A. **26**: 92, t. 11 fig. 1 a–j (1963).

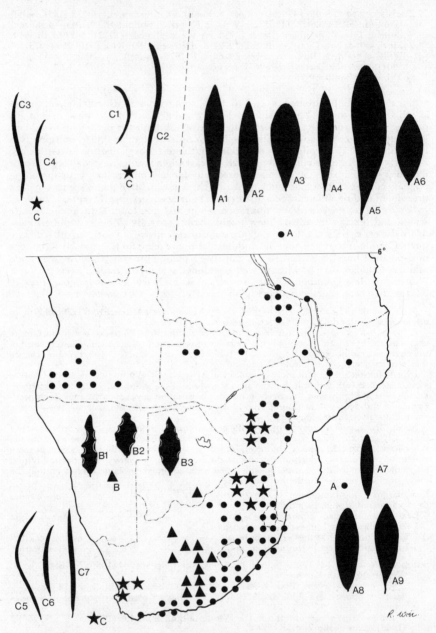

Tab. 67. Leaf-shape in EUCLEA CRISPA (all × ½). A. — Subsp. CRISPA. A1, *Menezes* 1854 (Angola); A2, *Menezes* 1250 (Angola); A3, *Santos* 2232 (Angola); A4, *Richards* 6981 (Zambia); A5, *Brenan* & *Greenway* 8145 (Zambia); A6, *Torre* & *Paiva* 16066 (Mozambique); A7, *White* 10073a (Zimbabwe); A8, *White* 10073b (Zimbabwe); A9, *Barrett* 45/55 (Zimbabwe). B. — Subsp. OVATA. B1, *Potts* 3827 (Orange Free State); B2, *Botha* R38 (Orange Free State); B3, *Acocks* 13517 (Orange Free State). C. — Subsp. LINEARIS. C1, *White* 10042c (Zimbabwe); C2, *White* 10015 (Zimbabwe); C3, *Keet* 6057 (Transvaal); C4, *Poynton* s.n. (Transvaal); C5, *White* 10701d (Cape Province); C6, *White* 10701c (Cape Province); C7, *White* 10701e (Cape Province).

Euclea lanceolata E. Mey. ex A.DC. Prodr. **8**: 218 (1844). — Hiern in Trans. Camb. Phil. Soc. **12**: 97 (1873), pro parte. — S. Moore in Journ. Linn. Soc., Bot. **40**: 134 (1911). — R.E. Fr., Wiss. Ergebn. Schwed. Rhod.-Kongo-Exped. **1**: 256 (1916). Type from South Africa.

   Gymnosporia ferruginea Baker in Kew Bull. **1897**: 247 (1897). Type: Malawi, Zomba Mt., ♂ fl. xii.1896, *Whyte* s.n. (K, holotype), synon. nov.

   Euclea dekindtii Gürke in Engl. Bot. Jahrb. **32**: 138 (1908). — F. White, F.F.N.R.: 333 (1962). Type from Angola, synon. nov.

   Euclea baumii Gürke in Warb. Kunene-Sambesi-Exped. Baum: 327 (1903). Type from Angola, synon. nov.

   Euclea bakerana Brenan in Mem. N.Y. Bot. Gard. **8**: 238 (1953). Type as for *Gymnosporia ferruginea*, synon. nov.

Rhizomatous suffrutex, shrub or small bushy tree up to 8 m. high. Leaves sometimes glaucous; lamina less than 7 times as long as broad, up to 8·5 cm. long and 3 cm. broad, usually straight, very rarely subfalcate, apex usually muticous, rarely shortly mucronate, margin not undulate, branchlets usually puberulous, very rarely subglabrous. *Chromosome number*: 2n = 30.

Zambia. N: Mbala Distr., top of Sunzu Hill, 2040 m, ♀ fl. 18.xi.1952, *White* 3704 (FHO; K). W: Kasempa, fr. 12.viii.1961, *Fanshawe*, 6672 (FHO). C: Katanino, ♂ fl. 3.xi.1955, *Fanshawe* 2568 (K). **Zimbabwe.** W: Matopos Hills, Besna Kobila, fl. 30.i.1973, *White* 10073a (FHO). C: 16 km. S. of Marandellas, 1370 m., fl. 16.xii.1962, *Moll* 317 (FHO; SRGH). E: Umtali Distr., Nyamaganu Peak, 1220 m., fl. 3.i.1960, *Chase* 7246 (BM; FHO; SRGH). S: Belingwe Distr., Mt. Buhwa 1250 m., fr. 1.v.1973, *Biegel* et al. 4249 (FHO; SRGH). **Malawi.** N: Mafingi Mts., 2380 m., ♂ fl. 22.xi.1952, *Angus* 836 (FHO). C: Lilongwe, ♀ fl. 15.xi.1951, *Jackson* 643 (BM; EA; K). S: Mt. Mulanje, Lukulezi valley, 1675–1830 m., ♂ fl. 10.x.1959, *Chapman* 457 (BM; FHO; PRE). **Mozambique.** N: Ribáuè Distr., Serra Mepáluè, 1700 m., fl. 11.xii.1967, *Torre & Correia* 16433 (COI; K; LISC; LMU). Z: Serra Patapane, 5 km. from Lioma, 1500 m., ♀ fl. 13.xi.1967, *Torre & Correia* 16066 (C; FHO; LISC; MO; SRGH; WAG). MS: Mount Umtereni, 1220 m., ♂ fl. 11.ix.1906, *Swynnerton* 1306 (K; SRGH).

Also in Angola and South Africa. In various types of woodland, especially in rocky places. Also on termite mounds and near dambos, and at edges of montane forest. 1220–2380 m. It can tolerate high concentrations of nickel in the soil.

In different parts of its range *E. crispa* subsp. *crispa* adopts a different habit. In Zambia it is almost invariably a suffrutex less than 0·5 m. tall, whereas on the mountains of Malawi and Mozambique its characteristic growth form is a 1–2 m. high shrub. In Zimbabwe it ranges from a 30 cm. high suffrutex to a tree 8 m. tall.

In leaf shape *E. crispa* subsp. *crispa* is sometimes very similar to *E. divinorum* and the two species have often been confused. The leaves of *E. crispa*, however, are usually crenulate, broadest above the middle and are cuneate, not slightly concave, at the base. The venation is also prominent on the upper surface and not reddish beneath. In addition, *E. crispa* normally has hairy branchlets, a shorter petiole and only 1 inflorescence in each leaf-axil.

Subsp. **linearis** (Zeyher ex Hiern) F. White in Bull. Jard. Bot. Nat. Belg. **50**: 396 (1980). TAB. **67**, fig. C. Type from South Africa.

   Euclea linearis Zeyher ex Hiern in Trans. Camb. Phil. Soc. **12**: 96 (1873). — de Winter in F.S.A. **26**: 96 (1963). — Wild in Kirkia, **5**: 63, 70, 71, 75, 83, phot. 5 (1965); op. cit., Suppl.: 29 (1970); op. cit. **9**: 213 (1974). — Palmer & Pitman, Trees of Southern Afr. **3**: 1777 cum tab. (1972). — Wild, Biegel & Mavi, Rhod. Bot. Dict., ed. 2: 163 (1972). — Drummond in Kirkia, **10**: 266 (1975). — K. Coates Palgrave, Trees of Southern Afr.: 738 (1977). Type as above.

   Euclea eylesii Hiern in Journ. Bot. Lond. **45**: 47 (1907). Type: Zimbabwe, Sebakwe, fl. nii.1904, *Eyles* 44 (K, holotype; BM).

Rhizomatous, sparsely branched shrub, usually 1–2 m. high and forming extensive clones. Older plants reach a height of 3 m. but are often leafless except towards the extremities. Leaves linear, more than 7 times as long as broad. *Chromosome number*: 2n = 30.

E. linearis cannot be upheld as a species since it is connected to *E. crispa* by innumerable and extensive populations which completely bridge the morphological gap between them. In Zimbabwe, however, the extreme variant in this complex (*E. crispa* subsp. *linearis* sens. strict., see below) is one of the most characteristic and conspicuous plants on serpentine soils in the southern half of the country (Wild, op. cit., White, pers. obs.). Here it is remarkably uniform over extensive areas. In general, in Zimbabwe, subsp. *linearis* sens. strict., is confined to serpentine and subsp. *linearis* sens. lat. characteristically occurs on more acidic soils.

Because variation within the *E. crispa/linearis* complex is continuous, the delimitation of the latter is somewhat arbitrary. However, nearly all the plants growing on the southern serpentines have leaves with a length/breadth ratio greater than 14. By contrast, the leaves of *E.*

crispa subsp. *crispa*, outside the area of overlap with subsp. *linearis*, are rarely more than 7 times as long as broad. Hence these figures commend themselves respectively for the circumscription of subsp. *linearis* sens. strict. and subsp. *crispa*.

The southern serpentines occupied by subsp. *linearis* sens. strict. represent an area of uniformity within the *E. crispa* complex, but elsewhere in Zimbabwe the pattern is very complex. In some places, specimens which, typologically at least, belong to subsp. *crispa* and subsp. *linearis* sens. strict. can occur in the same population, e.g. *White* 10025 (subsp. *crispa*) and *White* 10026 *bis* (subsp. *linearis* sens. strict.) from a serpentine outcrop at the "Terrace", south of Gwelo, and *White* 10032 & 10036 (subsp. *linearis* sens. strict.), which form part of a sample of 25 gatherings, otherwise consisting only of subsp. *crispa* and subsp. *linearis* sens. lat. This sample came from a population on granitic soils at Jacobsen's Farm, south of Gwelo.

Subsp. *linearis* sens. strict. also occurs in the Transvaal and the western part of Cape Province. It is somewhat variable. Nevertheless some individuals from all three areas are virtually identical. In the Transvaal it occurs both on serpentine and acid rocks (Waterberg Sandstone, Black Reef Quartzite). In Cape Province it occurs on Table Mountain Sandstone.

(a) subsp. **linearis** sens. strict.

Leaf-lamina 14 to 27 times as long as broad, up to 9·2 cm. long and 0·45 cm. broad, usually falcate, sometimes subfalcate, apex usually distinctly mucronate, sometimes slightly mucronate; branchlets usually glabrous, rarely puberulous (invisible to naked eye).

Zimbabwe. W: Umzingwane Distr., Essexvale Hills, st. 12.vi.1977, *Webber* s.n. (SRGH). C: Great Dyke, Silver Star Ranch, south of Ngezi Dam, fr. 27.i.1973, *White* 10016 (FHO; SRGH); Great Dyke, 24 km. S. of Selukwe, fr. 16.iii.1964, *Wild* 6365 (FHO; SRGH). S: Belingwe Distr., Shabani, fr. 28.i.1973, *White* 10042a (FHO).

Also in South Africa. Characteristically in open woodland and wooded grassland on serpentine. 1000–1450 m.

It is known as "Asbestos Bush" since asbestos is extensively mined in the serpentines of Zimbabwe.

(b) subsp. **linearis** sens. lat.

Leaf-lamina from 7–13·9 times as long as broad, up to 12·7 cm. long and 1·1 cm. broad, usually slightly falcate, apex usually slightly mucronate, rarely distinctly mucronate or muticous; branchlets usually puberulous, very rarely glabrous.

Zimbabwe. N: Lomagundi Distr., Trefonen farm, 19 km. NW. of Darwendale, 1350–1400 m., fr. 3.iii.1973, *Gordon* s.n. (SRGH). C: Salisbury, Prince Edward Dam road, fl. x.1948, *Armitage* 14/48 (SRGH); Gwelo to Lalepansi 24 km., imm. fr. 27.i.1973, *White* 10023 (FHO; SRGH). E: Inyanga Distr., Mt. Dombo, 1880 m., fr. 17.vii.1974, *Müller* 2175 (FHO; SRGH). S: Zimbabwe-Fort Victoria 14 km., imm. fr. 4.ii.1973, *White* 10122 (FHO).

Also in South Africa. In woodland and open woodland, especially on termite mounds and rocky outcrops 1150–1880 m.

3. **Euclea racemosa** Murr., Syst. Veg., ed. **13**: 747 (1774). — Hiern in Trans. Camb. Phil. Soc. **12**: 104 (1873). — O. B. Mill. in Journ. S. Afr. Bot. **18**: 68 (1952). — de Winter in F.S.A. **26**: 97 (1963). — Palmer & Pitman, Trees of Southern Africa, **3**: 1783 (1972). — K. Coates Palgrave, Trees of Southern Afr.: 739 (1977). — F. White in Bull. Jard. Bot. Nat. Belg. **50**: 398 (1980). TAB. **68**, fig. A–C. Type from South Africa.

Evergreen shrub or small tree up to 12 m. tall. Foliage dark green and glossy. Young shoots glabrous. *Leaves* alternate, subopposite, opposite or verticillate; petiole 0·1–0·3 cm. long; lamina very variable in shape and size, up to 11 × 4 cm.; lower surface glabrous; secondary nerves subprominent on upper surface, less so beneath, but conspicuous because of pink coloration, tertiary nerves and veins usually indistinct, especially on lower surface. Inflorescence 1·5–5 cm. long, unbranched, with glabrous axes. *Male flowers* up to 0·5 cm. long. Calyx 0·2 cm. long, saucer-shaped, glabrous, except sometimes for the minutely setulose margins, shallowly lobed or denticulate. Corolla lobed only in upper half, glabrous or sparsely strigulose along the median axis of the lobes. Stamens 10–20, mostly in pairs, filaments up to 0·13 cm. long; anthers up to 0·3 cm. long, strigulose, especially near apex. *Female flowers* similar to male but much smaller. Staminodes absent. Ovary glabrous or strigulose-tomentellous. *Fruit* 0·6–0·8 cm. in diameter. *Chromosome number*: 2n = 30.

From Egypt southwards to South Africa. From near sea-level to 1525 m.

Like *E. natalensis*, *E. racemosa* shows a complicated pattern of ecogeographical replacement, with most variation occurring in South Africa. Of the 6 South African subspecies, 2, subsp.

zuluensis and subsp. *sinuata*, enter our area only in southern Mozambique. The remaining subspecies, *schimperi*, which occurs north of the R. Limpopo is not very variable. However, some narrow-leaved specimens from the extreme south of its range approach subsp. *zuluensis*.

Leaves with strongly undulate margins, broadest near the apex - - - subsp. *sinuata*
Leaves usually not strongly undulate, broadest well below the apex:
    Leaves up to 4 times as long as broad, distinctly broadest in the upper half subsp. *schimperi*
    Leaves more than 5·5 times as long as broad, not or scarcely broadest above the middle
                                        subsp. *zuluensis*

Subsp. **sinuata** F. White in Bull. Jard. Bot. Nat. Belg. **50**: 399 (1980). TAB. **68**, fig. B. Type from Natal.

Leaves obovate, up to 5·2 × 2·4 cm. Differs from subsp. *schimperi* in its relatively broader leaves, which are more broadly rounded at the apex, much more attenuate at the base and have strongly undulate margins. Ovary stigulose-tomentellous. *Chromosome number*: 2n = 30.

Mozambique. SS: Rio das Pedras, ♂ fl. vii.1936, *Gomes e Šousa* 1790 (BR; K; LISC). LM: Inhaca Isl., fr. 8.vi.1970, *Correia & Marques* 1707 (LISC).
Also in Natal. Evergreen coastal thicket and scrub forest. From near sea-level to 200 m.

Subsp. **schimperi** (A.DC.) F. White in Bull. Jard. Bot. Nat. Belg. **50**: 399 (1980). TAB. **68**, fig. A. Type from Ethiopia.
    Kellaua schimperi A.DC. in Ann. Sci. Nat., Sér. 2, Bot. **18**: 209 (1842); in DC., Prodr. **8**: 290 (1844). Type as above.
    Euclea schimperi (A.DC.) Dandy in Andrews, Fl. Pl. Anglo-Egyptian Sudan, **2**: 370 (1952); in Kew Bull. **1953**: 108 (1953). — F. White, F.F.N.R.: 333, t. 61 (1962). — Drummond in Kirkia, **10**: 266 (1975). Type as above.
    Euclea kellau Hochst. in Flora, **26**: 83 (1843). — Hiern in Trans. Camb. Phil. Soc. **12**: 103 (1873). Steedman, Trees, etc. S. Rhod.: 63 (1933). Type from Ethiopia.
    Euclea mayottenis H. Perrier in Mém. Inst. Sci. Madag., Sér. B, **4**: 93, t. 1 fig. 1–4 (1952); Fl. Madag. Ebenaceae: 3, t. 1 fig. 1–4 (1954), synon. nov. Type from Comoro Isl.
    Euclea macrophylla sensu Wild, Guide Fl. Vict. Falls: 152 (1953).

Leaves thinly coriaceous, obovate or oblanceolate, up to 12 × 4 cm., up to 4 times as long as broad, distinctly broadest in the upper half. Ovary strigulose-tomentellous. *Chromosome number*: 2n = 30.

Zambia. N: Chienge, fr. 18.viii.1958, *Fanshawe* 4733 (K). W: Solwezi to Mwinilunga 130 km., ♂ fl. 16.ix.1952, *White* 3268 (BM; BR; EA; FHO; K). C: Lusaka Forest Nursery, ♂ fl. 4.iii.1952, *White* 2188 (BR; FHO; K). E: Lundazi, fr. 24.viii.1961, *Grout* 263 (FHO). S: Ngonga R., ♂ fl. 13.i.1960, *White* 6235 (FHO; K). **Zimbabwe.** N: Mazoe Distr., Chipoli, Maponongwe Drift, 820 m., fr. 20.ix.1958, *Moubray* 16 (K). W: Victoria Falls, fr. 17.xi.1949, *Wild* 3082 (FHO; K; SRGH). C: Rusape, ♀ fl. fr. 1952, *Dehn* 277a/53 (K). E: Nyagadza R., 8 km. S. of Chikore Mission 762 m., ♂ fl. iii.1962, *Goldsmith* 107/62 (FHO; SRGH). S: near Tokwe Dam, st. 4.ii.1973, *White* 10111 (FHO). **Malawi.** N: Mzimba R., 1370 m., ♀ fl. 26.ii.1959, *Robson* 1726 (BM; FHO; K; SRGH). C: Chongoni Forest, Bunda Hill, 1525 m., ♀ fl. 26.iii.1961, *Chapman* 1180 (FHO; SRGH). S: Lower Shire R. Valley, 92 m., st. v.1959, *Young* 43 (MAL). **Mozambique.** N: Malema, Mutuáli, base of Mt. Cucuteia, junction of R. Neuce, 650 m., ♀ fl. 16.iii.1964, *Torre & Paiva* 11205 (C; COI; FHO; K; LISC; LMU). Z: Alto Molócuè, 11 km. from Mocuba-Nauela crossroads, fr. 16.x.1949, *Barbosa & Carvalho* 4455 (K). T: monte de Zóbuè, fr. 21.x.1941, *Torre* 3686 (C; COI; FHO; K; LISC; LMU). MS: Marromeu, near Chupanga, ♂ fl. 12.ix.1942, *Mendonça* 197 (BR; EA; FHO; LISC; LUA; M; P; SRGH).
From Egypt, the Yemen and Oman southwards to our area. Also in east and southern Zaire and the Comoro Islands. Mostly in evergreen or semi-evergreen bushland, thicket and scrub forest, especially on termite mounds, on banks of watercourses and in rocky places. 90–1525 m.

Subsp. **zuluensis** F. White in Bull. Jard. Bot. Nat. Belg. **50**: 399 (1980). TAB. **68**, fig. C. Type from Natal.
    E. schimperi var *daphnoides* (Hiern) de Winter in F.S.A. **26**: 99 (1963) quoad relat. Natal et Transvaal tantum.

Leaves thickly coriaceous, narrowly oblong-elliptic or oblanceolate-elliptic, up to 7·2 × 1·3 cm., not or scarcely broadest above the middle. Ovary strigulose-tomentellous.

Mozambique. GI: Gaza, Caniçado, Mabalane (Vila Pinto Teixeira) to Meginge 23 km., fl. 22.viii.1969, *Correia & Marques* 1147 (LISC; SRGH). M: Bela Vista, Mazimiane Farm, base of Libombos Mts., immat. fr. 19.iii.1948, *Gomes e Sousa* 3701 (K; LISC).
Also in Swaziland and South Africa. In semi-evergreen bushland. 50–500 m.

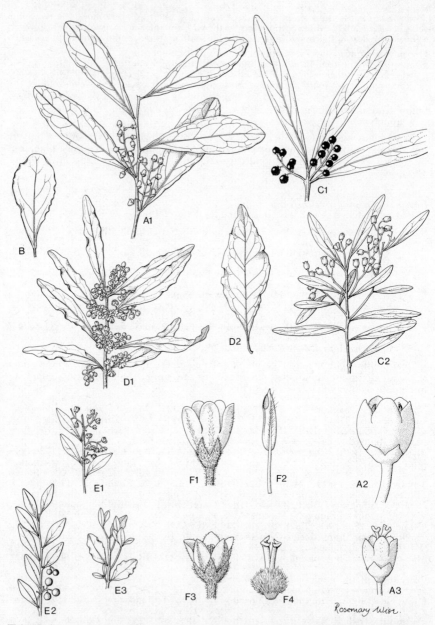

Rosemary Wise.

Tab. 68. A. — EUCLEA RACEMOSA subsp. SCHIMPERI. A1, flowering branchlet (× ½); *Goldsmith* 107/62; A2, male flower (× 4); *Goldsmith* 107/62; A3, female flower (× 4); *Apyettu* 27. B. — E. RACEMOSA subsp. SINUATA (× ½); *White* 10366. C. — E. RACEMOSA subsp. ZULUENSIS. C1, fruiting branchlet (× ½); *White* 10490; C2, flowering branchlet (× ½); *White* 10422. D. — E. DIVINORIUM. D1, flowering branchlet (× ½); *Angus* 1696; D2, leaf (× ½); *Angus* 71. E. — E. UNDULATA. E1, flowering branchlet (× ½); *de Beer* 679; E2, fruiting branchlet (× ½); *Yalala* 162; E3, leaves (× ½); *White* 10078 bis. F. — E. NATALENSIS. F1, male flower (× 4); *Graham* 2064; F2, stamen (× 10); *Graham* 2064; F3, female flower (× 4); *Codd* 1633; F4, pistil (× 8); *Codd* 1633.

Specimens from Natal, Swaziland and Mozambique form a very distinct taxon, which is sharply differentiated from subsp. *sinuata* which occurs in adjacent, more coastal, habitats. Specimens from the Transvaal, however, approach narrow-leaved variants of subsp. *schimperi* though nearly all lie outside the range of variation of the latter.

4. **Euclea divinorum** Hiern in Trans. Camb. Phil. Soc. **12**: 99 (1873). — Sim, For. Fl. Port. E. Afr.: 81 (1909). — R.E. Fr., Wiss. Ergebn. Schwed. Rhod.-Kongo-Exped. **1**: 256 (1916). — Codd, Trees & Shrubs Kruger Nat. Park: 144, t. 134 (1951). — O. B. Mill. in Journ. S. Afr. Bot. **18**: 67 (1952). — Wild, Guide Fl. Vict. Falls: 152 (1953); in Kirkia, **7**: 33, 34 (1968); in Portugaliae Acta Biol., B, **9**: 297 (1968); in Kirkia, **7**, Suppl.: 42 (1970); op. cit. **9**: 216, 218, 220, 243, 247, 249, 251, 252, 255, 269, 274 (1974). — F. White, F.F.N.R.: 333 (1962). — de Winter in F.S.A. **26**: 94 (1963). — de Winter & F. White, Prodr. Fl. SW. Afr. **107**: 6 (1967). — Palmer & Pitman, Trees of Southern Afr. **3**: 1779 cum tab & photogr. (1972). — Wild, Biegel & Mavi, Rhod. Bot. Dict., ed. 2: 163 (1972). — Drummond in Kirkia, **10**: 266 (1975). — K. Coates Palgrave, Trees of Southern Afr.: 737, t. 242 (1977). TAB. **68**, fig. D. Types: Zambia, Victoria Falls, ♂ fl. ix.1860, *Kirk* s.n. (K, lectotype, here designated); Mozambique, Delagoa Bay, ♂ fl., *Forbes* s.n. (K, paratype).

    Euclea lanceolata sensu Hiern, tom. cit.: 97 (1873) pro parte.
    Euclea katangensis De Wild. in Ann. Mus. Congo, Sér. 4 (Bot), Études Fl. Katanga: 222 (1903) synon. nov. Type from Zaire (Upper Shaba).
    Euclea huillensis Gürke in Warburg, Kunene-Sambesi-Exped.: 326 (1903) synon. nov. Type from Angola.
    Embelia oleifolia S. Moore in Journ. Bot. Lond. **47**: 297 (1909). Types: Zimbabwe, Bulawayo, fr. 1908, *Chubb* 31 (BM, syntype; SRGH); ♀ fl. vi. 1898, *Rand* 504 (BM, syntype).

Evergreen shrub or small tree up to 9 m. tall (elsewhere up to 15 m.). Bark grey-brown to black, usually rough and longitudinally fissured. Young shoots glabrous except for rusty, peltate scales. *Leaves* usually opposite or subopposite; petiole 0·4–0·6 cm. long; lamina up to 9 × 4 cm., variable in width, otherwise uniform, subrhombic or subtrullate, broadest at or below the middle, apex acute, the tip itself rounded, base attenuate and slightly concave, margin often strongly undulate; lower surface glabrous except for rusty, peltate scales; lateral nerves and main veins sometimes slightly raised above, otherwise inconspicuous. *Inflorescence* paired in at least some leaf-axils, unbranched, dense and contracted, up to 1·5 cm. long. *Male flowers* c. 0·35 cm. long. Calyx patelliform with short, broadly deltate teeth, glabrous except for rusty, peltate scales. Corolla deeply lobed, widely open at the throat, each lobe with more than 25 strigulose hairs along the mid-petaline line. Stamens c. 16, strigulose. Pistillode with 2 simple or bilobed stylodes. *Female flowers* without staminodes. Ovary densely strigulose. *Fruit* c. 0·7 cm. diameter. *Chromosome number*: 2n = 30.

Caprivi Strip: Linyanti to Katimo Mulilo, ♂ fl. xii.1958, *Killick & Leistner* 3173 (EA; M; SRGH). **Botswana.** N: Okavango R., Sepopa, ♀ fl. 25.ix.1954, *Story* 4737 (FHO; M). SE: Tuli Block, Majali R., ♂ fl. 9.viii.1962, *Yalala* 191 (BM; K; SRGH). **Zambia.** B: Mashi R., Shangombo, ♂ fl. 14.viii.1952, *Codd* 7555 (BM; EA; K; SRGH). N: Choma R., Kaputa, immat. fr. 18.x.1949, *Bullock* 1310 (K). C: Kapiri Mposhi, ♂ fl. 5.viii.1957, *Fanshawe* 3447 (FHO; K). S: Choma Distr., Siamambo Forest Reserve, ♂ fl. 8.ix.1957, *Angus* 1696 (BR; FHO; K). **Zimbabwe.** N: Bindura, Chipoli Farm, ♂ fl. 10.ix.1958, *Moubray* in GHS 88335 (K; SRGH). W: Wankie Distr., Kalala R., imm. fr. 24.x.1968, *Rushworth* 1236 (K; SRGH). C: Jacobsen's Farm, 10 km. S.E. of Gwelo, ♂ fl. 11.vi.1966, *Biegel* 1238 (FHO; SRGH). E: Umtali, Minini R., Darlington Commonage, ♀ fl. 23.ix.1953, *Chase* 5073 (BM; BR; FHO; K; SRGH). S: Lundi R., Chipinda Pools, immat. fr. 29.x.1960, *Goodier* 23 (FHO; K; SRGH). **Malawi.** S: Shire Valley, Lengwe, fr. v.1970, *Hall-Martin* 709 (FHO). **Mozambique.** N: Quissonga to Mipande 3 km., ♂ fl. vii.1953, *Pedro* 4092 (LMA). T: Zóbuè to Tete 50 km., 350 m., immat. fr. 12.iii.1964, *Torre & Paiva* 11166 (C; COI; FHO; K; LISC; LMA). MS: Sabi R., Maringa, st. 28.vi.1950, *Chase* 2564 (BM; FHO; LISC; SRGH). GI: Vilanculos, between Mabote and Zimane, ♂ fl. 1.ix.1944, *Mendonça* 1937 (EA; LISC; MO; SRGH; WAG), ♀ fl. 2.ix.1944, *Mendonça* 1954 (BR; LD; LISC; M; P; PRE). M: Magude, Ghobela, fr. 7.i.1948, *Torre* 7073 (C; FHO; LISC; WAG).

From Ethiopia and the Sudan Republic to South Africa. In various types of woodland, bushland and thicket, especially in rocky places, banks of rivers and on termite mounds. From near sea-level to 1400 m.

In the southern part of its range *E. divinorum* always has narrow leaves. A broad-leaved northern variant has been given specific rank (*E. huillensis*, *E. katangensis*, *E. keniensis*). Its distribution, however, is too sporadic to justify taxonomic recognition at any rank.

E. divinorum is moderately tolerant of heavy metals and often becomes locally abundant on metalliferous soils. It tolerates high concentrations of nickel, but only relatively low values of copper. It is particularly common on arsenical soils which are often associated with gold deposits or reefs, and hence it may have value as an indicator of the presence of gold. Caution needs to be observed, however, since it is often also equally abundant elsewhere. Although not a characteristic serpentine species, it is locally frequent where serpentine is associated with related rocks. It also occurs in lightly wooded grassland on the steep skeletal soils of graphitic quartzite outcrops in Mangula District, Zimbabwe.

5. **Euclea undulata** Thunb., Prodr.: 85 (1796). — Hiern in Trans. Camb. Phil. Soc. **12**: 105 (1873). — de Winter in F.S.A. **26**: 95 (1963). — de Winter & F. White, Prodr. Fl. SW. Afr. **107**: 6 (1967). — Wild in Kirkia, **7**: 29 (1968); op. cit. **9**: 247 (1974). — Palmer & Pitman, Trees of Southern Afr. **3**: 1781 cum tab. & photogr. (1972). Wild, Biegel & Mavi, Rhod. Bot. Dict., ed. **2**: 163 (1972). — Drummond in Kirkia, **10**: 266 (1975). — K. Coates Palgrave, Trees of Southern Afr.: 741 (1977). TAB. **68**, fig. E. Type from South Africa.
  Euclea undulata var *myrtina* (Burch.) Hiern, tom. cit.: 106 (1873). — de Winter, tom. cit.: 97 (1963). Type from South Africa, synon. nov.
  Euclea myrtina Burch., Trav. **1**: 465 (1822). Type as above.

Evergreen shrub or small tree up to 6 m. tall. Young shoots glabrous except for rusty peltate scales. *Leaves* usually opposite or subopposite; petiole up to 0·3 cm. long; lamina up to 4 cm. long and 1·5 cm. broad, mostly obovate, oblanceolate or oblanceolate-elliptic, apex obtuse to broadly rounded, base cuneate, but not concave, margin often strongly undulate; lower surface glabrous except for peltate scales; lateral nerves and veins almost invisible. *Inflorescence* solitary in leaf-axils, unbranched, c. 1 cm. long. *Male flowers* c. 0·3 cm. long. Calyx patelliform, denticulate, glabrous except for peltate scales. Corolla deeply lobed, widely open at the throat, each lobe with 0–20 (28) strigulose hairs. Stamens c. 16, glabrous or strigulose. Pistillode with 2 simple stylodes. *Female flowers* without staminodes. Ovary covered with whitish scales, otherwise glabrous. *Fruit* c. 0·7 cm. diameter. *Chromosome number:* 2n = 30.

Botswana. SE: Gaberones, Pharing, fl. fr. iii.1949, *Miller* B847 (K). **Zimbabwe.** W: Nyamandhlovu Research Station, ♀ fl. 9.iv.1962, *Denny* 363 (SRGH); Kezi to Sun Yat Sen 9 km., 31.i.1973, *White* 10078 bis (FHO). C: Que Que Distr., Silobela area, Vungu mine, st. 5.iii.1970, *Wild* 7789 (SRGH). **Mozambique.** M: Inhaca Isl., ♂ fl.-buds ix.1961, *Mogg* 31511 (J).
  Also in Namibia, Swaziland and South Africa. In bushland and scrub woodland. From near sea-level to 1200 m.

E. undulata tolerates low concentrations of copper in the soil. It is also locally dominant as dwarfed plants on soils with a high antimony and arsenic content.
E. undulata is very closely related to *E. divinorum*, which it replaces in large parts of South Africa and Namibia. In the narrow zone of overlap they sometimes occur side by side without any sign of intermediates.

108. OLEACEAE
By F. K. Kupicha

Trees, shrubs, climbers or suffrutices. Leaves opposite, rarely verticillate or alternate, exstipulate, simple or pinnate, sometimes with acarodomatia (small ± circular pits, often densely fringed with hairs) in axils of nerves on lower leaf-surface. Inflorescence cymose, often paniculate (thyrsoid), sometimes fasciculate, sometimes only one flower developing. Flowers actinomorphic, ☿ or rarely unisexual (not in F.Z. area), sometimes heterostylous. Calyx gamosepalous, lobes 4-many (obscure in *Schrebera*). Corolla gamopetalous, lobes 4-many. Stamens 2, epipetalous. Ovary superior, bicarpellate, bilocular, with 1, 2 or 4 ovules per loculus, axile, pendulous or ascending; style 1, stigma capitate or bifid. Fruit baccate, drupaceous or capsular. Seeds sometimes winged, usually endospermous.

A family of c. 25 genera and 400–500 species, with world-wide distribution.

Species of *Ligustrum* L., *Fraxinus* L. and *Forsythia* Vahl are sometimes cultivated in gardens and nurseries.

1. Corolla hypocrateriform; flowers conspicuous, heterostylous - - - - - 2
   - Corolla cup-shaped, lobes united only at base; flowers individually inconspicuous (*Olea* and *Chionanthus*) or attractive (*Menodora*), not heterostylous - - - - - 3
2. Corolla with patches of dark hairs at the mouth; trees or shrubs, not climbing; fruit a woody capsule containing 4 or 8 winged seeds - - - - - - - **1. Schrebera**
   - Corolla glabrous; plants usually sarmentose; fruit a deeply bilobed berry (or simple by abortion), each lobe containing one seed - - - - - - **2. Jasminum**
3. Scabrid suffrutex with trifid to pinnatifid, alternate leaves up to 2 cm. long; fruit borne on a strongly recurved pedicel, a membranous bilobed capsule with circumscissile dehiscence
   3. Menodora
   - Trees and shrubs with entire leaves 1·8–17 cm. long; fruit on erect pedicel, a drupe - 4
4. Corolla lobes oblong, cucullate; leaves usually with acarodomatia, without scales
   4. Chionanthus
   - Corolla lobes ovate, flat; leaves without acarodomatia, often with minute silvery scales on lower surface - - - - - - - - - - - - - **5. Olea**

1. SCHREBERA Roxb.

Schrebera Roxb., Pl. Corom. **2**: 1, t. 101 (1799), *nom. conserv.*
Nathusia Hochst. in Flora **24**: 671 (1841).

Trees and shrubs. Leaves opposite, simple or compound, petiolate, glabrous or hairy; acarodomatia absent. Inflorescences few- to many-flowered, cymose; flowers heterostylous, scented. Calyx campanulate, truncate or obscurely lobed. Corolla pale cream or pink, hypocrateriform, usually pilose within the tube; lobes 5–7, imbricate, each with a conspicuous patch of swollen brownish or purplish hairs at the base. Stamens 2, inserted at or near the mouth of the corolla tube; filaments short, anthers introrse. Ovary bilocular, with 4 ovules per loculus; style slender; stigma exserted or included, subcapitate or cylindrical. Fruit a woody bilocular capsule with loculicidal dehiscence. Seeds endospermous, with a long subapical wing, spinning during dispersal.

A small genus with 3 species in Africa, one (*S. swietenioides* Roxb.) in India and Burma, one (*S. kusnotoi* Kostermans) in Malaysia and one (*S. americana* (Zahlbr.) Gilg) in Peru.

Leaves simple; inflorescences few- to several-flowered, with pedicels 2–16 mm. long; area of dark hairs on corolla lobes extending almost to petal margins; anthers 2–3 mm. long; fruit pear-shaped, 4–6 cm. long, containing 4 winged seeds - - - - **1. *trichoclada***
Leaves imparipinnate, 5(7)-foliolate; inflorescences several- to many-flowered, with pedicels up to 2 mm. long; area of dark hairs on corolla lobes reaching only half-way from mouth of corolla tube to petal margin; anthers 1·25–2 mm. long; fruit clavate, 2·5–4·5 cm. long, containing 8 winged seeds - - - - - - - - - - **2. *alata***

1. **Schrebera trichoclada** Welw. in Trans. Linn. Soc. **27**: 41 (1869). — Baker in F.T.A. **4**: 15 (1902). — Turrill in F.T.E.A., Oleaceae: 2 (1952). — Topham, N.C.L. ed. 2: 72 (1958). — F. White, F.F.N.R.: 338 (1962). — Liben in Fl. d'Afr. Centr., Oleaceae: 4, t. 1 (1973). — Fanshawe, Check-list Woody Pl. Zamb.: 39 (1973). — R.B. Drumm. in Kirkia **10**: 267 (1975). TAB. **69**. Lectotype from Angola.
   Schrebera golungensis Welw., tom. cit.: 40, t. 15 (1869). — Baker tom. cit.: 14 (1902). — Topham, N.C.L. ed. 2: 72 (1958). Type from Angola.
   Nathusia trichoclada (Welw.) O. Kuntze, Rev. Gen. Pl. **2**: 412 (1891). Type as for *Schrebera trichoclada*.
   Nathusia golungensis (Welw.) O. Kuntze, loc. cit. Type as for *Schrebera golungensis*.
   Schrebera buchananii Baker in Kew Bull. **1895**: 95 (1895); loc. cit. — Topham, N.C.L. ed. 2: 72 (1958). Type: Malawi, Shire Highlands, *Buchanan* 418 (K, holotype).
   Schrebera affinis Lingelsh. apud Gilg & Schellenb. in Engl., Bot. Jahrb. **51**: 65 (1913). Type from Angola.
   Schrebera koiloneura Gilg apud Gilg & Schellenb., loc. cit. Type from Tanzania.
   Schrebera koiloneura var. *kakomensis* Lingelsh. apud Gilg & Schellenb., tom. cit.: 66 (1913). Type from Tanzania.
   Schrebera schellenbergii Lingelsh. in Engl., Pflanzenr. **IV**, 243: 98 (1920). — Topham, N.C.L. ed. 2: 72 (1958). Type from Angola.

Shrub or bushy tree 2–10 m. tall, the trunk with rough grey flaking bark. Young shoots whitish-pilose or less often glabrous. Leaves simple; petioles 5–15 mm. long, without articulation, usually softly hispid. Leaf lamina 4·5–14 × 2·5–7·5 cm., elliptic, oblong-elliptic or ovate, the apex rounded, acute, acuminate or apiculate, the base acute; leaves occasionally glabrous but usually upper surface at first pubescent, later glabrescent and lower surface softly hispid especially on main veins. Leaf

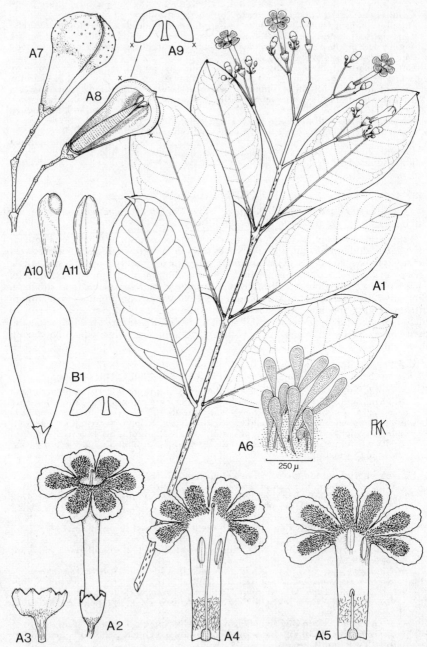

Tab. 69. A. — SCHREBERA TRICHOCLADA. A1, habit (× ⅔); A2, flower (× 2); A3, calyx (× 2); A4,
long-styled flower, dissection (× 2), A1–A4 from *Simão* 195/48; A5, short-styled flower,
dissection (× 2); A6, group of dilated hairs from upper surface of petal, A5 and A6 from
Torre & Correia 13750; A7, mature fruit beginning to dehisce (× ⅔); A8, one valve of fruit
(the front one removed) with seeds all fallen (× ⅔); A9, diagrammatic cross-section of this
valve; A10, seed (4 per fruit) (× ⅔); A11, sterile structure formed from 2 aborted seeds (2
per fruit) (× ⅔), A7–A11 from *Bingham* 688. B. — SCHREBERA ARBOREA. B1, outline and
cross-section of one valve of fruit (× ⅔) *Eggeling* 1343.

venation, including the fine tertiary reticulation, clearly visible; main nerves impressed on upper surface, prominent below, reticulation raised on both surfaces. Inflorescence a terminal few- to several-flowered cyme 3–6 cm. long; bracts and bracteoles inconspicuous and soon deciduous; axes glabrous to tomentose; pedicels 2–16 mm. long. Calyx 2–4 mm. long, glabrous or puberulous, truncate or irregularly 4–5-lobed at apex. Corolla cream-coloured or greenish; tube 13–21 mm. long, pilose inside near the base; lobes (5) 6 (7), 5–8 (11) × 3–4 (10) mm., unevenly obovate with apex obtuse or emarginate, each with a patch of dense maroon to dark brown dilated hairs extending from the corolla mouth almost to the petal margin. Anthers 2–3 mm. long, on filaments up to 1 mm. long; flowers dimorphic: stamens inserted near the top of the corolla so that the anthers either reach the corolla mouth or are 1.5–3 mm. distant from it. Style correspondingly either half as long as corolla tube or exserted. Fruit a woody bivalved pear-shaped capsule 4–6 cm. long, straw-coloured, often warty, splitting open at maturity; locules 2, each containing 2 winged seeds up to 5 cm. long and 2 under-developed seeds which remain united to form a woody compressed-elliptic structure.

Caprivi Strip. 16 km. W. of Katima Mulilo, fr. 28.viii.1967, *von Breitenbach* 1221 (PRE). **Botswana.** N: Chobe Distr., Kasane, 910 m., fl. 5.i.1966, *Mutakela* 1/66/8 (SRGH). **Zambia.** B: Zambesi Distr., fl., *Gilges* 321 (K; PRE; SRGH). N: Mporokoso Distr., Mwawe R., Mweru-wa-Ntipa, 880 m., fr. 6.viii.1962, *Tyrer* 322 (BM; SRGH). W: Mwinilunga, fl. 3.x.1937, *Milne-Redhead* 2534 (BM; K; PRE). E: Petauke Distr., 48 km. W. of Minga Forest Reserve, st. 24.v.1952, *White* 2885 (FHO; K). S: Namwala Distr., Ngoma, Kafue National Park, fl. buds 7.i.1963, *Mitchell* 17/4 (FHO; SRGH). **Zimbabwe.** N: Gokwe Distr., below northern escarpment of Charama Plateau, 910 m., st. 27.iv.1965, *Simon* 257 (SRGH). W: Wankie Distr., 48 km. from Kazungulu on Victoria Falls road, fl. 16.xi.1974, *Raymond* 287 (SRGH). C: Que Que Distr., Appleton road, c. 1275 m., st. 13.v.1976, *Biegel* 5303a (SRGH). E: Mubare Distr., Commonage, Macequece road, 980 m., fl. 31.i.1954, *Chase* 5191 (BM; K; LISC; PRE; SRGH). S: Lundi R., fr. 30.vi.1930, *Hutchinson & Gillett* 3301 (BM). **Malawi.** N: Mzimba Distr., Kasitu R. bridge, c. 1070 m., fl. 28.i.1975, *Pawek* 8996 (MAL; SRGH). C: Kasungu Distr., Kasungu National Park, 1010 m., fr. 15.ix.1970, *Hall-Martin* 1484 (K; SRGH). S: Mulanje Distr., Tuchila Experimental Station, fr. 28.v.1964, *Salubeni* 323 (BM; MAL). **Mozambique.** N: Meconta, c. 19 km. from Corrane to Liupo, c. 130 m., fl. 18.i.1964, *Torre & Paiva* 10058 (LISC). Z: between Nicuadala and Marral, fr. 26.vii.1942, *Torre* 4426 (BM; LISC). T: Boroma area, near Mágué, 270 m., fr. 23.vii.1950, *Chase* 2718 (BM; COI; LISC; SRGH). MS: Maringue, fr. vi.1973, *Bond* 9B 103 (SRGH).
Known also from Angola, Tanzania and Zaire. In thickets and savanna woodland, on rocky and sandy soils. Fanshawe (loc. cit.) records *S. trichoclada* from the central province of Zambia.

S. trichoclada is morphologically very close to the third African species of *Schrebera*, *S. arborea* A. Chev., and my delimitation of these two taxa is rather different from that of Liben (in Fl.d' Afr. Centr., Oleaceae: 2 (1973); note that *S. arborea* is called *S. golungensis*).

The characters used by Liben to separate *S. trichoclada* and *S. arborea* were as follows:

Lower leaf surface with main veins and reticulation in high relief; calyx strongly nerved; small trees of savanna and open forest - - - - - - - - - - - *trichoclada*
Lower leaf surface with main veins and reticulation scarcely raised; calyx smooth; tall trees of humid forest - - - - - - - - - - - - - *arborea*

I disagree that the leaf and calyx differences given above are diagnostic, because both types are found in *S. trichoclada* in the F.Z. area. Instead, variable characters which appear to correlate well with each other and with plant habit and habitat are pedicel length and fruit shape, as in the following key:

Pedicels of flowers at anthesis 2–16 mm. long; fruit pear-shaped, the two loculi of each valve constricted into deep crevices on either side of the septum (Tab. **69**, fig. A7–9); trees and shrubs up to 10 m. tall, of open woodland - - - - - - *trichoclada*
Pedicels of flowers at anthesis 1–3 mm. long; fruit obovoid, the two loculi of each valve wide, not constricted, on either side of the septum (Tab. **69**, fig. B1); tall forest trees of up to 40 m. - - - - - - - - - - - - - *arborea*

Using these characters, I would re-identify several of the Zaire collections listed by Liben under *S. arborea* (*golungensis*) as *S. trichoclada*: *Devred* 2394; *Quarré* 2602; *Thiebaud* 433 and *Toussaint* 122 & 2283.
The type of *S. golungensis*, *Welwitsch* 933 from Angola, is a sterile shoot, its leaves having inconspicuous venation. Because he relied on the leaf-venation character, Liben made *S. golungensis* synonymous with *S. arborea*, and the former, earlier, name took precedence over the latter. However, leaves exactly similar to those of *Welwitsch* 933 have been seen on undoubted *S. trichoclada* from Mozambique, and I have little hesitation in sinking *S. golungensis* into *S.*

trichoclada. I have seen specimens of *S. arborea* from W. Africa (Guinea-Bissau to Cameroon), Zaire, Uganda and the Sudan; as far as I know it does not occur in Angola.

2. **Schrebera alata** (Hochst.) Welw. in Trans. Linn. Soc. **27**: 41 (1869). — Baker in F.T.A. **4**: 15 (1902). — Turrill in F.T.E.A., Oleaceae: 4, fig. 1 (1952). — Verdoorn in Bothalia **6**: 550, fig. 1 (1956); in Fl. Southern Afr. **26**: 100 (1963). — F. White, F.F.N.R.: 338 (1962). — Chapman and White, Evergr. For. Malawi: 44 (1970). — Liben in Fl.d' Afr. Centr., Oleaceae: 2 (1973). — Fanshawe, Check-list Woody Pl. Zamb.: 39 (1973). — R.B. Drumm. in Kirkia **10**: 267 (1975). Type from Ethiopia.
   Nathusia alata Hochst. in Flora **24**, 1, Intelligenzbl.: 25 & 2: 672 (1841). Type as above.
   Schrebera saundersiae Harvey, Thes. Cap. **2**: 40, t. 163 (1859). — Wright in F.C. **4**: 483 (1907). Type from S. Africa (Natal).
   Schrebera alata var. *tomentella* Welw., tom. cit.: 42 (1869). Type from Angola.
   Nathusia holstii Engl. & Gilg in Engl., Pflanzenw. Ost-Afr. **C**: 308 (1895). Type from Tanzania.
   Schrebera goetzeana Gilg in Engl., Bot. Jahrb. **28**: 450 (1900). — Baker, tom. cit.: 16 (1902). — Turrill, tom. cit.: 5 (1952). Type from Tanzania.
   Schrebera obliquifoliolata Gilg, op. cit. **30**: 72 (1901). — Baker, tom. cit. Type from Kenya.
   Schrebera holstii (Engl. & Gilg) Gilg, loc. cit. — Baker, tom. cit.: 17 (1902). — Turrill, op. cit.: 6 (1952). Type as for *Nathusia holstii*.
   Schrebera welwitschii Gilg, tom. cit.: 73 (1901). — Baker, tom. cit.: 15 (1902). Syntypes from Angola.
   Schrebera latialata Gilg, loc. cit. Type from S. Africa (Natal).
   Schrebera tomentella (Welw.) Gilg, tom. cit.: 74 (1901). — Baker, tom. cit.: 16 (1902). Type as for *Schrebera alata* var. *tomentella*.
   Schrebera argyrotricha Gilg, loc. cit. — Wright in F.C. **4**: 483 (1907). — Verdoorn, tom. cit.: 553, fig. 2 (1956); op. cit.: 102 (1963). Type from S. Africa (Transvaal).
   Schrebera mazoensis S. Moore in Journ. Bot. **45**: 48 (1907). — Turrill, tom. cit.: 5 (1952). Type: Zimbabwe, Mazoe, *Eyles* 202 (BM, holotype).
   Schrebera saundersiae var. *latialata* (Gilg) Gilg & Schellenb. in Engl., op. cit. **51**: 66 (1913). Type as for *Schrebera latialata*.
   Schrebera nyassae Lingelsh. in Engl., Pflanzenr. **IV**, 243: 106 (1920). Type from Tanzania.
   Schrebera gilgiana Lingelsh. in Engl., tom. cit.: 108 (1920). Syntypes from S. Africa (Transvaal and Natal) and Zimbabwe, *Marloth* 3403 (not seen).
   Schrebera greenwayi Turrill in Kew Bull. **7**: 135 (1952); op. cit.: 5 (1952). Type from Tanzania.

A tree 4–15 m. tall, or sometimes a small bush; trunk slightly fluted; slash cream, becoming darker with exposure; bark pale brown or grey, fairly smooth, with short longitudinal fissures. Young branches glabrous or thinly to densely tomentose, older twigs with very conspicuous lenticels. Leaves imparipinnate with 5 (7) leaflets, the terminal leaflet and distal pair inserted together; leaf axis distinctly articulated at leaflet insertion; petiole and rhachis ± equal in length (in leaves with 5 leaflets). Total leaf length 6–26 cm.; terminal leaflet 3–14 × 1·2–5·5 cm., laterals 2–12 × 1–5 cm.; leaflets elliptic or obovate, rounded to acuminate at apex; terminal leaflet tapered at base, laterals acute or tapered. Petiole and rhachis often winged, glabrous or pubescent; lamina glabrous above, glabrous or sparsely pilose below especially on main veins. Tertiary vein reticulation conspicuous. Inflorescence a terminal several- to many-flowered cyme 4–9 cm. long with deciduous bracts and bracteoles and glabrous to tomentose branches; pedicels up to 2 mm. long. Flowers white to pink, sweet-scented, heterostylous with "pin" and "thrum" forms. Calyx 2–4 mm. long, minutely puberulous to pubescent, truncate or very unevenly dentate at apex. Corolla tube 8–12 mm. long in "pin" flowers, 11–16 mm. in "thrum" flowers, usually pilose inside; lobes (5) 6–7, 3–7 × 2–6 mm., unevenly obovate with apex usually bilobed, each with a triangular patch of brownish dilated hairs extending from the corolla mouth up to ½ the length of the petal. Stamens inserted near the top of the corolla tube ("pin" flower) or at the mouth ("thrum"); anthers 1·25–1·5 (2) mm. long, on filaments up to 1 mm. long. Style correspondingly of 2 lengths, either exserted from or slightly shorter than corolla tube. Fruit a woody bivalved capsule 2·5–4·5 cm. long, brown, ± clavate, with wrinkled surface, splitting open at maturity to reveal two locules divided by a papery dissepiment. Seeds 4 per locule, c. 2 cm. long, pendulous.

Zambia. N: Mpika Distr., Muchinga Escarpment, 47 km. N. of Mpika, 1830 m., fl. 29.xi.1952, *White* 3787 (FHO; K). W: Misaka Forest Reserve, st. 12.viii.1968, *Mutimushi* 2629

(K). E: Lundazi Distr., Nyika Plateau, near top of Kangampande Mt., 2130 m., fr. 2.v.1952, *White* 2556 (FHO; K). S: Choma, Siasikabole-Singane area, 3·2 km. S. of Singani's court, near Muzuma R., 1280 m., st. 11.viii.1961, *Bainbridge* 536 (FHO). **Zimbabwe.** N: Mazoe, 1340 m., fl. iv.1912, *Eyles* 7433 (BM). W: Bulawayo Distr., Old Gwanda road, fr. 23.iii.1968, *Bean* 16/68 (SRGH). C: Chilimanzi Distr., Shasha Kopjes, fr. v.1951, *Gibson* 31/51 (K; SRGH). E: Umtali Distr., Imbeza Valley, near Yakarra stream, 1280 m., fl. 15.i.1955, *Chase* 5443 (BM; COI; K; LISC; PRE; SRGH). S: Victoria, fr. 1909, *Monro* 978 (BM; K). **Malawi.** N: Nkhata Bay Distr., Viphya Plateau, 50 km. SW. of Mzuzu, 1770 m., fr. immat. 27.iii.1976, *Pawek* 10939 (MAL; SRGH). C: Ntchisi Distr., Ntchisi Mt., 1400 m., fr. 2.viii.1946, *Brass* 17109 (BM; K; PRE; SRGH). S: Likhubula Valley, c. 1830 m., fl. 7.ii.1958, *Chapman* 517 (FHO; K; MAL; PRE). **Mozambique.** N: Lichinga, Massangulo Mts., c. 1400 m., fr. immat. 25.ii.1964, *Torre & Paiva* 10818 (LISC). Z: Gúruè Mts., st. 24.ix.1944, *Mendonça* 2253 (LISC). T: near the Posto Zootécnico, Angónia, st. 13.v.1948, *Mendonça* 4224 (LISC). MS: Macuta Mts., 720 m., st. 2.vi.1971, *Müller & Gordon* 1783 (LISC; SRGH). M: Namaacha Falls, fl. 22.ii.1955, *E. M. & W.* 541 (LISC; SRGH).

Also known from S. Africa (Natal and Transvaal), Swaziland, Angola, Zaire, Rwanda, Burundi, Uganda, Tanzania, Kenya and Ethiopia. In fringing forest, in *Brachystegia* woodland, in evergreen forest in gullies. Fanshawe (loc. cit.) records *S. alata* from the Barotse Prov. of Zambia.

As the extensive synonymy implies, this widespread species is rather variable. Taxonomists have particularly emphasised the presence or absence of indumentum on young stems and leaves and the degree of winging of the petiole and leaf rhachis. In the F.Z. area the latter does not appear of any interest, but glabrous and pubescent forms occur in a well-marked geographical pattern. Thus all specimens from Malawi and from Mozambique (N, Z, T, MS) are glabrous, as are most specimens from M(M). All specimens from Zimbabwe (N, W, C, S) are pubescent. Collections from Zimbabwe (E) and Zambia may be either glabrous or hairy. The pattern does not seem to be continued outside the F.Z. area, most countries having a mixture of forms.

It is interesting that *Manilkara mochisia* (*Sapotaceae*) exhibits a very similar distribution of glabrous and hairy forms in the F.Z. area.

2. JASMINUM L.

Jasminum L., Sp. Pl. **1**: 7 (1753).

Shrubs, often with climbing habit. Leaves opposite, verticillate or alternate, simple or compound, glabrous or hairy; acarodomatia sometimes present in axils of lateral nerves on lower leaf surface. Inflorescences few- to many-flowered, cymose, or flowers solitary; flowers sweetly scented. Calyx with cup-shaped tube and (4) 5–15 teeth. Corolla white and often pink on reverse, or yellow, hypocrateriform, glabrous; lobes imbricate, (4) 5–13. Stamens 2, inserted in or at mouth of corolla tube; filaments short, anthers introrse. Ovary bilocular; style slender, either exceeding corolla tube or half as long (flowers heterostylous); stigma bilobed but lobes remaining together. Fruit a deeply bilobed 2-seeded berry, often one lobe aborting. Seeds without endosperm.

An Old World genus of 150 or more species, including 25–30 in Africa.

A number of exotic jasmines are grown for ornament in the F.Z. area; I have seen specimens of *J. grandiflorum* L., *J. humile* L., *J. mesnyi* Hance (syn. *J. primulinum* Hemsley), *J. multiflorum* (Burm. f.) Andrews, *J. polyanthum* Franchet and *J. sambac* (L.) Aiton. These and other members of *Jasminum* in cultivation are keyed and discussed by P. S. Green in Baileya **13**: 137–172 (1965).

1. Leaves compound - - - - - - - - - - - - - 2
 – Leaves simple (unifoliolate) - - - - - - - - - - - 5
2. Leaves alternate; leaflets 3–5; acarodomatia absent; flowers yellow
 4. *odoratissimum* subsp. *goetzeanum*
 – Leaves opposite; leaflets 3; acarodomatia present; flowers pink or white - - - 3
3. Flowers 12–17 mm. long (corolla tube 8·5–12 mm. long; lobes 5, up to 5 × 4 mm.); lobes of fruit ellipsoid - - - - - - - - - - - 3. *bakeri*
 – Flowers 20–40 mm. long; lobes of fruit subglobose - - - - - 4
4. Petiolules of lateral leaflets 1–3 (5) mm. long; corolla lobes (4) 5 (6), 5–10 × 4–8 mm.; inflorescences terminal and axillary - - - - - - 2. *abyssinicum*
 – Petiolules of lateral leaflets 4–18 mm. long; corolla lobes 6–9, 7·5–11 × 2·5–4 mm., inflorescences terminal only - - - - - - - 1. *fluminense*
5. Acarodomatia absent; lobes of fruit ellipsoid - - - - - 6
 – Acarodomatia present; lobes of fruit subglobose - - - - - 10

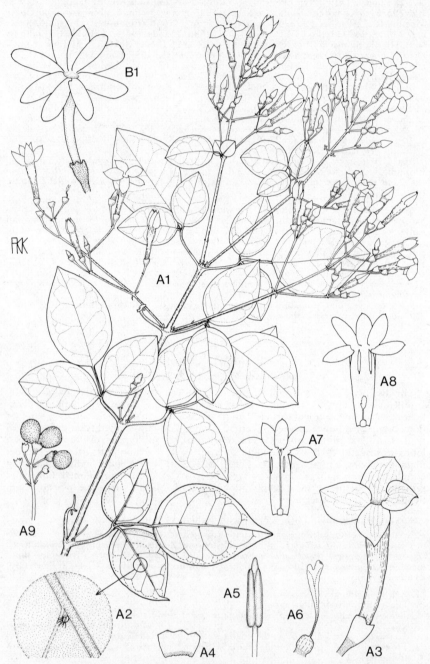

Tab. 70. A. — JASMINUM ABYSSINICUM. A1, habit (× ⅔); A2, detail of lower leaf surface showing
acarodomatium, A1 & A2 from *Pawek* 5619; A3, flower (× 2); A4, calyx (× 2); A5, stamen
(× 4); A6, gynoecium (× 4), A3–A6 from *Müller* 1593; A7, long-styled flower,
semidiagrammatic (× 1) *Lawton* 462; A8, short-styled flower (× 1) *Torre & Correia* 15654;
A9, part of infructescence (× ⅔) *Fanshawe* 7356. B. — JASMINUM FLUMINENSE. B1, flower
(× 2) *Gomes Pedro* s.n.

6. Leaves usually ternate; flowers borne in dense corymbs; calyx lobes c. 1 mm. long
   5. dichotomum
   – Leaves opposite; flowers solitary or in 3–11-flowered cymes; calyx lobes 2–10 mm. long
   7
7. Leaves pilose - - - - - - - - - - - 8. stenolobum
   – Leaves glabrous - - - - - - - - - - - 8
8. Flowers solitary or rarely in 3-flowered cymes; leaves up to 3 (4·5) × 1·4 (2) cm. (but usually much smaller), lanceolate, oblong-elliptic or narrowly ovate - - - 7. multipartitum
   – Flowers in 3–11-flowered cymes, rarely solitary; leaves either larger than above or less than twice as long as broad - - - - - - - - - - - 9
9. Leaves broadly ovate to rounded-triangular, thinly membranous and drying somewhat translucent; corolla tube 13–20 mm. long, lobes 10–15 mm.; anthers 2–4 mm. long
   6. brachyscyphum
   – Leaves elliptic to ovate, thicker-textured than above and drying opaque; corolla tube 19–33 mm. long, lobes 10–24 mm.; anthers 3·5–5 mm. long - - - 9. meyeri-johannis
10. Inflorescences borne in axils of leaves on the main stem - - - - 13. sp.A
    – Inflorescences borne terminally on lateral shoots - - - - - - - 11
11. Calyx tube 3 mm. long, lobes represented by minute thickened points or by conduplicate recurved teeth up to 2 mm. long; flowers present in axils of subterminal leaves
    12. breviflorum
    – Calyx tube 1–3 mm. long, lobes not conduplicate or recurved but with subulate part 2–9 mm. long; rarely calyx teeth very short, if so then flowers absent from axils of subterminal leaves - - - - - - - - - - - - - 12
12. Liane climbing to 6 m., sparsely branched; petioles 2–9 mm. long; flowers present in axils of subterminal leaves as well as in terminal cymes; corolla white inside, pink outside
    10. pauciflorum
    – Small shrub with twining stems and woody rootstock, richly branched, attaining 1·5 m.; petioles 1–3 mm. long; flowers only terminal, in cymes or solitary, not axillary; corolla creamy-white - - - - - - - - - - - 11. streptopus

1. **Jasminum fluminense** Vell., Fl. Flum.: 10 (1825); op. cit., Atl. **1**: t. 23 (1827). — Dandy in Kew Bull. **5**: 368 (1951). — Turrill in Kew Bull. **7**: 134 (1952); in F.T.E.A., Oleaceae: 19 (1952). — Verdoorn in Bothalia **6**: 562, t. 5 (1956); in Fl. Southern Afr. **26**: 107 (1963). — F. White, F.F.N.R.: 336 (1962). — P. S. Green in F.W.T.A. ed. 2, **2**: 50 (1963). — Friedrich-Holzhammer in Prodr. Fl. SW. Afr. **108**: 1 (1967). — Liben in Fl. d'Afr. Centr., Oleaceae: 10 (1973). — Fanshawe, Check-list Woody Pl. Zamb.: 24 (1973). — R.B. Drumm. in Kirkia **10**: 267 (1975). TAB. **70**.B. Type from Brazil.

   Jasminum mauritianum Bojer ex DC., Prodr. **8**: 310 (1844). — Baker in F.T.A. **4**: 10 (1902). — Harvey in F.C. **4**: 482 (1907). — Gilg & Schellenb. in Engl., Bot. Jahrb. **51**: 88 (1913). — O. B. Miller, B.C.L.: 48 (1948). Type from Mauritius.

   Jasminum tettense Klotzsch in Peters, Reise Mossamb., Bot. **1**: 284 (1862). Described from Mozambique "An Waldrändern und Hecken, in der Umgebung von Tette und Sena, *Peters*".

   Jasminum zanzibarense Bojer ex Klotzsch, tom. cit.: 283 (1862). Type from Zanzibar.

   Jasminum schroeterianum Schinz in Verh. Bot. Ver. Prov. Brand. **30**: 256 (1888). — Baker, tom. cit.: 12 (1902). Type from Namibia.

   Jasminum hildebrandtii ("*O.hildebrandtii*") Knobl. in Engl., op. cit. **17**: 538 (1893). — Baker, tom. cit.: 11 (1902). Type from Kenya.

   Jasminum megalosiphon Gilg in Engl., Pflanzenw. Ost-Afr. **C**: 309 (1895). — Baker, tom. cit.: 10 (1902). Syntypes from Tanzania.

   Jasminum holstii Gilg, loc. cit. — Baker, tom. cit.: 11 (1902). Syntypes from Tanzania.

   Jasminum pospischilii Gilg in Notizbl. Bot. Gart. Berl. **1**: 183 (1895). — Baker, tom. cit.: 10 (1902). Type from Kenya.

   Jasminum blandum S. Moore in Journ. Linn. Soc., Bot. **37**: 179 (1905). Type from Uganda.

   ?*Jasminum lanatum* Gilg & Schellenb. in Engl., Bot. Jahrb. **51**: 87 (1913) e descr. Type from Cameroon.

   Jasminum uhligii Gilg & Schellenb., tom. cit.: 85 (1913). Type from Tanzania.

   Jasminum rooseveltii De Wild., Pl. Bequaert. **1**: 533 (1922). Syntypes from Kenya.

   Jasminum fluminense subsp. *fluminense* var. *blandum* (S. Moore) Turrill in Kew Bull. **7**: 134 (1952); in F.T.E.A., Oleaceae: 21 (1952). Type as for *Jasminum blandum*.

   Jasminum fluminense subsp. *nairobiense* Turrill, tom. cit.: 133 (1952); loc. cit. Type from Kenya.

   Jasminum fluminense subsp. *mauritianum* (Bojer) Turrill, tom. cit.: 134 (1952); loc. cit. Type as for *Jasminum mauritianum*.

   Jasminum fluminense subsp. *holstii* (Gilg) Turrill, loc. cit.; tom. cit.: 23 (1952). Type as for *Jasminum holstii*.

Climber with long flexuous twining stems, attaining 6 m.; nodes of main branches well spaced. Young shoots, petioles and peduncles smooth and glabrous or more usually crispate-pilose or lanate, the indumentum pale greenish-yellow; bark of older branches at first smooth and brown, later cracking to reveal yellow streaks. Leaves trifoliolate, glabrous or more usually whitish- or yellowish-pilose on both surfaces. Petiole 0·5–2·5 cm. long; petiolules of lateral leaflets 0·4–1·8 cm. long, of terminal leaflets 1–2·2 cm. long. Leaflets ovate, elliptic or suborbicular, the apex acuminate, rounded or rarely emarginate, almost always with apiculum, the base rounded or acute; terminal leaflet 2·7–9 × 1·3–5 cm., lateral leaflets 1·2–6 × 0·8–3·8 cm. Acarodomatia present, usually only in axils of lower lateral nerves but occasionally occurring almost to leaflet apex. Flowers in terminal several- to many-flowered corymbs, strongly scented; pedicels up to 3 mm. long. Calyx glabrous or pubescent, tube 1·5–2 (2·5) mm. long, with 5–6 minute triangular teeth 0·25–0·5 mm. long. Corolla white, often tinged red on outer surface; tube 11–28 mm. long; lobes (5) 6–9, 7·5–11 × 2·5–4 mm., elliptic with rounded and minutely apiculate apex. Stamens inserted near the top of corolla tube; filaments up to 2·5 mm. long; anthers 3–4 mm. long, reaching to base of corolla lobes or up to 1 anther length distant. Style either exserted from or $\frac{1}{3}$–$\frac{1}{2}$ as long as tube. Fruit 1–2-lobed, lobes subglobose, 6–7 mm. long when dry.

Caprivi Strip. E. of the (Quando) Cuando R., 940 m., fl. x.1945, *Cusson* 1024 (PRE). **Botswana.** N: Okavango swamps, Moremi Wild Life Reserve, island on Gobega Lagoon, 940 m., fl. 5.iii.1972, *Biegel & Russell* 3849 (K; LISC; PRE; SRGH). **Zambia.** B: island in Kwando R. between Imasha and Sinjembele, fl. ix.1959, *Guy* in SRGH 98853 (SRGH). N: Mpika Distr., W. bank Luangwa R. between Kapamba R. and Mfuwe, fl. iii.1971, *Abel* 474 (SRGH). W: Kabompo R. Gorge, near Kabompo Boma, fr. 16.x.1952, *White* 3463 (FHO). C: Mt. Makulu Research Station, fl. & fr. 15.v.1956, *Angus* 1284 (COI; FHO; K; PRE; SRGH). E: Nsadzu Bridge, 900 m., fl. & fr. 27.xi.1958, *Robson* 747 (K; LISC; SRGH). S: Monze Distr., along Kondwa R. near Bweengwa, fl. & fr. 29.v.1963, *van Rensburg* 2242 (K; SRGH). **Zimbabwe.** N: Shamva Distr., Pote R., North Star Farm, c. 1000 m., fl. 3.xii.1972, *Simon* 2295 (LISC; PRE; SRGH). W: Wankie Distr., Gwaai R. c. 16 km. from Zambezi R., fl. & fr. 10.viii.1969, *Cannell* 78 (K; SRGH). C: Hartley Distr., Poole Farm, fl. & fr. 12.v.1950, *Hornby* 3173 (K; PRE; SRGH). E: Chipinga Distr., E. Sabi R., Sangwe crossing, c. 380 m., fl. & fr. 28.i.1957, *Phipps* 180 (SRGH). S: Chiredzi Distr., Sabi-Lundi junction, Chitsa's Kraal, fl. & fr. 5.vi.1950, *Chase* 2298 (BM; LISC; SRGH). **Malawi.** N: Karonga Distr., Kilupula V., fl. & fr. 24.viii.1956, *Banda* 331 (BM; SRGH). C: Kasungu, fl. & fr. 24.viii.1946, *Brass* 17406 (BM; K; PRE; SRGH). S: Mangochi Distr., c. 1·5 km. from Mangochi (Fort Johnston) on Monkey Bay road, fl. 12.vii.1963, *Salubeni* 61 (MAL; SRGH). **Mozambique.** N: Goa I., fl. & fr. 19.v.1961, *Leach & Rutherford-Smith* 10925 (K). Z: 7 km. from Namacurra to Vila da Maganja (Vila Maganja da Costa), c. 40 m., fl. & fr. 25.i.1966, *Torre & Correia* 14082 (LISC). T: Boroma area, Sisitso Station, Zambese R., 270 m., fl. 8.vii.1950, *Chase* 2605 (BM; SRGH). MS: Mringa area, Sabi R., 180 m., fl. & fr. 30.vi.1950, *Chase* 2484 (BM; COI; LISC; SRGH). GI: Inharrime, Mangorro, Malamba Experimental Station, fl. & fr. 2.iv.1954, *Barbosa & Balsinhas* 5519 (BM; LMA). M: between Bela Vista and Salamanga, fl. 11.xii.1961, *Lemos & Balsinhas* 289 (BM; COI; K; LISC; LMA; PRE).

Distributed throughout Africa, also in Arabia and the Seychelles; introduced into S. America. In riverine bush.

A very variable species which Turrill (loc. cit.) divided into several subspecies; these are not upheld in the F.Z. area. In the F.Z. area the great majority of specimens have pilose leaves and peduncles, but those from the two southern provinces of Mozambique (GI and M) are usually glabrous. This variation is not correlated with any other significant difference.

2. **Jasminum abyssinicum** Hochst. ex DC., Prodr. **8**: 311 (1844). — Baker in F.T.A. **4**: 11 (1902). — Gilg & Schellenb. in Engl., Bot. Jahrb. **51**: 84 (1913). — Turrill in F.T.E.A., Oleaceae: 18 (1952). — Verdoorn in Bothalia **6**: 563, t. 6 (1956); in Fl. Southern Afr. **26**: 107 (1963). — F. White, F.F.N.R.: 336 (1962). — Liben in Fl. d'Afr. Centr., Oleaceae: 11 (1973). — Fanshawe, Check-list Woody Pl. Zamb.: 24 (1973). — R.B. Drumm. in Kirkia **10**: 267 (1975). TAB. **70**.A. Type from Ethiopia.

Jasminum wyliei N.E. Br. in Kew Bull. **1909**: 419 (1909). Type from S. Africa (Natal).
Jasminum ruwenzoriense De Wild. in Rev. Zool. Afr. **9**, Suppl. Bot.: 87 (1921). Type from Zaire.
Jasminum butaguense De Wild., tom. cit.: 84 (1921). Type from Zaire.
Jasminum rutshuruense De Wild., tom. cit.: 85 (1921). Type from Zaire.
Jasminum mearnsii De Wild., Pl. Bequaert. **1**: 531 (1922). Type from Kenya.
Jasminum wittei Staner in De Wild. & Staner, Contrib. Fl. Katanga, Suppl. **4**: 80 (1932). Type from Zaire.

Climber with long flexuous twining stems attaining 5 m. or more; nodes of main branches well spaced. Young shoots and petioles smooth and glabrous or minutely puberulous, bark of older branches becoming rough. Leaves trifoliolate, glabrous except for the acarodomatia. Petiole 1–1·7 (3·7) cm. long; petiolules of lateral leaflets 1–3 (5) mm. long. Leaflets ovate, elliptic or suborbicular, the apex acute, acuminate or rounded, often apiculate, the base acute or rounded; terminal leaflet 4–5·7 (10) × 2·5–4·5 (5·3) cm., lateral leaflets 2–4·3 (8·5) × 1·2–2·9 (4·2) cm. Acarodomatia present, usually occurring in axils of lateral nerves almost up to leaflet apex. Flowers terminal and lateral, solitary or in few- to many-flowered loose cymes, scented; pedicels 2–8 mm. long; inflorescence axes usually puberulous. Calyx glabrous; tube 2–3 mm. long, with 5 (6) teeth which are sometimes shallowly triangular but often merely points on the calyx tube rim. Corolla white, sometimes tinged pink on the outside; tube 14–26 mm. long; lobes (4) 5 (6), 5–10 × 4–8 mm., elliptic or ovate-elliptic, with auricled base and rounded or cuspidate apex. Stamens inserted at or near the middle of the corolla tube; filaments very short or up to 3 mm. long; anthers (3) 4–5 (6) mm. long, never reaching to mouth of corolla tube. Style either very short or exserted from tube. Fruit 1–2-lobed, lobes globose, c. 7 mm. long when dry.

Zambia. N: Mafinga Mt., fl. 26.viii.1958, *Lawton* 462 (K). W: Mwinilunga Distr., Mwanamitowa R., fl. 9.viii.1930, *Milne-Redhead* 863 (K; PRE). E: Nyika, fr. 30.xii.1962, *Fanshawe* 7356 (K). **Zimbabwe.** E: Chirinda Forest, fl. 30.iv.1950, *Hack* 225/50 (SRGH). **Malawi.** N: Nkhata Bay Distr., S: Viphya near Chikangawa, c. 2000 m., fl. 11.ix.1970, *Müller* 1593 (K; MAL; SRGH). **Mozambique.** MS: Manica, Zuira Mts., Tsetsera, road to Chimoio (Vila Pery), c. 1800 m., fl. 3.iv.1966, *Torre & Correia* 15654 (LISC).

Also known from S. Africa (Natal and Transvaal), Zaire, Rwanda, Burundi, Uganda, Tanzania, Kenya and Ethiopia. In evergreen montane forest and mushitu.

Specimens of *J. abyssinicum* from the F.Z. area and S. Africa appear to be almost completely glabrous, but those from more tropical regions often have indumentum on stems and calyx and occasionally also on leaves.

It has perhaps not been recognised previously that floral characters are useful for separating *J. abyssinicum* from its relative *J. fluminense*: the former has broad corolla lobes, usually 5 in number, the latter narrower lobes, 6 or over. This difference is almost diagnostic, but a few specimens of *J. fluminense* outside the F.Z. area, notably some members of subsp. *nairobiensis* Turrill, have been seen which have only 5 corolla lobes.

3. **Jasminum bakeri** Scott Elliot in Journ. Linn. Soc., Bot. **30**: 86 (1894). — Baker in F.T.A. **4**: 12 (1902). — F. White, F.F.N.R.: 336 (1962). — P. S. Green in F.W.T.A. ed. 2, **2**: 50 (1963). — Liben in Fl. d'Afr. Centr., Oleaceae: 12, t. 3 (1973). — Fanshawe, Check-list Woody Pl. Zamb.: 24 (1973). Type from Sierra Leone.

Jasminum syringa S. Moore in Journ. Bot., Lond. **44**: 87 (1906). — Turrill in F.T.E.A., Oleaceae: 19 (1952). Type from Uganda.

Jasminum bequaertii De Wild. in Fedde, Repert. **11**: 539 (1913). Type from Zaire.

Climber with long flexuous twining stems; nodes of main branches well spaced. Young shoots and petioles smooth and glabrous; bark of older branches becoming corky. Leaves trifoliolate, glossy above, glabrous except for acarodomatia. Petiole 0·3–5 cm. long; petiolules of lateral leaflets 0·2–2·5 cm. long. Leaflets ovate, narrowly ovate, obovate or elliptic, the apex acute to acuminate, often minutely apiculate, the base obtuse or acute; terminal leaflet 3·5–10 × 1·5–7 cm., lateral leaflets 2·3–8·5 × 1–6 cm. Acarodomatia present, usually occurring in axils of lateral nerves almost up to leaflet apex. Inflorescence a neat many-flowered flat-topped or rounded corymb, terminal on side branches; peduncle and other axes densely whitish pubescent; flowers scented. Calyx puberulous, tube c. 2 mm. long, with 5 (6) shallowly triangular to shortly cuspidate-acuminate teeth up to 0·75 mm. long. Corolla white inside, greenish-white outside; tube 8·5–12 mm. long; lobes 5, 3–5 × 2–4 mm., broadly elliptic with rounded or cuspidate apex. Stamens inserted ± at the middle of the corolla tube; filaments up to 1 mm. long, anthers 3–4 mm. long. Style either reaching to mouth of corolla tube or half as long. Fruit 1–2-lobed, lobes c. 1 cm. long when dry, elliptic.

Zambia. N: Mansa (Fort Rosebery) Distr., Luapula R. — Mansa road 44 km. from Chembi Ferry, fl. 5.x.1947, *Brenan & Greenway* 8021 (FHO; K). W: Solwezi Distr., Solwezi R. gorge, fl. 13.ix.1952, *Angus* 442 (BM; FHO; K; SRGH).

W. Africa from Guinea to Cameroon, Central African Republic, Angola, Zaire, Uganda and Burundi. In mushitu and evergreen fringing forest.

According to Fanshawe, loc. cit., *J. bakeri* is also known from the Central Prov. of Zambia.

4. **Jasminum odoratissimum** L., Sp. Pl. **1**: 7 (1753).

Subsp. **goetzeanum** (Gilg) P. S. Green in Notes Roy. Bot. Gard. Edin. **23**: 375 (1961). Type from Tanzania.

> *Jasminum goetzeanum* Gilg in Engl., Bot. Jahrb. **28**: 451 (1900). — Baker in F.T.A. **4**: 12 (1902). — Turrill in F.T.E.A., Oleaceae: 18 (1952). — F. White, F.F.N.R.: 334 (1962). — Lisowski, Malaisse & Symoens in Bol. Soc. Brot., Sér. 2, **45**: 467 (1971). — Liben in Fl. d'Afr. Centr., Oleaceae: 8, t. 2 (1973). — Fanshawe, Check-list Woody Pl. Zamb.: 24 (1973). Type as above.

Straggling shrub up to 4 m. tall, all parts glabrous. Leaves alternate, 3–5 (7)-foliolate. Petiole 12–17 mm. long, rhachis (when present) 7–11 mm. long. Terminal leaflet up to 3·5 (6) × 1·3 (4) cm., elliptic or ovate-elliptic, acute at apex, tapered at base to a petiolule; lateral leaflets smaller then terminal, of similar shape but acute or rounded at base, not or scarcely petiolulate. Acarodomatia absent. Flowers in few- to several-flowered cymes, terminal on side shoots and in axils of penultimate leaves, overtopped by leaves; pedicels 3–12 mm. long. Calyx glabrous, tube 2·5–3 mm. long with (4) 5 (6) cuspidate teeth up to 0·75 mm. long. Corolla yellow; tube 11–18 mm. long; lobes 5 (–7), 4–9 × 3·5–5 mm., subcircular to elliptic, with adaxial surface often papillose. Stamens inserted either at the top of the corolla tube (anthers exserted) or at the middle; anthers subsessile, 3–4·5 mm. long. Style half as long as corolla tube or exserted. Fruit usually 1-lobed, lobe c. 9 mm. long when dry, elliptic.

Zambia. N: Mbala Distr., S. of Kasionolwa Hill, fl. x.1933, *Miller* D163 (FHO). **Malawi.** N: Rumphi Distr., Nyika Plateau, Chiwilamera Evergreen Forest Patch, Kasaramba road, fl. 10.xii.1965, *Banda* 798 (K; MAL; SRGH).
Also occurring in Tanzania, Kenya and Zaire. In evergreen forest.

J. odoratissimum differs florally from the other F.Z. species of *Jasminum* in several respects. First, it is the only yellow-flowered species. Second, a few specimens (including *Banda* 798) were found to have an additional flap of tissue, adnate to each petal at the mouth of the corolla tube, forming a kind of corona. Third, *J. odoratissimum* (or at least subsp. *goetzeanum*) seems to be truly heterostylous, having two distinct flower forms with the anthers and stigma borne at reciprocal levels. In all other F.Z. species long- and short-styled flowers occur but the variation in style length is not correlated with variation in stamen position; the latter varies a little, but apparently at random. Morphology suggests that there is an effective outbreeding mechanism in *J. odoratissimum* and that a similar system once operated in ancestors of the other species but has since degenerated. It would be interesting to know the genetic background and pollen-stigma reactions and morphology of these African jasmines.

5. **Jasminum dichotomum** Vahl, Enum. **1**: 26 (1804). — Baker in F.T.A. **4**: 9 (1902). — Gilg & Schellenb. in Engl., Bot. Jahrb. **51**: 89 (1913). — Turrill in F.T.E.A., Oleaceae: 23, fig. 7 (1952). — P. S. Green in F.W.T.A. ed. 2, **2**: 50 (1963). — Liben in Fl. d'Afr. Centr., Oleaceae: 14 (1973). — Fanshawe, Check-list Woody Pl. Zamb.: 24 (1973). Type from Guinea.

> *Jasminum noctiflorum* Afzel., Rem. Guin. Coll. **4**: 25 (1815). Type from Sierra Leone.
> *Jasminum guineense* G. Don, Gen. Syst. **4**: 60 (1838). Type from Dahomey.
> *Jasminum ternum* Knobl. in Engl., Bot. Jahrb. **17**: 535 (1893). Lectotype (of Turrill, op. cit.: 23, 1952) from Angola.
> *Jasminum brevipes* Baker in Kew Bull. **1895**: 93 (1895). Type from Angola.
> *Jasminum ternifolium* Baker, tom. cit.: 95 (1895); loc. cit. Type from the Sudan.
> *Jasminum bukobense* Gilg in Engl., Pflanzenw. Ost-Afr. **C**: 308 (1895). — Baker, loc. cit. (1902). Type from Tanzania.
> *Jasminum gardeniodorum* Gilg ex Baker in F.T.A. **4**: 8 (1902). — Gilg & Schellenb. in Engl., tom. cit.: 90 (1913). Type from Togo.
> *Jasminum mathildae* Chiov. in Ann. Bot. Roma **9**: 79 (1911). Syntypes from Ethiopia.
> *Jasminum gossweileri* Gilg & Schellenb., loc. cit. (1913). Type from Angola.

Scrambling or climbing shrub reaching 8 m. or more, with stiff straight branches. Young stems dark greenish or reddish-brown, minutely puberulous; older twigs with rough grey bark. Leaf insertion rather irregular but most often ternate, leaves thinly coriaceous, glabrous, glossy, bright green above and mat below. Petiole 8–10 mm. long, articulated near the base. Leaf lamina 2·3–6·5 × 1·5–3·2 cm., narrowly ovate or elliptic, the base acute, the apex rounded, acute or acuminate; main nerves impressed above, prominent below; one pair of lateral nerves particularly conspicuous, arising near leaf base and curving towards apex. Acarodomatia absent. Inflorescences terminal and axillary, the flowers numerous and densely arranged in ± flat-topped corymbs; pedicels 1–3 mm. long, glabrous. Calyx glabrous; tube 2–3 mm. long; teeth 5–6, up to 1 mm. long, narrowly triangular to subulate. Corolla pure white inside, pinkish-brown outside; tube 14–23 mm. long; lobes 6–9, 7–10 × 2–4·5 mm., elliptic with apex acute. Stamens inserted at the middle of the corolla tube or just above or below; filaments up to 2 mm. long; anthers 3–4 mm. long, never reaching mouth of corolla tube. Style equalling or c. ⅓ as long as tube. Fruit 1–2-lobed, lobes c. 1 cm. long when dry, elliptic.

Zambia. N: Nsama, fl. 12.ix.1958, *Fanshawe* 4812 (FHO; K). W: Ndola Distr., Kafue R. near Nkana Pump Station, c. 1225 m., fl. 14.vi.1961, *Linley* 156 (K; SRGH). E: Lundazi Distr., near Kalindi, Lukusuzi National Park, 1250 m., fl. buds 5.vii.1971, *Sayer* 1286 (SRGH). **?Malawi.** "Nyassaland", fl. vii.1935, *Smuts* 2100 (PRE).
Widely distributed in tropical Africa from Senegal east to the Sudan and Ethiopia and south to Angola, Zaire and Tanzania. In mushitu and riparian woodland.

According to P. S. Green, loc. cit., *J. dichotomum* occurs in Mozambique, but I have seen no material from that country. Fanshawe (loc. cit.) records this species from the Central Prov. of Zambia.

6. **Jasminum brachyscyphum** Baker in Kew Bull. **1895**: 93 (1895). Type: Malawi, Shire Highlands, Zambesi-land, *Buchanan* 224 (K, holotype).
   Jasminum meyeri-johannis — sensu Turrill in Kew Bull. **7**: 134 (1952) & in F.T.E.A., Oleaceae: 24 (1952) pro parte.
   Jasminum multipartitum — sensu R. B. Drumm. in Kirkia **10**: 267 (1975) pro parte quoad specim. *Eyles* 7938.

Small shrub, tree up to c. 6 m. or a liane climbing to 15 m. or more. Young shoots, petioles and inflorescences minutely puberulous, bark of older twigs yellowish- or greyish-brown, splitting longitudinally. Petioles 5–6 mm. long, articulated near the base or just below the middle. Leaves glossy green, glabrous, often darkening on drying; lamina 1·8–6 × 1–1·4 cm., not more than twice as long as broad, broadly ovate to rounded-triangular, the apex rounded, acute or acuminate with minute thickened point, the base broadly obtuse or cordate. Venation inconspicuous on upper leaf surface, visible below; lateral nerves several, the basal ones diverging at a wide angle from the midrib and not running up to the apex. Acarodomatia absent. Flowers terminal on short branches, in (1) 3–11-flowered cymes, sweetly scented; pedicels 3–9 mm. long. Calyx tube 2–3 mm. long, glabrous; lobes 5–8, 2–7 mm. long, subulate, glabrous or puberulous. Corolla white inside, reddish or purplish outside; tube 13–20 mm. long; lobes 7–10, 10–15 mm. long, oblong, the apex cuspidate or shortly acuminate. Stamens inserted towards top of corolla tube; filaments short; anthers 2–4 mm. long. Style either well exserted from tube or ¼–⅓ as long, with bifid stigma. Fruit usually bilobed, lobes 7–12 mm. long (when dry), ellipsoid.

Zimbabwe. C: Makoni Distr., Forest Hill Kop, 1460 m., fr. vii.1917, *Eyles* 755 (BM; K; SRGH). E: Mutare Distr., Murahwa's Hill, Christmas Pass, fl. 8.x.1948, *Chase* 1654 (BM; K; SRGH). **Malawi.** S: Mt. Mulanje, Likhubula Valley, 1460 m., fl. 17.x.1941, *Greenway* 6330 (K; PRE). **Mozambique.** Z: Gúruè Mts., near source of R. Malema, c. 1700 m., fr. 4.i.1968, *Torre & Correia* 16870 (LISC). MS: between Mavita and Valley of the Moçambize R., fl. 25.x.1944, *Mendonça* 2580 (BM; LISC).
Not known outside the F.Z. area. In fringing woodland and on granite kopjes.

The independent status of this species has been overlooked by earlier workers, who classified specimens of *J. brachyscyphum* as members of other species. Thus Turrill (1952) made *J. brachyscyphum* a synonym of *J. meyeri-johannis*. On the other hand, Verdoorn (Bothalia **6**: 569, 1956) found difficulty in distinguishing S. African *J. multipartitum* and specimens of "*J. meyeri-johannis*" (= *J. brachyscyphum*) from Zimbabwe and Mozambique. She suggested that these latter specimens should be classified as *J. multipartitum* but pointed out that more tropical specimens of *J. meyeri-johannis* appear to belong to a distinct species. Drummond (1975) cited *J. meyeri-johannis* in synonymy under *J. multipartitum*, and the specimen given as an example of this taxon, *Eyles* 7938, belongs to *J. brachyscyphum*.
Despite this prolonged taxonomic confusion, *J. meyeri-johannis*, *J. brachyscyphum* and *J. multipartitum* are well-defined species, separated by many differential characters and occupying individual habitats and geographical areas, and no intermediate specimens have been seen. See remarks under species 7, 8 and 9.

7. **Jasminum multipartitum** Hochst. in Flora **27**, 2: 825 (1844). — Harvey in F.C. **4**: 480 (1909) pro parte. — Verdoorn in Bothalia **6**: 567, t. 9 (1956); in Fl. Pl. Afr. **32**: t. 1272 (1958); in Fl. Southern Afr. **26**: 110 (1963). — R. B. Drumm. in Kirkia **10**: 267 (1975) pro parte excl. syn. *J. meyeri-johannis*. TAB. **71**. Type from S. Africa (Natal).
   Jasminum oleicarpum ("*oleaecarpum*") Baker in Kew Bull. **1895**: 95 (1895); in F.T.A. **4**: 8 (1902), pro parte. Syntypes: Mozambique, "Tette", fr. ii.1859, *Kirk* (K; lectotype of

Tab. 71. JASMINUM MULTIPARTITUM. 1, habit (× ⅔); 2–5, long-styled flower; 2, calyx (× 2); 3, corolla, opened out (× 2); 4, stamen (× 4); 5, gynoecium (× 4), 1–5 from *Mthalane* 1; 6–8, short-styled flower; 6, corolla, opened out, diagrammatic (× ⅔); 7, stamen (× 4); 8, gynoecium (× 4), 6–8 from *Pienaar & Meyer* s.n.; 9, fruiting twig (× ⅔); 10, seed (× ⅔), 9–10 from *Gardner* 21.

Turrill) and "opposite Senna", fr. i.1859 (K), both *J. multipartitum*; "Tette", fl. xi.1858, *Kirk* (K) and "Rovuma R., 30 miles up", 28.iii.1861, *Kirk* (K), both *J. stenolobum*.

Shrub up to 3 m. high, the branches sometimes spreading or sarmentose. Young shoots minutely puberulous, bark of older twigs pale fawn, splitting longitudinally. Petioles up to 7 mm. long, the articulation variously situated, from near the lamina to near the stem. Leaf lamina up to 3 (4·5) × 1·4 (2) cm., but usually much smaller, lanceolate, oblong-elliptic or narrowly ovate, the apex acute to obtuse with minute thickened tip, the base cuneate or tapered; both surfaces glabrous; venation inconspicuous above, midrib more prominent below, lateral nerves few, the basal pair often running almost to the leaf apex; margin of leaf narrowly thickened. Leaves often darkening on drying. Acarodomatia absent. Flowers terminal on short lateral branches, solitary or very rarely in 3-flowered cymes, very sweetly scented; pedicels 2–5 mm. long. Calyx tube 2–3 mm. long, glabrous; lobes 7–12, 2·5–5 mm. long, subulate, hispidulous. Corolla white inside, reddish outside; tube 17–32 mm. long; lobes (6) 9–12, (12·5) 15–18 mm. long, oblong, the apex cuspidate, shortly acuminate or trifid. Stamens inserted near the top of corolla tube; filaments short; anthers 2·5–4 mm. long, reaching to within 2–4 mm. of base of corolla lobes. Style either well exserted from tube or c. ⅓ as long, with bifid stigma of variable form. Fruit usually 2-lobed, 7–12 mm. long when dry, each lobe ellipsoid.

Zimbabwe. N: Mafungabusi Distr., st. 10.vii.1947, *McGregor* 24/47 (SRGH). W: Matobo Distr., Hope Fountain Mission, c. 1400 m., fl. 18.xi.1973, *Norrgrann* 402 (SRGH). C: Selukwe Distr., Wanderer's Valley 13 km. S. of Selukwe, fl. 8.xii.1966, *Biegel* 1538 (K; PRE; SRGH). E: Chipinga Distr., on Sabi-Tanganda Estate, fl. 18.xi.1959, *Goodier* 650 (K). S: Victoria Distr., Mushandike National Park, fl. 30.xi.1975, *Bezuidenhout* 270 (SRGH). **Mozambique.** T: near Tete, fl. 11.x.1943, *Torre* 6012 (BM; LISC). GI: between Inharrime and Chidenguele, fl. 10.xii.1944, *Mendonça* 3375 (BM; LISC). M: road to Goba border, fl. 13.ix.1961, *Lemos & Balsinhas* 186 (BM; COI; K; LISC; LMA; PRE; SRGH).

S. Africa (Cape Prov., Natal and Transvaal) and Swaziland. In dry habitats, on sandy or stony soils.

J. multipartitum is closely related to no. 6, *J. brachyscyphum*. These species can be distinguished by their different leaf shapes, by the normally solitary flowers of the former as against the (1) 3–11-flowered inflorescences of the latter, and by the difference in flower size (*J. multipartitum* has generally larger flowers).

J. multipartitum is cultivated as an ornamental in the F.Z. area and southern Africa.

8. **Jasminum stenolobum** Rolfe in Oates, Matabeleland ed. 2: 403 (1889). — Harvey in F.C. **4**: 481 (1907). — Gilg & Schellenb. in Engl., Bot. Jahrb. **51**: 91 (1913). — O. B. Miller, B.C.L.: 48 (1948). — Turrill in F.T.E.A., Oleaceae: 24 (1952). — Verdoorn in Bothalia **6**: 569, t. 10 (1956); in Fl. Southern Afr. **26**: 111 (1963). — F. White, F.F.N.R.: 336 (1962). — Fanshawe, Check-list Woody Pl. Zamb.: 24 (1973). — R. B. Drumm. in Kirkia **10**: 267 (1975). Type: Zimbabwe, Matabeleland, *Oates* (K, holotype).

Jasminum tomentosum Knobl. in Engl., Bot. Jahrb. **17**: 536 (1893). — Turrill, op. cit.: 29 (1952). Type from Kenya.

Jasminum oleicarpum ("*oleaecarpum*") Baker in Kew Bull. **1895**: 95 (1895); in F.T.A. **4**: 8 (1902), pro parte. Syntypes: Mozambique, "Tette", fr.ii.1859, *Kirk* (K, lectotype of Turrill) and "opposite Senna", fr. i.1859, *Kirk* (K) *J. multipartitum*; "Tette", fl. xi.1858, *Kirk* (K) and "Rovuma R., 30 miles up", 28.iii.1861, *Kirk* (K), both *J. stenolobum*.

Jasminum rotundatum Knobl. in Notizbl. Bot. Gart. Berl. **11**: 1078 (1934), emend. op. cit. **12**: 201 (1934). — Turrill, op. cit.: 30 (1952). Type from Tanzania.

Jasminum tomentosum var. *lutambense* Knobl., op. cit. **13**: 283 (1936). Type from Tanzania.

Suffrutex or low shrub 0·5–1·5 m. tall, rhizomatous, with much-branched sprawling stems. Young shoots, petioles and inflorescences crispate-pilose with white or yellowish silvery hairs, bark of older twigs grey or brownish, smooth. Petioles 5–13 mm. long, articulated in the middle or towards the base. Leaves pilose, not darkening on drying; lamina 0·9–6 × 0·6–2 cm., very variable in shape, elliptic-ovate, elliptic-oblong or narrowly lanceolate to suborbicular, the apex tapering, rounded or occasionally emarginate, with minute thickened point, the base acute or obtuse. Venation visible on both surfaces, main nerves raised below; lateral nerves several, the basal ones joining with others before reaching leaf apex. Acarodomatia absent. Flowers terminal on short branches, solitary or in 3 (5)-flowered cymes, sweetly scented; pedicels 3–10 mm. long. Calyx crispate-pilose; tube 2–3 mm. long; lobes 7–15, 2·5–9 mm. long, filiform. Corolla pure white; tube 16–33 mm. long; lobes

9–13, 12–22 mm. long, oblong or narrowly elliptic, the apex cuspidate or shortly acuminate. Stamens inserted near top of corolla tube; filaments short; anthers 2·5–4·5 mm. long, reaching to within $\frac{1}{2}$–$1\frac{1}{2}$ anther lengths of base of corolla lobes. Style either exserted from or $\frac{1}{3}$–$\frac{2}{3}$ as long as corolla tube, with bifid stigma of variable form. Fruits 1–2-lobed, lobes 1–1·5 cm. long when dry, ellipsoid.

Botswana. N: Chobe Distr., Nungwe Valley, 910 m., fl. 2.i.1967, *Henry* 49 (SRGH). SE: 8 km. E. of Lothlekane, fr. 23.iii.1965, *Wild & Drummond* 7256 (SRGH). Zambia. N: Mpika Distr., Luangwa Valley Game Reserve, c. 610 m., fl. 30.xii.1966, *Prince* 54 (K; SRGH). C: Concord Ranch 3 km. S. of Chisamba railway station, fl. 6.i.1958, *Benson* 215 (BM). E: Nsadzu to Chipata (Ft. Jameson) road near Chadiza turn-off, 900 m., fl. 25.xi.1958, *Robson* 708 (K; LISC; SRGH). S: Mazabuka Distr., near Lochinvar Ranch, fr. 3.i.1964, *van Rensburg* 2711 (K; SRGH). Zimbabwe. N: Mtoka Distr., Nyaderi Bridge, fl. 5.xii.1968, *Müller & Burrows* 958 (K; SRGH). W: Wankie Distr., S. of Main Camp, Wankie National Park, fl. 9.xii.1968, *Rushworth* 1327 (K; LISC; SRGH). C: Salisbury Distr., Twentydales, fl. 22.xi.1970, *Linley* 569 (SRGH). E: Umtali Commonage, fl. 9.xi.1948, *Chase* 1331 (BM; K; LISC; SRGH). S: Ndanga Distr., Gudus, Sangwe Reserve, 430 m., fl. i.1960, *Farrell* 119 (BM; PRE; SRGH). Malawi. N: Mzimba Distr., Mzambazi to Mperembe, 400 m., fl. 30.xii.1975, *Pawek* 10658 (K; MAL; PRE; SRGH). C: Kasungu National Park, 1010 m., fl. 5.xii.1970, *Hall-Martin* 1036 (SRGH). S: Ntcheu Distr., Livulezi R., fr. 31.i.1968, *Jeke* 147 (K; MAL; SRGH). Mozambique. N: near Nampula, fl. 1.xi.1942, *Mendonça* 1186 (BM; LISC). Z: Mocuba, Namagoa, 60–120 m., fl. xi.1942, *Faulkner* 262 (K; PRE; SRGH). T: Tete, Estima, fr. 29.i.1972, *Macêdo* 4739 (LISC; LMA; LMU). MS: Manica, Rotanda, 4 km. on road to Mavita, fl. 30.x.1965, *Torre & Pereira* 12611 (LISC). M: between Bela Vista and Porto Henrique, fl. 23.viii.1944, *Mendonça* 1858 (LISC).

Also known from Kenya, Tanzania and S. Africa (N. Natal, W. and N. Transvaal). In *Brachystegia* woodland, on sandy soils, often on termitaria. Fanshawe (loc. cit.) records *J. stenolobum* also from Barotseland (Zambia (B)).

J. stenolobum is closely related to *J. brachyscyphum* and *J. multipartitum* (nos. 6 & 7), from which it is most easily distinguished by its indumentum. Notes on specimens from the F.Z. area suggest that the flowers of *J. stenolobum* are pure white, not pink or maroon on the outside as in the other two species.

In S. Africa and the F.Z. area *J. stenolobum* usually has flowers solitary or in threes, but in Tanzania the inflorescences are often many-flowered.

9. **Jasminum meyeri-johannis** Engl. in Hochgebirgsflora Trop. Afr.: 334 (1892). — Turrill in Kew Bull. **7**: 134 (1952); in F.T.E.A., Oleaceae: 24 (1952), pro parte excl. syn. *J. brachyscyphum* et *J. oleicarpum*. Type from Tanzania.
  Jasminum engleri Gilg in Engl., Bot. Jahrb. **19**, Beibl. 47: 46 (1894). — Baker in F.T.A. **4**: 3 (1902). Type from Tanzania.
  Jasminum smithii Baker in Kew Bull. **1895**: 93 (1895); loc. cit. (1902). Type from Tanzania.
  Jasminum afu Gilg in Engl., Pflanzenw. Ost-Afr. **C**: 308 (1895). — Baker, op. cit.: 7 (1902). Syntypes from Tanzania.

Scandent shrub to 2 m. or more. Young stems glabrous or puberulous and soon glabrescent, older stems with smooth bark. Petioles 6–15 mm. long, articulated below the middle, sometimes with a line of pubescence on the upper surface. Leaves tough, glossy, glabrous, not darkening on drying; lamina 2·5–7·7 × 1·5–4·5 cm., elliptic to ovate, the base acute or rounded, the apex obtuse to acuminate. Venation inconspicuous. Acarodomatia absent. Flowers terminal on short lateral branches, solitary or in 3–11-flowered cymes, very sweetly scented; pedicels 3–12 mm. long. Calyx glabrous; tube 2–3 mm. long, lobes 5–7, 2–10 mm. long, subulate. Corolla white inside, pink or pale green outside; tube 19–33 mm. long, lobes 7–11, 10–24 mm. long, linear to narrowly elliptic, with apex acute or acuminate. Stamens inserted in the top half of corolla tube; filaments up to 3 mm. long, anthers 3·5–5 mm. long. Style either exserted from tube or $\frac{1}{3}$–$\frac{1}{2}$ as long, with bifid stigma. Fruit 1–2-lobed, lobes 10–14 mm. long when dry, ellipsoid.

Mozambique. GI: entrance to Zandamela and Zavala, fl. 10.x.1958, *Mogg* 32652 (LISC; SRGH); between Quissico and Mutote, fl. 24.ix.1947, *Pedro & Pedrogão* 1881 (LMA; LMU; PRE).
Eastern Kenya and Tanzania. In coastal savanna.

J. meyeri-johannis has been confused with no. 6, *J. brachyscyphum*, but in fact these species are easily separated on morphological characters, as the key shows, and their ranges are vicarious.

The flowers of *Mogg* 22652 are stated to be yellow. This is very unlikely to have been so, but the flowers of this species turn yellow when dry as in all African jasmines, and the observation may have been made some while after collection. (*Mogg* 28787, also from Mozambique (GI), an otherwise typical specimen of *J. fluminense*, is also said to have yellow flowers).

Jasminum angolense Welw. ex Baker (1895) (syn. *J. stenodon* Baker, 1895), of which I have seen a few specimens from Angola, is closely related to *J. meyeri-johannis*.

10. **Jasminum pauciflorum** Benth. in Hook., Niger Fl.: 443 (1849). — Baker in F.T.A. **4**: 6 (1902). — Turrill in F.T.E.A., Oleaceae: 28 (1952). — P. S. Green in F.W.T.A. ed. 2, **2**: 50 (1963). — Liben in Fl. d'Afr. Centr., Oleaceae: 17 (1973). Type from Ghana.
   Jasminum schweinfurthii Gilg in Notizbl. Bot. Gart. Berl. **1**: 72 (1895). — Baker, tom. cit.: 3 (1902). Type from Zaire.
   Jasminum welwitschii Baker in Kew Bull. **1895**: 94 (1895); in op. cit.: 7 (1902). Type from Angola.
   Jasminum obovatum Baker, loc. cit.; loc. cit. Type from Angola.
   Jasminum longipes ("*longpipes*") Baker, loc. cit.; loc. cit. Type from Angola.
   Jasminum bieleri De Wild., Ann. Mus. Congo, Bot., Sér. 5, **3**: 248 (1910). Type from Zaire.
   Jasminum talbotii Wernham in Rendle, Cat. Talbot Nig. Pl.: 58 (1913). Type from Nigeria.
   Jasminum angustilobum Gilg & Schellenb. in Engl., Bot. Jahrb. **51**: 95 (1913). Type from Nigeria.
   Jasminum umbellulatum Gilg & Schellenb., tom. cit.: 98 (1913). Type from Cameroon.
   Jasminum soyauxii Gilg & Schellenb., tom. cit.: 96 (1913). Type from Gabon.
   Jasminum callianthum Gilg & Schellenb., loc. cit. Syntypes from Nigeria and Togo.
   Jasminum dasyneurum Gilg & Schellenb., tom. cit.: 97 (1913). Syntypes from Cameroon.
   Jasminum cardiophyllum Gilg & Schellenb., tom. cit.: 102 (1913). Syntypes from Cameroon.
   Jasminum warneckei Gilg & Schellenb., tom. cit.: 103 (1913). Syntypes from Togo.
   Jasminum vanderystii De Wild., Pl. Bequaert. **4**: 350 (1928). Type from Zaire.
   Jasminum brieyi De Wild., op. cit. **5**: 471 (1932). Type from Zaire.
   Jasminum sp. near streptopus — F. White, F.F.N.R.: 337 (1962).

Liane climbing to 6 m., rather sparsely branched. Young stems glabrous or pilose and glabrescent, older stems with smooth bark. Petioles 2–9 mm. long, articulated at about the middle. Leaves drying papery, pilose or glabrous, with usually conspicuous venation; lamina 2·8–9 × 1·5–4·5 cm., ovate, oblong-ovate or obovate-oblong, the base cordate to obtuse, the apex shortly acuminate. Acarodomatia present in axils of most lateral veins, hispid-pilose. Flowers in terminal few-flowered cymes and also present, solitary or in few-flowered cymes, in axils of subterminal leaves, sweet-scented; peduncles up to 20 mm. long; pedicels 8–25 mm. long. Calyx glabrous to pilose; tube 2–3 mm. long, lobes 5–6, 2–7 mm. long, subulate. Corolla white inside, pink outside; tube 18–26 mm. long, lobes 6–8, 10–20 mm. long, linear, apex acute to acuminate. Stamens inserted near top of corolla tube; filaments c. 1 mm. long, anthers 2·5–4 mm. long. Style either exserted from tube or c. ½ as long, with bifid stigma. Fruit lobes c. 7 mm. long, subglobose.

Zambia. N: Kawambwa Distr., Chishinga Ranch, fl. 11.ix.1965, *Lawton* 1322 (K; SRGH). W: Ndola Distr., L. Ishibu, fl. 18.x.1953, *Fanshawe* 434 (K; SRGH).

W. tropical Africa from Guinea Bissau to Cameroon, Zaire, Angola, Tanzania, Kenya, Uganda, Burundi and the Sudan. In mushitu vegetation.

The illustration in F.W.T.A. (ed. 2, **2**: fig. 212, 1963) is atypical, as it shows *J. pauciflorum* with a very many-flowered inflorescence.

11. **Jasminum streptopus** E. Meyer, Comm. Pl. Afr. Austr. **1**: 173 (1838). — Harvey in F.C. **4**: 481 (1907). — Verdoorn in Bothalia **6**: 570 (1956); in Fl. Southern Afr. **26**: 111 (1963). — F. White, F.F.N.R.: 336 (1962). — Fanshawe, Check-list Woody Pl. Zamb.: 24 (1973). — R. B. Drumm. in Kirkia **10**: 267 (1975). Type from S. Africa (Natal).
   Jasminum bogosense Beccari ex Martelli in Fl. Bogos.: 51 (1886). Type from Ethiopia.
   Jasminum parvifolium Knobl. in Engl., Bot. Jahrb. **17**: 537 (1893). — Baker in F.T.A. **4**: 8 (1902). — Turrill in F.T.E.A., Oleaceae: 26 (1952). — Liben in Fl. d'Afr. Centr., Oleaceae: 18 (1973). Type from Kenya.
   Jasminum microphyllum Baker in Kew Bull. **1895**: 93 (1895). Type from Angola.
   Jasminum kirkii Baker, tom. cit.: 94 (1895); in tom. cit.: 6 (1902). Syntypes: Mozambique, Zambesi-land at Shiramba, *Kirk* (K) and between Lupata and Tette, *Kirk* (K).

Jasminum walleri Baker, tom. cit.: 95 (1895); tom. cit.: 9 (1902). Syntypes: Mozambique, Manganja Hills, *Waller* (K); Magomero Mission Station, 910 m., *Waller* (K).

Jasminum djuricum Gilg in Notizbl. Bot. Gart. Berl. **1**: 73 (1895). Type from the Sudan.

Jasminum dicranolepidiforme Gilg in Engl., Bot. Jahrb. **28**: 450 (1900). — Baker, tom. cit.: 8 (1902). Type from Tanzania.

Jasminum gerrardii Harvey ex C. H. Wright in F.C. **4**: 480 (1906). Lectotype from S. Africa (Natal).

Jasminum swynnertonii S. Moore in Journ. Linn. Soc., Bot. **40**: 135 (1911). — Turrill, tom. cit.: 27 (1952). Lectotype: Mozambique, Kurumadzi, R. Jihu, 610 m., 17.xi.1906, *Swynnerton* 180a (BM, holotype; K; SRGH, isotypes).

Jasminum hockii De Wild. in Bull. Jard. Bot. Brux. **3**: 279 (1911). Syntypes from Zaire.

?Jasminum viridescens Gilg & Schellenb. in Engl., Bot. Jahrb. **51**: 100 (1913), e descr. Type from Angola.

Jasminum bussei Gilg & Schellenb., tom. cit.: 101 (1913). — Turrill, tom. cit.: 29 (1952). Type from Tanzania.

Jasminum transvaalensis S. Moore in Journ. Bot., Lond. **56**: 10 (1918). Type from S. Africa (Transvaal).

?Jasminum punctulatum Chiov., Fl. Somala **2**: 279 (1932), e descr. Syntypes from Somalia.

Jasminum angustitubum Knobl. in Notizbl. Bot. Gart. Berl. **11**: 674 (1932); emend. tom. cit.: 1079 (1934). — Turrill, tom. cit.: 27 (1952). Type from Tanzania.

Jasminum biflorum Knobl., op. cit. **13**: 256 (1936). — Turrill, tom. cit.: 25 (1952). Type from Tanzania.

Jasminum stolzeanum Knobl., tom. cit.: 282 (1936). — Turrill, tom. cit.: 25 (1952). Type from Tanzania.

Jasminum ellipticum Knobl., loc. cit. — Turrill, tom. cit.: 26 (1952). Type from Tanzania.

Jasminum virgatum Knobl, tom. cit.: 283 (1938). Type from Tanzania.

Jasminum streptopus var. *transvaalensis* (S. Moore) Verdoorn in Bothalia **6**: 572, t. 12 (1956); in Fl. Southern Afr. **26**: 112 (1963). Type as for *Jasminum transvaalensis*.

Small shrub with woody rhizomatous rootstock and slender flexuous stems trailing or climbing up to 1·5 m. Young shoots yellowish crispate-pilose, bark of older twigs silvery grey. Petioles 1–3 mm. long, articulated above the middle, the joint often inconspicuous. Leaves dull green, often thin-textured, pilose on both surfaces, the hairs longest and most dense on midrib of lower surface, or rarely leaves glabrous except for acarodomatia. Leaf lamina 0·9–5 (7) × 0·6–3 (3·5) cm., elliptic, obovate-elliptic, ovate or lanceolate, the apex acuminate, acute, rounded or very rarely emarginate, with minute thickened point, the base obtuse or acute. Venation visible on both surfaces, veins level, raised or impressed above, raised below; lateral nerves several, the basal ones joining with others before reaching leaf apex; tertiary venation sometimes visible. Acarodomatia present in axils of lateral nerves up to or slightly above the middle of the leaf, rarely occurring also on lateral nerves. Inflorescences terminal on short shoots, with flowers solitary or in 2–3(7)-flowered cymes, almost never present in axils of subterminal leaves; pedicels 2–20 mm. long. Calyx glabrous to pilose; tube 1–2·5 mm. long; lobes 4–7, 0·6–5 (9) mm. long, triangular, subulate. Corolla creamy white; tube 13–26 mm. long; lobes 6–9, 8–18 mm. long, linear or narrowly elliptic, apex acute to acuminate. Stamens inserted in top half of corolla tube; filaments 1–2·5 mm. long; anthers 2–4 mm. long. Style either exserted from tube or c. ½ as long, with bifid stigma. Fruit with subglobose lobes 5–8 mm. long, black when ripe.

Zambia. B: Balovale, fl. xii.1953, *Gilges* 308 (PRE; SRGH). N: Kasama Distr., Chambeshe Flats, Kalungu R. levee, fl. 27.i.1962, *Astle* 1293 (K; SRGH). W: Mwinilunga Distr., near R. Kamwedzi, fl. 17.xii.1937, *Milne-Redhead* 3714 (K; PRE; SRGH). C: Mukulaikwa Agricultural Station c. 64 km. W. of Lusaka, fl. 6.i.1963, *Angus* 3472 (FHO; K). E: Chipata Distr., Mkhania area, c. 609 m., fl. 12.xii.1968, *Astle* 5382 (K; SRGH). S: Kafue National Park, Kalala turnoff, fl. 17.viii.1961, *Boughey* 12373 (SRGH). **Zimbabwe.** W: Wankie, fl. xi.1934, *Levy* 116 (PRE). C: Headlands Distr., Chigwani, fl. 18.xii.1951, *Greenhow* 86/51 (SRGH). E: Umtali Distr., Tsonzo Division, Kukwanisa, 1755 m., fl. 10.xii.1967, *Biegel* 2393 (K; PRE; SRGH). S: Victoria Distr., Kyle National Park Headquarters area, fl. 17.xi.1970, *Basera* 194 (SRGH). **Malawi.** N: Chitipa Distr., 72 km. down Nthalire road, c. 1750 m., fr. 18.iv.1975, *Pawek* 9353 (K; MAL; SRGH). C: Ntcheu Distr., near Njolomole village, fl. 12.i.1968, *Jeke* 143 (MAL). S: Zomba, near Likangala Bridge, Zomba to Jali road, fl. 17.xii.1963, *Salubeni* 181 (MAL; SRGH). **Mozambique.** N: Eráti, 2 km. from Alua to Mejuco, c. 450 m., fl. 13.xii.1963, *Torre & Paiva* 9545 (LISC). T: 23 km. from Tete to Changara, c. 300 m., fl. 21.xii.1965, *Torre & Correia* 13801 (LISC). MS: Madanda Forests, c. 120 m., fr. 5.xii.1906, *Swynnerton* 2114 (BM). M: Bela Vista, Ponta do Ouro, fl. 8.iv.1968, *Balsinhas* 1191 (LMA; PRE).

Also known from S. Africa, Angola, Burundi, Uganda, Kenya, Tanzania, Ethiopia, Somalia and Sudan. In scrub and woodland, often on termitaria.

The collection from Mozambique (M) cited above, *Balsinhas* 1191, differs from the rest of the F.Z. material in having leaves of somewhat stouter texture and large dimensions (up to 6 × 3 cm.) and 7-flowered cymes. It is exceptional in its habitat, having been found in dune vegetation. This collection belongs to *J. streptopus* var. *streptopus* sensu Verdoorn (in Bothalia **6**: 571, 1956), a taxon which is otherwise confined to the Durban area of Natal. Verdoorn treated the rest of S. African *J. streptopus* as var. *transvaalensis*. Tropical African collections of *J. streptopus* are variable in many characters, as the synonymy implies, and I have been unable to divide it into infraspecific groups. Many specimens from the F.Z. area agree with var. *transvaalensis* but these merge with the rest of the species without a noticeable discontinuity in the variation pattern.

J. streptopus is very similar to *J. pauciflorum* (no. 10), as noted by P. S. Green (in F.W.T.A. ed. 2, **2**: 50, 1963) and Verdoorn (in Fl. Southern Afr. **26**: 112, 1963). They can be separated by the morphological characters given in the key, and by their different leaf shapes; their habitats, too, are quite different.

12. **Jasminum breviflorum** Harvey ex C. H. Wright in F.C. **4**: 480 (1906). — Verdoorn in Bothalia **6**: 564, t. 7 (1956); op. cit. **7**: 15 (1958); in Fl. Southern Afr. **26**: 108 (1963). Type from S. Africa (Transvaal).
    Jasminum gerrardii Harvey ex C. H. Wright, loc. cit. (1906), pro parte excl. lectotyp.

Shrubby climber. Young stems pubescent, glabrescent, older stems with smooth bark. Petioles 2–5 mm. long, articulated at about the middle. Leaves coriaceous, sparsely pubescent to glabrous on both faces; lamina 1·6–4·5 × 0·8–3·4 cm., ovate to elliptic or narrowly elliptic, base acute to rounded, apex obtuse or acute, mucronate; venation slightly raised on both surfaces, inconspicuous below; lateral nerves several, the basal ones joining with others before reaching leaf apex. Acarodomatia present in axils of lateral nerves up to middle of leaf. Flowers borne on lateral branches in terminal few- (3–4)-flowered racemes and in the axils of upper leaf-pairs, sweet scented; pedicels 9–15 mm. long. Calyx glabrous or pubescent; tube 3 mm. long; lobes 4–5, represented by minute thickened points or by conduplicate, recurved teeth up to 2 mm. long. Corolla white, fleshy or thin-textured; tube 17–25 mm. long; lobes 4–7, 10–15 mm. long, elliptic, apex acute. Stamens inserted in the top half of corolla tube; filaments up to 2 mm. long; anthers 3–4 mm. long. Style either exserted from tube or c. ½ as long. Fruit with lobes subglobose, c. 7 mm. long.

Mozambique. M: Maputo, fl. 22.i.1947, *Hornby* 2559 (BM; K; LMA; PRE; SRGH); Maputo, between Catuane and L. Mandjene, fr. 14.iv.1949, *Myre & Balsinhas* 605 (PRE).
Also found in S. Africa (Transvaal, Natal and Cape Prov.). In evergreen thicket.

Verdoorn (loc. cit.) cites *J. gerrardii* as a synonym of *J. breviflorum*. I have seen both type collections of *J. gerrardii* (*Gerrard* 1477 (BM; K) and *Rehmann* 7706 (K), both from Natal), and while the latter is *J. breviflorum* the former is *J. streptopus* (var. *transvaalensis* (S. Moore) Verdoorn). *Gerrard* 1477 (K) is chosen as lectotype of *J. gerrardii*.

13. **Jasminum sp. A.**

Climber with flexuous glabrous stems. Leaves opposite or subopposite, simple; lamina 3·2–7·2 × 2–4·5 cm., ovate, glabrous (or very sparsely pilose) except for lanate acarodomatia in axils of main nerves on lower leaf face. Petioles 3–5 mm. long, articulated above the middle, pubescent on upper surface. Inflorescences 1–3 (-more?)-flowered, 1–2 (more?) cm. long, borne on main stems in leaf axils; axes pilose; bracts present, represented by a range of reduced leaves; pedicels (fruiting) c. 3 mm. long. Calyx tube c. 2 mm. long; lobes ?6, 2–3 mm. long, each folded longitudinally and curved outwards. Fruit with lobes subglobose, c. 8 mm. long.

Mozambique. Z: Mopeia, fr. 28.vii.1942, *Torre* 4438 (LISC).
Known only from this specimen. In dense scrub on river margin.

Within the F.Z. area, among species with simple leaves and acarodomatia, *J. sp. A.* is most similar to *J. breviflorum* in its rather tough leaf texture and conduplicate recurved calyx lobes. It differs from *J. breviflorum* in having compact inflorescences borne in the axils of leaves on the main stem. Taking into account species from outside the F.Z. area, *J. sp. A.* agrees completely in technical characters with *J. schimperi* Vatke, a widespread and variable species found in Zaire, Rwanda, Burundi, Ethiopia, Uganda, Kenya and Tanzania. The geographically closest population of *J. schimperi* is in Tanzania (T6, F.T.E.A.). Its synonymy is given below.

Jasminum schimperi Vatke in Linnaea **40**: 210 (1876). Type from Ethiopia.

> *Jasminum eminii* Gilg in Engl., Pflanzenw. Ost-Afr. **C**: 309 (1895). — Baker in F.T.A. **4**: 6 (1902). — Turrill in F.T.E.A., Oleaceae: 28 (1952). — Liben in Fl. d'Afr. Centr.: 16, t. 4 (1973). Type from Uganda.
> *Jasminum radcliffei* S. Moore in Journ. Linn. Soc., Bot. **37**: 178 (1905). Type from Uganda.
> *Jasminum pulvilliferum* S. Moore in Journ. Bot. Lond. **44**: 24 (1906). Type from Kenya.
> *Jasminum mildbraedii* Gilg & Schellenb. in Mildbr., Wiss. Ergebn. Deutsch. Zentr.-Afr.-Exped. 1907–08, **2**: 529 (1913). Type from Tanzania.
> *Jasminum buchananii* S. Moore, op. cit. **54**: 288 (1916). Type from Kenya.
> *Jasminum albidum* De Wild., Pl. Bequaert. **1**: 528 (1922). Type from Zaire.

Although *J. schimperi* occasionally has lateral inflorescences like those of *sp. A*, this is not the typical form. Usually it has compact, several- to many-flowered inflorescences with short pedicels (2–5(10) ·mm.) borne terminally on lateral branches, with additional flowers or cymes in axils of subterminal leaves.

It is probable that *Torre* 4438 belongs to *J. schimperi*, but further collecting of better specimens is needed to confirm the presence of this species in Mozambique.

3. MENODORA Humb. & Bonpl.

Menodora Humb. & Bonpl., Pl. Aequin. **2**: 98, t. 110 (1809). — Steyermark in Ann. Mo. Bot. Gard. **19**: 87 (1932).
> *Bolivaria* Cham. & Schlecht. in Linnaea **1**: 207 (1826).
> *Calyptrospermum* A. Dietr. in L., Sp. Pl., ed. 6, **1**: 226 (1831).

Low shrubs or suffrutices, glabrous or hairy, rarely spinose (not in Africa). Leaves alternate or opposite, entire or lobed, sessile or petiolate; acarodomatia absent. Inflorescence 1 to many-flowered, cymose; bracts and bracteoles absent; pedicels erect or recurved in fruit. Calyx 5–15-lobed, segments entire or divided. Corolla yellow or white, usually subrotate to shortly infundibuliform, rarely (not in Africa) hypocrateriform, 5 (6)-lobed, often pubescent within the tube. Stamens 2, inserted on the corolla tube, usually with relatively long filaments; anthers introrse. Ovary bilobed and bilocular, with axile placentation and 2 or 4 ovules per loculus; style long and slender, stigma capitate. Fruit a bilocular capsule forming 2 cocci with membranous pericarp; dehiscence usually circumscissile. Seeds angular, reticulate; endosperm absent.

A genus of c. 17 species, all occurring in arid or semi-arid environments, found in three disjunct regions: the SE. states of the U.S.A. and Mexico; temperate S. America; southern Africa.

Menodora heterophylla Moric. ex A.DC. in DC., Prodr. **8**: 316 (1844). Type from Mexico.

Var. **australis** Steyermark in Ann. Mo. Bot. Gard. **19**: 127, t. 6 fig. 2 (1932). — Verdoorn in Bothalia **6**: 606, t. 29 (1956); in Fl. Southern Afr. **26**: 128 (1963). TAB. **72**. Type from S. Africa (Transvaal).
> *Menodora heterophylla* sensu Oliver in Hook., Ic. Pl.: t. 1459 (1884). — sensu C. H. Wright in F.C. **4**: 484 (1907).

Suffrutex 5–24 cm. high with woody taproot and caudex bearing many erect sparsely branched herbaceous stems; branches strongly ribbed, scabrid. Leaves 7–19 mm. long, alternate or subopposite, subcoriaceous, subsessile, almost entire to irregularly trifid or pinnatifid with ± elliptic lobes; lamina minutely pitted, scabrid on the margins and abaxial midrib; margins revolute. Inflorescences terminal and axillary, cymose, 1–few-flowered. Flowers not scented. Calyx scabrid; tube c. 2 mm. long, lobes 10 (15?), 5–8 mm. long, usually linear, acute at apex, rarely divided. Corolla yellow, shortly infundibuliform, delicate; tube 2–3 mm. long, puberulous within, lobes up to 12 × 7 mm., broadly elliptic, rounded or mucronate at apex. Stamens inserted at mouth of corolla tube; anthers 2–2·75 mm. long, free part of filaments 3–4 mm. long. Ovary 2-lobed; style c. 1 cm. long, slender, terete, with capitate, obscurely bilobed stigma. Fruiting pedicel nodding; mature capsule c. 6 mm. long, with circumscissile dehiscence; seeds 1–4 in each coccus, 6–9 × 4–6 mm., with reticulate testa.

Botswana. SE: c. 10 km. NE. of Gaberones, 1010 m., fl. 2.xii.1954, *Codd* 8937 (K; PRE; SRGH).

Also found in S. Africa (Transvaal). In open woodland on sandy soil.

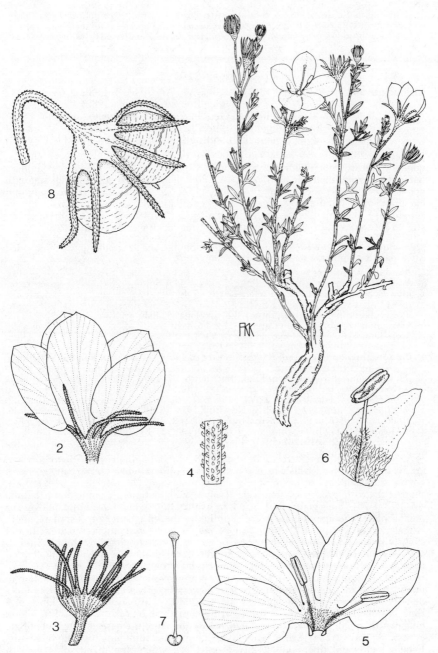

Tab. 72. MENODORA HETEROPHYLLA var. AUSTRALIS. 1, habit (×1) *Rogers* 6225; 2, flower (×⅔); 3, calyx opened out (×⅔); 4, detail of one calyx segment (×10); 5, corolla and androecium (×⅔); 6, stamen, showing level to which filament and corolla are adnate (×4); 7, gynoecium (×⅔), 2–7 from *Harbor* 17012; 8, fruit (×4) *Pegler* 950.

This species has a remarkably disjunct distribution, var. *heterophylla* being native to Texas and Mexico.

According to Verdoorn, in Fl. Southern Afr. **26**: 127 (1963), *Menodora africana* is also recorded from Botswana, but I have seen no material of this species from the F.Z. area. It is distinguished from *M. heterophylla* by its bipinnatifid leaves and divided calyx segments.

4. CHIONANTHUS L.

Chionanthus L., Sp. Pl. **1**: 8 (1753). — Stearn in Ann. Mo. Bot. Gard. **63**: 355–357 (1977); in Journ. Linn. Soc., Bot. **80**: 191–206 (1980).
Linociera Swartz in Schreber, Gen. Pl. **2**: 784 (1791), *nom. conserv*.
Dekindtia Gilg in Engl., Bot. Jahrb. **32**: 139 (1902).
Campanolea Gilg & Schellenb. in Engl., Bot. Jahrb. **51**: 73 (1914).

Trees and shrubs, evergreen (in F.Z. area) or rarely deciduous. Leaves simple, opposite, entire, not lepidote; acarodomatia usually present. Flowers in axillary paniculate cymes or fascicles, 4-merous. Calyx deeply lobed. Corolla valvate, with very short tube and linear lobes, these strongly concave, with incurved margins and cucullate apex. Stamens 2, inserted on corolla tube; anthers introrse. Ovary subglobose, bilocular, with 2 ventral or pendent ovules per loculus; style short, stigma subcapitate. Fruit a 1-seeded drupe. Seed lacking endosperm.

A genus of c. 100 species found in America, Africa, Asia and Australia, mostly tropical and subtropical but with a few temperate representatives.

1. Acarodomatia present - - - - - - - - - - - - - 2
 – Acarodomatia absent - - - - - - - - - - - 4. *richardsiae*
2. Flowers borne in dense sessile axillary clusters; calyx 2·5–3·5 mm. long, densely appressed pilose - - - - - - - - - - - - - 1. *battiscombei*
 – Flowers borne in loose cymes; calyx 1–1·5 mm. long, ± glabrous - - - - 3
3. Acarodomatia glabrous; leaves large (6–17 × 2·5–8 cm.) - - - - - 2. *niloticus*
 – Acarodomatia pubescent; leaves smaller (4–9 × 1·5–3·5 cm.) - - - 3. *foveolatus*

1. **Chionanthus battiscombei** (Hutch.) Stearn in Journ. Linn. Soc., Bot. **80**: 197 (1980). TAB. **73**. Type from Kenya.
 Dekindtia africana Gilg in Engl., Bot. Jahrb. **32**: 139 (1902) non *Mayepea africana* Knobl. (1893) nec *Chionanthus africanus* (Knobl.) Stearn (1980). — Baker in F.T.A. **4**: 588 (1904). — Turrill in F.T.E.A., Oleaceae: 16, fig. 5 (1952). Lectotype from Angola.
 Linociera battiscombei Hutch. in Kew Bull. **1914**: 17 (1914). — Verdoorn in Bothalia **6**: 600, t. 26 (1956); in Fl. Southern Afr. **26**: 124 (1963). — Chapman & White, Evergr. For. Malawi: 44 (1970). — Fanshawe, Check-list Woody Pl. Zamb.: 26 (1973). — R. B. Drumm. in Kirkia **10**: 267 (1975). Type as for *Chionanthus battiscombei*.

A shrub or small to tall tree, 2·5–30 m. high, with smooth grey or brown bole and cream slash turning reddish-brown on exposure. Twigs characteristically straight; young shoots glabrous or minutely puberulous and glabrescent; bark smooth, sometimes with prominent lenticels. Leaves with petiole 3–7 mm. long; lamina 4·5–15 × 1·75–6 cm., glabrous, lanceolate, ovate, elliptic, oblong-elliptic, obovate or oblanceolate, the apex apiculate to acuminate, the base acute or cuspidate; main lateral nerves 4–7-paired; upper leaf surface glossy, with impressed midrib and slightly raised lateral nerves; lower surface mat, minutely pitted, with both midrib and lateral nerves raised, and pubescent acarodomatia in axils of most of the lateral nerves (very rarely acarodomatia absent); leaf margin slightly thickened, not revolute. Inflorescence glomerulate, comprising sessile axillary clusters of few to many flowers and their subtending bracts; flowers sweetly scented; bracts and calices densely appressed-pilose. Calyx 2·5–3·5 mm. long, campanulate, the 4 lobes in 2 slightly unequal opposite pairs joined at base into a short tube. Corolla 5–6·5 mm. long, white or pale yellow, the 4 lobes separate almost to the base, strongly cucullate at apex. Anthers 1·25–1·5 mm. long. Gynoecium 1·5–2 mm. long, the stigma sessile. Fruit an ellipsoid drupe c. 1·5 cm. long, black when ripe. Seed solitary, with basilateral attachment.

Zambia. W: Kitwe, by Kafue R., fl. 14.x.1972, *Fanshawe* 11495 (K). E: Lundazi Distr., Nyika Plateau, Kangampande Mt., 2130 m., st. 8.v.1952, *White* 2797 (FHO). **Zimbabwe.** N: Mazoe Distr., Mazoe, below dam, st. vii.1920, *Henkel* s.n. (SRGH). W: Matobo, Besna Kobila, 1340 m., fl. buds & fr. iv.1953, *Miller* 1779 (PRE; SRGH). C: Wedza Distr., Wedza Mt., fr. 19.ii.1963, *Wild* 5996 (K; LISC; PRE; SRGH). E: Mutare, in gully S. of Circular Drive,

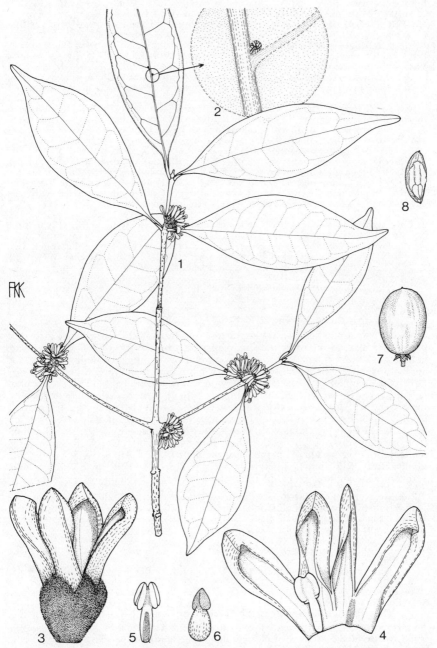

Tab. 73. CHIONANTHUS BATTISCOMBEI. 1, habit (× ⅔); 2, detail of lower leaf surface showing an acarodomatium in the angle between midrib and lateral nerve; 3, flower (× 6); 4, corolla, opened out, one stamen removed (× 6); 5, stamen (× 6); 6, gynoecium (× 6), 1–6 from *Chase* 7903; 7, fruit (× 1) *Barbosa* 1586; 8, seed (× 1) *Müller* 1536.

Commonage, fl. 30.xi.1962, *Chase* 7903 (BM; K; LISC; PRE; SRGH). S: Chibi Distr., Nyoni Range, 900 m., st. 17.vii.1973, *Müller* 2083 (SRGH). **Malawi.** N: Viphya, headwaters of Lonjoswa stream, 1830 m., fr. 24.x.1964, *Chapman* 2272 (FHO). C: Ntchisi Mt., fr. 25.vii.1946, *Brass* 16938 (K; PRE; SRGH). S: Mulanje Distr., Litchenya Crater, c. 1200 m., fr. 4.ix.1970, *Müller* 1536 (K; PRE; SRGH). **Mozambique.** N: Ribáuè, Chinga Mts., c. 1200 m., fr. immat. 12.xii.1967, *Torre & Correia* 16482 (LISC). Z: Morumbala Mts., 700 m., fl. 12.xii.1971, *Müller & Pope* 2001 (K; LISC; SRGH). MS: Chimoio, between Chimoio (Vila Pery) and Revuè, near Zemba Mt., fl. buds & fr. 27.iv.1948, *Barbosa* 1586 (LISC). M: Bela Vista, Zitundo, near Malongane, fl. buds & fr. immat. 3.iii.1970, *Balsinhas* 1605 (MA).

Also known from Kenya, Tanzania, Angola and S. Africa (Transvaal). In evergreen forest in gullies. Fanshawe (loc. cit.) records the species also from the northern province of Zambia.

In the F.Z. area *Chionanthus battiscombei* is apparently confined to the kloof forest habitat; the only exception seen was the collection *Balsinhas* 1605 from Mozambique (M), found in dune scrub associated with *Mimusops caffra*. In the F.Z. area the species flowers in November and December (the beginning of the rainy season) but flower buds are often found on specimens collected from April onwards (i.e. in the dry season). This suggests that the characteristic inflorescence of *C. battiscombei*, with its close-packed structure and dense indumentum, may have evolved as an adaptation protecting the flower buds throughout the months of drought.

It is interesting, however, that the species behaves differently further north: in Kenya and Tanzania it is found in open grassland as well as in gully forest, and specimens with open flowers have been collected in March, May, July, August and November.

2. **Chionanthus niloticus** (Oliver) Stearn in Journ. Linn. Soc., Bot. **80**: 202 (1980). Type from Uganda.

> *Linociera nilotica* Oliver in Trans. Linn. Soc. **29**: 106, t. 117 (1875). — Baker in F.T.A. **4**: 19 (1902). — Turrill in F.T.E.A., Oleaceae: 14 (1952). — P. S. Green in F.W.T.A. ed. 2, **2**: 48 (1963). — Liben in Fl. d'Afr. Centr., Oleaceae: 30 (1973). — Fanshawe, Check-list Woody Pl. Zamb.: 26 (1973). Type as above.
> *Mayepea nilotica* (Oliver) Knobl. in Engl., Bot. Jahrb. **17**: 528 (1893). Type as above.
> *Linociera holtzii* Gilg & Schellenb. in Engl., Bot. Jahrb. **51**: 73 (1913). Lectotype (Turrill) from Tanzania.
> *Linociera sp. 2.* — F. White, F.F.N.R.: 337 (1962).

Shrub, sometimes semi-scandent, or tree up to 9 m. tall. Young twigs glabrous; bark pallid, flaking; stems brittle. Leaves stiffly coriaceous; petiole 5–10 mm. long; lamina 6–17 × 2·5–8 cm., glabrous, oblong-elliptic, elliptic or oblong-obovate, the apex obtusely apiculate, the base acute to acuminate; main lateral nerves usually 8–10-paired; upper leaf surface glossy with impressed midrib and scarcely raised lateral nerves; lower surface mat, minutely pitted, with raised midrib and lateral nerves, and with glabrous acarodomatia in axils of most of lateral nerves; leaf margin slightly thickened, not revolute. Inflorescences few- to several-flowered axillary cymes 0·5–1·5 (6) cm. long with axes appressed-pubescent. Calyx 1·25–1·5 mm. long, sparsely minutely ciliate, the lobes obtusely triangular. Corolla 5–7 mm. long, white, lobes separate almost to the base, cucullate at apex. Anthers 1·5–2 mm. long. Gynoecium 1·5–2 mm. long, style sparsely pubescent. Fruit up to 2·5 cm. long, subglobose or compressed ovoid, blue-black when ripe. Seed up to 1·5 cm. long.

Zambia. N: Kawambwa Distr., near Kafulwe Mission, fl. 3.xi.1952, *White* 3587 (FHO; K).
Widely distributed in tropical Africa, from Mali to Gabon and eastwards to Angola, Zaire, Kenya, Uganda, Tanzania and the Sudan. In riparian mushitu. Fanshawe (loc. cit.) records the species also from the western province of Zambia.

3. **Chionanthus foveolatus** (E. Meyer) Stearn in Journ. Linn. Soc., Bot. **80**: 198 (1980). Type from S. Africa (Cape Prov.).

> *Olea foveolata* E. Meyer, Comm. Pl. Afr. Austr. **1**: 176 (1838). — Harvey in F.C. **4**: 485 (1907). Type as above.
> *Linociera foveolata* (E. Meyer) Knobl. in Fedde, Repert. **41**: 151 (1937). — Verdoorn in Fl. Southern Afr. **26**: 120 (1963). Type as above.
> *Linociera marlothii* Knobl., loc. cit. Type from S. Africa (Natal).

Shrub or small tree; leaves obtuse at apex, with petioles 2–4 mm. long -   subsp. *foveolatus*
Tall forest tree; leaves caudate (cuspidate-acuminate) at apex, with petioles 5–13 mm. long
subsp. *major*

Subsp. **foveolatus**

> *Linociera foveolata* subsp. *foveolata* — Verdoorn in Bothalia **6**: 595, t. 21 & 22 (1956); in Fl. Southern Afr. **26**: 121, fig. 13, 1 (1963). Type as above.

Shrub c. 3–6 m. tall. Twigs short and frequently branched; young growth glabrous, bark pale grey, fairly smooth. Leaves with petioles 2–4 mm. long; lamina

5·5–9 × 1·5–3·5 cm., glabrous, elliptic, the apex obtuse, the base acute; main lateral nerves 4–6-paired; upper leaf surface glossy with faintly impressed midrib and raised lateral nerves, lower surface mat, minutely pitted, with raised midrib and lateral nerves and with pubescent/acarodomatia in axils of most of lateral nerves; leaf margin slightly thickened, not or slightly revolute. Inflorescences few- to several-flowered axillary cymes up to 2 (3) cm. long; axes ± glabrous; flowers scented. Calyx 1–1·5 mm. long, sometimes with lobes sparsely ciliate. Corolla 5–7 mm. long, white, lobes free almost to the base, strongly or moderately cucullate at apex. Anthers 1·25–2 mm. long. Gynoecium 2 mm. long. Fruit 1·5–2 mm. long, subglobose.

Mozambique. M: Namaacha, Goba, near the Libombos fountain, fl. 18.viii.1967, *Gomes e Sousa & Balsinhas* 4933 (COI; LMA; PRE; SRGH).
Distributed widely in S. Africa from the Cape to the Transvaal. In gallery forest.

This subspecies is variable in habit, leaf shape and habitat, and the above description applies to representatives found in the F.Z. area.

Subsp. **major** (Verdoorn) Stearn in Journ. Linn. Soc., Bot. **80**: 199 (1980). Type from S. Africa (Transvaal).
    Linociera foveolata subsp. *major* Verdoorn in Bothalia **6**: 598, fig. 13, t. 24 (1956); in Fl. Southern Afr. **26**: 122, fig. 13, 3 (1963). — R. B. Drumm. in Kirkia **10**: 267 (1975). Type as above.

Differs from subsp. *foveolatus* in the following characters. Tall forest trees, up to 40 m. high. Leaves with petiole 5–13 mm. long, lamina 4–8 × 2–3·5 cm., oblong or obovate, with apex caudate and base acute to acuminate. Fruit up to 2·7 cm. long.

Zimbabwe. E: Inyanga Distr., S. slopes of Inyangani, c. 1800 m., fl. buds 8.xi.1967, *Müller* 700 (K; PRE; SRGH). **Mozambique.** MS: Tsetsera area, near road to Mavita, c. 1700 m., fl. 28.xi.1966, *Müller* 491 (K; LISC; PRE; SRGH).
Also known from S. Africa (Transvaal). In evergreen forest.

The third subspecies of *C. foveolatus*, subsp. *tomentellus* (Verdoorn) Stearn, occurs in S. Africa (Natal and Cape Prov.).

4. **Chionanthus richardsiae** Stearn in Journ. Linn. Soc., Bot. **80**: 204, figs. 4, 6 (1980).
    Type: Zambia, road to Inono, 1520 m., fl. 18.i.1955, *Richards* 4144 (K, holotype).
    Linociera sp. 1 — F. White, F.F.N.R.: 337 (1962).

Shrub or small tree 0·6–4 m. tall. Young twigs appressed-pubescent, glabrescent, bark rough, grey, flaking. Leaves glabrous; petiole c. 3 mm. long; leaf lamina 3–7·5 × 1·25–4·25, stiffly coriaceous, elliptic to obovate, with apex obtuse to apiculate and base acute; upper surface dark dull green, with midrib impressed and lateral nerves scarcely visible; lower surface paler, minutely punctate, midrib raised, lateral nerves slightly impressed; margin slightly revolute; acarodomatia absent. Inflorescences 3–5-flowered axillary cymes up to 1·5 cm. long, with axes appressed-pubescent; flowers sweetly scented. Calyx spreading, 1·5 mm. long with sparsely ciliate lobes. Corolla 6·5–8·5 mm. long, white, lobes separate almost to the base, cucullate at apex. Anthers 1·25–2 mm. long. Gynoecium 1·5–2 mm. long, narrowly conical. Fruit unknown.

Zambia. N: Mbala (Abercorn) Distr., Kalambo Falls, 1520 m., fl. 6.i.1955, *Bock* 154 (PRE).
Unknown outside Zambia. In open bush, often on steep slopes, in sandy and stony soil.

5. OLEA L.

Olea L., Sp. Pl. **1**: 8 (1753).

Trees and shrubs, often with very hard wood. Leaves opposite and decussate, simple, often lepidote; acarodomatia absent. Flowers small, borne in paniculate cymes, 4-merous. Calyx lobes shallow. Corolla valvate, with very short tube and ovate petals, these not cucullate. Stamens 2, inserted on corolla tube; anthers introrse. Ovary subglobose, bilocular with 2 pendent ovules per loculus; style short, stigma capitate (obscurely biloped). Fruit a drupe with hard thick endocarp, usually 1-seeded. Seed containing endosperm.

An Old World genus of c. 20 species, with 6 native to mainland Africa.

1. Inflorescences terminal and lateral; lower leaf surface silvery or golden with dense covering
   of minute circular scales   -   -   -   -   -   -   -   - 1. *europaea* subsp. *africana*
  - Inflorescence terminal (includes upper two leaf pairs); lower leaf surface glabrous or with
   scattered scales, not metallic   -   -   -   -   -   -   -   -   -   -   -   - 2
2. Tree 10–40 m. high; leaves 4·5–9 (11) × 1·2–4 cm., with petiole 8–15 mm. long - 2. *capensis*
  - Shrub or small tree up to 3 m. high; leaves 2·4–7·5 × 0·5–1·4 cm., with petiole 4–6 mm. long
                                                                         3. *chimanimani*

1. **Olea europaea** L., Sp. Pl. **1**: 8 (1753). Described from Spain, Italy and France.

Subsp. **africana** (Miller) P. S. Green in Kew Bull. **34**: 69 (1979). Type a cultivated specimen
originally from S. Africa (Cape Prov.).
> *Olea africana* Miller, Gard. Dict. ed. 8, Olea no. 4 (1768). — O. B. Miller, B.C.L.: 48
> (1948). — Verdoorn in Bothalia **6**: 573, fig. 3, t. 13 (1956); in Fl. Southern Afr. **26**: 113
> (1963). — F. White, F.F.N.R.: 337 (1962). — Friedrich-Holzhammer in Prodr. Fl. SW.
> Afr. **108**: 2 (1967). — F. White in Chapman and White, Evergr. For. Malawi: 44 (1970).
> — Fanshawe, Check-list Woody Pl. Zamb.: 31 (1973). — R. B. Drumm. in Kirkia **10**: 267
> (1975). Type as above.
> *Olea chrysophylla* Lam., Tabl. Encycl. Méth. Bot. **1**: 29 (1791); Encycl. Méth. Bot. **4**:
> 544 (1798). — Baker in F.T.A. **4**: 18 (1902). — Chevalier in Rev. Int. Bot. Appl. Agric.
> Trop. **28**: 4 (1948). — Turrill in F.T.E.A., Oleaceae: 9 (1952). — Liben in Fl. d'Afr.
> Centr., Oleaceae: 22 (1973). Type from Réunion.

For fuller synonymy, see P. S. Green, loc. cit.

Evergreen shrub or tree 2–16 m. tall. Young stems glabrous, square in cross-
section with 4 ridges running down each internode from the pair of leaves inserted
above; older branches with moderately smooth grey bark. Leaves coriaceous; petiole
1–13 mm. long; lamina 1·8–9·5 × 0·8–2·4 cm., lanceolate, narrowly elliptic or elliptic,
the apex acute or obtuse with midrib excurrent into a small point, base acute; upper
leaf surface glossy dark green, obscurely minutely punctate, midrib impressed,
lateral nerves obscure, joining into a wavy submarginal vein; lower surface densely
covered with small circular scales (just visible with hand lens), giving a dull or
metallic, golden or silvery appearance, sometimes scales more sparsely distributed;
margin slightly revolute, not undulate. Inflorescences many-flowered paniculate
cymes, terminal and lateral, usually shorter than the subtending leaf; flowers sweet-
scented; bracts 1–2 mm. long, deciduous. Calyx c. 0·75 mm. long, cup-shaped,
subentire with 4-toothed rim. Corolla c. 2·5 mm. long, white, globose in bud, the
lobes soon spreading; lobes elliptic-acute, c. 4–5 times as long as tube. Anthers c. 1·5
mm. long. Gynoecium c. 1·5 mm. long. Fruit 0·5–1 cm. long (dry), ellipsoid.

Botswana. SW: Ootsi, 1070 m., fl. buds xi.1940, *Miller* B231 (PRE). SE: 31 km. WNW. of
Lobatsi, 1280 m.., fl. 18.i.1960, *Leach & Noel* 183 (SRGH). **Zambia.** E: Nyika, fr. 31. xii.
1962, *Fanshawe* 7384 (FHO; K). S: Livingstone Distr., Candelabra Gorge, Victoria Falls, fr.
2.iii.1963, *Mitchell* 17/89 (SRGH). **Zimbabwe.** N: Gokwe Distr., Umi R. gorge c. 19 km. N.
of Gokwe, fl. 18.i.1964, *Bingham* 1054 (LMU; SRGH). W: Matopos, fl. buds xi.1925, *Eyles*
5581 (SRGH). C: Selukwe Distr., Umbetekwe R., fl. 10.ix.1975, *Wild* 8009 (SRGH). E:
Melsetter Distr., Umvumvumvu R., 760 m., fl. 29.ix.1955, *Chase* 5810 (BM; COI; K; LISC;
PRE; SRGH). S: Victoria Distr., Acropolis, fl. 4.x.1949, *Wild* 3037 (K; LISC; SRGH).
Malawi. N: Viphya, Chikangawa, MacDonalds Camp, st. 15.vi.1954, *Jackson* 1349 (FHO; K;
MAL). **Mozambique.** MS: Cheringoma, near Cundui, st. 24.vii.1946, *Simão* 811 (LMA;
PRE). GI: Inhambane, between Nhacoongo and Inharrime, fr. 6.iv.1959, *Barbosa & Lemos*
8513 (COI; LISC; LMA; PRE; SRGH). M: Namaacha Falls, fr. 22.ii.1955, *E. M. & W.* 543
(BM; LISC; SRGH).
Widely distributed from Southern Africa through E. tropical Africa to Arabia, and in the
Mascarenes and Madagascar. In bush vegetation, almost always near running water.

2. **Olea capensis** L., Sp. Pl. **1**: 8 (1753). — Harvey ex C. H. Wright in F.C. **4**: 487 (1907). —
Verdoorn in Bothalia **6**: 582, fig. 6–10, t. 16–20 (1956); in Fl. Southern Afr. **26**: 116, fig. 12
(1963). — F. White, F.F.N.R.: 337 (1962); in Chapman and White, Evergr. For. Malawi:
44 (1970). — Fanshawe, Check-list Woody Pl. Zamb.: 31 (1973). Type from S. Africa
(Cape Prov.).
> *Olea laurifolia* Lam., Tabl. Encycl. Méth. Bot. **1**: 29 (1791). — Harvey ex C. H.
> Wright, loc. cit. Type from S. Africa (Cape Prov.).
> *Olea concolor* E. Meyer, Comm. Pl. Afr. Austr.: 176 (1837). Type from S. Africa (Cape
> Prov.).
> *Mayepea welwitschii* Knobl. in Engl., Bot. Jahrb. **17**: 530 (1893). Type from Angola.
> *Linociera welwitschii* (Knobl.) Knobl. in Bot. Centralbl. **61**: 129 (1895). Type as above.
> *Linociera urophylla* Gilg in Engl., Bot. Jahrb. **30**: 373 (1901). Type from Tanzania.
> *Olea hochstetteri* Baker in F.T.A. **4**: 17 (1902). — Turrill in F.T.E.A., Oleaceae: 10

(1952). — P. S. Green in F.W.T.A. ed. 2, **2**: 49 (1963). — Liben in Fl. d'Afr. Centr., Oleaceae: 23 (1973). Type from Ethiopia.

Olea enervis Harvey ex C. H. Wright, op. cit.: 488 (1907). Type from S. Africa (Natal).

Olea macrocarpa C. H. Wright in F.C. **4**: 1129 (1909); in Kew Bull. **1909**: 186 (1909). Lectotype from S. Africa (Transvaal).

Olea urophylla (Gilg) Gilg & Schellenb. in Engl., Bot. Jahrb. **51**: 75 (1913). Type as for *Linociera urophylla*.

Olea welwitschii (Knbol.) Gilg & Schellenb. in Engl., tom. cit.: 76 (1913). — Turrill, tom. cit.: 12, fig. 3 (1952). — Liben, tom. cit.: 24, t. 6 (1973). Type as for *Mayepea welwitschii*.

Olea schliebenii Knobl. in Notizbl. Bot. Gart. Berl. **12**: 199 (1934). — Turrill, tom. cit.: 10 (1952). Type from Tanzania.

Olea guineensis Hutch. & C. A. Smith in Kew Bull. **1937**: 336 (1937). Type from Ivory Coast.

Olea mussolinii Chiov. in Atti R. Accad. Ital., Mem. Clas. Sci. **11**: 48 (1940). Syntypes from Ethiopia.

Olea capensis subsp. *macrocarpa* (C. H. Wright) Verdoorn, tom. cit.: 590, fig. 10, t. 20 (1956); in tom. cit.: 119, fig. 12, 2 (1963). — R. B. Drumm. in Kirkia **10**: 267 (1975). Type as for *Olea macrocarpa*.

Olea capensis subsp. *enervis* (Harvey ex C. H. Wright) Verdoorn, tom. cit.: 588, fig. 9, t. 19 (1956); in op. cit. **7**: 15 (1958); in tom. cit.: 117, fig. 12, 3 (1963). Type as for *Olea enervis*.

Tree 10–40 m. high; bark pale grey, longitudinally fissured; slash reportedly of various colours: orange-brown, cream or rich green; wood exceedingly hard and durable. Young stems glabrous, ± terete, bark grey, fairly smooth. Leaves coriaceous; petiole 8–15 mm. long, pale to conspicuously dark brown; lamina 4·5–9 (11) × 1·2–4 cm., elliptic, ovate-elliptic or oblong-elliptic, with apex acute to acuminate and tipped with a small hard point, and base cuspidate to acuminate; upper surface glossy, with midrib impressed or level and the 5–6 pairs of lateral nerves faintly raised, joining inconspicuously in loops; lower surface mat, and paler, with prominent midrib and faintly raised lateral nerves; both surfaces punctate with scattered circular scales just visible with hand lens, these sometimes difficult to see on upper surface, lower surface never with metallic sheen; margin slightly revolute, often somewhat undulate. Inflorescence a many-flowered paniculate cyme, terminal and in axils of upper two leaf-pairs; bracts 0·3–1 mm. long, subulate. Calyx c. 1 mm. long, cup-shaped, subentire with 4-toothed rim. Corolla c. 3 mm. long, white, globose in bud and remaining thus for a prolonged period, ultimately opening, the elliptic-acute lobes then reflexed; tube c. ⅓ as long as lobes. Anthers c. 2 mm. long. Gynoecium c. 1·5 mm. long. Fruit up to 1·7 cm. long (dry), ellipsoid, with very thick woody endocarp. Seed not seen.

Zambia. N: Lake Young, Shiwa Ngandu, fl. buds 20.ix.1938, *Greenway* 5752 (FHO; K; SRGH). W: Mwinilunga, fl. buds 26.x.1955, *Holmes* 1289 (K). **Zimbabwe.** C: Wedza Distr., Wedza Mt., fl. buds 14.v.1964, *Wild* 6559 (FHO; SRGH). E: Inyanga Distr., gorge of Pungwe tributary on S. slopes of Inyangani, c. 1800 m., fl. buds 10.xi.1967, *Müller* 718 (K; PRE; SRGH). **Malawi.** N: Nyika Plateau, immat. fr. xi.1965, *Cottrell* 52 (K; SRGH). C: Ntchisi Forest, immat. fr. 8.v.1961, *Chapman* 1292 (BM; K; LISC; MAL; PRE; SRGH). S: Mangoche Mt., fl. buds 1935, *Clements* 537 (FHO; MAL). **Mozambique.** N: area of Caiaia, R. Neôce, st. 5.vii.1967, *Macedo* 4479 (LMA). Z: Gúruè, near source of R. Malema, c. 1700 m., fl. 5.i.1968, *Torre & Correia* 16958 (LISC). MS: Tsetsera area, near road to Mavita, c. 1900 m., st. 29.xi.1966, *Müller* 506 (K; LISC; SRGH).

Widely distributed in tropical and southern Africa. In evergreen or mixed evergreen forest. Fanshawe (loc. cit.) records *O. capensis* also from the Central Prov. of Zambia.

Apart from the differences given in the key, *O. capensis* can be distinguished from *O. europaea* subsp. *africana* by the larger fruits of the former. Moreover, in leaves of *O. capensis* the lateral nerves join in a series of loops some distance from the leaf margin, whereas in *O. europaea* subsp. *africana* there is an almost straight submarginal vein formed from these lateral nerves.

O. capensis in, Southern Africa has been treated in great detail by Verdoorn, loc. cit., who identified three subspecies in her area. I agree that these taxa can be recognised, and find that the F.Z. material belongs to subsp. *macrocarpa*. I have not used this infraspecific group because the situation becomes more complex in tropical Africa to the north and west of the F.Z. area. Here authors have mainly recognised two species within what should be called *O. capensis*: *O. hochstetteri* and *O. welwitschii*. They are distinguished by leaf shape, viz. absolute length of lamina and petiole and length: breadth ratio of lamina; *O. hochstetteri* has relatively short, broad leaves and *O. welwitschii* long, narrow ones. A survey of many specimens throughout the entire species range shows that while plants with very different leaf shape have sometimes been collected from the same locality (e.g. Mt. Meru in Tanzania), leading observers to be convinced of their specific distinctness, the range of variation is continuous when leaves from many places are considered together.

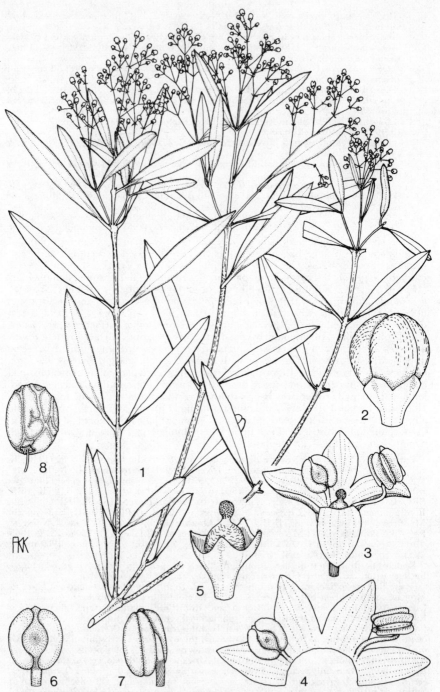

Tab. 74. OLEA CHIMANIMANI. 1, habit (× 1) *Bamps, Symoens & Vanden Berghen* 830; 2, flower bud (× 12); 3, flower, with one anther before and one after dehiscence (× 10); 4, corolla opened out (× 10); 5, calyx and gynoecium (× 12); 6–7, stamen; 6, adaxial view; 7, lateral view (× 15), 2–7 from *Goodier & Phipps* 308; 8, fruit (× 3) *Wild* 4575. From Kew Bull. **34**: 72 (1979).

When leaf length was plotted against length/breadth, scoring specimens from the whole species range, the type specimens of *O. hochstetteri* and *O. welwitschii* fell near together, both within the main cluster of specimens. It was found that in general leaf length tends to increase from south to north in *O. capensis*, and that the longest leaves tend to be proportionately narrowest and to have the longest petioles.

Within the F.Z. area, *O. capensis* is not widespread but occurs in isolated patches. Specimens from each locality have a characteristic leaf shape. Collections from the Gúruè (Z) and Tsetsera area of Mozambique (MS) have conspicuously narrow leaves, like those of the type specimen of *O. urophylla* from southern Tanzania. These specimens, being from the southern half of the species' range, have leaves rather too small to fit the concept of *O. welwitschii*, although their proportions are correct. The F.Z. specimens, in general, agree with Tanzanian "*O. hochstetteri*".

These notes are given to justify my broad treatment of *O. capensis*. The species still needs to be thoroughly investigated. Wood anatomy and fruit shape should be taken into account, since collectors have suggested that significant differences exist in these features too.

3. **Olea chimanimani** Kupicha in Kew Bull. **34**: 71, fig. 1 & map 1 (1979). TAB. **74**. Type: Zimbabwe, Chimanimani Mts., Stonehenge, c. 0·4 km. W. of Mountain Hut, 1710 m., fl. 28.xii.1959, *Goodier & Phipps* 308 (K, holotype; PRE; SRGH, isotypes).

Shrub or small tree 2–3 m. high. Young stems glabrous, ± terete, bark grey, fairly smooth. Leaves thinly coriaceous; petiole 4–6 mm. long; lamina 2·4–7·5 × 0·5–1·4 cm., oblanceolate or narrowly elliptic, with apex obtuse, acute or acuminate and tipped with a small hard point, base acuminate; upper surface glossy dark green, with midrib level or slightly raised and lateral nerves faintly visible or obscure; lower surface mat, paler, with prominent midrib and lateral nerves completely obscure; both surfaces punctate, scales absent; margin not or slightly revolute, not undulate. Inflorescence a many-flowered paniculate cyme, terminal and in axils of upper 2 (3) leaf-pairs, slightly exceeding subtending leaf; bracts 0·5–1·5 mm. long, subulate. Calyx c. 0·75 mm. long, cup-shaped, deeply 4-lobed, lobes triangular. Corolla c. 2 mm. long, whitish or cream-coloured; lobes ovate, obtuse, shortly united at base. Anthers c. 1–1·5 mm. long. Gynoecium c. 1·5 mm. long. Fruit 6–8 × 5–6 mm. (when dry), oblong-ellipsoid or subglobose. Seed solitary, pendulous.

Zimbabwe. E: Chimanimani Mts., 1560 m., fl. 14.i.1974, *Bamps, Symoens & Vanden Berghen* 830 (LISC; PRE; SRGH). **Mozambique.** MS: Chimanimani Mts., fr. immat. 1973, *Dutton* 77 (LMA).

Not known outside this limited area. In scrub vegetation among quartzite crags.

109. LOGANIACEAE

By A. J. M. Leeuwenberg

Woody or less often herbaceous plants. Leaves mostly opposite, less often ternate (often in *Nuxia*, sometimes in *Strychnos*), occasionally quaternate (sometimes in *Nuxia*), or sometimes subopposite or alternate (sometimes in *Buddlejeae*, especially in *Buddleja* and *Nuxia*). Stipules true or false, present, reduced to lines connecting the petiole bases, or absent; in some cases leaves connate-perfoliate (in some *Buddleja* spp. with opposite leaves); lamina simple, variously shaped, pinnately veined, entire, incised, or less often lobed; sometimes with 1–3 pairs of basal secondary veins larger and curved along the margin, rendering the leaves seemingly 3–7-veined (most *Strychnos* spp.). Inflorescence usually thyrsoid or otherwise cymose, or 1-flowered, rarely a raceme (*Gomphostigma*). Flowers regular, mostly hermaphrodite, homo- or (in *Mostuea*) heterostylous, mostly 4- or 5-merous (corolla and androecium 8–16-merous in *Anthocleista*), mostly actinomorphic but then often with subequal or unequal sepals, less often subactinomorphic or zygomorphic. Sepals usually green, free or united, usually persistent, imbricate, valvate, or apert in bud. Corolla usually coloured, sympetalous, variously shaped; lobes valvate, imbricate, or contorted in bud. Stamens as many as corolla lobes and alternating with them or sometimes less; filaments free from each other, from much shorter to much longer than the anthers; anthers basifix, often versatile; cells 2 or (outside F.Z. area) 4, discrete or confluent at the apex, parallel or divergent at the base, dehiscent throughout by a longitudinal slit. Pistil simple; ovary superior or sometimes slightly inferior, mostly 2-celled, but sometimes 1-, or 4-celled; style simple, terminal, persistent or deciduous; stigma simple or sometimes branched; ovules 2-many, on an axile placenta attached to the septum or in a unicellular ovary to the bottom (*Strychnos spinosa*) or parietal. Fruit a capsule or a berry, 1-many-seeded. Seeds variously shaped, small or large, sometimes winged, with fleshy, starchy, or horny (*Mostuea, Strychnos*) endosperm surrounding a rather small or large straight embryo.

A family distributed throughout the warm-temperate and tropical regions of the world. Represented in the F.Z. area by 6 genera and 38 species, 4 of which are cultivated.

1. Corolla lobes 8–16; inflorescence terminal, dichasal or nearly so, usually large; plant brittle when dry; fruit a berry; trees - - - - - - - - - - **-1. Anthocleista**
- Corolla lobes 4–5; if fruit a berry leaves mostly triplinerved (*Strychnos*) - - - 2
2. Fruit a capsule (a berry only in the cult. *Buddleja madagascariensis*); leaves not triplinerved 3
- Fruit a berry, often large and thick-walled; leaves mostly triplinerved; often huge climbers with hooked tendrils; aesitvation valvate - - - - - - - **6. Strychnos**
3. Stigma twice dichotomously branched; corolla mostly white with a yellow base, infundibuliform; capsule mostly bilobed - - - - - - **4. Mostuea**
- Stigma capitate or nearly so; capsule ellipsoid or nearly so - - - - - - 4
4. Anther cells confluent; stamens exserted; corolla tube cylindrical, included in the tubular calyx which is about as long as the capsule - - - - - - - **5. Nuxia**
- Anther cells discrete; if corolla tube cylindrical, it is much longer than the calyx and the stamens are included - - - - - - - - - - - - - 5
5. Inflorescence racemose; corolla subrotate - - - - - **3. Gomphostigma**
- Inflorescence cymose, much branched if corolla subrotate - - - **2. Buddleja**

1. ANTHOCLEISTA Afzel. ex R. Br.

Anthocleista Afzel. ex R. Br., in Tuckey, Narrat. Exped. R. Zaire, App. **5**: 449 (1818). — Martius, Nov. Gen. **2**: 91 (1827). — Leeuwenberg, Acta Bot. Neerl. **10**: 1 (1961).

Entirely glabrous trees, 1–35 m. high, shrubs, or lianas. Leaves opposite, those of a pair equal or unequal, petiolate or sessile; bases or petioles joined, often auriculate, and, especially in young plants more or less conspicuously ligulate at the base; lamina soft and brittle or coriaceous when living, membranaceous or papyraceous and often brittle, or coriaceous when dry, entire or minutely crenate; margin recurved or not; secondary veins conspicuous or not. Stipules intrapetiolar (ligular). Inflorescence terminal, erect, sometimes pendulous when in fruit, almost dichasial, 1–5 times branched, easily breaking at the nodes when dry. Lower bracts foliaceous, the others usually very small, triangular or ovate. In continental species mostly only one flower of each inflorescence open at a time. Sepals 4, green, creamy, or occasionally partially orange, free, or sometimes connate at the base (*A. laxiflora*), orbicular or nearly so, concave, decussate, appressed to the corolla tube and later to the fruit, usually rounded at the apex, entire, the 2 inner ones usually becoming retuse or torn by the development of the corolla, often spreading when dry, often enlarged under the fruit. Corolla white, creamy, violet, violet-blue, mauve, or sometimes pale yellow, the limb often paler than the tube which is sometimes green, actinomorphic, tubular, usually not contracted when mature, thick, fleshy, brittle, also when living, often sweet-scented; tube approximately cylindrical, more or less gradually widened towards the throat; lobes 8–16, contorted in bud usually turned to the right, spreading or recurved, elliptic to lanceolate, usually obtuse, entire. Stamens as many as the corolla lobes and alternating with them, exserted, equal; filaments short or very short, mostly shorter than the anthers, entirely connate into a short tube or occasionally connate for two-thirds of their length, inserted near the apex of the corolla tube; anthers white or creamy, often partially green, sometimes brownish, lanceolate, obtuse or sometimes acute at the apex, usually sagittate at the base; cells 2, parallel and if sagittate only at the base divergent, dehiscent throughout by a longitudinal split. Ovary superior, ovoid-conic, cylindric, or obovoid-cylindric, 4-celled; style thick, about as long as the corolla tube, persisting during a short period after the corolla is shed; stigma large, usually obovoid-cylindric and apically bilobed, often slightly laterally compressed. In each cell one large bilobed placenta with numerous ovules on both sides. Fruit a berry, dark green or yellow, hard, globose or ellipsoid (irregular or regular indentations or furrowings are due to shrinkage and are always artificial, rounded at the apex, sometimes apiculate; wall usually thick; septa thin. Seeds obliquely ovate-circular or irregularly polyhedral in outline, flattened, medium to dark brown, slightly verrucose, faveolate. Small colleters in one rank in the axils of the leaves, bracts, and sepals.

A small genus comprising 14 species in tropical Africa, Madagascar, and the Comores.

1. Branches armed with short usually paired spines; flower buds uniformly rounded or subtruncate at apex; corolla tube 1·25–2 × as long as the calyx; swamps; mostly with stilt-roots - - - - - - - - - - - - - - - - - 4. *vogelii*
- Branches unarmed; or — if with incipient or occasional spines then flower buds not rounded, but tapering at apex - - - - - - - - - - - - - - - 2
2. Leaves sessile or subsessile, — if petiolate — then buds not rounded but tapering at apex
3
- Leaves petiolate and buds usually uniformly rounded at apex; blade narrowly obovate; calyx definitely constricted at mouth; river banks - - - - - - 2. *liebrechtsiana*
3. Sepals drying smooth, clasping the corolla tube at anthesis, only spreading under the mature fruit; leaves, at least the upper ones, usually petiolate, lower ones often sessile or subsessile, usually subcoriaceous, and with inconspicuous tertiary veins; corolla tube about 1–1·5 × as long as the lobes which are large and reflexed; berry never shrivelled when dry
3. *schweinfurthii*
- Sepals drying rugulose, outer pair at least more or less spreading, not closely clasping the corolla tube at anthesis, widely spreading under the fruit; leaves sessile or sometimes petiolate, usually membranaceous, and with conspicuous tertiary veins; corolla tube 1·25–2·5 × as long as the lobes; berry when dry irregularly shrivelled - 1. *grandiflora*

1. **Anthocleista grandiflora** Gilg in Engl., Bot. Jahrb. **17**: 582 (1893). — Leeuwenberg in Acta. Bot. Neerl. **10** (1961) 1–53: 28, fig. 13, map 9. Type from the Comores.
    A. zambesiaca Bak. in Kew Bull. **1895**: 99 (April 1895). — F.T.A. **4**(1): 540 (1903). — Prain & Cummins in Fl. Cap. **4**, 1: 1049 (1909). — Bruce in Kew Bull. **10**: 54 (1955). — Bruce & Lewis in F.T.E.A., Loganiaceae: 10, f. 2 (1960). Type from Malawi: Shire Highlands, *Buchanan* 84 (K, holotype; isotypes: E, K).
    A. insignis Galpin in Kew Bull. **1895**: 150 (vi–vii 1895). Type from Swaziland.

Tree, 5–35 m. high, unarmed. Leaves sessile, those of larger trees sometimes petiolate; lamina medium to dark green above, paler beneath, when dry greenish, medium to dark brown, paler beneath, brittle, often thinly papyraceous to coriaceous, narrowly or very narrowly obovate, in young plants usually narrower, 1·75–3·5 (in young plants up to 5) times as long as wide, 15–70 × 7–25 cm., up to 135 × 50 cm. in young plants or low-levelled branches, narrowed to the auricles or long-decurrent into the petiole; veins conspicuous; margin not recurved. Sepals green, rounded, usually spreading when dry, the outer ones circular or broadly ovate, 5–8 × 5–8 mm., when dry rugulose outside and often pointed, the inner ones usually slightly larger. Corolla in the young bud rounded or tapering, in the mature bud 5–10 as long as the calyx, 35–60(70) mm. long, tapering at the apex, white, the limb paler than the tube which is slightly greenish outside; tube 3·8–6·5 × as long as the calyx, 1·25–2·5 × as long as the lobes; lobes 11–13, very narrowly elliptic, spreading or recurved. Berry ellipsoid, when dry irregularly shrivelled, conspicuously rugulose, acuminate.

Zimbabwe. E: Inyanga, near Nyamingura R., *Phipps* 1250 (BR; EA) S: Bikita District, *Wild* 4379 (K; MO). **Malawi** C: Ntchisi, Ntchisi Distr., *Brass* 17072 (BM; BR; EA; K; MO). S: Thyolo Mt., Cholo District, *Brass* 17854 (BR; K; MO). **Mozambique**. N: Macondes Distr. between Mueda and Chomba, *Gomes Pedro* 5352 (LISU). Z: Gurué, *Andrada* 1842 (COI; LISC). MS: Bárué, Choa Mts., Catandica (Vila Gouveia), *Mendonça* 287 (LISC). GI: Maruma Mt., *Swynnerton* 27 (BM; K).
Also in East Africa, from Uganda and Kenya to Transvaal, in Zanzibar and in the Comores. In open often swampy places, in rain forests, or in gallery forests, mostly in mountains, 0–2300 m.

2. **Anthocleista liebrechtsiana** De Wild. et Dur., Compt. Rend. Soc. Bot. Belg. **38** (2): 96 (1899). — Baker in F.T.A. **4**, 1: 540 (1903). — Leeuwenberg in Acta. Bot. Neerl. **10**, 1–53: 22, fig. 11, map 7. (1961). TAB. **75**. Type from Zaire.

Tree or few-stemmed shrub, 1·50–12 m. high, without spines. Leaves petiolate; lamina dark green above, pale greyish-green beneath, drying greenish-brown and coriaceous, very narrowly obovate to linear or sometimes narrowly ovate, 2½–10 (usually about 4–7) × as long as wide, 11–75 × 3–15 (usually about 15–40 × 3–8) cm.,

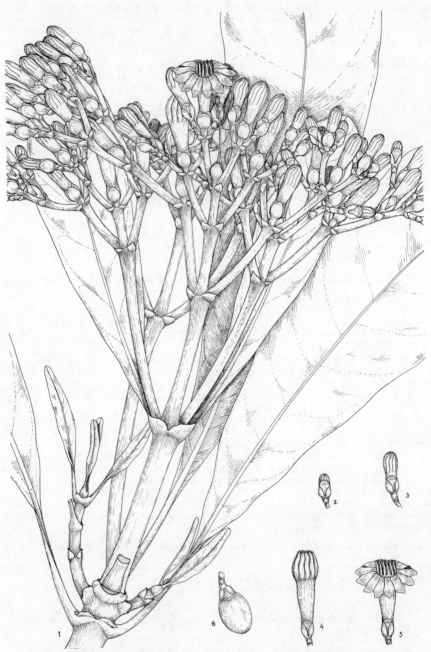

Tab. 75. ANTHOCLEISTA LIEBRECHTSIANA. 1, branch (×½); 2–3, young flower buds (×½), 1–3 from *Lebrun* 1672; 4, expanding bud (×½); 5, flower (×½), 4–5 from *Leemans* 219; 6, fruit (×½), from *J. Leonard* 675.

long-decurrent into the petiole; costa prominent and acutely triangular beneath; secondary veins rather inconspicuous; margin neither recurved nor revolute. Sepals pale green, rounded, also when dry strongly appressed to the base of the corolla tube, which therefore is slightly contacted, and later to the fruit, smooth and shining, the outer ones broadly ovate or orbicular, 4–8 × 4–8 mm., the inner ones slightly larger and becoming torn by the development of the corolla. Corolla in the mature bud 4·5–7 times as long as the calyx, 32–54 mm. long, and as in the young bud rounded or somewhat tapering at the apex, white, the limb paler than the tube which is usually greenish-white; tube 3–5 times as long as the calyx, 1·8–3 times as long as the lobes; lobes 10–12, narrowly elliptic, spreading. Berry globose or ovoid, when dry irregularly shrivelled, thin-walled; wall about 1 mm. thick.

Zambia. B: North of Chavuma, Zambesi [Balovale] District, *Angus* 642 (BM; BR; K), 642A (K) S: Katombora, Southern Livingstone, *Brenan & Morze* 7740 (EA; K).
Also from Ghana to Angola. In open places in swamps or in water and in usually periodically inundated forests. Alt. 0–400 m.

3. **Anthocleista schweinfurthii** Gilg in Engl. Bot. Jahrb. **17**: 579 (1893). — Baker in F.T.A. **4**, 1: 541 (1903). — Bruce, Kew Bull. **10**: 51 (1955). — Bruce & Lewis in F.T.E.A., Loganiaceae 11. (1960). — Leeuwenberg in Acta Bot. Neerl. **10**, 1–53: 24, fig. 12, map 8 (1961). TAB. **76**. Lectotype from Zaire.
A. nobilis Baker, 1 c. p. 539, quoad spec. *Schweinfurth* 3037 et 3726, non G. Don.

Tree 3–30 (usually 4–10) m. high. Twigs without or occasionally, especially in young plants, with short paired partially united spines, often with small broadly conical cushions. Leaves usually petiolate but often sessile in young plants or on low-level branches; lamina dark green and usually (?) glossy above, paler beneath, when dry medium to dark brown above, paler beneath, papyraceous to coriaceous, narrowly to very narrowly ovate, in young plants usually narrower, 1·75–3·5 times as long as wide, 7·45 × 3·5–18 cm., in young plants up to 100 × 30 cm. or more (?), cuneate at the base; costa more or less acute beneath; tertiary veins inconspicuous; margin not curved, but often revolute. Sepals green, rounded, when dry usually smooth, especially the outer ones, in flower uually appressed to the base of the corolla, in fruit often spreading, the outer ones orbicular or slightly broader than long, 8–13 × 9–13 mm., the inner ones usually slightly larger, becoming retuse by the development of the corolla. Corolla in the mature bud 5·5–7 times as long as the calyx, 55–61 mm. long, tapering at the apex as in the young bud, white or creamy, tube darker than the lobes, often greenish-white; tube about 3–4 times as long as the calyx, about 1–1·5 times as long as the lobes; lobes 10–11, very narrowly elliptic, reflexed. Berry globose or ellipsoid rounded or apiculate at the apex, never shrivelled when dry; thick-walled.

Zambia. N: near Samfya, Lake Bangweulu, Mansa [Fort Roseberry] District, *Angus* 276 (BR; K, with fr. in spirit coll.), 291 (BM; BR; K, with fr. in spirit coll.). W: Ndola District, *Fanshawe* 1684 (BR, K).
Also in Central Africa, from Nigeria to the Sudan in the north and Angola and Tanzania in the south. In secondary or gallery forests, in thickets, or sometimes in savannas or rain forests, usually not in moist places, 0–1800 m.

4. **Anthocleista vogelii** Planch. in Hook., Icon. Pl. **8**: tt. 793–794 (1848). — Hook., Niger Fl.: 459, t. 43–44 (1849). — Hutch. & Dalz., F.W.T.A. **2**: 18 (1931). — Bruce, Kew Bull. **10**: 48 (1955). — Bruce & Lewis, in F.T.E.A., Loganiaceae: 8 (1960). — Leeuwenberg in Acta Bot. Neerl. **10**, 1–53: 16. Fig. 7; Map 5 (1961). Type from Nigeria.
A. nobilis Baker, F.T.A. **4**: 538 (1903), quoad spec. *Vogel* 51, non G. Don.

Tree, 6–20 m. high or more (?). Twigs with 2(4) spines, which are divergent and confluent at the base, or occasionally unarmed. Leaves sessile or very shortly petiolate; lamina dark green and often glossy above, pale glaucous beneath, when dry dark brown above, paler beneath, brittle, papyraceous to coriaceous, narrowly to very narrowly obovate, in young plants usually narrower, 1·75–3·5 (usually about 2, in young plants up to 4 times as long as wide, 15–45 × 6–24 cm., up to 150 × 45 cm. in young plants, narrowed to the auricles or decurrent into the petiole, if petiolate cordate at the very base; margin usually recurved. Sepals pale green, occasionally partially orange, rounded, when dry more or less rugulose and somewhat spreading or the outer ones circular or broader than long, 1–1½ × as broad as long, 4–12 × 7–15 mm., the inner ones usually larger, up to about 2 × as long as the others, often

Tab. 76. ANTHOCLEISTA SCHWEINFURTHII. 1, branch (× ½); 2–3 young flower buds (× ½), 1–3 from *Schmitz* 3332; 4, mature bud (× ½), from *Tisserant* 1934; 5, flower (× ½), from *Schmitz* 3332; 6, fruit (× ½), from *Louis* 3100.

partially torn by the development of the corolla. Corolla in the young bud at the apex as in the mature bud rounded or sometimes obtuse; in the mature bud 2, 5–4 × as long as the calyx, 23–37 mm. long, and rounded or subtruncate at the apex, creamy or sometimes pale yellow; the tube darker than the lobes; tube 1·25–2 × as long as the calyx, 0·9–1·5(1·7) × as long as the lobes; lobes 13–16, narrowly elliptic, spreading. Berry thick-walled, globose or ellipsoid, rounded at the apex, when dry and mature occasionally apiculate, but not shrivelled, occasionally so when dry and immature.

Zambia. W: near the headwaters of the Lunga R., Mwinilunga-Kolwezi Road, Mwinilunga District, *Angus* 574 (BM; BR; K; SRGH).

Also in Tropical Africa, from Sierra Leone to Uganda in the north and Angola in the south. Usually in moist places, in swamps, in Raphia groves, on river banks; in primary rain or secondary forests. Alt. 0–1500 m.

2. BUDDLEJA L.

Buddleja L., Sp. Pl. **1**: 112. 1753. Leeuwenberg in Meded. Landb. Wag. **79**-6: 1–163 (1979).

Shrubs, less often trees or suffrutescent herbs, 0·25–30 m. high, mostly with a white or pale grey indumentum of stellate hairs which dries usually rusty on the branchlets, the lower side of the leaves, and the inflorescences. Colleters absent. Bark rough, fibrous, longitudinally fissured. Branchlets terete to quadrangular, in the latter case sometimes narrowly winged. Leaves opposite and those of a pair equal, less often subopposite, or in a few species alternate, petiolate, sessile, or — if opposite — sometimes connate-perfoliate. Stipules leafy, reduced to a line, or absent. Lamina circular to very narrowly elliptic, entire, crenate to dentate, or sometimes evenlobed.Inflorescence terminal and/or axillary, thyrsoid, paniculate, or variously reduced even to a single globose head. Flowers 4-merous, actinomorphic. Calyx green, campanulate or nearly so, less often cup-shaped or obconical; lobes usually subequal, from slightly longer to much shorter than the tube, erect, entire. Corolla cup-shaped, campanulate, funnel- or salver-shaped, variously coloured, from white to orange or to dark violet, or purple, often with an orange throat, sometimes darker at anthesis, outside with an indumentum of stellate and/or glandular hairs, but glabrous at the base, which is included in the calyx tube, and at the apices of the lobes, less often entirely glabrous or hairy on the entire outer side of the lobes; lobes usually shorter than the tube, imbricate in African species, elliptic or suborbicular, at the apex rounded, obtuse, or emarginate, entire to crenate, spreading or erect. Stamens inserted in the corolla, from the base of the tube to the mouth, with a certain degree of variation in a single species; filaments mostly short or very short, long only if stamens well-exserted (section *Chilianthus*); anthers oblong or less often subcircular in outline, mostly glabrous, apiculate to emarginate at the apex, mostly deeply cordate at the base, introrse; cells parallel or sometimes slightly divergent at the base, discrete, longitudinally dehiscent throughout. Pistil: ovary superior, usually 2-celled. Style short or long, included or exserted; stigma often large, clavate, capitate, or less often bilobate. In each cell one axile placenta with several to many ovules. Fruit a capsule or sometimes a berry, subglobose to narrowly ellipsoid, surrounded or subtended by the persistent calyx and often at the same time by the persistent corolla. Capsule bivalved, septicidal; valves mostly splitting up to the middle. Seeds pale to dark brown, obliquely fusiform or obliquely polyhedral, winged or not, laterally compressed or not, reticulate or less often smooth or ridged.

About 90 species in the tropics and subtropics of America, Africa, and Asia.

1. Inflorescence paniculate; plants indigenous - - - - - - - - 2
– Inflorescence thyrsoid or spicate; plants cultivated - - - - - - 6
2. Flowers small, up to 3 mm. long; stamens exserted; corolla tube short, up to 1 × as long as the lobes - - - - - - - - - - - - - - 3
– Flowers larger, 4–9 mm. long; stamens included; corolla tube 2·5–6·5 × as long as the lobes 4
3. Leaves almost triangular to ovate, irregularly dentate to crenate - - 4. *dysophylla*
– Leaves narrowly elliptic, entire or obscurely sinuate - - - - 8. *saligna*
4. Leaves deeply cordate to auriculate at the base, narrowly ovate or narrowly elliptic, bullate, often sessile, crenate; corolla tube 2·5–3·2 × as long as the calyx, 4–7 mm. long
9. *salviifolia*
– Leaves decurrent, cuneate, or rounded at the base, ovate or elliptic or narrowly so, not or

sometimes slightly bullate, petiolate, entire, serrate, or lobed; corolla tube 3·5–6·5 × as long
as the calyx, 5·5–9 mm. long- - - - - - - - - - - 5
5. Leaves mostly serrate, never lobed; venation conspicuous, reticulate, much impressed
above; stipules often present - - - - - - - - 2. *auriculata*
– Leaves often lobed and furthermore entire; venation inconspicuous; stipules none
7. *pulchella*
6. Corolla orange; fruit baccate- - - - - - - - 6. *S. madagascariensis*
– Corolla white to dark violet; fruit capsular - - - - - - - 7
7. Corolla white, inflorescence spicate; corolla tube 2·5–5 mm. long - - 1. *asiatica*
– Corolla mostly lilac to violet; — if inflorescence spicate — corolla tube 12–17 mm. long 8
8. Corolla tube 12–17 mm. long, often curved; inflorescence mostly spicate; leaves largely
sinuate-dentate to entire - - - - - - - - - 5. *lindleyana*
– Corolla tube 6–11 mm. long, straight; inflorescence thyrsoid; leaves serrate to subentire
3. *davidii*

1. **Buddleja asiatica** Lour., Fl. cochin.: 72 (1790). — Bentham in De Candolle., Prod. **10**: 446
(1846). — F. White, F.F.N.R.: 339 (1962). — Leenhouts in Fl. Males. ser. 1. **6**: 337, fig. 24
(1963). — Leeuwenberg, Vidal & Galibert, Fl. Camb. Laos Viet. **13**: 92, fig. 15. 1–8
(1972). — Leeuwenberg in Meded. Landb. Wag. **79**–6: 92, fig. 22 (1979). Type from
Vietnam.

Shrub, undershrub or sometimes small tree, 0·80–7 m. high. Branchlets terete or
nearly so, densely stellate-pubescent or -wooly with white, grey or fulvous hairs.
Leaves opposite, those in the inflorescence often more or less alternate, shortly
petiolate; petiole 2–15 mm. long; blade narrowly to very narrowly elliptic or ovate,
3–8 times as long as wide, 3–30 × 0·5–7 cm., long-acuminate at the apex.
Inflorescence terminal and/or axillary, thyrsoid, spiciform. Pedicels very short,
0·2–2 mm. long; flowers fragrant, crowded or more or less remote, (sub)sessile, in
1–3-(rarely more) flowered cymes; each cyme in the axil of a linear bract. Calyx
campanulate, 1·3–4·5 mm. long, outside stellate-pubescent or -tomentose; lobes
0·4–3 times as long as the tube, subequal. Corolla white, sometimes pale violet or
greenish, with erect lobes 1·3–3 times as long as the calyx and 3–6 mm. long, outside
densely or less often sparsely stellate-tomentose; tube nearly cylindrical, 1·4–2·4 × as
long as the calyx, 1·7–4 × as long as the lobes, 2·5–4·8 × 1·2–1·5 mm.; lobes orbicular
or nearly so, 1–1·7 × 1–1·5 mm., rounded, entire or crenate, spreading. Stamens
included. Ovary glabrous or lepidote. Capsule ellipsoid, 3–5 × 1·5–3 × 1·5–3
mm.; valves often torn at the apex. Seeds pale brown, reticulate, winged at
both ends, 0·8–1 × 0·3–0·4 × 0·2 mm.; grain ellipsoid, apiculate at both ends,
0·3–0·4 × 0·3 × 0·2 mm.

Zambia. C: Lusaka Forest Nursery, *F. White* 3034 (BR; FHO; K).
Nepal, India, Bangladesh, Burma, Thailand, Laos, Cambodia, Vietnam, Malaysia,
Indonesia, New Guinea, Philippines, China (Yünnan, Kwangsi, Szechwan, Fukien,
Kwantung, Hainan), Hong-Kong, Taiwan. In open places or light forests. Alt. 200–2000 m.

2. **Buddleja auriculata** Benth. in Hooker, Comp. Bot. Mag. **2**: 60 (1837). — Prain &
Cummins in Fl. Cap. **4**, 1: 1047 (1909). — Verdoorn in Fl. S. Afr. **26**: 162, fig. 22, 2 (1963).
— Leeuwenberg in Meded. Landb. Wag. **79**–6: 20, fig. 2; phot. 1; map 1 (1979). TAB. **77**.
Type from S. Africa.
B. auriculata var. *euryfolia* Prain & Cummins, tom. cit.: 1048. Type from S. Africa.

Shrub 0·50–3·50 m. high, semiscandent, sometimes a small tree. Branchlets
stellate-tomentose with short whitish hairs, subangular or terete. Leaves opposite,
petiolate; petiole 3–11 mm. long, stellate-tomentose; stipules foliate, sometimes
reduced to a line, elliptic or ovate or narrowly so, 2–5 times as long as wide,
2–13 × 0·5–5 cm., acuminate at the apex, cuneate or rounded at the base, serrate or
sometimes entire and revolute at the margin, dark green, shiny, glabrous, almost
bullate above, whitish-tomentose with short stellate hairs (often drying rusty)
beneath; venation conspicuous, reticulate, much impressed above. Inflorescence
terminal and axillary, large, paniculate, many-flowered, 3–25 × 2–20 cm., rather lax.
Flowers sweet-scented, shortly pedicellate or subsessile. Calyx green, campanulate,
1·5–2·2 mm. long, outside whitish-tomentose with short stellate hairs; lobes 1·2–4
times as long as the lobes; lobes subequal, often broadly triangular, 0·5–1 × 0·8–1
mm., acute or subacute, entire. Corolla creamy, white with orange throat, orange-
yellow, lilac, or salmon, with erect lobes 4–5 times as long as the calyx, 7–11 mm.
long, outside whitish-tomentose with short stellate hairs, sometimes partially
glabrescent, inside pilose with simple hairs from near the apex of the ovary to just

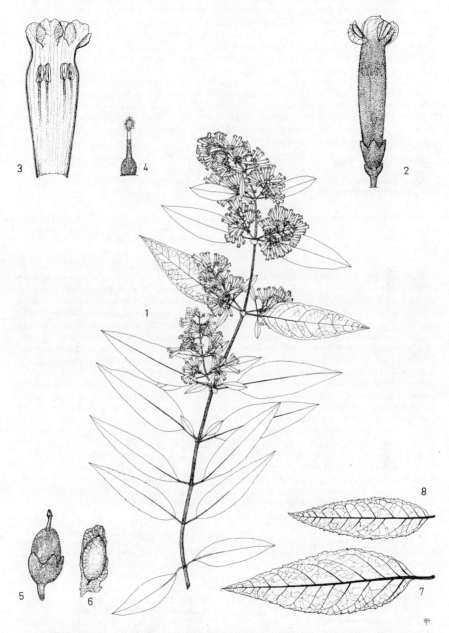

Tab. 77. BUDDLEJA AURICULATA. 1, flowering branch (× ½); 2, flower (× 5); 3, opened corolla
(× 5); 4, pistil (× 5), 1–4 from *Bayliss* 6411; 5, fruit (× 5); 6, seed (× 15), 5–6 from *Devenish*
1263; 7–8, leaves (× ½), 7 from *Compton* 27871; 8 from *Rogers* 20167.

below the insertion of the stamens; tube nearly cylindrical, 3·3–4·5 times as long as
the calyx, 3·5–6 times as long as the lobes, 5·5–9 × 1·2–2 mm.; lobes oblong or
subcircular, mostly slightly longer than wide, 1·5–2 × 1·2–1·8 mm., rounded, entire,
spreading. Stamens included, inserted at 2–2·5(–3·5) mm. from the corolla mouth;
filaments short, 0·3–1 mm. long; anthers oblong or nearly so, 0·9–1·2 × 0·3–0·4 mm.,
rounded or obtuse at the apex, deeply cordate at the base, glabrous; cells parallel.
Pistil much shorter than the corolla tube, 2·2–4·3 mm. long; ovary subglobose or
ovoid, laterally compressed, 0·9–1·5 × 0·7–1 × 0·6–0·9 mm., shortly stellate-
tomentose, 2-celled; style included, stellate-tomentose at the base, 0·5–1·7 mm. long;
stigma large, clavate or sometimes capitate, 0·3–1·2 × 0·3–0·5 mm. In each cell one
axile placenta with 10–15 ovules. Capsule ellipsoid, 2–4 × 1·5–2 × 1·2–1·5 mm.,
laterally compressed half exserted from the calyx, hairy like the ovary but less
densely. Seed medium brown, obliquely polyhedral, 1·2–1·5 × 0·6–0·7 × 0·3–0·4
mm., reticulate, narrowly winged.

   Zimbabwe. E: Pungwe R. source, fl. iv, *Chase* 4930 (BM; COI; LISC; MO; PRE; SRGH).
Mozambique. MS: Tsetsera fl. vi, *Biegel* 3962 (K; SRGH).
   Also in Swaziland, South Africa (Transvaal, Natal, Transkei, Cape Province). In montane
forests or thickets, often in gullies. Alt. 600–2000 m.

3. **Buddleja davidii** Franch., Pl. David. ex. Sinarum Imp. in Nouv. Arch. Mus. Paris Sér. 2.
      10: 103 (1888) (as *Budleia davidi*). — F. White, F.F.N.R.: 339 (1962). — Leenhouts in Fl.
      Males. **1**. 6: 340 (1963). — Leeuwenberg in Meded. Landb. Wag. **79**–6: 113, fig. 29
      (1979). Type from China.
         B. variabilis Hemsl., Journ. Linn. Soc., Bot. **26**: 120 (1889). Type from China.

   Shrub 0·50–3 m. high, often sarmentose. Branchlets subquadrangular, stellate-
tomentose when young, glabrescent. Leaves opposite or sometimes subopposite,
shortly petiolate; stipules foliate, often only present on main branches; long; lamina
narrowly ovate or narrowly elliptic, 3–6 times as long as wide, 4–20 × 1–7 cm.,
acuminate at the apex, cuneate at the base, serrate to subentire. Inflorescence
terminal, sometimes also lateral, thyrsoid, long and rather narrow, 10–30 × 3–5 cm.,
rarely smaller, composed of mostly short-stalked, lax, many-flowered cymes.
Flowers all shortly pedicellate or some sessile. Calyx narrowly campanulate,
2–3·5 × 1–1·5 mm., outside stellate-pubescent to glabrous, tube (1·5)2–5 times as
long as the lobes; lobes subequal, usually narrowly triangular, 0·5–2 × 0·5–0·8 mm.,
acute or acuminate, entire. Corolla violet or lilac (white not seen in specimens in the
field), orange-yellow within the throat, with erect lobes 3–5 times as long as the calyx,
7·5–14 mm. long, outside glabrous or stellate-pubescent and/or with minute
glandular hairs; tube nearly cylindrical, 2–4 times as long as the calyx, 3–5·5 times as
long as the lobes, 6–11·5 mm. long, 0·9–1·5 mm. wide below the throat and from
there abruptly widened; lobes circular or slightly longer than wide, 1–3 × 1–3 mm.,
rounded, entire to crenate, spreading. Stamens included; filaments very short,
mostly inserted in the middle of the corolla tube. Ovary oblong or nearly so, laterally
compressed, glabrous, minutely pubescent, or sometimes with glandular hairs.
Capsule narrowly ellipsoid or narrowly ovoid, (3)5–9 × 1·2–2 mm.

   Zambia. N: Kawambwa, *F. White* 3641 (FHO; K); E: Chipata (Fort Jameson), *F. White*
2454 (FHO; K). **Zimbabwe.** C: Ruwa, Goromonzi District, *Biegel* 4422 (PRE; SRGH).
   Indigenous in China (Tibet, Yünnan, Hunan, Szechwan, Kweichow, Kansu, Kiangsu,
Kwangsi, Hupeh) and Japan, but cultivated and often naturalized all over the world. In
mountains and thickets. Alt. 600–3000 m. In Africa cultivated in gardens in the mountains and
sometimes naturalized.

4. **Buddleja dysophylla** (Benth.) Radlk., Abh. Nat. Ver. Bremen **8**: 410 (1883). — Phillips in
      Journ. S. Afr. Bot. **12**: 114 (1946), comb. illegit. — Bruce & Lewis, F.T.E.A.,
      Loganiaceae: 40 (1960). — Verdoorn in Fl. S. Afr. **26**: 167. fig. 23.4.(1963). — Brummitt
      Wye Coll. Malawi Proj. Rep.: 67 (1973) and in Kew Bull. **31**: 173 (1976). — Leeuwenberg
      in Meded. Landb. Wag. **79**–6: 37 (1979). TAB. **78**. Type from S. Africa.
         Nuxia dysophylla Benth. in Hooker, Comp. Bot. Mag. **2**: 60 (1837). Type as above.
         Chilianthus dysophyllus (Benth.) A.DC. in DC., Prodr. **10**: 436 (1846). Type as above.

   Shrub, straggling or scandent (erect if growing isolated), 1–10 m. high,
divaricately branched. Bark pale brown. Branchlets terete to quadrangular, striate,
tawny or rusty-pubescent to densely tomentose (also in living specimen), hairs rather
long and often stellate. Leaves opposite, petiolate; petiole hairy as branchlets, 0·3–3

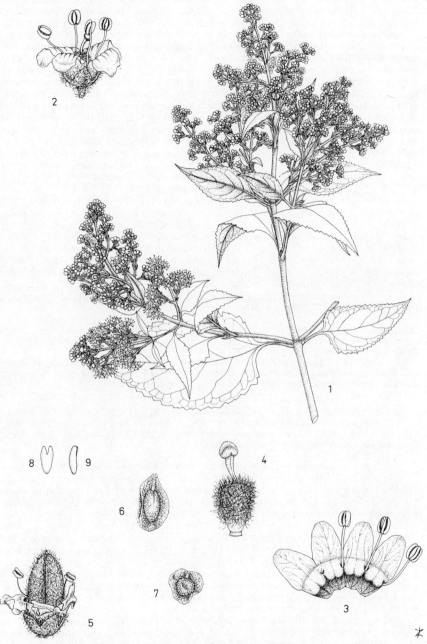

Tab. 78. BUDDLEJA DYSOPHYLLA. 1, flowering branch (× ½); 2, flower (× 5); 3, opened corolla (× 5); 4, pistil (× 10), 1–4 from *Salubeni* 1476; 5, fruit (× 5); 6–7 seeds (× 15); 8–9 embryos (× 45), 5–9 from *Pegler* 1154.

cm. long; lamina almost triangular to ovate, 1·5–2(3) times as long as wide, 1–10 × 0·7–7·5 cm., with the base truncate, subcordate or wedge-shaped, often decurrent into the petiole, acuminate to rounded at the apex, irregularly dentate to crenate, sparsely to densely stellate-tomentose and glabrescent above, persistently sparsely to densely stellate-pubescent or -tomentose beneath; veins impressed above. Inflorescence terminal, paniculate, large, lax or less often more or less congested in ultimate branchings, many-flowered, 4 × 4 to 20 × 20 cm. Flowers all at least shortly pedicellate, occasionally sessile, sweet-scented. Sepals up to one third connate, sometimes two halfway united, equal or subequal, ovate or approximately so, 1·5–3 times as long as wide, 0·8–2 × 0·5–1 mm., obtuse to acuminate, entire, tawny- or rusty-tomentose outside, with a few minute glandular hairs or (less often) glabrous inside. Corolla white, greenish, creamy, or yellowish or sometimes pale mauve, with the throat maroon-coloured; lobes erect 2–3 times as long as the calyx and 2·5–5 mm. long, outside mostly at least on lobes papillose-pubescent and sometimes also pilose on tube on which both types of hairs may occur, inside papillose-pubescent on lobes, glabrous within tube; tube cup-shaped, often slightly contracted at the mouth, slightly longer than the calyx, about as long as the lobes, 1·2–2·4 × 1·2–2·2 mm., rounded, entire, suberect at first, later recurved. Stamens mostly well-exserted; filaments long, often curved, 1–3 mm. long, inserted 0·1–0·5 mm. below the corolla mouth; anthers subcircular in outline, 0·3–0·8 mm. long, glabrous, cordate at the base; cells slightly divergent at the base. Pistil 2–4 mm. long; ovary obovoid or ellipsoid, laterally compressed, 1·2–2 × 0·9–1·3 × 0·5–1 mm., tomentose except for the glabrous disk-like base; style glabrous, broadened at the apex, with large 0·8–2 mm. long stigma; stigma up to half as long as the style, more or less peltate or clavate. Capsule ovoid or ellipsoid, 2·2–3·4 × 2–2·3 × 1·2–1·5 mm., tomentose, about 3 times as long as the calyx, apiculate. Seed pale brown, flattened, winged, often obscurely angular, circular, obliquely elliptic or ovate, often angular, 0·6–0·7 × 0·4–0·5 × 0·2 mm., pale brown, minutely reticulate.

Malawi. N: N. end of Nyika Plateau, *Brummitt, Munthali & Synge* WC 248 (K); C: Dedza Mt. bud vi, *Brummitt & Salubeni* 11691 (WAG) S: Zomba Plateau, fl. viii, *Brummitt* 12400 (K; PRE; WAG; SRGH).
Also in S. Africa, Transkei and Swaziland and in southern tropical Africa from Zaire (Shaba) and Tanzania to Malawi. Forest edges or scrub. Alt. 0–2600 m. (0–1100 m. in Swaziland, Transkei, and S. Africa; 2000–2600 m. in Zaire, Tanzania and Malawi).

5. **Buddleja lindleyana** Fortune in Lindley, Bot. Reg. **30**. Misc. 25 (1844); **32**. t. 4 (1846). — Prain & Cummins in Fl. Cap. **4**, 1: 1048 (1909). — Leeuwenberg in Meded. Landb. Wag. 79–6: 129 (1979). Type a cultivated specimen from the Cambridge Bot. Gard., Great Britain, raised from seeds sent from China.

Shrub or undershrub, 1–3 m. high. Branchlets quadrangular or approximately so, rusty-pubescent with stellate and/or glandular hairs, less often almost glabrous. Leaves opposite or subopposite, very variable in shape and size, shortly petiolate; stellate-pubescent, 1–7 mm. long; lamina sparsely pubescent to glabrous above, minutely to densely pubescent with often stellate hairs beneath, membranaceous when dry, ovate, elliptic, or narrowly elliptic, 2–3(–5) times as long as wide, 2–11 × 0·7–5 cm. or sometimes smaller, long-acuminate at the apex, cuneate at the base, entire to coarsely sinuate-dentate; secondary veins conspicuous. Inflorescence terminal, almost spicate, more or less interrupted, 4–20 × 2–4 cm. Flowers shortly pedicellate. Calyx campanulate or urceolate, 2–3·5 mm. long, outside densely pubescent with glandular and often also some stellate hairs; tube 3–10 × as long as the lobes; lobes often dentate, subequal or unequal, mostly broadly triangular, 0·2–1 × 0·5–1 mm. Corolla purple, usually curved, with erect lobes 4–10 times as long as the calyx, 13–20 mm. long, pubescent with glandular and often also some stellate hairs outside; tube nearly cylindrical, gradually widened from somewhat above the insertion of the stamens, 3·5–8 times as long as the calyx, 5–6 times as long as the lobes, 11–17 mm. long, lobes suborbicular, about as long as wide or wider, 2–3 mm. long, stamens included; filaments inserted 3–6 mm. above the base of the corolla.

Zimbabwe. C: Makoni, Rusapi District, *Wild* 4646 (BR; K; PRE; SRGH).
China (Szechwan, Hupeh, Kiangsu, Anhwei, Shanghai, Chekiang, Fukien, Kiangsi, Hunan,

Kwangsi, Kwangtung, Yünnan), Macao, Hong Kong, Japan. Bush on road sides, mostly in the mountains. Alt. 300–1800 m.
Most probably escaped from cultivation.

6. **Buddleja madagascariensis** Lam., Enc. Méth., Bot. **1**: 513 (1785); Tab. Encycl. Méth., Bot. **1**: 291, t. 69, fig. 3 (1792). — Vahl, Symb. **3**: 14 (1794). — Hook., Bot. Mag. **55**: t. 2824 (1828). — Leenhouts in Fl. Males. **1**. 6: 340 (1963). — Leeuwenberg in Meded. Landb. Wag. **79**–6:59 (1979), fig. 14, map 6. Type from Madagascar.
   Nicodemia madagascariensis (Lam.) R.N. Parker, For. Fl. Pujab. 2nd ed.: 357 (1924). — Bruce & Lewis in F.T.E.A., Loganiaceae: 35, 36, fig. 7, 3 (1960). — F. White, F.F.N.R.: 339 (1962). Type from Madagascar.
   Adenoplea madagascariensis (Lam.) Eastw., Leafl. West. Bot. **1**: 197 (1936). Type from Madagascar.

Sarmentose shrub, 2–4 m. high, or 8–10 m. long climber. Branchlets terete, white-tomentose with stellate hairs, drying rusty. Leaves opposite or sometimes subopposite, petiolate; petiole 5–20 mm. long, stellate-tomentose; lamina narrowly ovate or elliptic, 2–3(4) times as long as wide, 4–14 × 1·5–7 cm., acuminate at the apex, rounded, cuneate, or rarely subcordate at the base, entire, dark green, with impressed reticulate venation, and glabrous or nearly so above (only tomentose when young), white-tomentose with stellate hairs (drying rusty) beneath. Inflorescence thyrsoid or paniculate, 5–25 × 2–15 cm. Flowers sessile or shortly pedicellate. Calyx campanulate or urceolate, 2–3·5 × 1·5–2·5 mm., acute or obtuse, entire. Corolla dark yellow, orange, or salmon, with erect lobes 3–6 times as long as the calyx, 9·5–13 mm. long, outside stellate-tomentose, inside with a 3–5 mm. wide pilose ring about 2 mm. above the base; tube nearly cylindrical, 2·5–4·5 times as long as the calyx, 2·1–3·4 times as long as the lobes, 7·3–10 mm. long, slightly widened towards the throat or not, in the middle 1·5–2 mm. wide; lobes subcircular, up to 1·5 times as long as wide, 2·2–4 × 2–3 mm., rounded, entire, spreading. Stamens barely included; filaments very short, 0·1–0·5 times as long as the anthers, glabrous, inserted just below the corolla mouth; anthers oblong, 1–1·4 × 0·3–0·8 mm., deeply cordate at the base, rounded or retuse at the apex, glabrous; cells parallel. Pistil 5–8 mm. long; ovary subglobose, usually laterally compressed, 1–1·4 × 0·9–1·2 × 0·7–1 mm., stellate-tomentose at the apex or occasionally entirely glabrous, 4-celled; style with stigma 4–6·8 m. long. Each cell with one axile oblong bilobed placenta with about 30–40 ovules. Berry blue-violet, globose or nearly so, 2·5–5 mm. in diam., many-seeded. Seed medium brown, obliquely ovoid or ellipsoid, 0·6–0·9 × 0·4–0·6 × 0·4–0·6 mm., not winged, minutely reticulate.

Botswana. S.E.: Gaberones fl. viii, *O.B. Miller* B918 (PRE). **Zambia.** W: Kitwe fl., 12.viii.1967, *Fanshawe* s.n. (SRGH); C: 16 km. S. of Lusaka, fl. ix, *Coxe* 58 (K); E: Fort Jameson fl. iv, *F. White* 2459 (FHO; K). **Mozambique.** M: Maputo, fl. ix, *Balsinhas* 1932 (K; PRE).
   Indigenous in Madagascar, but cultivated and often naturalized all over the world in tropical and subtropical regions. Bush in the mountains. Alt. 600–2000 m.

7. **Buddleja pulchella** N.E. Brown in Kew Bull. **1894**: 389 (1894). — Prain & Cummins in Fl. Cap. **4**, 1: 1048 (1909). — Bruce & Lewis, F.T.E.A., Loganiaceae: 38, fig. 6. 7. (1960). — Verdoorn in Fl. S. Afr. **26**: 163 (1963). — Palmer & Pitman, Trees S. Afr. **3**: 1883 (1973). — Leeuwenberg in Meded. Land. Wag. **79**–6: 67, fig. 16, map 8 (1979). TAB. **79**. Type a cultivated specimen from S. Africa.

Shrub, often climbing, 1–10 m. high and up to at least 20 m. long. Branchlets terete, white- or pale grey-tomentose with stellate hairs; indumentum drying rusty. Leaves opposite or subopposite, petiolate; petiole 5–20 mm. long; lamina variable in shape and size, ovate, triangular, oblong or narrowly oblong, or narrowly obovate, 1·2–3 times as long as wide, 1·8–10(–15) × 1–5(–7) cm., acuminate to obtuse at the apex, gradually or abruptly narrowed at the base and at the same time mostly decurrent into the petiole, sometimes with 1–2 large lobes near the base, furthermore entire, pale grey-stellate-tomentose, glabrescent, and with costa (and often main veins) impressed above, stellate-tomentose to pilose beneath; venation inconspicuous. Stipules none. Inflorescence paniculate, very variable in size, usually lax; 3–25 × 2–25 cm., several times branched. Flowers in rather loose clusters, shortly pedicellate or sessile, sweet-scented or malodorous. Calyx cylindrical or nearly so, 3–5 × 1·2–3 mm., white-stellate-tomentose outside (indumentum usually drying rusty), inside glabrous; tube 4–8 times as long as the lobes; lobes subequal, broadly to

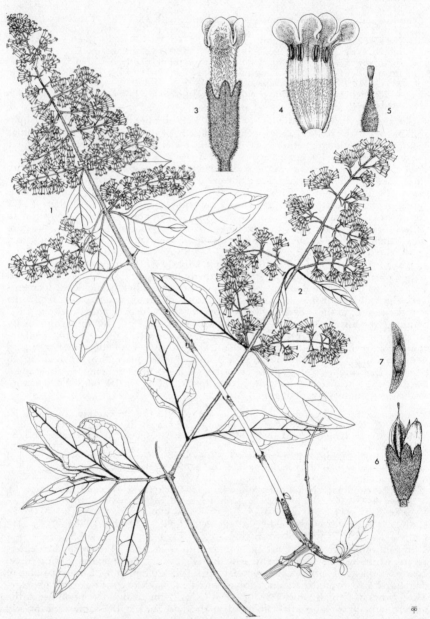

Tab. 79. BUDDLEJA PULCHELLA. 1–2 flowering branches (×½), 1 from *Rudatis* 1054; 2 from *Wood* 574; 3, flower (×4½); 4, opened corolla (×4½); 5, pistil (×4½); 3–5 from *Drummond* & *Hemesley* 4328; 6, fruit (×4½); 7, seed (×6), 6–7 from *Rehmann* 7563.

very narrowly triangular, 0·5–1 × 0·5–1 mm., acute or obtuse, entire. Corolla white, pale orange, or yellow to whitish with orange or yellow throat, with 7–12 mm. long erect lobes, outside stellate-tomentose, inside hirto-pilose in a 2–3 mm. wide ring which has its lower edge at 1–2 mm. above the base; tube nearly cylindrical, 1·5–2·5 times as long as the calyx, 3·7–6·5 times as long as the lobes, 5·5–9 mm. long, usually slightly widened towards the 1–1·8 mm. wide throat; lobes subcircular to oblong, 1–3 × 1–2·5 mm., broadly rounded, entire, spreading to reflexed. Stamens included; anthers sessile with the apex 0–1 mm. below the corolla mouth, oblong, 2–5 times as long as wide, 0·8–1·2 × 0·2–0·5 mm., deeply cordate at the base, rounded to retuse at the apex, glabrous; cells parallel. Pistil 2·5–5·2 mm. long, stellate-tomentose, often glabrous at the apex of the style; ovary ovoid or narrowly ovoid, often laterally compressed, 1–2 × 0·4 to 1·2 × 0·4–1 mm.; style rather short; stigma clavate to subcapitate, decurrent into the style, 0·8–1·2 mm. long. Each cell with one axile oblong placenta with 1–15 ovules outside. Capsule narrowly ellipsoid, 4·5–6 × 1·2 to 1·8 × 1·2 to 1·5 mm., 1·4–2 times as long as the calyx, acute at the apex, hairy like the ovary but less densely; valves often split at the apex. Seed pale brown, oblong, flat, 2·5–2·8 × 0·5–0·8 × 0·2–0·3 mm., narrowly winged at the edges or sometimes not winged; wings acuminate at both ends, minutely reticulate all over.

Zambia. N: Nyika, fr. xii, *Fanshawe* 7296 (FHO; K). **Zimbabwe.** C: Diana's Vow farm, Makoni District, *Chase* 8563 (FHO; K; PRE; SRGH). E: Umtali Commonage, fl. vi, *Chase* 770 (BM; K; LISC; S; SRGH). **Mozambique.** Z: Quelimane, *Sim* 20530 (PRE).
Also in eastern tropical Africa and eastern S. Africa. In tropical Africa in woodland or light forests in the mountains; alt. 1200–2000 m.; in southern Africa in forest, mostly at edges, or in open places. Alt. 300–1100 m.

8. **Buddleja saligna** Willd., Enum. Hort. Berol. **1**: 159 (1809). — Verdoorn in Fl. S. Afr. **26**: 163, f. 23. 1. (1963). — Palmer & Pitman, Trees S. Afr. **3**: 1883 c. phot. (1973) — Leeuwenberg in Meded. Landb. Wag. **79**–6: 72, fig. 17, phot. 7, map 9 (1979). TAB. **80**.
Type a cultivated specimen from Hort. Schönbrunn, Vienna, Austria.
B. salicifolia Jacq., Hort. Schoenbr. **1**: 12, t. 29 (1797), non Vahl (1794). — Phillips, Journ. S. Afr. Bot. **12**: 114 (1946).
Nuxia saligna (Willd.) Benth. in Hook., Comp. Bot. Mag. **2**: 59 (1836).
Scoparia arborea L.f., Suppl.: 125 (1781), not *Buddleja arborea* Meyen (1834–1835).
Type from S. Africa.
Chilianthus oleaceus Burchell., Trav. **1**: 94 (1822).

Shrub or small tree, 0·5–12 m. high, much branched. Branchlets terete or quadrangular and with 4 ridges or narrow wings, lepidote especially at the apex. Leaves decussate, those of a pair equal, often shortly petiolate: petiole lepidote, glabrescent, 1–20 mm. long; lamina subcoriaceous, narrowly elliptic to linear, variable in shape and size, 4–12 times as long as wide, 1·2–15 × 0·2–2(–3) cm. acute, acuminate, or less often rounded at the apex, cuneate at the base or decurrent into the petiole, entire or obscurely sinuate at the revolute margin, above lepidote-scaly when young, soon glabrous, much paler and densely grey- or tawny-pubescent with stellate hairs beneath; venation reticulate, impressed above, forming a line more or less parallel to the margin. Inflorescence terminal and often also in the axils of the upper leaves, paniculate, large, many-flowered, lax or rather so, 3 × 3 to 18 × 16 cm. Flowers subsessile in 3-flowered cymes, sweet-scented. Calyx cup-shaped or nearly so, 0·8–1·4 mm. long, outside densely lepidote-scaly; tube 0·7–4 × as long as the lobes; lobes equal or subequal, mostly broadly triangular, 0·5–1 times as long as wide, 0·2–0·6 × 0·4–0·6 mm., acute or obtuse, entire. Corolla white or creamy, with erect lobes 1·8–2·2 times as long as the calyx, 1·8–3 mm. long, inside pilose with rather stiff hairs in the throat; tube cup-shaped or nearly so, mostly slightly shorter than the calyx, 0·5–1·2 × 0·7–1·2 mm.; lobes 1·2–2·5 times as long as the tube, suborbicular or elliptic, 1–1·5 times as long as wide, 1–1·8 × 0·8–1·2 mm., rounded, entire, recurved. Stamens well-exserted; filaments 1–2·5 mm. long, inserted at or just above the middle of the corolla tube; anthers subcircular, 0·3–0·4 mm. long, glabrous; cells parallel or slightly divergent at the base. Pistil densely lepidote all over, 1–2 mm. long; ovary ovoid or nearly so, sometimes laterally compressed, 0·5–1·2 × 0·4–0·8 × 0·3–0·8 mm., 2-celled; style with stigma about as long as the ovary, 0·5–0·8 mm. long; each cell with one subcircular peltate placenta with 4–6 ovules outside. Capsule oblong, often laterally· compressed, 1·5–2·5 × 0·8–1·2 × 0·8–1 mm., sparsely lepidote, about twice as long as the calyx, 2-

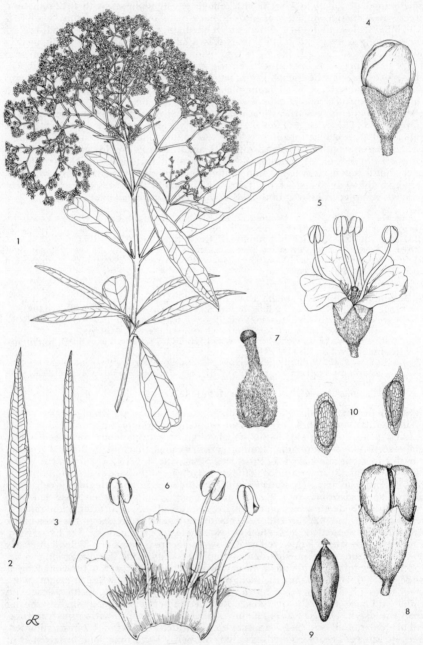

Tab. 80. BUDDLEJA SALIGNA. 1, flowering branch (× ½), from *Schlechter* 1939; 2–3 leaves (× ½), 2 from *Wood* 4647, 3 from *Ecklon & Zeyher* 94.12; 4, flower bud (× 15); 5, flower (× 10); 6, opened corolla (× 15); 7, pistil (× 20), 4–7 from *Schlechter* 1939; 8, fruit (× 15); 9, placenta (× 15); 10, seeds (× 15), 8–10 from *Ecklon & Zeyher* 94.12.

(later 4-)valved. Seeds medium brown, obliquely tetrahedral or nearly fusiform, 0·9–1·3 × 0·4–0·6 × 0·2–0·4 mm., obscurely winged or not, minutely reticulate.

Zimbabwe. W: Essexvale, *Borle* 102 (A; BR; PRE). S: Umzingwane, Fern Kloof, *Queen Victoria Mem. Mus.* 7119 (SRGH).
Also in S. Africa. Dry hillsides, mixed scrub bushveld, mountain slopes, wooded valleys, forest edges, along rivers and in coastal bush at river mouths. Alt. 0–2000 m.

9. **Buddleja salviifolia** (L.) Lam., Encycl. Méth., Bot. **1**: 513 (1785); (as *Budleja salvifolia*). — Baker in F.T.A. **4**, 1: 516 (1903). — Marquand in Kew Bull. **1930**: 198 (1930). — Bruce & Lewis in F.T.E.A., Loganiaceae: 38, fig. 6, 6 (1960). — Verdoorn in Fl. S. Afr. **26**: 160, fig. 22, 1. (1963). — Leeuwenberg in Meded. Landb. Wag. **79**–6: fig. 18, phot. 8, map 10 (1979). Type from S. Africa.
Lantana salviifolia L., Syst. Nat. ed., 10, **2**: 1116. (1759) (as *L. alvifolia*). Type as above.

Shrub or sometimes small tree, 1–8 m. high. Branchlets white- or, (especially when dry) rusty-tomentose with stellate hairs as the leaves beneath, the peduncles, bracts beneath, and calyx and corolla outside. Leaves opposite, sessile or shortly petiolate; petiole — if present — up to 7 mm. long; lamina narrowly ovate to narrowly oblong, (2–)4–6·5 times as long as wide, 4–17 × 0·8–4·5 cm., sometimes a little smaller, long-acuminate, with the apex acute, deeply cordate to auriculate at the base, crenate, bullate, finely and distinctly rugose and glabrescent above; venation conspicuous, reticulate. Stipules foliate. Inflorescence paniculate, very variable in size, 2–15 × 1–10 cm. (30 × 25 cm.); ultimate branches 3-flowered, more or less sessile and clustered. Flowers sessile, sweet-scented. Calyx campanulate, 2–3 mm. long; tube 1–3 times as long as the lobes; lobes subequal, broadly to rather narrowly triangular, 0·8–1·5 times as long as wide, 0·5–1·5 × 0·5–1 mm., acute, entire. Corolla white or lilac to purple, with a deep orange throat, with erect lobes (2–)2·5–3·5 times as long as the calyx, 6–9 mm. long, outside tomentose, inside hirsute on the base of the lobes and within the tube, except for the glabrous base; tube nearly cylindrical, (1·4)2–2·7 times as long as the calyx, 2·5–3·2 times as long as the lobes, 4·2–6·8 mm. long, slightly widened towards the throat and there about 2 mm. wide; lobes suborbicular to oblong, 1·8–2·5 × 1·2–1·8 mm., rounded, entire, spreading. Stamens included; filaments very short, inserted 3–4 mm. above the corolla base; anthers oblong, 0·8–1·5 × 0·3–0·5 mm., glabrous; cells parallel. Pistil 2·8–5 mm. long; ovary subglobose, laterally compressed, 1–1·2 × 1 × 0·8–1 mm., hirto-pubescent with stellate hairs, bilocular; style with stigma 2–3 mm. long, glabrescent; stigma large, clavate, often larger than the style. In each locule one axile circular peltate placenta with about 20 ovules outside. Capsule 3–4·5 × 2·4 × 1·4–2 mm., ellipsoid, exserted by about half from the calyx, apiculate, hairy as the ovary but less densely. Seed medium brown, obliquely tetrahedral, 0·8–1·5 × 0·4–0·6 × 0·3 mm. narrowly winged at the edges; both wings and testa minutely reticulate.

Zambia. N: Nyika Plateau, fl. viiii, *Coxe* 32 (SRGH); ibid. fl. x, *Robson & Angus* 481 (EA; K). **Zimbabwe.** E: Mt. Nuza, fl. vi, *Gilliland* 390 (BM, FHO; K; PRE; SRGH), 2055 (BM). **Malawi.** N: Nyika Plateau, fl. viii, *Brass* 17166 (A; BM; BR; EA; FHO; K; MO; NY; PRE; SRGH; US). S: Mulanje Mt., Litchenya Plateau, fl. vii, *Brass* 16646 (A; BM; BR; K; MO; NY; PRE; SRGH), 16665 (BR; K; MO; NY; SRGH; US; WAG). **Mozambique.** Z: Gurué, Pico Namuli, fl. viii, *Mendonça* 2245 (LISC).
Southern and eastern Africa, southwards from Angola and Kenya. Forest edges, rocky slopes, along water courses, and in montane grassland. Alt. in the tropics 1200–2500 m., near the Cape from 150 m.

3. GOMPHOSTIGMA Turcz.

Gomphostigma Turcz. in Bull Soc. Nat. Mosc. **16**: 53 (1843). — Benth in DC., Prodr. **10**: 434 (1846); in Journ. Linn. Soc., Bot. **1**: 95 (1857); in Benth & Hook. f., Gen. Pl. **2**: 792 (1876). — Leeuwenberg in Acta Bot. Neerl. **16**: 143 (1967); op. cit. **20**: 682 (1971); in Meded. Landb. Wag. **77**, 8: 15–30 (1977).

Undershrubs or herbs with woody base, glabrous or bearing stellate hairs. Leaves opposite, sessile. Inflorescence a terminal raceme. Lower bracts leafy, decreasing towards the apex. Pedicels suberect, bibracteate. Flowers 4-merous. Sepals equal, green, glabrous on both sides or scaly outside, united for 0·2–0·4 of their length, obtuse, rounded or nearly so at the apex, entire, persistent. Corolla cup-shaped, mostly glabrous outside, inside pillose, at least in the tube; lobes imbricate, suborbicular, rounded, entire, spreading. Stamens exserted, inserted just below the corolla mouth or somewhat lower; anthers shorter or longer than the filaments, oblong or nearly so, deeply cordate at the base, rounded to apiculate at the apex, glabrous, introse; cells parallel, discrete, dehiscent throughout by a longitudinal slit. Pistil glabrous; ovary narrowly ovoid to oblong, laterally compressed, rounded or shortly bilobed at the apex, with a disk-like base, 2-celled; style persistent; stigma capitate. Capsule oblong, laterally compressed or not, fairly bilobed, with an indented line along the line of dehiscence, bivalved; valves torn at the apex, with a revolute margin being the torn septum, glabrous inside; dry placentae with seeds in valves.

Two species occuring in Southern Africa.

Gomphostigma virgatum (L.f.) Baillon, Hist. Pl. **9**: 348 (1888). — O. Kuntze, Rev. Gen. Pl. **3**, 3: 201 (1898) comb. illegit. — N.E. Br. in Kew Bull. **1929**: 143 (1929). — Phillips, Gen. S. Afr. Fl. Pl. ed. 2: 576 (1951). — Verdoorn in Fl. S. Afr. **26**: 169, fig. 24, 1 (1963). — Leeuwenberg in Meded. Landb. Wag. **77** — 8: 20, fig. 5, photos. 6–8, map 2 (1977). TAB. **81**. Type from S. Africa (Cape Prov.).

Buddleja virgata L.f., Suppl. Pl.: 123 (1781) as *"Budleia"*. Type as above.
Gomphostigma scorpoides Turcz. in Bull. Soc. Nat. Mosc. **16**: 53 (1843). — Benth in DC., Prodr. **10**: 434 (1846). — Solereder in Engl. & Prantl, Nat. Pflanzenfam. **4**, 2: 46, fig. 26C–E (1892). — Prain & Cummins in F.C. **4**, 1: 1038 (1909). Type from S. Africa (Cape Prov.).
Sopubia eenei S. Moore, Journ. Bot. London **38**: 462 (1900). Type from Namibia.
Sopubia leposa S. Moore, tom. cit.: 468 (1900). Type: Zimbabwe, Salisbury, *Rand* 158 (BM, holotype).

Plant erect, 0·50–2 m. high, glabrous or more often with a silvery indumentum of stellate scales on the branchlets, both sides of the leaves, the inflorescences, and the calyx outside. Bark shallowly longitudinally fissured. Leaves 5–80 × 1–5 mm., opposite with a connecting ridge, narrowly or very narrowly oblong, rarely very narrowly obovate, acute, acutish, or sometimes rounded at the apex, at the margin entire or sometimes remotely and obscurely toothed, usually revolute. Inflorescence a raceme, 3–15 cm. long, terminal. Pedicels 2–12 mm. long, suberect, bibratate. Flowers scented. Sepals (1·5)2–4·5 mm. long, oblong, glabrous inside, obtuse, rounded or mucronate at the apex. Corolla (3·5)5–10 mm. long, when lobes erect 1·6–3·2 × as long as the calyx, white or exceptionally blue or pink, subpersistent, accrescent, outside glabrous, inside pilose in the tube and at the base of the lobes; tube 1·5–4 mm. long, mostly slightly shorter than the calyx; limb 6–15 mm. in diam.; lobes (2)2·5–6 mm. in diam. Stamens: filaments 1–2 mm. long, glabrous; anthers 1–2 × 0·3–1 mm., dark brown at the edges, turning completely dark brown at the anthesis. Pistil 4–7 mm. long, glabrous, mostly curved; ovary 2–3·8 × 1–1·8 × 0·6–1·2 mm., narrowly ovoid to oblong, laterally compressed; style 1·6–3·7 mm. long; stigma 0·4–0·7 × 0·4–0·8 mm. In each cell one axile placenta with c. 80–100 ovules. Capsule 3–8·5 × 1·5–4 × 1·5–3 mm. Seed 1–1·2 × 0·8–1 × 0·6–1 mm., medium brown, shining, obliquely polyhedral or nearly so, minutely reticulate, not winged. Embryo 0·6 mm. in diam., straight, white; rootlet 0·2 mm. in diam., obtuse; cotyledons 0·3 × 0·25 mm., oblong, rounded at the apex. Endosperm mealy, white, not copious.

Botswana. SE: Kanye Distr., Mogoclune valley, fl. & fr. v., *Miller* B/363 (BR; K; M; PRE; Z). **Zambia.** W: above Mutanda Bridge, 37 km. Solwezi-Kasempa Road, fl. & fr. ix, *Angus* 460 (BM; K). S: Ngongo R., 48 km. Choma-Namwala road, fr. vi, *F. White* 2960 (K). **Zimbabwe.** N: Nsutevi & Futu R. junction, Urungwe Distr., fl. & fr. iv, *Goodier* 236 (K; PRE; SRGH). W: 8 km. S. of Plumtree fl. iii, *Davies* 324 (K; PRE; SRGH). C: near Poole Farm, Hartley Distr., fr. viii, *Hornby* 3309 (PRE; SRGH). E: Pounsley, Umtali Distr., fl. & fr., *Phipps* 717 (BR; SRGH). S: Victoria Distr., Mtilikwe R. bed, fr. viii, *Chase* 5738 (BM; COI; LISC; PRE; SRGH).

Also in Southern Africa from southern Angola and southern Zaire. Along and in water-courses and rivers, only in running water among boulders or along the banks in sand and mud; 600–2000 m.

4. MOSTUEA Didr.

Mostuea Didr., Vidensk. Medd. Kjoeb. **1853**: 86 (1854); Leeuwenberg, Meded. Landouwb. Wag. **61** (1): 1 (1961).

Shrubs, undershrubs, or sometimes lianas, with simple hairs or nearly glabrous. Stems much branched; twigs usually thin. Leaves opposite, those of a pair equal or

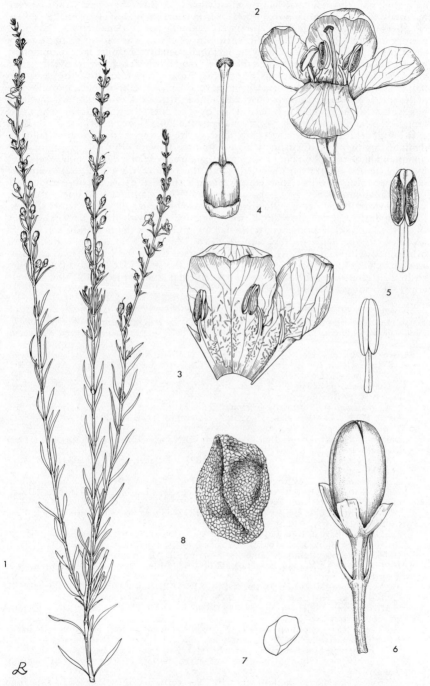

Tab. 81. GOMPHOSTIGMA VIRGATUM. 1, branch (× $\frac{1}{2}$); 2, flower (× 4); 3, portion of corolla with stamens (× 5); 4, pistil (× 5); stamen, both sides (× 5), 1–5 from *Norlindh* & *Weimark* 4499; 6, fruit (× 4); 7, seed (× 10); 8, seed (× 20), 6–8 from *Wilms* 1079.

subequal, shortly petiolate; lamina mostly papyraceous when dry, ovate to very narrowly elliptic, variable in shape and size, on lateral branches often smaller, entire or obscurely sinuate-dentate, pinnately veined; veins conspicuous. Stipules membranaceous, usually triangular, entire, those of a pair united, or all fused into a short ocrea. Inflorescence axillary or terminal, usually on short lateral branches, mostly obliquely and incompletely dichasial, 1-many-flowered. Bracts mostly small and sepal-like. Flowers 5-merous, dimorphic. Sepals green, connate at the base up to half their length, equal or unequal, ovate to very narrowly elliptic, obtuse to subulate at the apex, entire or sometimes obscurely sinuate-dentate. Corolla white, sometimes pale yellow, orange, or red, yellow at the base or not, infundibuliform, 2·5–9 × as long as the calyx; tube about 3–5 × as long as the lobes; lobes imbricate in the bud, spreading, orbicular or nearly so, rounded, entire or sometimes obscurely sinuate-dentate. Stamens included or in long-staminate flowers often exserted, equal or unequal; filaments pubescent or sometimes glabrous (*M. microphylla, M. brunonis*), inserted at one-quarter to one-third from the base of the corolla tube; anthers orbicular or oblong; cells 2, discrete, divergent at the base, dehiscent throughout by a longitudinal slit. Ovary superior, ovoid, usually with 2 impressed lines, 2-celled, with 2 ovules in each cell; style simple, shorter or longer than the stamens, when longer slightly exserted, deciduous, minutely pubescent with glandular hairs; stigma twice dichotomously branched, lobes narrow, pubescent (hairs glandular). Ovules attached to the base of the septum. Fruit capsular, obcordate, bilobed, or occasionally ellipsoid, flattened, with an impressed line in the middle, loculicidal, 4-valved; valves hinging on the septum; cells with 1–2 seeds. Seeds plano-convex (not so if fruit ellipsoid), obliquely ovate to orbicular, pale brown, dull and densely appressed-pilose (African species).

Eight species; one in northern South America, the others in tropical Africa.

Sepals obtuse, shorter than the ovary which has brush-like hairs at the apex -   2. *microphylla*
Sepals acute to subulate, usually longer than the variously hairy or glabrous ovary- 1. *brunonis*

1. **Mostuea brunonis** Didr., Vidensk. Meddel. Dansk. Naturk. Foren. Kjoeb. **1853**: 87 (1854). — Hiern, Cat. Welw. Afr. Pl. **1**: 699 (1898). — Baker in Fl. Trop. Afr. **4**, 1: 505. 1903. Type from Lower Congo.
    Leptocladus thomsonii Oliv., Journ. Linn. Soc. **8**: 160, t. 12. 3rd. fig. (1865). — Hiern, loc. cit. Type from S. Nigeria.
    M. thomsonii (Oliv.) Benth. in Hook., Icon. Pl. **12**: 83 (1876). — Gilg in Engl., Bot. Jahrb. **17**: 561 (1893). — Baker, loc. cit. — Hutchinson & Dalziel, F.W.T.A. **2**: 20 (1931). Type as above.
    M. madagascarica Baill., Bull. Mens. Soc. Linn. Paris, **1**: 245 (1880). Lectotype from Madagascar.
    M. schumanniana Gilg in tom. cit.: 560 (1893). — Baker, tom. cit.: 508 (1903). Type from Moyen Congo.
    M. grandiflora Gilg ex Engl., Abh. Preuss. Akad. Wiss. **51**: 69 (1894) nomen. — Gilg in Engl., Pflanzenw. Ost-Afr. **C**: 310 (1895); in Engl. Bot. Jahrb. **23**: 198 (1896). — Baker, tom. cit.: 507. Type from Tanzania.
    M. fuchsiifolia Bak. in Kew Bull. **1895**: 96 (1895); in tom. cit.: 507 Hiern, loc. cit. Type from Angola.
    M. walleri Bak. in Kew Bull. **1895**: 96 (Apr.–May 1895); in tom. cit.: 507 (1903). — Bruce in Kew Bull. **11**: 159 (1956). — Bruce & Lewis in F.T.E.A., Loganiaceae: 4, fig. 1. (1960). Type: Mozambique: Zambezia, Morrumbala Mt., *Waller* s.n. (K, holotype).
    M. zenkeri Gilg, Notizbl. Bot. Gart. Berlin **1**: 73 (5 June 1895). — Baker, tom. cit.: 507. Type from Cameroun.
    M. densiflora Gilg in Engl., Bot. Jahrb. **23**: 198 (1896). — Baker, tom. cit.: 508. — Cavaco in Bull. Hist. Nat. Paris Ser. 2. **29**: 513 (1957). Type from Zaire.
    M. ulugurensis Gilg in op. cit. **23**: 198 (1896); in op. cit. **28**: 117 (1899). — Baker, tom. cit.: 506. Type from Tanzania.
    M. camporum Gilg in op. cit. **28**: 177 (1899). — Baker in tom. cit.: 509. — Bruce & Lewis in tom. cit.: 5. Type from Tanzania.
    M. megaphylla Good in Journ. Bot. **67**: 100 (1929). Type from Angola.
    M. gracilipes Mildbr. in Notizbl. Bot. Gart. Berlin, **11**: 675 (1932). Type from Tanganyika.
    M. gossweileri Cavaco in Bull. Mus. Hist. Nat. Paris, Sér. 2, **29**: 513 (1957). Type from Angola.
    M. lundensis Cavaco, in loc. cit. Type from Angola.

Shrub, undershrub, or occasionally liana, 0·30–7 m. high, usually much branched. Stems erect or overhanging, with spreading branches. Twigs near the apex variously hairy or glabrous. Leaf blade dull when dry, extremely variable in shape and size, 1·2–6 × as long as wide, variously hairy or glabrous, beneath often with domatia in the angles of some secondary veins. Peduncle, branches, and pedicels short or long, densely appressed-pubescent to glabrous, sometimes slightly strigillose, all more or less independently variable. Bracts very small, triangular, approximately sepal-like. Sepals connate at the base up to half their length, equal or unequal, ovate, ovate-lanceolate, or ovate-linear (the first two forms and the last two usually together in a single flower!), 1·5–6 × as long as wide, 1–4(5) × 0·3–1 mm., acute to subulate at the apex, hairy or glabrous outside, usually longer than the ovary. Stamens included or in long staminate flowers slightly exserted; filaments usually pubescent, those of short stamens often widened near the anther when young, inserted at about one-third from the base of the corolla tube; anthers 0·7–1·2 × 0·6–1 mm. Ovary about 1·2–2 × as long as wide, 1–1·5 × 0·6–1·2 mm., glabrous or appressed-pubescent near the apex (varying independently from the indumentum and the shape of the sepals). Capsule mostly medium to dark brown when dry, usually bilobed, glabrous or hairy, dull.

Throughout tropical Africa, also in Madagascar. Various habitats. Alt. 0–2000 m.

Var. **brunonis**

Leaf blade 0·6–15(28) × 0·3–8(12·5) cm. Corolla white, lilac, or pink, mostly with a yellow or orange base and throat (sometimes entirely yellow, orange, or red, or red with yellow), 6–18 mm. long; sometimes amply infundibuliform, usually glabrous outside; tube 5–13 mm. long, 1–1·5 mm. wide at the base, 2–9 mm. at the throat; lobes 1–5 mm. in diam. Capsule mostly 1·6–2·4 × as wide as long, 4–7(13) × 8–14 mm., composed of two oblique laterally compressed-globose or ovoid cells, with an impressed line in the middle, apically emarginate to bilobed, often with an apiculate tip, often pink or red, mostly medium to dark brown when dry, sparsely appressed-pubescent to glabrous or appressed-piloso-pubescent, mostly truncate to subcordate at the base. Other characters see species description.

Zambia. N: Mbala (Abercorn), Lunzuwa Falls, fl. fr. (juv.) Dec. 1953, *E. Nash* 31 (BM). **Zimbabwe.** E: Vumba "Elephant" Forest, forest edge, Umtali Distr., fl. 8.xi.1957, alt. 5500', *Chase* 6745 (COI; K). **Malawi.** N: Willindi Forest, fl. Sept. 1954, *Chapman* 243 (K; BR; MO). S: Big Ruo Gorge, Mulanje Mt., fl. vii.1958. *Chapman* 605 (FHO; K). **Mozambique.** N: Cabo Delgado, Túnguè, Nangade road to Mued, *Mendonça* 974 (LISC). MS: Manica E Gorungosa Mts., near Gogogo, fl. 10.x.44, *Mendonça* 2438 (LISC).
Distribution as for the species. In moist or dry places, in secondary, gallery, or rain forests. Alt. 0–2000 m.

This taxon is extremely variable and could be separated into several "forms" none of which should, however, be given a definite taxonomic rank at this stage of our knowledge. Those "forms" have been discussed by Leeuwenberg in his above cited revision from which this treatment has been compiled.

2. **Mostuea microphylla** Gilg in Engl., Pflanzenw. Ost-Afr. C: 310 (1895). — Baker in Fl. Trop. Afr. **4**, 1: 505 (1903). — Bruce in Kew Bull. **11**: 160 (1956). — Leeuwenberg in Meded. Landb. Wag. **61–4**, 13, fig. 2; map 5 (1961). Type from Zanzibar. TAB. **82**.
    M. amabilis Turrill in Kew Bull. **1920**: 25 (1920). Type from Mozambique: near the mouth of the Msalu R., *C.E.F. Allan* 90 (K, 2 sheets, holo- and isotype).

Shrub or small liana, about 1–1·5 m. high. Twigs pubescent and usually with a bare stripe above the leaves when young, glabrescent. Leaf lamina ovate, narrowly ovate, elliptic, or narrowly elliptic, variable in shape and size, 1·25–2·5 × as long as wide, 6–35 × 5–23 mm., sometimes smaller, obtuse, rounded, or occasionally acute at the apex, cuneate, rounded, or subcordate at the base, above sparsely pubescent on the costa, beneath less so on the costa and veins to glabrous. Inflorescence terminal on short lateral branches or axillary, usually much longer than the leaves, 1–5·5 cm. long, 1-many-flowered. Flowers small. Sepals equal, connate at the base, narrowly ovate, about 2 × as long as wide, 1–1·2 × 0·5–0·6 mm., obtuse, often minutely ciliate, glabrous on both sides; after the shedding of the corolla seemingly topped by the brush-like apical hairs of the ovary which is slightly longer. Corolla white or orange, 4–7 × as long as the calyx, 4–8 mm. long. Capsule medium brown, sparsely appressed-pubescent or -pilose.

Mozambique: near mouth of Msalu R., *C.E.F. Allen* 90 (BM; K, type of *M. amabilis*).
Also in Zaire, Somalia, Kenya and Tanzania. Coastal evergreen bushland, gallery, or rain forests. Alt. up to 350 m.

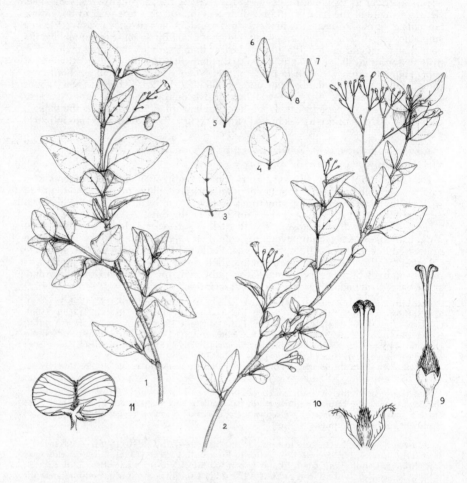

Tab. 82. MOSTUEA MICROPHYLLA. 1–2, branches ($\times \frac{1}{2}$), 1 from *Adamson* 318, 2 from *Peter* 45423; 3–8, leaves ($\times \frac{1}{2}$), 3–5 from *Adamson* 318, 6 from *Senni* 104, 7–8 from *Schlieben* 5747; 9, *Calyx* with *pistil* ($\times 6$). MOSTUEA BRUNONIS. 10, calyx with pistil ($\times 6$), from *Gossweiler* 8973; 11, fruit ($\times 2$), from *Callens* 2695.

5. NUXIA Lam.

Nuxia Lam., Tab. Encycl. Méth. Bot. **1**: 295 (1792). — Leeuwenberg in Meded. Landb. Wag. **75**–8: 1 (1975).
Lachnopylis Hochst. in Flora **26**: 77 (1843). — De Candolle, Prod. **9**: 22 (1845). — Hutchinson & Moss in F.W.T.A. **2**: 20 (1931). — Phillips, Gen. S. Afr. Fl. Pl., 2nd ed.: 575 (1951).

Shrubs or trees. Trunk often fluted. Branches unarmed, with fissured bark, not lenticellate. Leaves opposite, ternate, quaternate, or in some lateral branches occasionally alternate, petiolate (or only in *N. gracilis* often sessile). Inflorescence terminal, thyrsoid, but often dichotomously branched. Ultimate branches topped by solitary flowers or by heads of three or more flowers which may be globose. Lower bracts mostly leafy, the others small, often scale-like. Flowers 4-merous, actinomorphic except for the more or less irregularly lobed calyx. Calyx campanulate to cylindrical, 4-lobed; lobes much shorter than the tube, triangular, acute or rarely obtuse or rounded, entire or sometimes tridenate (only observed in *N. floribunda* and *N. gracilis*), erect or suberect, sometimes two or two pairs coherent or partially united; flowers with almost regularly 4-lobed calyx or the irregularly lobed calyx in a single inflorescence, with minute glandular hairs and often also simple or branched hairs; inner side mostly sericeous to appressed-pubescent. Corolla mostly white or creamy, circumscissile, outside with some minute glandular hairs at the base of the lobes and the apex of the tube; inside with a pilose or villose ring of mostly recurved hairs in the throat and with minute glandular hairs in the tube except for its glabrous base (hairs in corolla throat or in tube may be absent on a few occasions, but in those cases never both kinds simultaneously); tube slightly shorter or less often longer than the calyx, cylindrical; lobes oblong, recurved from above the base and there convex and often with a minute callosity, concave and rounded to acute at the apex, entire. Stamens well exserted, inserted at the mouth of the corolla tube; filaments only at the very base vilose in a line with the group of hairs on the corolla throat or slightly above or entirely glabrous, straight or curved, geniculate at the base only in *N. floribunda*; anthers glabrous, 0·5–2 mm. long; cells 2, parallel when young, conspicuously divergent after the pollen has been shed, confluent at the apex, dehiscent throughout by a longitudinal slit. Pistil: ovary superior, obcordoid, ovoid, or ellipsoid, mostly slightly laterally compressed, appressedly pubescent or hirto-pubescent, glabrous only in *N. floribunda*, at the very base, however, always glabrous, surrounded by a narrow disk-like ring formed by the base of the shed corolla, 2-celled; style well exserted, glabrous; stigma small, capitate or occasionally slightly bilobed. In each cell one axile peltate placenta with many ovules attached to the middle of the septum. Capsule about as long as and included in the persisting calyx or up to about one-third longer, with about the same indumentum as the ovary, bivalved, septicidal, retuse or rounded at the apex; often later longitudinally cleft, the capsule becoming 4-valved. Seeds small, medium brown, fusiform, longitudinally striate or reticulate-striate.

15 species in southern Arabia, tropical Africa (inclusive of Madagascar, the Comoro Islands, and the Mascarenes), and South Africa.

1. Leaves opposite -   -   -   -   -   -   -   -   -   -   -   -   - 2
 - Leaves ternate or sometimes quaternate -   -   -   -   -   -   -   - 3
2. Leaves 2·5–4(6) × as long as wide, acuminate or acute at the apex; petiole mostly over 10 mm. long; ovary and capsule glabrous -   -   -   -   -   -   - 2. *floribunda*
 - Leaves (2–)4–8 × as long as wide, obtuse or rounded at the apex; petiole 2–10 mm. long; ovary and capsule appressed-pubescent -   -   -   -   -   - 3. *oppositifolia*
3. Corolla lobes at or near the apex mostly with recurved pubescence outside; ovary pubescent at the apex; calyx sericeous inside; petiole 3–20 mm. long -   -   -   - 1. *congesta*
 - Corolla with some minute glandular hairs outside; ovary glabrous; calyx inside sparsely and/or minutely appressed-pubescent; petiole (3)10–55 mm. long -   -   2. *floribunda*

1. **Nuxia congesta** R. Br. ex Fresen. in Flora **21**: 606 (1838). — R. Br. in Salt, Abyss., App.: 63 (1814), nomen. — Bentham in De Candolle, Prodr. **10**: 435 (1846). — Baker in F.T.A. **4**: 512 (1903). — Prain & Cummins in Fl. Cap. **4**, 1: 1041, 1042 (1909). — P. Jovet in Bull. Hist. Nat. Toulouse **82**: 18 (1948); Bruce & Lewis in F.T.E.A., Loganiaceae: 44, fig. 8. 7–8 (1960). — Verdoorn in Fl. S. Afr. **26**: 156, figs. 19. 2, 20. 21 (1963). — Onochie & Leeuwenberg in F.W.T.A., 2nd ed., **2**: 46, f. 211 (1963). — Leeuwenberg, Fl. Cam. **12** (= Fl. Gabon **19**): 38, pls. 12 and 13 (1972). — Palmer & Pitman, Trees S. Afr. **3**: 1895, c. photographs (1973). — Leeuwenberg in Meded. Landb. Wag. **75**–8: 12, figs. 3 and 4; phots. 1–6; map 3 (1975). Type from Ethiopia.

L. ternifolia Hochst. in Flora **26**: 7 (1843). — De Candolle, Prodr. **9**: 23 (1845). Type from Ethiopia.

N. sambesina Gilg. in Engl., Pflanzenw. Ost-Afr. **C**: 312 (1895). — Baker in F.T.A. **4**, 1: 514. — Eyles in Trans. Roy. Soc. S. Afr. **5**, 4: 442 (1916). Type from Malawi: Zomba, *Kirk* s.n. (holotype, B†; lectotype: K).

N. viscosa Gibbs in Journ. Linn. Soc., Lond. **37**: 454 (1906). — Eyles, loc. cit. Type from Zimbabwe: Matopo Hills, near the American Mission and on the Silozi, *Gibbs* 246 (BM, holotype).

Evergreen or sometimes in Ethiopia deciduous tree or shrub, 2–25 m. high, often gnarled. Trunk in larger trees fluted and irregularly ridged, 5–100(200) cm. in diam.; bark pale grey-brown, fissured. Branchlets glabrous to densely pubescent, not lenticellate, with 6(8) raised lines and often sulcate when dry. Leaves ternate (occasionally 4-nate), petiolate; petiole glabrous to densely pubescent, 3–20 mm. long; lamina coriaceous or subcoriaceous, very variable in shape and size, elliptic, narrowly elliptic, obovate, nearly rhomboid, or suborbicular, 1·3–3(5) times as long as wide, (1)2–15 × (0·3)1·2–7·5 cm., acuminate to emarginate at the apex, cuneate or less often rounded at the base or decurrent into the petiole, entire or sometimes more or less distinctly serrate-dentate, or crenate, glabrous to subtomentose with stellate hairs on both sides. Leaves on sucker shoots (mostly in the shade) darker green, herbaceous, serrate, and more hairy in tropical countries, but thicker in southern Africa. Inflorescence variable in shape and size, seemingly umbellate to paniculate, congested or rather lax, 3–15 × 3–15 cm., 3–7 times branched. Peduncle and branches hairy like the branchlets. Pedicels short or obsolete. Flowers solitary or three together, fragrant. Calyx green, often viscid, 3–8 × 1·5–2·2 mm., minutely to manifestly pubescent with glandular and mostly also ordinary hairs outside, inside sericeous; lobes variable in size, mostly about as long as wide or wider, 0·5–2 × 1–1·5 mm., acute or less often obtuse, entire, valvate, often two coherent or partially united. Corolla white, with recurved pubescence on the lobes which may be reduced to scattered hairs or is rarely absent and with minute glandular hairs at the base of the lobes and the apex of the tube (the latter sometimes on the middle of the tube or on nearly the complete tube too), inside with curved hairs in the throat and sometimes also at the apices of the lobes and with minute glandular hairs in the tube except for its glabrous base (hairs in throat or hairs in tube are occasionally absent, although never both simultaneously); tube slightly shorter or sometimes — when calyx very short only — longer than the calyx; lobes oblong, 2–3 × as long as wide, 2–5 × 1–2 mm., obtuse or subacute at the apex. Stamens: anthers glabrous, 1·2–2 mm. long. Pistil (4·5–)7·5–13 mm. long; ovary ovoid or ellipsoid, slightly laterally compressed, 1·5–2·5 × 1·2–2 × 1–1·5 mm., appressedly hirto-pubescent, glabrous at the base. Capsule 0·5–1·5 mm. longer than the calyx, appressedly hirto-pubescent.

Botswana. SE: Lobasti (fl. x) *Rogers* 6213 (BM; BOL; K; PRE; Z). **Zambia.** E: Nyika Plateau, fl. 25.x.1958, *Robson* 369 (SRGH). **Zimbabwe.** N: Mazoe, near the Dam, fl. vii, *Gililand* 133 (BM); Korakora stream, S. of Farm Absent, fl. ix, *O. B. Miller* 1788 (PRE; SRGH). C: Chilimanzi District, *Gibson* 29/51 (K; SRGH). E: Gungunyana For. Res., fl. ix, *Goldsmith* 67/61 (K; LISC; M; MO; PRE; SRGH). **Malawi.** N: Nyika Plateau, Sangule Hill, Rumphi District, fl. viii, *Adlard* 303 (FHO; K; PRE; SRGH). C: Ntchisi Mt., fl. viii, *Brass* 17061 (K; MO; SRGH; US) 17618 (K; MO; PRE; SRGH; US). S: Zomba Plateau, bud, fr. v, *Brass* 16143 (MO; SRGH; US). **Mozambique.** Z: Mt. Tumbini, fl. viii, *Andrada* 1814 (COI; LISC) and 1814 b (LISC).

Also in southern Arabia, tropical and southern continental Africa and islands in the Gulf of Guinea. In tropical Africa: Mostly in light montane forest, there often dominant or codominant, less often scrub, grassland, or bamboo zone; in Ethiopia and Eastern Africa often in Juniperus forest; alt. (460)1100–3000 m. On high mountains on the slopes from about 2000 m. to the timber-line, on lower mountains near or at the summit. In southern Africa: Often at lower altitudes, even at sea level. Dry rocky crests and wooded stony slopes, montane forests or gallery forest, and towards the south in rocky gorges near the coast.

2. **Nuxia floribunda** Benth. in Hook., Comp. Bot. Mag. **2**: 59 (1836). — Walpers, Rep. **3**: 331 (1845). — Bentham in De Candolle, Prodr. **10**: 435 (1846). — Baker in F.T.A. **4**, 1: 515 (1903). — Prain & Cummins in Fl. Cap. **4**, 1: 1039 (1909). — Marloth, Fl. S. Afr. **3**, 1: 48, fig. 16. 1–8, fig. 18. (1932). — Bruce & Lewis, in F.T.E.A., Loganiaceae: 43, f. 8. 1–3 (1960), p.p., excl. syn. *N. siebenlistii.* — Verdoorn in Fl. S. Afr. **26**: 156 (1963). — Leeuwenberg in Meded. Landb. Wag. **75**–8: 38, fig. 6, map 5 (1975). TAB. **83**. Type from S. Africa.

N. polyantha Gilg in Engl., Bot. Jahrb. **30**: 376 (1910). — Baker, tom. cit.: 513. Type from Tanzania.

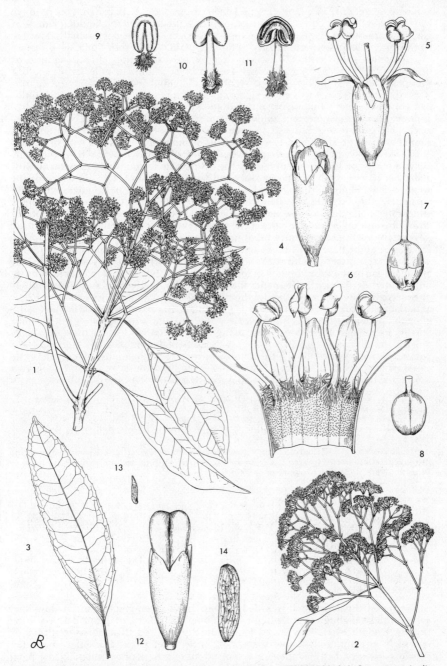

Tab. 83. NUXIA FLORIBUNDA. 1, flowering branch ($\times\frac{1}{2}$), from *Wild* 6593; 2, inflorescence ($\times\frac{1}{2}$), from *Lewalle* 815; 3, leaf ($\times\frac{1}{2}$), from *Donis* 3936; 4, opening bud ($\times$7), from *Torre & Correira*; 5, flower ($\times$7); 6, opened corolla ($\times$10); 7, pistil with corolla base ($\times$10); 8, ovary ($\times$10), 5–8 from *Wild* 6593; 9, young stamen ($\times$10), from *Torre & Correira* 14853; 10, 11, dorsal and ventral side of stamen ($\times$10), 10–11 from *Greenway* 3520.

Shrub or tree, 2–25 m. high. Trunk 10–60 cm. in diam. Bark rough, grey or grey-brown, fissured. Branchlets glabrous or minutely pubescent with glandular hairs and glabrescent, sulcate. Leaves ternate, opposite, decussate, or occasionally in a few lateral branchlets alternate, mostly petiolate; petiole 3–55 mm. long, glabrous or minutely pubescent; lamina coriaceous or papyraceous, elliptic, narrowly elliptic, or nearly so, 2·5–4(6) × as long as wide, 4–16 × 1–7 cm., acuminate or less often acute at the apex, cuneate at the base or decurrent into the petiole which is sometimes partly winged, dentate, crenate, or entire, glabrous or with very minute scattered glandular hairs on both sides; venation often reticulate, more or less prominent on both sides. Inflorescence lax or sometimes only in the ultimate ramifications congested, mostly large, paniculate or seemingly umbellate, 4–32 × 4–25 cm., 4–9 × branched. Branches often greenish or creamy, glabrous or minutely pubescent with glandular hairs. Flowers fragrant, pedicellate or sessile, solitary or in clusters of three together. Calyx creamy or green, often viscid, 2·5–3·8 mm. long, minutely pubescent with glandular hairs outside, inside sparsely and/or minutely appressed-pubescent; tube about 3–5 × as long as the lobes which are equal or unequal, triangular or broadly triangular, 0·6–1·2 × 0·6–1·2 mm., acute or less often obtuse, entire or sometimes tridentate; sometimes two lobes coherent or partly united. Corolla white, outside with some minute glandular hairs at the base of the lobes and at the apex of the tube (less often on almost the entire tube), inside with a ring of slightly recurved rather thick hairs at the base of the lobes and pubescent with glandular hairs in the tube of which extreme base is glabrous; tube slightly shorter than the calyx; lobes oblong, about as long as the tube, 1·8–3·3 as long as wide, 1·8–2·2, 0·6–1 mm., acute, obtuse, or rounded at the apex. Stamens: filaments white, geniculate and there pilose-pubescent just above the base, and at the extreme base pubescent with glandular hairs, as the latter is situated within the zone of glandular hairs at the inner side of the corolla, after the pollen is shed 2–3·8 mm. long; anthers white, turning pale brown, about 0·5–0·6 mm. long. Pistil 3–6·5 mm. long; ovary subglobose or nearly obcordoid, mostly laterally compressed, 1 × 0·7 × 0·5–1·5 × 1·3 × 1 mm., glabrous or occasionally with a few hairs at the apex. Capsule brown, mostly conspicuously longer than the calyx, nearly obovoid, 3–5 × 1·3–2 × 1·3–2 mm., glabrous, retuse and with two impressed lines of dehiscence at the apex. Seeds 0·8–0·9 × 0·2 mm.

Zambia. N: Nyika Plateau, Lundazi District, *F. White* 2737 (FHO; K). **Zimbabwe.** E: Vumba Mts., fl. v, *Chase* 4525 (BM; BR; COI; K; LISC; MO; S; SRGH; UPS). S: Salisbury, fr. iv, *Brain* 10837 (SRGH). **Malawi.** N: Viphya Escarpment, Kawendama, fl. vii, *Chapman* 1666 (FHO; K; SRGH). **Mozambique.** Z: Gúruè, Serra do Gúruè, fl. ii, *Torre & Correia* 14853 (LISC).

Also in eastern and southern Africa. Forests (where often on riverbanks), scrub, savanna, or heath vegetation; in tropical regions from 800 to 2400 m. altitude and there often dominant in forest; in Natal and Cape province at 0–1000 m. altitude.

3. **Nuxia oppositifolia** (Hochst.) Benth in De Candolle, Prodr. **10**: 435 (1846). — O. Schwartz, Fl. Trop. Arab.: 184 (1939). — P. Jovet in Bull. Hist. Nat. Toulouse **82**: 17 (1948). — Bruce & Lewis in F.T.E.A., Loganiaceae: 43, f. 8. 4–6 (1960). — F. White, F.F.N.R.: 340 (1962). — Verdoorn in Fl. S. Afr. **26**: 153 (1963). — Leeuwenberg in Meded. Landb. Wag. **75**–8: 52, fig. 10, map 7 (1975). Type from Ethiopia.
Lachnopylis oppositifolia Hochst. in Flora **26**: 77 (1843). Type as for the species above.
Nuxia dentata R. Br. ex Benth., loc. cit. — Baker in F.T.A. **4**, 1: 513 (1903). — Prain & Cummins in Fl. Cap. 4, 1: 1040 (1909). — Eyles in Trans. Roy. Soc. S. Afr. **5**, 4: 442 (1916). Type from Ethiopia.

Evergreen shrub or small tree, 1–15(20) m. high, often gnarled and willow-like, sometimes straggling. Trunk often rather fluted, 10–60 cm. in diam. Bark reddish-brown, smooth or shallowly and longitudinally fissured. Branchlets glabrous or sparingly and minutely puberulous, not lenticellate, slightly angular and often sulcate when dry. Leaves opposite, less often subopposite, or occasionally alternate, petiolate; petiole glabrous or hairy as the branchlets, 2–12 mm. long; lamina not much paler beneath and subcoriaceous or coriaceous when dry, narrowly to very narrowly elliptic, (2–)4–8 × as long as wide, (1·5)3–13 × 0·4–3(5) cm., obtuse or rounded and often mucronulate at the apex, cuneate at the base or decurrent into the petiole, bluntly serrate to entire, glabrous or with minute glandular hairs on both sides; secondary veins not very conspicuous. Inflorescence rather congested, 1·5–7 × 2–8 cm., 2–5 × branched. Peduncle, branches, and pedicels glabrous or scarcely hairy. Ultimate branches and pedicels very short or obsolete. Flower slightly fragrant, 1–3(8) together. Calyx green, 3·5–5 × 1·2–2 mm., minutely pubescent with glandular and often also ordinary hairs outside, when dry seemingly glabrous and often shining, inside sericeous; tube 4–8 × as long as the lobes; lobes triangular, about as long as wide or wider, 0·5–1·2(1·4) × 0·7–1·2 mm., acute or subacute, entire, sometimes two coherent or partially united. Corolla white, creamy, or sometimes pale lilac, outside with some minute glandular hairs at the base of the lobes, inside with a villose ring of recurved hairs in the throat and with minute glandular hairs in the tube except for its glabrous base; tube slightly shorter or less often slightly longer than the calyx; lobes oblong, 1·5–3 × as long as wide, 1·8–3 × 0·8–2 mm., obtuse or rounded at the apex. Stamens: anthers about 1·2 mm. long. Pistil: ovary obcordoid or ovoid, slightly laterally compressed, 1–1·6 × 0·8–1·2 × 0·7–1 mm., appressed-pubescent, glabrous at the base. Capsule about as long as the calyx, appressed-pubescent. Seeds 0·5–0·8 × 0·15–0·2 mm.

Zambia. N: Tanganyika Plateau, *Whyte* s.n., vii.1896 (K). W: 65 km. S. of Mwinilunga, fl. viii, *Milne-Redhead* 924 (BR; K; PRE; WAG). C: Mt. Makulu, 18 km. S. of Lusaka, fl. viii, *Angus* 1280 (BR; FHO; K). E: Chipata [Fort Jameson], fr., fl. x, *Mutimushi* 2328 (K). S: Victoria Falls, *Greenway* 6261 (EA; K; PRE). **Zimbabwe.** N: Lomagundi, fl. x, *Eyles* 2671 (A; K; NBG; PRE; SRGH). W: Masene Springs, fl. x, *Whellan* 399 (K; SRGH). C: Lalapanzi, Gwelo District, fl. x, *Walters* 2439 (K; SRGH). E: Umvumvumu R., fl. xii–i, *Chase* 488 (BM; COI; K; LISC; PRE; SRGH). S: Lundi R., Nuanetsi District, fl. xii, *Davies* 2266 (BR; K; PRE; SRGH). **Malawi.** C: Kapeni Stream, Dzonze, Dedza District, *Chapman* 1354 (FHO; K; MO; SRGH). S: Nawadzi R., Thyolo [Cholo] District, fl. ix, *Brass* 17850 (BM; BR; K; MO; SRGH; US). **Mozambique.** Z: Lumba R. bank, along road to Morrumbala, fl. x, *Andrada* 1945 (COI; LISC). T: Kabankangywe Kraal, Mazoe R., fl. ix, *Wild* 2593 (BR; K; S; SRGH). MS: Inhacoro, *Surcouf*. 20.ix.1929 (BR; P). M: Maputo, Goba, fl. xi–xii, *Mendonça* 3076 (LISC).

Also in southern Arabia, eastern, south central, and southern Africa, and Madagascar. Montane scrub on river banks or gallery forest; alt. 500–2400 m.; in Madagascar and in Natal at lower elevation, even at 50 m. alt.

6. STRYCHNOS L.

Strychnos L., Sp. Pl. **1**: 189 (1753)

Erect or climbing shrubs, lianas, or trees. Lianas with curled tendrils, solitary or arranged in 1–3 pairs above each other on short branches, in the axils of small bracts or — only if solitary — sometimes in the axils of ordinary leaves. Trees: usually less than 10 m. high (savanna spp.), 10–35(40) m. (forest spp.); trunk 10–100 cm. in diam.; without tendrils, often with arching branches (*S. usambarensis, S. xantha*). Wood mostly hard, usually with bark-islets; bark mostly thin, smooth or less often rough, in lianas often with large lenticels, sometimes thick and corky (S. cocculoides). Branches armed with axillary or sometimes terminal simple straight or slightly recurved spines, unarmed, often lenticellate, rarely corky; branchlets terete, quadrangular, sometimes sulcate, especially when dry. Stipules mostly reduced to an often ciliate and straight rim connecting the petiole-bases. Leaves opposite, sometimes decussate, or on the main axis sometimes ternate, those of a pair or whorl equal or subequal, petiolate or sometimes subsessile, mostly inserted on distinct leaf-cushions; blade variously shaped, orbicular to narrowly elliptic, mostly coriaceous, in the shade thinner, often larger, and more acute at the apex, entire; with 1–2(3) pairs of distinct secondary veins from or from above the base curved along the margin, usually not fully reaching the apex, anastomosing with the other veins or less often pinnately veined. Inflorescence terminal, axillary, or both, thyrsoid, 1-many-flowered, shorter or longer than the leaves, lax or congested, sometimes sessile. Flowers 4–5-merous, actinomorphic or with only sepals unequal. Calyx: sepals green or coloured approximately like the corolla, free or connate up to one-half of their length, equal, subequal, or sometimes unequal, imbricate, orbicular to linear, outside hairy or glabrous. Corolla rotate to salver-shaped, white to yellowish, greenish, pale green, or rarely orange or ochraceous, thin at the base, always more or less thickened towards the lobes, on both sides variously hairy, papillose, or glabrous, but inside at the base always glabrous, sometimes with a corona at the mouth; lobes valvate in the bud, triangular to oblong, acute or subacute, entire, erect to recurved. Stamens exserted or included, inserted at the mouth of or in the corolla tube; filaments glabrous or sometimes hairy; anthers introrse, orbicular to narrowly oblong, cordate, deeply so, or less often sagittate at the base. Pistil: ovary 2- or rarely 1-celled (*S. mellodora, S. spinosa*); stigma capitate, less often obscurely bilobed, or occasionally conical. In each cell of a 2-celled ovary one axial placenta with 2 to about 50 ovules, attached to the middle of the septum. In a one-celled ovary one basal placenta which is mostly globose, with few (*S. mellodora*) or many ovules (*S. spinosa*). Fruit a berry, 1–2-celled, globose or nearly so, mostly yellow to red, less often green when mature, sometimes blue-black (S. potatorum), immature often glaucous, glabrous, subtended by the persisting calyx, 0·8–18 cm. in diameter, 1–45-seeded. Wall thin and soft in small fruits, thicker and brittle in (mature) larger (very hard in nearly mature!!). Pulp juicy, fleshy, often edible (*S. cocculoides, S. madagascariensis, S. pungens, S. spinosa*, and also in some small-fruited species). Seeds large, 0·5–3(5) cm. long, variably shaped, generally disc-shaped to subglobose. Testa thick and osseous to very thin and membranaceous, rough and scabrid-pubescent (only when thick) or smooth and sericeous to glabrous. Endosperm horny, slightly diaphanous. Sometimes some colleters above the bracts, on the base of the sepals, and near the base of the petioles.

A circumtropical genus of about 170 species.

Key to flowering material

1. Inflorescence terminal - - - - - - - - - - - 2
 - Inflorescence axillary and sometimes also terminal at the same time - - - 6
2. Sepals narrow, 2 × as long as wide or more, acuminate to subulate - - - - 3
 - Sepals orbicular to triangular, rounded to acute; fruits small, up to 3 cm. in diameter 4
3. Sepals outside, mostly, at least apically glabrous, never with an even indumentum; branchlets usually glabrous; branches and bark not corky; ovary 1-celled - 17. *spinosa*
 - Sepals outside with an even pubescence; branchlets usually pubescent; branches and bark corky; ovary 2-celled - - - - - - - - - - 2. *cocculoides*
4. Pistil pilose, 3·5–4 mm. long; corolla-lobes recurved; stamens well-exserted; climber with paired tendrils - - - - - - - - - - - 14. *panganensis*
 - Pistil glabrous up to 2·8 mm. long; stamens included or covered by the not recurved corolla-lobes; shrubs or climbers with solitary tendrils - - - - - 5
5. Corolla pubescent outside; inflorescence dense, c. 15–50-flowered; sepals outside at least at the base pubescent; climber with solitary tendrils - - - - - 10. *matopensis*
 - Corolla glabrous outside; inflorescence lax, 3–10 flowered; sepals mostly glabrous outside; shrubs - - - - - - - - - - - 13. *myrtoides*
6. Pistil pilose or pubescent - - - - - - - - - - - 7
 - Pistil glabrous - - - - - - - - - - - - 12
7. Leaves with a sharp apex; savanna tree or shrub with densely lenticillate branches and large globose fruits - - - - - - - - - - 16. *pungens*
 - Leaves not with sharp apex - - - - - - - - - - 8
8. Corolla with a bush-like ring in the throat; sepals always rounded or obtuse - - 9
 - Corolla pilose or villose inside - - - - - - - - - 11
9. Flowers 4-merous; savanna shrub or tree with mostly rounded leaves - - - 10
 - Flowers 5-merous; forest climber with acuminate leaves and paired tendrils - 8. *lucens*
10. Leaves glaucous, not or hardly paler beneath, mat or dull with mostly pale green reticulate prominent venation on both sides - - - - - - - 6. *innocua*
 - Leaves shiny and dark green above, more distinctly paler beneath; tertiary venation reticulate, not or hardly prominent above - - - - 9. *madagascariensis*
11. All flowers distinctly pedicellate; leaves rather thin, glabrous above; mature bud 3·5–5 mm. long; pistil 3·5–4 mm. long - - - - - - - - 14. *panganensis*
 - At least some flowers nearly sessile; leaves coreaceous, pubescent on main veins above, mature bud and pistil (4–)5–7 mm. long - - - - - 7. *kasengaensis*
12. Branchlets pubescent, often sparsley so - - - - - - - 13
 - Branchlets glabrous - - - - - - - - - - - 16
13. Branches lenticellate; petioles not rugulose - - - - - - - 14
 - Branches not lenticellate; petioles rugulose; inflorescence sometimes transformed into rose-like galls - - - - - - - - - - 1. *angolensis*
14. Stamens inserted in corolla-tube; anthers bearded; inflorescence congested; mature bud 3·5–4 mm. long; forest tree - - - - - - - - 12. *mitis*
 - Stamens at mouth of corolla-tube; anthers glabrous; inflorescence often lax - - 15
15. Mature bud 4–6 mm. and pistil 4–5 mm. long; inflorescence lax; leaves obtuse rounded or broadly and bluntly acuminate - - - - - - - 3. *decussata*
 - Mature bud 2–3·5 mm. and pistil 1–2·5 mm. long; inflorescence congested or not; leaves acuminate - - - - - - - - - 18. *usambarensis*

16. Branches lenticellate - - - - - - - - - - - 17
 - Branches not lenticellate - - - - - - - - - - 20
17. Inflorescences in axils of scales at base of branchlets; branches dichotomously branched
                                                                    15. *potatorum*
 - Inflorescences in axils of ordinary leaves and sometimes also terminal - - - 18
18. Mature flower-bud 2·5–3 mm. long; pistil 1·1–2·5 mm. long - - - - - 19
 - Mature flower-bud 4–6 mm. long; pistil (3–)4–6 mm. long (see also 12. *mitis*) 3. *decussata*
19. Inflorescence lax, many-flowered; flowers 4-merous: corolla lobes 4–5 × as long as tube;
    anthers glabrous - - - - - - - - - - 11. *mellodora*
 - Inflorescence more or less congested; flowers 4–5-merous; corolla lobes 0·8–1·6 × as long as
    tube; anthers bearded - - - - - - - - - - 12. *mitis*
20. Mature flower-bud 1·6–4 mm. long; pistil 1·1–2·2(–3) mm. long; trees or shrubs without
    tendrils - - - - - - - - - - - - - 21
 - Mature flower-bud 4·5–6·7 mm. long; pistil (3·5–)4–6 mm. long; climbers or small trees
    with lianescent branches with tendrils in 1–3 pairs above each other or small fire resistant
    savanna shrubs 10–50 cm. high - - - - - - - - - 22
21. Inflorescence 1–2 cm. long, congested, flowers 5-merous - - - 5. *henningsii*
 - Inflorescence 3–11 cm. long, lax; flowers 4-merous - - - 11. *mellodora*
22. Leaves with the apex rounded, acute, or sometimes shortly acuminate, drying dark brown
    to pale greenish-brown - - - - - - - - - 4. *gossweileri*
 - Leaves acuminate, usually drying yellow - - - - - - 19. *xantha*

Key to fruiting material

1. Fruit 4–15 cm. in diameter; wall 5–8 mm. thick, rarely less; seeds (10–)30–100 - 2
 - Fruit c. 1–3 cm. in diameter; wall 1–3 mm. thick; seeds 1–2(12) - - - - 5
2. Infructescence terminal; branches often spiny; sepals narrow; fruits strictly globose
                                                                turn to 3 of first key
 - Infructescence axillary; branches without spines - - - - - - 3
3. Leaves with sharp apex; savanna tree or shrub - - - - 16. *pungens*
 - Leaves without a sharp apex - - - - - - - - - 4
4. Savanna shrubs or trees; leaves mostly rounded at apex - - turn to 10 of first key
 - Climber with paired tendrils, in forests - - - - - - 8. *lucens*
5. Fruit blue-black, with one ellipsoid sericeous seed; branches repeatedly dichotomously
    divided; shrub or tree - - - - - - - - 15. *potatorum*
 - Fruit orange, yellow, red or dark brown - - - - - - - 6
6. Seed not dented, sometimes grooved and then glabrous; pistil hairy or glabrous - 7
 - Seed with a deep pit at one side, covered by false papillae simulated by curved hairs; pistil
    glabrous - - - - - - - - - - turn to 22 of first key
7. Seeds ellipsoid or subglobose, not or hardly flattened, smooth - - - - 8
 - Seeds obliquely elliptic or tetrahedral, flattened, often rough - - - - 12
8. Seed with a deep groove, glabrous; savanna shrub or tree; leaf (especially when old) with
    veins prominent on both sides - - - - - - - 5. *henningsii*
 - Seed without groove, pubescent or glabrous; trees or climbers; veins of leaf not or not very
    prominent - - - - - - - - - - - - 9
9. Petiole rugulose, pubescent like branchlets; branches not lenticellate - 1. *angolensis*
 - Petiole not rugulose, glabrous or hairy; branches lenticeliate - - - - 10
10. Tree or shrub, the latter only in savannas - - - - - - - 11
 - Climber with solitary tendrils or sometimes tree - - - 18. *usambarensis*
11. Forest tree; leaves mostly acuminate; pistil 2·3–3 mm. long; seed glabrous - 12. *mitis*
 - Savanna tree or shrub; leaves rounded, obtuse or broadly and bluntly acuminate; pistil 4–5
    mm. long; seed minutely pubescent - - - - - - 3. *decussata*
12. Seed smooth, minutely pubescent or glabrous; montane forest tree; peduncle and pedicels
    pubescent; infructescence rather large and lax - - - - 11. *mellodora*
 - Seed scabrid; testa rather thick; climbers, shrubs, or small trees, mostly in the savannas;
    infructescence small - - - - - - - - - - 13
13. Infructescence terminal - - - - - - - - - - 14
 - Infructescence axillary and often at the same time terminal - - - - 16
14. Infructescence congested, with many dried-up flowers; branchlets appressed-pubescent;
    leaves often circular - - - - - - - - - 10. *matopensis*
 - Infructescence lax, without or with few dried-up flowers - - - - 15
15. Pistil 3·5–4 mm. long; pilose; fruit 8–18 mm. in diameter; climber with paired tendrils
                                                                    14. *panganensis*
 - Pistil 0·9–1·4 mm. long, glabrous; fruit 6–9 mm. in diameter; shrub or small tree
                                                                turn to 5 of first key
16. Savanna tree or shrub; leaves mostly rounded at apex - - - 13. *myrtoides*
 - Climber with paired tendrils; leaves acuminate or nearly so - - - - 17
17. Leaves thickly coriaceous, mostly glabrous on both sides; margin revolute; tertiary
    venation prominent on both sides - - - - - - - 8. *lucens*
 - Leaves coriaceous or papery, glabrous or hairy; margin not revolute - - - 18

18. Leaves coriaceous, pubescent on main veins above; fruit about 30 mm. in diameter. about
       10–20-seeded; pistil (4)5–7 mm. long - - - - - - - 7. *kasengaensis*
— Leaves rather thin, glabrous above; fruit 8–18 mm. in diameter, (1)3–6-seeded; pistil 3·5–4
       mm. long - - - - - - - - - - - - 14. *panganensis*

1. **Strychnos angolensis** Gilg in Engl., Bot. Jahrb. **17**: 571 (1893). — Hiern, Cat. Afr. Pl.
   Welw. **1**: 703 (1898). — Baker in F.T.A. **4**, 1: 522 (1903). — Bruce & Lewis in F.T.E.A.,
   Loganiaceae: 29 (1960). — Onochie & Leeuwenberg in F.W.T.A., 2nd ed., **2**: 43 (1963).
   — Leeuwenberg in Meded. Landb. Wag. **69**–1: 57, fig. 4, map 3 (1969). Type from
   Angola.

Climbing shrub or liana, 3–30 m. high, climbing over shrubs and in trees, or small
semiscandent tree, 5–12 m. high. Trunk 4–20 cm. in diam. Bark smooth, thin, pale
brown, not or slightly lenticellate, in séction pale yellow; wood pale yellow. Branches
not lenticellate, medium to dark brown when dry, usually hairy like the branchlets;
branchlets ochraceous-pubescent, green, terete, not sulcate when dry. Tendrils
solitary, in the axils of small triangular bracts. Leaves: petiole ochraceous-
pubescent, mostly transversely rugose when dry, 1–5(6) mm. long; lamina shining or
mat and medium to dark green above, paler and often with dark veins beneath,
papyraceous, coriaceous, or subcoriaceous, thinner when growing in shade, also
when fresh, very variable in shape and size, ovate, elliptic or narrowly ovate,
1–3·5 × as long as wide, broadest on the main axis and at the base of the branchlets,
1·8–7(10) × 1–4(5) cm., usually obtuse and apiculate, but often acute, sometimes
obtusely acuminate or emarginate at the apex, rounded, cuneate, or sometimes
subcordate at the base, glabrous or ochraceous-pubescent beneath on the costa and
main secondary veins, especially at the base, and above on the costa near the base; one
or two pairs of secondary veins from or from above the base curved along the margin;
tertiary venation inconspicuous above, prominent beneath. Inflorescence axillary
and sometimes also terminal, solitary, lax, few-flowered, about 0·5–1 × as long as the
leaves, sometimes transformed into rose-like galls. Lateral branches usually 3-
flowered. Peduncle and branches slender, ochraceous-pubescent as the very short
pedicels. Bracts narrowly triangular or sepal-like. Flowers 4–5-merous, even in a
single inflorescence. Sepals pale green, equal or sometimes unequal, connate at the
base up to one-third of their length, broadly ovate, 0·7–1·2 × as long as wide,
(0·5)0·8–1·2 × (0·5)0·7–1 mm., acute or obtuse at the apex, glabrous on both sides,
ciliate, without colleters. Corolla in the mature bud 2·5–3 × as long as the calyx,
2–2·5 mm. long, white or yellow, fading from white to yellow, subrotate, 4–4·5 mm.
in diam., glabrous or minutely papillose-pubescent outside, inside on the base of the
lobes densely pilose with dirty white hairs (sometimes complete lobes pilose); tube
very short, shorter than the calyx, 0·4–0·6 mm. long; lobes ovate to triangular,
2·7–4·5 × as long as the tube, comparatively wider in 4-merous flowers, 1·2–1·7 × as
long as wide, 1·5–2 × 1–1·4 mm., acute, spreading. Stamens just exserted; filaments
about as long as the anthers, inserted at the mouth of the corolla tube; anthers yellow,
glabrous, ovate or suborbicular, 0·5–1 × 0·5–0·8 mm., deeply cordate at the base.
Pistil glabrous, 1·2–1·6 mm. long; ovary globose or broadly ovoid, 0·8–1·2 × 0·6–0·9
mm.; style short or very short, 0·1–0·7 mm. long; stigma capitate. In each cell 4–6
ovules. Fruit orange or red, immature glaucous, small, soft, ellipsoid or globose,
12 × 12–22 × 18 mm., 1-seeded, with smooth skin, slightly shining or mat, obliquely
pedicellate. Wall thin, 1 mm. thick. Seed shining when living, dark brown, smooth,
approximately   bean-shaped   with   an   impressed   hilum   or   ellipsoid,
8·5 × 6 × 5–15 × 11 × 9 mm., glabrous; testa very thin, easily rubbed off.

Zambia. B: 20 km. S. of Kwambwa Boma fr. x, F. White 3530 (FHO; K). N: Samfya, fr. xi,
Mutimushi 1165 (K); W: Kitwe, fl. ii–iii, *Fanshawe* 2105 (BR; EA; K; LISC; SRGH), 2122 (BR;
K), 2387 (BR; EA; K; LISC). **Zimbabwe.** E.: Pungwe Gorge, Inyanga District, ix, *Wild* 4613
(K; LISC; SRGH). **Mozambique.** Z: 42 km. from Lioma, between Lioma and Gúruè, fl. ix,
Barbosa & Carvalho 4996 (K; PRE).
    Also in Nigeria, Central and East Africa. On river banks in rain forests or in gallery forests;
alt. 0–1500 m.

2. **Strychnos cocculoides** Bak. in Kew Bull. **1895**: 98 (1895); in F.T.A. **4**, 1: 533 (1903). —
   Hiern, Cat. Afr. Pl. Welw. **1**: 704 (1898). — Bruce & Lewis in F.T.E.A., Loganiaceae: 16,
   fig. 3, 1–5 (1960). — Verdoorn in Fl. S. Afr. **26**: 149 (1963). — Leeuwenberg in Meded.
   Landb. Wag. **69**–1: 86, fig. 12, map 10 (1969). Type from Angola.
       S. suberosa T.R. Sim, Forest Fl. Port E. Afr.: 90, pl. 76 (1909). Type from
   Mozambique: Zambézia Maganja da Costa, near Maquebela, *Sim* 6013 (PRE, holotype).

Deciduous shrub or small tree, (0·30)1–6(8) m. high, sometimes flowering on one-year shoots with small leaves on old fire-cut stumps. Trunk 4–25 cm. in diam., branched from low down; bark pale grey to pale brown, thick, corky, longitudinally ridged, not scaly, nor lenticellate; wood hard, whitish, without bark-islets. Branches pale to dark brown when dry, more or less fissured and corky by which often irregularly thickened, not lenticellate, often with recurved or sometimes with straight spines, sometimes terminating in a straight spine; branchlets yellow-green and often tinged with red, when living greenish brown, dark brown, reddish-brown, or purplish-brown and often sulcate when dry, pubescent or sometimes glabrous, sometimes (especially when older) fissured like branches. Leaves: petiole short, pubescent or less often glabrous, 2–8 mm. long; lamina mat or shining and pale to dark green above, mat and pale beneath, veins often pale on both sides, coriaceous, young thinner and papyraceous when dry, very variable in shape and size, orbicular, ovate, elliptic, or sometimes narrowly elliptic, usually comparatively wider on flowering branches, smaller (as in *S. spinosa*) on the very spiny young shoots on fire cut stumps, 1–1·5(2·2) times as long as wide, 2–6(10) × 1–5(9) cm., emarginate, rounded to acute and apiculate, or sometimes acuminate at the apex, rounded, subcordate, or cuneate at the base, often slightly bullate or subpapillose, often with hair-pockets in the angles of the main veins beneath, pubescent or less often glabrous on both sides; 1–3 pairs of secondary veins from or from above the base curved along the margin. Inflorescence terminal, seemingly umbellate, congested, 2·5–4 × 2–5 cm., mostly many-flowered. Flowers 5-merous. Sepals pale green, connate at the base, narrowly triangular, equal, subequal, or sometimes unequal, (2)2·5–5 × as long as wide, 2–5 × 0·7–1·3 mm., acuminate or subulate at the apex, not distinctly ciliate, outside pubescent with an even indumentum of mostly erect hairs, inside pubescent at the apex and glabrous at the base or sometimes entirely pubescent, without colleters. Corolla when mature 1–2 × as long as the calyx, (3·4)4–5 mm. long, pale green, greenish, white, or greenish-yellow, sparsely pubescent or glabrous outside, inside with a narrow entire white-penicillate corona at the mouth of the tube 1–2 × as long as the lobes, (2)2·5–3 × 2–3 mm., urceolate or campanulate, often somewhat contracted at the throat; lobes triangular, 1–2 × as long as wide, 1·3–2·5 × 1–1·5 mm., acute, erect or suberect. Stamens included, inserted 0·4–1 mm. from the base of the corolla; anthers oblong or elliptic, 1·2–1·5 × 0·8–1·2 mm., ciliate all around, cells parallel. Pistil pubescent, 1·8–2·8 mm. long; ovary ovoid, broadly ovoid or globose, 0·8–1·3 times as long as wide, 1–1·6 × 1–1·4 mm., acuminate or rounded at the apex, 2-celled; style mostly very short; stigma oblong, large, up to 0·9 mm. long. In each cell about (30)50 ovules. Fruit large, hard, resembling an orange, globose, yellow or orange, often speckled with green, nearly mature dark green and mottled with pale green, when immature often blue-green, often slightly shining, with granular skin, 6–11 cm. in diam., with about 10–100 seeds. Wall rather thick, 2–5 mm. thick, thickened inside above the pedicel. Pulp edible. Seeds ochraceous, flattened, more or less plano-convex, obliquely ovate or elliptic, usually irregularly curved, about 1·2–1·6 times as long as wide, 12 × 8 × 2·5–22 × 15 × 4 mm., slightly rough, very shortly pubescent.

Botswana. N: Ngamiland, fl. xii–i, *Curson* 3311 (PRE), ibid., *O. B. Miller* B/434 (FHO). SW: N. Kalahari, *Schoenfelder* 4 (PRE). **Zambia.** B: near Senenga, Barotseland, fl. vii, *Codd* 7221 (BM; BR; COI; EA; K; L; PRE; SRGH; UPS). N: Sunzu Hills, Mbale [Abercorn] District, *Burtt* 6345 (BM; BR; K). W: Ndola, *Fanshawe* 826 (K), ibid., fl. v fr. ii, *R. G. Miller* 170 (BM; FHO), 300a (FHO, NY). C: Chilanga, *Rogers* 8594, partly (Z). S: Livingstone, fl. x, *Morze* 21 (FHO). **Zimbabwe.** N: Gokwe District, *Vincent* 78 (MO; SRGH). W: Matobo District, *O.B. Miller* 1904 (SRGH). C: Salisbury, fl. x, *Brain* 6238 (MO; SRGH), ibid., fl. x–xi fr. xii, *Eyles* 3694 (BOL; K; NBG; SRGH). S: Victoria District *Monro* 555 (BM; SRGH). **Malawi.** C: Kasungu, fr. viii, *Brass* 17433 (K; MO; NY; SRGH; US). **Mozambique.** N: km. 11 of Montepuez-Namuno road, fr. xii, *Torre & Paiva* 9712 (LISC). Z: between Alto Ligonha and Alto Molócuè, *Barbosa & Carvalho* 4406 (K; SRGH). T: Moatize, between Zóbuè and Metengobalama, *Correia* 397 (LISC). MS: Gorongosa, National Game Park *Torre & Paiva* 9136 (LISC). SS: Cubine, *Le Testu* 617 (P). M: Maganja da Costa, near Maquebela, *Sim* 6013 (PRE, type of *S. suberosa* Sim, non De Wild).
  Also in Central, East, and northern South Africa. Woodland; alt. 400–2000 m.

3. **Strychnos decussata** (Pappe) Gilg in Engl., Bot. Jahrb. **28**: 121 (1899). — Bruce & Lewis in F.T.E.A., Loganiaceae: 29 (1960). — Verdoorn in Fl. S. Afr. **26**: 139 (1963). — Leeuwenberg in Meded. Land. Wag. **69**–1: 100, fig. 18, map 14 (1969). TAB. **84**. Type from S. Africa (Cape Prov.).

Tab. 84. 1–7 STRYCHNOS DECUSSATA. 1, branch (×½); 2, flower (×2); 3, portion of corolla with stamen (×2); 4, hair of the corolla (×15), 1–4 from *Rudatis* 1226; 5, fruit (×½), from *Bally* 1936; 6, seed (×1), from *Horby* 2485; 7, leaves (×½), from *Bally* 1936, *Ecklon & Zeyher* 3368, *Pole Evans* 1635 and *Rudatis* 1226. 8–14 STRYCHNOS HENNINGSII. 8–10, branches and leaves (×½), 8 from *Busse* 2511a, 9 from *Flanagan* 1102 and 10 from *Greenway* 6562; 11, flower (×3); 12, portion of corolla with stamen and filament (×3); 13, fruit (×½), 11–13 *Flanagan* 1102; 14, seed (×2), from *Gossweiler* 12608.

Atherstonea decussata Pappe, Silv. Cap. 2nd ed. **29** (1862). Type as for the species above.

Strychnos atherstonei Harv., Thes. Cap. **2**: 41, t. 164 (1863) with *S. baculum* Harv. in syn. Type from S. Africa.

Shrub or small tree, 2–12(17) m. high. Trunk 8–30(45) cm. in diam. Bark smooth, pale to dark grey; wood hard. Branches lenticellate, pale brown or pale grey to black when dry; branchlets lenticellate or not, not or obscurely sulcate, and terete or nearly so when dry. Tendrils none. Leaves: petiole glabrous, 2–7(10) mm. long lamina shining and dark green above, paler beneath, pale greenish- or yellowish-brown when dry, subcoriaceous, also when fresh, very variable in shape and size, obovate, elliptic, narrowly elliptic, narrowly obovate, ovate, or narrowly ovate, (1)1·5–3(4) times as long as wide, (1)2–5(11·5) × (0·4)0·8–3(6·5) cm., rounded, obtuse, or broadly and bluntly acuminate at the apex, cuneate or rounded at the base, glabrous on both sides; one or two pairs of secondary veins from or from somewhat above the base curved along the margin and often a faint submarginal pair; margin often slightly involute. Inflorescences axillary and sometimes also terminal, or ramiflorous, often appearing before the season's new leaves, often several together, lax, several flowered, 1·5 × 1·5–3 × 2·5 cm., 1–2 times branched. Peduncle, branches, and pedicels slender, minutely or conspicuously papillose-pubescent or less often glabrous. Flower (4)5-merous, fragrant. Sepals connate up to one-quarter of their length, equal, subequal, or occasionally unequal, broadly ovate to triangular, 1–1·5(1·7) times as long as wide, 1–1·3 × 0·6–1 mm. (when unequal smallest 0·6 × 0·4 mm.), acute or obtuse at the apex, minutely ciliate, glabrous on both sides or like the pedicels minutely or conspicuously papillose-pubescent outside, without colleters. Corolla in the mature bud 3·5–5 times as long as the calyx, 4·6 mm. long, and rounded at the apex, white, creamy, or yellow, outside glabrous or minutely papillose-pubescent, inside densely pilose except for the glabrous apex of the lobes and base of the tube; tube short-campanulate or nearly cylindrical, often much widened towards the throat, very variable in length, 1–2·5 times as long as the calyx, 1–3·2 mm. long, at the throat (1·5)2–3 mm. wide; lobes oblong or narrowly triangular, 0·6–3 times as long as the tube, 1·7–2·5 times as long as wide, 1·7–3·2 × 1–1·6 mm., acute, spreading or sometimes recurved. Stamens exserted; filaments short or long, very variable in length and elongate at anthesis, longer in flowers with shorter corolla tube (like in *S. potatorum*), 0·5–2 times as long as the anthers, glabrous, inserted at the mouth of the corolla tube; anthers oblong, 0·8–1·3 × (0·4)0·5–1 mm., glabrous. Pistil glabrous, slender, 3·8–5·2 mm. long; ovary ovoid or narrowly ovoid, 1–2 × 0·8–1·2 mm., gradually narrowed into the style, 2-celled; style 1·8–3·9 mm. long; stigma capitate or sometimes bilobed, often rather small. In each cell 5–10 ovules. Fruit orange, nearly mature glaucous or green, small, soft, ellipsoid or globose, sometimes stipitate, 1·5–2 × 1·5–1·9 mm., 1(–2)-seeded, with somewhat granular skin, slightly shining. Wall thin, dry, about 0·3 mm. thick. Pulp edible. Seeds pale grey-brown, not or hardly flattened, ellipsoid, slightly longer than wide and wider than thick, 10 × 10 × 7–16 × 12 × 9 mm., densely and shortly appressed-pubescent, smooth; testa sticking to the pulp, often wrinkled in dry fruits.

Zambia. S: Lusito, fr. v, *Fanshawe* 6599 (K). **Zimbabwe.** N: bank Manora R., Urungwe District, *Phipps* 927 (K; LISC; SRGH). E: Odzi R., Hot Springs, Melsetter District, fl. xi, *Chase* 4704 (BM; BR; COI; K; LISC; MO; PRE; SRGH; WAG). S: Sabi and Lundi Rs. Junction, fr. vii, *Wild* 3494 (BR; EK; K; LISJC; NY; PRE; SRGH). **Malawi.** S: Npatamanga Gorge, Blantyre District, fl. xii, *Chapman* 1070 (FHO; PRE; SRGH). **Mozambique.** MS: Cheringoma Distr. Chiniziua, fr. iv, *Gomes e Sousa* 4346 (K). T: Sisitso, Zumbo area, Zambesi R., fr. vii, *Chase* 2740 (BM; LISC; MO; SRGH). MS: Gorongosa, *Torre & Paiva* 9072 (LISC). GI: Between Limpopo and Nuanetsi Rs., fl. vii, *Smuts* 337 (BM; K; PRE). M: Moamba, fl. xii, *Torre* 2178 (LISC).

Also in East and South Africa and Madagascar. In woodlands, often near rocks, sometimes on river banks, in Madagascar often in dry forests and bushes on limestone; alt. 0–1500 m.

4. **Strychnos gossweileri** Exell in Journ. Bot., Lond. **67**. Suppl. 2: 102 (1929). —
Leeuwenberg in Meded. Land. Wag. **69**–1: 124, fig. 8, map 5 (1969). Type from Angola.
S. caespitosa Good in Journ. Bot., Lond. **67**. Suppl. Gamopet.: 104 (1929). Type from Angola.

Small fire resistant savanna shrub, 10–50 cm. high, or climbing shrub or liana, 2–5 m. high climbing and about 10–20 m. long or more. Trunk in shrub very thin, about

0·5 cm. in diam., in liana up to 12 cm. in diam. Bark dark brown, shallowly fissured, with large tuberculate lenticels; wood pale yellow. Branches usually quadrangular, glabrous, not or hardly sulcate when dry. Tendrils in 1–3 pairs above each other. Leaves with a glabrous, 2–5 mm. long petiole; lamina mat or slightly shining and dark green above, dull or slightly shining and paler and often glaucous beneath, drying dark brown to pale greenish-brown, in the sun coriaceous, in the shade subcoriaceous, elliptic, obovate, or on the main axis (climbers) subcircular, (1)1·5–5·5 times as long as wide, 2–7(11) × 1–4(5) cm., acute, apiculate, or shortly acuminate at the apex, cuneate or less often rounded at the base, glabrous on both sides; one pair of secondary veins from or from above the base curved along the margin and often accompanied by a faint submarginal pair; costa impressed above; tertiary venation reticulate, often prominent on both sides. Inflorescence axillary, solitary, more or less congested, usually few-flowered, much shorter than the leaves, 1–2 times branched. Peduncle, branches, and pedicels usually short, glabrous. Flowers 4(–5)-merous. Sepals connate at the base, equal or subequal, subcircular, 1·2 times as long as wide, 1·2–1·4 × 1–1·2 mm., obtuse or subacute at the apex, glabrous on both sides, often very minutely ciliate, without colleters. Corolla in the mature bud (3·5)4·5–5·5 times as long as the calyx, (5·2)5·7–6·7 mm. long, slightly contracted somewhat beneath the throat, and tapering at the apex, white with often pale green tips at the lobes, outside glabrous or sometimes minutely papillose-pubescent, inside densely pilose to glabrous on the lobes and sometimes also in the tube; tube cylindrical or nearly so, (2)3–3·5 times as long as the calyx, (1·4)1·7–2 times as long as the lobes, (3)3·5–4·5 mm. long, at the throat 1–1·4(2) mm. wide; lobes oblong, 2·2–2·5 mm., acute, recurved. Stamens slightly exserted; filaments glabrous, usually short, inserted at the mouth of the corolla tube; anthers oblong, 0·7–1 × 0·3–0·7 mm., glabrous. Pistil glabrous, (3·5)5–6 mm. long; ovary ovoid, 1–1·5(2) × 0·7–1(1·2) mm., 2-celled; style slender, (1·5)5–6 mm. long; stigma small, capitate or obscurely bilobed. Each cell with 5–9 ovules. Fruit orange-yellow or yellowish, immature glaucous, small, soft, obovoid or ellipsoid, 1·5 × 1–2 × 1·5 cm., apiculate at the apex, often slightly obliquely pedicellate, 1-seeded, with a smooth skin. Wall thin. Seed flattened, obliquely ellipsoid, 13 × 8 × 4 to 16 × 11 × 6 mm., at one side with a deep pit at the other with a bulge surrounded by a shallow groove, seemingly papillose; false papillae simulated by curved hairs.

Zambia. B: Sikongo For. Res., Kalabo District, *F. White* 2080 (BM; BR; FHO; K; PRE). N: Kamwedzi, Mfumbwe, Dongwe, Kasempa District, fr. vi, *Holmes* 1099 (FHO; K). S: Samau, near Livingstone (?), *J. D. Martin* 738 (FHO).

Also widespread in tropical Africa. Gallery forests, groves on granitic rocks, or in woodland; alt. 0–700 m.

5. **Strychnos henningsii** Gilg in Engl., Bot. Jahrb. **17**: 569 (1893). — Bruce & Lewis in F.T.E.A., Loganiaceae: 32 (1960). — Verdoorn in Fl. S. Afr. **26**: 140, f. 18. 3 (1963). — Leeuwenberg in Meded. Land. Wag. **69**–1: 126, fig. 18; map 19 (1969). TAB. **84**. Type from S. Africa (Cape Province).
    S. holstii Gilg in Engl., Abh. Preuss. Akad. Wiss. **36**: 36 (1894); in Engl., Pflanzenw. Ost-Afr. **C**: 310 (1895). — Baker in F.T.A. **4**, 1: (1903). Type from Tanzania.
    S. sennensis Bak. in Kew Bull. **1895**: 97 (1895) and in F.T.A. **4**, 1: 529 (1903). Type from Mozambique; Valley of Zambesi R., opposite Senna, *Kirk* s.n. (K, holotype).
    S. pauciflora Gilg in Engl., Bot. Jahrb. **28**: 121 (1899). — Prain & Cummins in Fl. Cap. **4**, 1: 1053 (1909). Type from Mozambique: Maputo, *Schlechter* 11682 (holotype B†; BM lectotype G; K; Z isotypes).
    S. procera Gilg & Busse in Engl., Bot. Jahrb. **36**: 97, fig. 1 (1905). Type from Tanzania.

Shrub or small tree, 2–10(20) m. high with spreading rounded crown. Trunk 40–50 cm. in diam. (more or less); bark pale grey or pale brown, rough. Branches pale grey or pale brown, sometimes shallowly fissured, not lenticellate; branchlets conspicuously sulcate when dry, glabrous. Leaves petiole short, glabrous 1–3 mm. long; lamina pale to dark to dark green and shining above, less shining and paler beneath, coriaceous, very variable in shape and size even in a single branchlet, elliptic, oblong, narrowly elliptic, or ovate, (1·2)1·5–3(3·5) × as long wide, (1·5)2–6(10) × (0·6)1–3(6) cm., rounded to acuminate at the apex, cuneate, rounded, or sometimes on main axis subcordate (and then comparatively wider); glabrous on both sides; one pair of secondary veins from or from above the base curved along the margin and one rarely two faint submarginal pairs; tertiary venation reticulate, prominent on both sides, especially in thick leaves. Inflorescence axillary and sometimes also terminal, much shorter than the leaves, 1 × 1–2 × 2 cm., congested, few- or many-flowered. Peduncle often very short, sparsely pubescent to glabrous like the branches and pedicels. Bracts small, upper sepal-like, lower larger, sparsely pubescent to glabrous beneath, often with colleters in the axils. Flowers fragrant, 5-merous, sessile or subsessile. Sepals pale green, connate at the base, broadly orbicular or nearly so, 1–1·4 × 1–1·4 mm., rounded or obtuse at the apex, minutely ciliate, glabrous on both sides, without colleters. Corolla in the mature bud 2·3–3 × as long as the calyx, 2·8–4 mm. long, greenish-yellow, creamy, or white, subrotate and 4–5 mm. in diam. when open, glabrous outside, inside pilose or villose at the base of the lobes or sometimes entirely glabrous; tube short, 0·7–1·2 × as long as the calyx, 0·8–1·5 mm. wide; lobes thick, triangular to ovate, 1·7–2·5 × as long as the tube, 1·7–2 × as long as wide, 2–2·5 × 1–1·6 mm., acute, spreading. Stamens just exserted; inserted at the mouth of the corolla tube; anthers elliptic, 0·8–1 × 0·5–0·8 mm., glabrous. Pistil glabrous, 1·6–2·2(3) mm. long; ovary globose, depressed-globose, or sometimes ovoid, 0·8–1·2(1·5) × 0·9–1·4 mm., rounded or sometimes acuminate at the apex, 2-celled; style 0·7–1 mm. long; stigma capitate. In each cell 8–12 ovules. Fruit yellow, orange, or red, ellipsoid, 1 × 0·8–2 × 1·5 cm., one-seeded. Wall thin. Seed pale brown, ellipsoid, not flattened, 0·8 × 0·5 × 0·5–1·2 × 0·7 × 0·7 cm., glabrous, smooth, with a deep closed groove at one side (like coffee-bean), very minutely fovelate.

Zambia. N: Molwe, Mbale [Abercorn] District, *Glover in Bredo* 6423 (BR; EA; K; LISC; P; WAG). W: Chingola, fl. xii, *Fanshawe* 2405 (BR; EA; K). **Zimbabwe.** E: Mangazi R. Valley, Melsetter District, iii, *Goldsmith* 82/62 (K; LISC; PRE; SRGH). S: near Chibilia Falls, Sabi R., Ndanga District, *Chase* 2350 (BM; BR; COI; K; LISC; SRGH); Sabi-Lundi Rs. Junction, Ndanga District, *Wild* 3363 (BR; K; LISJC; PRE; S; UC). **Malawi.** S: Tangazi R. Valley, Chiromo, *Topham* 692 (FHO). **Mozambique.** N: Meconta, *Torre* 1017 (COI; LISC). Z: Namagoa, Mocuba District, viii, *Faulkner* 55 (COI; EA; K; PRE). MS: Inhamitanga, fl. vii, *Simão* 62 (LISC), 1336 (K; PRE). GI: Near Chidenguele, fl. viii, *Pedro & Pedrogão* 1818 (PRE). M: Maputo, fl. xii, *Schlechter* 11682 (BM; G; GRA; K; Z, type of *S. pauciflora*).

Also in Zaire, Angola, East Africa and South Africa, and Madagascar. Woodlands or sometimes light forests; alt. 0–2000 m.

6. **Strychnos innocua** Del., Cent. Pl. Méroë: 53 (1826) and in Calliaud, Voyage 'à Méroé **4**: 343 (1827). — Bruce & Lewis in F.T.E.A., Loganiaceae: 25 (1960), *innocua* pro parte quoad subsp. *innocua*. — Onochie & Leeuwenberg in F.W.T.A. 2nd ed. **2**: 496 (1963. — Leeuwenberg in Meded. Land. Wag. **69**–1: 138, fig. 20, phot. 3, map 21 (1969). Type from Ethiopia.

S. unguacha A. Rich., Voy. Abyss. Bot. Atlas: t. 73 (1847); Tent. Fl. Abyss. **2**: 52 (1851). — Baker in F.T.A. **4**, 1: 534 (1903). The holotype of this is the neotype of the preceding name.

Deciduous shrub or small often much branched tree, 2–12(18) m. high. Trunk 7–40 cm. in diam., branched from low down. Bark pale grey, grey brown, or sometimes dark grey-brown, smooth, somewhat powdery, flaking in small rounded or square scales near base of trunk; wood hard, pale yellowish. Branches pale grey-brown, powdery or not, lenticellate or not, not sulcate, terete; branchlets glabrous or pubescent. Leaves: petiole often short, glabrous or pubescent, 2–7 mm. long; lamina mat or dull, glaucous, and with mostly pale green reticulate veins on both sides, beneath slightly paler, coriaceous (living and dry) or papyraceous (dry), elliptic, narrowly elliptic, obovate, or narrowly obovate, (1)1·5–3(3·5) × as long as wide, (2)4–10(20) × (1)2–7(13·5) cm., rounded (or in sucker shoots often acute) at the apex, cuneate or less often rounded at the base, glabrous or pubescent on both sides; one or two pairs of distinct secondary veins from or from above the base curved along the margin and often a faint submarginal pair; tertiary venation reticulate and distinctly prominent on both sides. Inflorescences axillary or ramiflorous, usually several together, very short and nearly fasciculate, 1 × 1–1·5 × 1·5 cm., few-flowered. Peduncle, branches, and pedicels short or very short. Flowers 4-merous. Sepals pale green, free, subequal, the inner slightly smaller, ovate, broadly ovate, or suborbicular, 0·8–1·5 × as long as wide, 1·7–3·5 × 1·5–2·5 mm., rounded at the apex, ciliate, glabrous or pubescent outside, glabrous inside, without colleters. Corolla in the mature bud 2·2–4 × as long as the calyx and (6)6·5–9(10·5) mm. long, creamy or greenish-yellow, glabrous outside, inside with a brush-like ring of white lanate hairs in the throat and just on the bae of the lobes; tube cylindrical or nearly so, 1·6–2·5 × as long as the calyx, 1–1·7 × as long as the lobes, (3)3·5–5·5(6) mm. long, 1·5–2·5(3) mm., wide at the throat; lobes thick, narrowly triangular 1·7–2·3 × as long as wide, 3–4(4·5) × 1·3–2(2·3) mm., acute or subacute, spreading. Stamens hardly exserted; filaments extremely short, glabrous, inserted at the mouth of the corolla tube; anthers oblong, about twice as long as wide, 1·2–2 × 0·6–1 mm., glabrous. Pistil hirto-pilose in the middle, (4)5–7·5 mm. long; ovary narrowly ovoid or oblong, 1·5–3 × 1–1·5 mm., hirto-pilose at the very apex, further glabrous, often with a disk-like base, 2-celled; style thick, (2)3–4·5 mm. long, at the base hairy like the ovary at the apex; stigma capitate. In each cell (14)18–30 ovules. Fruit orange or yellow, nearly mature bluish-green, large, hard, globose, (2·5)4–7·5(9·5) cm. in diam., with (3–8–50 seeds, with somewhat granular skin, slightly shining. Wall thick, 2·5–5(6) mm. thick, thicker above the pedicel, brittle in mature fruits, hard and not broken by hand when nearly mature and/or dry. Pulp orange, edible. Seeds pale ochraceous, flattened or not, often more or less plano-convex, obliquely ovate, elliptic or tetrahedral, usually irregularly curved, 1–1·5 × as long as wide, 17–21 × 13–20 × 5–8 mm., with thick very short erect hairs, rather rough.

Zambia. B: Nega Nega Hills, *Burtt-Davy* 20788 (FHO). N: Nsisi, near Mbala [Abercorn], *R. E. Fries* 1324 (UPS). W: Ndola, x, Duff 232/37 (FHO; K), ibid., fl. fr. xi, *Miller* 205 (FHO), 298 (FHO; NY), ibid., fl. fr. ix, *Trapnell* 1994 (BR; COI; EA; K; S; SRGH). **Zimbabwe.** N: Chipoli Mazoe District, *Moubray* SRGH 89321 (BR; K; LISC; S; SRGH). C: Salisbury, Spelonken, fl. 6.i.1966, *Dale* S.K.F. 160 (SRGH). S: Victoria, *Monro* annis 1909–1912 (BM). **Malawi.** N: 30 km. NE. of Rumphi, *Langdale-Brown* 97 (EA). C: Kasungu, fr. viii, *Brass* 17434 (K; MO; NY). **Mozambique.** Z: Namagoa Estate, Mocuba District, xi, *Faulkner* 214 (BR; EA; K; LD; LISC; PRE; S; SRGH; UPS).

Also in tropical Africa. Deciduous woodlands; alt. 0–1600 m.

7. **Strychnos kasengaensis** De Wild. in Bull. Jard. Bot. Brux. **5**: 46 (1915). — Bruce & Lewis in F.T.E.A., Loganiaceae: 30 (1960). — Leeuwenberg in Meded. Landb. Wag. **69**–1: 150, fig. 29, map 23 (1969). Type from Zaire.

Climbing shrub or liana, at least up to 12 m. long and climbing 6 m. high. Branches not lenticellate, medium brown when dry, pubescent; branchlets brown-pubescent, and like the branches not or hardly sulcate when dry. Tendrils paired. Leaves with a pubescent, 1–4 mm. long petiole; lamina coriaceous, not or slightly shining when dry, elliptic, narrowly elliptic, ovate, or narrowly ovate, 1·5–3 times as long as wide, 2·5–9 × 1·5–4 cm., on main axis and at the bases of the branches comparatively wider and often smaller, acuminate at the apex, cuneate, rounded, or — if ovate — often subcordate at the base, minutely pubescent on main veins above, sparsely pubescent beneath, especially on the veins; one pair of secondary veins from or from somewhat above the base curved along the margin, and accompanied by a faint submarginal pair; tertiary venation reticulate, prominent beneath. Inflorescence axillary, congested, 1 × 1 to 2·5 × 2 cm., 2–4 times branched, several- to many-flowered, usually much shorter than the leaves. Flowers 5-merous. Sepals connate at the base, circular or nearly so, 1·2–1·5 mm. long, obtuse or rarely acute at the apex, ciliate, mostly sparsely pubescent outside, glabrous inside, without colleters. Corolla in the mature bud (3·5)4·7–6 times as long as the calyx, (4)6–7 mm. long, and more or less tapering at the apex, white, outside sparsely pubescent or glabrous, inside densely pilose in the tube, except for the glabrous base; tube (1·3)2·3–3·3 times as long as the calyx, (0·8)1–1·3 times as long as the lobes, (1·8)3–4 mm. long, widely cylindrical; lobes oblong, 2·5–3 times as long as wide (2·4)3–4 × 1–1·2 mm., acute, spreading. Stamens well-exserted; filaments as long as the anthers or slightly longer, up to 1·5 mm. long, glabrous, inserted at the mouth of the corolla tube; anthers oblong, 1·2–1·5 × 0·6–0·9 mm., bearded with few curled hairs at the base. Pistil pilose, (4)5–7 mm. long; ovary ovoid, 2–2·5 × 1·2–1·5 mm., glabrous at the very base, 2-celled; style (1·5)3–4·5 mm. long, only at the base pilose; stigma capitate. In each cell 15–20 ovules. Fruit bright yellow, rather small, rather soft, subglobose, about 3 cm. in diam., about 10–20-seeded, with somewhat granular skin, slightly shining. Wall 1 mm. thick when dry. Seed pale brown, flattened, more or less plano-convex, obliquely ovate, elliptic, or trullate, usually irregularly curved, 12–14 × 10–11 × 4–5 mm., with thick very short erect hairs, rough.

Zambia. Z: Mbala (Abercorn) District, *Gerstner* PRE 29240 (PRE); ibid., Kalambo Gorges Escarpment, *H. M. Richards* 19655 (K; WAG).

Also in Zaire, Burundi, Tanzania, Zambia. Riverine forests or in crevices of granitic rocks in woodlands; alt. 0–1500 m.

8. **Strychnos lucens** Bak. in Kew Bull. **1895**: 97 (1895); in F.T.A. **4**, 1: 524 (1903). — Bruce & Lewis, in F.T.E.A., Loganiaceae: 23, f. 4. 1–2 and 5. 1–3 (1960). — Leeuwenberg in Meded. Land. Wag. **69**–1: 157, fig. 22, map 25 (1969). Type from Angola.

  S. milneredheadii Duvign. et Staquet in Bull. Soc. Roy. Bot. Belg. **84**: 69 (1951). Type from Zambia: Matonchi R., Mwinilunga District, *Milne-Redhead* 2947 (BR, holotype, not seen; K; PRE, isotypes seen).

Climbing shrub or liana, 3–20 m. high climbing, up to at least 30 m. long. Trunk 2–5 cm. in diam., or more; bark dark grey, lenticellate, pale grey-brown to black; branchlets glabrous or occasionally pubescent. Tendrils paired. Leaves: petiole glabrous or with scattered hairs, 2–7 mm. long; lamina shining and dark green above, paler and mat or slightly shining beneath, often thickly coriaceous (thinner in the shade) also when fresh, elliptic, narrowly elliptic, or narrowly ovate, 1·5–3 times as long as wide, extremely variable in size, (2)4–10(16) × (1)2·5–5(8·5) cm., acute, shortly acuminate, or sometimes rounded or subtruncate at the apex, rounded or cuneate at the base; glabrous or rarely with few hairs near the base beneath; one pair of secondary veins from above the base curved along the margin and often also with a faint submarginal pair; tertiary venation reticulate, prominent on both sides in dry leaves. Inflorescence axillary, 3–20-flowered, rather congested or lax, pubescent with brown hairs or sometimes glabrous, shortly pedunculate, 1·5–5 × 1·5–4 cm. Flowers 5-merous, sessile or shortly pedicellate. Sepals green, connate at the very base, usually subequal, ovate or broadly ovate, 1·2–1·5 times as long as wide, (1·7)2–3 × 1·5–2·2 mm., rounded or sometimes obtuse at the apex, ciliate, glabrous on both sides, without colleters. Corolla in the mature bud 2·2–2·5 times as long as the calyx, 5·5–7·5 mm. long, and rounded at the apex, pale green-yellow, glabrous outside, inside with a white densely sericeous-villose ring in the throat; tube nearly cylindrical, 1·2–1·4 times as long as the calyx, 0·9–1·5 times as long as the lobes, at the base 1·5–1·8 mm., at the throat 1·5–2·5 mm. wide; lobes thicker towards the apex, oblong, 2–3·3 × as long as wide, 2·4 × 1–1·2 mm., acute at the apex, spreading. Stamens slightly exserted; filaments glabrous, very short, inserted at the mouth of the corolla tube; anthers oblong, 1–2 × 0·5–0·6 mm., sagittate at the base, glabrous; cells parallel. Pistil pilose or hirto-pilose in the middle, 4·5–6 mm. long; ovary ovoid, 1·5–2·5 × 1–1·5 mm., glabrous at the base, usually gradually narrowed into the style, 2-celled; style glabrous at the apex, 3–4 mm. long; stigma capitate. In each cell 15–25(30) ovules. Fruit orange or yellow, nearly mature glaucous, small or rather large, globose, 1·5–7 cm. in diam., hard when not small, (1)10–30(45)-seeded. Wall 1–3 mm. thick when dry. Pulp orange or yellow, slimy. Seed pale brown or ochraceous, obliquely elliptic, ovate, ellipsoid, or more or less tetrahedral, 1 × 7·5 × 4–21 × 14 × 5 mm., mostly somewhat thickened at one side in the middle, rough, minutely scabrid-pubescent.

Zambia. N: Muchinga escarpment, Mpika District, xi, *F. White* 3786 (BM; BR; FHO; K). W: Ndola, fl., fr. x, *Fanshawe* 620 (BR; EA; K; LISC), 1641 (BR; EA; K; SRGH). C: Kabwe (Broken Hill), fr. ix, *Mutimushi* 953 (K; P). S: Chibila R., *Trapnell* 1120 (K). **Zimbabwe.** E: Chisenga For. Res., fl. fr. xii, *Armitage* 34/55 (K; SRGH), 174/55 (FHO; K; LISC; P; PRE; SRGH); near Drumfad, Umtali District, x–xi, *Chase* 4686 (BM; BR; COI; LISC; MO; PRE; SRGH; UPS). **Malawi.** N: Nyika Plateau, *Robinson* 3094 (BR; K; M; PRE; SRGH). S: Mulanje Mt., *Chapman* 377 bis (FHO; K; MO). **Mozambique.** Z: Milange, fr. i, *Correia* 454 (LISC). MS: Chimoio (Vila Pery), *Esselen* PRE 29239 (PRE).

  Also in Zaire, Angola, and East Africa. Riverine forests, gallery forests, or on rocky hills in bushlands or woodlands; alt. 0–1700 m.

9. **Strychnos madagascariensis** Poir. in Lam., Encycl. Méth., Bot. **8**: 696 (1808). — Leeuwenberg in Meded. Landb. Wag. **69**–1: 160, fig. 23: map 26 (1969). TAB. **85**. Type from Madagascar.

  S. dysophylla Benth. in Journ. Linn. Soc. **1**: 103 (1856). — Baker in F.T.A. **4**, 1: 533 (1903). — Bruce & Lewis in Kew Bull. **11**: 273 (1956); in F.T.E.A., Loganiaceae: 27 (1960). Type from Mozambique: Delagoa Bay, *Forbes* 62 (K, holotype; P, isotype).

  S. randiaeformis Baill. in Bull. Mens. Soc. Linn. Paris **1**: 246 (1880). The holotype of this synonym is the isotype of the preceding.

  S. quaqua Gilg in Engl., Bot. Jahrb. **17**: 567 (1893) and **32**: 176 (1902) and **36**: 101 (1905). — Bak. in F.T.A. **4**, 1: 531 (1903). Type from Mozambique: Quelimane, *Stuhlmann* 1041 (B holotype †, photographs seen in FHO and K; HBG, lectotype).

  S. burtoni Bak. in Kew Bull. **1895**: 98 (1895); in F.T.A. **4**, 1: 533 (1903). Type from Mozambique: Manica e Sofala, Chupanga, *Kirk* 368 (K, lectotype, designated by Bruce & Lewis).

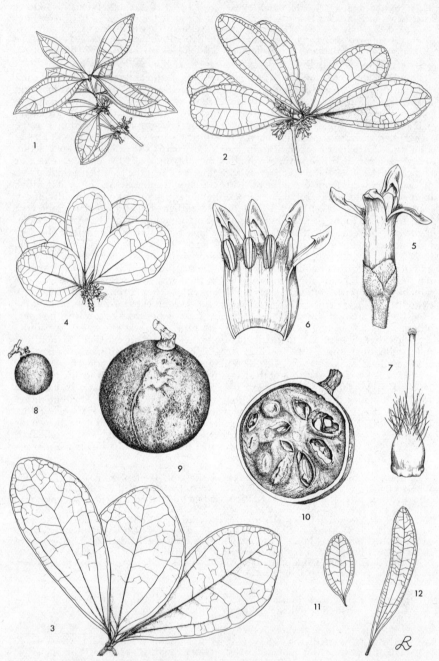

Tab. 85. STRYCHNOS MADAGASCARIENSIS. 1–4 flowering branches (× ½), 1 from *Capuron* 892, 2
Barbosa 2596, 3 *Ward* 4046; 5, flower (× 4); 6, opened corolla with stamens (× 5); 7, pistil
(× 8), 4–7 from *Mendonça* 3048; 8–9 fruit (× ½), 8 from *Peter* 13345, 9 from *Haerdi* 25; 10,
dried slice of nearly mature fruit, testae adhering to the pulp (× ½), from *Ward* 4092; 11–12
leaves (× ½), 11 from *Welch* 414, 12 from *B. D. Burtt* 5366.

S. innocua subsp. *burtonii* (Bak.) Bruce & Lewis in Kew Bull. **11**: 272 (1956); in
F.T.E.A., Loganiaceae: 26 (1960). — Verdoorn in Fl. S. Afr. **26**: 145 (1963). —
Leeuwenberg, Act. Bot. Neerl. **14**: 219 (1965). Lectotype as above.

Often many-stemmed and much-branched shrub or small tree, 1·50–10(20) m.
high, deciduous. Trunk 7–60 cm. in diam., mostly branched from low down. Bark
mostly pale grey or greyish-white, smooth; wood whitish. Branches pale grey to dark
brown, powdery or not, lenticellate or not, smooth; branchlets glabrous or
pubescent, terete. Leaves: petiole glabrous or pubescent, often short, 1–5 mm. long;
lamina shining and dark green above, paler beneath, coriaceous, very variable in
shape and size, suborbicular, elliptic, narrowly elliptic, obovate or narrowly obovate,
(1)1·5–4(5) times as long as wide, 2–10(15·5) × 1–4(5·5) cm., rounded, or especially
in Madagascar and near the coast in Moçambique and Natal and furthermore
elsewhere in shade branches acute or acuminate at the apex, cuneate or rounded at
the base, glabrous or pubescent on both sides; one or two pairs of distinct secondary
veins from or from above the base curved along the margin and often a faint
submarginal pair; tertiary venation reticulate, not or slightly prominent above.
Inflorescences axillary or ramiflorous, usually several together, very short and nearly
fasciculate, 1 × 1–2 × 1·5 cm., 1–2 × branched, few-flowered. Peduncle, branches,
and pedicels short or very short. Flowers 4-merous. Sepals pale green, free or nearly
so, subequal, the inner slightly smaller, ovate, broadly ovate or suborbicular, 1–1·6
times as long as wide, 1·7–2·5(3) × 1·5–2·2 mm., rounded or obtuse at the apex,
ciliate, glabrous or with some minute appressed hairs at the base inside, without
colleters. Corolla in the mature bud (1·8)2·8–3·3(3·6) times as long as the calyx,
(4·5)5–8 mm. long, white or greenish-yellow, glabrous outside, inside with a brush-
like ring of white lanate hairs in the throat and just on the base of the lobes; tube
cylindrical or nearly so (1)1·2–1·8(2·2) times as long as the calyx, 0·7–1·7 × as long as
the lobes, 2·5–4·5(5) mm. long, 1·5–2·8 mm. wide at the throat; lobes thick, narrowly
triangular, 1·5–3 times as long as wide, 2–3·6 × 1–1·9 mm., acute or subacute,
spreading. Stamens hardly exserted; filaments extremely short, inserted at the
mouth of the corolla tube; anthers oblong, about twice as long as wide,
1·2–1·8 × 0·6–0·8 mm., glabrous. Pistil hirto-pilose in the middle, (3·5)4–6 mm.
long; ovary narrowly ovoid or oblong, 1·5–3 × 0·8–1·5 mm., hirto-pilose at the very
apex, further glabrous, often with a disc-like base, gradually narrowed into the style,
2-celled; style thick (2)2·5–4 mm. long, at the base hairy like the ovary at the apex;
stigma capitate. In each cell 5–30 ovules. Fruit orange or yellow, nearly mature
bluish-green, large or sometimes rather small, hard when not small, globose, 2–8(10)
cm. in diam., with about 5–50 seeds, with somewhat granular skin, slightly shining.
Wall mostly thick, (1)2–4 mm. thick, thicker above the pedicel, brittle in mature
fruits, hard and not broken by hand when nearly mature and/or dry. Pulp orange,
slimy, edible. Seeds pale ochraceous, flattened or not, often more or less plano-
convex, obliquely ovate, elliptic, or tetrahedral, usually irregularly curved, 1·4–1·8
times as long as wide, 11–25 × 6–18 × 5–8 mm., with thick very short erect hairs,
rather rough.

Botswana. N: Chobe R., *van Son* TRV 28860 (BM; F; K; PRE). **Zambia.** B: Sasheke, *F.
White* 1990 (BR; FHO; K). S: Gwembe, *Bainbridge* 201/55 (FHO; K). **Zimbabwe.** N: Mtoko
Distr., Suskwe, fl. 4.xii.1953, *Phelps* 82 (SRGH). W: Matobo (July), *Hodgson* 8/52 (K; MO;
PRE; SRGH). C: Hartley Distr., Gowe, *Whellan* 429 (SRGH). E: Hot Springs, Melsetter
District (Nov), *Chase* 4705 (BM; COI; LISC; MO; PRE; SRGH). S: Ft. Victoria District,
Wild 3016 (BR; K; PRE; S). **Malawi.** N: Karonga District, *Jackson* 1259 (BR; FHO; K). C:
Lake Coast, near Salima, *Burtt* 6075 (BM; BR; K). S: Mulange Mt., *Chapman* 379 (BM; BR;
FHO; K; PRE). **Mozambique.** N: between Quissanga and Ingoane, *Barbosa* 2050 (LISC;
PRE). Z: km. 55 of Alto Molócuè-Gilé road, *Barbosa & Carvalho* 4440 (K; SRGH). T: Sisitso,
Zumbo area, *Chase* 2778 (BM). MS: Dombe, Luite R., fl. x, *Gomes Pedro* 4434 (BR; K; PRE).
GI: between Mabalane and Mabote, *Barbosa & Lemos* 8635 (K; LISC; PRE). M: Catembe, xii,
Schlechter 11615 (BM; BOL; BR; COI; E; EA; G; GRO; HBG; K; LE; P; PR; Z).
Also in East and south Africa, and in Madagascar.

10. **Strychnos matopensis** S. Moore in Journ. Bot. **43**: 48 (1905). — Bruce & Lewis in
F.T.E.A., Loganiaceae: 21 (1960). — Leeuwenberg in Meded. Landb. Wag. **69**–1: 177,
fig. 25, map 27 (1969). Type from Zimbabwe: Matobo Hills, *Eyles* 1182 (BM, holotype;
SRGH, isotype).

Climber or scandent shrub, about 4 m. high, forming a dense tangle. Branches brown or grey-brown, more or less terete, not sulcate when dry, not or hardly lenticellate, often hairy like the branchlets; branchlets with a pale grey- or brown appressed pubescent indument. Tendrils solitary in the axils of ordinary leaves. Leaves with an appressed-pubescent, short, 1–3(4) mm. long petiole; lamina shining and dark green above, paler and less shining beneath, coriaceous, ovate, orbicular, or sometimes elliptic, 1–2(3) times as long as wide, 1·8–5(7) × 1–3(5) cm., sub-acuminate, acute, or rarely rounded and mucronulate at the apex, rounded to cuneate at the base, above glabrous or sparingly appressed-pubescent on the impressed costa or less often glabrous, often minutely ciliate; with one or two pairs of secondary veins from or from above the base curved along the margin and often with a faint submarginal pair; tertiary venation inconspicuous. Inflorescence terminal and (occasionally) also axillary, congested, many-flowered, 0·7 × 1–2 × 2·5 cm. Flowers 5-merous, sessile. Sepals one-third connate, equal or subequal, ovate or broadly ovate, 1·3–1·8 × 1–1·4 mm., acute at the apex, ciliate, outside appressed-pubescent, especially at the base, inside glabrous, without colleters. Corolla in the mature bud 2·4–2·5 × as long as the calyx, 3·3–4·2 mm. long, white or pale yellow, campanulate, outside glabrous or puberulous, inside white-penicillate in the throat; tube wide, 1·3–1·5 times as long as the calyx, 1–1·5 times as long as the lobes, 2–2·2 × 2–3 mm.; lobes triangular, 1·3–1·7 × as long as wide, 1·3–2 × 1–1·2 mm., thick, acute, suberect. Stamens included; filaments very short, 0·7 times as long as the anthers, glabrous, inserted at one-half to two-thirds from the base of the corolla tube; anthers nearly cordate, 0·9–1·2 × 0·6–0·8 mm., bearded at the base. Pistil glabrous, 2–2·8 mm. long; ovary broadly ovoid, 0·8–1·2 times as long as wide, 1–1·2 × 1–1·2 mm., 2-celled; style 1–1·8 mm. long, rather thick; stigma capitate. Fruit orange or yellow, soft, ellipsoid, slightly longer than wide, 12 × 9–15 × 12 mm., 1(2)-seeded, with smooth skin, slightly shining. Wall thin. Seed dark brown, flattened, more or less plano-convex, elliptic, 9–10 × 7–9 × 2 mm., with thick very short erect hairs, rather rough.

Zambia. N: Kawambwa, xi, *Fanshawe* 4058 (BR; FHO; K). **Zimbabwe.** W: Matobo Hills, xi, *Eyles* 1182 (BM; SRGH, type). W: Umzingwane District, x, *Davies* 2587 (M; SRGH). **Mozambique.** T: Chicoa, Song o Hills, xii, *Torre & Correia* 13972 (LISC).
Also in East Africa. Semi-evergreen bushland or Brachystegia-woodland, on rocky hills, termite-mounds and gallery forest; alt. 900–1600 m.

11. **Strychnos mellodora** S. Moore in Journ. — Leeuwenberg in Meded Landb. Wag. **69**–1: 180, fig. 26, map 29 (1969). Linn. Soc. **40**: 147 (1911). — Bruce & Lewis in F.T.E.A., Loganiaceae: 23 (1960). Type from Zimbabwe: Chirinda Forest, *Swynnerton* 101 (BM, holotype; K; NBG; SRGH; US; Z, isotypes).

Tree, 20–35 m. high. Trunk c. 45 cm. in diam. Bark thin. Branches not or obscurely lenticellate, medium to dark brown and not sulcate when dry; branchlets glabrous, usually slightly sulcate when dry, mostly darker than in *S. mitis*. Leaves with glabrous, 3–8 mm. long petioles, those of a pair joined at the base by a distinct stipular line; lamina dull, drying pale greenish-brown with dark brown, paler beneath, coriaceous, elliptic or narrowly elliptic, 2–3·5 times as long as wide, 5–12 × 2–5 cm., acuminate at the apex, cuneate at the base, glabrous on both sides; secondary veins 5–8 on each side; tertiary venation not very conspicuous, slightly prominent on both sides. Inflorescence axillary, usually also in the axils of the apical leaves and therefore seemingly terminal, lax, many-flowered, shorter or longer than the leaves, 3–11 × 2·5–8 cm., 3–4 times branched. Flowers 4-merous. Sepals free, equal, broadly ovate, about 1–1·5 times as long as wide, 0·7–1·2 × 0·7–1 mm., rounded at the apex, ciliate, outside glabrous or nearly so, inside glabrous, without colleters. Corolla in the mature bud 2·5–4·5 × as long as the calyx 2·5–3 mm. long, and rounded at the apex, white, outside glabrous, inside densely pilose on the lobes (of which the apex sometimes glabrous); tube very short, 0·5–0·7 × as long as the calyx, 0·5–0·7 mm. long; lobes oblong, 4–5 × as long as the tube, 1·7–2 × as long as wide, 2–2·5 × 1·1–1·5 mm., acute, spreading. Stamens exserted; filaments about 1–1·5 × as long as the anthers, inserted at the mouth of the corolla tube; anthers orbicular or nearly so, 0·5–0·6 × 0·5 mm., glabrous. Pistil glabrous, 1·1–1·2 mm. long; ovary depressed-globose or -ovoid, 0·7 × 0·7 mm., 1-celled; style short, 0·4–0·5 mm. long; stigma capitate. One basal placenta with 2–9 ovules. Fruit small, globose or nearly so, 12 × 10–18 × 18 mm., one-seeded. Wall thin. Seed flattened, ellipsoid, 8–9 × 6–7 × 3 mm., pubescent (?). Testa thin sticking to the pulp.

Zimbabwe. E: Chirinda Forest, x, *Swynnerton* 101 (BM; K; NBG; SRGH; US; Z, type).
Mozambique. MS: Chimoio, Garuso, xi, *Simão* 637 (LISC).
Also in Tanzania. Montane rain forests; alt. 800–1200 m.

12. **Strychnos mitis** S. Moore in Journ. Linn. Soc. **40**: 146 (1911). — Bruce & Lewis in
F.T.E.A., Loganiaceae: 21 (1960). — Verdoorn in Fl. S. Afr. **26**: 141, fig. 17, 2 (1963). —
Leeuwenberg in Meded. Landb. Wag. **69**–1: 190, fig. 26; map 30 (1969). Type from
Zimbabwe: Chirinda forest, *Swynnerton.*17a (BM, lectotype; K, Z isotypes).
  S. adolphi-frederici Gilg in Mildbraed, Wiss. Ergebn. Deutsch. Zentr.-Afr. Exped.
1907–1908, **2**: 531, t. 74 F/J (1914). TAB. **86**. Type from Zaire.

Evergreen tree with a rounded crown, 6–35(40) m. high. Trunk 20–100 cm. in
diam.; bark smooth, grey, grey-green, or grey-brown; wood whitish, hard. Branches
pale grey or buff, mostly paler than in *S. mellodora*, lenticellate, ascending, with a
fissured bark; branchlets glabrous or sometimes pubescent, sulcate when dry. Leaves
with a 2–5 mm. long, glabrous or sometimes (if branchlets hairy) pubescent petiole;
lamina shining and dark green above, paler beneath, coriaceous (not thick), narrowly
elliptic, oblong, or sometimes ovate or narrowly ovate, 1·7–3·5 times as long as wide,
4–11·5 × 1·5–5 cm., usually comparatively narrower towards the apices of the
branchlets, mostly acuminate, more rarely acute or obtuse at the apex, cuneate or
rounded at the base or decurrent into the petiole, glabrous or sometimes pubescent
on both sides, with one pair of distinct secondary veins from about 1 cm. above the
base and a faint submarginal pair from the base; tertiary venation spreading, not very
prominent. Inflorescence axillary and terminal, usually dense, much shorter than the
leaves, 1–3(4) × 1–2·5 cm. Peduncle pubescent, usually short. Flowers 4–5-merous
even in a single inflorescence, sessile. Sepals pale green, connate at the very base,
broadly ovate or orbicular, about as long as wide, 1·5–1·8 × 1·5–1·8 mm., obtusely
acute at the apex, ciliate, pubescent to glabrous outside, glabrous and with colleters
at the base inside. Corolla in the mature bud 2–2·7 times as long as the calyx, 3·5–4
mm. long, and rounded at the apex, creamy, yellow, or greenish, outside glabrous,
inside densely villose at the base of the lobes; tube campanulate, 0·9–1·3(1·5) × as
long as the calyx, 0·6–1·2 times as long as the lobes, 1·5–2·2 mm. long; lobes thick at
the apex, triangular to ovate, 1·1–1·7 times as long as wide, 1·5–2·5 × 1·1–1·8 mm.,
acute at the apex, often slightly widened above the base, spreading or sometimes
recurved. Stamens just exserted; filaments glabrous, short, inserted at two-thirds
from the base of the corolla tube; anthers cordate-circular when young, becoming
oblong after the pollen is shed, 1–1·2 × 0·6–0·7 mm., bearded at the base. Pistil
glabrous, 2·3–3 mm. long; ovary ovoid or ovoid-conical, about 1·5 times as long as
wide, 1–2 × 0·7–1·3 mm., 2-celled; style 0·8–1·5 mm. long; stigma capitate. In each
cell about 10–15 ovules. Fruit yellow or orange, small, subglobose, 1–2 cm. in diam.
Wall thin. Seeds 1(–2), pale ochraceous, subellipsoid, usually flattened at one side,
1–1·2 × 0·8–1 × 0·5–0·6 cm., smooth, glabrous, not grooved, very minutely
foveolate; hilum in the middle of the mostly flattened side. Testa thin.

Zimbabwe. C: Wedza Mt., *Wild* 6563 (FHO; K; SRGH). E: Chirinda Forest, Chipinga
District, *Goldsmith* 1/65 (WAG). S: Benga, Bikita District (ii) *O. West* 6327 (WAG).
Mozambique. MS: km. 15 of Mavita-Chamoio (Vila Pery) road, fr. vi, *Torre* 4348 (LISC).
Maputo, Gobo, fl. x, fr. xii, *Barbosa* 742 (LISC).
  Also in Angola, East Africa from the Sudan and Ethiopia to Kentani in South Africa, and the
Comoro Islands. Upland and lowland rain forests and gallery forests; alt. 0–2100 m.

13. **Strychnos myrtoides** Gilg et Busse in Engl., Bot. Jahrb. **32**: 178 c.tab. (indument in fig. C
incorrect) (1902) and **36**: 100 (1905). — Baker in F.T.A. **4**, 1: 531 (1903). — Bruce &
Lewis in F.T.E.A., Loganiaceae: 15 (1960). — Leeuwenberg in Meded. Landb. Wag.
69–1: 199, fig. 30, map 32 (1969). Type from Tanzania.

Much branched densely leafy shrub or small tree, 1–5 m. high. Branches pale
brown, not lenticellate, with the bark peeling off, not sulcate, often dichotomously
branched; branchlets slender, pubescent with ascending brown hairs, terete or
nearly so, not sulcate when dry. Leaves small or very small, subsessile or shortly
petiolate; petiole 0·2–1(2) mm. long; pubescent with brown ascending hairs, lamina
paler beneath, subcoriaceous, very variable in shape and size, ovate, narrowly ovate,
elliptic, narrowly elliptic, obovate, or sometimes suborbicular, (1)1·5–4(5) times as
long as wide, (4)6–20(27) × 3–8(11) mm. in the continent, rounded to acute or
broadly and bluntly acuminate at the apex, cuneate or rounded at the base, glabrous
or with ranked appressed short hairs on the costa above, beneath sparsely pubescent

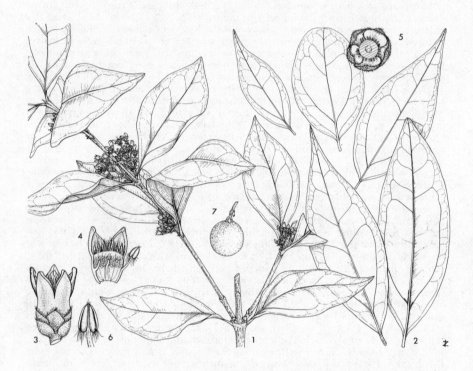

Tab. 86. 8–14 STRYCHNOS MITIS. 1, branch (×½) from *Eggeling* 6482; 2, leaves (×½), from *Barbosa* 742, *Fegen* PRF 2944, *Gossweiler* 13083 and *Swynnerton* 17; 3, flower (×4); 4, portion of corolla with stamen (×4); 5, calyx above (×5); 6, anther (×8), 4–6 from *Eggeling* 6482; 7, fruit (×½) from *Swynnerton* 17.

or with some hairs at the base and often also on the costa, often minutely ciliate above; with one pair of not very conspicuous secondary veins from or from above the base curved along the margin and often with a faint submarginal pair; margin more or less revolute, especially in old leaves. Inflorescence terminal and (occasionally) also axillary, lax or rather lax, few-(sometimes 1)-flowered, 5 × 3–15 × 10 mm., 0–2 × times branched. Peduncle and branches pubescent with brown ascending hairs, especially at the base. Flowers (4–)5-merous. Sepals connate at the base up to one-third of their length, equal or subequal, broadly ovate to suborbicular, 1–1·2 times as long as wide, 0·8–1·2 × 0·8–1·2 mm., obtuse to acute at the apex, distinctly ciliate, glabrous on both sides, without colleters. Corolla in the mature bud 2·8–3·3 times as long as the calyx, 2·2–4 mm. long, white, outside glabrous or minutely papillose-pubescent, inside white-penicillate in the throat; tube campanulate to nearly cylindrical, 1·2–1·9 times as long as the calyx, 0·8–1·4 times as long as the lobes, at the throat 1·2–1·5 mm. wide; lobes narrowly triangular, 1·5–2 times as long as wide, 1·2–1·8 × 0·8–1 mm., acute, erect or suberect. Stamens included; filaments very short, inserted on the middle of the corolla tube; anthers circular or oblong, 0·5–0·9 × 0·5–0·7 mm., ciliate at the base or sometimes all around. Pistil glabrous, very short, 0·9–1·4 mm. long; ovary ovoid, 0·8–1 × 0·6 mm., 2-celled; style very short, up to 0·6 mm. long, or none; stigma capitate. In each cell 2–10 ovules. Fruit red or orange, small, soft, globose or laterally compressed (?), about 8–9 mm. in diam., 1-seeded, with somewhat granular skin, slightly shining. Wall thin. Pulp red or orange. Seed ochraceous-brown, flattened, disk-like, elliptic, 8 × 7 × 1·5 mm., rough, with thick very short erect hairs, slightly dented at one side.

Mozambique. N: Palma, *Gomes Pedro* 5418 (EA). Z: between Mocuba and Namacurra, viii, *Barbosa & Carvalho* 3824 (K; SRGH). M: Maputo, *Sim* 20934 (PRE).
Also in South-eastern Tanzania and Madagascar. Brachystegia-woodlands or light forests; alt. 0–600 m.

14. **Strychnos panganensis** Gilg in Engl., Pflanzenw. Ost-Afr. **C**: 311 (1895): in Engl., Bot. Jahrb. **36**: 94 (1905). — Baker in F.T.A. **4**, 1: 526 (1903). — Bruce & Lewis in F.T.E.A., Loganiaceae: 28 (1960). — Leeuwenberg in Meded. Landb. Wag. **69**–1: 208, fig. 32, map 33 (1969). Type from Tanzania.
   S. guerkeana Gilg, loc. cit. (1895) tom. cit.: 521 (as *"guerckeana"*). — Duvign. in Bull. Soc. Roy. Bot. Belg. **85**: 27 (1952). Type from Tanzania.
   S. bicirrifera Dunkley in Kew Bull. **1935**: 263 (1935). — Duvign., loc. cit., fig. 8B. Type from Kenya.

Scandent or scrambling evergreen shrub or liana, 3–20 m. high. Branches lenticellate, twining, terete, pale to dark brown and not sulcate when dry; branchlets hirto-pubescent, terete, often slightly sulcate when dry. Tendrils paired. Leaves subsessile or shortly petiolate; petiole hirto-pubescent, up to 3 mm. long; lamina shining and dark green above, paler beneath, coriaceous also when fresh, variable in shape and size, ovate, narrowly ovate, elliptic, narrowly elliptic, cordate, or subcircular, comparatively wider on the main axis and at the bases of the branchlets, 0·75–3 times as long as wide, 8–50(67) × 6–30 mm., acute, or shortly acuminate, often apiculate, rarely obtuse at the apex, cordate, rounded, or cuneate at the base, glabrous, except for a few hairs on the costa beneath and sometimes above, with one or two pairs of secondary veins from or from just above the base curved along the margin, outer pair usually fainter; tertiary venation not very conspicuous, prominent on both sides in dry leaves; costa impressed above. Inflorescence terminal and at the same time axillary, rather lax, few- to many-flowered, 1·5 × 1·5–3 × 3 cm., 2–3 times branched. Peduncle, branches, and pedicels slender. Pedicels 2–4 mm. long. Flowers 5-merous. Sepals subequal, connate at the base, ovate or broadly ovate, 1–1·5 × as long as wide 1–1·5 × (0·6)0·8–1 mm., obtuse at the apex, ciliate, glabrous on both sides (or outside at the base occasionally with a single minute hair), without colleters. Corolla in the mature bud 3–4 × as long as the calyx, 3·5–5 mm. long, white, greenish, or creamy, outside glabrous, inside pilose in the throat and at the base of the lobes; tube campanulate, short, 0·8–1·2 × as long as the calyx, 1–1·7 mm. long; lobes thick 2–3 × as long as the tube, oblong, 2·3–3 × as long as wide, 2·5–3·5 × 1–1·4 mm., acute, recurved. Stamens well-exserted; filaments 1·2–2 mm. long, inserted at the mouth of the corolla tube; anthers oblong, 1·2–1·5 × 0·6–0·9 mm., glabrous or bearded with a single pilose hair at base. Pistil pilose in the middle, 3·5–4 mm. long; ovary ovoid, pilose at the apex, 1–1·5 × 0·9–1 mm., 2-celled; style thick, glabrous at the very apex, 2·3–3 mm. long; stigma capitate. In each cell 10–12 ovules. Fruit yellow, small, soft, subglobose, 8–18 mm. in diam., with (1)3–6 seeds, with a smooth skin, slightly shining, often mucronate by the style. Wall thin. Seed pale brown or pale ochraceous, flattened, more or less plano-convex, obliquely elliptic or subtetrahedral, about 1·2 × as long as wide, 7 × 5 × 1·5–10 × 7 × 3 mm., rough, with scabrid short hairs. Testa not sticking to the pulp.

Mozambique. MS: Macuti, 6 km. from Beira, fr. iii, *Wild & Leach* 5188 (K; PRE; SRGH). Z: Gobene Forest, 45 km. from Maganja da Costa, *Torre & Correia* 14626 (LISC). GI: Panda, fr. ii, *Exell, Mendonça & Wild* 610 (BM; LISC; SRGH).; Manjacaze, fr. iii, *Torre* 2546 (LISC).

Also in East Africa and Madagascar. Lowland rain forests, gallery forests, or thickets in coastal evergreen bushland; alt. 0–500 m.

15. **Strychnos potatorum** L. f., Suppl.: 148 (1781). — J. S. Gamble, Fl. Madras: 868 (1921). — H. H. Haines, Botany Bihar and Orissa **2**: 572 (1922), reprinted 1961. p. 592). — Leeuwenberg in Meded. Landb. Wag. **69**–1: 218, fig. 35, map 34 (1969). Type from India.
    S. stuhlmannii Gilg in Engl., Bot. Jahrb. **17**: 570 (1893). — Baker in F.T.A. **4**, 1: 529 (1903). — Bruce & Lewis in F.T.E.A., Loganiaceae: 33 (1960). — Verdoorn in Fl. S. Afr. **26**: 143, fig. 17. 3 (1963). Type from Mozambique: Tete Province, Zambesi R., opposite Chiramba (Shinamba), viii-1859, *Kirk* s.n. (K, lectotype).

Deciduous much branched small or medium sized tree or sometimes shrub, (2)4–18 m. high trunk 20–100 cm. d.b.h. Bark pale grey or grey-brown, smooth, lenticellate, thin, in section green on the periphery and orange or buff towards the centre; wood pale yellowish-brown, with bark-islets. Branches often repeatedly dichotomously branched, pale to very dark brown, distinctly lenticellate, terete, not or slightly sulcate when dry, with protruding persistent cup-like petiole-bases; branchlets glabrous. Apex of branchlet modified into a spine-like, 1–3 mm. long apex. Leaves with a short, glabrous, 1–7 mm. long petiole; lamina dark green above, mat or dull and paler beneath, papyraceous or thinly coriaceous, membranaceous when young (in flowering branches), variable in shape and size, elliptic, narrowly elliptic, ovate, or narrowly ovate, 1·5–3(3·5) times as long as wide, 6–15 × 3–9 cm., in the shade up to 19·5 × 11 cm., acute, acuminate, or sometimes, especially in young shoots, subcordate at the base, glabrous on both sides; two pairs of distinct mostly pale green secondary veins from or from above the base curved along the margin, not reaching the apex; tertiary venation mostly pale green, reticulate. Inflorescence in the axils of small scales at the bases of the branchlets, solitary, lax or rather congested, 1·5–2·5 × 1–2 × branched. Peduncle, branches, and pedicels slender, glabrous. Flowers (4–)5-merous, variable in shape and size, appearing before or with the young leaves. Sepals dark green, connate at the base, subequal, variable in shape and size, ovate, broadly ovate, or sometimes oblong (1)1·3–2 times as long as wide, 1–2·2(2·5) × 0·7–1·5 mm., acute or less often acuminate at the apex, not ciliate, glabrous on both sides, without colleters. Corolla in the mature bud (3)3·5–5(6) times as long as the calyx, (4)4·5–7·5 mm. long, white, creamy, or yellow, outside glabrous, inside pilose with white hairs on the base of the lobes and often also in the throat; tube very variable in length and width, (0·8)1–2·5(3·2) times as long as the calyx, 1–3·5 mm. long, at the throat 1·5–3 mm. wide; lobes 1–3 times as long as the tube, oblong, 2–4·3 × as long as wide, 3–4·5 × (0·8)1–2·2 mm., acute, spreading. Stamens exserted; filaments glabrous, longer in flowers with shorter corolla tube (like in *S. decussata*), 0·5–1·7 times as long as the anthers, 1–2·5 mm. long, inserted at the mouth of the corolla tube; anthers oblong, mostly about twice as long as wide, 1·1–2 × (0·3)0·6–1 mm., glabrous. Pistil glabrous, (3·5)4·4–6 mm. long; ovary ovoid or conical, 1·2–1·5 times as long as wide, 1–2 × 0·8–1·4 mm., 2-celled; style rather thick, (2·5)2·7–4·3 mm. long; stigma often small, capitate or occasionally obscurely bilobed. In each cell 5–13 ovules. Fruit blue-black, small, cherry-like, soft, globose, (10)15–25 mm. in diam., 1-seeded, with smooth skin, shining. Wall thin, dry about 0·3 mm. thick. Pulp purplish. Seed slightly glossy, pale brown, depressed-globose or ellipsoid, 10 × 9 × 6–13 × 12·5 × 9 mm., with an obscurely angular line all around, densely sericeous, smooth.

Botswana. N: Serondela, Chobe R., (fr. vii), *Robertson & Elffers* 58 (K; PRE; SRGH). **Zambia.** B: Nangweshi, Barotseland, *Codd* 7126 (K; PRE). W: Luanshya, vii, *Fanshawe* 1397 (BR; FHO; K). C: Mt. Makulu Research Station, 18 km. S. of Lusaka, *Angus* 1286 (FHO; K).

E: Lundazi District, *Feedy* 117 (K). S: Katombora, *Greenway & Trapnell* 7952 (BR; EA; FHO; K; PRE). **Zimbabwe.** N: Binga Hill, Sebungwe District, xi, *Phipps* 1417 (BR; K; PRE; SRGH). W: Victoria Falls, Wankie District, *Wild* 3091 (BR; K; SRGH). C: Lower Sabi R., Chibuwe R., *Wild* 2328 (BR; K; S; SRGH). E: Above railroad to Beira, Umtali District, *Chase* 554 (BM; BR; K; S; SRGH). S: E. Sabi, Sangwe Crossing, *Phipps* 51 (BR; COI; K; LISC; SRGH). **Malawi** N: near Rumphi Boma, *F. White* 2542 (FHO; K).

Also in East and northern South Africa and in India, Ceylon, and Burma. In gallery forest, in Brachystegia-woodland, semi-evergreen bushland, often on river banks, on banks of dry riverbeds, or on termitaries; alt. 0–1600 m.

16. **Strychnos pungens** Solered. in Engl. & Prantl, Nat. Pflanzenf. **4**, 2: 40 (1892); in Engl., Bot. Jahrb. **17**: 554 (1893). — Hiern in Cat. Afr. Pl. Welw. **1**: 704 (1898). — Gilg in Engl., Bot. Jahrb. **32**: 176 (1902). — Baker in F.T.A. **4**, 1: 530 (1903). — Bruce & Lewis in F.T.E.A., Loganiaceae: 24 (1960). — Verdoorn in Fl. S. Afr. **26**: 144 (1963). — Leeuwenberg in Meded. Landb. Wag. **69**–1: 224, fig. 36, map 35 (1969). Type from Tanzania.

Deciduous tree or shrub (0·30)2–8(16) m. high. Trunk 10–20 cm. in diam. or more. Bark grey or brown, rough, closely and shallowly reticulate, not corky, smooth and grey higher up or in young trees; inner bark yellow; wood yellowish, with large bark-islets. Branches pale or dark brown, conspicuously and densely lenticellate, not sulcate; branchlets glabrous or occasionally with few hairs. Leaves: petiole short, glabrous, 1–4 mm. long; lamina shining and dark green above, hardly or not paler and less shining beneath, coriaceous, rigid, elliptic, narrowly elliptic, obovate, or occasionally orbicular, (1)2–4(5) × as long as wide, (2)3–8(10) × 1–3·5(4·5) cm., acute or rounded and sharply pointed at the apex, cuneate or rounded at the base, glabrous on both sides or occasionally partly pubescent beneath; one pair of secondary veins from or from above the base curved along the margin and often a faint submarginal pair; tertiary venation reticulate. Inflorescences axillary or ramiflorous, mostly several together, usually very short and about 0·25 × as long as the leaves, 1 × 1–2 × 2(4 × 5) cm., usually congested and subsessile, rarely lax. Pedicels very short. Flowers 5-merous. Sepals green, nearly free, subequal, the inner slightly smaller, ovate, broadly ovate, or sometimes orbicular, 1–1·5 × as long as wide, 2–4 × 2–3 mm., acute, obtuse, or rounded at the mostly slightly keeled apex, conspicuously ciliate, glabrous on both sides or inside minutely appressed-pubescent at the base, without colleters. Corolla in the mature bud 2·4–3·2 × as long as the calyx, 7–9·5 mm. long, greenish-creamy, or -yellow, glabrous outside, inside with a brush-like ring of white lanate hairs in the throat and just on the base of the lobes; tube cylindrical or nearly so, 1·4–2 × as long as the calyx, 1–1·7 × as long as the lobes, 4–5·5 mm. long, 1·5–3 mm. wide at the base, 2–3·5 mm. at the throat; lobes thick, narrowly triangular, 2–2·7 × as long as wide, 3–4 × 1·3–2 mm., acute, spreading. Stamens slightly exserted; filaments 0·1–0·2 × as long as the anthers, inserted at the mouth of the corolla tube, glabrous; anthers oblong, 1–2 × 0·5–1 mm., glabrous. Pistil pilose in the middle, 5–7·5 mm. long; ovary ovoid or oblong, 1–3 × 0·7–2 mm., 2-celled, apically pilose, often with a disk-like base; style thick, 2·5–4·5 mm. long, at the base hairy like the ovary at the apex; stigma capitate. In each cell 25–60 ovules. Fruit orange or yellow, nearly mature bluish-green, large, hard, globose, 5–12(15) cm. in diam., with about 20–100 seeds, with somewhat granular skin, slightly shining. Wall thick, (2)4–6 mm. thick, thicker above pedicel, woody when dry. Pulp sweet-tasting. Seeds ochraceous, flattened, more or less plano-convex, obliquely ovate, elliptic, or trullate, usually irregularly curved, 20–24 × 12–7 × 5–10 mm., with thick very short erect hairs, rather rough.

Botswana. N: Ngamiland, *O. B. Miller* B/420 (PRE). SE: Kanye District, *Yalala* 171 (K; WAG). **Zambia.** N: Mbala (Abercorn), ix, *Bullock* 101 (BR; EA; K; S). W: Mwinilunga District, fl., et fr. ix, *F. White* 3355 (BM; BR; FHO; K). C: E. of Lusaka, xi, *Robinson* 5807 (EA, K). **Zimbabwe.** N: Sebungwe District, x, *Davies* 1565 (COI; SRGH), W: Near Victoria Falls, *Kirk* July 1860 (K). C: Salisbury Distr., Salisbury, fl. 26.xii.1924, *Eyles* 4544 (SRGH). **Malawi.** C: Kasungu, *Brass* 17432 (K; MO; NY; SRGH; US).

Also in Central, East and northern South Africa. Woodland, often with Brachystegia; alt. 0–2000 m.

17. **Strychnos spinosa** Lam., Tab. Encycl. Méth. Bot. **2**: 38 (1794). — Baker in F.T.A. **4**, 1: 536 (1903). pro parte excl. specim. Loandensis *Welwitsch* lecta — Hutchinson & Dalziel, F.W.T.A. **2**: 22 (1931). — Aubréville, Fl. Soud-Guin.: 438, pl. 96 1–4 (1950). — Bruce & Lewis in F.T.E.A., Loganiaceae: 17, fig. 3. 1–16 (1960), partly (excl. syn. *S. madagascariensis*). — Verdoorn in Fl. S. Afr. **26**: 147, fig. 17. 4, 18. 2 (1963). —

Leeuwenberg in Meded. Landb. Wag. **69**–1: 239, figs. 39, 40, phot. 4, map 38 (1969). Type from Madagascar.

S. tonga Gilg in Engl., Bot. Jahib. **17**: 575 (1893); in Engl., Pflanzenw. Ost-Afr. **C**: 311, t. 38 B/K (1895). — Baker, in tom. cit.: 527. Type: Mozambique, Quilimane, *Stuhlmann* 103 (B†; HBG, lectotype).

S. volkensii Gilg in Abh. Kön. Akhad. Wiss. Berlin **1894**: 25 (1894); in Engl., Pflanzenw. Ost-Afr. **C**: 311 (1895); in Notizbl. Bot. Gart. Berl. **1**: 76. (1895). — Hiern., Cat. Af. Pl. Welw. **1**: 702 (1898). — Type from Tanzania.

S. carvalhoi Gilg in Engl., Bot. Jahrb. **28**: 23 (1899). Type from Mozambique: between Mussoril and Cabeceira, *Carvalho* s.n. (1884)–'85 (B†; COI lectotype).

S. cardiophylla Gilg Busse in Engl., Bot. Jahrb. **36**: 110 (1905). Type from Tanzania.

S. cuneifolia Gilg Busse, tom. cit.: 109, f. 20. Type from Tanzania.

S. harmsii Gilg Busse, loc. cit. Type from Tanzania.

S. leiosepala Gilg Busse tom. cit.: 111. Type from Angola.

S. radiosperma Gilg Busse tom. cit.: 108, f. 2C. Type from Tanzania.

S. rhombifolia Gilg Busse tom. cit.: 107. Type from Sudan.

Deciduous shrub or small tree, (0·30)1–6(10) m. high, sometimes flowering on first year shoots (with small leaves) on old fire-cut stumps. Trunk 4–15 cm. in diam. (sometimes fluted), branched from below; bark pale to dark grey or brown, shallowly fissured, not or sometimes thinly corky, often scaly, sometimes blackened by bush fire, not lenticellate; wood hard, whitish, without bark-islets. Branches sometimes deeply ringed at nodes, grey or brown, not lenticellate, often with curved or straight spines; branchlets glabrous or pubescent, green or yellow-green when fresh, green or brown and often sulcate when dry, sometimes terminating in a straight spine. Leaves on main axis sometimes ternate; petiole 2–10 mm. long, glabrous or pubescent; lamina coriaceous also when living, young thinner and papyraceous when dry, very variable in shape and size, orbicular, elliptic, narrowly elliptic, ovate, or obovate, 1–2(3) × as long as wide, sometimes comparatively wider or narrower on main axis, usually comparatively wider on flowering branches, 1·4–9·5 × 1·2–7·5 cm., rarely larger, up to 13·5 × 7·5 cm., emarginate, rounded to acute and often apiculate, sometimes acuminate (mostly in trees along water courses), or rarely abruptly acuminate at the apex, cuneate, less often rounded, or sometimes on main axis subcordate at the base, often with hair pockets in the angles of the main veins beneath, glabrous or pubescent on both sides (in sucker shoots and seedlings of plants with few hairs more hairy, smaller and cordate); 1–3 pairs of distinct secondary veins curved from above the base curved along the margin. Inflorescence terminal, seemingly umbellate, congested or less often not, 1·5 × 1·5–5 × 3·5(6·5 × 7) cm., mostly many-flowered. Flowers 5-merous. Sepals pale green, connate at the base, narrowly triangular to linear, equal, subequal, or sometimes unequal, 2–6(10) × as long as wide, 1·5–6 × (0·3)0·5–1·2 mm., acuminate or subulate at the apex, minutely ciliate at the base or not, outside mostly sparsely pubescent at the base and glabrous at the apex or entirely glabrous, never with an even indumentum all over, inside glabrous and usually with some hairs at the base or sometimes minutely pubescent, without colleters. Corolla when mature 0·8–2·7 × as long as the calyx, 4–5(3·8–6) mm. long, pale green, greenish, or less often white, creamy, or yellow, glabrous or sparsely pubescent outside, inside with a narrow entire white-penicillate corona at the mouth of the tube; tube 1·2–2·7 × as long as the lobes, 2·5–3(2·2–4·4) × 2–3(3·5) mm., urceolate or campanulate, often somewhat contracted in the throat; lobes triangular, 1·2–1·8 × as long as wide, 1·2–2 × 1–1·5 mm., acute, erect or suberect. Stamens included; filaments glabrous, 0·7–1·5 × as long as the anthers, inserted at 0·4–0·8 mm. from the base of the corolla; anthers oblong or elliptic, 1·2–1·4 × 0·8–1 mm., ciliate with a villose or sometimes pilose indumentum all around by which they are coherent. Pistil pubescent, rarely obscurely so, glabrous just below the stigma, 1·8–2·4(3) mm. long; ovary ovoid, broadly ovoid, or globose, (1)1·2–1·6(2) × 1–1·3 mms., 1-celled; stigma subsessile, oblong. One large globose basal placenta with about 60–120 ovules. Fruit yellow or when nearly mature yellow-green or green, large, hard, resembling an orange or a grapefruit, globose, often slightly shining, with granular skin, 7–11(15) cm. in diam., with about 10–100 seeds, immature often slightly pear-shaped. Wall rather thick, hard, 2–5(8) mm., thick, thickened inside above the pedicel. Pulp yellow, edible. Seed ochraceous in fruit, very pale brown when dried up, darkening in rotten fruits, obliquely ovate or elliptic, flattened, more or less plano-convex, usually irregularly curved, about 1–2 × as long as wide, 11 × 6 × 2–23 × 18 × 5 mm., mostly smooth, very shortly pubescent. Testa thick.

Zambia. N: Kasama District, x, *Robinson* 3953 (EA; K; M; SRGH). C: 20 km. S. of Lusaka, x, *Angus* 1414 (BM; BR; FHO; K; PRE). S: Gwembe District, fr. iv, *F. White* 2619 (BR; FHO). **Zimbabwe.** C: Salisbury, xi, *Brain* 10835 (SRGH). E: Melsetter District, x, *R. Goodier & Phipps* 303 (EA; K; M; PRE; S; SRGH). **Malawi.** C: Kasungu, fr. viii, *Brass* 17445 (K; MO; NY; SRGH; US). S: Blantyre, vii, *Buchanan* 34 (E; K). **Mozambique.** N: Mogincual, fr. iii, *Torre & Paiva* 11444 (LISC). Z: Mocuba, xii, *Torre* 4780 A (LISC). T: Zóbuè Mts., x, *Torre* 3694 (LISC). MS: near Catandica (Vila Gouveia), xii, *Torre & Correia* 13413 (LISC). M: Santaca, Maputo, viii, *Gomes e Sousa* 3788 (C; COI; K; PRE).

Also in Tropical and South Africa. Woodlands, bushlands, or sometimes gallery forests; alt. 0–2200 m.

18. **Strychnos usambarensis** Gilg in Abh. Preuss. Akad. Wiss. **1894**: 36 (1894); in Engl., Pflanzenw. Ost-Afr. **C**: 311 (1895). — Baker in F.T.A. **4**, 1: 526 (1903). — Bruce & Lewis in F.T.E.A., Loganiaceae: 34 (1960). — Verdoorn in Fl. S. Afr. **26**: 140, fig. 18, 5 (1963). — Onochie & Leeuwenberg in F.W.T.A., 2nd ed. **2**: 44 (1963). — Leeuwenberg in Meded. Landb. Wag. **69**–1: 267, fig. 44, map 44 (1969). Type from Tanzania.

S. micans S. Morre, Journ. Linn. Soc. **40**: 146 (1911). — Verdoorn in Bothalia **3**: 587–588, fig. 3 (1939). Type from Zimbabwe: Chirinda Forest, *Swynnerton* 125 (BM, holotype; isotypes K, Z).

Shrub, climbing shrub, large liana or much branched small tree (as shrub 1·50–5 m. high, as liana 2–20 m. high or more, climbing over shrubs and in trees and then up to 70 m. long, as tree 3–10(15) m. high). Liana trunk 2–9 cm. in diam. (or more ?); bark dark brown, thin, with many lenticels; wood pale yellow. Tree trunk 15–25 cm. in diam.; bark pale or dark grey or pale grey-brown with darker patches, smooth, in section orange; wood pale yellow. Branches lenticellate, usually very dark brown, often covered with a pale skin which splits and peels off, not sulcate when dry; branchlets glabrous or shortly pubescent, not or slightly sulcate when dry. Tendrils solitary, only present in climbing shrubs or lianas. Leaves: petiole glabrous, 2–6 mm. long; lamina slightly shining and dark green above, less shining to dull and paler beneath, coriaceous or thinly coriaceous in the sun, in the shade thinner and thinly coriaceous or papyraceous, also when living, ovate, narrowly ovate, elliptic, or narrowly elliptic, 1·5–3 times as long as wide, 3–8 × 1·2–3·5 cm., in lianas in the shade up to 16 × 7 cm., distinctly acuminate or in the shade mostly caudate at the apex, usually mucronate at the very apex, cuneate, rounded, or occasionally subcordate at the base, glabrous on both sides; two pairs of secondary veins from or from above the base curved along the margin, outer usually fainter; tertiary venation reticulate, inconspicuous. Inflorescences axillary, solitary or several together, lax or congested, usually much shorter than the leaves, 1–2·5 × 1–1·5 cm., few-flowered, 1–3 times branched. Flowers 4–5-merous. Sepals pale green, connate at the base, subequal ovate, broadly ovate, or triangular, 1–1·3 times as long as wide, 0·6–1 × 0·6–1 mm., acute or obtuse, very minutely ciliate, glabrous on both sides or shortly pubescent outside, without colleters. Corolla in the mature bud 3–4 times as long as the calyx, 2–3·5 mm. long, and rounded at the apex, white, yellow, or sometimes orange, glabrous or minutely papillose-pubescent outside, inside with a ring of hairs in the throat and often, when outside so, minutely papillose-pubescent on the lobes; tube short, urceolate, 1–1·4(1·7) times as long as the calyx, 1–1·4 mm. long; at the throat 1·4–2·2 mm. wide; lobes 1·4–1·6(2·7) times as long as the tube, oblong, 1·3–2·2 times as long as wide, 1·2–2·2 × 0·8–1·3 mm., acute, recurved from somewhat below the middle. Stamens exserted; filaments 0·5–1·2(2) times as long as the anthers, inserted at the mouth of the corolla tube; anthers suborbicular or oblong, 0·6–0·9 × 0·3–0·7 mm., glabrous. Pistil glabrous, 1·1–2·5 mm. long; ovary ovoid, slightly longer than wide, 0·7–1 × 0·6–0·8 mm., 2-celled; style short, 0·4–1·5 mm. long; stigma capitate or sometimes obscurely bilobed. In each cell 4–7 ovules. Fruit orange or orange-yellow, immature pale green and often with a glaucous bloom, or glacous, small, soft, globose or subglobose, often laterally compressed, sometimes shortly stipitate within the calyx, 11–18 × 10–18 × 10–18 mm., with smooth skin, 1-seeded, mat, mostly oblique pedicellate. Wall thin, about 0·3 mm. thick. Pulp orange. Seed pale brown, depressed-globose or ellipsoid, 9 × 7 × 5–12 × 11 × 8 mm., shortly and densely pubescent, smooth, with a central hilum at one side. Testa thin.

Zambia. N: Lunzua R., near Mbala [Abercorn], *R. E. Fries* 1234 (UPS). S: Kalomo, *Mitchell* 15/30 (K). **Zimbabwe.** W: Matobo District, *Guy* 1/58 (K; SRGH); ibid., fr. ix, *Plowes* 1465 (BR; MO; PRE; SRGH). C: Selukwe District, *Biegel* 1543 (WAG). E: Umtali District, fr. viii, *Chase* 553 (BM; PRE; SRGH), 1656 (BM; LISC; SRGH), 4146 (BM; BR; COI; MO; NY; PRE; SRGH). S: km. 45 of Uvuma-Fort Victoria road, fl., fr. v, *Grout* 22/47 (FHO; SRGH). **Mozambique.** MS: Inhamitanga, xii, *Simão* 28 (LISC) 1282 (PRE). M: Maputo, i, *Hornby* 2547 (BM; PRE; SRGH).

Also in West, Central, East and northern South Africa. Upland and lowland rain forests and secondary forests, especially on river banks, gallery forests, semi-evergreen, and coastal evergreen bushlands; alt. 0–2000 m.

19. **Strychnos xantha** Leeuwenberg in Meded. Landb. Wag. **69**–1: 274, fig. 45, map 45 (1969). Type from Zambia: Kitwe District, *Fanshawe* 3185 (K, holotype; BR; EA; PRE; SRGH; WAG, isotypes).

Climbing shrub or liana, at least 6–10 m. high, climbing over shrubs or in trees, not climbing shrub or small tree with short thick trunk and often liana-like branches, 2–12 m. high. Trunk short (in trees), 50 cm. in diam. Bark slightly rough, grey or black, in section faded yellow; sapwood white; wood faded yellow. Branches ochraceous or black-brown, often spotted, not lenticellate, terete, not or hardly sulcate when dry; branchlets glabrous, terete, ochraceous and often sulcate when dry. Tendrils — if present — paired. Leaves with a glabrous, 5–8 mm. long petiole; lamina hardly shining and pale green, not or slightly paler beneath, usually yellow when dry, often stiffly coriaceous when fresh, elliptic, narrowly elliptic, ovate or narrowly ovate, (1·5)2–3 × as long as wide, 4–9(11·5) × 1·5–4·5(7) cm., acuminate at the apex, cuneate or rounded at the base, glabrous on both sides; one or two pairs of secondary veins from or from above the base curved along the margin and often a faint submarginal pair; tertiary venation reticulate, conspicuous, in dry leaves prominent on both sides, in living beneath only. Inflorescence axillary and occasionally also terminal, rather congested, few-flowered, much shorter than the leaves, 1·5 × 1–2·5 × 1·5 cm. Peduncle, branched, and pedicels short, thin, glabrous. Corolla in the mature bud 3·5–5 × as long as the calyx, 4·5–6·5 mm. long and rounded at the apex, white, outside glabrous or minutely papillose-pubescent, inside densely pilose on the lobes and in the throat, less pilose in the tube, glabrous at the base; tube cylindrical or nearly so, 1·7–3 × as long as the calyx, 1–1·3 × as long as the lobes, 2·4–3·5 mm. long, at the throat 1·8–2·2 mm. wide; lobes oblong, 1·7–2·5 × as long as wide, 2·4–3·3 × 1·2–1·5 mm., acute, recurved. Stamens exserted; filaments 0·8–1·5 × as long as the anthers, elongate at anthesis, inserted at the mouth of the corolla tube; anthers oblong, 0·7–1 × 0·4–0·6 mm., glabrous. Pistil glabrous, 4–5·5 mm. long; ovary ovoid or nearly so, 1–1·5 × 0·8–1 mm., 2-celled; style slender, 2·8–4·2 mm. long; stigma obscurely bilobed or less often capitate. In each cell 6–8 ovules. Fruit orange or yellow, nearly mature pale green, with paler dots, small, soft, ellipsoid, 15 × 10–20 × 18 mm., often mucronate, often obliquely pedicellate, with smooth skin, 1–2-seeded. Wall thin, when dry 0·3–0·5 mm. thick. Pulp orange or yellow. Seed ochraceous, often obliquely ellipsoid, flattened, 10 × 8 × 3–14 × 11 × 4 mm., at one side with a deep pit at the other with a bulge surrounded by a shallow groove; seemingly papillose; false papillae simulated by short curved hairs.

Zambia. N: Kawambwa, *Fanshawe* 3827 (BR; K; LISC; SRGH). W: Kitwe District, iv, *Fanshawe* 3185 (BR; EA; K; PRE; SRGH; WAG, type). **Mozambique.** N: Nampula, *Torre & Paiva* 11594 (LISC). MS: Maringua, Save R., *Chase* 2534 (BM; LISC; SRGH).

Also in Eastern Zaire and Tanzania. In gallery forests or riverine thickets; alt. 0–1800 m.

110. SALVADORACEAE

By A. R. Vickery

Trees or shrubs, unarmed or with axillary spines. Leaves usually coriaceous, opposite, simple. Stipules minute or absent. Inflorescence of dense axillary or terminal fascicles or panicles. Flowers mostly 4-merous, bisexual or dioecious, actinomorphic. Calyx-lobes 2–4. Petals 4, imbricate, free or partially joined.

Stamens 4, epipetalous or inserted at base of petals and alternate with them. Anthers 2-thecous; thecae opening by longitudinal slits. Ovary 1–2-locular; style short; ovules 1–2, erect. Fruit a berry or drupe. Seeds without endosperm; embryo with thick cordate cotyledons.

Family of 3 genera, palaeotropical, often forming an important part of vegetation in arid areas. The fibrillated stems of *Salvadora*, and, less frequently, *Dobera*, are widely used as toothbrushes.

1. Shrubs bearing axillary spines - - - - - - - - - - **1. Azima**
 - Unarmed shrubs or trees - - - - - - - - - - - - 2
2. Petals free; stamens hypogynous - - - - - - - - - **2. Dobera**
 - Petals shortly united at base; stamens epipetalous - - - - - **3. Salvadora**

1. AZIMA Lam.

Azima Lam., Encycl. Méth., Bot. **1**: 343 (1783).

Glabrous or pubescent dioecious shrubs or small trees; stems much branched, bearing paired or solitary axillary spines. Leaves opposite, simple. Flowers small, axillary. Calyx campanulate, shortly 4-toothed or 2-partite. Petals 4, free, lanceolate. Male flowers: stamens 4, free, slightly exserted, alternating with the petals; anthers oblong, nearly equalling the filaments; ovary rudimentary. Female flowers: staminoides 4, alternating with the petals; ovary ovoid, 2-locular; ovules basal, 1 (or more rarely 2) per locule; stigma sessile, bilobed hairy. Fruit a berry 1-(more rarely 2-) seeded.

A genus of 3–4 species occurring from South and tropical Africa and Madagascar to India, Malaya and the Philippines.

Azima tetracantha Lam., Encycl. Méth. Bot. **1**: 343 (1783) and Illust., t. 807 (1799). — Bak. in Oliv., F.T.A. **4**, 1: 22 (1902). — Verdoorn in Fl. Southern Afr. **26**: 129 (1963). — Friedrich-Holzhammer in Merxm., Prodr. Fl. SW. Afr. **78**: 2 (1968). — Verdc. in F.T.E.A. Salvadoraceae: 3 (1968). — Palmer & Pitman, Trees of S. Afr. **3**: 1837 (1972). — Diniz in Rev. Ciênc. Biol., Sér. A, **15**: 68 (1972). TAB. **87**. Syntypes from East Indies and plants cultivated in France.

Dioecious shrubs to 3(5) m. high, stems usually much branched, sometimes scrambling, bearing paired or solitary axillary straight spines. Young stems green, sometimes 4-angled, glabrous or pubescent. Leaf-blades elliptic or oblong to suborbicular, 1·3–5·0 cm. long, 0·7–4·5 cm. wide, with stiff mucro at apex. Flowers small, in axillary clusters, often elongating, into interrupted spikes near ends of twigs. Calyx campanulate, 4-toothed, 2–4 mm. long. Corolla greenish, petals 4, 2–5 mm. long. Male flowers: stamens inserted at base of rudimentary ovary, alternating with petals, anthers exserted. Female flowers: staminoides with short filaments bearing sagittate sterile anthers; ovary up to 4·5 mm. long; 2-locular with 1–2 ovules per locule; stigma sessile. Berries green, turning white, 0·6–0·8 cm. diameter, stigma persistent. Seeds discoidal, dark brown.

Zambia. C: Feira, ♂ fl. immat. 8.ix.1964, *Fanshawe* 48892 (K; LISC). **Zimbabwe.** E: Melsetter Distr., Birchenough Bridge, 520 m., st. 26.vii.1958, *Chase* 6968 (K; LISC). S: Gwanda, Chikwarakwara, 210 m., ♂ fl. xi.1956, *Davies* 2165 (K; LISC). **Malawi.** S: Chikwawa, 200 m., ♂ fl. 1.x.1946, *Brass* 17892 (K). **Mozambique.** T: between Tete and Boroma, 22 km. from Tete, fr. 18.x.1965, *Rosa* 50 (LISC). MS: Gorongosa, Chitengo, fr. 23.x.1965, *Balsinhas* 990 (COI). GI: Gaza, Chibuto, near Maniquenique, Moaze, ♀ fl. 10.x.1957, *Barbosa & Lemos* 7986 (COI; K; LISC). M: Costa do Sol, fr. 6.v.1964, *Balsinhas* 720 (LISC).

Widespread from Arabia and the Somali Republic through east and central Africa to South Africa and Namibia, Madagascar, Aldabra, Comoro Is., and extending to India and Ceylon and the Philippines. In scrub, especially on saline or alluvial soils near rivers and near coast.

2. DOBERA Juss.

Dobera Juss., Gen. Pl.: 425 (1789).
Platymitium Warb. in Pflanzenw. Ost. Afr. **C**: 279, t. 31 (1895).

Glabrous or slightly pubescent unarmed trees or shrubs. Flowers bisexual, sessile in axillary or terminal panicles. Calyx ovoid, 3–4(5) toothed, sometimes ± bilobed. Petals 4(5) free, elliptic, oblong or linear-oblong, sometimes ± spathulate,

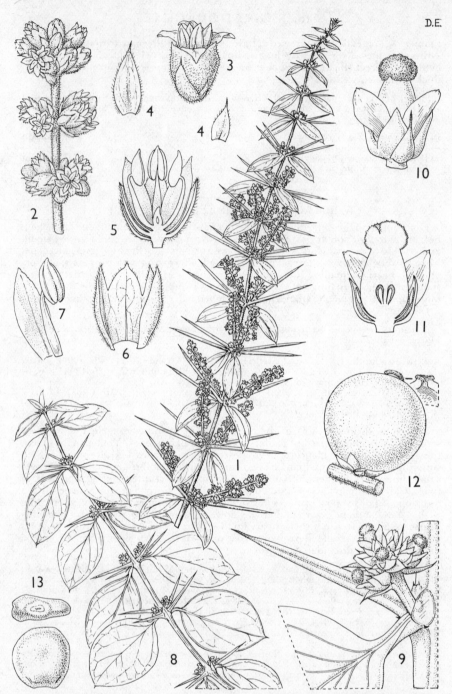

D.E.

Tab. 87. AZIMA TETRACANTHA. 1, part of male flowering branch (× ⅔); 2, male inflorescence (× 2); 3, male flower (× 6); 4, bracts (× 8); 5, male flower in longitudinal section (× 6); 6, calyx (× 8); 7, petal and stamen (× 8), 1–7 from *Milne-Redhead & Taylor* 7152; 8, part of female flowering branch (× ⅔); 9, part of node from same (× 3); 10, female flower (× 8); 11, same in longitudinal section (× 8), 8–11 from *Faulkner* 1877; 12, fruit showing attachment and separate detail of apex in section (× 4) *Verdcourt* 3113; 13, seed, two aspects (× 3) *Bogdan* 4358. From F.T.E.A.

imbricate. Stamens 4(5), hypogynous; filaments dilated basally, shortly connate forming a tube, free and filiform above; anthers ovate. Disk composed of 4 glands each at the base of a petal and alternating with the stamens. Ovary 1-locular; style short; stigma obtuse or truncate; ovules 1–2. Fruit a subglobose or ellipsoidal 1-seeded drupe.

A genus of 2 species, both occurring in tropical Africa, one extending to southern Arabia and India.

Dobera loranthifolia (Warb.) Harms in Engl. & Prantl., Nat. Pflanzenfam. Nachtr. **2–4**: 282 (1897). — Bak. in Oliv., F.T.A. **4**, 1: 22 (1902). — Verdc. in Kew. Bull. **19**: 157 (1964); in F.T.E.A. Salvadoraceae: 4 (1968). — Diniz in Rev. Ciênc. Biol., Sér. A, **5**: 70 (1972). TAB. **88**. Syntypes from Tanzania.
    Platymitium loranthifolium Warb. in Pflanzenw. Ost. Afr. **C**: 279, t. 31 (1895). Types as above.
    Dobera allenii N.E. Br. in Kew Bull. **1914**: 80 (1914); in Hook., Ic. Pl. **31**: t. 3017 (1915). Type: Mozambique, *Allen* 95 (K).

Much-branched evergreen tree, 5–15 m. high, unarmed. Bark variable, black flaky, pale grey rough and fissured, or grey and smooth. Leaf-blades subcoriaceous or thin, elliptic or obovate, 1·5–9·5 cm. long, 1·2–4·5(–7) cm. wide, mostly acute, almost glabrous, glaucous. Petioles 4–5 mm. long. Flowers small, greenish white, ± 3 mm. long, in axillary or terminal panicles. Ovary containing two ovules, but fruit always 1-seeded. Fruit ellipsoid, 1–1·5 cm. long, 7–10 mm. wide.

Mozambique. N: Cabo Delgado, Macondes, between Mueda and Nairoto, fr. 20.ix.1948, *Barbosa* 2224 (LISC). Z: Morrumbala, fl. 15.viii.1953, *Gomes Pedro* 4156 (LMA).

Also in the Somali Republic, Kenya and Tanzania. In scrub and wooded grassland, usually on alluvial or sandy soils.

The type of *Dobera allenii* is cited as having been collected at "Antari." This locality cannot be traced, and the comparison of its label with other labels by the same collector leads one to suspect that "Antari" is a vernacular name.

3. SALVADORA L.

Salvadora L., Sp. Pl. **1**: 122 (1753).

Glabrous or pubescent, unarmed shrubs or small trees, often scrambling. Flowers bisexual, small, in axillary or terminal panicles. Calyx campanulate, 4-toothed. Corolla campanulate; lobes imbricate, shortly joined at base, elliptic, obtuse. Stamens 4, inserted at base or middle of corolla tube. Disk of 4 glands alternating with the stamens, or absent. Ovary 1-locular; style short, columnar; stigma subpeltate or broadly truncate; ovule erect, solitary. Fruit a globose drupe.

A genus of about 4 species occurring from south Africa through Arabia and India to China.

Petioles distinct; leaf-blades ovate - - - - - - - - - 1. *persica*
Petioles indistinct; leaf-blades linear-oblong - - - - - - 2. *australis*

1. **Salvadora persica** L., Sp. Pl. **1**: 122 (1753). — Bak. in F.T.A. **4**, 1: 23 (1902). — Verdoorn in Fl. Southern Afr. **26**: 132 (1963). — Verdc. in Kew Bull. **19**: 147 (1964). — Friedrich-Holzhammer in Merxm., Prodr., Fl. SW. Afr. **78**: 2 (1968). — Verdc. in F.T.E.A. Salvadoraceae: 7 (1968). — Palmer & Pitman, Trees of S. Afr. **3**: 1839 (1968). — Diniz in Rev. Ciênc. Biol., Sér. A, **5**: 71 (1972). TAB. **89**. Lectotype: Linnean Herbarium 164. 1 (LINN).

Much branched shrubs or small trees to 6 m. high, unarmed. Branches long, often pendulous or semiscandent, glabrous or pubescent. Leaves subsucculent; blades coriaceous, landeolate to elliptic, occasionally orbicular, 1·3–10 cm. long, 1·2–3 cm. wide, rounded to acute at apex, cuneate to subcordate at base. Flowers small, greenish-white in lateral and terminal panicles up to 10 cm. long. Petals (1)–3 mm. long. Drupes red or dark red purple when ripe.

1. Plant glabrous - - - - - - - - - - - 2
 – Plant pubescent - - - - - - - - - var. *pubescens*
2. Anthers 0·75–1·00 mm. long - - - - - - var. *persica*
 – Anthers 0·25 mm. long - - - - - - - var. *parviflora*

Var. **persica**

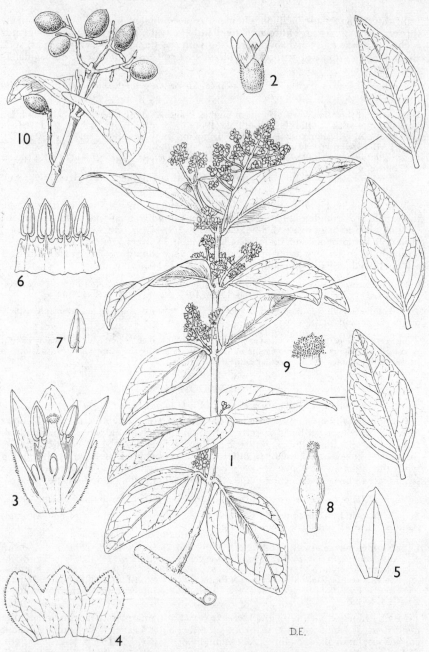

D.E.

Tab. 88. DOBERA LORANTHIFOLIA. 1, part of flowering branch (×⅔); 2, flower (×4); 3, longitudinal section of same (×8); 4, calyx opened out, viewed from inside (×8); 5, petal (×8); 6, stamens, with tube opened out, viewed from inside (×8); 7, anther, outer aspect (×8); 8, gynoecium (×8); 9, style and stigma (×20), 1–9 from *Greenway* 10853; 10, part of fruiting branchlet (×⅔), from Bally 2191. From F.T.E.A.

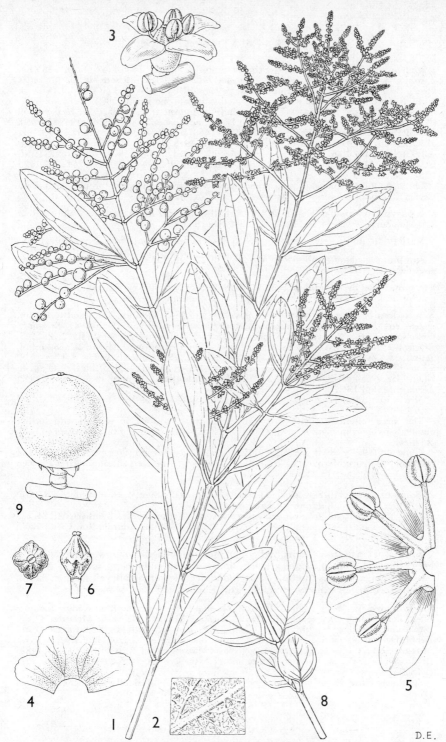

Tab. 89. SALVADORA PERSICA var. PERSICA. 1, flowering branch (×⅔); 2, detail from lower surface of leaf (×8); 3, flower (with pedicel and small segment of rhachis) (×6); 4, calyx opened out (×8); 5, corolla and androecium opened out (×8); 6, gynoecium and pedicel, side view (×8); 7, same viewed from above (×8); 8, fruiting branch (×⅔); 9, fruit (×6), all from *Verdcourt* 3581. From F.T.E.A.

Botswana. N: Ngamiland, near Sehitwa Store, fl. immat. 18.vi.1937, *Erens* 236 (K). **Zambia.** S: Mambova, fl. 6.ix.1962, *Fanshawe* 7022 (K; LISC). **Rhodesia.** E: Melsetter Distr., Sabi R., Birchenough Bridge, fl. & fr. 12.ix.1949, *Chase* 1745 (BM; K; LISC). S: Nuanetsi Distr., banks of Lundi R., near junction of Sabi and Lundi, 300 m., fl. 13.ix.1967, *Müller* 660 (K; LISC). **Malawi.** S: Mangochi (Fort Johnston), fl. 10.viii.1954, *Jackson* 1359 (BM). **Mozambique.** N: Memba, between Lúrio and Chaonde, fl. & fr. 13.x.1948, *Barbosa* 2386 (LISC). Z: Mopeia, fl. & fr. 14.x.1941, *Torre* 3650 (LISC). T: Boroma, fl. 6.ix.1941, *Torre* 3370 (LISC). MS: Gorongosa, National Game Park, 40 m., fr. 5.xi.163, *Torre & Paiva* 9060 (LISC). GI: Gaza, Caniçado, Lagoa Nova, Massingir road, fr. 10.xi.1970, *Correia* 1878 (BM; COI; LISC). M: Marracuene, Costa do Sol, fl. 10.viii.1959, *Barbosa & Lemos* 8645 (COI: K; LISC; SRGH).

Throughout dry regions of tropical and north-east Africa, extending through Arabia and Ceylon. In coastal or riverine scrub on saline, sandy or alluvial soils.

Var. **parviflora** Verdc. in Kew Bull. **19**: 152 (1964). Type: Zambia, banks of Zambesi R. at Mambova, 29.v.1954, *Munro* 16 (EA, holotype; K; PRE).

Variety distinguished by its very small flowers; known only from type collection.

Zambia. S: banks of Zambesi R. at Mambova, fl. 29.v.1954, *Munro* 16 (K). Not known from elsewhere. On river banks.

Var. **pubescens** Brenan in Kew Bull. **4**: 90 (1949). — Verdc. in Kew Bull. **19**: 153 (1964). — Verdc. in F.T.E.A. Salvadoraceae: 9 (1968). Type from Tanzania.

Botswana. N: Khwai R., fl. vii.1971, *Smith* 163 (K; LISC). **Zambia.** C: Katondwe, 395 m., fr. 6.x.1964, *Fanshawe* 8947 (K). S: Livingstone, near Katombora, fl. 26.viii.1947, *Brenan & Greenway* 7755 (K). **Zimbabwe.** N: Sipolilo Distr., Mashumbi, Hunyani R., fl. 12.viii.1965, *Bingham* 1596 (K; LISC). W: Setungwe Distr., Kavira Hotsprings, Mlibizi R., 560 m., fl. 31.ix.1958, *Lovemore* 549 (K; LISC). E: Melsetter Distr., Birchenough Bridge, fl. 12.ix.1949, *Chase* 1777 (BM; K: LISC). S: Bukera Distr., W. of Birchenough Bridge, 520 m., fl. 26.vii.1958, *Chase* 6967 (K; LISC; SRGH). **Mozambique.** T: Chioco, fl. & fr. 26.ix.1942, *Mendonça* 456 (LISC).

Also in Kenya and Tanzania.

The three varieties recognised here do not appear to differ in ecology, and further observations on living material are necessary to establish the validity of var. *parviflora* and var. *pubescens*.

Verdcourt (1964) notes that *Bolus* 9701 (K) from Delagoa Bay and other specimens from near Maputol (Lourenço Marques) have uniformly long slender pedicels and approach the Indian var. *wightiana* (Planch. ex Thwaites) Verdc.

2. **Salvadora australis** Schweick. in Bothalia **3**: 248 (1937). Syntypes: Mozambique, Mapai, Guijá, *Lea* 5 (K) and from the Transvaal.
    Salvadora angustifolia var. *australis* (Schweick.) Verdoorn in Fl. Southern Afr. **26**: 133 (1963). Palmer & Pitman, Trees of S. Afr. **3**: 1841 (1968). — Diniz in Rev. Ciênc. Biol., Sér. A, **5**: 73 (1972). Types as above.

Shrubs or small trees to 5 m. high, often much branched. Young stems terete, with grey pubescence. Leaf blades greyish, linear-oblong, pubescent, rounded at apex, narrowly cuneate at base, petiole short, indistinct. Flowers small, yellowish, c. 4 mm. long, in terminal or axillary panicles. Fruit globose, c. 5 mm. diameter.

Zimbabwe. W: Wankie Distr., Wankie National Park, near road from Robins Camp to Sinamatella Camp, 1000 m., fr. 4.xi.1970, *Müller* 1703 (K; SRGH). E: Melsetter Distr., Birchenough Bridge, fr. 12.ix.1949, *Chase* 1755 (BM; K; SRGH). S: Gwanda Distr., Beitbridge, junction of Shashi and Limpopo, fr. 28.viii.1958, *West* 3709 (K; LISC; SRGH). **Mozambique.** GI: Gaza, Caniçado, Lagoa Nova, Massingir road, fr. 10.xi.1970, *Correia* 1881 (BM; COI; LISC). M: between Moamba and Mavabaze, fr. 28.xi.1944, *Mendonça* 3111 (LISC).

South-east Africa from Zimbabwe to northern Zululand. Alluvial soils, river banks and flood plains.

INDEX TO BOTANICAL NAMES